T0327695

FIBONACCI AND LUCAS NUMBERS WITH APPLICATIONS

FIBONACCI AND LUCAS NUMBERS WITH APPLICATIONS

Volume One

Second Edition

THOMAS KOSHY
Framingham State University

WILEY

Published by John Wiley & Sons, Inc., Hoboken, New Jersey
Published simultaneously in Canada

For general information on our other products and services or for technical support, please contact our Customer Care Department within the United States at (800) 762-2974, outside the United States at (317) 572-3993 or fax (317) 572-4002.

Wiley also publishes its books in a variety of electronic formats. Some content that appears in print may not be available in electronic formats. For more information about Wiley products, visit our web site at www.wiley.com.

Library of Congress Cataloging-in-Publication Data:

Names: Koshy, Thomas.
Title: Fibonacci and Lucas numbers with applications / Thomas Koshy,
 Framingham State University.
Description: Second edition. | Hoboken, New Jersey : John Wiley & Sons, Inc.,
 [2017]- | Series: Pure and applied mathematics: a Wiley series of texts,
 monographs, and tracts | Includes bibliographical references and index.
Identifiers: LCCN 2016018243 | ISBN 9781118742129 (cloth : v. 1)
Subjects: LCSH: Fibonacci numbers. | Lucas numbers.
Classification: LCC QA246.5 .K67 2017 | DDC 512.7/2–dc23 LC record available at
 https://lccn.loc.gov/2016018243

Set in 10/12pt, TimesNewRomanMTStd by SPi Global, Chennai, India

Printed in the United States of America

10 9 8 7 6 5 4 3 2 1

Dedicated to
Suresh, Sheeba, Neethu, and Vinod

CONTENTS

LIST OF SYMBOLS

Symbol	Meaning
$\mathbb{N}$	set of positive integers 1, 2, 3, 4, ...
W	set of whole numbers 0, 1, 2, 3, ...
$\mathbb{Z}$	set of integers $\dots, -2, -1, 0, 1, 2, \dots$
$\mathbb{R}$	set of real numbers
$\{s_n\}_1^\infty = \{s_n\}$	sequence with general term s_n
$\displaystyle\sum_{i=k}^{i=m} a_i = \sum_{i=k}^{m} a_i = \sum_{k}^{m} a_i$	$a_k + a_{k+1} + \cdots + a_m$
$\displaystyle\sum_{i \in I} a_i$	sum of the values of a_i as i runs over the values in I
$\displaystyle\sum_{P} a_i$	sum of the values of a_i, where i satisfies properties P
$\displaystyle\sum_{i,j} a_{ij}$	$\sum_i(\sum_j a_{ij})$
$\displaystyle\prod_{i=k}^{i=m} a_i = \prod_{i=k}^{m} a_i = \prod_{k}^{m} a_i$	$a_k a_{k+1} \cdots a_m$
$n!$ (n factorial)	$n(n-1)\cdots 3 \cdot 2 \cdot 1$, where $0! = 1$
$\lvert x \rvert$	absolute value of x
$\lfloor x \rfloor$ (the floor of x)	the greatest integer $\leq x$
$\lceil x \rceil$ (the ceiling of x)	the least integer $\geq x$
PMI	principle of mathematical induction

Symbol	Meaning
a div b	quotient when a is divided by b
a mod b	remainder when a is divided by b
$a\|b$	a is a factor of b
$a \nmid b$	a is not a factor of b
$\{x, y, z\}$	set consisting of the elements x, y, and z
$\{x\|P(x)\}$	set of elements x with property $P(x)$
$\|A\|$	number of elements in set A
$A \cup B$	union of sets A and B
$A \cap B$	intersection of sets A and B
(a, b)	greatest common factor of a and b
$[a, b]$	least common factor of a and b
$A = (a_{ij})_{m \times n}$	$m \times n$ matrix A whose ijth element is a_{ij}
$\|A\|$	determinant of square matrix A
$\in$	belongs to
$\approx$	is approximately equal to
$\equiv$	is congruent to
∞	infinity symbol
■	end of a proof, solution, example, theorem, lemma, or corollary
$\overline{AB}$	line segment with end points A and B
AB	length of line segment AB
$\overleftrightarrow{AB}$	line containing the points A and B
$\overrightarrow{AB}$	ray AB
$\angle ABC$	angle ABC
$\overleftrightarrow{AB} \parallel \overleftrightarrow{CD}$	lines $\overleftrightarrow{AB}$ and $\overleftrightarrow{CD}$ are parallel
$\overleftrightarrow{AB} \perp \overleftrightarrow{CD}$	lines $\overleftrightarrow{AB}$ and $\overleftrightarrow{CD}$ are perpendicular
RHS	right-hand side
LHS	left-hand side
cis θ	$\cos \theta + i \sin \theta$

PREFACE

Man has the faculty of becoming completely absorbed in one subject, no matter how trivial and no subject is so trivial that it will not assume infinite proportions if one's entire attention is devoted to it.

—Tolstoy, *War and Peace*

THE TWIN SHINING STARS

The Fibonacci sequence and the Lucas sequence are two very bright shining stars in the vast array of integer sequences. They have fascinated amateurs, and professional architects, artists, biologists, musicians, painters, photographers, and mathematicians for centuries; and they continue to charm and enlighten us with their beauty, their abundant applications, and their ubiquitous habit of occurring in totally surprising and unrelated places. They continue to be a fertile ground for creative amateurs and mathematicians alike, and speak volumes about the vitality of this growing field.

This book originally grew out of my fascination with the intriguing beauty and rich applications of the twin sequences. It has been my long-cherished dream to study and to assemble the myriad properties of both sequences, developed over the centuries, and to catalog their applications to various disciplines in an orderly and enjoyable fashion. As the cryptanalyst Sophie Neveu in Dan Brown's bestseller *The Da Vinci Code* claims, "the [Fibonacci] sequence ... happens to be one of the most famous mathematical progressions in history."

An enormous wealth of information is available in the mathematical literature on Fibonacci and Lucas numbers; but, unfortunately, most of it continues to be widely scattered in numerous journals, so it is not easily accessible to many, especially to non-professionals. The first edition was the end-product of materials collected and presented from a wide range of sources over the years; and to the best of my knowledge, it was the largest comprehensive study of this beautiful area of human endeavor.

So why this new edition? Since the publication of the original volume, I have had the advantage and fortune of hearing from a number of Fibonacci enthusiasts from around the globe, including students. Their enthusiasm, support, and encouragement were really overwhelming. Some opened my eyes to new sources and some to new charming properties; and some even pointed out some inexcusable typos, which eluded my own eyes. The second edition is the byproduct of their ardent enthusiasm, coupled with my own.

Many Fibonacci enthusiasts and amateurs know the basics of Fibonacci and Lucas numbers. But there are a multitude of beautiful properties and applications that may be less familiar. Fibonacci and Lucas numbers are a source of great fun; teachers and professors often use them to generate excitement among students, who find them stimulating their intellectual curiosity and sharpening their mathematical skills, such as pattern recognition, conjecturing, proof techniques, and problem-solving. In the process, they invariably appreciate and enjoy the beauty, power, and ubiquity of the Fibonacci family.

AUDIENCE

As can be predicted, this book is intended for a wide audience, not necessarily of professional mathematicians. College undergraduate and graduate students often opt to study Fibonacci and Lucas numbers because they find them challenging and exciting. Often many students propose new and interesting problems in periodicals. It is certainly delightful and rewarding that they often pursue Fibonacci and Lucas numbers for their senior and master's thesis. In short, it is well-suited for projects, seminars, group discussions, proposing and solving problems, and extending known results.

High School students have enjoyed exploring this material for a number of years. Using Fibonacci and Lucas numbers, students at Framingham High School in Massachusetts, for example, have published many of their discoveries in *Mathematics Teacher*.

As in the first edition, I have included a large array of advanced material and exercises to challenge mathematically sophisticated enthusiasts and professionals in such diverse fields as art, biology, chemistry, electrical engineering, neurophysiology, physics, music, and the stock market. It is my sincere hope that this edition will also serve them as a valuable resource in exploring new applications and discoveries, and advance the frontiers of mathematical knowledge, experiencing a lot of satisfaction and joy in the process.

MAJOR CHANGES

In the interest of brevity and aestheticism, I have consolidated several closely-related chapters, resulting in fewer chapters in the new edition. I also have rearranged some chapters for a better flow of the development of topics. A number of new and charming properties, exercises, and applications have been added; so are a number of direct references to Fibonacci numbers, the golden ratio, and the pentagram to D. Brown's *The Da Vinci Code*. The chapters on *Combinatorial Models I* (Chapter 14) and *Graph-theoretic Models I* (Chapter 21) present spectacular opportunities to interpret Fibonacci and Lucas numbers combinatorially; so does the section on *Fibonacci Walks* (Section 4.6). I also have added a new way of looking at and studying them geometrically (Chapter 6).

Again, in the interest of brevity, the chapters on Fibonacci, Lucas, Jacobsthal, and Morgan-Voyce polynomials have been dropped from this edition; but they will be treated extensively in the forthcoming Volume Two. The chapters on tribonacci numbers and polynomials also will appear in the new volume.

ORGANIZATION

In the interest of manageability, the book is divided into 30 chapters. Nearly all of them are well within reach of many users. Most chapters conclude with a substantial number of interesting and challenging exercises for Fibonacci enthusiasts to explore, conjecture, and confirm. I hope, the numerous examples and exercises are as exciting for readers as they are for me. Where the omission can be made without sacrificing the essence of development or focus, I have omitted some of the long, tedious proofs of theorems. Abbreviated solutions to all odd-numbered exercises are given in the back of the book.

SALIENT FEATURES

Salient features of this edition remain the same as its predecessor: a user-friendly, historical approach; a non-intimidating style; a wealth of applications, exercises, and identities of varying degrees of difficulty and sophistication; links to combinatorics, graph theory, matrices, geometry, and trigonometry; the stock market; and relationships to everyday life. For example, works of art are discussed vis-à-vis the *golden ratio (phi)*, one of the most intriguing irrational numbers. It is no wonder that Langdon in *The Da Vinci Code* claims that "PHI is the most beautiful number in the universe."

APPLICATIONS

This volume contains numerous and fascinating applications to a wide spectrum of disciplines and endeavors. They include art, architecture, biology, chemistry, chess, electrical engineering, geometry, graph theory, music, origami, poetry, physics, physiology, neurophysiology, sewage/water treatment, snow plowing, stock market trading, and trigonometry. Most of the applications are well within the reach of mathematically sophisticated amateurs, although they vary in difficulty and sophistication.

HISTORICAL PERSPECTIVE

Throughout, I have tried to present historical background for the material, and to humanize the discourse by giving the name and affiliation of every contributor to the field, as well as the year of contribution. I have included photographs of a number of mathematicians, who have contributed significantly to this exciting field. My apologies to any discoverers whose names or affiliations are still missing; I would be pleased to hear of any such inadvertent omissions.

NUMERIC AND GEOMETRIC PUZZLES

This volume contains several numeric and geometric puzzles based on Fibonacci and Lucas numbers. They are certainly a source of fun, excitement, and surprise for every one. They also provide opportunities for further exploration.

LIST OF SYMBOLS

An updated List of symbols appears between the Contents and Preface. Although they are all standard symbols, they will come in handy for those not familiar with them.

APPENDIX

The Appendix contains a short list of the fundamental properties from the theory of numbers and the theory of matrices. It is a good idea to review them as needed. Those who are curious about their proofs will find them in *Elementary Number Theory with Applications* by the author.

The Appendix also contains a list of the first 100 Fibonacci and Lucas numbers, and their prime factorizations. They all should come in handy for computations.

A WORK IN PROGRESS

A polynomial approach to Fibonacci and Lucas numbers creates new opportunities for optimism, creativity, and elegance. It acts like a thread unifying Fibonacci, Lucas, Pell, Pell-Lucas, Chebyshev, and Vieta polynomials. Such polynomials, and their combinatorial and graph-theoretic models, among other topics, will be studied in detail in a successor volume.

ACKNOWLEDGMENTS

It is my great pleasure and joy to express my sincere gratitude to a number of people who have helped me to improve the manuscript of both editions with their constructive suggestions, comments, and support, and to those who sent in the inexcusable typos in the first edition. To begin with, I am deeply grateful to the following reviewers of the first or second edition for their boundless enthusiasm and input:

Napoleon Gauthier	The Royal Military College, Kingston, Ontario, Canada
Henry W. Gould	West Virginia University, Morgantown, West Virginia
Marjorie Bicknell-Johnson	Santa Clara, California
Thomas E. Moore	Bridgewater State University, Bridgewater, Massachusetts
Carl Pomerance	University of Georgia, Athens, Georgia
M.N.S. Swamy	Concordia University, Montreal, Quebec, Canada
Monte J. Zerger	Adams State University, Alamosa, Colorado
Gerald Alexanderson	Santa Clara University, Santa Clara, California
Anonymous Reviewer (2nd edition)	

Thanks to (the late) Angelo DiDomenico of Framingham High School, who read an early version of the first edition and offered valuable suggestions; to Margarite Landry for her superb editorial assistance; to Kevin Jackson-Mead for preparing the canonical prime factorizations of the Lucas numbers L_{51} through L_{100}, and for proofreading the entire work of the first edition; to (the late) Thomas Moore for the graphs in Figures 5.10, 17.41, and 17.42; to Marjorie Bicknell-Johnson and Krishnaswami Alladi for their quotes; and to the staff at Wiley, especially, Susanne Steitz-Filler, Allison McGinniss, and Kathleen Pagliaro for their enthusiasm, cooperation, support, and confidence in this huge endeavor.

Finally, I would be delighted to hear from Fibonacci enthusiasts about any possible elusive errors. If you should have any questions, or should come across or discover any additional properties or applications, I would be delighted to hear about them.

Thomas Koshy
tkoshy@emeriti.framingham.edu
Framingham, Massachusetts
August, 2017

If I have been able to see farther, it was only
because I stood on the shoulders of giants.
– Sir Isaac Newton (1643–1727)

The author has provided a lucid and comprehensive treatment of Fibonacci and Lucas numbers. He has emphasized the beauty of the identities they satisfy, indicated the settings in mathematics and in nature where they occur, and discussed several applications. The book is easily readable and will be useful to experts and non-experts alike.

–Krishnaswami Alladi, University of Florida

LEONARDO FIBONACCI

Leonardo Fibonacci, also called Leonardo Pisano or Leonard of Pisa, was the most outstanding mathematician of the European Middle Ages. Little is known about his life except for the few facts he gives in his mathematical writings. Ironically, none of his contemporaries mention him in any document that survives.

Fibonacci was born around 1170 into the Bonacci family of Pisa, a prosperous mercantile center. ("Fibonacci" is a contraction of "Filius Bonacci," son of Bonacci.) His father Guglielmo (William) was a successful merchant, who wanted his son to follow his trade.

Around 1190 when Guglielmo was appointed collector of customs in the Algerian city of Bugia (now called Bougie), he brought Leonardo there to learn the art of computation. In Bougie, Fibonacci received his early education from a Muslim schoolmaster, who introduced him to the Indian numeration system and Indian computational techniques. He also introduced Fibonacci to a book on algebra, *Hisâb al-jabr w'almuqabâlah*, written by the Persian mathematician al-Khowarizmi (ca. 825). (The word *algebra* is derived from the title of this book.)

As an adult, Fibonacci made frequent business trips to Egypt, Syria, Greece, France, and Constantinople, where he studied the various systems of arithmetic then in use, and exchanged views with native scholars. He also lived for a time at the court of the Roman Emperor, Frederick II (1194–1250), and engaged in scientific debates with the Emperor and his philosophers.

Fibonacci and Lucas Numbers with Applications, Volume One, Second Edition. Thomas Koshy.
© 2018 John Wiley & Sons, Inc. Published 2018 by John Wiley & Sons, Inc.

Fibonacci*

Around 1200, at the age of 30, Fibonacci returned home to Pisa. He was con-
vinced of the elegance and practical superiority of the Indian numeration system
over the Roman system then in use in Italy. In 1202 Fibonacci published his
pioneering work, *Liber Abaci* (*The Book of the Abacus*). (The word *abaci* here
does not refer to the hand calculator called an abacus, but to computation in
general.) *Liber Abaci* was devoted to arithmetic and elementary algebra; it introduced the Indian numeration system and arithmetic algorithms to Europe. In
fact, Fibonacci demonstrated in his book the power of the Indian numeration system more vigorously than in any mathematical work up to that time. *Liber Abaci*'s
15 chapters explain the major contributions to algebra by al-Khowarizmi and
Abu Kamil (ca. 900), another Persian mathematician. Six years later, Fibonacci

revised *Liber Abaci* and dedicated the second edition to Michael Scott, the most famous philosopher and astrologer at the court of Frederick II.

After *Liber Abaci*, Fibonacci wrote three other influential books. *Practica Geometriae* (*Practice of Geometry*), published in 1220, is divided into eight chapters and is dedicated to Master Domonique, about whom little is known. This book skillfully presents geometry and trigonometry with Euclidean rigor and some originality. Fibonacci employs algebra to solve geometric problems and geometry to solve algebraic problems, a radical approach for the Europe of his day.

The next two books, the *Flos* (*Blossom* or *Flower*) and the *Liber Quadratorum* (*The Book of Square Numbers*) were published in 1225. Although both deal with number theory, *Liber Quadratorum* earned Fibonacci his modern reputation as a major number theorist, ranked with the Greek mathematician Diophantus (ca. 250 A.D.) and the French mathematician Pierre de Fermat (1601–1665). Both *Flos* and *Liber Quadratorum* exemplify Fibonacci's brilliance and originality of thought, which outshine the abilities of most scholars of his time.

In 1225, Frederick II wanted to test Fibonacci's talents, so he invited Fibonacci to his court for a mathematical tournament. The contest consisted of three problems, prepared by Johannes of Palumbo, who was on the Emperor's staff. The first was to find a rational number x such that both $x^2 - 5$ and $x^2 + 5$ are squares of rational numbers[†]. Fibonacci gave the correct answer 41/12: $(41/12)^2 - 5 = (31/12)^2$ and $(41/12)^2 + 5 = (49/12)^2$.

The second problem was to find a solution to the cubic equation $x^3 + 2x^2 + 10x - 20 = 0$. Fibonacci showed geometrically that it has no solutions of the form $\sqrt{a + \sqrt{b}}$, but gave an approximate solution, 1.3688081075, which is correct to nine decimal places. This answer appears in the *Flos* without any explanation.

The third problem, also recorded in *Flos*, was to solve the following:

Three people share 1/2, 1/3, and 1/6 of a pile of money. Each takes some money from the pile until nothing is left. The first person then returns one-half of what he took, the second one-third, and the third one-sixth. When the total thus returned is divided among them equally, each possesses his correct share. How much money was in the original pile? How much did each person take from the pile?

Fibonacci established that the problem is indeterminate and gave 47 as the smallest answer. None of Fibonacci's competitors in the contest could solve any of these problems.

The Emperor recognized Fibonacci's contributions to the city of Pisa, both as a teacher and as a citizen. Today, a statue of Fibonacci stands in the Camposanto Monumentale at Piazza dei Miracoli, near the Cathedral and the Leaning Tower of Pisa. Until 1990, it had been at a garden across the Arno River for some years.

[†]A solution to the problem appears in *The Mathematics Teacher*, Vol. 45 (1952), 605–606. R.A. Laird of New Orleans, Louisiana, reproduced it in *The Fibonacci Quarterly* 3 (1965), pp. 121–122. The general solution to the problem that both $x^2 - m$ and $x^2 + m$ be rational squares appears in O. Ore, *Number Theory and its History*, McGraw-Hill, New York, 1948, pp. 188–193.

Head of the Statue#

Statue of Fibonacci*

Not long after Fibonacci's death in 1240, Italian merchants began to appreciate the beauty and power of the Indian numeration system, and gradually adopted it for business transactions. By the end of the sixteenth century, most of Europe had accepted it. *Liber Abaci* remained the European standard for more than two centuries, and played a significant role in displacing the unwieldy Roman numeration system, thereby spreading the more efficient Indian number system to the rest of world.

FIBONACCI NUMBERS

It may be hard to define mathematical beauty,
but that is true of beauty of any kind.
–G.H. Hardy (1877–1947), *A Mathematician's Apology*

2.1 FIBONACCI'S RABBITS

Fibonacci's Classic work, *Liber Abaci*, contains many elementary problems, including the following famous problem about rabbits:

Suppose there are two newborn rabbits, one male and the other female. Find the number of rabbits produced in a year if:

- Each pair takes one month to become mature;
- Each pair produces a mixed pair every month, beginning with the second month; and
- Rabbits are immortal.

Suppose, for convenience, that the original pair of rabbits was born on January 1. They take a month to become mature, so there is still only one pair on February 1. On March 1, they are two months old and produce a new mixed pair, a total of two pairs. Continuing like this, there will be three pairs on April 1, five pairs on May 1, and so on; see the last row of Table 2.1.

Fibonacci and Lucas Numbers with Applications, Volume One, Second Edition. Thomas Koshy.
© 2018 John Wiley & Sons, Inc. Published 2018 by John Wiley & Sons, Inc.

TABLE 2.1.   Growth of the Rabbit Population

Number of Pairs	Jan	Feb	Mar	Apr	May	Jun	Jul	Aug
Adults	0	1	1	2	3	5	8	13
Babies	1	0	1	1	2	3	5	8
Total	1	1	2	3	5	8	13	21

2.2   FIBONACCI NUMBERS

The numbers in the bottom row are called *Fibonacci numbers*, and the sequence
1, 2, 3, 5, 8, … is the *Fibonacci sequence*. Table A.2 in the Appendix lists the first
100 Fibonacci numbers.

The sequence was given its name in May, 1876, by the outstanding French
mathematician François Édouard Anatole Lucas, who had originally called it
"the series of Lamé," after the French mathematician Gabriel Lamé (1795–1870).
It is a bit ironic that despite Fibonacci's numerous mathematical contributions,
he is primarily remembered for the sequence that bears his name.

François Édouard Anatole Lucas* was born
in Amiens, France, in 1842. After completing
his studies at the École Normale in Amiens,
he worked as an assistant at the Paris Obser-
vatory. He served as an artillery officer in the
Franco-Prussian war and then became professor of mathematics at the Lycée Saint-Louis
and Lycée Charlemagne, both in Paris. He was
a gifted and entertaining teacher. Lucas died
of a freak accident at a banquet; his cheek was
gashed by a shard from a plate that was accidentally dropped; he died from the infection
within a few days, on October 3, 1891.

Lucas loved computing and developed plans for a computer, but it never
materialized. Besides his contributions to number theory, he is known for his
four-volume classic on recreational mathematics, *Récréations mathématiques*
(1891–1894). Best known among the problems he developed is the *Tower of
Brahma* (or *Tower of Hanoi*).

The Fibonacci sequence is one of the most intriguing number sequences. It
continues to provide ample opportunities for professional mathematicians and
amateurs to make conjectures and expand their mathematical horizon.

The sequence is so important and beautiful that *The Fibonacci Association*,
an organization of mathematicians, has been formed for the study of Fibonacci
and related integer sequences. The association was co-founded in 1963 by Verner
E. Hoggatt, Jr. of San Jose State College (now San Jose State University),

Figure source: https://en.wikipedia.org/wiki/File:Elucas_1.png.

California, Brother Alfred Brousseau of St. Mary's College in California, and I. Dale Ruggles of San Jose State College. The association publishes *The Fibonacci Quarterly*, devoted to articles related to integer sequence.

Verner Emil Hoggatt, Jr.* (1921–1980) received his Ph.D. in 1955 from Oregon State University. His "life was marked by dedication to the study of the Fibonacci sequence. His production and creativity were [truly] astounding" [44], according to Marjorie Bicknell-Johnson, who has written extensively in *The Fibonacci Quarterly*.

Hoggatt was the founding editor of the *Quarterly*. He authored or co-authored more than 150 research articles. In addition, he wrote the book *Fibonacci Numbers* in 1969, and edited three other books. In short, Hoggatt had a brilliant and productive professional life.

Brother Alfred Brousseau[#] (1907–1988) began teaching at St. Mary's College, Moraga, California, in 1930. While there, he continued his studies in physics and earned his Ph.D. in 1937 from the University of California. Four years later, he became Principal of Sacred Heart High School. In 1937, Br. Alfred returned to St. Mary's College.

The April 4, 1969 issue of *Time* [170] featured Hoggatt and Br. Alfred in the article "The Fibonacci Numbers."

Br. Alfred Brousseau later became Br. U. Alfred, when the Catholic brotherhood changed the way brothers were named; see Chapter 5.

* *Figure source*: http://faculty.evansville.edu/ck6/bstud/hoggatt.html. Reproduced with permission of Marjorie Bicknell-Johnson.
[#] *Figure source*: Reproduced with permission of Marjorie Bicknell-Johnson.

The Fibonacci sequence has a fascinating property: every Fibonacci number, except the first two, is the sum of the two immediately preceding Fibonacci numbers. (At the given rate, there will be 144 pairs of rabbits on December 1. This can be confirmed by extending Table 2.1 through December.)

RECURSIVE DEFINITION

This observation yields the following recursive definition of the nth Fibonacci number F_n:

$$F_1 = F_2 = 1 \qquad \leftarrow \quad \text{initial conditions}$$

$$F_n = F_{n-1} + F_{n-2}, \qquad \leftarrow \quad \text{recurrence} \qquad (2.1)$$

where $n \geq 3$. We will formally confirm the validity of this recurrence shortly.

It is not known whether Fibonacci knew of the recurrence. If he did, we have no record to that effect. The first written confirmation of the recurrence appeared three centuries later, when the great German astronomer and mathematician Johannes Kepler (1571–1630) wrote that Fibonacci must have surely noticed this recursive relationship. In any case, it was first noted in the west by the Dutch mathematician Albert Girard (1595–1632).

Numerous scholars cite the Fibonacci sequence in Sanskrit. Susantha Goonatilake attributes its discovery to the Indian writer Pingala (200 B.C.?). Parmanand Singh of Raj Narain College, Hajipur, Bihar, India, writes that what we call Fibonacci numbers and the recursive formulation were known in India several centuries before Fibonacci proposed the problem; they were given by Virahanka (between 600 and 800 A.D.), Gopala (prior to 1135 A.D.), and the Jain scholar Acharya Hemachandra (about 1150 A.D.). Fibonacci numbers occur as a special case of a formula established by Narayana Pandita (1156 A.D.).

The growth of the rabbit population can be displayed nicely in a tree diagram, as Figure 2.1 shows. Each new branch of the "tree" becomes an adult branch in one month and each adult branch, including the trunk, produces a new branch every month.

Table 2.1 shows several interesting relationships among the numbers of adult pairs, baby pairs, and total pairs. To see them, let A_n denote the number of adult pairs and B_n the number of baby pairs in month n, where $n \geq 1$. Clearly, $A_1 = 0$, and $A_2 = 1 = B_1$.

Suppose $n \geq 3$. Since each adult pair produces a mixed baby pair in month n, the number of baby pairs in month n equals that of adult pairs in the preceding month; that is, $B_n = A_{n-1}$. Then

$$\begin{pmatrix} \text{Number of pairs} \\ \text{in month } n \end{pmatrix} = \begin{pmatrix} \text{Number of adult} \\ \text{pairs in month } n-1 \end{pmatrix} + \begin{pmatrix} \text{Number of baby} \\ \text{pairs in month } n-1 \end{pmatrix}$$

$$A_n = A_{n-1} + B_{n-1}$$

$$= A_{n-1} + A_{n-2}, \quad n \geq 3.$$

Month   Total number of branches

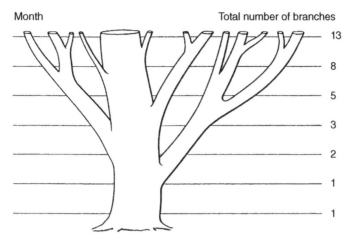

Figure 2.1. A Fibonacci tree.

Thus A_n satisfies the Fibonacci recurrence, where $A_2 = 1 = A_3$. Consequently, $F_n = A_{n+1}$, where $n \geq 1$.

Notice that

$$\begin{pmatrix} \text{Total number of} \\ \text{pairs in month } n \end{pmatrix} = \begin{pmatrix} \text{Number of adult} \\ \text{pairs in month } n \end{pmatrix} + \begin{pmatrix} \text{Number of baby} \\ \text{pairs in month } n \end{pmatrix}.$$

That is, $F_n = A_n + B_n = A_n + A_{n-1} = F_{n-1} + F_{n-2}$, where $n \geq 3$. This establishes the Fibonacci recurrence observed earlier.

Since $F_n = A_{n+1}$, it follows that the ith element in row 1 is F_{i-1}, where $i \geq 2$. Likewise, since $B_n = A_{n-1} = F_{n-2}$, the ith element in row 2 is F_{i-2}, where $i \geq 3$.

The tree diagram in Figure 2.2 illustrates the recursive computing of F_5, where each dot represents an addition.

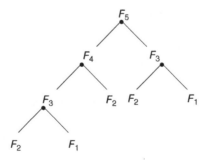

Figure 2.2. Tree diagram for recursive computing of F_5.

Using Fibonacci recurrence, we can assign a meaningful value to F_0. Since $F_2 = F_1 + F_0$, it follows that $F_0 = 0$. This will come in handy in our pursuit of Fibonacci properties later.

Fibonacci recurrence has an immediate consequence to geometry. To see this, consider a nontrivial triangle. By the triangle inequality, the sum of the lengths of any two sides is greater than the length of the third side. Consequently, it follows by the Fibonacci recurrence that *no three consecutive Fibonacci numbers can be the lengths of the sides of a nontrivial triangle.*

Next we introduce another integer family.

LUCAS NUMBERS

The Fibonacci recurrence, coupled with different initial conditions, can be used to construct new number sequences. For instance, let L_n be the nth term of a sequence with $L_1 = 1, L_2 = 3$ and $L_n = L_{n-1} + L_{n-2}$, where $n \geq 3$. The resulting sequence $1, 3, 4, 7, 11, \ldots$ is the *Lucas sequence*, named after Lucas. Table A.2 also lists the first 100 Lucas numbers.

In later chapters, we will see that the Fibonacci and Lucas families are very closely related, and hence the title of this book. For instance, both F_n and L_n satisfy the same recurrence.

2.3   FIBONACCI AND LUCAS CURIOSITIES

Next we present some quick and interesting characteristics of the two families.

FIBONACCI AND LUCAS SQUARES AND CUBES

Of the infinitely many Fibonacci numbers, some have special qualities. For example, only two distinct Fibonacci numbers are squares, namely 1 and 144. This was established in 1964 by J.H.E. Cohn of the University of London [126]. In the same year, he also established that 1 and 4 are the only Lucas squares (see Chapter 23).

In 1969, H. London of McGill University and R. Finkelstein of Bowling Green State University, Bowling Green, Ohio, proved that there are only two distinct Fibonacci cubes, namely, 1 and 8, and that the only Lucas cube is 1 [416].

A UBIQUITOUS FIBONACCI NUMBER AND ITS LUCAS COMPANION

A Fibonacci number that appears to be ubiquitous is 89. Let us see why.

- Since 1/89 is a rational number, its decimal expansion is periodic:

$$\frac{1}{89} = 0.\overline{0112359550561797752808(89)887640449438202247191}.$$
↑

The period is 44, and a surprising number occurs in the middle of a repeating block.

- It is the eleventh Fibonacci number, and both 11 (the fifth Lucas number) and 89 are primes. While 89 can be viewed as the $(8 + 3)$rd Fibonacci number, it can also be looked at as the $(8 \cdot 3)$rd prime.

- Concatenating 11 and 89 gives 1189. Since $1189^2 = (1 + 2 + \cdots + 1681)/2$, it is also a triangular number. Interestingly, there are 1189 chapters in the Bible, of which 89 are in the four gospels.

- Eighty-nine is the smallest number to stubbornly resist being transformed into a palindrome by the familiar "reverse the digits and then add" method. In this case, it takes 24 steps to produce a palindrome, namely, 8813200023188.

- $8 + 9$ is the sum of the four primes preceding 11, and $8 \cdot 9$ is the sum of the four primes following it: $17 = 2 + 3 + 5 + 7$ and $72 = 13 + 17 + 19 + 23$.

- The most recent year divisible by 89 is 1958: $1958 = 2 \cdot 11 \cdot 9$. Notice the prominent appearance of 11 again.

- The next year divisible by 89 is $2047 = 2^{11} - 1$. Again, 11 makes a conspicuous appearance. It is, in fact, the smallest number of the form $2^p - 1$ which is not a prime. Primes of the form $2^p - 1$ are called *Mersenne primes*, after the French Franciscan priest Marin Mersenne (1588–1648). The smallest Mersenne number that is not prime is 2047 [369].

- On the other hand, $2^{89} - 1$ is a Mersenne prime. It is the tenth Mersenne prime, discovered in 1911 by R.E. Powers. Its decimal value contains 27 digits and looks like this:
$$2^{89} - 1 = 6189700196\ldots11.$$

The first three digits are significant because they are the first three digits of an intriguing irrational number we will encounter in Chapters 16–20. Once again, note the occurrence of 11 at the end.

- Multiply the two digits of 89; add the sum of the digits to the product; the sum is again 89: $(8 \cdot 9) + (8 + 9) = 89$. (It would be interesting to check if there are other numbers that exhibit this remarkable behavior.) Also, $8/9 \approx 0.89$.

- There are only two consecutive positive integers, one of which is a square and the other a cube: $8 = 2^3$ and $9 = 3^2$.

- Square the digits of 89, and add them to obtain 145. Add the squares of its digits again. Continue like this. After eight iterations, we return to 89:
$$89 \rightarrow 145 \rightarrow 42 \rightarrow 20 \rightarrow 4 \rightarrow 16 \rightarrow 37 \rightarrow 58 \rightarrow 89.$$

In fact, if we apply this algorithm to any number, we will eventually arrive at 89 or 1.

- On 8/9 in 1974, an unprecedented event occurred in the history of the United States – the resignation of President Richard M. Nixon. Strangely enough, if we swap the digits of 89, we get the date on which Nixon was pardoned by his successor, President Gerald R. Ford.

All these fascinating observations about 11 and 89 were made in 1966 by M.J. Zerger of Adams State College, Colorado [619].

Soon after these Fibonacci curiosities appeared in *Mathematics Teacher*, G.J. Greenbury of England (private communication, 2000) contacted Zerger with two of his own curiosities involving the decimal expansions of two primes:

$$\frac{1}{29} = 0.\overline{0344827586206(89)6551724137931}$$

$$\frac{1}{59} = 0.\overline{016949152542372881355932202203(89)830508474576271186440677966}1.$$

Curiously enough, 89 makes its appearance in the repeating block of each expansion.

R.K. Guy of the University of Calgary, Alberta, Canada, in his fascinating book, *Unsolved Problems in Number Theory* [254], presents an interesting number sequence $\{x_n\}$. It has a quite remarkable relationship with 89, although it is not obvious. The sequence is defined recursively as follows:

$$x_0 = 1$$

$$x_n = \frac{1 + x_0^3 + x_1^3 + \cdots + x_{n-1}^3}{n}.$$

For example, $x_0 = 1$, $x_1 = (1 + 1^3)/1 = 2$, and $x_2 = (1 + 1^3 + 2^3)/2 = 5$.

Surprisingly enough, x_n is integral for $0 \le n < 89$, but x_{89} is *not*.

FIBONACCI PRIMES

Zerger also observed that the product $F_6 F_7 F_8 F_9$ is the product of the first seven prime numbers: $F_6 F_7 F_8 F_9 = 13 \cdot 21 \cdot 34 \cdot 55 = 510,510 = 2 \cdot 3 \cdot 5 \cdot 7 \cdot 11 \cdot 13 \cdot 17$. Interestingly enough, 510 is the Dewy Decimal Classification Number for Mathematics.

FIBONACCI AND LUCAS PRIMES

Many Fibonacci and Lucas numbers are indeed primes. For example, the Fibonacci numbers 2, 3, 5, 13, 89, 233, and 1597 are primes; and so are the Lucas numbers 3, 7, 11, 29, 47, 199, and 521. Although it is widely believed that there are infinitely many Fibonacci and Lucas primes, their proofs still remain elusive.

The largest known Fibonacci prime is F_{81839}. Discovered in 2001 by Walter Broadhurst and Bouk de Water, it is 17,103 digits long. The largest known

Fibonacci prime with two distinct digits is 233, discovered by Patrick Capelle. The largest known Lucas prime is L_{56003}, discovered by Broadhurst and Sean A. Irvine in 2006. It has 11,704 decimal digits. (In Chapter 5, we will discuss a method for computing the number of digits in both F_n and L_n.)

Table A.3 lists the canonical prime factorizations of the first 100 Fibonacci numbers. Lucas had found the prime factorizations of the first 60 Fibonacci numbers before March 1877 and most likely even earlier. For instance, the largest prime among the first 100 Fibonacci numbers is F_{83}.

Table A.4 gives the complete prime factorizations of the first 100 Lucas numbers.

CUNNINGHAM CHAINS

A *Cunningham chain*, named after the British Army officer Lt. Col. Allan J.C. Cunningham (1842–1928), is a sequence of primes in which each element is one more than twice its predecessor. Interestingly, the smallest six-element chain begins with 89: 89–179–359–719–1439–2879.

Are there Fibonacci and Lucas numbers that are one more than or one less than a square? Less than a cube? We will find the answers shortly.

FIBONACCI AND LUCAS NUMBERS $w^2 \pm 1, w \geq 0$

In 1973, Finkelstein established yet another curiosity: The only Fibonacci numbers of the form $w^2 + 1$, where $w \geq 0$, are 1, 2, and 5: $1 = 0^2 + 1, 2 = 1^2 + 1$, and $5 = 2^2 + 1$ [174].

Two years later, he proved that the only Lucas numbers of the same form are 1 and 2: $1 = 0^2 + 1$ and $2 = 1^2 + 1$ [175].

In 1981, N.R. Robbins of Bernard M. Baruch College, New York, proved that the only Fibonacci numbers of the form $w^2 - 1$, where $w \geq 0$, are 3, and 8: $3 = 2^2 - 1$ and $8 = 3^2 - 1$ [500]. The only such Lucas number is 3.

FIBONACCI AND LUCAS NUMBERS $w^3 \pm 1, w \geq 0$

In the same year, Robbins also determined all Fibonacci and Lucas numbers of the form $w^3 \pm 1$, where $w \geq 0$ [500]. There are two Fibonacci numbers of the form $w^3 + 1$, namely, 1 and 2: $1 = 0^3 + 1$ and $2 = 1^3 + 1$. There are two such Lucas numbers: 1 and 2.

There are *no* Fibonacci numbers of the form $w^3 - 1$, where $w \geq 0$. But there is exactly one such Lucas number, namely, 7: $7 = 2^3 - 1$.

FIBONACCI NUMBERS $(a^3 \pm b^3)/2$

Certain Fibonacci numbers can be expressed as one-half of the sum or difference of two cubes. For example, $1 = (1^3 + 1^3)/2, 8 = (2^3 + 2^3)/2$, and $13 = (3^3 - 1^3)/2$. In fact, at the 1969 Summer Institute of Number Theory at Stony Brook, New York, H.M. Stark of the University of Michigan at Ann Arbor asked: *Which Fibonacci numbers have this distinct property?* This problem is linked to the finding of all complex quadratic fields with class 2. In 1983, J.A. Antoniadis tied such fields to solutions of certain diophantine equations.

FIBONACCI NUMBERS THAT ARE LUCAS

There are exactly three distinct Fibonacci numbers that are also Lucas numbers: $1 = L_1, 2 = L_0$, and $3 = L_2$; see Example 5.5. S. Kravitz of Dover, New Jersey, established this in 1965 [373].

FIBONACCI NUMBERS IN ARITHMETIC PROGRESSION

Are there four distinct positive Fibonacci or Lucas numbers that are in arithmetic progression? Unfortunately, the answer is *no*; see Exercise 2.21.

FIBONACCI AND LUCAS TRIANGULAR NUMBERS

A *triangular number* is a positive integer of the form $n(n + 1)/2$. The first five triangular numbers are 1, 3, 6, 10, and 15; they can be represented geometrically, as Figure 2.3 shows.

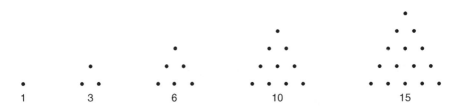

1   3   6   10   15

Figure 2.3. The first five triangular numbers.

In 1963, M.H. Tallman of Brooklyn, New York, observed that the Fibonacci numbers 1, 3, 21, and 55 are triangular numbers: $1 = (1 \cdot 2)/2, 3 = (2 \cdot 3)/2$,

$21 = (6 \cdot 7)/2$, and $55 = (10 \cdot 11)/2$ [555]. He asked if there were any other Fibonacci numbers that are also triangular [555].

Twenty-two years later, C.R. Wall of Trident Technical College, South Carolina, established that there are *no* other Fibonacci numbers in the first one billion Fibonacci numbers. In fact, he conjectured there are no other such Fibonacci numbers [584].

In 1976, Finkelstein proved that 1, 3, 21, and 55 are the only triangular Fibonacci numbers of the form F_{2n} [176].

Eleven years later, L. Ming of Chongqing Teachers' College, China, established conclusively that 1, 3, 21, and 55 are the only triangular Fibonacci numbers [442]. This result is a byproduct of the two following results by Ming:

- $8F_n + 1$ is a square if and only if $n = 0, \pm 1, 2, 4, 8, 10$.
- F_n is triangular if and only if $n = \pm 1, 2, 4, 8, 10$.

Are there Lucas numbers that are also triangular? Obviously, 1 and 3 are. In fact, in 1990 Ming also established that 1, 3, and 5778 are the only such Lucas numbers: $1 = (1 \cdot 2)/2, 3 = (2 \cdot 3)/2$, and $5778 = (107 \cdot 108)/2$.

FIBONACCI AND LUCAS EVEN PERFECT NUMBERS

Ming's results have an interesting byproduct to the study of even perfect numbers $2^{p-1}(2^p - 1)$, where both p and $2^p - 1$ are prime [369]. Every perfect number is triangular and the first four even perfect numbers are 6, 28, 496, and 8128. Since the triangular numbers 1, 3, 21, 55, and 5778 are not even perfect numbers, it follows that *no* Fibonacci or Lucas numbers are even perfect numbers.

Wall reached the same conclusion in 1968 using congruences [583].

FIBONACCI AND THE BEASTLY NUMBER

In 1989, S. Singh of St. Laurent's University in Quebec, Canada, discovered some intriguing relationships between the infamous *beastly number* 666 and Fibonacci numbers [528]:

- $666 = F_{15} + F_{11} - F_9 + F_1$, where $15 + 11 - 9 + 1 = 6 + 6 + 6$.

- $666 = F_1^3 + F_2^3 + F_4^3 + F_5^3 + F_6^3$, where the sum of the subscripts equals $1 + 2 + 4 + 5 + 6 = 6 + 6 + 6$.

- $666 = \left[F_1^3 + (F_2 + F_3 + F_4 + F_5)^3 \right] / 2$.

FIBONACCI AND *THE DA VINCI CODE*

To the delight of math lovers everywhere, the deranged sequence 13–3–2–21–1–1–8–5 of the first eight Fibonacci numbers plays an important role in D. Brown's bestseller, *The Da Vinci Code* [71]. The sequence is one of the clues left behind by the murdered museum curator Jacques Saunière for his granddaughter Sophie Neveu (nicknamed Princess Sophie), a cryptanalyst:

> 13-3-2-21-1-1-8-5
> O, Draconian devil!
> Oh, lame saint!

This scrambled sequence appears three times in the novel, and the correct version twice, Fibonacci numbers in both cases. The correct version happens to be the key that opens Saunière's mysterious safe-deposit box.

The term *Fibonacci sequence* occurs 12 times in the book and *Fibonacci numbers* four times. Sophie says the Fibonacci sequence "happens to be one of the most famous mathematical progressions in history," and describes the Fibonacci recurrence. The beastly number appears twice in the book.

FIBONACCI AND *EUREKA*

Eureka is a fantastic show produced by Chamber Theatre Productions of Boston, Massachusetts, which plays every year around the U.S. It is a delightful production that promotes, among other topics, the beauty of Fibonacci numbers, "the golden ratio, and the unlikely bond between pineapples, pinecones, and rabbit hutches." Although *Eureka* is primarily aimed at teenagers, even non-mathematically inclined adults will find it entertaining. Visit www.chambertheatre.com for details.

ENDS OF FIBONACCI AND LUCAS NUMBERS

Infinitely many Fibonacci numbers end in 0; they are of the form F_{50n}. But *not* a single Lucas number ends in 0; see Chapter 23.

MATHEMATICAL TOURIST ATTRACTIONS

Finally, take a good look at Figures 2.4 and 2.5. Both exemplify the popularity of Fibonacci numbers, and are tourist attractions, especially for the mathematically inclined. Figure 2.4 is a chimney at the Turun Energia Power Plant, Turku, Finland. Figure 2.5 shows the first 29 Fibonacci numbers on a building in Sweden.

Figure 2.4. The Fibonacci chimney at the Turun Energia Power Plant. *Source*: https://en.wikipedia. org/wiki/Fibonacci_numbers_in_popular_culture#/media/File:Fibonacci.JPG.

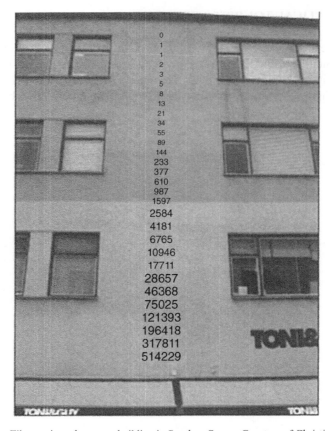

Figure 2.5. Fibonacci numbers on a building in Sweden. *Source*: Courtesy of Christian Liljeberg.

EXERCISES 2

1. Determine the value of L_0.
2. Using Fibonacci recurrence, compute the value of F_{-n}, where $1 \le n \le 10$.
3. Using Exercise 2.2, predict the value of F_{-n} in terms of F_n.
4. Compute the value of L_{-n}, where $1 \le n \le 10$.
5. Using Exercise 2.4, predict the value of L_{-n} in terms of L_n.

To commemorate the publication of the maiden issue of the *Journal of Recreational Mathematics*, L. Bankoff of Los Angeles published his discovery that $F_{20} - F_{19} - F_{15} - F_5 - F_1 = F_{17} + F_{13} + F_{11} + F_9 + F_7 + F_3$, and that each sum gives the year.

6. Find the year in which the journal was first published.

7. Verify that the sums of the subscripts of the Fibonacci numbers on both sides are equal.

Let c be an integer ≥ 2. Define a sequence $\{g_n\}_{n\geq 0}$ recursively by $g_{n+1} = g_n g_{n+1}$, where $g_0 = 1$ and $g_1 = c$.

8. Predict an explicit formula for g_n.

9. Prove the formula in Exercise 2.8.

Predict a formula for each sum.

10. $\displaystyle\sum_{i=1}^{n} F_i$.

11. $\displaystyle\sum_{i=1}^{n} L_i$.

12. $\displaystyle\sum_{i=1}^{n} F_i^2$.

13. $\displaystyle\sum_{i=1}^{n} L_i^2$.

14. Verify that $F_{2n} = F_n L_n$ for $n = 3$ and $n = 8$.

15. Verify that $L_n = F_{n-1} + F_{n+1}$ for $n = 4$ and $n = 7$.

16. Prove that $nL_{n+1} > (n+1)L_n$, where $n \geq 3$ (Shannon, 1969 [521]).

17. Prove that $F_{n+2} < 2^n$, where $n \geq 3$ (Fuchs, 1964 [202]).

Let a_n denote the number of additions needed to compute F_n recursively.

18. Define a_n recursively.

19. Show that $a_n = F_n - 1$, where $n \geq 1$.

20. Prove that $F_n < 1.75^n$ for every positive integer n (LeVeque, 1962 [389]).

21. Show that there are *no* four distinct Fibonacci numbers in arithmetic progression (Silverman, 1964 [525]).

22. Let $I_n = \int_0^1 x^{I_n-1} dx$, where $n \geq 2$, and $I_1 = \int_0^1 x\, dx$. Evaluate I_n (Lind, 1965 [394]).

23. If $F_n < x < F_{n+1} < y < F_{n+2}$, then $x + y$ cannot be a Fibonacci number (Hoggatt, 1980 [304]).

Suppose we introduce a mixed pair of 1-month-old rabbits into a large enclosure on the first day of a certain month. By the end of each month, the rabbits become mature and each pair produces $k - 1$ mixed pairs of offspring at the beginning of the following month, where $k \geq 2$. For example, at the beginning of the second month, there is one pair of 2-month-old rabbits, and $k - 1$ pairs of 0-month-olds; and at the beginning of the third month, there is one pair of 3-month-olds, $k - 1$ pairs of 1-month-olds, and $k(k - 1)$ pairs of 0-month-olds. Assume the rabbits are immortal. Let a_n denote the average age of the rabbit pairs at the beginning of the nth month (Filipponi and Singmaster, 1990 [172]).

*24. Define a_n recursively.

*25. Predict an explicit formula for a_n.

*26. Prove the formula in Exercise 2.25.

*27. Find $\lim\limits_{n \to \infty} a_n$.

FIBONACCI NUMBERS IN NATURE

Come forth into the light of things,
let Nature be your teacher.
–William Wordsworth (1770–1850)

The fabulous Fibonacci numbers occur in nature in quite unexpected places. In this chapter, we will cite a number of such occurrences to pique your curiosity. You may come across others as well.

FIBONACCI AND THE EARTH

Do Fibonacci numbers appear anywhere? Zerger observed that the equatorial diameter of the Earth in miles is approximately the product of two alternate Fibonacci numbers, and this, in kilometers, is approximately the product of two consecutive Fibonacci numbers: $55 \cdot 144 = 7920$ miles and $89 \cdot 144 = 12,816$ kilometers. In addition, the Earth's diameter, according to *The 2000 World Almanac and Book of Facts*, is 7928 miles or 12,756 kilometers; the polar diameter is 7901 miles. The diameter of Jupiter, the largest planet, is 11 times that of the Earth.

Fibonacci and Lucas Numbers with Applications, Volume One, Second Edition. Thomas Koshy.
© 2018 John Wiley & Sons, Inc. Published 2018 by John Wiley & Sons, Inc.

FIBONACCI AND ILLINOIS

In 1992, Zerger discovered some astonishing occurrences of Fibonacci numbers in relation to the state of Illinois [618]:

- Illinois was admitted to the Union on the 3rd of December.
- Illinois was the 5th largest state by population, according to the 1990 census.
- Illinois' name consists of 8 letters.
- Illinois is the 13th state, when the states are arranged alphabetically.
- Illinois was the 21st state admitted to the Union. The postal abbreviation IL is formed with the ninth and twelfth letters: $9 + 12 = 21$.
- Interstate 55 begins in Chicago and roughly follows the 89th parallel to New Orleans.

3.1 FIBONACCI, FLOWERS, AND TREES

The number of petals in many flowers is often a Fibonacci number. For instance, count the number of petals in the flowers pictured in Figure 3.1. Enchanter's nightshade has two petals, iris and trillium three, wild rose five, and delphinium and cosmos eight. Most daisies have 13, 21, or 34 petals; there are even daisies

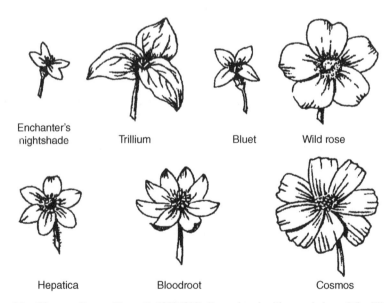

Figure 3.1. Flowers. *Source*: Hoggatt, 1979 [303]. Reproduced with permission of the Fibonacci Association.

TABLE 3.1. An Assortment of Flowers

Plant	Number of petals
Enchanter's nightshade	2
Iris, lily	3
Buttercup, columbine, delphinium, larkspur, wall lettuce	5
Celandine, delphinium, field senecio, squalid senecio	8
Chamomile, cineraria, corn marigold, double delphinium, globeflower	13
Aster, black-eyed Susan, chicory, doronicum, helenium, hawkbit	21
Daisy, gailliardia, plantain, pyrethrum, hawkweed	34

with 55 and 89 petals. Table 3.1 lists the Fibonacci number of petals in an assortment of flowers. Although some plants, such as buttercup and iris, always display the same number of petals, some do not. For example, delphinium blossoms sometimes have 5 petals and sometimes have 8 petals; and some Michaelmas daisies have 55 petals, while some have 89 petals.

The cross section of an apple reveals a pentagonal shape with five pods. The starfish, with five limbs, also exhibits a Fibonacci number; see Figure 3.2. Interestingly, the most common number of petals in flowers is five; Figure 3.3 shows an assortment of such flowers.

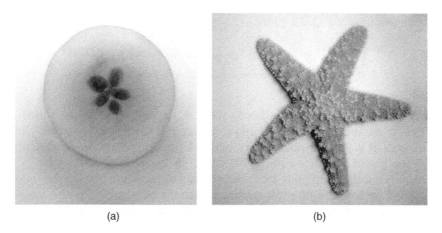

(a)   (b)

Figure 3.2. (a) Cross section of an apple. (b) Starfish.

FIBONACCI AND TREES

Fibonacci numbers are also found in some spiral arrangements of leaves on the twigs of plants and trees. From any leaf on a branch, count up the number of

Figure 3.3. An assortment of flowers with five petals.

leaves until you reach the leaf directly above it; the number of leaves is often a Fibonacci number. On basswood and elm trees, this number is 2; on beech and hazel trees, it is 3; on apricot, cherry, and oak trees it is 5; on pear and poplar trees, it is 8; and on almond and willow trees, it is 13; see Figure 3.4.

Here is another intriguing fact. The number of turns, clockwise or counter-clockwise, we can take from the starting leaf to the terminal leaf is also usually a Fibonacci number. For example, on basswood and elm trees, it takes one turn; for beech and hazel trees, it is also 1; for apricot, cherry, and oak trees, it is 2; for pear and poplar trees, it is 3; and for almond and willow trees it is 5.

The arrangement of leaves on the branches of trees is *phyllotaxis*[†]. Accordingly, the ratio of the number of turns to that of leaves is called the *phyllotactic ratio* of the tree. Thus the phyllotactic ratio of basswood and elm is 1/2; for beech

[†]The word *phyllotaxis* is derived from the Greek words *phyllon*, meaning *leaf*, and *taxis*, meaning *arrangement*.

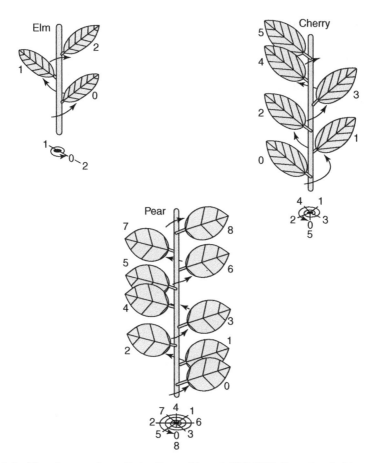

Figure 3.4. Elm, cherry, and pear limbs. *Source*: Hoggatt, 1979 [303]. Reproduced with permission of the Fibonacci Association.

and hazel, it is 1/3; for apricot, cherry, and oak, it is 2/5; for pear and poplar, it is 3/8; and for almond and willow, it is 5/13. These data are summarized in Table 3.2. As an example, it takes 3/8 of a full turn to reach from one leaf to the next leaf on a pear tree.

TABLE 3.2. Phyllotactic Ratios

Tree	Number of turns	Number of leaves	Phyllotactic ratio
Basswood, elm	1	2	1/2
Beech, hazel	1	3	1/3
Apricot, cherry, oak	2	5	2/5
Pear, poplar	3	8	3/8
Almond, willow	5	13	5/13

FIBONACCI AND SUNFLOWERS

Mature sunflowers display Fibonacci numbers in a unique and remarkable way. The seeds of the flower are tightly packed in two distinct spirals, emanating from the center of the head to the outer edge; see Figures 3.5 and 3.6. One goes clockwise, and the other counterclockwise. Studies have shown that although there are

Figure 3.5. A sunflower. *Source*: Reproduced with permission of the Fibonacci Association.

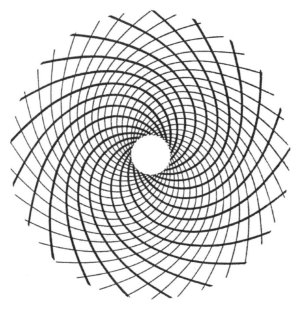

Figure 3.6. The spiral pattern in a sunflower. *Source*: Huntley, 1970 [344]. Reproduced with permission of Dover Publications.

exceptions, the numbers of spirals, by and large, are adjacent Fibonacci numbers; usually, they are 34 and 55. Hoggatt reported a large sunflower with 89 spirals in the clockwise direction and 55 in the opposite direction, and a gigantic one with 144 spirals clockwise and 89 counterclockwise.

It is interesting to note that Br. Brousseau once gave Hoggatt a sunflower with 123 clockwise spirals and 76 counterclockwise spirals, two adjacent Lucas spirals.

In 1951, John C. Pierce of Goddard College, Beverly, Massachusetts reported in *The Scientific Monthly* [484] that the Russians had grown a sunflower head with 89 and 144 spirals. After reading his article on Fibonacci numbers, Margaret K. O'Connell and Daniel T. O'Connell of South Londonderry, Vermont, examined their sunflowers, grown from Burpee's seeds [458]. They found heads with 55 and 89 spirals, some with 89 and 144 spirals, and one giant one with 144 and 233 spirals. The latter seems to be a world record.

FIBONACCI, PINECONES, ARTICHOKES, AND SUNFLOWERS

The scale patterns on pinecones, artichokes, and pineapples provide additional visual occurrences of Fibonacci numbers. The scales are modified leaves closely packed on short stems, and they form two sets of spirals, called *parastichies*[†]. Some spirals run clockwise and the rest counterclockwise, as on a sunflower. Spiral numbers are often adjacent Fibonacci numbers. Some cones have 3 clockwise spirals and 5 counterclockwise spirals; some have 5 and 8; and some 8 and 13. Figures 3.7 and 3.8 show the scale patterns on two pinecones.

Interestingly, some pinecones display three different patterns. Their numbers, as you would expect, are also adjacent Fibonacci numbers.

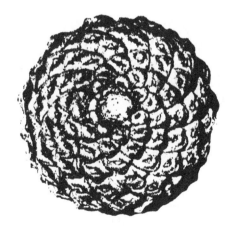

Figure 3.7. Pinecone. *Source*: Reproduced with permission of American Museum of Natural History Library.

[†]The word *parastichies* is derived from the Greek words *para*, meaning *beside*, and *stichos*, meaning *row*.

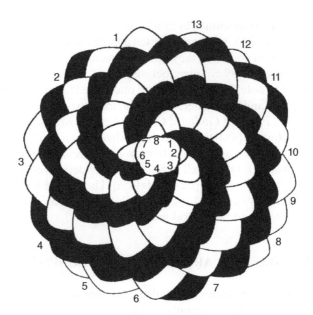

Figure 3.8. Scale patterns (8 clockwise spirals, 13 counterclockwise spirals). *Source*: Hoggatt, 1979 [303]. Reproduced with permission of the Fibonacci Association.

Artichokes show a similar pattern: the number of spirals in the two directions often are adjacent Fibonacci numbers. Usually, there are 3 clockwise and 5 counterclockwise spirals, or 5 clockwise and 8 counterclockwise spirals. Figure 3.9 shows two artichokes of the latter type.

Figure 3.9. Artichokes. *Source*: Garland, 1987 [213]. Reproduced with permission of Pearson Education, Inc.

The scales on pineapples are nearly hexagonal in shape; see Figure 3.10. Since hexagons tessellate a plane perfectly and beautifully (see Figure 3.11), the scales form three different spiral patterns. Once again, the numbers of spirals are the adjacent Fibonacci numbers 8, 13, and 21. According to the *1977 Yearbook of Science and the Future*, a careful study of 2000 pineapples confirmed this unusual Fibonacci pattern.

Figure 3.10. Pineapple.

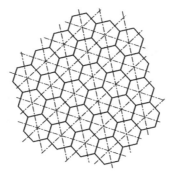

Figure 3.11. Hexagonal scale patterns. *Source*: Hoggatt, 1979 [303]. Reproduced with permission of the Fibonacci Association.

3.2   FIBONACCI AND MALE BEES

Male bees come from unfertilized eggs, so a male bee (M) has a mother but no father. A female bee (F), on the other hand, is developed from a fertilized egg, so it has both parents. Figure 3.12 shows the genealogical tree of a drone for seven generations. Count the total number of bees at each level, that is, in each generation. It is a Fibonacci number, as Table 3.3 demonstrates. Notice that it looks very much like Table 2.1.

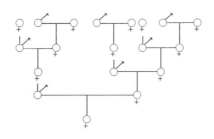

♀ Female

♂ Male

Figure 3.12.   The family tree of a male bee.

TABLE 3.3.   Number of Bees per Generation

Generation	1	2	3	4	5	6	7	8
Number of female bees	0	1	1	2	3	5	8	13
Number of male bees	1	0	1	1	2	3	5	8
Total number of bees	1	1	2	3	5	8	13	21

Let a_n, b_n, and t_n denote the number of female bees, the number of male bees, and the total number of bees in generation n, respectively, where $n \geq 1$. Clearly, $a_1 = 0$ and $b_1 = 1$. Since drones have exactly one parent, it follows that $b_n = a_{n-1}$, $a_n = a_{n-1} + b_{n-1}$, and $t_n = a_n + b_n$.

Since $a_1 = 0$ and $a_2 = 1$, and $a_n = a_{n-1} + a_{n-2}$, it follows that $a_n = F_{n-1}$. Then $t_n = a_n + b_n = a_n + a_{n-1} = a_{n+1}$, where $t_1 = a_2 = 1$ and $t_2 = a_3 = 1$; so $t_n = F_n$. Thus the number of ancestors of the drone in generation n is F_n. This fascinating relationship, originally presented by W. Hope-Jones in 1921 [322], is examined further in Chapter 18.

FIBONACCI AND BEES

Consider two adjacent rows of cells in an infinite beehive, as pictured in Figure 3.13. We would like to find the number of paths the bee can take to crawl from

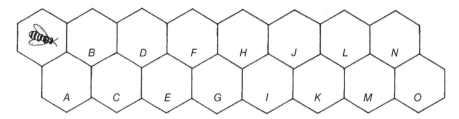

Figure 3.13.

one cell to an adjacent one. It can move in only one general direction, namely, to the right.

Let b_n denote the number of different paths to the nth cell. Since there is exactly one path to cell $A, b_1 = 1$; see Figure 3.14. There are two distinct paths to cell B, as Figure 3.15 shows. So $b_2 = 2$. There are three distinct paths the bee can take to cell C; so $b_3 = 3$; see Figure 3.16. There are five distinct paths the bee can take to cell D; see Figure 3.17. Consequently, $b_4 = 5$. Likewise, $b_5 = 8$.

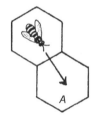

Figure 3.14.

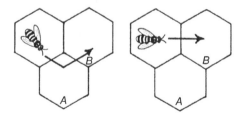

Figure 3.15.

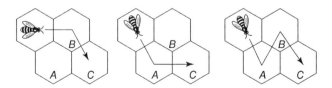

Figure 3.16.

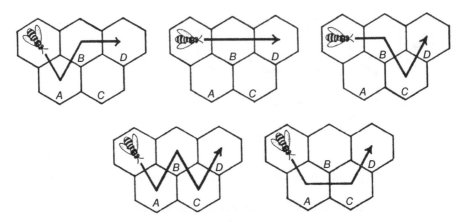

Figure 3.17.

Clearly, a pattern emerges, as Table 3.4 shows. It follows inductively that there are $b_n = F_{n+1}$ distinct paths for the bee to crawl to cell n; see Exercise 3.1.

TABLE 3.4.   Number of Paths

n	1	2	3	4	5	...	n
b_n	1	2	3	5	8	...	?

The next application was conceived in 1972 by L. Carlitz of Duke University, Durham, North Carolina [114].

3.3   FIBONACCI, LUCAS, AND SUBSETS

Example 3.1. *Find the number of subsets, including the null set, of a set of n points such that consecutive points are* not *allowed if the points lie on*

1) *a line; and*
2) *a circle.*

Solution.

1) We will denote the n points by $1, 2, 3, \ldots, n$. Suppose they are linear. Let A_n denote the number of subsets. It follows from Figure 3.18 that $A_1 = 2$ and $A_2 = 3$.

          1   1   2

Subsets: ∅, {1}   Subsets: ∅, {1}, {2}

Figure 3.18.

Let $n \geq 3$. Let n be the extreme point, so it has just one neighbor; see Figure 3.19. By definition, there are A_{n-1} subsets that do not contain n, and A_{n-2} subsets that contain n. Thus, by the addition principle, $A_n = A_{n-1} + A_{n-2}$. This, coupled with the initial conditions, implies that $A_n = F_{n+2}$, where $n \geq 1$. (Notice the similarity between this example and Example 4.1.)

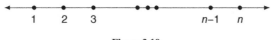

Figure 3.19.

2) Suppose the n points lie on a circle. Let B_n denote the number of subsets that do *not* contain consecutive points. It follows by Figure 3.20 that $B_1 = 2$, $B_2 = 3$, and $B_3 = 4$.

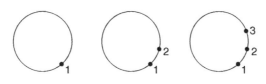

Figure 3.20.

Now let $n \geq 4$. Consider the point n; see Figure 3.21. There are A_{n-1} subsets that do not contain n, and A_{n-3} subsets that contain n. So $B_n = A_{n-1} + A_{n-3} = F_{n+1} + F_{n-1} = L_n$, where $n \geq 2$.

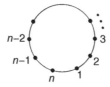

Figure 3.21.

■

3.4 FIBONACCI AND SEWAGE TREATMENT

The next example deals with sewage treatment. It was studied by R.A. Deninger of the University of Michigan at Ann Arbor in 1972 [143].

Example 3.2. *There are n towns on the bank of a river. They discharge untreated sewage into the stream and pollute the water, so the towns would like to build treatment plants to control pollution. It is economically advantageous to build one or more central treatment plants along the main sewers, and then send the wastewater from each town to another one. It is not economical to split the sewage of a town* *between two adjacent towns, since this would require the building of two sewers for the same town. Find the number of economically feasible solutions.*

Solution. Let $f(n)$ denote the number of economic solutions. Clearly, $f(1) = 1$. Suppose $n = 2$. Then there are three possible solutions: each town has its own plant, one plant in town 1, or one plant in town 2; see Figure 3.22. Thus $f(2) = 3$.

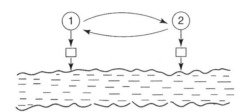

Figure 3.22.

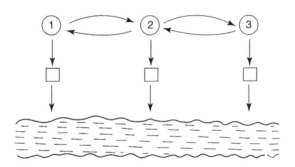

Figure 3.23.

Suppose there are three towns; see Figure 3.23. Since there is no transfer of sewage between adjacent towns, each town can build its own treatment plan, send the sewage upstream ($\rightarrow$), or send it downstream ($\leftarrow$). Let 0 denote no transport between neighboring towns, 1 upstream transport, and 2 downstream transport. Figure 3.24 shows the various economic solutions. They can be represented symbolically as follows: 00 01 02 10 11 12 20 2̸1̸ 22, where 21 is *not* a solution, since a town cannot simultaneously transfer waste both upstream and downstream. Thus $f(3) = 8 = 3f(2) - 1$.

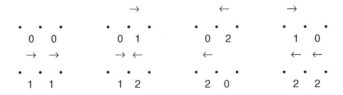

Figure 3.24.

With $n = 4$ towns, there are $f(4) = 21$ solutions:

000	001	002	010	011	012	0~~2~~1	022
100	101	102	110	111	112	1~~2~~1	122
200	201	202	210	211	212	2~~2~~1	222

Notice that $f(4) = 3f(3) - f(2)$, since there are three "words" that end in 21.

More generally, consider $n + 1$ towns with $f(n + 1)$ solutions. Adding one town increases the number of solutions to $3f(n)$. From this we must subtract the number of words ending in 21, namely, $f(n - 1)$. Thus $f(n + 1) = 3f(n) - f(n - 1)$, where $n \geq 2$.

Using this recurrence, the value of $f(n)$ can be computed for various values of n. It appears from Table 3.5 that $f(n) = F_{2n}$.

TABLE 3.5. Number of Economic Solutions

n	1	2	3	4	5	6
$f(n)$	1	3	8	21	55	144

To confirm this formula, notice that $f(1) = 1 = F_2$ and $f(2) = 4 = F_4$. So it suffices to show that F_{2n} satisfies the Fibonacci recurrence:

$$3f(n) - f(n - 1) = 3F_{2n} - F_{2n-2}$$
$$= 2F_{2n} + F_{2n-1}$$
$$= F_{2n} + F_{2n+1}$$
$$= F_{2n+2}$$
$$= f(n + 1).$$

Thus $f(n) = F_{2n}$, where $n \geq 1$, as desired. ∎

3.5 FIBONACCI AND ATOMS

The *atomic number* Z of an atom is the number of protons in it. The periodic table shows an interesting relationship between the atomic numbers of inert gas and Fibonacci numbers; see Table 3.6.

There are six inert gases – helium, neon, argon, krypton, xenon, and radon – and they are exceptionally stable chemically. With the exception of helium, their atomic numbers are approximately the same as the Fibonacci numbers F_7 through F_{11}, as Table 3.6 shows. Compute $\lfloor Z/18 + 1/2 \rfloor$ for each gas. It follows from column 4 that each is a Fibonacci number.

TABLE 3.6. Atomic Number and Fibonacci Number

Inert Gas	Atomic Number Z	Corresponding Fibonacci Number	$\lfloor Z/18 + 1/2 \rfloor$
Helium	2	8	0
Neon	10	13	1
Argon	18	21	1
Krypton	36	34	2
Xenon	54	55	3
Radon	86	89	5

The nucleus of an atom consists of two kinds of particles: protons and neutrons. A proton has a charge equal but opposite to that of an electron, while a neutron is neutral. Let N denote the number of neutrons in the nucleus. Nuclei having the values 2, 8, 14, 20, 28, 50, 82, or 126 for N or Z are considered more stable than others. (The origin of these numbers is a mystery.) Compute $\lfloor N/10 + 1/2 \rfloor$ for each N. Surprisingly, each is again a Fibonacci number; see Table 3.7.

TABLE 3.7.

N	2	8	14	20	28	50	82	126
$\lfloor N/10 + 1/2 \rfloor$	0	1	1	2	3	5	8	13

FIBONACCI AND THE BALMER SERIES

In 1885, the Swiss schoolteacher Johann Jacob Balmer (1825–1898) discovered that the wavelengths (in angstroms) of four lines in the hydrogen spectrum (now known as the *Balmer series*) can be expressed as the product of the constant 364.5 (in nanometers) and a fraction:

$$656 = \frac{9}{5} \times 364.5$$

$$486 = \frac{4}{3} \times 364.5$$

$$434 = \frac{25}{21} \times 364.5$$

$$410 = \frac{9}{8} \times 364.5.$$

Notice that the denominators of the fractions are Fibonacci numbers. J. Wlodarski of Germany made this observation in 1973 [599].

3.6 FIBONACCI AND REFLECTIONS

Fibonacci numbers have made a real world appearance in *optics*, the branch of physics that deals with light and vision. To see this, consider two glass plates placed face to face. Such a stack has four reflective faces, as Figure 3.25 shows.

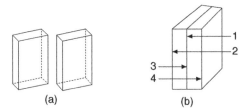

Figure 3.25. (a) Two separate glass plates. (b) The stack has four reflective faces, labeled 1–4.

Suppose a ray of light falls on the stack. We would like to determine the number of distinct reflective paths a_n made with n reflections, where $n \geq 0$ (L. Moser and W. Wyman studied this problem in 1963) [72, 451]. To this end, we first collect some data.

When $n = 0$, there are no reflections. So the ray just passes through the glass plates, as Figure 3.26 shows. So $a_0 = 1$.

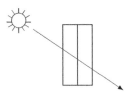

Figure 3.26. Stacked glass plates with no reflections.

Suppose the ray undergoes one reflection. Then there are two distinct possible paths, so $a_1 = 2$; see Figure 3.27.

Figure 3.27. Stacked glass plates with one reflection.

If the ray is reflected twice, three possible paths can emerge; so $a_2 = 3$; see Figure 3.28. If it is reflected thrice, there are five possible reflecting patterns; so $a_3 = 5$; see Figure 3.29. Likewise, $a_4 = 8$; see Figure 3.30.

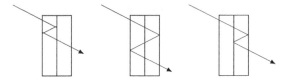

Figure 3.28.   Stacked glass plates with two reflections.

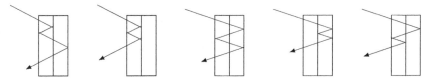

Figure 3.29.   Stacked glass plates with three reflections.

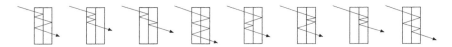

Figure 3.30.   Stacked glass plates with four reflections.

More generally, suppose the ray is reflected n times, so the last reflection occurs at face 1 or 3. Then the previous reflection must have occurred on face 2 or 4; see Figure 3.31. The number of paths with the nth reflection on face 1 equals the number of paths reaching face 1 after $n - 1$ reflections. There are a_{n-1} such paths.

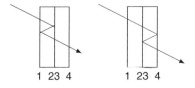

Figure 3.31.

Suppose the nth reflection takes place on face 3. Then the $(n - 1)$st reflection must have occurred on face 4. Such a ray must already have had $n - 2$ reflections before reaching face 4. By definition, there are a_{n-2} such paths.

Thus, by the addition principle, $a_n = a_{n-1} + a_{n-2}$, where $a_1 = 2$ and $a_2 = 3$. Consequently, $a_n = F_{n+2}$.   ∎

Next we will investigate the occurrences of Fibonacci and Lucas numbers in the study of hydrocarbon molecules. To do this, first we will introduce the concept of a graph.

3.7 PARAFFINS AND CYCLOPARAFFINS

Graph theory is a relatively new branch of mathematics. A *graph* is a finite, nonempty set of *vertices* and *edges* (arcs or line segments) [368]. Figures 3.32 and 3.33 depict two graphs.

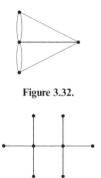

Figure 3.32.

Figure 3.33. Ethane molecule C_2H_6.

Graphs are useful in the study of hydrocarbon molecules. The English mathematician Arthur Cayley (1821–1895) was the first to employ graphs to examine hydrocarbon isomers.

A hydrocarbon molecule consists of hydrogen and carbon atoms. Each hydrogen atom (H) is bonded to a single carbon atom (C), whereas a carbon atom bonds to two, three, or four atoms, which can be carbon or hydrogen. But carbon atoms in saturated hydrogen molecules, such as ethane, contain only single bonds; see Figure 3.33.

Deleting hydrogen atoms from the structural formulas of saturated hydrocarbons yields graphs consisting of carbon atoms and edges between two adjacent vertices. The *topological index* of such a graph with n vertices is the total number of different ways the graph can be partitioned into disjoint subgraphs containing exactly k edges, where $k \geq 0$ [331].

For example, Figure 3.34 shows the carbon atom skeleton of the paraffin pentane, C_5H_{12}; and Figure 3.35 shows its various possible partitionings. Notice that the topological index of pentane is $1 + 4 + 3 = 8 = F_6$.

Figure 3.34.

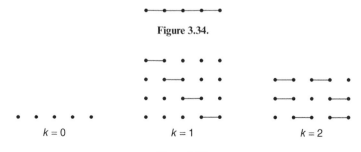

$k = 0$ $k = 1$ $k = 2$

Figure 3.35.

Cayley*

Table 3.8 shows the carbon atom graphs G_n and their topological indices of ten paraffins C_nH_{2n+2}, where $n \geq 1$. For a graph consisting of a single vertex, the index is defined as one. It appears from the table that the index of G_n is F_{n+1}.

To confirm this observation, let t_n denote the topological index of the carbon atom graph G_n of a paraffin with n vertices; see Figure 3.36.

Figure 3.36.

Figure source: https://en.wikipedia.org/wiki/Arthur_Cayley#/media/File:Arthur_Cayley.jpg.

TABLE 3.8. Topological Indices of Paraffins C_nH_{2n+2}

Paraffin	n	Graph	k						Total
			0	1	2	3	4	5	
Methane	1	•	1						1
Ethane	2	•—•	1	1					2
Propane	3	•—•—•	1	2					3
Butane	4	•—•—•—•	1	3	1				5
Pentane	5	•—•—•—•—•	1	4	3				8
Hexane	6	•—•—•—•—•—•	1	5	6	1			13
Heptane	7	•—•—•—•—•—•—•	1	6	10	4			21
Octane	8	•—•—•—•—•—•—•—•	1	7	15	10	1		34
Nonane	9	•—•—•—•—•—•—•—•—•	1	8	21	20	5		55
Decane	10	•—•—•—•—•—•—•—•—•—•	1	9	28	35	15	1	89

$$\uparrow$$
$$F_{n+1}$$

Case 1. Suppose the edge $v_{n-1}-v_n$ is *not* included in the partitioning. Then the edge $v_{n-2}-v_{n-1}$ may or may not be included. Consequently, the topological index of the remaining graph G_{n-1} is t_{n-1}.

Case 2. Suppose the edge $v_{n-1}-v_n$ is included in the partitioning. Then the edge $v_{n-2}-v_{n-1}$ is not included. This yields the graph G_{n-2}, and its index is t_{n-2}.

Thus, by the addition principle, $t_n = t_{n-1} + t_{n-2}$, where $t_1 = 1$ and $t_2 = 2$. So $t_n = F_{n+1}$.

Next we investigate cycloparaffins.

CYCLOPARAFFINS

Table 3.9 shows the carbon atom skeleton A_n of ten cycloparaffins C_nH_{2n}, and the corresponding indices. A similar argument shows that the index of A_n = index of G_n + index of $G_{n-2} = F_{n+1} + F_{n-1}$, where $n \geq 3$. Notice that the index of A_n is in fact the Lucas number L_n. (We will confirm in Chapter 5 that $F_{n+1} + F_{n-1} = L_n$.)

We will revisit the triangular arrays in Tables 3.8 and 3.9 in Chapter 12.

TABLE 3.9. Topological Indices of Cycloparaffins C_nH_{2n}

Cycloparaffin	n	Graph	k						Total
			0	1	2	3	4	5	
•	1	•	1						1
•	2	•—•	1	2					3
Cyclopropane	3	△	1	3					4
Cyclobutane	4	▭	1	4	2				7
Cyclopentane	5	⬠	1	5	5				11
Cyclohexane	6	⬡	1	6	9	2			18
Cycloheptane	7	⬡	1	7	14	7			29
Cyclooctane	8	⯃	1	8	20	16	2		47
Cyclononane	9	⯃	1	9	27	30	9		76
Cyclodecane	10	⯃	1	10	35	50	25	2	123

$$\uparrow$$
$$L_n$$

3.8 FIBONACCI AND MUSIC

Fibonacci numbers occur in music also. Such links were observed by Zerger [618].

- The word MUSIC begins with the 13th and 21st letters of the alphabet. With the 8th, 13th, and 21st letters, we can form the word HUM.
- The Library of Congress Classification Number for Music is M, the 13th letter of the alphabet.
- The Dewey Decimal Classification Number for Music is 780, where $780 = 1 \cdot 2 \cdot 3 \cdot 5 \cdot 13$, a product of Fibonacci numbers.
- Pianos are often tuned to a standard of 440 cycles per second, where $440 = 8 \cdot 55$.

The keyboard of a piano provides a fascinating visual illustration of the link between Fibonacci numbers and music. An octave on a keyboard represents a musical interval between two notes, one higher than the other. The frequency of the higher note is twice that of the lower. On the keyboard, the octave is divided into 5 black and 8 white keys, a total of 13 keys; see Figure 3.37. The five black keys form two groups, one of two keys and the other of three keys.

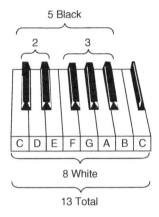

Figure 3.37. Fibonacci numbers in the octave of a piano keyboard. *Source*: Garland, 1987 [213]. Reproduced with permission of Pearson Education, Inc.

The 13 notes in an octave form the chromatic scale, the most popular scale in Western music. The chromatic scale was preceded by two other scales, the 5-note pentatonic scale and the 8-note diatonic scale. Popular tunes such as "Mary had a Little Lamb" and "Amazing Grace" can be played using the pentatonic scale, while melodies such as "Row, Row, Row Your Boat" use the diatonic scale.

The major sixth and the minor sixth (six tones apart and $5\frac{1}{2}$ tones apart, respectively) are the two musical intervals most pleasing to the ear. A major sixth, for example, consists of the notes C and A; they make 264 and 440 vibrations per second, respectively; see Figure 3.38. Notice that $264/440 = 3/5$, a Fibonacci ratio.

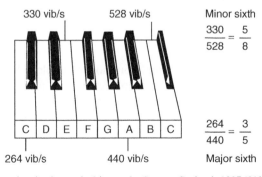

Figure 3.38. Fibonacci ratios in musical intervals. *Source*: Garland, 1987 [213]. Reproduced with permission of Pearson Education, Inc.

A minor sixth interval, for instance, consists of the notes E and C, making 330 and 528 vibrations a second. Their ratio is also a Fibonacci ratio: 330/528 = 5/8.

We will study the ratios of consecutive Fibonacci (and Lucas) numbers in detail in Chapter 16.

Fibonacci numbers have found their way into the art of poetry also. We will explore such occurrences now.

3.9 FIBONACCI AND POETRY

A limerick, according to Webster's dictionary, is a nonsensical poem of 5 lines, of which the first, second, and fifth have 3 beats, and the other two have 2 beats and rhyme. The following limerick [213], for example, is made up of 5 lines; they contain 2 groups of 2 beats and 3 groups of 3 beats, a total of 13 beats. Once again, all numbers involved are Fibonacci numbers.

A fly and a flea in a flue	3 beats
Were imprisoned, so what could they do?	3 beats
Said the fly, "Let us flee!"	2 beats
"Let us fly!" said the flea,	2 beats
So they fled through a flaw in the flue.	3 beats
Total =	13 beats.

In the 1960s, G.E. Duckworth of Princeton University, New Jersey, analyzed the *Aeneid*, an epic poem written in Latin about 20 B.C. by Virgil (70–19 B.C.), the "greatest poet of ancient Rome and one of the outstanding poets of the world" [151, 603]. Duckworth discovered frequent occurrences of the Fibonacci numbers and several variations in this masterpiece:

$$1, 3, 4, 7, 11, \ldots \quad \leftarrow \text{Lucas sequence}$$
$$1, 4, 5, 9, 14, \ldots$$
$$1, 5, 6, 11, 17, \ldots$$
$$1, 6, 7, 13, 20, \ldots$$
$$2, 3, 5, 8, 13, \ldots$$
$$3, 7, 10, 17, 27, \ldots$$
$$4, 9, 13, 22, 35, \ldots$$
$$6, 13, 19, 32, 61, \ldots$$

The mathematical symmetry Virgil consciously employed in composing the *Aeneid* brings the harmony and aesthetic balance of music to the ear, since ancient poetry was written to be read out loud.

According to Duckworth's investigations into Virgil's structural patterns and proportions [151], there is evidence that even Virgil's contemporary poets, such as Catullus, Lucretius, Horace, and Lucan, used the Fibonacci sequence in the structure of their poems. Duckworth's study lends credibility to the theory that the Fibonacci sequence and the Golden Section (see Chapter 16) were known to the ancient Greeks and Romans, although no such mention of it exists.

3.10 FIBONACCI AND NEUROPHYSIOLOGY

In 1976, Kurt Fischer of the University of Regensburg, Germany, studied a model of the physiology of nerves, and discovered yet another occurrence of Fibonacci numbers [177].

Impulses traveling along nerve fibers originate from sodium or potassium ions, and flow through identical transmembrane pores consisting of $n \geq 2$ cells. Tiny quantities of calcium ions, Ca^{2+}, can enter the pores and stop the flow of sodium ions, Na^{+}, in these pores. They can occupy one or two cells, except at the entrance of the pore. These two states are denoted by 1 and 2, respectively. Figure 3.39 shows a typical pore, where 0 denotes an empty cell.

2	0	1	2

Figure 3.39. A sample pore.

Suppose that sodium can enter or leave at either end of a pore, whereas calcium can do so only at the left side of the pore. Consequently, calcium ions within a pore impede the flow of sodium through this pore.

This Markovian stochastic process, named after the Russian mathematician Andrey Andreyevich Markov (1856–1922), can be depicted by a tree structure; the vertices of the tree represent the possible states of a pore and its edges represent the possible transitions between states. Figure 3.40, for example, shows the various possible states of a pore with five nonempty cells.

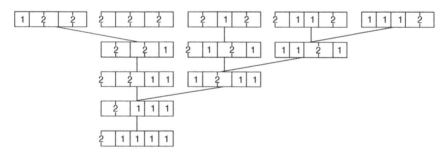

Figure 3.40. Tree of states of a pore with 5 cells.

Notice that the tree consists of two kinds of vertices, those with a 1 in the far right cell or a 2 in the middle of the two right cells. Every state in level five has the latter property, and shows that the translation of sodium ions to the right is no longer feasible because of the presence of calcium on the right side of each state.

Figure 3.41 depicts a tree-skeleton of Figure 3.40, which very much resembles the Fibonacci tree in Figure 2.1. It follows from either figure that a pore with five nonempty cells has $5 = F_5$ states at level 5.

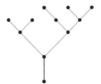

Figure 3.41.   A tree diagram of Figure 3.40.

More generally, a pore with n nonempty cells has F_n states at level n. This follows from the fact that the number of states at level n satisfies the Fibonacci recurrence.

In 1963, S.L. Basin of then San Jose State College, San Jose, California, wrote that "even those people interested in electrical networks cannot escape from our friend Fibonacci" [23]. We will now show how Fibonacci numbers occur in the study of electrical networks.

3.11   ELECTRICAL NETWORKS

Consider a network of n resistors, arranged in the shape of a ladder; see Figure 3.42. We will show that the resistance $Z_o(n)$ (*output impedance*) across the output terminals C and D, the resistance $Z_i(n)$ (*input impedance*) between the input terminals A and B, and the *attenuation* $A(n) = Z_o/Z_i$ are all very closely related to Fibonacci numbers in an unusually special case.

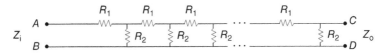

Figure 3.42.   Resistor network: n ladder sections.

First, consider two resistors, R_1 and R_2, arranged in series. Let V denote the voltage drop across a resistor R due to current I; see Figure 3.43. Then $V = IR = I(R_1 + R_2)$. So $R = R_1 + R_2$. On the other hand, suppose the resistors are connected in parallel, as in Figure 3.44. Then $V = I_1 R_1 = I_2 R_2 = (I_1 + I_2)R$, so

$$\frac{1}{R} = \frac{I_1 + I_2}{V} = \frac{I_1}{V} + \frac{I_2}{V} = \frac{1}{R_1} + \frac{1}{R_2}.$$

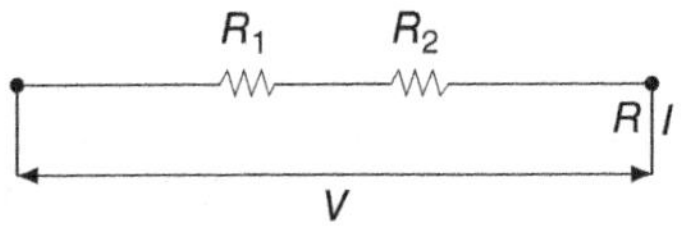

Figure 3.43. 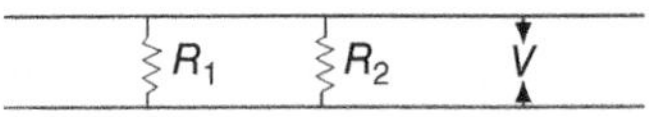   **Figure 3.44.**

Thus, if R_1 and R_2 are connected in parallel, the resultant resistance R is given by

$$\frac{1}{R} = \frac{1}{R_1} + \frac{1}{R_2}.$$

We are now ready to tackle the ladder network problem step by step. Suppose $n = 1$; that is, the network consists of one section; see Figure 3.45. Then $Z_o(1) = R_2$ and $Z_i(1) = R_1 + R_2$, so

$$A(1) = \frac{Z_i(1)}{Z_o(1)} = \frac{R_1}{R_2} + 1. \tag{3.1}$$

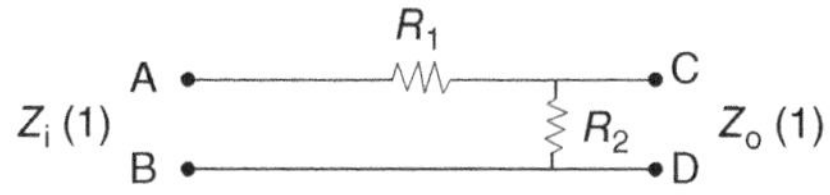

Figure 3.45.

Suppose $n = 2$. The corresponding circuit is obtained by adding a section to the one in Figure 3.45; see Figure 3.46. Since the resistors R_1 and R_2 in the extension are connected in series, they can be replaced by a resistor $R_3 = R_1 + R_2$; this yields the equivalent network in Figure 3.47.

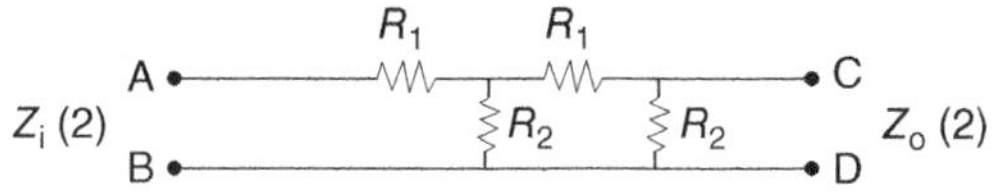

Figure 3.46.

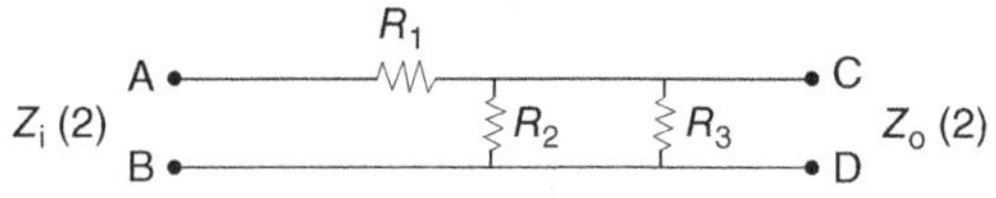

Figure 3.47.

Now R_2 and R_3 are connected in parallel, so they can be replaced by a resistor R_4; see Figure 3.48. Then

$$\frac{1}{R_4} = \frac{1}{R_2} + \frac{1}{R_3}$$

$$= \frac{1}{R_2} + \frac{1}{R_1 + R_2}$$

$$= \frac{R_1 + 2R_2}{R_2(R_1 + R_2)}$$

$$R_4 = \frac{R_2(R_1 + R_2)}{R_1 + 2R_2}.$$

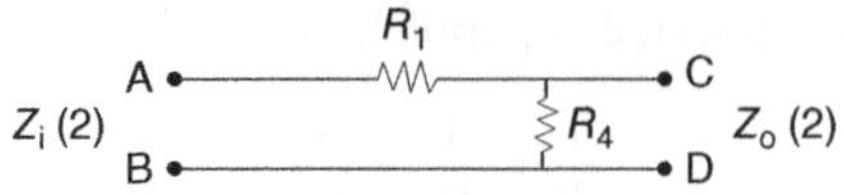

Figure 3.48.

Since the resistors R_1 and R_4 are connected in series, we have

$$
\begin{aligned}
Z_i(2) &= R_1 + R_4 \\
&= R_1 + \frac{R_2(R_1 + R_2)}{R_1 + 2R_2} \\
&= \frac{R_1(R_1 + 2R_2) + R_2(R_1 + R_2)}{R_1 + 2R_2}.
\end{aligned}
\tag{3.2}
$$

To compute the output impedance Z_o of the circuit in Figure 3.46, we traverse it in the opposite direction, that is, from left to right. The first resistor R_1 plays no role in its computation, so we simply ignore it; see Figure 3.49. The resistors R_1 and R_2 are in series, so they can be replaced by a resistor $R_3 = R_1 + R_2$; see Figure 3.50.

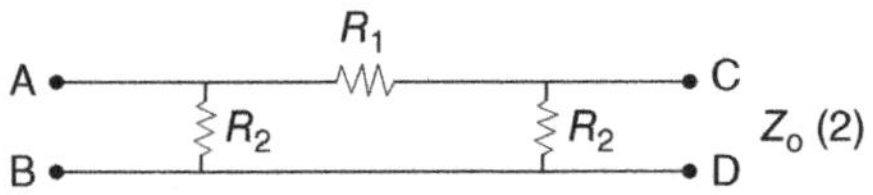

Figure 3.49.

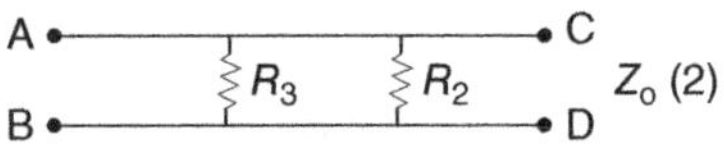

Figure 3.50.

This yields a circuit with two parallel resistors R_3 and R_2, so

$$
\begin{aligned}
\frac{1}{Z_o(2)} &= \frac{1}{R_3} + \frac{1}{R_2} \\
&= \frac{1}{R_1 + R_2} + \frac{1}{R_2} \\
&= \frac{R_1 + 2R_2}{R_2(R_1 + R_2)} \\
Z_o(2) &= \frac{R_2(R_1 + R_2)}{R_1 + 2R_2};
\end{aligned}
\tag{3.3}
$$

see Figure 3.51.

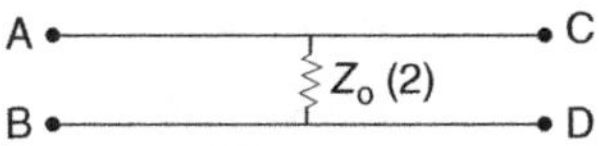

Figure 3.51.

Then

$$A(2) = \frac{Z_o(2)}{Z_i(2)}$$

$$= \frac{R_2(R_1 + R_2)}{R_1(R_1 + 2R_2) + R_2(R_1 + R_2)}. \tag{3.4}$$

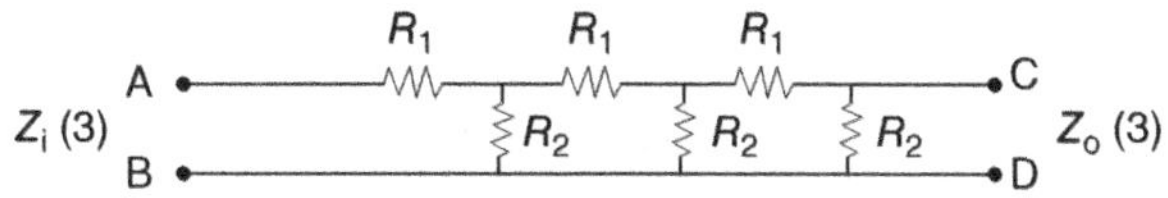

Figure 3.52.

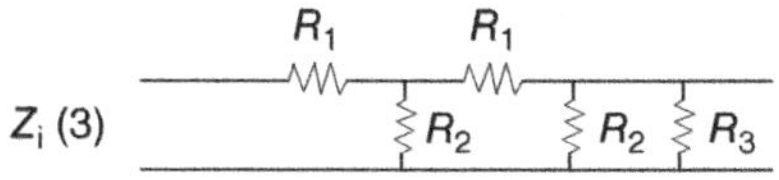

Figure 3.53.

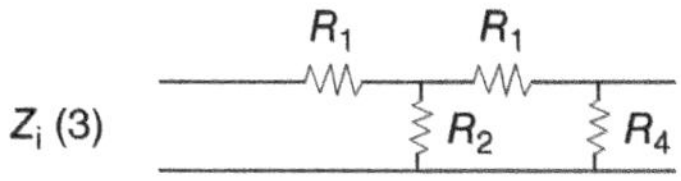

Figure 3.54.

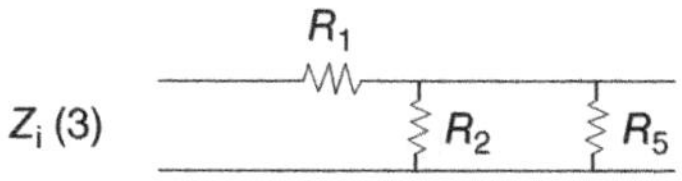

Figure 3.55.

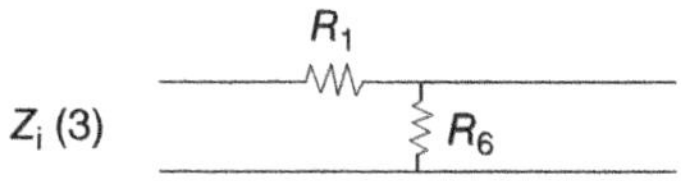

Figure 3.56.

Now consider a ladder with $n = 3$ sections; see Figure 3.52. Using Figures 3.53–3.56, we have

$$R_3 = R_1 + R_2$$

$$R_4 = \frac{R_2(R_1 + R_2)}{R_1 + 2R_2}$$

$$R_5 = R_1 + R_4$$

$$= R_1 + \frac{R_2(R_1 + R_2)}{R_1 + 2R_2}$$

$$= \frac{R_1(R_1 + 2R_2) + R_2(R_1 + R_2)}{R_1 + 2R_2}$$

$$\frac{1}{R_6} = \frac{1}{R_2} + \frac{1}{R_5}$$

$$= \frac{1}{R_2} + \frac{R_1 + 2R_2}{R_1(R_1 + 2R_2) + R_2(R_1 + R_2)}$$

$$R_6 = \frac{R_2(R_1^2 + 3R_1 R_2 + R_2^2)}{R_1^2 + 4R_1 R_2 + 3R_2^2}$$

$$Z_i(3) = R_1 + R_6$$

$$= R_1 + \frac{R_2(R_1^2 + 3R_1 R_2 + R_2^2)}{R_1^2 + 4R_1 R_2 + 3R_2^2}$$

$$= \frac{R_1^3 + 5R_1^2 R_2 + 6R_1 R_2^2 + R_2^3}{R_1^2 + 4R_1 R_2 + 3R_2^2}. \tag{3.5}$$

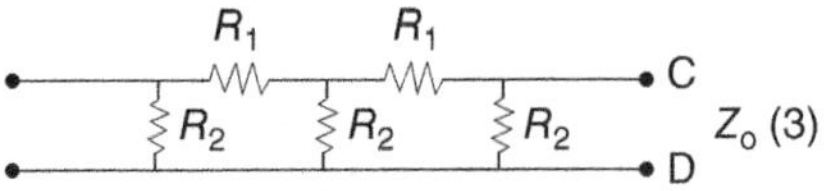

Figure 3.57.

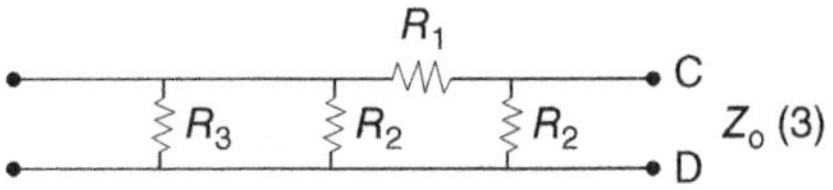

Figure 3.58.

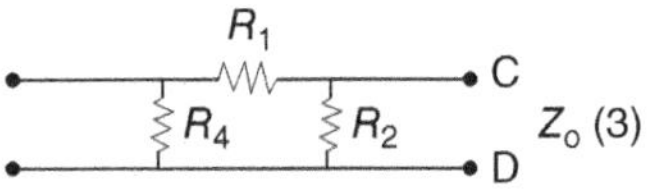

Figure 3.59.

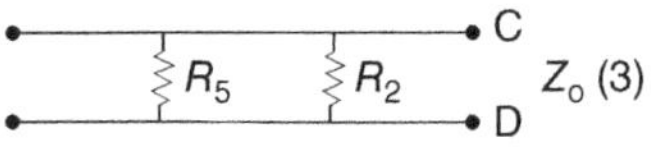

Figure 3.60.

Using the same technique employed for the case $n = 2$, and Figures 3.57–3.60, we get

$$R_3 = R_1 + R_2$$

$$\frac{1}{R_4} = \frac{1}{R_3} + \frac{1}{R_2}$$

$$= \frac{1}{R_1 + R_2} + \frac{1}{R_2}$$

$$R_4 = \frac{R_2(R_1 + R_2)}{R_1 + 2R_2}$$

$$R_5 = R_1 + R_4$$

$$= R_1 + \frac{R_2(R_1 + R_2)}{R_1 + 2R_2}$$

$$= \frac{R_1(R_1 + 2R_2) + R_2(R_1 + R_2)}{R_1 + 2R_2}$$

$$\frac{1}{Z_o(3)} = \frac{1}{R_5} + \frac{1}{R_2}$$

$$= \frac{R_1(R_1 + 2R_2)}{R_1(R_1 + 2R_2) + R_2(R_1 + R_2)} + \frac{1}{R_2}$$

$$= \frac{R_1^2 + 4R_1R_2 + 3R_2^2}{R_2(R_1^2 + 3R_1R_2 + R_2^2)}$$

$$Z_o(3) = \frac{R_2(R_1^2 + 3R_1R_2 + R_1^2)}{R_1^2 + 4R_1R_2 + 3R_2^2}. \tag{3.6}$$

So

$$A(3) = \frac{R_2(R_1^2 + 3R_1R_2 + R_2^2)}{R_1^3 + 5R_1^2R_2 + 6R_1R_2^2 + R_2^3}. \tag{3.7}$$

In particular, let $R_1 = R_2 = 1$ ohm. Then equations (3.1) through (3.7) yield the following:

$$Z_o(1) = \frac{1}{1} \qquad Z_i(1) = \frac{2}{1} \qquad A(1) = \frac{2}{1}$$

$$Z_o(2) = \frac{2}{3} \qquad Z_i(2) = \frac{5}{3} \qquad A(2) = \frac{5}{2}$$

$$Z_o(3) = \frac{5}{8} \qquad Z_i(3) = \frac{13}{8} \qquad A(3) = \frac{13}{5}.$$

More generally, we predict that

$$Z_o(n) = \frac{F_{2n-1}}{F_{2n}}, \quad Z_i(n) = \frac{F_{2n+1}}{F_{2n}}, \quad \text{and} \quad A(n) = \frac{F_{2n+1}}{F_{2n-1}},$$

where $n \geq 1$. This can be confirmed using PMI[†].

To prove that $Z_o(n) = \frac{F_{2n-1}}{F_{2n}}$:

Since $Z_o(1) = 1/1 = F_1/F_1$, the result is true when $n = 1$. Assume it is true for an arbitrary integer $k \geq 1$: $Z_o(k) = F_{2k-1}/F_{2k}$.

Consider a ladder network with $k + 1$ resistors; see Figure 3.61. By the inductive hypothesis, the first k sections can be replaced by a resistor with resistance $Z_o(k)$. This yields the circuit in Figure 3.62.

[†]PMI is an abbreviation of the Principle of Mathematical Induction.

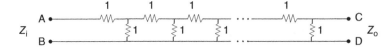

Figure 3.61.   Resistor network: $k + 1$ ladder sections.

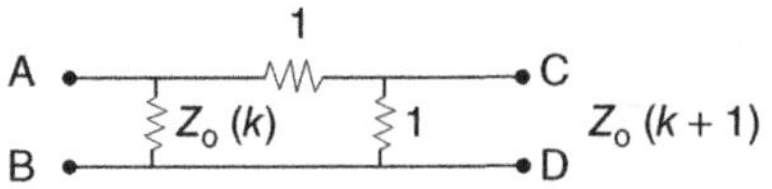

Figure 3.62. 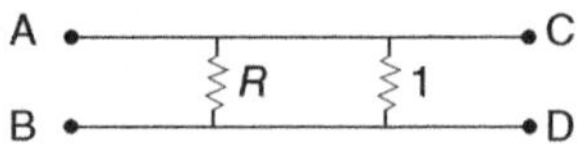   **Figure 3.63.**

Using Figures 3.62 and 3.63, we have

$$R = Z_o(k) + 1$$

$$= \frac{F_{2k-1}}{F_{2k}} + 1$$

$$= \frac{F_{2k+1}}{F_{2k}}$$

$$\frac{1}{Z_o(k+1)} = \frac{1}{R} + \frac{1}{1}$$

$$= \frac{F_{2k}}{F_{2k+1}} + 1$$

$$Z_o(k+1) = \frac{F_{2k+1}}{F_{2k+2}}.$$

So the formula works for $n = k + 1$ also. Thus, by PMI, it works for all $n \geq 1$.   ∎

It can similarly be shown that $Z_i(n) = F_{2n+1}/F_{2n}$ and hence $A(n) = F_{2n+1}/F_{2n-1}$, where $n \geq 1$.

EXERCISES 3

1. Let b_n denote the number of distinct paths the bee in Figure 3.13 can take to cell n. Show that $b_n = F_{n+1}$, where $n \geq 1$.

Exercises 3.2–3.10 require a basic knowledge of binary trees.

The *Fibonacci tree* B_n is a binary tree, defined recursively as follows: Both B_1 and B_2 consist of a single vertex; when $n \geq 3$, B_n has a root, a left subtree B_{n-1}, and a right subtree B_{n-2}.

2. Draw the first five Fibonacci trees.

3. Is B_n a full binary tree?

4. Is B_6 a balanced binary tree?

5. Is B_5 a complete tree?

6. For what values of n is B_n a complete binary tree?

7. How many leaves l_n does B_n have?

Use B_n to find the following.

8. The number of internal vertices i_n.

9. The number of vertices v_n.

10. The height h_n.

Let $f(n, k)$ denote the element in row n and column k of the triangular array in Table 3.8, where $n \geq 1$ and $k \geq 0$.

11. Find $f(7, 2)$ and $f(10, 4)$.

12. Define $f(n, k)$ recursively.

Let $g(n, k)$ denote the element in row n and column k of the triangular array in Table 3.9, where $n \geq 1$ and $k \geq 0$.

13. Find $g(7, 2)$ and $g(10, 4)$.

14. Define $g(n, k)$ recursively.

ADDITIONAL FIBONACCI AND LUCAS OCCURRENCES

A thing of beauty is a joy forever:
Its loveliness increases; it will never
Pass into nothingness.

–John Keats (1795–1821)

Fibonacci and Lucas numbers appear in many other unexpected places. For example, index cards come in size 2×3 or 3×5; most rugs come in five different sizes: 2×3, 3×5, 4×6, 6×9, or 9×12. In the first two cases, the dimensions are adjacent Fibonacci numbers; in the third and fourth cases, the ratio $4:6 = 6:9 = 2:3$, a Fibonacci ratio; and in the last case, $9:12 = 3:4$, the ratio of two adjacent Lucas numbers. We will study such ratios in Chapter 16.

4.1 FIBONACCI OCCURRENCES

Before turning to our first example, we will make a few formal definitions. A *word* is an ordered arrangement of symbols; it does *not* need to have a meaning. For example, *abc* is a word using the letters of the English alphabet, whereas 001101 is a *binary word*. A *bit* is a 0 (zero) or a 1 (one). The word *bit* is a contraction of the phrase *bi*nary dig*it*. The *length* of a word is the number of symbols in it. The *empty word* is of length zero, denoted by the Greek letter λ (lowercase lambda).

We are now ready for a number of interesting examples; always look for similarities.

Fibonacci and Lucas Numbers with Applications, Volume One, Second Edition. Thomas Koshy.
© 2018 John Wiley & Sons, Inc. Published 2018 by John Wiley & Sons, Inc.

Example 4.1. *Let a_n denote the number of n-bit words that do* not *contain two consecutive 1s. Find a_n.*

Solution. First, we collect such *n*-bit words corresponding to $n = 1, 2, 3$, and 4; see Table 4.1. It follows from the table that $a_0 = 1$, $a_1 = 2$, $a_2 = 3$, $a_3 = 5$, and $a_4 = 8$.

TABLE 4.1.

n	*n*-bit Words							Count
0	λ							1
1	0	1						2
2	00	01	10					3
3	000	010	100	001	101			5
4	0000	0100	1000	0010	1010	0001	0101	1001	8

$$\uparrow$$
$$F_{n+2}$$

Consider an arbitrary *n*-bit word *w*, where $n \geq 2$.

Case 1. Suppose *w* ends in 0: $\underbrace{\cdots \quad 0}_{n \text{ bits}}$. The $(n-1)$st bit can be a 0 or

1: $\underbrace{\begin{array}{c}\cdots \quad 0\,0 \\ \cdots \quad 1\,0\end{array}}_{n \text{ bits}}$. So there are no restrictions on the $(n-1)$st bit. Thus there are

a_{n-1} *n*-bit words ending in 0, but not containing consecutive 1s.

Case 2. Suppose *w* ends in 1: $\underbrace{\cdots \quad 1}_{n \text{ bits}}$. The $(n-1)$st bit must be 0: $\underbrace{\cdots \quad 01}_{n \text{ bits}}$.

There are no restrictions on the $(n-2)$nd bit; it can be 0 or 1: $\underbrace{\begin{array}{c}\cdots \quad 0\,0\,1 \\ \cdots \quad 1\,0\,1\end{array}}_{n-2 \text{ bits}}$.

There are a_{n-2} *n*-bit words ending in 0, but not containing consecutive 1s.

Since the two cases are mutually exclusive, by the addition principle, $a_n = a_{n-1} + a_{n-2}$, where $a_0 = 1$ and $a_1 = 2$. Thus $a_n = F_{n+2}$, where $n \geq 0$. ∎

This example does not provide a constructive algorithm for systematically listing all *n*-bit words with the desired property; see Exercise 4.1.

Example 4.1 has a delightful byproduct. Suppose *n* coins are flipped sequentially at random. The total number of outcomes such that no two consecutive coins fall heads is F_{n+2}. Consequently, the probability that *no* two adjacent coins fall heads is $F_{n+2}/2^n$.

The next example is very closely related to Example 4.1.

Example 4.2. *An n-storied apartment building needs to be painted green or yellow in such a way that* no *two adjacent floors can be painted yellow, where $n \geq 1$. Find the number of different ways b_n of painting the building.*

Solution. Figure 4.1 shows the various ways of painting the building when $n = 1, 2, 3$, or 4. It follows from the figure that $b_1 = 2, b_2 = 3, b_3 = 5$, and $b_4 = 8$.

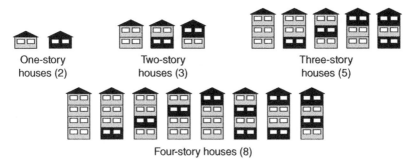

One-story   Two-story   Three-story
houses (2)   houses (3)   houses (5)

Four-story houses (8)

Figure 4.1. Possible ways of painting the building.

This is essentially the same as Example 4.1. With green = 0 and yellow = 1, every *n*-bit word in Example 4.1 represents a possible way of painting the building, and vice versa. Thus $b_n = F_{n+2}$, where $n \geq 1$. ∎

The next example deals with tessellating a $2 \times n$ rectangular area with 1×2 dominoes.

Example 4.3. *Let a_n denote the number of ways of tessellating a $2 \times n$ rectangular area with 1×2 dominoes. Find a_n.*

Solution. When $n = 1, a_1 = 1$; see Figure 4.2. When $n = 2$, there are two different ways of covering a 2×2 area, so $a_2 = 2$; see Figure 4.3.

Figure 4.2.   **Figure 4.3.**

More generally, let $n \geq 3$. Since the pattern can begin with one vertical domino or two horizontal dominoes, it follows that $a_n = a_{n-1} + a_{n-2}$. This Fibonacci recurrence, coupled with the initial conditions, implies that $a_n = F_{n+1}$, where $n \geq 1$. ∎

The next example deals with the family of subsets of a set with a special property. Irving Kaplansky (1917–2006) of the University of Chicago studied it [125].

Example 4.4. *Let a_n denote the number of subsets of the set $S_n = \{1, 2, \ldots, n\}$ that do* not *contain consecutive integers, where $n \geq 0$. We define $S_0 = \emptyset$. Find a_n.*

Solution. To get some idea about a_n, we first find its value for $0 \leq n \leq 4$; see Table 4.2. It appears from the table that $a_n = F_{n+2}$. We will now confirm it by finding a recurrence for a_n.

TABLE 4.2.

n	Subsets That Do Not Contain Consecutive Integers	a_n
0	$\emptyset$	1
1	$\emptyset, \{1\}$	2
2	$\emptyset, \{1\}, \{2\}$	3
3	$\emptyset, \{1\}, \{2\}, \{3\}, \{1, 3\}$	5
4	$\emptyset, \{1\}, \{2\}, \{3\}, \{4\}, \{1, 3\}, \{1, 4\}, \{2, 4\}$	8

$$\uparrow$$
$$F_{n+2}$$

Clearly, $a_0 = 1$ and $a_1 = 2$. Let $n \geq 2$. Let A be an arbitrary subset of S_n that does *not* contain consecutive integers. Then either $n \in A$ or $n \notin A$.

Case 1. Suppose $n \in A$. Then $n - 1 \notin A$. By definition, S_{n-2} has a_{n-2} subsets containing no consecutive integers. Now add n to each of them. The resulting subsets have the desired property, so S_n has a_{n-2} such subsets.

Case 2. Suppose $n \notin A$. By definition, there are a_{n-1} such subsets of S_n having the required property.

These two cases are mutually exclusive; so by the addition principle, $a_n = a_{n-1} + a_{n-2}$, where $a_0 = 1$ and $a_1 = 2$. So $a_n = F_{n+2}$.   ∎

Although Examples 4.1 and 4.4 look different, they are basically the same. For example, the subsets of $S_3 = \{1, 2, 3\}$ that do not contain consecutive integers are $\emptyset, \{1\}, \{2\}, \{3\}$, and $\{1, 3\}$. Using the correspondence $\emptyset \leftrightarrow 000, \{1\} \leftrightarrow 100, \{2\} \leftrightarrow 010, \{3\} \leftrightarrow 001$, and $\{1, 3\} \leftrightarrow 101$, we can recover all 3-bit words that do not contain consecutive 1s.

More generally, let A be a subset of S_n that does not contain consecutive integers. The corresponding n-bit word has a 1 in position i if and only if $i \in A$, where $1 \leq i \leq n$.

The next example, studied by the French mathematician Olry Terquem (1782–1862), deals with another special class of subsets of the set

$S_n = \{1, 2, \ldots, n\}$ [125]. A subset of S_n is *alternating* if its elements, when arranged in ascending order, follow the pattern odd, even, odd, even, and so on.

For example, $\{3\}$, $\{1, 2, 5\}$, and $\{3, 4\}$ are alternating subsets of the set S_4, whereas $\{1, 3, 4\}$ and $\{2, 3, 4, 5\}$ are not; $\emptyset$ is considered alternating.

Example 4.5. *Let a_n denote the number of alternating subsets of the set S_n. Prove that $a_n = F_{n+2}$, where $n \geq 0$.*

Proof. First, we collect some data on a_n to get a feel for it; see Table 4.3. It appears from the table that $a_n = F_{n+2}$, where $n \geq 0$. We will now confirm this.

TABLE 4.3.

n	Alternating Subsets of S_n	a_n
0	$\emptyset$	1
1	$\emptyset$, $\{1\}$	2
2	$\emptyset$, $\{1\}$, $\{1, 2\}$	3
3	$\emptyset$, $\{1\}$, $\{3\}$, $\{1, 2\}$, $\{1, 2, 3\}$	5
4	$\emptyset$, $\{1\}$, $\{3\}$, $\{1, 2\}$, $\{1, 4\}$, $\{3, 4\}$, $\{1, 2, 3\}$, $\{1, 2, 3, 4\}$	8

$$\uparrow$$
$$F_{n+2}$$

Let A be any alternating subset of S_n.

Case 1. Suppose $1 \in A$. Let $A' = A - \{1\}$. The smallest (nonzero) element in A is even. Subtracting 1 from each element in A' yields an alternating subset of S_{n-1}. By definition, there are a_{n-1} alternating subsets of $S_n - \{1\}$. Consequently, there are a_{n-1} alternating subsets A of S_n.

Case 2. Suppose $1 \notin A$. So $2 \notin A$. This leaves $n - 2$ elements in S_n. By definition, they can be used to form a_{n-2} alternating subsets of S_n.

Thus $a_n = a_{n-1} + a_{n-2}$, where $a_0 = 1$ and $a_1 = 2$. So $a_n = F_{n+2}$, where $n \geq 0$. ∎

In the next example, we will study a combinatorial problem, originally investigated by D. Lind of the University of Virginia, Charlottesville, Virginia [408]. We will solve it by developing a constructive algorithm.

Example 4.6. *Find the number of n-bit words a_n that do not contain the subword 000 or 111.*

Solution. Table 4.4 shows such acceptable words of length n, where $1 \leq n \leq 4$.

The table reveals two interesting patterns:

1) The number of admissible *n*-bit words is $2F_{n+1}$, where $1 \leq n \leq 4$.
2) The acceptable 4-bit words can be constructed from such 2-bit and 3-bit words.

TABLE 4.4.

n	Acceptable n-bit Words	Number of Such Words
1	0, 1	2
2	00, 01, 10, 11	4
3	001, 010, 011, 100, 101, 110	6
4	0010, 0011, 0100, 0101, 0110	10
	1001, 1010, 1011, 1100, 1101	

$$\uparrow$$
$$2F_{n+1}$$

This in turn gives the desired algorithm.

Step 1. Let x be an arbitrary acceptable n-bit word. Suppose it ends in 0. Then appending a 1 to it yields an acceptable $(n + 1)$-bit word $x1$. On the other hand, suppose x ends in 1. Then appending a 0 produces a distinct acceptable word $x0$ of length $n + 1$. There are a_n such $(n + 1)$-bit words.

Step 2. Let y be an acceptable $(n - 1)$-bit word. Suppose it ends in 0. Then appending a 11 gives an acceptable $(n + 1)$-bit word $y11$. If y ends in 1, then append a 00; this creates an acceptable $(n + 1)$-bit word $y00$. There are a_{n-1} such $(n - 1)$-bit words.

Clearly, Steps 1 and 2 produce only acceptable $(n + 1)$-bit words. Since there are no duplicates, they generate a total of $a_n + a_{n-1}$ acceptable $(n + 1)$-bit words.

Is this algorithm reversible? To answer this, consider an arbitrary acceptable $(n + 1)$-bit word $w = w_1 w_2 \ldots w_{n+1}$. Suppose $w_{n+1} = 0$. If $w_n = 0$, then $w_{n-1} = 1$. So $w_1 \ldots w_{n-1}$ is a valid $(n - 1)$-bit word. On the other hand, if $w_n = 1$, then w_{n-1} can be 0 or 1. So $w_1 \ldots w_n$ is an admissible n-bit word.

Suppose $w_{n+1} = 1$. If $w_n = 1$, then $w_{n-1} = 0$. Then $w_1 \ldots w_{n-1}$ is an acceptable $(n - 1)$-bit word. But if $w_n = 0$, then w_{n-1} can be 0 or 1. So $w_1 \ldots w_n$ is a valid n-bit word.

Thus reversing the algorithm yields acceptable n- and $(n - 1)$-bit words. So the algorithm is reversible. Consequently, there is a bijection between the set of acceptable $(n + 1)$-bit words, and the set of n- or $(n - 1)$-bit acceptable words. Thus, by the addition principle, $a_{n+1} = a_n + a_{n-1}$, where $a_1 = 2 = 2F_2$ and $a_2 = 4 = 2F_3$.

Suppose $a_n = 2F_{n+1}$ and $a_{n-1} = 2F_n$. It then follows by the recurrence that $a_{n+1} = 2F_{n+1} + 2F_n = 2F_{n+2}$. Thus, by PMI, $a_n = 2F_{n+1}$ for every $n \geq 1$. ∎

Next we will study the different ways a positive integer can be expressed as ordered sums of 1s and 2s.

4.2   FIBONACCI AND COMPOSITIONS

A *composition* of a positive integer n is an ordered sum of 1s and 2s. In the summer of 1974, Krishnaswami Alladi of Vivekananda College, India, and Hoggatt studied the compositions of positive integers [10].

For example, 3 has three distinct such compositions, and 4 has five; see Table 4.5. Notice that $1 + 2$ and $2 + 1$ are considered distinct compositions; so order matters. It appears from the table that the number of distinct compositions C_n of n is a Fibonacci number. The next theorem does indeed confirm this observation.

TABLE 4.5.

n	Compositions	C_n
1	1	1
2	$1 + 1, 2$	2
3	$1 + 1 + 1, 1 + 2, 2 + 1$	3
4	$1 + 1 + 1 + 1, 1 + 1 + 2, 1 + 2 + 1, 2 + 1 + 1, 2 + 2$	5
5	$1 + 1 + 1 + 1 + 1, 1 + 1 + 1 + 2, 1 + 1 + 2 + 1, 1 + 2 + 1 + 1,$ $2 + 1 + 1 + 1, 1 + 2 + 2, 2 + 1 + 2, 2 + 2 + 1$	8

$$\uparrow$$
$$F_{n+1}$$

Theorem 4.1 (Alladi and Hoggatt, 1974 [10]).   *The number of distinct compositions C_n of a positive integer n with 1s and 2s is F_{n+1}, where $n \geq 1$.*

Proof. Let $C_n(1)$ and $C_n(2)$ denote the number of compositions of n that end in 1 and 2, respectively. Clearly, $C_1(1) = 1$ and $C_1(2) = 0$; so $C_1 = C_1(1) + C_1(2) = 1$. Likewise, $C_2 = C_2(1) + C_2(2) = 1 + 1 = 2$.

Now consider an arbitrary composition of n, where $n \geq 3$.

Case 1.   Suppose the composition ends in 1. Deleting this 1 yields a composition of $n - 1$. On the other hand, adding a 1 at the end of a composition of $n - 1$ yields a composition of n that ends in 1. Thus $C_n(1) = C_{n-1}$.

Case 2.   Suppose the composition ends in 2. Deleting this 2, we get a composition of $n - 2$. On the other hand, adding a 2 or two 1s at the end, we get a composition of n. But the latter has already been counted in Case 1, so $C_n(2) = C_{n-2}$.

Thus $C_n = C_n(1) + C_n(2) = C_{n-1} + C_{n-2}$. This recurrence, coupled with the initial conditions, gives the desired result.   ∎

We will re-confirm this fact in Chapter 14 by an alternate method.

The next two results were also discovered in 1974 by Alladi and Hoggatt, where $f(n)$ denotes the total number of 1s in the compositions of n; and $g(n)$ that of 2s. For example, $f(3) = 5$ and $g(3) = 2$; see Table 4.5.

Theorem 4.2. *Let $n \geq 3$. Then*

1) $f(n) = f(n-1) + f(n-2) + F_n$
2) $g(n) = g(n-1) + g(n-2) + F_{n-1}$.

Proof.

1) As in the preceding proof, we have $C_n = C_n(1) + C_n(2)$. Since $C_n(2) = C_{n-2}$, there are C_{n-2} compositions of n that end in 2. But C_{n-2} denotes the number of compositions of $n-2$. By definition, there is total of $f(n-2)$ 1s in the compositions of $n-2$.

   Since $C_n(1) = C_{n-1}$, there are C_{n-1} compositions of n that end in 1. Excluding this 1, they contain $f(n-1)$ 1s. Since each of the C_{n-1} compositions contains a 1 as the final addend, they contain a total of $f(n-1) + C_{n-1} = f(n-1) + F_n$ 1s. Thus $f(n) = f(n-1) + f(n-2) + F_n$, where $n \geq 3$.

2) Similarly, $g(n) = g(n-1) + g(n-2) + F_{n-1}$, where $n \geq 3$. ∎

   For example,

$$f(5) = 20 = 10 + 5 + 5 = f(4) + f(3) + F_5$$
$$g(5) = 10 = 5 + 2 + 3 = g(4) + g(3) + F_4.$$

Theorem 4.3. $f(n) = g(n+1)$, *where $n \geq 1$*.

Proof. (We will establish this using PMI.) Since $f(1) = 1 = g(2)$, and $f(2) = 2 = g(3)$, the result is true when $n = 1$ and $n = 2$.

Now assume it is true for all positive integers $< n$. Then $f(n-1) = g(n)$ and $f(n-2) = g(n-1)$. By Theorem 4.2, we then have $f(n) = f(n-1) + f(n-2) + F_n = g(n) + g(n-1) + F_n = g(n+1)$. So the given result is true for n also. Thus, by PMI, it is true for all positive integers. ∎

For example, $f(3) = 5 = g(4)$, and $f(4) = 10 = g(5)$.

COMPOSITIONS WITH ODD SUMMANDS

In the next example, we will study compositions involving sums of odd positive integers.

Example 4.7. *Let a_n denote the number of ways n can be written as an ordered sum of odd positive integers, where $n \geq 1$. Find a_n.*

Solution. Table 4.6 shows the various possibilities for $1 \leq n \leq 5$. It appears from the table that $a_n = F_n$. In fact, $a_n = a_{n-1} + a_{n-2}$, where $a_1 = 1 = a_2$; so the conjecture is true.

TABLE 4.6.

n	Compositions	a_n
1	1	1
2	1 + 1	1
3	1 + 1 + 1, 3	2
4	1 + 1 + 1 + 1, 1 + 3, 3 + 1	3
5	1 + 1 + 1 + 1 + 1, 1 + 1 + 3, 1 + 3 + 1, 3 + 1 + 1, 5	5

$$\uparrow$$
$$F_n$$

∎

COMPOSITIONS WITH SUMMANDS GREATER THAN 1

Next we explore compositions with summands greater than 1. Eugen Otto Erwin Netto (1848–1919) of Germany studied this problem in 1901.

Example 4.8. *Let b_n denote the number of compositions of a positive integer n using summands greater than 1. Find b_n.*

Solution. Table 4.7 shows the possible compositions for integers 1–7. Once again, it appears from the table that $b_n = F_{n-1}$. We will now confirm this conjecture.

TABLE 4.7.

n	Compositions	b_n
1	–	0
2	2	1
3	3	1
4	2 + 2, 4	2
5	2 + 3, 3 + 2, 5	3
6	2 + 2 + 2, 2 + 4, 3 + 3, 4 + 2, 6	5
7	2 + 2 + 3, 2 + 3 + 2, 2 + 5, 3 + 2 + 2, 3 + 4, 4 + 3, 5 + 2, 7	8

$$\uparrow$$
$$F_{n-1}$$

Clearly, $b_1 = 0$ and $b_2 = 1$. So let $n \geq 3$. Notice, for example, that three compositions of 7 can be obtained by adding 2 as a summand to every composition of 5: $2 + 3 + 2$, $3 + 2 + 2$, and $5 + 2$; the other five compositions can be obtained by adding 1 to the last summand of every composition of 6: $2 + 2 + 3$, $2 + 5$, $3 + 4$, $4 + 3$, and 7.

More generally, every composition of n can be obtained from the compositions of $n - 2$ by inserting 2 as a summand; and by adding a 1 to the last summand of every composition of $n - 1$. Since there is no overlapping between the two procedures, it follows that $b_n = b_{n-1} + b_{n-2}$, where $b_1 = 0$ and $b_2 = 1$. Thus $b_n = F_{n-1}$, as conjectured.   ∎

The next example deals with another special class of n-bit words.

Example 4.9. *Let b_n denote the number of n-bit words $w_1 w_2 w_3 \ldots w_n$, where $w_1 \leq w_2, w_2 \geq w_3, w_3 \leq w_4, w_4 \geq w_5, \ldots$. Find b_n.*

Solution. There are three such 2-bit words: 00, 01, and 11. Table 4.8 shows such n-bit words for $1 \leq n \leq 4$. Using the table, we conjecture that $b_n = F_{n+2}$, where $n \geq 1$. In fact, this is true.

TABLE 4.8.

n	n-bit Words							Count
1	0	1						2
2	00	01	11					3
3	000	010	011	110	111			5
4	0000	0100	0001	0101	0111	1100	1101	1111	8

$$\uparrow$$
$$F_{n+2}$$ ∎

4.3 FIBONACCI AND PERMUTATIONS

The next example shows the occurrence of Fibonacci numbers in combinatorics, specifically, in permutations. A *permutation* on a set S is a bijective function $f: S \to S$. In words, it is nothing but a rearrangement of the elements of S.

Example 4.10. *Let p_n denote the number of permutations f of the set $S_n = \{1, 2, 3, \ldots, n\}$ such that $|i - f(i)| \leq 1$ for all $1 \leq i \leq n$, where $n \geq 1$. In words, p_n counts the number of permutations that move each element no more than one position from its natural position. Find p_n.*

Solution. Figure 4.4 shows the possible permutations for $1 \leq n \leq 4$; and Table 4.9 summarizes the data. Using the table, we conjecture that $p_n = F_{n+1}$. We will now confirm this.

TABLE 4.9.   Number of Permutations of S_n

n	1	2	3	4	$\cdots$	n
p_n	1	2	3	5	$\cdots$	?

Case 1. Let $f(n) = n$. Then the remaining $n - 1$ elements can be used to form p_{n-1} permutations such that $|i - f(i)| \leq 1$ for all i.

Case 2. Let $f(n) \neq n$. Then $f(n) = n - 1$ and $f(n - 1) = n$. The remaining $n - 2$ elements can be employed to form p_{n-2} permutations with the desired property.

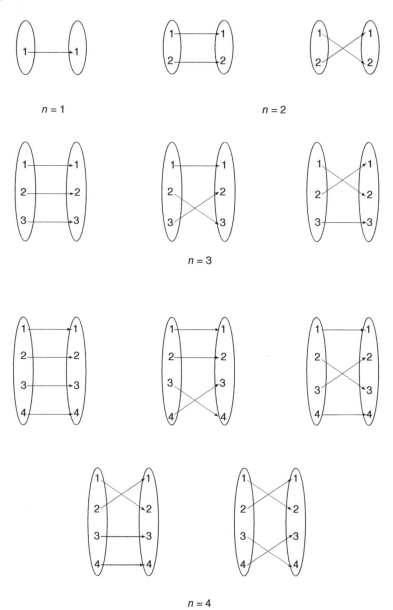

Figure 4.4. Desired permutations of S_n, where $1 \leq n \leq 4$.

By invoking the addition principle, we then have $p_n = p_{n-1} + p_{n-2}$, where $p_1 = 1$ and $p_2 = 2$. Thus $p_n = F_{n+1}$, as conjectured.   ■

Since there are $n!$ permutations of the elements of S_n, it follows by this example that there are $n! - F_{n+1}$ permutations f of S_n such that $|i - f(i)| > 1$, where $1 \leq i \leq n$. In other words, there are $n! - F_{n+1}$ permutations of S_n that move at least one element by two spaces or more from its natural position.

In particular, there are $3! - F_4 = 3$ such permutations of the set $\{1, 2, 3\}$; see Figure 4.5.

Figure 4.5.

4.4   FIBONACCI AND GENERATING SETS

Next we show that Fibonacci numbers occur in the study of generating sets. To begin with, let $[n] = \{1, 2, 3, \dots, n\}$, where $n \geq 1$. Let S be a nonempty subset of $[n]$. Let $S + 1 = \{s + 1 | s \in S\}$. A nonempty subset S is said to *generate* $[n + 1]$ if $S \cup (S + 1) = [n + 1]$.

For example, let $n = 7$ and $S = \{1, 3, 5, 7\}$. Then $S + 1 = \{2, 4, 6, 8\}$ and $S \cup (S + 1) = \{1, 2, 3, 4, 5, 6, 7, 8\} = [8]$; so S generates the set $[8]$, as does the set $\{1, 3, 4, 5, 7\}$, but *not* $\{1, 3, 4, 7\}$.

Example 4.11.   *Let s_n denote the number of subsets of $[n]$ that generate $[n + 1]$. Find s_n.*

Solution. Clearly, $s_1 = 1 = s_2$. There are two subsets of $[3]$ that generate $[4]$: $\{1, 3\}$ and $\{1, 2, 3\}$; so $s_3 = 2$. There are three subsets of $[4]$ that generate $[5]$: they are $\{1, 2, 4\}$, $\{1, 3, 4\}$, and $\{1, 2, 3, 4\}$; so $s_4 = 3$.

There are $s_5 = 5$ subsets of $[5]$ that generate $[6]$. Three of these subsets can be obtained by inserting the element 5 in each of the subsets $\{1, 2, 4\}$, $\{1, 3, 4\}$, and $\{1, 2, 3, 4\}$: $\{1, 2, 4, 5\}$, $\{1, 3, 4, 5\}$, and $\{1, 2, 3, 4, 5\}$. The remaining two can be obtained by inserting 5 in each of the subsets $\{1, 3\}$ and $\{1, 2, 3\}$: $\{1, 3, 5\}$ and $\{1, 2, 3, 5\}$. Thus s_5 equals the number of subsets of $[4]$ that generate $[5]$, plus that of $[3]$ that generate $[4]$.

More generally, the subsets of $[n]$ that generate $[n + 1]$ can be obtained by inserting n in each of the subsets of $[n - 1]$ that generate $[n]$; and by inserting n

in each of the subsets of $[n-2]$ that generate $[n-1]$. Thus

$$s_n = \text{number of subsets of } [n-1] \text{ that generate } [n]$$
$$+ \text{ number of subsets of } [n-2] \text{ that generate } [n-1]$$
$$= s_{n-1} + s_{n-2},$$

where $s_1 = 1 = s_2$. Thus $s_n = F_n$. ∎

4.5 FIBONACCI AND GRAPH THEORY

The next example establishes a bridge between Fibonacci numbers and graph theory. To see this, we present additional graph-theoretic terminology [368].

Recall that a *graph* $G = (V, E)$ consists of a set V of points, called *vertices*, and a set of arcs or line segments, called *edges*, joining them. An edge connecting vertices v and w is denoted by $v-w$. A vertex v is *adjacent* to vertex w if there is an edge connecting them.

For example, the graph in Figure 4.6 has four vertices – $A, B, C,$ and D – and seven edges. Vertex A is adjacent to B, but *not* to C.

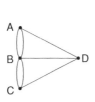

Figure 4.6.

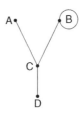

Figure 4.7.

An edge emanating from and terminating at the same vertex is a *loop*. *Parallel edges* have the same vertices. A loop-free graph that contains no parallel edges is a *simple graph*.

For example, the graph in Figure 4.7 has a loop at B. The graph in Figure 4.6 has parallel edges, while the graph in Figure 4.8 contains no loops or parallel edges, making it a simple graph.

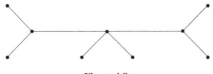

Figure 4.8.

A *subgraph* $H = (V', E')$ is a graph such that $V' \subseteq V$ and $E' \subseteq E$. A *path* between two vertices v_1 and v_n is a sequence $v_1 - v_2 - \cdots - v_n$ of vertices and edges connecting them; its *length* is $n - 1$. A graph is *connected* if there a path between every two distinct vertices. For instance, consider the graph in Figure 4.6. The length of the path $A - B$ is one, and that of $A - B - A - B - C$ is four.

INDEPENDENT SUBSET OF THE VERTEX SET

Let V denote the set of vertices of a graph. A subset I of V is *independent* if *no* two vertices in I are adjacent. In other words, I is independent if and only if edge $v - w$ does *not* exist for any $v, w \in I$.

For example, consider the graph in Figure 4.9. Then $\{a, c, e\}$ and $\{b, d\}$ are independent, whereas $\{a, c, d, f\}$ is *not*.

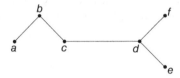

Figure 4.9.

We are now ready for the graph-theoretic example.

Example 4.12. *Let P_n denote the simple path graph $v_1 - v_2 - \cdots - v_n$ connecting the vertices $v_1, v_2, \ldots, v_n$. Let a_n denote the number of independent subsets of the vertex set V_n. Find a_n.*

Solution. The path P_1 consists of a single point; so there are two possible independent subsets: $\emptyset, \{v_1\}$. The path P_2 is $v_1 - v_2$. There are three independent subsets of $\{v_1, v_2\}$: $\emptyset, \{v_1\}$, and $\{v_2\}$.

The path P_3 contains three vertices: v_1, v_2, and v_3. Accordingly, there are five independent subsets: $\emptyset, \{v_1\}, \{v_2\}, \{v_3\}$, and $\{v_1, v_3\}$.

These data are summarized in Table 4.10. Clearly, a pattern emerges. We conjecture that $a_n = F_{n+2}$, where $n \geq 1$.

TABLE 4.10.

n	Path	Independent Subsets	a_n
1	•	$\emptyset, \{v_1\}$	2
2	•—•	$\emptyset, \{v_1\}, \{v_2\}$	3
3	•—•—•	$\emptyset, \{v_1\}, \{v_2\}, \{v_3\}, \{v_1, v_3\}$	5
4	•—•—•—•	$\emptyset, \{v_1\}, \{v_2\}, \{v_3\}, \{v_4\}, \{v_1, v_3\}, \{v_1, v_4\}, \{v_2, v_4\}$	8

$$\uparrow$$
$$F_{n+2}$$

Let W be any independent subset of V_n. Suppose $v_n \in W$. Then $v_{n-1} \notin W$. Since v_{n-2} may or may not be in W, by definition, there are a_{n-2} such independent subsets.

On the other hand, suppose $v_n \notin W$. It follows by a similar argument that there are a_{n-1} such independent subsets.

Thus $a_n = a_{n-1} + a_{n-2}$, where $a_1 = 2$ and $a_2 = 3$, and $n \geq 3$. This implies that $a_n = F_{n+2}$, as predicted. ∎

Interestingly, Examples 4.1 and 4.12 are closely related. To see this, let S be an independent subset of the set of vertices $v_n = \{v_1, v_2, \ldots, v_n\}$ on P_n. We will now employ S to construct an n-bit word $w = w_1 w_2 \ldots w_n$ containing no two consecutive 1s: $w_i = 0$ if and only if $v_i \in S$. Since $v_i, v_{i+1} \notin S$, this definition guarantees that $w_i \neq 0$ and $w_{i+1} \neq 0$; so w cannot contain two adjacent 1s.

This algorithm is clearly reversible. So it establishes a bijection between the set of independent subsets of V and the set of n-bit words with no consecutive 1s. Thus the number of independent subsets of V equals the number of n-bit words containing no consecutive 1s.

For example, let $n = 4$. Then $V = \{v_1, v_2, v_3, v_4\}$, and $\{v_1, v_4\}$ is an independent subset of V on path P_3. The corresponding binary word is 0110. The 4-bit word corresponding to $\emptyset$ is 1111. Going in the reverse order, consider the 4-bit word 1010; the corresponding independent subset is $\{v_2, v_4\}$.

Table 4.11 shows the independent subsets of the set of vertices $\{v_1, v_2, v_3\}$ on path P_2, and the corresponding 3-bit words.

TABLE 4.11.

Independent Subsets	3-bit Words
$\emptyset$	111
$\{v_1\}$	011
$\{v_2\}$	101
$\{v_3\}$	110
$\{v_1, v_3\}$	010

4.6 FIBONACCI WALKS

Next we will study another occurrence of Fibonacci numbers in combinatorics. To this end, we first introduce the concept of a lattice point and lattice path.

A *lattice point* on the Cartesian plane is a point with integral coordinates. For example, $(2, -3)$ and $(5, -13)$ are lattice points, whereas $(0, \sqrt{2})$ and $(-3, \pi)$ are *not*. A *lattice path* is a sequence of connected horizontal, vertical, or diagonal unit steps $\overline{X_k X_{k+1}}$, where X_k and X_{k+1} are lattice points. The *length* of a lattice path is the number of unit steps in the path.

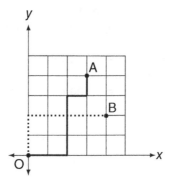

Figure 4.10.   Lattice paths.

For example, Figure 4.10 shows two lattice paths. The solid path from O to A has length 7. It contains two unit steps in the easterly direction (E), followed by three in the northerly direction (N), one in the easterly direction (E), and one in the northerly direction (N). It is denoted by the word *EENNNEN*. The dotted path *NNEEEE* from O to B has length 6.

We will now study horizontal lattice paths of length n, where $n \geq 0$. We can use E-steps, or D-steps, which are double E-steps (length 2) [370, 457]. We call such horizontal paths *Fibonacci walks* (or *paths*).

Figure 4.11 shows Fibonacci walks of length n, where $0 \leq n \leq 4$, where the thick dot indicates the origin. It appears from Table 4.12 that the number of Fibonacci paths b_n of length n is a Fibonacci number: $b_n = F_{n+1}$.

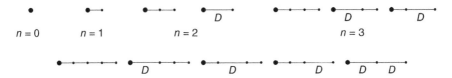

Figure 4.11.   Fibonacci walks of length ≤ 4.

TABLE 4.12.   Number of Fibonacci Walks

n	0	1	2	3	4
b_n	1	1	2	3	5

We will now confirm this conjecture. Clearly, $b_0 = 1 = F_1$ and $b_1 = 1 = F_2$. So it remains to show that b_n satisfies the Fibonacci recurrence. To this end, consider an arbitrary Fibonacci path of length n, where $n \geq 2$.

Suppose the walk ends in D: ⏝D. By definition, there are b_{n-2} such
length $n-2$

Fibonacci paths of length n. On the other hand, suppose the walk ends in
E: ⏝E. There are b_{n-1} such paths. These two are mutually exclusive cases.
length $n-1$

So, by the addition principle, $b_n = b_{n-1} + b_{n-2}$.

Thus b_n satisfies the Fibonacci recurrence with $b_0 = F_1$ and $b_1 = F_2$. So
$b_n = F_{n+1}$, where $n \geq 0$. ∎

4.7 FIBONACCI TREES

In order to study Fibonacci trees, we will need to build up some additional
graph-theoretic vocabulary.

A *cycle* is a path with the same endpoints; it contains no repeated vertices.
A graph is *acyclic* if it contains no cycles. A connected, acyclic graph is a *tree*. A
tree with n vertices has exactly $n - 1$ edges.

For example, the graph in Figure 4.12 shows the family tree of the Bernoullis of
Switzerland, a distinguished family of mathematicians. The graph in Figure 4.13
is *not* a tree because it is cyclic. The tree in Figure 4.12 contains a specially des-
ignated vertex, called the *root*. Its root is Nicolaus. The basic terminology of
(rooted) trees reflects that of a family tree.

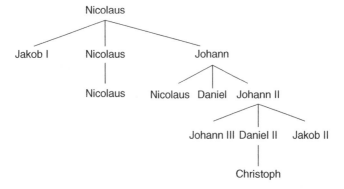

Figure 4.12. The Bernoulli family tree.

Figure 4.13.

PARENT, CHILD, SIBLING, ANCESTOR, DESCENDANT, AND SUBTREE

Let T be a tree with root v_0. Let $v_0 - v_1 - \cdots - v_{n-1} - v_n$ be the path from v_0 to v_n. Then

- v_{i-1} is the *parent* of v_i.
- v_i is a *child* of v_{i-1}.
- The vertices $v_0, v_1, \ldots, v_{n-1}$ are ancestors of v_n.
- The descendants of a vertex v are those vertices for which v is an ancestor.
- A vertex with no children is a *leaf* or a *terminal vertex*.
- A vertex that is not a leaf is an *internal vertex*.
- The subtree rooted at v consists of v, its descendants, and all it edges.

For example, consider the tree in Figure 4.14. It is rooted at a. Vertex b is the parent of both e and f, so e and f are the children of b. Vertices a, b, and e are ancestors of i. Vertices b and e are descendants of a. Vertex f has no children, so it is a leaf. Vertices b and d have at least one child, so both are internal vertices. Figure 4.15 displays the subtree rooted at b.

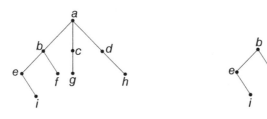

Figure 4.14.   Figure 4.15.

BINARY TREES

An *ordered rooted tree* is a rooted tree in which the vertices at each level are ordered as the first, second, third, and so on. Such a tree is a *binary tree* if every vertex has at most two children.

For example, the tree in Figure 4.15 is a binary tree. Its *left subtree* is the binary tree rooted at e, and its *right subtree* is the binary tree rooted at f.

We are now ready to explore the close relationships between Fibonacci trees and Fibonacci numbers.

FIBONACCI TREES

The *n*th *Fibonacci tree* T_n is a binary tree, defined recursively as follows, where $n \geq 1$:

- Both T_1 and T_2 are binary trees with exactly one vertex each.
- Let $n \geq 3$. Then T_n is a binary tree whose left subtree is T_{n-1} and whose right subtree is T_{n-2}.

Figure 4.16 shows the Fibonacci trees T_1 through T_6.

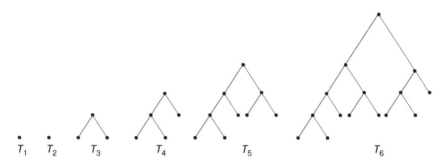

Figure 4.16. Fibonacci trees.

We would like to determine the number of vertices v_n, the number of leaves l_n, the number of internal vertices i_n, and the number of edges e_n of T_n. To facilitate our study, first we will collect some data from Figure 4.16, and summarize them in a table; see Table 4.13.

TABLE 4.13.

n	1	2	3	4	5	6	$\cdots$	n
v_n	1	1	3	5	9	15	$\cdots$	?
l_n	1	1	2	3	5	8	$\cdots$	?
i_n	0	0	1	2	4	7	$\cdots$	?
e_n	0	0	2	4	8	14	$\cdots$	?

Using the table, we conjecture that $v_n = 2F_n - 1, l_n = F_n, i_n = l_n - 1 = F_n - 1$, and $e_n = v_n - 1 = 2F_n - 2$. Using recursion, it is easy to establish these results, as the next theorem shows.

Theorem 4.4. *Let v_n, l_n, i_n, and e_n denote the number of vertices, leaves, internal vertices, and edges of a Fibonacci tree T_n, respectively, where $n \geq 1$. Then $v_n = 2F_n - 1, l_n = F_n, i_n = F_n - 1$, and $e_n = 2F_n - 2$.*

Proof.

1) Clearly, the formula works when $n = 1$ and $n = 2$. Suppose $n \geq 3$. Since T_n has T_{n-1} as a left subtree and T_{n-2} as its right subtree, it follows that $v_n = v_{n-1} + v_{n-2} + 1$, where $v_1 = 1 = v_2$. Let $b_n = v_n + 1$. Then $b_n = b_{n-1} + b_{n-2}$,

where $b_1 = 2 = b_2$. So $b_n = 2F_n$, and hence $v_n = 2F_n - 1$. Thus the formula works for $n \geq 1$.

2) It follows by the recursive definition of T_n that $l_n = l_{n-1} + l_{n-2}$, where $l_1 = 1 = l_2$. Thus $l_n = F_n$, where $n \geq 1$.

3) Clearly, $i_n = v_n - l_n = (2F_n - 1) - F_n = F_n - 1$.

4) Again, by the recursive definition of T_n, $e_n = e_{n-1} + e_{n-2} + 2$, where $e_1 = 0 = e_2$. Let $c_n = e_n + 2$. Then $c_n = c_{n-1} + c_{n-2}$, where $c_1 = 2 = c_2$. Thus $c_n = 2F_n$, so $e_n = 2F_n - 2$, where $n \geq 1$.   ∎

4.8   PARTITIONS

Now we will study how Fibonacci and Lucas numbers can be used to split the set of positive integers $\mathbb{N}$ into two disjoint subsets. To this end, first we will introduce the concept of a partition.

PARTITION OF A SET

Let $P = \{S_i | i \in \mathbb{N}\}$ be a family of subsets of a non-empty set S. Then P is a *partition* of S if:

- each S_i is nonempty;
- $\bigcup_{i \in \mathbb{N}} S_i = S$; and
- $S_i \cap S_j = \emptyset$ if $i \neq j$.

For example, the set of positive even integers and that of positive odd integers form a partition of $\mathbb{N}$.

PARTITIONS OF $\mathbb{N}$

Around 1976, D.L. Silverman of Los Angeles, California, proposed an interesting problem: prove that the set of positive integers $\mathbb{N}$ can be partitioned into two disjoint subsets A and B such that the sum of any two distinct elements from the same subset is *not* a Fibonacci number [526]. So every non-Fibonacci number is the sum of two distinct elements from the same subset.

Suppose $F_i, F_{i+1} \in X$, where $X = A$ or B. Since $F_i + F_{i+1} = F_{i+2}$, it follows that both F_i and F_{i+1} *cannot* belong to X. But $F_i, F_{i+2} \in X$. Consequently, $F_2, F_4, F_6, F_8, \ldots$ must belong to the same subset, say, A; and $F_3, F_5, F_7, F_9, \ldots$ must belong to the same subset B; that is,

$$1, 3, 8, 21, \ldots \in A \quad \text{and} \quad 2, 5, 13, 34, \ldots \in B.$$

Using the fact that no two distinct elements in X can add up to a Fibonacci number, we can find as many elements in X as we please. Thus

$$A = \{\mathbf{1}, \mathbf{3}, 6, \mathbf{8}, 9, 11, 14, 16, 17, 19, \mathbf{21}, \ldots\}$$

$$B = \{\mathbf{2}, 4, \mathbf{5}, 7, 10, 12, \mathbf{13}, 15, 18, 20, \ldots\},$$

where we have boldfaced Fibonacci numbers. We have $A \cup B = \mathbb{N}$, $A \cap B = \emptyset$, and the sum of no two distinct elements in A or B is a Fibonacci number. See [129] for a proof.

A GRAPH-THEORETIC INTERPRETATION

The subsets A and B can be used to construct an infinite graph G with $\mathbb{N}$ as its vertex set. Two vertices a and b in G are adjacent if $a + b$ is a Fibonacci number. Figure 4.17 shows the resulting graph. It is a *bipartite* graph, meaning every edge is incident with a vertex in A and a vertex in B.

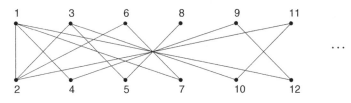

Figure 4.17.

Lucas numbers also enjoy a similar property:

$$C = \{\mathbf{1}, \mathbf{4}, 5, 8, 9, \mathbf{11}, 12, 15, 16, 19, \ldots\}$$

$$D = \{2, \mathbf{3}, 6, \mathbf{7}, 10, 13, 14, 17, \mathbf{18}, 20, 21, \ldots\},$$

where Lucas numbers are boldfaced, $C \cup D = \mathbb{N}$, $C \cap D = \emptyset$, and no two distinct elements in C or D add up to a Lucas number. Figure 4.18 shows the corresponding graph; it is also bipartite.

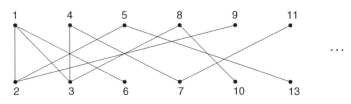

Figure 4.18.

More generally, K. Alladi, P. Erdös, and Hoggatt proved that $\mathbb{N}$ can be uniquely partitioned using any sequence $\{u_n\}$ that satisfies the Fibonacci recurrence, where $u_1 = 1$ and $u_2 \geq 1$, as the following theorem shows. In the interest of brevity, we omit its proof [11].

Theorem 4.5 (Alladi, Erdös, and Hoggatt, 1978 [11]).   *Let* $U = \{u_n \mid u_{n+2} = u_{n+1} + u_n, u_1 = 1, u_2 > 1, \text{and } n \geq 1\}$. *Then* $\mathbb{N}$ *can be partitioned into two disjoint subsets such that no sum of any two distinct elements in the same subset belongs to U.*   ∎

4.9   FIBONACCI AND THE STOCK MARKET

In the 1930s, Ralph Nelson Elliott, an engineer, made an extensive study of the fluctuations in the U.S. stock market [486]. The Dow Jones Industrials Average (DJIA), an indicator based on stocks of 30 top companies, is often used as a measure of stock market activity, and hence of the health of the economy. The DJIA varies according to market conditions, which reflect investors' optimism and pessimism. Nevertheless, according to Prechter and Frost [486], Elliott discovered, based on his study and observations, "that the ever-changing stock market tended to reflect a basic harmony found in nature and from this discovery developed a rational system of stock market analysis."

In 1939, Elliott expressed his analysis as a theoretical principle, which has since been called the *Elliott Wave Principle*. In practice, the wave principle corresponds to the performance of the DJIA.

Elliott observed that the stock market unfolds according to a fundamental pattern comprising a complete cycle of eight waves. Each cycle consists of two

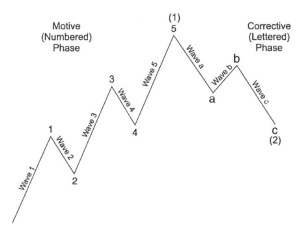

Figure 4.19.   The fundamental behavior of the Elliott Wave Principle. *Source*: Prechter and Frost, 1978/2014 [486]. Copyright Robert R. Prechter, Jr. Reproduced with permission of New Classics Library.

phases, the *numbered phase* and the *lettered phase*; see Figure 4.19. The numbered phase consists of eight waves: five upward waves and three downward waves. The upward waves 1, 3, and 5 are *impulse waves*, and they reflect optimism in the stock market; the downward waves 2 and 4 are *corrective waves*, and they are corrections to those impulses, indicating pessimism. Wave 2 corrects wave 1, and wave 4 corrects wave 3.

The upward trend, depicted by the sequence 1–2–3–4–5, is then corrected by the downward trend, namely, the lettered phase *a–b–c*; the downward trend is, in fact, made up of two downward waves, *a* and *c*, and one upward wave *b*. The five-wave sequence 1–2–3–4–5 indicates a *bull market*, whereas the corrective sequence *a–b–c* indicates a *bear market*. Thus one complete cycle consists of the sequence 1–2–3–4–5–*a–b–c*. According to the wave principle, this cycle of upward and downward turns continues.

Now the numbered phase can be considered a wave, say, wave ①, and the lettered phase wave ②. Thus there are waves within waves; see Figure 4.20. Counting ① and ② separately, we get two waves. The pattern (1)–(2)–(3)–(4)–(5) consists of 21 smaller waves, and the downward trend (*a*)–(*b*)–(*c*) consists of 13 smaller waves; so the pattern (1)–(2)–(3)–(4)–(5)–(*a*)–(*b*)–(*c*) consists of 34 smaller waves. Thus we have the following pattern:

$$①–② = 2 \text{ waves}$$

$$(1)–(2)–(3)–(4)–(5)–(a)–(b)–(c) = 8 \text{ waves}$$

$$1–2–3–4–5–\cdots–1–2–3–4–5 = 34 \text{ waves.}$$

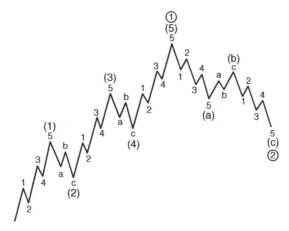

Figure 4.20. Waves within waves. *Source*: Prechter and Frost, 1978/2014 [486]. Copyright Robert R. Prechter, Jr. Reproduced with permission of New Classics Library.

That is, a wave of a large degree can be split into two waves of lower degree. These two waves can be divided into eight waves of next lower degree, and they in turn can be subdivided into 34 waves of even lower degree. This subdividing pattern also implies that waves can be combined to form waves of higher degrees. Whether waves are divided or combined, the underlying behavior remains invariant; see Figure 4.21.

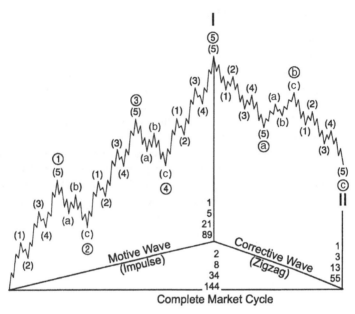

Figure 4.21.   Forming large waves. *Source*: Prechter and Frost, 1978/2014 [486]. Copyright Robert R. Prechter, Jr. Reproduced with permission of New Classics Library.

The complete cycle in Figure 4.21 comprises a bull market and a bear market. The bull market cycle consists of five primary waves, which can be subdivided into 21 intermediate waves, and they in turn can be re-subdivided into 89 minor waves. The corresponding figures for the bear market are 1, 3, 13, and 55, respectively. This observation yields interesting dividends, as Table 4.14 shows. Odd as it may seem, according to the Elliott Wave Principle, this Fibonacci rhythmic pattern continues indefinitely.

TABLE 4.14.

Waves	Bull Market Cycle	Bear Market Cycle	Total Waves
Cycle waves	1	1	2
Primary waves	5	3	8
Intermediate waves	21	13	34
Minor waves	89	55	144

EXERCISES 4

1. An n-bit word containing no consecutive 1s can be constructed recursively as follows: Append a 0 to such $(n-1)$-bit words or append a 01 to such $(n-2)$-bit words. Using this procedure, construct all 5-bit words containing no consecutive 1s. There are 13 such words.

2. Let a_n denote the number of n-bit words that do not contain the pattern 111. Define a_n recursively.

Let a_n denote the number of ways a person can climb up a ladder with n rungs. At each step, he/she can climb one or two rungs (Cohen, 1978 [125]).

3. Define a_n recursively.

4. Find an explicit formula for a_n.

Let b_n denote the number of ways of forming a sum of n (integral) dollars using only one- and two-dollar bills, taking order into consideration (Moser, 1963 [449]).

5. Define b_n recursively.

6. Find an explicit formula for b_n.

Let b_n denote the number of compositions of a positive integer n using 1, 2, and 3 as summands (Netto, 1901).

7. Find b_3 and b_4.

8. Define b_n recursively.

9. The *secant method* is one of the ancient techniques of solving the equation $f(x) = 0$. (It is the inverse of linear interpolation.) Let x_1 and x_2 denote two initial approximations of a solution. Then the $(n+1)$st approximation is given by the iterative formula

$$x_{n+1} = \frac{x_{n-1}f(x_n) - x_n f(x_{n-1})}{f(x_n) - f(x_{n-1})}.$$

   Let $f(x) = x^2$, and $x_1 = 1 = x_2$. Prove that $x_n = 1/F_n$ (Thoro, 1963 [558]).

10. Let b_n denote the number of different ways the pattern of n boxes in Figure 4.22 can be filled with 0s and 1s such that *no* box with a 0 points to one with a 1. We let $b_0 = 1$. Figure 4.23 shows the possible configurations with n boxes, where $0 \leq n \leq 3$. Prove that $b_n = F_{n+2}$, where $n \geq 0$ (Carlitz; see Cohen, 1978 [125]).

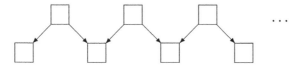

Figure 4.22.

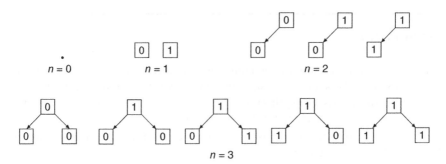

Figure 4.23.

A set of integers A is *fat* if each of its elements is $\geq |A|$, where $|A|$ denotes the cardinality of the set A. For example, $\{5, 7, 91\}$ is a fat set, whereas $\{3, 7, 36, 41\}$ is not. $\emptyset$ is considered a fat set. Let a_n denote the number of fat subsets of the set $\{1, 2, 3, \ldots, n\}$ (G.E. Andrews; see Cohen, 1978 [125]).

11. Define a_n recursively.

12. Find an explicit formula for a_n.

An ordered pair of subsets (A, B) of the set $S_n = \{1, 2, \ldots, n\}$ is *admissible* if $a > |B|$ for every $a \in A$ and $b > |A|$ for every $\in B$, where $|X|$ denotes the cardinality of the set X. For example, $(\{2, 3\}, \{4\})$ is an admissible pair of subsets of S_4.

13. Find all admissible pairs of subsets of the sets S_0, S_1, and S_2.

*14. Predict the number of admissible ordered pairs of subsets of S_n.

*15. Let S_n denote the sum of the elements in the nth term of the sequence of sets of Fibonacci numbers $\{1\}, \{1, 2\}, \{3, 5, 8\}, \{13, 21, 34, 55\}, \ldots$. Find a formula for S_n.

FIBONACCI AND LUCAS IDENTITIES

Mathematics is the music of reason.

–James Joseph Sylvester (1814–1897)

Both Fibonacci and Lucas numbers satisfy numerous identities. Many were discovered centuries ago. In this chapter we will explore several fundamental identities.

We will begin with a few summation formulas. Exercise 2.10, for example, required that we conjecture a formula for the sum $\sum_{i=1}^{n} F_i$. In doing so, we notice an interesting pattern:

$$F_1 = 1 = 2 - 1 = F_3 - 1$$
$$F_1 + F_2 = 2 = 3 - 1 = F_4 - 1$$
$$F_1 + F_2 + F_3 = 4 = 5 - 1 = F_5 - 1$$
$$F_1 + F_2 + F_3 + F_4 = 7 = 8 - 1 = F_6 - 1$$
$$F_1 + F_2 + F_3 + F_4 + F_5 = 12 = 13 - 1 = F_7 - 1.$$

Following this pattern, we conjecture that $\sum_{i=1}^{n} F_i = F_{n+2} - 1$.

Fibonacci and Lucas Numbers with Applications, Volume One, Second Edition. Thomas Koshy.
© 2018 John Wiley & Sons, Inc. Published 2018 by John Wiley & Sons, Inc.

We will now state this as a theorem, and then establish it in two different ways. See Exercise 5.21 for a third method.

A popular technique that is useful in dealing with finite sums is a *telescoping sum*. A sum of the form $S = \sum_{i=m+1}^{n} (a_i - a_{i-1})$ is a telescoping sum. It is easy to show that $S = a_n - a_m$. This will come in handy in establishing the conjecture.

Theorem 5.1 (Lucas, 1876).

$$\sum_{i=1}^{n} F_i = F_{n+2} - 1. \tag{5.1}$$

Proof. By the Fibonacci recurrence we have

$$\sum_{i=2}^{n+1} F_{i-1} = \sum_{i=2}^{n+1} (F_{i+1} - F_i)$$

$$\sum_{i=1}^{n} F_i = F_{n+2} - F_2$$

$$= F_{n+2} - 1,$$

as conjectured.   ∎

AN ALTERNATE METHOD BY PMI

Since $F_1 = 1 = F_3 - 1$, the formula works when $n = 1$.

Now assume it works for an arbitrary positive integer k. Then

$$\sum_{i=1}^{k+1} F_i = \sum_{i=1}^{k} F_i + F_{k+1}$$

$$= (F_{k+2} - 1) + F_{k+1}$$

$$= (F_{k+2} + F_{k+1}) - 1$$

$$= F_{k+3} - 1.$$

So the formula works for $k + 1$ also.

Thus it is true for all positive integers n.   ∎

For example, $\sum_{i=1}^{20} F_i = F_{22} - 1 = 17{,}711 - 1 = 17{,}710$. You can verify this by direct computation.

Theorem 5.1 has a simple, but nice, geometric interpretation. Suppose n line segments $\overline{A_i B_i}$ are placed end to end, where $A_i B_i = F_i$ and $1 \leq i \leq n$. Then $A_1 B_n = F_{n+2} - 1$.

Theorem 5.1 is also the basis of an interesting puzzle, conceived by W.H. Huff:

Add up any finite number of consecutive Fibonacci numbers. Now add the second term to this sum. The resulting sum is a Fibonacci number.

The next example justifies the validity of this puzzle.

Example 5.1. *Prove that* $\sum_{j=0}^{k} F_{i+j} + F_{i+1} = F_{i+k+2}.$

Proof. By Theorem 5.1, we have

$$\sum_{j=0}^{k} F_{i+j} + F_{i+1} = \sum_{r=1}^{i+k} F_r - \sum_{r=1}^{i-1} F_r + F_{i+1}$$

$$= (F_{i+k+2} - 1) - (F_{i+1} - 1) + F_{i+1}$$

$$= F_{i+k+2},$$

as desired. ∎

This example, in fact, identifies the Fibonacci number that is the final sum in Huff's puzzle. Obviously, this example and therefore the puzzle can be extended to the generalized Fibonacci sequence; see Exercise 7.12.

Using the technique employed in Theorem 5.1, we can derive a formula for the first n odd-numbered Fibonacci numbers. We omit the proof in the interest of brevity; see Exercise 5.1.

Theorem 5.2 (Lucas, 1876).

$$\sum_{i=1}^{n} F_{2i-1} = F_{2n}. \tag{5.2}$$

∎

For example, $\sum_{i=1}^{10} F_{2i-1} = 6765 = F_{20}.$

The next result now follows by Theorems 5.1 and 5.2. Again, we omit the proof for brevity; see Exercise 5.2.

Corollary 5.1 (Lucas, 1876).

$$\sum_{i=1}^{n} F_{2i} = F_{2n+1} - 1. \tag{5.3}$$

∎

Corollary 5.1 has a wonderful application to graph theory. Before we present it, we need two definitions.

5.1   SPANNING TREE OF A CONNECTED GRAPH

A *spanning tree* of a connected graph G is a subgraph that is a tree containing every vertex of G. The *complexity* $k(G)$ of a graph is the number of distinct spanning trees of the graph.

For example, the graph in Figure 5.1 has three distinct spanning trees; so its complexity is three; see Figure 5.2.

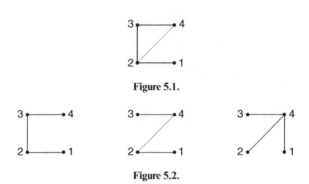

Figure 5.1.

Figure 5.2.

FAN GRAPH

A *fan graph* (or simply a *fan*) G_1 of order 1 consists of two vertices, 0 and 1, and exactly one edge between them. A fan G_n of order n is obtained by adding a vertex n to a fan G_{n-1}, and then connecting vertex n to vertices 0 and $n-1$, where $n \geq 2$. Figure 5.3 shows fans of orders 1 through 4.

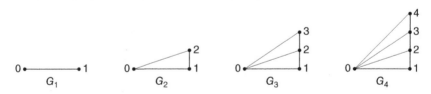

Figure 5.3.   Fan graphs.

Next we look for the number s_n of spanning trees of a fan G_n. Fan G_1 has one spanning tree, so $s_1 = 1$; see Figure 5.4. Fan G_2 has three spanning trees, so $s_2 = 3$; see Figure 5.5. G_3 has eight spanning trees; so $s_3 = 8$; see Figure 5.6. Thus $s_1 = 1 = F_2, s_2 = 3 = F_4$ and $s_3 = 8 = F_6$. So we predict that $s_4 = F_8$.

Figure 5.4.   Spanning tree of G_1.

Figure 5.5.   Spanning trees of G_2.

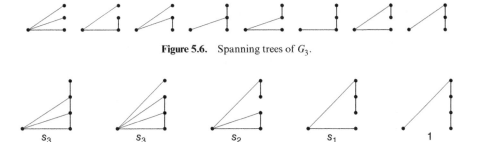

Figure 5.6. Spanning trees of G_3.

s_3 $\qquad$ s_3 $\qquad$ s_2 $\qquad$ s_1 $\qquad$ 1

Figure 5.7. Possible ways of having vertex 4 in a spanning tree of G_4.

To confirm this observation, consider the possible ways of having vertex 4 in a spanning tree of G_4. It follows from Figure 5.7 that

$$s_4 = s_3 + \sum_{i=1}^{3} s_i + 1$$
$$= F_6 + (F_6 + F_4 + F_2) + 1$$
$$= F_6 + (F_7 - 1) + 1$$
$$= F_8.$$

More generally,

$$s_n = s_{n-1} + \sum_{i=1}^{n-1} s_i + 1$$
$$= F_{2n-2} + \sum_{i=1}^{n-1} F_{2i} + 1$$
$$= F_{2n-2} + (F_{2n-1} - 1) + 1$$
$$= F_{2n}.$$

For example, a fan of order 5 has $F_{10} = 55$ spanning trees.

Before we state the next property, let us study the following delightful pattern involving three consecutive Fibonacci numbers:

$$F_1 F_3 - F_2^2 = 1 \cdot 2 - 1^2 = (-1)^2$$
$$F_2 F_4 - F_3^2 = 1 \cdot 3 - 2^2 = (-1)^3$$
$$F_3 F_5 - F_4^2 = 2 \cdot 5 - 3^2 = (-1)^4$$
$$F_4 F_6 - F_5^2 = 3 \cdot 8 - 5^2 = (-1)^5$$
$$\vdots$$

Clearly, a pattern emerges. So we conjecture that $F_{n-1}F_{n+1} - F_n^2 = (-1)^n$, where $n \geq 1$. This leads us to the next result, which was discovered in 1680 by the Italian-born French astronomer and mathematician Giovanni Domenico Cassini (1625–1712). It was discovered independently in 1753 by the Scottish mathematician Robert Simson (1687–1768) of the University of Glasgow.

Cassini[*]   Simson[#]

There are several ways of proving this. For now, we will employ PMI.

Theorem 5.3 (Cassini's Formula).   *Let $n \geq 1$. Then*

$$F_{n-1}F_{n+1} - F_n^2 = (-1)^n. \tag{5.4}$$

Proof. Since $F_0F_2 - F_1^2 = 0 \cdot 1 - 1 = (-1)^1$, the given result is true when $n = 1$. Now assume it is true for an arbitrary positive integer k. Then

$$\begin{aligned}
F_kF_{k+2} - F_{k+1}^2 &= (F_{k+1} - F_{k-1})(F_k + F_{k+1}) - F_{k+1}^2 \\
&= F_kF_{k+1} - F_kF_{k-1} - \left[F_k^2 + (-1)^k \right] \\
&= F_kF_{k+1} - F_k \left(F_{k-1} + F_k \right) + (-1)^{k+1} \\
&= F_kF_{k+1} - F_kF_{k+1} + (-1)^{k+1} \\
&= (-1)^{k+1}.
\end{aligned}$$

So the formula works for $n = k + 1$. Thus, by PMI, it works for all $n \geq 1$.   ∎

[*] *Figure source*: https://en.wikipedia.org/wiki/Giovanni_Domenico_Cassini#/media/File:Giovanni_Cassini.jpg.
[#] *Figure source*: https://en.wikipedia.org/wiki/Robert_Simson#/media/File:Robert_Simson.jpg

We will encounter Cassini's formula numerous times in later chapters. For example, it is the basis of an interesting geometric paradox in Chapter 6.

The following result is an immediate consequence of Cassini's formula.

Corollary 5.2. *Any two consecutive Fibonacci numbers are relatively prime; that is, $(F_n, F_{n+1}) = 1$, where (a, b) denotes the greatest common divisor of the integers a and b.*

Proof. Let p be a prime factor of both F_n and F_{n+1}. Then, by Cassini's formula, $p | \pm 1$, which is a contradiction. Thus $(F_n, F_{n+1}) = 1$, as desired. ∎

Substituting for F_{n+1}, Cassini's formula can be rewritten as $F_{n-1}^2 + F_n F_{n-1} - F_n^2 = (-1)^n$. This implies that the Diophantine equation $x^2 + xy - y^2 = (-1)^n$ has infinitely many solutions $(x, y) = (F_{n-1}, F_n)$.

Cassini's formula has another consequence: we can express the product $F_{2n} F_{2n+2} F_{2n+4}$ as a product of three consecutive integers; see Exercise 5.79 [412]. For example, $F_{10} F_{12} F_{14} = 55 \cdot 144 \cdot 377 = 2,985,840 = 143 \cdot 144 \cdot 145$.

Now we will investigate the sum of the squares of the first n Fibonacci numbers. Once again, let us look for a pattern: $F_1^2 + F_2^2 = 2 = F_2 F_3$. This result has a nice geometric interpretation: The sum of the areas of an $F_1 \times F_1$ square and an $F_2 \times F_2$ square equals the area of an $F_2 \times F_3$ rectangle; see Figure 5.8.

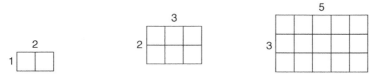

Figure 5.8.   **Figure 5.9.**

Likewise, $F_1^2 + F_2^2 + F_3^2 = 6 = 2 \cdot 3 = F_3 F_4$, and $F_1^2 + F_2^2 + F_3^2 + F_4^2 = 15 = 3 \cdot 5 = F_4 F_5$. These results also can be interpreted geometrically; see Figure 5.9.

More generally, we have the following result. Its proof also follows by PMI; see Exercise 5.8.

Theorem 5.4 (Lucas, 1876).

$$\sum_{i=1}^{n} F_i^2 = F_n F_{n+1}. \tag{5.5}$$

∎

For example,

$$\sum_{i=1}^{25} F_i^2 = F_{25} F_{26} = 75,025 \cdot 121,393 = 9,107,509,825.$$

LUCAS COUNTERPARTS

Interestingly, summation formulas (5.1) through (5.5) have analogous results for the Lucas family:

$$\sum_{i=1}^{n} L_i = L_{n+2} - 3 \tag{5.6}$$

$$\sum_{i=1}^{n} L_{2i-1} = L_{2n} - 2 \tag{5.7}$$

$$\sum_{i=1}^{n} L_{2i} = L_{2n+1} - 1 \tag{5.8}$$

$$L_{n-1}L_{n+1} - L_n^2 = 5(-1)^{n-1} \tag{5.9}$$

$$\sum_{i=1}^{n} L_i^2 = L_n L_{n+1} - 2. \tag{5.10}$$

These formulas also can be established using PMI; see Exercises 5.3–5.7. In addition, we can establish formula (5.10) using the Euclidean algorithm; see Exercise 9.19.

Identity (5.10) also can be interpreted geometrically: The sum of the areas of the n squares $L_i \times L_i$ equals the area of the rectangle $L_n \times L_{n+1}$ minus 2, where $1 \le i \le n$.

5.2   BINET'S FORMULAS

To derive new identities, we now present explicit formulas for both F_n and L_n. To this end, let α and β be the solutions of the quadratic equation $x^2 - x - 1 = 0$; so $\alpha = (1 + \sqrt{5})/2$ and $\beta = (1 - \sqrt{5})/2$; see Figure 5.10. (The choice of the equation will become clear in Chapter 13.)

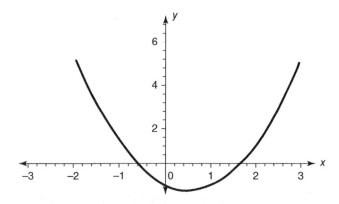

Figure 5.10.   Graph of $y = x^2 - x - 1$.

So $\alpha + \beta = 1$ and $\alpha\beta = -1$. Then $\alpha^2 = \alpha + 1$, $\alpha^3 = \alpha^2 + \alpha = 2\alpha + 1$, and $\alpha^4 = 2\alpha^2 + \alpha = 2(\alpha + 1) + \alpha = 3\alpha + 2$. Thus we have

$$\alpha = 1\alpha + 0$$
$$\alpha^2 = 1\alpha + 1$$
$$\alpha^3 = 2\alpha + 1$$
$$\alpha^4 = 3\alpha + 2.$$

Clearly, an interesting pattern emerges: The constant terms and the coefficients of α on the RHS[†] appear to be adjacent Fibonacci numbers. This observation leads us to the following result.

Lemma 5.1. $\alpha^n = \alpha F_n + F_{n-1}$, where $n \geq 0$. ∎

This also follows by PMI; see Exercise 5.52.
Changing $\sqrt{5}$ to $-\sqrt{5}$ yields the following corollary.

Corollary 5.3. $\beta^n = \beta F_n + F_{n-1}$, where $n \geq 0$. ∎

Lemma 5.1 and Corollary 5.3 now pave the way for an explicit formula for F_n. To this end, we let $u_n = (\alpha^n - \beta^n)/\sqrt{5}$, where $n \geq 1$. Then

$$u_1 = \frac{\alpha - \beta}{\sqrt{5}} = \frac{\sqrt{5}}{\sqrt{5}} = 1 \quad \text{and} \quad u_2 = \frac{\alpha^2 - \beta^2}{\sqrt{5}} = \frac{(\alpha + \beta)(\alpha - \beta)}{\sqrt{5}} = 1.$$

Suppose $n \geq 3$. Then

$$u_{n-1} + u_{n-2} = \frac{\alpha^{n-1} - \beta^{n-1}}{\sqrt{5}} + \frac{\alpha^{n-2} - \beta^{n-2}}{\sqrt{5}}$$

$$= \frac{\alpha^{n-2}(\alpha + 1) - \beta^{n-2}(\beta + 1)}{\sqrt{5}}$$

$$= \frac{\alpha^{n-2} \cdot \alpha^2 - \beta^{n-2} \cdot \beta^2}{\sqrt{5}}$$

$$= \frac{\alpha^n - \beta^n}{\sqrt{5}}$$

$$= u_n.$$

[†]RHS and LHS are abbreviations of right-hand side and left-hand side, respectively.

Thus u_n satisfies the Fibonacci recurrence and the same two initial conditions. Consequently, $u_n = F_n$.

This gives an explicit formula for F_n.

Theorem 5.5 (Binet's Formula). *Let α be the positive root of the quadratic equation $x^2 - x - 1 = 0$ and β its negative root. Then*

$$F_n = \frac{\alpha^n - \beta^n}{\alpha - \beta},$$

where $n \geq 1$. ∎

This explicit formula for F_n is called *Binet's formula,* after the French mathematician Jacques-Marie Binet (1786–1856), who found it in 1843. In fact, it was discovered in 1718 by the French mathematician Abraham De Moivre (1667–1754) using generating functions (see Chapter 13). It was independently derived in 1844 by the French engineer and mathematician Gabriel Lamé (1795–1870).

Binet[*]

De Moivre[#]

In any case, we will derive the formula in two other ways in later chapters. As we can predict, it can be employed to derive a myriad of Fibonacci identities, as we will see in several chapters.

The next two results, for example, follow by Binet's formula. Again, we omit their proofs for brevity; see Exercises 5.28 and 5.29.

[*]*Figure source*: https://en.wikipedia.org/wiki/Jacques_Philippe_Marie_Binet#/media/File:Jacques_Binet.jpg.
[#]*Figure source*: https://en.wikipedia.org/wiki/Abraham_de_Moivre#/media/File:Abraham_de_moivre.jpg.

Lamé*

Corollary 5.4 (Lucas, 1876).

$$F_{n+1}^2 + F_n^2 = F_{2n+1} \tag{5.11}$$

$$F_{n+1}^2 - F_{n-1}^2 = F_{2n}. \tag{5.12}$$

∎

For example, $F_8^2 + F_7^2 = 441 + 169 = 610 = F_{15}$, and $F_{11}^2 - F_9^2 = 7921 - 1156 = 6765 = F_{20}$.

Interestingly, both Cassini's formula and identity (5.11) are special cases of the identity

$$F_m F_{n+k} - F_k F_{n+m} = (-1)^k F_{m-k} F_n; \tag{5.13}$$

see Exercise 5.77. Larry Taylor of Rego Park, New York, discovered this formula in 1981 [556].

For example, when $n = 1$ and $m = k + 1$, it gives

$$F_{k+1}^2 - F_k F_{k+2} = (-1)^k,$$

which is Cassini's formula. Identity (5.11) can be deduced similarly; see Exercise 5.78.

Next we will present an interesting occurrence of Fibonacci numbers. It appeared as a problem in the 68th Annual William Lowell Putnam Mathematical Competition in 2007 [596]. The featured proof employs Binet's formula and PMI, and the properties $\alpha + \beta = 1, \alpha\beta = -1$, and $\beta^2 = \beta + 1$.

---

*Figure source: https://en.wikipedia.org/wiki/Gabriel_Lam%C3%A9#/media/File:Gabriel-Lam%C3%A9.jpeg.

Example 5.2. *Find an explicit formula for x_n, where $x_{n+1} = 3x_n + \lfloor \sqrt{5}x_n \rfloor$, where $x_0 = 1, n \geq 1$, and $\lfloor x \rfloor$ denotes the floor of the real number x.*

Solution. First, we compute a few initial values of x_n, and then look for a pattern:

$$x_0 = \quad 1 = 1 \cdot 1$$

$$x_1 = \quad 5 = 1 \cdot 5$$

$$x_2 = \quad 26 = 2 \cdot 13$$

$$x_3 = 136 = 4 \cdot 34$$

$$x_4 = 712 = 8 \cdot 89$$

$$\vdots \qquad \uparrow \ \uparrow$$

Powers of 2   Fibonacci numbers

Each x_n is a product of a power of 2 and a Fibonacci number. A closer examination helps us to make a conjecture: $x_n = 2^{n-1}F_{2n+3}$, where $n \geq 0$. We will now establish the validity of this conjecture using PMI.

Clearly, the formula works when $n = 0$. Now assume it works for an arbitrary integer $n \geq 0$. Then

$$x_{n+1} = 3x_n + \lfloor \sqrt{5}x_n \rfloor$$

$$= \lfloor (3 + \sqrt{5})x_n \rfloor$$

$$= \lfloor 2\alpha^2 x_n \rfloor$$

$$= \lfloor 2^n \alpha^2 F_{2n+3} \rfloor$$

$$= \left\lfloor 2^n \alpha^2 \cdot \frac{\alpha^{2n+3} - \beta^{2n+3}}{\alpha - \beta} \right\rfloor$$

$$= \left\lfloor \frac{2^n}{\alpha - \beta} \left[(\alpha^{2n+5} - \beta^{2n+5}) + (\beta^{2n+5} - \beta^{2n+1}) \right] \right\rfloor$$

$$= 2^n F_{2n+5} + \left\lfloor \frac{2^n}{\alpha - \beta} \beta^{2n+3}(\beta^2 - \alpha^2) \right\rfloor$$

$$= 2^n F_{2n+5} + \lfloor -2^n \beta^{2n+3} \rfloor .$$

Since $\beta^2 = \dfrac{3-\sqrt{5}}{2}$, $2^n \beta^{2n+2} = \dfrac{(3-\sqrt{5})^{n+1}}{2}$. Then $-2^n \beta^{2n+3} = \dfrac{\sqrt{5}-1}{4}(3-\sqrt{5})^{n+1}$.

Clearly, $0 < -2^n \beta^{2n+3} < 1$, so $\lfloor -2^n \beta^{2n+3} \rfloor = 0$. Consequently, $x_{n+1} = 2^n F_{2n+5}$.
Thus, by PMI, the conjecture is true for all $n \geq 0$.   ∎

In particular, $x_{11} = 2^{10}F_{25} = 2^{10} \cdot 75025 = 76,825,600$.

In Chapter 23, we will prove that there are only two distinct Fibonacci squares, namely, 1 and 144. Consequently, identity (5.11) has a nice geometric interpretation: *No two consecutive Fibonacci numbers can be the lengths of the legs of a right triangle with integral sides.*

The next theorem provides a link between four consecutive Fibonacci numbers and the lengths of the sides of a Pythagorean triangle, as established in 1948 by C.W. Raine [490].

Theorem 5.6. *Let ABC be a triangle with $AC = F_k F_{k+3}, BC = 2F_{k+1}F_{k+2},$ and $AB = F_{2k+3}$. Then $\triangle ABC$ is a Pythagorean triangle, right-angled at C.* ∎

It suffices to show that $AB^2 = AC^2 + BC^2$, so we leave its proof as an exercise.

For example, let $AC = F_7 F_{10} = 13 \cdot 55 = 715, BC = 2F_8 F_9 = 2 \cdot 21 \cdot 34 = 1428$, and $AB = F_{17} = 1597$. Then $AC^2 + BC^2 = 715^2 + 1428^2 = 2,550,409 = 1597^2 = AB^2$; so $\triangle ABC$ is a right triangle, right-angled at C.

Corresponding to Binet's formula for F_n, there is one for L_n, as the next theorem shows. We invite you to confirm it; see Exercise 5.14.

Theorem 5.7. *Let $n \geq 0$. Then $L_n = \alpha^n + \beta^n$.* ∎

The two Binet formulas can be used in tandem to derive an array of identities. The following corollary gives five such identities.

Corollary 5.5.

$$F_n L_n = F_{2n} \tag{5.14}$$

$$F_{n+1} + F_{n-1} = L_n \tag{5.15}$$

$$F_{n+2} + F_{n-2} = 3F_n \tag{5.16}$$

$$F_{n+2} - F_{n-2} = L_n \tag{5.17}$$

$$L_{n+1} + L_{n-1} = 5F_n. \tag{5.18}$$

∎

For example, $F_{20} = 6765 = 55 \cdot 123 = F_{10}L_{10}, F_{11} + F_{13} = 89 + 233 = 322 = L_{12}, F_{11} - F_7 = 89 - 13 = 76 = L_9,$ and $L_{10} + L_{12} = 123 + 322 = 445 = 5 \cdot 89 = 5F_{11}$.

Identity (5.14) implies that every Fibonacci number F_{2n} has nontrivial factors, when $n \geq 3$. It has an additional byproduct. To see this, let $n = 2^m$, where $m \geq 1$. Then

$$
\begin{aligned}
F_{2^m} &= L_{2^{m-1}} F_{2^{m-1}} \\
&= L_{2^{m-1}}(L_{2^{m-2}} F_{2^{m-2}}) = L_{2^{m-1}} L_{2^{m-2}} F_{2^{m-2}} \\
&= L_{2^{m-1}} L_{2^{m-2}}(L_{2^{m-3}} F_{2^{m-3}}) = L_{2^{m-1}} L_{2^{m-2}} L_{2^{m-3}} F_{2^{m-3}} \\
&\;\;\vdots \\
&= L_{2^{m-1}} L_{2^{m-2}} \cdots L_8 L_4 L_2 L_1.
\end{aligned}
$$

For example, $F_{32} = L_{16}L_8L_4L_2L_1 = 2207 \cdot 47 \cdot 7 \cdot 3 \cdot 1 = 2{,}178{,}309$.

It follows from identity (5.15) that the sum of any two Fibonacci numbers that are two units apart is a Lucas number. Likewise, by identity (5.17), the difference of any two Fibonacci numbers that lie four units away is also a Lucas number.

The next example is an interesting confluence of Fibonacci and Lucas numbers, and the well-known *Pell's equation* $x^2 - dy^2 = (-1)^n$, where x, y, d, and n are positive integers, and d is nonsquare [370, 454].

Example 5.3. *Solve the Pell's equation* $x^2 - 5y^2 = 1$.

Solution. Clearly, $(x_1, y_1) = (9, 4)$ is a solution: $9^2 - 5 \cdot 4^2 = 1$. So is $(161, 72)$. The solution $(9, 4)$ is the *fundamental (minimal) solution*.

It is well known that the general solution (x_n, y_n) of $x^2 - dy^2 = 1$ is given by

$$x_n = \frac{1}{2}\left[(x_1 + y_1\sqrt{d})^n + (x_1 - y_1\sqrt{d})^n\right]$$

$$y_n = \frac{1}{2\sqrt{d}}\left[(x_1 + y_1\sqrt{d})^n - (x_1 - y_1\sqrt{d})^n\right],$$

where (x_1, y_1) is the fundamental solution and $n \geq 2$.

Consequently,

$$x_n = \frac{1}{2}\left[(9 + 4\sqrt{5})^n + (9 - 4\sqrt{5})^n\right]$$

$$y_n = \frac{1}{2\sqrt{5}}\left[(9 + 4\sqrt{5})^n - (9 - 4\sqrt{5})^n\right].$$

We now make an interesting observation. By Lemma 5.1, we have

$$\alpha^6 = \alpha F_6 + F_5$$

$$= 8\alpha + 5$$

$$= 9 + 4\sqrt{5},$$

and hence $\beta^6 = 9 - 4\sqrt{5}$.

Thus

$$x_n = \frac{\alpha^{6n} + \beta^{6n}}{2} = \frac{1}{2}L_{6n}$$

and

$$y_n = \frac{\alpha^{6n} - \beta^{6n}}{2\sqrt{5}} = \frac{1}{2}F_{6n}.$$

Consequently, $(x_n, y_n) = (\frac{1}{2}L_{6n}, \frac{1}{2}F_{6n})$.   ■

In particular, $(x_2, y_2) = (\frac{1}{2}L_{12}, \frac{1}{2}F_{12}) = (\frac{322}{2}, \frac{144}{2}) = (161, 72)$, as expected.

Since $(x_n, y_n) = (\frac{1}{2}L_{6n}, \frac{1}{2}F_{6n})$ is the general solution of the equation $x^2 - 5y^2 = 1$, it follows that $L_{6n}^2 - 5F_{6n}^2 = 4$; see Exercise 5.37. It also follows that both L_{6n} and F_{6n} are even integers.

We will make one more observation. Since $0 < |\beta| < |\alpha|$, we have

$$\lim_{n \to \infty} \frac{x_{n+1}}{x_n} = \lim_{n \to \infty} \frac{\alpha^{6n+6} + \beta^{6n+6}}{\alpha^{6n} + \beta^{6n}}$$

$$= \alpha^6 \cdot \lim_{n \to \infty} \frac{1 + (\beta/\alpha)^{6n+6}}{1 + (\beta/\alpha)^{6n}}$$

$$= \alpha^6$$

$$= \lim_{n \to \infty} \frac{y_{n+1}}{y_n}.$$

For example, $\dfrac{x_4}{x_3} = \dfrac{51841}{2889} \approx 17.94427137$ and $\alpha^6 = 9 + 4\sqrt{5} \approx 17.94427191$.

The next example is an interesting application of Theorem 5.7 and the summation formula (5.6). Br. U. Alfred studied it in 1965 [8]. (Br. U. Alfred *was* Br. Alfred Brousseau; see p. 7.)

Example 5.4. *Compute the sum* $\displaystyle\sum_{k=1}^{n} \lfloor \alpha^k \rfloor$, *where* $\lfloor x \rfloor$ *denotes the floor of the real number* x.

Solution. By Binet's formula for L_k, $\alpha^k = L_k - \beta^k$, where $\beta < 0$ and $|\beta| < 1$. So $\lfloor \alpha^k \rfloor = L_k$ if k is odd; and $\lfloor \alpha^k \rfloor = L_k - 1$ if k is even. Thus

$$\sum_{k=1}^{n} \lfloor \alpha^k \rfloor = \sum_{k=1}^{n} L_k - \lfloor n/2 \rfloor$$

$$= L_{n+2} - \lfloor n/2 \rfloor - 3. \qquad \blacksquare$$

The next result is an application of Cassini's formula, the two Binet formulas, and identity (5.15); in addition, it requires a basic knowledge of extension fields in abstract algebra. Ira Gessel of Harvard University established it in 1972 [217]. But M.J. Wasteels proved it 70 years earlier [586].

Theorem 5.8. *A positive integer n is a Fibonacci number if and only if $5n^2 \pm 4$ is a square.*

Proof. Using Exercise 5.30 and Cassini's formula, we have

$$L_r = F_{r+1} + F_{r-1}$$

$$F_r^2 + (-1)^r = F_{r+1}F_{r-1}$$

$$L_r^2 - 4[F_r^2 + (-1)^r] = (F_{r+1} + F_{r-1})^2 - 4F_{r+1}F_{r-1}$$

$$= (F_{r+1} - F_{r-1})^2$$

$$= F_r^2$$

$$L_r^2 = 5F_r^2 + 4(-1)^r.$$

Consequently, if n is a Fibonacci number, then $5n^2 \pm 4$ is a square.

Conversely, let suppose $5n^2 \pm 4$ is a square m^2. Then

$$m^2 - 5n^2 = \pm 4$$

$$\frac{m + n\sqrt{5}}{2} \cdot \frac{m - n\sqrt{5}}{2} = \pm 1.$$

Since m and n have the same *parity* (both odd or both even), both $(m + n\sqrt{5})/2$ and $(m - n\sqrt{5})/2$ are integers in the extension field $Q(\sqrt{5}) = \{x + y\sqrt{5} | x, y \in Q\}$, where Q denotes the field of rational numbers. Since their product is ± 1, they must be units in the field. But the only integral units in $Q(\sqrt{5})$ are of the form $\pm \alpha^{\pm i}$. Then

$$\frac{m + n\sqrt{5}}{2} = \alpha^i$$

$$= \frac{1}{2}\left[(\alpha^i + \beta^i) + (\alpha^i - \beta^i)\right]$$

$$= \frac{L_i + F_i\sqrt{5}}{2}.$$

Thus $n = F_i$, a Fibonacci number.   ∎

SCHUB'S IDENTITY

P. Schub of the University of Pennsylvania discovered the identity $L_n^2 - 5F_n^2 = 4(-1)^n$ in 1950 [512]; see Exercise 5.37. We encountered this identity in the proof of Theorem 5.8. This identity implies that the Pell's equation $x^2 - 5y^2 = 4(-1)^n$ has infinitely many solutions $(x, y) = (L_n, F_n)$ [370].

Schub's Identity has an additional application. To see this, let $n = 2m + 1$. The resulting square, $L_{2m+1}^2 = 5F_{2m+1}^2 - 4$, is the discriminant of the quadratic equation $(F_{2m+1} \pm 1)x^2 - F_{2m+1}x - (F_{2m+1} \mp 1) = 0$. Consequently, its solutions are rational.

For example,

$$1x^2 - 2x - 3 = (1x + 1)(1x - 3)$$
$$6x^2 - 5x - 4 = (2x + 1)(3x - 4)$$
$$12x^2 - 13x - 14 = (3x + 2)(4x - 7)$$
$$35x^2 - 34x - 33 = (5x + 3)(7x - 11).$$

More generally,

$$[F_{2m+1} + (-1)^m]x^2 - F_{2m+1}x - [F_{2m+1} - (-1)^m] = (F_{m+1}x + F_m)(L_m x - L_{m+1}).$$

The truth of this rests on the following facts:

$$F_m F_{m-1} = F_m^2 - F_{m-1}^2 + (-1)^m$$
$$F_{m-1}F_{m+1} = F_m^2 + (-1)^m$$
$$F_m^2 + F_{m+1}^2 = F_{2m+1}.$$

A. Struyk made these observations in 1959 [546].

The next three examples are also interesting applications of identity (5.15).

Are there Fibonacci numbers other than 1, 2, and 3 that are also Lucas numbers? Unfortunately, there aren't any, as S. Kravitz established in 1965 [374].

Example 5.5. *Prove that no Fibonacci number other than 1, 2, and 3 is a Lucas number.*

Proof. Clearly, $F_1 = F_2 = 1 = L_1, F_3 = 2 = L_0$, and $F_4 = 3 = L_2$. (If we include Fibonacci numbers with negative subscripts, then $F_{-1} = 1 = L_1$ and $F_{-3} = 2 = L_0$.) Suppose $F_n = L_k$ for some integer $k \geq 3$, where $n \geq 5$. Then, by identity (5.15), $F_n = F_{k+1} + F_{k-1}$. Since $k \geq 3, F_n > F_{k+1}$; so $n > k + 1$. Then $F_n = F_{k+1} + F_{k-1} < F_{k+1} + F_k = F_{k+2} \leq F_n$, which is a contradiction. This gives the desired result. ∎

5.3 CYCLIC PERMUTATIONS AND LUCAS NUMBERS

Example 5.6. *Consider the permutation problem in Example 4.10. This time we arrange the numbers around a circle; see Figure 5.11. Let q_n denote the number of cyclic permutations g that move no element more than one position from its natural position on the circle, where $n \geq 3$. Find q_n.*

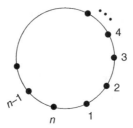

Figure 5.11.

Proof. Figure 5.12 shows such cyclic permutations for $n = 3$; so $q_3 = 4$. Notice that with 1 and 3 swapped, there is just one permutation; otherwise, there are three, for a total of four cyclic permutations.

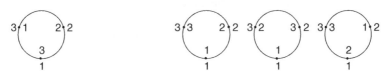

1 and 3 swapped   1 and 3 not swapped

Figure 5.12.

More generally, let $g(n) = 1$. The remaining $n - 2$ elements can be rearranged in p_{n-2} ways such that no element is moved by more than one space from its natural position. On the other hand, let $g(n) \neq 1$. This case is precisely the same as the permutation problem for the linear case; so there are p_n such cyclic permutations. Thus $q_n = p_{n-2} + p_n = F_{n-1} + F_{n+1} = L_n$. ∎

Example 5.7. *Recall from Example 4.1 that there are $a_n = F_{n+2}$ n-bit words that do not contain consecutive 1s. This time, in lieu of arranging the bits linearly, we arrange them around a circle such that no two adjacent bits are 1s. Let b_n denote the number of such cyclic arrangements, where $n \geq 2$. Prove that $b_n = L_n$.*

Proof. Table 5.1 shows the possible such n-bit words, where $2 \leq n \leq 5$. It appears from the table that $b_n = L_n$, where $n \geq 2$.

To confirm the observation, consider an arbitrary cyclic n-bit word $w = w_1 w_2 \ldots w_n$. Suppose $w_n = 1$: $w = \underbrace{\cdots \quad 1}_{n \text{ bits}}$. Then $w_1 = 0$ and $w_{n-1} = 0$;

so $w = 0 \underbrace{\cdots \quad 01}_{n-3 \text{ bits}}$. There are no restrictions on w_2 or w_{n-2}. By Example 4.1,

there are $F_{(n-3)+2} = F_{n-1}$ such cyclic permutations with the desired property.

TABLE 5.1. Cyclic *n*-bit Words With No Consecutive 1s

n	n-bit Words										b_n
2	01	10									3
3	000	001	010	100							4
4	0000	0001	0010	0100	1000	0101	1010				7
5	00000	00001	00010	00100	01000	10000	01010	01001	10100	10010	00101	11

$$\uparrow$$
$$L_n$$

On the other hand, suppose $w_n = 0$; $w = \underbrace{\quad \cdots \quad}_{n \text{ bits}} 0$. Then both w_1 and w_{n-1}

can be 0 or 1: $w = \underbrace{\begin{matrix} 0 & \cdots & 00 \\ 1 & \cdots & 10 \end{matrix}}_{n-1 \text{ bits}}$. Again, by Example 4.1, there are $F_{(n-1)+2} =$

F_{n+1} such cyclic permutations with the required property.

Thus $b_n = F_{n-1} + F_{n+1} = L_n$, as desired.   ∎

GRAPH-THEORETIC BRIDGES

As we can predict, this example and cycle graphs are very closely related. To see this, we make another graph-theoretic definition. A simple graph with vertices $v_1, v_2, \ldots, v_n$ is a *cycle graph* C_n if v_i is adjacent to v_{i+1}, and v_n to v_1, where $1 \leq i \leq n-1$.

We will now show that there is a bijection between the independent subsets S of the vertex set $V_n = \{v_1, v_2, \ldots, v_n\}$ of C_n, and the set B_n of circular n-bit words with *no* consecutive 1s. To this end, we define $w_i = 1$ if and only if $v_i \in S$. Since $v_i, v_{i+1} \notin S$, it follows that $w_i \neq 1$ and $w_{i+1} \neq 1$; so the binary word w has the desired property.

Since the process is reversible, this matching between the family of independent subsets of V_n and B_n is bijective. So the number of such independent subsets of V_n equals the number of circular binary sequences with no adjacent 1s, namely, L_n.

For example, the 5-bit word corresponding to the independent subset $\{v_1, v_4\}$ of V_5 is 10010; see Figure 5.13, where the vertices v_1 and v_4 are circled.

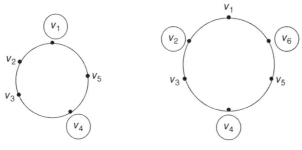

Figure 5.13.   Figure 5.14.

Figure 5.14 shows the independent subset of V_6 corresponding to the 6-bit word 010101.

A COMBINATORIAL INTERPRETATION OF $L_n = F_{n+1} + F_{n-1}$

We are now in a position to offer a combinatorial interpretation of the identity $L_n = F_{n+1} + F_{n-1}$, where $n \geq 3$. The essence of the combinatorial approach lies in the *Fubini principle,* named after the Italian mathematician Guido Fubini (1879–1943): *Counting the same set of objects in two different ways yields the same result.*

Let $V_n = \{v_1, v_2, \ldots, v_n\}$ denote the vertex set of the cycle graph C_n. The number of independent subsets of V_n is L_n.

We will now compute it in a different way. Let W be an arbitrary independent subset of V_n. Suppose $v_n \notin W$. Then v_{n-1} may or may not be in W. Since a path P_k has F_{k+2} independent subsets of its vertex set, it follows that there are F_{n+1} such independent subsets W. On the other hand, suppose $v_n \in W$. Then $v_{n-1}, v_1 \notin W$. There are F_{n-1} such independent subsets W. Combining the two cases, there are a total of $F_{n+1} + F_{n-1}$ independent subsets of V_n.

Combining the two counts, we get $L_n = F_{n+1} + F_{n-1}$, where $n \geq 3$.

5.4   COMPOSITIONS REVISITED

The next example presents another occurrence of Lucas numbers, in the study of compositions.

Example 5.8.   *Find the number of distinct compositions C_n of the positive integer n such that they do* not *begin with or end in 2, where $n \geq 2$.*

Solution.   Table 5.2 shows the possible compositions for $2 \leq n \leq 5$. Using these data, we conjecture that $C_n = L_{n-1}$, where $n \geq 2$. We will now confirm this.

TABLE 5.2.   **Compositions That Do Not Begin With or End in 2**

n	Desired Compositions	Count
2	$1 + 1$	1
3	$1 + 1 + 1,\ 1 + 2,\ 2 + 1$	3
4	$1 + 1 + 1 + 1,\ 1 + 1 + 2,\ 1 + 2 + 1,\ 2 + 1 + 1$	4
5	$1 + 1 + 1 + 1 + 1,\ 1 + 1 + 1 + 2,\ 1 + 1 + 2 + 1,\ 1 + 2 + 1 + 1,$	7
	$2 + 1 + 1 + 1,\ 1 + 2 + 2,\ 2 + 2 + 1$	

$$\uparrow$$
$$L_{n-1}$$

Since $C_2 = L_1$, we let $n \geq 3$. Consider an arbitrary composition of n.

Case 1. Suppose it ends in 1. Then it can begin with a 1 or 2; and the summands, excluding 1, must add up to $n - 1$: $\underbrace{\cdots}_{\text{sum} = n-1} + 1$. By Theorem 4.1, there are F_n such compositions.

Case 2. On the other hand, suppose the composition ends in 2. Then the composition must begin with a 1, and the remaining summands must add up to $n - 2$: $1 + \underbrace{\cdots}_{\text{sum} = n-3} + 2$. Again, by Theorem 4.1, there are F_{n-2} such compositions.

Thus $C_n = F_n + F_{n-2} = L_{n-1}$, as expected. ∎

5.5 NUMBER OF DIGITS IN F_n AND L_n

Binet's formulas can be employed successfully to predetermine the number of digits in both F_n and L_n. To achieve this, we rewrite Binet's formula for F_n as

$$F_n = \frac{\alpha^n}{\sqrt{5}} \left[1 - (\beta/\alpha)^n \right].$$

Since $|\beta| < |\alpha|$, $(\beta/\alpha)^n \to 0$ as $n \to \infty$. Therefore, when n is sufficiently large,

$$F_n \approx \frac{\alpha^n}{\sqrt{5}}$$

$$\log F_n \approx n \log \alpha - (\log 5)/2.$$

So

$$\text{Number of digits in } F_n = 1 + \text{characteristic of } \log F_n$$
$$= \lceil \log F_n \rceil$$
$$= \lceil n \log \alpha - (\log 5)/2 \rceil$$
$$= \lceil n[\log(1 + \sqrt{5}) - \log 2] - (\log 5)/2 \rceil,$$

where $\lceil x \rceil$ denotes the ceiling of the real number x.

For example, the number of digits in F_{30} is given by

$$\lceil 30[\log(1 + \sqrt{5}) - \log 2] - (\log 5)/2 \rceil = \lceil 5.92014420533 \rceil = 6.$$

Notice that $F_{30} = 832,040$ does indeed consist of six digits. Likewise, F_{45} is ten digits long; see Table A.2.

Since $L_n = \alpha^n + \beta^n$, it follows that $L_n \approx \alpha^n$, when n is sufficiently large; so $\log L_n \approx n \log \alpha$. Thus L_n consists of exactly

$$\lceil \log L_n \rceil = \lceil n \log \alpha \rceil = \lceil n[\log(1 + \sqrt{5}) - \log 2] \rceil$$

digits.

For example, L_{39} is $\lceil 39[\log(1 + \sqrt{5}) - \log 2] \rceil = 9$ digits long; and L_{50} is $\lceil 50[\log(1 + \sqrt{5}) - \log 2] \rceil = 11$ digits long.

5.6   THEOREM 5.8 REVISITED

Theorem 5.8 has an interesting counterpart for Lucas numbers. It was developed in 1974 by G. Wulczyn of Bucknell University, Lewisburg, Pennsylvania [605].

Theorem 5.9.   *A positive integer n is a Lucas number if and only if $5n^2 \pm 20$ is a square.*

Proof. Let $n = L_{2m+1}$. By Binet's formula, we then have

$$
\begin{aligned}
5n^2 + 20 &= 5(\alpha^{2m+1} + \beta^{2m+1})^2 + 20 \\
&= 5\left[\alpha^{4m+2} + \beta^{4m+2} + 2(\alpha\beta)^{2m+1}\right] + 20 \\
&= 5\left[\alpha^{4m+2} + \beta^{4m+2} - 2(\alpha\beta)^{2m+1}\right] \\
&= 5(\alpha^{2m+1} - \beta^{2m+1})^2 \\
&= 5\left(\sqrt{5}F_{2m+1}\right)^2 \\
&= 25F_{2m+1}^2.
\end{aligned}
$$

On the other hand, let $n = L_{2m}$. Then

$$
\begin{aligned}
5n^2 - 20 &= 5(\alpha^{2m} + \beta^{2m})^2 - 20 \\
&= 5\left[\alpha^{4m} + \beta^{4m} + 2(\alpha\beta)^{2m}\right] - 20 \\
&= 5\left[\alpha^{4m} + \beta^{4m} - 2(\alpha\beta)^{2m}\right] \\
&= 5(\alpha^{2m} - \beta^{2m})^2 \\
&= 5\left(\sqrt{5}F_{2m}\right)^2 \\
&= 25F_{2m}^2.
\end{aligned}
$$

Thus, if n is a Lucas number, then $5n^2 \pm 20$ is a square.

The proof of the converse is a bit complicated, so we omit it.   ∎

For example, let $n = 199 = L_{11}$. Then $5n^2 + 20 = 5 \cdot 199^2 + 20 = 198{,}025 = 445^2$, a square. On the other hand, let $n = 843 = L_{14}$. Then $5n^2 - 20 = 5 \cdot 843^2 - 20 = 3{,}553{,}225 = 1885^2$, again a square.

Theorem 5.8 has an interesting byproduct: The Diophantine equation $25y^2 - 5x^2 = 20(-1)^{n+1}$ has infinitely many solutions $(x, y) = (L_n, F_n)$; that is, the Pell's equation $x^2 - 5y^2 = 4(-1)^n$ has infinitely many solutions $(x, y) = (L_n, F_n)$. Consequently, $L_n^2 - 5F_n^2 = 4(-1)^n$. We can confirm this identity independently; see Exercise 5.37.

With Binet's formulas at our disposal, we now extend Fibonacci and Lucas numbers to negative subscripts.

FIBONACCI AND LUCAS NUMBERS WITH NEGATIVE SUBSCRIPTS

If we apply Fibonacci recurrence to the negative side, we get

	F_{-4}	F_{-3}	F_{-2}	F_{-1}	F_0	F_1	F_2	F_3	F_4	
$\cdots$	-3	2	-1	1	0	1	1	2	3	$\cdots$

So it appears that $F_{-n} = (-1)^{n+1}F_n$, where $n \geq 1$.

To confirm this, assume Binet's formula works for negative exponents. Then

$$
\begin{aligned}
F_{-n} &= \frac{\alpha^{-n} - \beta^{-n}}{\alpha - \beta} \\
&= \frac{(-\beta)^{-n} - (-\alpha)^{-n}}{\alpha - \beta} \\
&= \frac{(-1)^{n+1}(\alpha^n - \beta^n)}{\alpha - \beta} \\
&= (-1)^{n+1}F_n.
\end{aligned}
\tag{5.19}
$$

Likewise,

$$
L_{-n} = (-1)^n L_n.
\tag{5.20}
$$

Thus $F_{-n} = F_n$ if and only if n is odd; and $L_{-n} = L_n$ if and only if n is even. We can now extend Lemma 5.1 to negative exponents.

$$
\begin{aligned}
\alpha^{-n} &= \alpha F_{-n} + F_{-n-1} \\
&= (-1)^{n+1}\left(\alpha F_n - F_{n+1}\right) \\
&= \begin{cases} \alpha F_n - F_{n+1} & \text{if } n \text{ is odd} \\ F_{n+1} - \alpha F_n & \text{otherwise.} \end{cases}
\end{aligned}
\tag{5.21}
$$

For example, $\alpha^{-12} = F_{11} - \alpha F_{12} = 89 - 144\alpha$.

Formula (5.21) can also be established by PMI, or by showing that $\alpha^n(\alpha F_n - F_{n+1}) = (-1)^{n+1}$.

It follows from formula (5.21) that

$$
\beta^{-n} = \begin{cases} \beta F_n - F_{n+1} & \text{if } n \text{ is odd} \\ F_{n+1} - \beta F_n & \text{otherwise.} \end{cases} \tag{5.22}
$$

Two intriguing patterns emerge from the negative powers of α:

$$
\begin{aligned}
\alpha^{-1} &= 1 \cdot \alpha - 1 \\
\alpha^{-2} &= 2 - 1 \cdot \alpha \\
\alpha^{-3} &= 2 \cdot \alpha - 3 \\
\alpha^{-4} &= 5 - 3 \cdot \alpha \\
\alpha^{-5} &= 5 \cdot \alpha - 8 \\
&\;\;\vdots
\end{aligned}
$$

They are indicated by the criss-crossing arrows. The absolute values of the coefficients of α are consecutive Fibonacci numbers; and so are the absolute values of the various constants.

The summation formulas (5.1) through (5.3) are special cases of the generalized formula in the next theorem. It has an array of interesting byproducts. Its proof is a consequence of Binet's formula for Fibonacci numbers and the geometric sum

$$
\sum_{i=0}^{n} r^i = \frac{r^{n+1} - 1}{r - 1},
$$

where $r \neq 1$.

Theorem 5.10.

$$
\sum_{i=0}^{n} F_{ki+j} = \frac{F_{nk+k+j} - (-1)^k F_{nk+j} - F_j - (-1)^j F_{k-j}}{L_k - (-1)^k - 1}. \tag{5.23}
$$

Proof.

$$
\sqrt{5} \sum_{i=0}^{n} F_{ki+j} = \sum_{i=0}^{n} (\alpha^{ki+j} - \beta^{ki+j})
$$

$$
= \alpha^j \sum_{i=0}^{n} \alpha^{ki} - \beta^j \sum_{i=0}^{n} \beta^{ki}
$$

$$= \alpha^j \cdot \frac{\alpha^{nk+k} - 1}{\alpha^k - 1} - \beta^j \cdot \frac{\beta^{nk+k} - 1}{\beta^k - 1}$$

$$= \frac{(\alpha^{nk+k+j} - \alpha^j)(\beta^k - 1) - (\beta^{nk+k+j} - \beta^j)(\alpha^k - 1)}{(\alpha\beta)^k - (\alpha^k + \beta^k) + 1}$$

$$\sum_{i=0}^{n} F_{ki+j} = \frac{-F_{nk+k+j} + (-1)^k F_{nk+j} + F_j - (\alpha^k \beta^j - \alpha^j \beta^k)/\sqrt{5}}{(-1)^k - L_k + 1}.$$

Since

$$\alpha^k \beta^j - \alpha^j \beta^k = (\alpha\beta)^j (\alpha^{k-j} - \beta^{k-j}) = (-1)^j \sqrt{5} F_{k-j},$$

this yields the desired result. ∎

Letting $j = 0$, formula (5.23) yields the following result.

Corollary 5.6.

$$\sum_{i=0}^{n} F_{ki} = \frac{F_{nk+k} - (-1)^k F_{nk} - F_k}{L_k - (-1)^k - 1}. \tag{5.24}$$

∎

In particular, this gives the summation formulas (5.1) through (5.3).

Corollary 5.7.

$$\sum_{i=1}^{n} F_i = F_{n+2} - 1$$

$$\sum_{i=1}^{n} F_{2i-1} = F_{2n}$$

$$\sum_{i=1}^{n} F_{2i} = F_{2n+1} - 1.$$

∎

It follows from formula (5.24) that

$$\sum_{i=0}^{n} F_{3i} = \frac{F_{3n+2} - 1}{2}.$$

For example,

$$\sum_{i=0}^{5} F_{3i} = \frac{F_{17} - 1}{2} = \frac{1597 - 1}{2} = 798.$$

Identity (5.23) also implies the formula in Example 5.1:

$$\sum_{i=0}^{n} F_{i+j} = F_{n+j+2} - F_{j+1}.$$

For example,

$$\sum_{i=0}^{8} F_{i-5} = F_5 - F_{-4} = 5 - (-3) = 8.$$

5.7   CATALAN'S IDENTITY

Using Binet's formula for F_n, we can generalize Cassini's identity, as the next theorem shows. It was established in 1879 by the Belgian mathematician Eugene Charles Catalan (1814–1894). The ubiquitous Catalan numbers[†] are named after him.

Theorem 5.11 (Catalan, 1879).   *Let k be a positive integer and $n \geq k$. Then*

$$F_{n+k}F_{n-k} - F_n^2 = (-1)^{n+k+1}F_k^2. \tag{5.25}$$

Proof. Using Binet's formula and the identity $L_{2n} = 5F_n^2 + 2(-1)^n$ (see Exercise 5.40), we have

$$
\begin{aligned}
5(\text{LHS}) &= (\alpha^{n+k} - \beta^{n+k})(\alpha^{n-k} - \beta^{n-k}) - (\alpha^n - \beta^n)^2 \\
&= (\alpha\beta)^n(\alpha^k\beta^{-k} - \alpha^{-k}\beta^k) + 2(\alpha\beta)^n \\
&= (-1)^n(\alpha^k\beta^{-k} - \alpha^{-k}\beta^k) + 2(-1)^n \\
&= 2(-1)^n - (-1)^{n+k}(\alpha^{2k} + \beta^{2k}) \\
&= 2(-1)^n + (-1)^{n+k+1}[5F_k^2 + 2(-1)^k] \\
&= 5(-1)^{n+k+1}F_k^2 + 2(-1)^n + 2(-1)^{n+2k+1} \\
&= 5(-1)^{n+k+1}F_k^2 \\
\text{LHS} &= (-1)^{n+k+1}F_k^2,
\end{aligned}
$$

as desired.   ∎

For example, let $n = 10$ and $k = 3$. Then $F_{13}F_7 - F_{10}^2 = 233 \cdot 13 - 55^2 = 4 = (-1)^{14}F_3^2$.

Notice that the identity $F_{n+1}^2 + F_n^2 = F_{2n+1}$ follows from Catalan's identity (5.25).

[†]For a detailed discussion of Catalan numbers, see the author's *Catalan Numbers with Applications*, Oxford University Press, New York, 2009.

GELIN–CESÀRO IDENTITY

Catalan's identity has a beautiful byproduct. It can be employed to develop a spectacular identity involving the difference of two fourth-order Fibonacci products, namely, the *Gelin–Cesàro identity*

$$F_{n+2}F_{n+1}F_{n-1}F_{n-2} - F_n^4 = -1. \tag{5.26}$$

E. Gelin originally stated this identity, and the Italian mathematician Ernesto Cesàro (1859–1906) proved it in 1880 [147, 437].

We can easily establish this using Catalan's identity:

$$\begin{aligned}
\text{LHS} &= (F_{n+2}F_{n-2})(F_{n+1}F_{n-1}) - F_n^4 \\
&= \left[F_n^2 - (-1)^n\right]\left[F_n^2 + (-1)^n\right] - F_n^4 \\
&= -1 \\
&= \text{RHS}.
\end{aligned}$$

Catalan's identity has a Lucas counterpart:

$$L_{n+k}L_{n-k} - L_n^2 = 5(-1)^{n+k}F_k^2.$$

The proof follows along the same lines; see Exercise 5.36. It follows from this Catalan-like identity that $L_{2n+1} = L_{n+1}^2 - 5F_n^2 = 5F_{n+1}^2 - L_n^2$; so $L_{n+1}^2 + L_n^2 = 5F_{2n+1}$.

As can be predicted, we can use this Cassini-like identity to develop the pleasing Gelin–Cesàro-like identity for Lucas numbers:

$$L_{n+2}L_{n+1}L_{n-1}L_{n-2} - L_n^4 = -25. \tag{5.27}$$

The proof follows along the same lines as in the case of identity (5.26).

For example, $L_7 L_6 L_4 L_3 - L_5^4 = 29 \cdot 18 \cdot 7 \cdot 4 - 11^4 = -25$.

MELHAM'S IDENTITY

There is a corresponding formula for the difference of two third-order Fibonacci products:

$$F_{n+1}F_{n+2}F_{n+6} - F_{n+3}^3 = (-1)^n F_n. \tag{5.28}$$

R.S. Melham of the University of Technology, Sydney, Australia, discovered this attractive identity in 2003 [437].

We will now establish this, omitting some basic algebra. Since $F_{n+6} = 5F_{n+2} + 3F_{n+1}$, we have

$$\begin{aligned}
\text{LHS} &= F_{n+1}F_{n+2}(5F_{n+2} + 3F_{n+1}) - (F_{n+2} + F_{n+1})^3 \\
&= 2F_{n+2}^2 F_{n+1} - F_{n+2}^3 - F_{n+1}^3 \\
&= F_{n+2}^2(2F_{n+1} - F_{n+2}) - F_{n+1}^3 \\
&= (F_{n+1} + F_n)^2(F_{n+1} - F_n) - F_{n+1}^3 \\
&= (F_{n+1} + F_n)(F_{n+1}^2 - F_n^2) - F_n^3 \\
&= F_{n+1}F_n(F_{n+1} - F_n) - F_n^3 \\
&= F_n(F_{n+1}F_{n-1}) - F_n^3 \\
&= F_n\left[F_n^2 + (-1)^n\right] - F_n^3 \\
&= (-1)^n F_n \\
&= \text{RHS}. \quad\blacksquare
\end{aligned}$$

Melham's identity has a Lucas counterpart:

$$L_{n+1}L_{n+2}L_{n+6} - L_{n+3}^3 = 5(-1)^n L_n. \tag{5.29}$$

The proof follows along similar lines.

For example, $L_6 L_7 L_{11} - L_8^3 = 18 \cdot 29 \cdot 199 - 47^3 = 55 = 5(-1)^4 L_5$.

We can even generalize Catalan's identity; see identities 2 and 19 in the following section.

5.8   ADDITIONAL FIBONACCI AND LUCAS IDENTITIES

Over the centuries, a vast array of Fibonacci and Lucas identities have been developed. The following list shows some of them. It would be a good exercise to confirm the validity of each.

1. $F_{n+4}^3 = 3F_{n+3}^3 + 6F_{n+2}^3 - 3F_{n+1}^3 - F_n^3$ (Zeitlin and Parker, 1963 [617]).
2. $F_m F_n - F_{m+k}F_{n-k} = (-1)^{n-k}F_{m+k-n}F_k$.
3. $5\sum_{i=1}^{n} F_{i-2}F_i = L_{2n-1} + 3v + 1$, where $v = \begin{cases} 1 & \text{if } n \text{ is odd} \\ 0 & \text{otherwise.} \end{cases}$
4. $\alpha^m F_{n-m+1} + \alpha^{m-1}F_{n-m} = \alpha^n$.
5. $\beta^m F_{n-m+1} + \beta^{m-1}F_{n-m} = \beta^n$.
6. $F_n = F_m F_{n-m+1} + F_{m-1}F_{n-m}$.
7. $L_n = L_m F_{n-m+1} + L_{m-1}F_{n-m}$.
8. $\left(\dfrac{L_n + \sqrt{5}F_n}{2}\right)^m = \dfrac{L_{mn} + \sqrt{5}F_{mn}}{2}$ (Fisk, 1963 [180]).

9. $\alpha^{-n} = (-1)^{n+1}(\alpha F_n - F_{n+1})$.

10. $L_{2m}L_{2n} = L_{m+n}^2 + 5F_{m-n}^2$ (Wall, 1964 [576]).

11. $L_{2m}L_{2n} = 5F_{m+n}^2 + L_{m-n}^2$ (Wall, 1964 [576]).

12. $L_{2m}L_{2n} = L_{m+n}^2 + L_{m-n}^2 - 4(-1)^{m+n}$ (Hoggatt and Lind, 1967 [314]).

13. $L_{2m+2n} + L_{2m-2n} = L_{2m}L_{2n}$.

14. $L_{2m+2n} - L_{2m-2n} = 5F_{2m}F_{2n}$.

15. $L_{4n} = 5F_{2n}^2 + 2$ (Lucas, 1876).

16. $L_{4n+2} = 5F_{2n+1}^2 - 2$ (Lucas, 1876).

17. $2(F_n^4 + F_{n+1}^4 + F_{n+2}^4) = (F_n^2 + F_{n+1}^2 + F_{n+2}^2)^2$ (Candido, 1951 [97]).

18. $(F_nF_{n+3})^2 + (2F_{n+1}F_{n+2})^2 = F_{2n+3}^2$ (Raine, 1948 [490]).

19. $F_{n+h}F_{n+k} - F_nF_{n+h+k} = (-1)^nF_hF_k$ (Everman et al., 1960 [162]).

20. $\displaystyle\sum_{i=1}^{n}(-1)^{i-1}F_{i+1} = (-1)^{n-1}F_n$.

21. $\displaystyle 5\sum_{i=1}^{n} F_{2i}^2 = 3F_{2n+1}^2 + 2F_{2n+2}^2 - 6F_{2n}F_{2n+2} - 2n - 5$ (Rao, 1953 [493]).

22. $\displaystyle 5\sum_{i=1}^{n} F_{2i-1}^2 = 3F_{2n}^2 + 2F_{2n-1}^2 - 4F_{2n-2}F_{2n} + 2n - 2$ (Rao, 1953 [493]).

23. $\displaystyle 5\sum_{i=1}^{n} F_{2i-1}F_{2i+1} = 2F_{2n+2}^2 - 3F_{2n+1}^2 + 3n + 1$ (Rao, 1953 [493]).

24. $\displaystyle 5\sum_{i=1}^{2n} F_{2i}F_{2i+2} = 3F_{2n+2}^2 - 2F_{2n+1}^2 - 3n - 1$ (Rao, 1953 [493]).

25. $\displaystyle \sum_{i=1}^{2n} F_iF_{i+2} = F_{2n+1}F_{2n+2} - 1$ (Rao, 1953 [493]).

26. $\displaystyle 5\sum_{i=1}^{n} F_{2i}F_{2i+1} = F_{2n+2}^2 + F_{2n+1}^2 - n - 2$ (Rao, 1953 [493]).

27. $\displaystyle 5\sum_{i=1}^{n} F_{2i-1}F_{2i} = 4F_{2n+1}^2 - F_{2n+2}^2 + n - 3$ (Rao, 1953 [493]).

28. $\displaystyle \sum_{i=1}^{2n} F_iF_{i+1} = F_{2n+1}^2 - 1$ (Rao, 1953 [493]).

29. $\displaystyle 5\sum_{i=1}^{n} F_{2i-1}F_{2i+3} = 3F_{2n+2}^2 - 2F_{2n+1}^2 + 7n - 1$ (Rao, 1953 [493]).

30. $\displaystyle 5\sum_{i=1}^{n} F_{2i}F_{2i+4} = 2F_{2n+4}^2 - 3F_{2n+3}^2 - 7n - 6$ (Rao, 1953 [493]).

31. $\displaystyle 2\sum_{i=1}^{2n} F_iF_{i+4} = F_{4n+6} - F_{2n+3}^2 - 4$ (Rao, 1953 [493]).

32. $\displaystyle 4\sum_{i=1}^{n} F_{2i}F_{2i+1}F_{2i+2} = F_{2n+2}^3 - F_{2n+2}$ (Rao, 1953 [493]).

33. $\displaystyle 4\sum_{i=1}^{n} F_{2i-1}F_{2i}F_{2i+1} = F_{2n+1}^3 + F_{2n+1} - 2$ (Rao, 1953 [493]).

34. $4 \sum_{i=1}^{n} F_{2i-1}^3 = F_{2n}^3 + 3F_{2n}$ (Rao, 1953 [493]).

35. $4 \sum_{i=1}^{n} F_{2i}^3 = F_{2n+1}^3 - 3F_{2n+1} + 2$ (Rao, 1953 [493]).

36. $4 \sum_{i=1}^{n} F_i^3 = F_{n+2}^3 - 3F_{n+1}^3 + 3(-1)^n F_n + 2$ (Rao, 1953 [493]).

37. $4 \sum_{i=1}^{n} F_{2i-1} F_{2i+1} F_{2i+3} = F_{2n+2}^3 + 7F_{2n+2} - 8$ (Rao, 1953 [493]).

38. $4 \sum_{i=1}^{n} F_{2i} F_{2i+2} F_{2i+4} = F_{2n+2}^3 - 7F_{2n+3} + 6$ (Rao, 1953 [493]).

39. $10 \sum_{i=1}^{n} F_i F_{i+2} F_{i+4} = F_{3n+8} - 16(-1)^n F_{n+1} - 5$ (Rao, 1953 [493]).

40. $F_n F_{n+1} F_{n+2} = F_{n+1}^3 - (-1)^n F_{n+1}$.

41. $F_{3n} = 4F_{3n-3} + F_{3n-6}$.

42. $F_m F_{2m} F_{3n} = F_{n+m}^3 - (-1)^m L_m F_n^3 + (-1)^m F_{n-m}^3$ (Halton, 1965 [255]).

43. $F_{m+n} = F_{m+1} F_{n+1} - F_{m-1} F_{n-1}$ (Mana, 1968 [426]).

44. $F_{r+s+t} = F_{r+1} F_{s+1} F_{t+1} + F_r F_s F_t - F_{r-1} F_{s-1} F_{t-1}$.

45. $F_{3n} = 5F_n^3 + 3(-1)^n F_n$ (Halton, 1965 [255]).

46. $F_r F_{m+n} = F_{m+r} F_n - (-1)^r F_m F_{n-r}$ (Halton, 1965 [255]).

47. $[5F_m^2 + 2(-1)^m]F_{2r} = F_{m+r+1}^2 - F_{m+r-1}^2 + F_{m-r+1}^2 - F_{m-r-1}^2$ (Halton, 1965 [255]).

48. $[5F_m^2 + 2(-1)^m]F_r^2 + 2(-1)^r F_m^2 = F_{m+r}^2 + F_{m-r}^2$ (Halton, 1965 [255]).

49. $F_m F_{2m} F_{3r} = F_{m+r}^3 + (-1)^r F_{m-r}^3 - (-1)^m L_m F_r^3$ (Halton, 1965 [255]).

50. $F_{2m+1} F_{2n+1} = F_{m+n+1}^2 + F_{m-n}^2$ (Tadlock, 1965 [554]).

51. $L_{2m+1} L_{2n+1} = L_{m+n+1}^2 - L_{m-n}^2 + 4(-1)^{m-n}$ (Tadlock, 1965 [554]).

52. $F_n^2 + F_{n+2k}^2 = F_{n+2k-2} F_{n+2k+1} + F_{2k-1} F_{2n+2k-1}$ (Sharpe, 1965 [523]).

53. $F_n^2 + F_{n+2k+1}^2 = F_{2k+1} F_{2n+2k+1}$ (Sharpe, 1965 [523]).

54. $F_{n+2k}^2 - F_n^2 = F_{2k} F_{2n+2k}$ (Sharpe, 1965 [523]).

55. $F_{n+2k+1}^2 - F_n^2 = F_{n-1} F_{n+2} - F_{2k} F_{2n+2k+2}$ (Sharpe, 1965 [523]).

56. $2 \sum_{i=1}^{n} i F_i^2 = 2nF_n F_{n+1} - 2F_n^2 + [1 - (-1)^n]$.

57. $2 \sum_{i=1}^{n} (n - i + 1) F_i^2 = 2F_n F_{n+2} - [1 - (-1)^n]$.

58. $\sum_{i=1}^{n} (2n - i) F_i^2 = F_{2n}^2$ (Hoggatt, 1964 [272]).

59. $2 \sum_{i=1}^{n} F_i F_{i+1} = 2F_{n+1}^2 - [1 + (-1)^n]$.

60. $8 \sum_{i=0}^{n} \sum_{j=0}^{i} \sum_{k=0}^{j} \sum_{l=0}^{k} F_l^2 = 8F_{n+2}^2 - 2n^2 - 8n - 11 + 3(-1)^n$ (Graham, 1965 [240]).

61. $L_n L_{n+1} = L_{2n+1} + (-1)^n$ (Hoggatt, 1968 [278]).

62. $F_n^2 + F_{n+4}^2 = F_{n+1}^2 + F_{n+3}^2 + 4F_{n+2}^2$ (Swamy, 1967 [548]).

63. $F_{n+1}^5 - F_n^5 - F_{n-1}^5 = 5F_{n+1}F_nF_{n-1}[2F_n^2 + (-1)^n]$ (Carlitz, 1969 [109]).

64. $L_{n+1}^5 - L_n^5 - L_{n-1}^5 = 5L_{n+1}L_nL_{n-1}[2L_n^2 - 5(-1)^n]$ (Carlitz, 1969 [109]).

65. $F_{n+1}^7 - F_n^7 - F_{n-1}^7 = 7F_{n+1}F_nF_{n-1}[2F_n^2 + (-1)^n]^2$ (Carlitz, 1969 [109]).

66. $L_{n+1}^7 - L_n^7 - L_{n-1}^7 = 7L_{n+1}L_nL_{n-1}[2L_n^2 - 5(-1)^n]^2$ (Carlitz, 1969 [109]).

67. $L_{2n} = 1 + \lfloor \sqrt{5}F_{2n} \rfloor$ (Seamons, 1967 [516]).

68. $F_{4n+1} - 1 = L_{2n+1}F_{2n}$ (Dudley and Tucker, 1971 [152]).

69. $F_{4n+3} - 1 = L_{2n+1}F_{2n+2}$ (Dudley and Tucker, 1971 [152]).

70. $F_{n+1}L_{n+2} - F_{n+2}L_n = F_{2n+1}$ (Carlitz, 1967 [105]).

71. $F_nL_{n+r} - F_{n+r}L_{n-r} = (F_{2r} - F_r)F_{2n-r+1} + (F_{2r-1} - F_{r-1})F_{2n-r} - (-1)^n[F_r + (-1)^rF_{2r}]$.

72. $F_nL_{n+r} - L_nL_{n-r} = F_{2n+r} - L_{2n-r} - (-1)^n[F_r + (-1)^rL_r]$.

73. $L_{n+r}^2 + L_{n-r}^2 = L_{2n}L_{2r} + 4(-1)^{n+r}$.

74. $L_{10n} = [(L_{4n} - 3)^2 + (5F_{2n})^2]L_{2n}$ (Jarden, 1967 [352]).

75. $5(F_{n+r}^2 + F_{n-r}^2) = L_{2n}L_{2r} - 4(-1)^{n+r}$.

76. $5(F_{m+r}F_{m+r+1} + F_{m-r}F_{m-r+1}) = L_{2m+1}L_{2r} - 2(-1)^{m+r}$.

77. $F_{m+r}F_{m+r-2} + F_{m-r}F_{m-r-2} = L_{2m-2}L_{2r} - 6(-1)^{m+r}$.

78. $L_{m+r}L_{m+r+1} + L_{m-r}L_{m-r+1} = L_{2m+2r+1} + L_{2m-2r+1} + 2(-1)^{m+r}$.

79. $L_{m+r}L_{m+r+1} + L_{m-r}L_{m-r+1} = L_{2m+1}L_{2r} + 2(-1)^{m+r}$.

80. $\dfrac{L_{m+n} + L_{m-n}}{F_{m+n} + F_{m-n}} = \begin{cases} 5F_m/L_m & \text{if } n \text{ is odd} \\ L_m/F_m & \text{otherwise (Wall, 1965 [580]).} \end{cases}$

81. $2F_{m+n} = F_mL_n + L_mF_n$ (Ferns, 1967 [168]).

82. $2L_{m+n} = L_mL_n + 5F_mF_n$ (Ferns, 1967 [168]).

83. $L_{m+n} + L_{m-n} = \begin{cases} 5F_mF_n & \text{if } n \text{ is odd} \\ L_mL_n & \text{otherwise.} \end{cases}$

84. $L_{m+n} - L_{m-n} = \begin{cases} L_mL_n & \text{if } n \text{ is odd} \\ 5F_mF_n & \text{otherwise.} \end{cases}$

85. $L_{m+n}^2 - L_{m-n}^2 = 5F_{2m}F_{2n}$.

86. $(F_nF_{n+1} - F_{n+2}F_{n+3})^2 = (F_nF_{n+3})^2 + (2F_{n+1}F_{n+2})^2$ (Umansky and Tallman, 1968 [567]).

87. $(L_nL_{n+1} - L_{n+2}L_{n+3})^2 = (L_nL_{n+3})^2 + (2L_{n+1}L_{n+2})^2$ (Umansky and Tallman, 1968 [567]).

88. $F_{n+1}^3 - F_n^3 - F_{n-1}^3 = 3F_{n+1}F_nF_{n-1}$ (Carlitz, 1969 [109]).

89. $L_{n+1}^3 - L_n^3 - L_{n-1}^3 = 3L_{n+1}L_nL_{n-1}$ (Carlitz, 1969 [109]).

90. $L_n^2 - F_n^2 = 4F_{n-1}F_{n+1}$ (Hoggatt, 1969 [282]).

91. $L_n L_{n+2} + 4(-1)^n = 5F_{n-1}F_{n+3}$ (Hoggatt, 1969 [281]).

92. $\sum_{i=1}^{n} F_i F_{3i} = F_n F_{n+1} F_{2n+1}$ (Recke, 1968 [498]).

93. $F_{n-1}^4 + F_n^4 + F_{n+1}^4 = 2[2F_n^2 + (-1)^n]^2$ (Hunter, 1966 [339]).

94. $L_{n-1}^4 + L_n^4 + L_{n+1}^4 = 2[2L_n^2 - 5(-1)^n]^2$ (Carlitz and Hunter, 1969 [118]).

95. $F_{n-1}^6 + F_n^6 + F_{n+1}^6 = 2[2F_n^2 + (-1)^n]^3 + 3F_{n-1}^2 F_n^2 F_{n+1}^2$.

96. $F_{n-1}^8 + F_n^8 + F_{n+1}^8 = 2[2F_n^2 + (-1)^n]^4 + 8F_{n-1}^2 F_n^2(F_{n-1}^4 + F_n^4 + 4F_{n-1}^2 F_n^2 + 3F_{n-1}F_n F_{2n-1})$.

97. $F_{m+n}^2 L_{m+n}^2 - F_m^2 L_m^2 = F_{2n} F_{4m+2n}$ (Hunter, 1966 [339]).

98. $75 \sum_{i=1}^{n-1} \sum_{j=1}^{i} \sum_{k=0}^{j} F_{2k-1}^2 = 3F_{4n} + n(5n^2 - 14)$ (Swamy, 1970 [551]).

99. $25 \sum_{i=1}^{n} \sum_{j=1}^{i} F_{2j-1}^2 = L_{4n+2} + 5n(n+1) - 3$ (Peck, 1970 [474]).

100. $F_{n+2}^3 - F_{n-1}^3 - 3F_n F_{n+1} F_{n+2} = F_{3n}$ (Padilla, 1969 [470]).

101. $L_{n+1}^3 + L_n^3 - L_{n-1}^3 = 5L_{3n}$.

102. $(F_n F_{n+3})^2 + (2F_{n+1} F_{n+2})^2 = F_{2n+3}^2$ (Anglin, 1970 [14]).

103. $F_{m+n} = F_m L_n - (-1)^n F_{m-n}$ (Ruggles, 1963 [504]).

104. $L_{5n} = L_n[L_{2n}^2 - (-1)^n L_{2n} - 1]$ (Carlitz, 1971 [113]).

105. $L_{5n} = L_n\{[L_{2n} - 3(-1)^n]^2 + 25(-1)^n F_n^2\}$ (Carlitz, 1972 [114]).

106. $F_{n+3}^2 = 2F_{n+2}^2 + 2F_{n+1}^2 - F_n^2$.

107. $F_{3n} = L_n F_{2n} - (-1)^n F_n$ (Cheves, 1970 [121]).

108. $F_{3n} = F_n[L_{2n} + (-1)^n]$.

109. $F_n^2 + F_{n+3}^2 = 2(F_{n+1}^2 + F_{n+2}^2)$ (Thompson, 1929).

110. $(F_n + F_{n+6})F_k + (F_{n+2} + F_{n+4})F_{k+1} = L_{n+3}L_{k+1}$ (Blank, 1956 [49]).

111. $\sum_{j=0}^{13} F_{i+j} = 29F_{i+8}$ (Heath, 1950 [261]).

112. $L_n^3 = 2F_{n-1}^3 + F_n^3 + 6F_{n-1}F_{n+1}^2$ (Barley, 1972 [21]).

113. $5F_{2n+3}F_{2n-3} = L_{4n} + 18$ (Blazej, 1975 [51]).

114. $1 + 4F_{2n+1}F_{2n+2}^2 F_{2n+3} = (2F_{2n+2}^2 + 1)^2$ (Hoggatt and Bergum, 1977 [307]).

115. $1 + 4F_{2n+1}F_{2n+2}F_{2n+3}F_{2n+4} = (2F_{2n+2}F_{2n+3} + 1)^2$ (Hoggatt and Bergum, 1977 [307]).

116. $F_{8n} = L_{2n} \sum_{k=1}^{n} L_{2n+4k-2}$ (Higgins, 1976 [263]).

117. $F_n F_{n+3}^2 - F_{n+2}^3 = (-1)^{n+1} F_{n+1}$ (Hoggatt and Bergum, 1977 [307]).

118. $F_{n+3}F_n^2 - F_{n+1}^3 = (-1)^{n+1} F_{n+2}$ (Hoggatt and Bergum, 1977 [307]).

119. $F_n F_{n+3}^2 - F_{n+4} F_{n+1}^2 = (-1)^{n+1} L_{n+2}$ (Hoggatt and Bergum, 1977 [307]).

120. $F_n L_{n+3}^2 - F_{n+4} L_{n+1}^2 = (-1)^{n+1} L_{n+2}$ (Hoggatt and Bergum, 1977 [307]).

121. $7F_{n+2}^3 - F_{n+1}^3 - F_n^3 = 3L_{n+1}F_{n+2}F_{n+3}$ (Barley, 1973 [22]).

122. $F_{3n} = \prod_{k=0}^{n-1}(L_{2\cdot 3^k} - 1)$ (Usiskin, 1974 [568]).

123. $L_{3n} = \prod_{k=0}^{n-1}(L_{2\cdot 3^k} + 1)$ (Usiskin, 1974 [569]).

124. $F_{mn} = L_m F_{m(n-1)} + (-1)^{m+1} F_{m(n-2)}$ (Cheves, 1975 [122]).

125. $L_{(2m+1)(4n+1)} - L_{2m+1} = 5F_{2n(2m+1)} F_{(2m+1)(2n+1)}$.

126. $L_n^2 + L_{n+1}^2 = L_{2n} + L_{2n+2}$.

127. $F_{n+k}^3 + (-1)^k F_{n-k}(F_{n-k}^2 + 3F_{n+k}F_n L_k) = L_k^3 F_n^3$ (Mana, 1978 [430]).

128. $F_{n+k}^3 - L_{3k}F_n^3 + (-1)^k F_{n-k}^3 = 3(-1)^n F_n F_k F_{2k}$ (Wulczyn, 1978 [607]).

129. $F_{n+10}^4 = 55(F_{n+8}^4 - F_{n+2}^4) - 385(F_{n+6}^4 - F_{n+4}^4) + F_n^4$ (Wulczyn, 1979 [608]).

130. $F_k F_{n+j} - F_j F_{n+k} = (-1)^j F_{k-j} F_n$ (Taylor, 1981 [556]).

131. $F_k L_{n+j} - F_j L_{n+k} = (-1)^j F_{k-j} L_n$ (Taylor, 1981 [556]).

Additional identities can be found in the exercises and in the following chapters.

5.9 FERMAT AND FIBONACCI

A judge by profession, the French mathematician Pierre de Fermat (1601–1665) observed that the numbers 1, 3, 8, and 120 have a fascinating property: *One more than the product of any two of them is a square.*

Fermat*

Figure source: https://en.wikipedia.org/wiki/Pierre_de_Fermat#/media/File:Pierre_de_Fermat.jpg.

For example,

$$1 + 1 \cdot 3 = 2^2 \qquad 1 + 1 \cdot 8 = 3^2 \qquad 1 + 1 \cdot 120 = 11^2$$
$$1 + 3 \cdot 8 = 5^2 \qquad 1 + 3 \cdot 120 = 19^2 \qquad 1 + 8 \cdot 120 = 31^2.$$

In 1969, Alan Baker and Harold Davenport of Trinity College, Cambridge University, proved that if 1, 3, 8, and x have this property, then x must be 120 [18].

Intriguingly enough, notice that $1 = F_2$, $3 = F_4$, $8 = F_6$, and $120 = 4 \cdot 2 \cdot 3 \cdot 5 = 4F_3F_4F_5$. Accordingly, eight years later, Hoggatt and G.E. Bergum of South Dakota State University picked up on this observation and established the following generalization [307].

Theorem 5.12 (Hoggatt and Bergum, 1977 [307]). *The numbers F_{2n}, F_{2n+2}, F_{2n+4}, and $4F_{2n+1}F_{2n+2}F_{2n+3}$ have the property that one more than the product of any two of them is a square.*

Proof. By Cassini's formula, $1 + F_{2n}F_{2n+2} = F_{2n+1}^2$, $1 + F_{2n+1}F_{2n+3} = F_{2n+2}^2$, and $1 + F_{2n+2}F_{2n+4} = F_{2n+3}^2$. Then

$$
\begin{aligned}
1 + F_{2n}(4F_{2n+1}F_{2n+2}F_{2n+3}) &= 1 + 4(F_{2n}F_{2n+2})(F_{2n+1}F_{2n+3}) \\
&= 1 + 4(F_{2n+1}^2 - 1)(F_{2n+2}^2 + 1) \\
&= 4F_{2n+1}^2 F_{2n+2}^4 - 4(F_{2n+2}^2 - F_{2n+1}^2) - 3 \\
&= 4F_{2n+1}^2 F_{2n+2}^2 - 4F_{2n+3}F_{2n} - 3 \\
&= 4F_{2n+1}^2 F_{2n+2}^2 - 4F_{2n+3}(F_{2n+2} - F_{2n+1}) - 3 \\
&= 4F_{2n+1}^2 F_{2n+2}^2 - 4F_{2n+3}F_{2n+2} + 4F_{2n+1}F_{2n+3} - 3 \\
&= 4F_{2n+1}^2 F_{2n+2}^2 - 4F_{2n+3}F_{2n+2} + 4(F_{2n+2}^2 + 1) - 3 \\
&= 4F_{2n+1}^2 F_{2n+2}^2 - 4F_{2n+2}(F_{2n+3} - F_{2n+2}) + 1 \\
&= 4F_{2n+1}^2 F_{2n+2}^2 - 4F_{2n+1}F_{2n+2} + 1 \\
&= (2F_{2n+1}F_{2n+2} - 1)^2.
\end{aligned}
$$

Similarly, it can be shown that $1 + F_{2n+2}(4F_{2n+1}F_{2n+2}F_{2n+3}) = (2F_{2n+2}^2 + 1)^2$ and $1 + F_{2n+4}(4F_{2n+1}F_{2n+2}F_{2n+3}) = (2F_{2n+2}F_{2n+3} + 1)^2$. Thus one more than the product of any two of the numbers is a square. ∎

For example, $n = 1$ yields the Fermat quadruple $(F_2, F_4, F_6, 4F_3F_4F_5) = (1, 3, 8, 120)$; $n = 2$ yields the quadruple $(F_4, F_6, F_8, 4F_5F_6F_7) = (3, 8, 21, 2080)$; and $n = 3$ yields the quadruple $(F_6, F_8, F_{10}, 4F_7F_8F_9) = (8, 21, 55, 37128)$.

Hoggatt and Bergum also proved the following theorem for Fibonacci numbers with consecutive subscripts.

Theorem 5.13 (Hoggatt and Bergum, 1977 [307]). *Let $x = 4F_{2n+2}F_{2n+3}F_{2n+4}$. Then*

1) $1 + xF_{2n+1} = (2F_{2n+2}F_{2n+3} + 1)^2$
2) $1 + xF_{2n+3} = (2F_{2n+3} - 1)^2$
3) $1 + xF_{2n+5} = (2F_{2n+3}F_{2n+4} - 1)^2$. ∎

For example, let $n = 3$. Then $x = 4F_8 F_9 F_{10} = 4 \cdot 21 \cdot 34 \cdot 55 = 157{,}080$. Then

$$1 + xF_7 = 1 + 157080 \cdot 13 = 1429^2 = (2 \cdot 21 \cdot 34 + 1)^2 = (2F_8 F_9 + 1)^2$$

$$1 + xF_9 = 1 + 157080 \cdot 34 = 2311^2 = (2 \cdot 34^2 - 1)^2 \quad = (2F_9^2 - 1)^2$$

$$1 + xF_{11} = 1 + 157080 \cdot 89 = 3739^2 = (2 \cdot 34 \cdot 55 - 1)^2 = (2F_9 F_{10} - 1)^2.$$

An unusual relationship exists between the well-known geometric constant π and Fibonacci numbers.

5.10   FIBONACCI AND π

In 1985, Yuri V. Matiyasevich of St. Petersburg, Russia, developed a wonderful formula for π in terms of Fibonacci numbers:

$$\pi = \lim_{n \to \infty} \sqrt{\frac{6 \log F_1 F_2 \cdots F_n}{\log[F_1, F_2, \ldots, F_n]}},$$

where $[x, y]$ denotes the least common multiple (lcm) of the integers x and y. A proof of this formula, using some sophisticated number theory, appeared in *The American Mathematical Monthly* the following year [431].

For example,

$$\sqrt{\frac{6 \log F_1 F_2 \cdots F_{10}}{\log[F_1, F_2, \ldots, F_{10}]}} \approx 2.7732249039$$

$$\sqrt{\frac{6 \log F_1 F_2 \cdots F_{12}}{\log[F_1, F_2, \ldots, F_{12}]}} \approx 2.8454900617.$$

It would be an interesting exercise to determine the value of n for which the formula gives a desired approximation of π. Additionally, does there exist a corresponding formula for Lucas numbers?

Finally, we turn to two simple, but interesting, Fibonacci puzzles.

TWO SIMPLE FIBONACCI PUZZLES

1. Think of two positive integers. Add them to get a third number. Add the second and third numbers to get a fourth number. Add the third and fourth numbers to get a fifth number. Continue like this by adding the last two numbers until there are ten numbers. Now add all ten numbers. The resulting sum is 11 times the seventh number. R.V. Heath discovered this puzzle in 1950 [261]. (See Exercise 7.11 for a justification of the puzzle.)
2. Write down four consecutive Fibonacci numbers. The (positive) difference of the squares of the two middle numbers equals the product of the other two. (See Exercise 5.33.)

EXERCISES 5

Prove each.
  1. Theorem 5.2.
  2. Corollary 5.1.
3–7. Identities (5.6) through (5.10).
  8. Theorem 5.4 using PMI.

Disprove each.
  9. $L_{n+1}L_{n-1} - L_n^2 = (-1)^n$.
  10. $L_{n+1}(L_n + L_{n+2}) = L_{2n+2}$.

Let $v_n = \alpha^n + \beta^n$, where $n \geq 1$. Verify each.
  11. $v_1 = 1$ and $v_2 = 3$.
  12. $v_n = v_{n-1} + v_{n-2}$, where $n \geq 3$.
(Exercises 5.11 and 5.12 confirm that $v_n = L_n$.)

Prove each using PMI.
  13. Binet's formula for F_n.
  14. Binet's formula for L_n.
  15. Prove that $F_n = L_n$ if and only if $n = 1$.

Find a quadratic equation with the given roots, where k is a real number.
  16. α^n, β^n.
  17. $\alpha^n + k, \beta^n + k$.
  18. α^n, α^{-n}, where n is odd.

Solve each equation, where $n \geq 2$.
  19. $F_{n-1}x^2 - F_n x - F_{n+1} = 0$ (Umansky, 1972 [566]).
  20. $L_{n-1}x^2 - L_n x - L_{n+1} = 0$.

Using Lemma 5.1, prove both.

21. Formula (5.1).
22. Formula (5.6).

23. Prove that $F_{n+5} = 5F_{n+1} + 3F_n$, where $n \geq 0$.
24. Prove that $5 \mid F_{5n}$ for every $n \geq 0$.
25. Using Binet's formula, establish Cassini's formula (5.4).
26. Solve the recurrence $D_{n+1} = D_n + L_{2n} - 1$, where $D_0 = 0$ (Hoggatt, 1973 [296]).

Establish each identity, where $m, n \geq 1$.

27. $F_{2n} = F_n L_n$.
28. $F_{n+1}^2 + F_n^2 = F_{2n+1}$.
29. $F_{n+1}^2 - F_{n-1}^2 = F_{2n}$.
30. $F_{n+1} + F_{n-1} = L_n$.
31. $F_{n+2} - F_{n-2} = L_n$.
32. $L_{n+1} + L_{n-1} = 5F_n$.
33. $F_{n+1}^2 - F_n^2 = F_{n-1} F_{n+2}$.
34. $5(F_n^2 + F_{n-2}^2) = 3L_{2n-2} - 4(-1)^n$.
35. $L_{n+1}^2 + L_n^2 = 5F_{2n+1}$.
36. $L_{n+k} L_{n-k} - L_n^2 = 5(-1)^{n+k} F_k^2$.
37. $L_n^2 - 5F_n^2 = 4(-1)^n$.
38. $L_n^2 - F_n^2 = 4F_{n+1} F_{n-1}$.
39. $L_n^2 = L_{2n} + 2(-1)^n$.
40. $L_{2n} = 5F_n^2 + 2(-1)^n$.
41. $L_{n+2} - L_{n-2} = 5F_n$.
42. $L_{n+1}^2 - L_n^2 = L_{n-1} L_{n+2}$.
43. $L_n + F_n = 2F_{n+1}$.
44. $L_n - F_n = 2F_{n-1}$.
45. $F_{2n-2} < F_n^2 < F_{2n-1}$, where $n \geq 2$ (Hoggatt, 1963 [267]).
46. $F_{2n-1} < L_{n-1}^2 < F_{2n}$, where $n \geq 2$ (Hoggatt, 1963 [267]).
47. $F_{n+2} < 2^n$, where $n \geq 3$ (Fuchs, 1964 [202]).
48. $\alpha^n > 2F_n$, where $n \geq 2$ (Jarden, 1964 [351]).
49. $6\alpha^n > 5L_n$, where $n \geq 2$ (Jarden, 1964 [351]).
50. $\alpha^{n-1} < F_{n+1} < \alpha^n$, where $n \geq 2$ (Niven and Zuckerman, 1980 [456]).
51. $L_{-n} = (-1)^n L_n$.
52. $\alpha^n = \alpha F_n + F_{n-1}$.
53. $1 + \alpha^{2n} = \begin{cases} \sqrt{5} F_n \alpha^n & \text{if } n \text{ is odd} \\ L_n \alpha^n & \text{otherwise.} \end{cases}$

54. $1 + \beta^{2n} = \begin{cases} -\sqrt{5}F_n\beta^n & \text{if } n \text{ is odd} \\ L_n\beta^n & \text{otherwise.} \end{cases}$

55. $L_{2m+n} - (-1)^m L_n = 5F_m F_{m+n}.$

56. $F_{2m+n} - (-1)^m F_n = F_m L_{m+n}.$

57. $F_{2m+n} + (-1)^m F_n = L_m F_{m+n}.$

58. $F_{m+n} + F_{m-n} = \begin{cases} L_m F_n & \text{if } n \text{ is odd} \\ F_m L_n & \text{otherwise.} \end{cases}$

59. $F_{m+n} - F_{m-n} = \begin{cases} F_m L_n & \text{if } n \text{ is odd} \\ L_m F_n & \text{otherwise.} \end{cases}$

60. $F_{m+n}^2 - F_{m-n}^2 = F_{2m}F_{2n}.$

61. $L_{3n} = L_n[L_{2n} - (-1)^n].$

62. $L_{2n} = 1 + \lfloor \sqrt{5}F_{2n} \rfloor$, where $n > 0$ (Seamons, 1967 [516]).

63. $F_n F_{n+1} - F_{n-1}F_{n-2} = F_{2n-1}$ (Lucas, 1876).

64. $3F_n + L_n = 2F_{n+2}.$

65. $F_{n+1}L_{n+2} - F_{n+2}L_n = F_{2n+1}$ (Carlitz, 1967 [107]).

66. $L_{n+1}F_{n+2} - L_{n+2}F_n = F_{2n+1}.$

67. $F_{n+2}^2 + F_n^2 - 3F_{n+2}F_n = (-1)^n.$

68. $2F_{n-1}^3 + F_n^3 + 6F_{n-1}F_{n+1}^2 = L_n^3$ (Barley, 1973 [22]).

69. $2L_{n-1}^3 + L_n^3 + 6L_{n-1}L_{n+1}^2 = 125F_n^3.$

70. $5L_{3n} = L_{n+1}^3 + L_n^3 - L_{n-1}^3$ (Long, 1986 [418]).

71. $5L_{3n} = L_{n+2}^3 - L_{n-1}^3 - 3L_{n+2}L_{n+1}L_n.$

72. $F_{n+3}^2 = 2F_{n+2}^2 + 2F_{n+1}^2 - F_n^2$ (Gould, 1963 [231]).

73. $F_{n+1}^3 + F_n^3 - F_{n-1}^3 = F_{3n}$ (Lucas, 1876).

74. $F_{n+2}^3 - 3F_n^3 + F_{n-2}^3 = 3F_{3n}$ (Ginsburg, 1953 [221]).

75. $\alpha^m L_n + \alpha^{m-1}L_{n-1} = \alpha^n L_m + \alpha^{n-1}L_{m-1}$ (Edwards, 2004 [158]).

76. $\beta^m L_n + \beta^{m-1}L_{n-1} = \beta^n L_m + \beta^{n-1}L_{m-1}.$

77. $F_m F_{n+k} - F_k F_{n+m} = (-1)^k F_{m-k}F_n$ (Taylor, 1981 [556]).

78. Deduce the identity $F_n^2 + F_{n-1}^2 = F_{2n-1}$ from Exercise 5.77.

79. $F_{2n}F_{2n+2}F_{2n+4}$ can be expressed as the product of three consecutive integers (Lindstrom, 1976 [412]).

80. $L_n^4 = L_{4n} + 4(-1)^n L_{2n} + 2.$

81. $F_{n+4}^3 - 3F_{n+3}^3 - 6F_{n+2}^3 + 3F_{n+1}^3 + F_n^3 = 0.$

82. $(\alpha F_n + F_{n-1})^{1/n} + (-1)^{n+1}(F_{n+1} - \alpha F_n)^{1/n} = 1$ (Sofo, 1999 [534]).

83. $\displaystyle\sum_{k=1}^{2n-1} (2n-k)F_k^2 = F_{2n}^2$ (Hoggatt, 1964 [272]).

84. $\displaystyle\sum_{k=1}^{2n-1} (2n-k)L_k^2 = L_{2n}^2 - 4(n+1).$

85. $\displaystyle\sum_{k=1}^{n} F_k F_{k+1} = \begin{cases} F_{n+1}^2 & \text{if } n \text{ is odd} \\ F_{n+1}^2 - 1 & \text{otherwise.} \end{cases}$

86. $\displaystyle\sum_{k=1}^{n} L_k L_{k+1} = \begin{cases} L_{n+1}^2 - 4 & \text{if } n \text{ is odd} \\ L_{n+1}^2 + 1 & \text{otherwise.} \end{cases}$

87. $x^2 - x - 1$ is a factor of $x^{2n} - L_n x^n + (-1)^n$, where $n \geq 1$ (Mana, 1972 [429]).

88. $[L_{2n} + 3(-1)^n]/5$ is the product of two Fibonacci numbers (Freitag, 1974 [188]).

89. $L_{2n} - 3(-1)^n$ is the product of two Lucas numbers (Freitag, 1974 [189]).

90. Let $R_m = \displaystyle\sum_{i=0}^{m} F_{i+1} L_{m-i}$. Then $S_m = 10 R_m/(m+2)$ is the sum of two Lucas numbers (Freitag, 1975 [191]).

91. $2L_{n-1}^3 + L_n^3 + 6L_{n+1}^2 L_{n-1}$ is a cube (Wulczyn, 1976 [606]).

92. The sum of any $2n$ consecutive Fibonacci numbers is divisible by F_n, where n is even (Lind, 1964 [391]).

93. Let $n \geq 1$ and $(1 + \sqrt{5})^n = a_n + b_n \sqrt{5}$, where a_n and b_n are positive integers. Then $2^{n-1}|a_n$ and $2^{n-1}|b_n$ (Mana, 1969 [427]).

94. Let $t_n = n(n+1)/2$. Then $L(n) = (-1)^{t_n - 1}(L_{t_{n-1}} L_{t_n} - L_{n^2})$ (Freitag, 1982 [199]).

95. If $2F_{2n-1}F_{2n+1} - 1$ is prime, then so are $2F_n^2 + 1$ and $F_{2n}^2 + F_{2n-1}F_{2n+1}$ (Guillotte, 1972 [253]).

96. Find a formula for $K_n = (K_1 + K_2 + \cdots + K_{n-1}) + F_{2n-1}$, where $K_1 = 1$ (Hoggatt, 1972 [295]).

97. Let $\{g_n\}$ be a number sequence. Then $\displaystyle\sum_{k=1}^{n} (g_{k+2} + g_{k+1} - g_k)F_k = g_{n+2}F_n + g_{n+1}F_{n+1} - g_1$ (Recke, 1969 [499]).

Let f be a function defined by

$$f(n) = \begin{cases} f((n+1)/2) + f((n-1)/2) & \text{if } n \text{ is odd} \\ f(n/2) & \text{otherwise,} \end{cases}$$

where $f(1) = 1$. Then prove both.

98. $f([2^{n+1} + (-1)^n]/3) = F_{n+1}$ (Lind, 1970 [410]).

99. $f([7 \cdot 2^{n-1} + (-1)^n]/3) = L_n$ (Lind, 1970 [410]).

100. Evaluate the sum $\displaystyle\sum_{i=1}^{n} F_i P_i$, where $P_{n+2} = 2P_{n+1} + P_n, P_1 = 1$, and $P_2 = 2$. (Mead, 1965 [435]).

101. Let S_n denote the sum of the numbers in row n of the triangular array of Fibonacci numbers in Figure 5.15. Derive a formula for S_n.

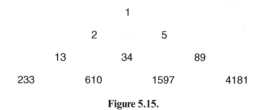

Figure 5.15.

102. Redo Exercise 5.101 using the array in Figure 5.16.

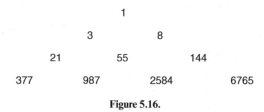

Figure 5.16.

103. Prove that the area of the trapezoid with bases F_{n+1} and F_{n-1}, and sides F_n, is $\sqrt{3}F_{2n}/4$ (Calendar Problems, 1993 [94]).

104. Evaluate $\lim\limits_{n\to\infty} \dfrac{F_{n+k}}{L_n}$, where $n, k > 0$ (Dence, 1968 [142]).

105. Prove that $\displaystyle\sum_{n=0}^{\infty} \dfrac{F_{n+1}}{2^n} = 4$ (Butchart, 1968 [92]).

Evaluate each infinite product.

106. $\displaystyle\prod_{k=1}^{\infty} \dfrac{F_{2k}F_{2k+2} + F_{2k-1}F_{2k+2}}{F_{2k}F_{2k+2} + F_{2k}F_{2k+1}}$ (Edwards, 2003 [157]).

107. $\displaystyle\prod_{k=1}^{\infty} \dfrac{L_{2k}L_{2k+2} + L_{2k-1}L_{2k+2}}{L_{2k}L_{2k+2} + L_{2k}L_{2k+1}}$.

6

GEOMETRIC ILLUSTRATIONS AND PARADOXES

A picture is worth a thousand words.

−A Chinese proverb

In the previous chapter, we encountered numerous Fibonacci and Lucas identities, and summation formulas. In this chapter, we add a geometric dimension to some of them, and to some new ones, by illustrating them pictorially.

To begin with, consider the identities

$$F_{n+1}^2 = 4F_n F_{n-1} + F_{n-2}^2 \tag{6.1}$$

$$F_{n+1}^2 = 2F_n^2 + 2F_{n-1}^2 - F_{n-2}^2 \tag{6.2}$$

$$F_{n+1}^2 = 4F_{n-1}^2 + 4F_{n-1}F_{n-2} + F_{n-2}^2 \tag{6.3}$$

$$F_{n+1}^2 = 4F_n^2 - 4F_{n-1}F_{n-2} - 3F_{n-2}^2. \tag{6.4}$$

These can be confirmed easily; see Exercises 6.1 through 6.4. We can obtain their Lucas counterparts by changing F_k to L_k. The techniques work for any sequence that satisfies the Fibonacci recurrence, but always adjust the least value as required.

6.1 GEOMETRIC ILLUSTRATIONS

Interestingly, we can illustrate geometrically these Fibonacci identities (and hence their Lucas counterparts), as Br. A. Brousseau did, among others, in 1972 [69].

Fibonacci and Lucas Numbers with Applications, Volume One, Second Edition. Thomas Koshy.
© 2018 John Wiley & Sons, Inc. Published 2018 by John Wiley & Sons, Inc.

For example, consider identity (6.1). Draw an $F_{n+1} \times F_{n+1}$ square. Since $F_{n+1} = F_n + F_{n-1}$, we divide the square into four $F_n \times F_{n-1}$ rectangles with one at each corner, and an $F_{n-2} \times F_{n-2}$ square in the middle; see Figure 6.1. Adding the five areas we get $F_{n+1}^2 = 4F_n F_{n-1} + F_{n-2}^2$.

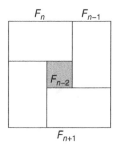

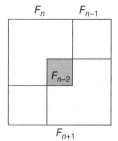

Figure 6.1. $F_{n+1}^2 = 4F_n F_{n-1} + F_{n-2}^2$.   **Figure 6.2.** $F_{n+1}^2 = 2F_n^2 + 2F_{n-1}^2 - F_{n-2}^2$.

On the other hand, suppose we mark off two $F_n \times F_n$ squares at opposite corners of the square. This produces two $F_{n-1} \times F_{n-1}$ squares at the remaining corners, and an overlapping $F_{n-2} \times F_{n-2}$ square again in the middle; see Figure 6.2. When we add up the areas and account for the overlapping area, we get $F_{n+1}^2 = 2F_n^2 + 2F_{n-1}^2 - F_{n-2}^2$.

We can similarly interpret identities (6.3) and (6.4); see Figures 6.3 and 6.4.

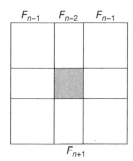

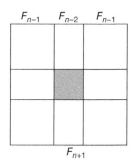

Figure 6.3. $F_{n+1}^2 = 4F_{n-1}^2 + 4F_{n-1}F_{n-2} + F_{n-2}^2$.   **Figure 6.4.** $F_{n+1}^2 = 4F_n^2 - 4F_{n-1}F_{n-2} - 3F_{n-2}^2$.

We will represent a few more identities pictorially and leave some as exercises.

Example 6.1.   *Represent pictorially the identity $L_n^2 = 8F_n F_{n-1} + F_{n-3}^2$.*

Solution. Notice that $L_n = F_{n+1} + F_{n-1} = F_n + 2F_{n-1}$. So we place a $2F_{n-1} \times F_n$ rectangle at each corner of an $L_n \times L_n$ square; see Figure 6.5. This leaves a square with a side of length $L_n - 2F_n = F_{n-3}$. Adding up the areas, we get the desired identity. See Exercise 6.5 for an algebraic proof.

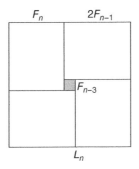

Figure 6.5. $L_n^2 = 8F_n F_{n-1} + F_{n-3}^2.$

∎

Similar techniques can be employed to interpret geometrically the following identities:

$$F_{n+1}^2 = 8F_{n-1}^2 + 2F_{n-2}^2 - L_{n-2}^2 \tag{6.5}$$

$$L_n^2 = 2F_{n+1}^2 + 2F_{n-1}^2 - F_n^2 \tag{6.6}$$

$$L_n^2 = 8F_{n-1}^2 + 2F_n^2 - F_{n-3}^2 \tag{6.7}$$

$$F_{n+1}^2 = 4F_{n-1}^2 + 4F_{n-1}F_{n-2} + F_{n-2}^2 \tag{6.8}$$

$$F_{n+1}^2 = 4F_n^2 - 4F_{n-1}F_{n-2} - 3F_{n-2}^2. \tag{6.9}$$

Here are some useful hints in order to pursue these:

1) Choose an $F_{n+1} \times F_{n+1}$ square, where $F_{n+1} = 2F_{n-1} + F_{n-2}$.
2) Draw an $L_n \times L_n$ square, where $L_n = F_{n+1} + F_{n-1}$.
3) Construct an $L_n \times L_n$ square, where $L_n = 2F_{n-1} + F_n$.
4) Draw an $F_{n+1} \times F_{n+1}$ square, where $F_{n+1} = F_{n-1} + F_{n-2} + F_{n-1}$.
5) Construct an $F_{n+1} \times F_{n+1}$ square, where $F_{n+1} = F_n + F_{n-1}$.

We will do two more identities. They both generate beautiful geometric patterns.

Example 6.2. *Represent geometrically the identity* $L_n^2 = 4F_n^2 + 4F_{n-1}^2 - 4F_{n-2}^2 + F_{n-3}^2.$

Solution. We begin with an $L_n \times L_n$ square. Since $L_n = F_{n-1} + F_n + F_{n-1}$, divide each side into segments of length F_{n-1}, F_n, and F_{n-1} in that order. Using the middle segments, complete the four $F_n \times F_n$ squares. They overlap in four $F_{n-2} \times F_{n-2}$ squares; and the side of the central square has length $L_n - 2F_n = (F_{n+1} + F_{n-1}) - 2F_n = 2F_{n-1} - F_n = F_{n-1} - F_{n-2} = F_{n-3}$; see Figure 6.6.

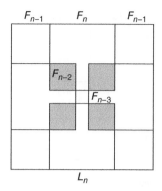

Figure 6.6. $L_n^2 = 4F_n^2 + 4F_{n-1}^2 - 4F_{n-2}^2 + F_{n-3}^2.$

Adding up all the areas in the original square and accounting for all duplicate areas, we get the given identity. See Exercise 6.7 for an algebraic proof. ■

We will now study a similar identity.

Example 6.3. *Interpret geometrically the identity* $F_{n+1}^2 = 4F_{n-1}^2 + 4F_{n-2}^2 + L_{n-2}^2 - 4F_{n-3}^2.$

Solution. We begin with an $F_{n+1} \times F_{n+1}$ square. Since $F_{n+1} = F_{n-1} + F_{n-2} + F_{n-1}$, divide each side into segments of length F_{n-1}, F_{n-2}, and F_{n-1} in that order. Complete the four $F_{n-1} \times F_{n-1}$ squares at the corners. Since $L_{n-2} = F_{n-1} + F_{n-3} = F_{n-2} + 2F_{n-3}$, locate centrally an $L_{n-2} \times L_{n-2}$ square; see Figure 6.7. This square overlaps with the four $F_{n-1} \times F_{n-1}$ squares; each overlapping square is $F_{n-3} \times F_{n-3}$.

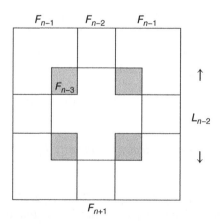

Figure 6.7. $F_{n+1}^2 = 4F_{n-1}^2 + 4F_{n-2}^2 + L_{n-2}^2 - 4F_{n-3}^2.$

Now add up all the areas inside the original square, and subtract the overlapping areas. This gives the given identity; see Exercise 6.8 for an algebraic proof. ∎

Figures 6.8 through 6.11 are four additional illustrations of Fibonacci identities; R.L. Ollerton of the University of Western Sydney developed them in 2008 [464].

These identities can take different forms depending on how we visualize the given areas. For example,

$$\begin{aligned}
F_{n+1}^2 &= F_n^2 + F_{n-1}^2 + 2F_n F_{n-1} \\
&= F_{n+1}F_n + F_n F_{n-1} + F_{n-1}^2 \\
&= 2F_{n+1}F_{n-1} + F_n^2 - F_{n-1}^2.
\end{aligned}$$

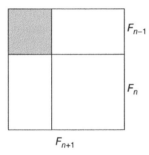

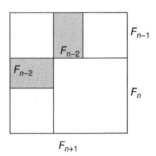

Figure 6.8. $F_{n+1}^2 = F_n^2 + F_{n-1}^2 + 2F_n F_{n-1}.$ **Figure 6.9.** $F_{n+1}^2 = F_n^2 + 3F_{n-1}^2 + 2F_{n-1}F_{n-2}.$

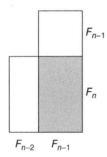

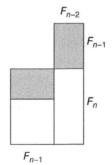

Figure 6.10. $F_n^2 = F_{n+1}F_{n-1} + F_n F_{n-2} - F_{n-1}^2.$ **Figure 6.11.** $F_n^2 = F_{n+1}F_{n-2} + F_{n-1}^2.$

Obviously, the technique of demonstrating Fibonacci and Lucas identities becomes more complicated as their powers increase. For example, we need to use volumes to illustrate the third-degree Fibonacci identity $F_{n+1}^3 = F_n^3 + F_{n-1}^3 + 3F_{n+1}F_nF_{n-1}$; see identity 88 in Chapter 5. Figure 6.12 gives a simple visualization of this identity [12].

Figure 6.12. $F_{n+1}^3 = F_n^3 + F_{n-1}^3 + 3F_{n-1}F_nF_{n+1}$.

This geometric pursuit is an opportunity for further enrichment activities.

Next we turn to Candido's algebraic identity. It has a nice Fibonacci (and Lucas) byproduct; we encountered it in Chapter 5.

6.2 CANDIDO'S IDENTITY

The *Candido identity*, named after the Italian mathematician, Giacomo Candido (1871–1941),

$$\left[x^2 + y^2 + (x+y)^2\right]^2 = 2\left[x^4 + y^4 + (x+y)^4\right],$$

provides an interesting application to Fibonacci (and Lucas) numbers, where x and y are arbitrary real numbers. In particular, let $x = F_n$ and $y = F_{n+1}$. Then we get the Fibonacci identity $(F_n^2 + F_{n+1}^2 + F_{n+2}^2)^2 = 2(F_n^4 + F_{n+1}^4 + F_{n+2}^4)$. As can be predicted, this result also has an interesting geometric interpretation.

To see this, consider a line segment $\overline{AD}$ such that $AB = F_n^2$, $BC = F_{n+1}^2$, and $CD = F_{n+2}^2$. Now complete the square $ADEF$; see Figure 6.13. Then

$$\text{Area } ADEF = (F_n^2 + F_{n+1}^2 + F_{n+2}^2)^2$$
$$= 2(F_n^4 + F_{n+1}^4 + F_{n+2}^4)$$
$$= 2(\text{sum of the three squares}).$$

See Figure 6.14 also.

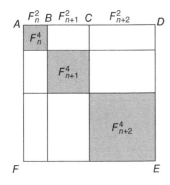

Figure 6.13.

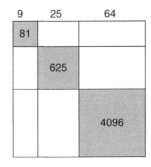

Figure 6.14.

Candido's identity works for any sequence $\{G_n\}$ that satisfies the Fibonacci recurrence.; see Figure 6.15.

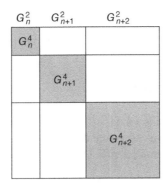

Figure 6.15.

Next we will investigate tilings with Fibonacci square tiles. In the process, we will revisit identity (5.5).

6.3 FIBONACCI TESSELLATIONS

Suppose we would like to tile a 13×8 floor with *distinct* $F_k \times F_k$ tiles. (Assume that $F_1 \times F_1$ and $F_2 \times F_2$ tiles are distinct.) Always use the largest possible size first at each step.

The 13×8 area can be tiled with one 8×8, one 5×5, one 3×3, one 2×2, and two 1×1 tiles with no gaps or overlapping; see Figure 6.16.

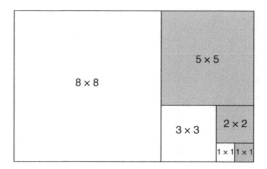

Figure 6.16. A Fibonacci Tessellation.

Since the total area equals the sum of the areas of the square tiles, it follows that $\sum_{i=1}^{6} F_i^2 = F_6 F_7$, a special case of the summation formula (5.5).

More generally, it follows by identity (5.5) that every $F_{n+1} \times F_n$ floor can be tiled using each $F_k \times F_k$ tile exactly once, where $1 \le k \le n$.

Next we will explore tiling a $L_{n+1} \times L_n$ floor with distinct Lucas square tiles. This will take us back to the summation formula (5.10).

6.4 LUCAS TESSELLATIONS

Suppose we would like to tile an 11×7 floor with *distinct $L_k \times L_k$* tiles. Beginning with a 7×7 tile, we will need one 4×4, one 3×3, and one 1×1 tiles; see Figure 6.17. There are no smaller tiles left. This leaves a 2×1 area that cannot be filled. Thus

$$\text{Tiled area} = \sum_{k=1}^{4} L_k^2 = 11 \times 7 - 2,$$

a special case of formula (5.10).

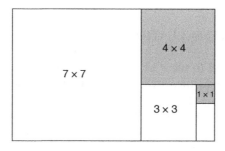

Figure 6.17.

Since

$$\sum_{k=1}^{n} L_k^2 = L_n L_{n+1} - 2,$$

it follows that *no* $L_{n+1} \times L_n$ floor can be tiled with distinct Lucas square tiles, where $n \geq 2$.

We will now turn to some interesting geometric paradoxes.

6.5 GEOMETRIC PARADOXES

Cassini's identity $F_{n+1}F_{n-1} - F_n^2 = (-1)^n$ is the cornerstone of two classes of fascinating geometric paradoxes. When n is even, say, $n = 2k$, the identity yields $F_{2k}^2 - F_{2k+1}F_{2k-1} = -1$; the first paradox is based on this fact. When $n = 2k + 1$, we have $F_{2k+1}^2 - F_{2k+2}F_{2k} = 1$; this leads to the second paradox.

The first paradox was a favorite of the famous English puzzlist, Charles Lutwidge Dodgson (1832–1898), better known by his pseudonym, Lewis Carroll. This puzzle, first proposed in 1774 by William Hooper in his *Rational Recreations*, reappeared in a mathematics periodical in Leipzig, Germany, in 1868, 666 years (again the beastly number) after Fibonacci published his rabbit-breeding problem.

W.W. Rouse Ball claims in his *Mathematical Recreations and Essays* (Macmillan, New York, 1973), which is a jewel in recreational mathematics, that 1868 was the earliest date he could find for the first appearance of this puzzle. Although the origin of the puzzle is still a mystery, the elder Sam Loyd, a puzzle columnist for the *Brooklyn Daily Eagle*, claimed that he presented it to the American Chess Congress in 1858.

6.6 CASSINI-BASED PARADOXES

To begin with, consider an 8×8 square; cut it up into four pieces, A, B, C, and D, as in Figure 6.18. Now rearrange the pieces to form a 13×5 rectangle; see Figure 6.19. While the area of the square is 64 square units, that of the rectangle is 65 square units. In other words, by re-assembling the pieces of the original square, we have gained one square unit:

Area of the parallelogram = area of the rectangle − area of the square

$$= 13 \times 5 - 8^2$$

$$= F_7 F_5 - F_6^2$$

$$= 1.$$

This is certainly paradoxical.

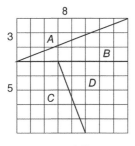

Figure 6.18.

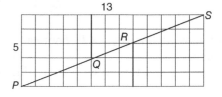

Figure 6.19.

How is this possible? In Figure 6.19, it appears that the "diagonal" $PQRS$ of the rectangle is a line (segment). In fact, this appearance is deceptive. The points P, Q, R, and S are in fact the vertices of a very narrow parallelogram; see Figure 6.20.

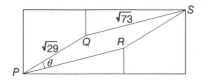

Figure 6.20.

Its sides are $\sqrt{29}$ and $\sqrt{73}$ units long, and the diagonal is $\sqrt{194}$ units long. Let θ be the acute angle between the adjacent sides of the parallelogram. Then, by the law of cosines in trigonometry,

$$\cos\theta/2 = \frac{194 + 29 - 73}{2\sqrt{29 \cdot 194}}$$

$$\approx 0.763898460833$$

$$\theta \approx 1°31'40''.$$

This explains why it is a very narrow parallelogram.

There is nothing sacred about the choice of the size of the square, except that $8 = F_6$ is a Fibonacci number with an even subscript, and $13 = F_7$ and $5 = F_5$ are its adjacent neighbors.

Since $F_{n+1}F_{n-1} - F_n^2 = 1$, when n is even, the puzzle can be extended to any $F_n \times F_n$ square: cut it up into four figures, two triangles and two trapezoids, as in Figure 6.21; and then rearrange the pieces to form the deceptive looking $F_{n+1} \times F_{n-1}$ rectangle, as Figure 6.22 shows.

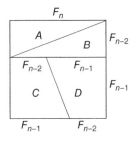

Figure 6.21.

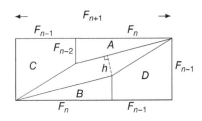

Figure 6.22.

The parallelogram, magnified in Figure 6.22, has an area of one square unit. So we can determine its height h (dotted):

$$\text{Area of the parallelogram} = \text{height} \times \text{base}$$

$$1 = h\sqrt{F_n^2 + F_{n-2}^2}$$

$$h = \frac{1}{\sqrt{F_n^2 + F_{n-2}^2}}.$$

Thus, as the size of the original square increases, the parallelogram becomes narrower and narrower, and the gap becomes less and less noticeable.

Sam Loyd's son was the first person to discover that the four pieces in Figure 6.18 can be arranged to form an area of 63 square units, as Figure 6.23 shows. The son adopted his father's name and inherited his father's puzzle column in the *Daily Eagle*.

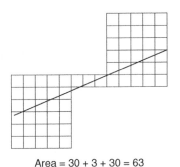

Area = 30 + 3 + 30 = 63

Figure 6.23.

To illustrate a paradox of the second kind, consider a 5×5 square, and cut into four pieces as before; see Figure 6.24. Now re-assemble the pieces to form an 8×3 "rectangle" in Figure 6.25. The area of the square is square 25 units, whereas that of the rectangle is only 24 square units; so we have lost one square unit area. The overlap along the diagonal accounts for the missing area. Notice that the area of the square $= F_5^2 = F_4F_6 + 1 =$ area of the rectangle $+ 1$.

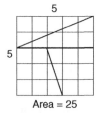

Figure 6.24.

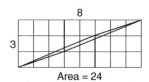

Figure 6.25.

More generally, let n be odd. Suppose an $F_n \times F_n$ square is cut into four pieces as in Figure 6.26; and then assemble the pieces to form an $F_{n+1} \times F_{n-1}$ rectangle, as in Figure 6.27. Then we would be missing an area of one square unit: $F_{n+1}F_{n-1} - F_n^2 = -1$.

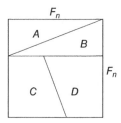

Figure 6.26.

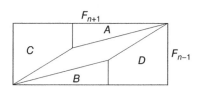

Figure 6.27.

In 1962, A.F. Horadam of the University of New England, Australia, derived the formula for $\tan\theta_n$, where θ_n denotes the acute angle between the adjacent sides of the parallelogram [323]. To derive the formula, we first consider the case n even, where $n \geq 4$; see Figure 6.28.

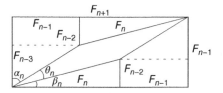

Figure 6.28.

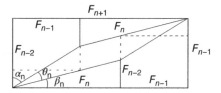

Figure 6.29.

Using Figure 6.28, identity (5.11), and the trigonometric identity $\tan^{-1} x + \tan^{-1} 1/x = \pi/2$, we have

$$\theta_n = \pi/2 - (\alpha_n + \beta_n)$$

$$= \frac{\pi}{2} - \tan^{-1} \frac{F_{n-1}}{F_{n-3}} - \tan^{-1} \frac{F_{n-2}}{F_n}$$

$$= \tan^{-1} \frac{F_{n-3}}{F_{n-1}} - \tan^{-1} \frac{F_{n-2}}{F_n}$$

$$\tan \theta_n = \frac{(F_{n-3}/F_{n-1}) - (F_{n-2}/F_n)}{1 + (F_{n-3}/F_{n-1}) \cdot (F_{n-2}/F_n)}$$

$$= \frac{F_{n-3}F_n - F_{n-1}F_{n-2}}{F_{n-1}F_n + F_{n-3}F_{n-2}}$$

$$= \frac{F_{n-3}(F_{n-1} + F_{n-2}) - F_{n-2}(F_{n-2} + F_{n-3})}{F_{n-1}(F_{n-1} + F_{n-2}) + F_{n-3}F_{n-2}}$$

$$= \frac{F_{n-1}F_{n-3} - F_{n-2}^2}{F_{n-1}^2 + F_{n-2}(F_{n-2} + F_{n-3}) + F_{n-3}F_{n-2}}$$

$$= \frac{(-1)^{n-2}}{F_{n-1}^2 + F_{n-2}^2 + 2F_{n-3}F_{n-2}}$$

$$= \frac{(-1)^n}{F_{2n-3} + 2F_{n-3}F_{n-2}}$$

$$= \frac{1}{F_{2n-3} + 2F_{n-3}F_{n-2}}.$$

Now let n be odd. Then there is an overlap; see Figure 6.29. It follows from the figure that

$$\theta_n = (\alpha_n + \beta_n) - \pi/2$$

$$= \tan^{-1} \frac{F_{n-1}}{F_{n-3}} + \tan^{-1} \frac{F_{n-2}}{F_n} - \frac{\pi}{2}$$

$$= \tan^{-1} \frac{F_{n-2}}{F_n} - \tan^{-1} \frac{F_{n-3}}{F_{n-1}}.$$

As before, this leads to

$$\tan \theta_n = \frac{(-1)^{n-1}}{F_{2n-3} + 2F_{n-3}F_{n-2}}$$

$$= \frac{1}{F_{2n-3} + 2F_{n-3}F_{n-2}}.$$

Thus both cases yield the same value for $\tan \theta_n$:

$$\tan \theta_n = \frac{1}{F_{2n-3} + 2F_{n-3}F_{n-2}},$$

where $n \geq 4$.

Table 6.1 shows the values of θ_n for the first few Fibonacci triplets (F_{n-1}, F_n, F_{n+1}). It follows from the table that, as n increases slowly, $\theta_n \to 0$ rapidly; thus $\theta_n \to 0$ as $n \to \infty$.

TABLE 6.1.

n	Fibonacci Triplets			θ_n
	F_{n-1}	F_n	F_{n+1}	
4	2	3	5	$\tan^{-1} \frac{1}{7} \approx 8°7'48''$
5	3	5	8	$\tan^{-1} \frac{1}{17} \approx 3°21'59''$
6	5	8	13	$\tan^{-1} \frac{1}{46} \approx 1°14'43''$
7	8	13	21	$\tan^{-1} \frac{1}{119} \approx 28'53''$
8	13	21	34	$\tan^{-1} \frac{1}{313} \approx 10'59''$
9	21	34	55	$\tan^{-1} \frac{1}{818} \approx 4'12''$
10	34	55	89	$\tan^{-1} \frac{1}{2143} \approx 1'36''$
11	55	89	144	$\tan^{-1} \frac{1}{5609} \approx 37''$
12	89	144	233	$\tan^{-1} \frac{1}{14686} \approx 14''$

6.7   ADDITIONAL PARADOXES

There are many delightful puzzles in which Fibonacci-based rectangles can be cut into several pieces, and the pieces rearranged to form a rectangle of larger or smaller area. One such paradox is Langman's paradox, developed by H. Langman of New York City.

LANGMAN'S PARADOX

Cut a 13×8 rectangle into four pieces, as Figure 6.30 shows. Now rearrange the pieces to form a 21×5 rectangle; see Figure 6.31. We gain one unit square.

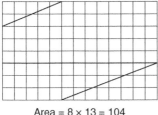

Area = 8 × 13 = 104

Figure 6.30.

Area = 5 × 21 = 105

Figure 6.31.

ANOTHER VERSION OF LANGMAN'S PARADOX

Another version of Langman's paradox involves gaining two square units when the pieces of the 21 × 8 rectangle in Figure 6.32 are re-assembled. Cut out the shaded area and place it on top of the unshaded area in such a way that the diagonal cuts form one long diagonal; now switch pieces *A* and *B*. The resulting area is 170 square units.

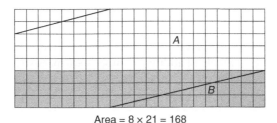

Area = 8 × 21 = 168

Figure 6.32.

The next paradox was developed in 1953 by Paul Curry, an amateur magician from New York City. It involves two alternate Fibonacci numbers.

CURRY'S PARADOX

Swap the positions of the triangles *B* and *C* in Figure 6.33 to form the 13 × 5 rectangle in Figure 6.34. This results in an apparent loss of one square unit. In fact, the loss occurs in the shaded area. Figure 6.33 contains 15 shaded cells,

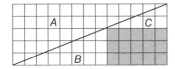

Figure 6.33.

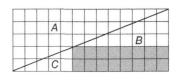

Figure 6.34.

whereas Figure 6.34 requires 16 cells to complete the 13×5 rectangle. In other words, we seem to have lost one square area in the process.

AN INTRIGUING SEQUENCE

Finally, suppose we construct a number sequence beginning with two arbitrary real numbers a and b, and then use the Fibonacci recurrence to construct the remaining elements. All such sequences, except one, can be used to develop the preceding puzzles. So which sequence will *not* produce a puzzle? In other words, under what conditions will the square and rectangle have exactly the same area?

To answer this, we must consider the following additive number sequence with $a = 1$ and $b = \alpha$:

$$1, \alpha, \alpha + 1, 2\alpha + 1, 3\alpha + 2, \ldots, s_n, \ldots .$$

Suppose we pick any three consecutive terms: s_{n-2}, s_{n-1}, and s_n. Using Lemma 5.1, then $s_{n+1}s_{n-1} = \alpha^{n-1} \cdot \alpha^{n+1} = \alpha^{2n} = s_n^2$; so the area of the square indeed equals that of the rectangle!

Interestingly, $\{s_n\}$ is the only additive number sequence with this striking behavior. The ratio of any two consecutive terms of the sequence is a magic constant: $s_{n+1}/s_n = \alpha$. Martin Gardner (1914–2010), who wrote a popular column called *Mathematical Games* in *Scientific American*, referred to the sequence $\{\alpha^n\}$ as the "golden series," which all additive number sequences struggle to become.

EXERCISES 6

Prove each algebraically.

1–4. Identities (6.1) through (6.4).

5. $L_n^2 = 8F_nF_{n-1} + F_{n-3}^2$.

6. $F_{n+1}^2 = 2F_n^2 + 2F_{n-1}^2 - F_{n-2}^2$.

7. $L_n^2 = 4F_n^2 + 4F_{n-1}^2 - 4F_{n-2}^2 + F_{n-3}^2$.

8. $F_{n+1}^2 = 4F_{n-1}^2 + 4F_{n-2}^2 + L_{n-2}^2 - 4F_{n-3}^2$.

9. $F_{n+1}^3 = F_n^3 + F_{n-1}^3 + 3F_{n+1}F_nF_{n-1}$.

GIBONACCI NUMBERS

We can study properties shared by Fibonacci and Lucas numbers by investigating a number sequence that satisfies the Fibonacci recurrence, but with arbitrary initial conditions. This is our focus in this chapter.

7.1 GIBONACCI NUMBERS

Consider the sequence $\{G_n\}$, where $G_n = G_{n-1} + G_{n-2}$, $G_1 = a$, $G_2 = b$, and $n \geq 3$. The ensuing sequence

$$a, b, a + b, a + 2b, 2a + 3b, 3a + 5b, \ldots$$

is the *gibonacci sequence* (generalized Fibonacci sequence); G_n is the *n*th *gibonacci number*. (A.T. Benjamin and J.J. Quinn coined the term *gibonacci* [35].)

Take a good look at the coefficients of a and b in the various terms of this sequence. They follow the same pattern we saw in Chapter 5 for powers of α: The coefficients of a and b are Fibonacci numbers. We will now pinpoint these two coefficients in the following theorem.

Theorem 7.1. *Let G_n be the nth gibonacci number. Then $G_n = aF_{n-2} + bF_{n-1}$, where $n \geq 3$.*

Proof. We will establish this using PMI. Since $G_1 = a = aF_{-1} + bF_0$ and $G_2 = b = aF_0 + bF_1$, the result is true when $n = 1$ and $n = 2$.

Fibonacci and Lucas Numbers with Applications, Volume One, Second Edition. Thomas Koshy.
© 2018 John Wiley & Sons, Inc. Published 2018 by John Wiley & Sons, Inc.

Suppose it is true for all positive integers $\leq k$. Then

$$
\begin{aligned}
G_{k+1} &= G_k + G_{k-1} \\
&= (aF_{k-2} + bF_{k-1}) + (aF_{k-3} + bF_{k-2}) \\
&= a(F_{k-2} + F_{k-3}) + b(F_{k-1} + F_{k-2}) \\
&= aF_{k-1} + bF_k.
\end{aligned}
$$

Thus, by the strong version of PMI, the formula works for all positive integers n. ■

GIBONACCI NUMBERS AND BEES

Interestingly, gibonacci numbers occur in the study of a bee colony. To see this, suppose we start the colony with a male and b female bees. Table 7.1 shows their genealogical growth for five generations. It follows from the table that the drone has a total of $G_{n+2} = aF_n + bF_{n-1}$ descendants in generation n.

TABLE 7.1.   Gibonacci Growth of Bees

Generation	1	2	3	4	5
Number of female bees	b	$a+b$	$a+2b$	$2a+3b$	$3a+5b$
Number of male bees	a	b	$a+b$	$a+2b$	$2a+3b$
Total number of bees	$a+b$	$a+2b$	$2a+3b$	$3a+5b$	$5a+8b$

The Fibonacci identities from Chapter 5 can be extended to gibonacci numbers. We will study a few in the following theorems.

Theorem 7.2.

$$
\sum_{i=1}^{n} G_{k+i} = G_{n+k+2} - G_{k+2}.
$$

Proof. By Theorems 5.1 and 7.1, we have

$$
\begin{aligned}
\sum_{i=1}^{n} G_{k+i} &= a \sum_{i=1}^{n} F_{k+i-2} + b \sum_{i=1}^{n} F_{k+i-1} \\
&= a(F_{n+k} - F_k) + b(F_{n+k+1} - F_{k+1}) \\
&= (aF_{n+k} + bF_{n+k+1}) - (aF_k + bF_{k+1}) \\
&= G_{n+k+2} - G_{k+2},
\end{aligned}
$$

as desired. ■

Notice that formulas (5.1) and (5.6) follow from this theorem.

The next summation formula follows from Theorem 7.1. Its proof is straightforward, so we leave it as an exercise; see Exercise 7.26.

Theorem 7.3.

$$5\sum_{i=1}^{n} G_i G_{i+1} = 5a^2(F_{n-2}^2 - v) + 5b^2(F_{n-1}^2 - v + 1) + ab(L_{2n-1} + 5F_n F_{n-1} + v + 1),$$

$$\text{where } v = \begin{cases} 1 & \text{if } n \text{ is odd} \\ 0 & \text{otherwise.} \end{cases} \qquad \blacksquare$$

Theorem 7.1 can be used to develop the Binet-like formula for G_n, as the following theorem shows.

Theorem 7.4 (Binet-like formula). *Let* $c = a + (a - b)\beta$ *and* $d = a + (a - b)\alpha$. *Then*

$$G_n = \frac{c\alpha^n - d\beta^n}{\alpha - \beta}.$$

Proof. By Theorem 7.1, we have

$$G_n = aF_{n-2} + bF_{n-1}$$
$$(\alpha - \beta)G_n = a(\alpha^{n-2} - \beta^{n-2}) + b(\alpha^{n-1} - \beta^{n-1})$$
$$= \alpha^n \left(\frac{a}{\alpha^2} + \frac{b}{\alpha} \right) - \beta^n \left(\frac{a}{\beta^2} + \frac{b}{\beta} \right)$$
$$= \alpha^n(a\beta^2 - b\beta) - \beta^n(a\alpha^2 - b\alpha)$$
$$= [a + (a - b)\beta]\alpha^n - [a + (a - b)\alpha]\beta^n$$
$$= c\alpha^n - d\beta^n.$$

This yields the desired formula. $\qquad \blacksquare$

Notice that

$$cd = [a + (a - b)\beta][a + (a - b)\alpha]$$
$$= a^2 + (a - b)^2\alpha\beta + a(a - b)(\alpha + \beta)$$
$$= a^2 - (a - b)^2 + a(a - b)$$
$$= a^2 + ab - b^2.$$

This constant occurs in many of the formulas for gibonacci numbers. It is called the *characteristic* of the gibonacci sequence. We will denote it by the Greek letter μ (mu): $\mu = a^2 + ab - b^2$.

The characteristic of the Fibonacci sequence is 1 and that of the Lucas sequence is -5.

Binet's formula for G_n opens the door for a myriad of formulas for the gibonacci family. The next theorem, for instance, is one such generalization of Cassini's formula. Additional formulas can be found in the exercises.

Theorem 7.5.
$$G_{n+1}G_{n-1} - G_n^2 = \mu(-1)^n.$$

Proof. By Theorem 7.4, we have

$$
\begin{aligned}
(\alpha - \beta)(G_{n+1}G_{n-1} - G_n^2) &= (c\alpha^{n+1} - d\beta^{n+1})(c\alpha^{n-1} - d\beta^{n-1}) - (c\alpha^n - d\beta^n)^2 \\
&= -cd(\alpha^{n+1}\beta^{n-1} + \alpha^{n-1}\beta^{n+1}) + 2cd(\alpha\beta)^n \\
&= -\mu(\alpha\beta)^{n-1}(\alpha^2 + \beta^2) + 2\mu(\alpha\beta)^n \\
&= (\alpha - \beta)\mu(-1)^n.
\end{aligned}
$$

This gives the desired generalization.   ∎

In particular, $L_{n+1}L_{n-1} - L_n^2 = 5(-1)^{n-1}$.

The Catalan-like identity $G_{n+k}G_{n-k} - G_n^2 = (-1)^{n+k+1}\mu F_k^2$ (see Exercise 7.20) can be used to generalize the Gelin–Cesàro identity (5.26):

$$G_{n+2}G_{n+1}G_{n-1}G_{n-2} - G_n^4 = -\mu^2. \tag{7.1}$$

This is true since $(G_{n+2}G_{n-2})(G_{n+1}G_{n-1}) = \left[G_n^2 - (-1)^n\mu\right]\left[G_n^2 + (-1)^n\mu\right] = -\mu^2$.

MELHAM'S IDENTITY REVISITED

Melham's identity (5.28) can be generalized to gibonacci numbers:

$$G_{n+1}G_{n+2}G_{n+6} - G_{n+3}^3 = \mu(-1)^n G_n. \tag{7.2}$$

The proof follows the same steps as in the proof of identity (5.28); so we just give the key steps in its proof. Clearly, $G_{n+6} = 5G_{n+2} + 3G_{n+1}$. Consequently,

$$
\begin{aligned}
\text{LHS} &= G_{n+1}G_{n+2}(5G_{n+2} + 3G_{n+1}) - (G_{n+2} + G_{n+1})^3 \\
&= G_{n+2}^2(2G_{n+1} - G_{n+2}) - G_{n+1}^3 \\
&= (G_{n+1} + G_n)^2(G_{n+1} - G_n) - G_{n+1}^3 \\
&= (G_{n+1} + G_n)(G_{n+1}^2 - G_n^2) - G_n^3 \\
&= G_{n+1}G_n(G_{n+1} - G_n) - G_n^3 \\
&= G_n(G_{n+1}G_{n-1}) - G_n^3
\end{aligned}
$$

$$= G_n \left[G_n^2 + \mu(-1)^n \right] - G_n^3$$

$$= \mu(-1)^n G_n$$

$$= \text{RHS}.$$

Clearly, identities (5.28) and (5.29) follow from identity (7.2).

In 1956, H.L. Umansky of Emerson High School in Union City, New Jersey, extended Raine's result in Theorem 5.6, as the next theorem shows [563].

Theorem 7.6. *Let ABC be a triangle with* $AC = G_k G_{k+3}$, $BC = 2G_{k+1} G_{k+2}$, *and* $AB = G_{2k+3}$. *Then* $\triangle ABC$ *is a right triangle with hypotenuse* $\overline{AB}$. ∎

In Chapter 5, we found that the sum of any ten consecutive Fibonacci numbers is 11 times the seventh number of the sequence. Is this true for gibonacci numbers?

To answer this, consider the first 10 gibonacci numbers $a, b, a + b, a + 2b, 2a + 3b, 3a + 5b, 5a + 8b, 8a + 13b, 13a + 21b$, and $21a + 34b$. Their sum is $55a + 88b$, which is clearly 11 times the seventh term $5a + 8b$. Interestingly enough, $11 = L_5$. Thus

$$\sum_{i=1}^{10} G_i = L_5 G_7,$$

where $L_5 = (55, 88) = (55, 89 - 1) = (F_{10}, F_{11} - 1)$.

More generally, is $\sum\limits_{i=1}^{n} G_i$ a multiple of some Lucas number L_m? To answer this, recall from Theorem 7.1 that $G_i = aF_{i-2} + bF_{i-1}$. So

$$\sum_{i=1}^{n} G_i = a \sum_{i=1}^{n} F_{i-2} + b \sum_{i=1}^{n} F_{i-1}$$

$$= aF_n + b(F_{n+1} - 1).$$

When $n = 10$, this sum is divisible by L_5, as we just observed. Consequently, let us look for a way to factor the sum on the RHS. Since a and b are arbitrary, we look for the common factors of F_n and $F_{n+1} - 1$. [Although $(F_n, F_{n+1}) = 1$, F_n and $F_{n+1} - 1$ need *not* be relatively prime.]

Table 7.2 shows a few specific values of $F_n, F_{n+1} - 1$, and their factorizations; we have omitted those cases where $(F_n, F_{n+1} - 1) = 1$.

It is apparent from the table that, when n is of the form $4k + 2$, $(F_n, F_{n+1} - 1)$ is a Lucas number and the various quotients are consecutive Fibonacci numbers; and when n is of the form $4k$, $(F_n, F_{n+1} - 1)$ is a Fibonacci number and the various quotients are consecutive Lucas numbers.

To confirm these two observations, we will need the following facts: $F_{4n+1} - 1 = L_{2n+1} F_{2n}$ and $F_{4n+3} - 1 = L_{2n+1} F_{2n+2}$. These follow by Exercise 5.58; see Exercises 7.65 and 7.67 also.

TABLE 7.2.

n	F_n	$F_{n+1} - 1$	Factorization	
6	8	–	$4 \cdot \boxed{2}$	← Fibonacci numbers
		12	$4 \cdot \boxed{3}$	
8	21	–	$3 \cdot \boxed{7}$	← Lucas numbers
		33	$3 \cdot \boxed{11}$	
10	55	–	$11 \cdot \boxed{5}$	← Fibonacci numbers
		88	$11 \cdot \boxed{8}$	
12	144	–	$8 \cdot \boxed{18}$	← Lucas numbers
		232	$8 \cdot \boxed{29}$	
14	377	–	$29 \cdot \boxed{13}$	← Fibonacci numbers
		609	$29 \cdot \boxed{21}$	
16	987	–	$21 \cdot \boxed{47}$	← Lucas numbers
		1596	$21 \cdot \boxed{76}$	
18	2584	–	$76 \cdot \boxed{34}$	← Fibonacci numbers
		4180	$76 \cdot \boxed{55}$	
20	6765	–	$55 \cdot \boxed{123}$	← Lucas numbers
		10945	$55 \cdot \boxed{199}$	

Proof.

Case 1. Let $n = 4k + 2$. Then

$$\sum_{i=1}^{4k+2} G_i = aF_{4k+2} + b(F_{4k+3} - 1)$$
$$= aL_{2k+1}F_{2k+1} + bL_{2k+1}F_{2k+2}$$
$$= L_{2k+1}(aF_{2k+2} + bF_{2k+2})$$
$$= L_{2k+1}G_{2k+3}.$$

Case 2. Let $n = 4k$. Then

$$\sum_{i=1}^{4k} G_i = aF_{4k} + b(F_{4k+1} - 1)$$
$$= aL_{2k}F_{2k} + bL_{2k+1}F_{2k}$$
$$= F_{2k}(aL_{2k} + bF_{2k+1})$$
$$= F_{2k}[a(F_{2k-1} + F_{2k+1}) + b(F_{2k} + F_{2k+2})]$$
$$= F_{2k}[(aF_{2k-1} + bF_{2k}) + (aF_{2k+1} + bF_{2k+2})]$$
$$= F_{2k}(G_{2k+1} + G_{2k+3}). \qquad \blacksquare$$

In particular, $\sum_{i=1}^{10} G_i = L_5 G_7 = 11 G_7$, as observed earlier. This is an interesting case, since multiplication by 11 is remarkably easy. Likewise, we can compute $\sum_{i=1}^{30} G_i$ by multiplying G_{17} by $L_{15} = 1364$.

It follows from Case 2 that we can compute $\sum_{i=1}^{4k} G_i$ by multiplying the sum $G_{2k+1} + G_{2k+3}$ by F_{2k}. For instance, $\sum_{i=1}^{20} G_i = 55(G_{11} + G_{13})$.

7.2   GERMAIN'S IDENTITY

Finally, we pursue an interesting consequence to the gibonacci family of an algebraic identity, developed by the French mathematician Sophie Germain (1776–1831), who used to correspond with many mathematicians using the name M. Leblanc: $4a^4 + b^4 = [a^2 + (a+b)^2][a^2 + (a-b)^2]$.

It follows from this identity that

$$4G_n^4 + G_{n-1}^4 = (G_n^2 + G_{n+1}^2)(G_n^2 + G_{n-2}^2)$$
$$4G_{n+1}^4 + G_{n-1}^4 = [G_{n+1}^2 + (G_{n+1} + G_{n-1})^2](G_{n+1}^2 + G_n^2).$$

These two gibonacci identities imply that

$$4F_n^4 + F_{n-1}^4 = (F_n^2 + F_{n-2}^2)F_{2n+1}$$
$$4L_n^4 + L_{n-1}^4 = 5(L_n^2 + L_{n-2}^2)F_{2n+1}$$
$$4F_{n+1}^4 + F_{n-1}^4 = (L_n^2 + F_{n+1}^2)F_{2n+1}$$
$$4L_{n+1}^4 + L_{n-1}^4 = 5(L_{n+1}^2 + 25F_n^2)F_{2n+1}.$$

For example, $4F_6^4 + F_5^4 = 17{,}009 = (F_6^2 + F_4^2)F_{13}$; and $4L_7^4 + L_5^4 = 2{,}843{,}765 = 5(L_7^2 + 25F_6^2)F_{13}$.

EXERCISES 7

1. Let $\{A_n\}$ be a sequence such that $A_1 = 2, A_2 = 3$, and $A_n = A_{n-1} + A_{n-2}$, where $n \geq 3$. Find an explicit formula for A_n (Jackson, 1969 [348]).

Let $c = a + (a - b)\beta$ and $d = a + (a - b)\alpha$. Evaluate each.

2. $c + d$.

3. $c - d$.

4. $\displaystyle\lim_{n \to \infty} \frac{G_n}{F_n}$.

5. $\displaystyle\lim_{n \to \infty} \frac{G_n}{L_n}$.

6. Solve the equation $G_{n-1}x^2 - G_n x - G_{n+1} = 0$ (Umansky, 1972 [566]).

Prove each, where $n \geq k \geq 0$.

7. $\displaystyle\sum_{i=1}^{n} G_i = G_{n+2} - b$.

8. $\displaystyle\sum_{i=1}^{n} G_{2i-1} = G_{2n} + a - b$.

9. $\displaystyle\sum_{i=1}^{n} G_{2i} = G_{2n+1} - a$.

10. $\displaystyle\sum_{i=1}^{n} G_i^2 = G_n G_{n+1} + a(a - b)$.

11. $\displaystyle\sum_{i=1}^{10} G_{k+i} = 11G_{k+7}$ (Hoggatt, 1963 [268]).

12. $\displaystyle\sum_{i=0}^{n} G_{k+i} = G_{n+k+2} - G_{k+1}$ (Huff).

13. $\displaystyle\sum_{i=1}^{n} iG_i = nG_{n+2} - G_{n+3} + a + b$ (Wall, 1964 [577]).

14. $\displaystyle\sum_{i=1}^{n} (n - i + 1)G_i = nG_{n+4} - a - (n + 2)b$ (Wall, 1965 [580]).

15. $\displaystyle\sum_{i=1}^{n} F_i G_{3i} = F_n F_{n+1} G_{2n+1}$ (Krishna, 1972 [376]).

16. $\displaystyle\sum_{i=1}^{n} (-2)^i \binom{n}{i} G_i = 5^{(n-1)/2}[c(-1)^n - d]$.

17. $\displaystyle\sum_{i+j+k=n} \frac{(-1)^k G_{j+2k}}{i!j!k!} = 0$ (Brady, 1974 [55]).

18. $G_{-n} = (-1)^{n+1}(aF_{n+2} - bF_{n+1})$.

19. $5G_{n+k}G_{n-k} = 5L_{2n} - (-1)^{n-k}\mu L_{2k}$.

20. $G_{n+k}G_{n-k} - G_n^2 = (-1)^{n+k+1}\mu F_k^2$ (Tagiuri, 1901; see Dickson, 1952 [147]).

21. $G_n^2 + G_{n-1}^2 = (3a - b)G_{2n-1} - \mu F_{2n-1}$.

22. $G_n = G_m F_{n-m+1} + G_{m-1}F_{n-m}$ (Ruggles, 1963 [504]).

23. $G_{m+n} = G_m F_{n+1} + G_{m-1}F_n$.

24. $G_{m-n} = (-1)^n(G_m F_{n-1} - G_{m-1}F_n)$.

25. $G_{m+k}G_{n-k} - G_m G_n = (-1)^{n-k+1}\mu F_k F_{m+k-n}$ (Tagiuri, 1901; see Dickson, 1952 [147]).

26. Theorem 7.3.

27. $G_n^2 = G_{n-3}^2 + 4G_{n-1}G_{n-2}$ (Umansky, 1956 [564]).

28. $G_n^4 + G_{n-1}^4 = 2G_n^2 G_{n-1}^2 + G_{n+1}^2 G_{n-2}^2$ (Umansky, 1956 [564]).

29. $G_n^2 + G_{n+3}^2 = 2(G_{n+1}^2 + G_{n+2}^2)$ (Horadam, 1971 [326]).

30. $G_{n+2}^2 - 3G_{n+1}^2 + G_n^2 = 2\mu(-1)^{n+1}$ (Zeitlin, 1965 [614]).

31. $(2G_m G_n)^2 + (G_m^2 - G_n^2)^2 = (G_m^2 + G_n^2)^2$.

32. $(G_n^2 + G_{n+1}^2 + G_{n+2}^2)^2 = 2(G_n^4 + G_{n+1}^4 + G_{n+2}^4)$.

33. $[G_m^2 + G_n^2 + (G_m + G_n)^2]^2 = 2[G_m^4 + G_n^4 + (G_m + G_n)^4]$.

34. $5(G_{n+r}^2 + G_{n-r}^2) = (a^2 L_{2n-4} + 2ab L_{2n-3} + b^2 L_{2n-2})L_{2r} - 4\mu(-1)^{n+r}$.

35. $5(G_{n+r}G_{n+r+1} + G_{n-r}G_{n-r+1}) = (a^2 L_{2n-3} + 2ab L_{2n-2} + b^2 L_{2n-1})L_{2r} - 2\mu(-1)^{n+r}$.

36. $G_{n+r}G_{n+r+1} + G_{n-r}G_{n-r+1} = (a^2 F_{2n-3} + 2ab F_{2n-2} + b^2 F_{2n-1})F_{2r}$.

37. $G_n^2 G_{n+3}^2 + 4G_{n+1}^2 G_{n+2}^2 = (G_{2n+1}^2 + G_{2n+2}^2)^2$.

38. $G_{n+1}^3 - G_n^3 - G_{n-1}^3 = 3G_{n+1}G_n G_{n-1}$.

39. $G_{n+1}^5 - G_n^5 - G_{n-1}^5 = 5G_{n+1}G_n G_{n-1}[2G_n^2 + \mu(-1)^n]$.

40. $G_{n+1}^7 - G_n^7 - G_{n-1}^7 = 7G_{n+1}G_n G_{n-1}[2G_n^2 + \mu(-1)^n]^2$.

41. $G_{n+1}^4 + G_n^4 + G_{n-1}^4 = 2[2G_n^2 + \mu(-1)^n]^2$.

42. $G_{n+1}^6 + G_n^6 + G_{n-1}^6 = 2[2G_n^2 + \mu(-1)^n]^3 + 3G_{n+1}^2 G_n^2 G_{n-1}^2$.

43. $G_{n+1}^8 + G_n^8 + G_{n-1}^8 = 2[2G_n^2 + \mu(-1)^n]^4 + G_n^4 + G_{n-1}^4 + 8G_n^2 G_{n-1}^2\{G_n^4 + G_{n-1}^4 + 4G_n^2 G_{n-1}^2 + 3G_n G_{n-1}[(3a - b)G_{2n-1} - \mu F_{2n-1}]\}$.

44. $G_{m-1}G_n - G_m G_{n-1} = \mu(-1)^{n-1}F_{m-n}$.

Let $A_n = \left(\sum\limits_{k=1}^{n} kF_k\right)\bigg/\left(\sum\limits_{k=1}^{n} F_k\right)$. Verify each (Ledin, 1966 [383]).

45. $\lim\limits_{n\to\infty}(A_{n+1} - A_n) = 1$.

46. $\lim\limits_{n\to\infty}(A_{n+1}/A_n) = 1$.

47. Let $\{H_n\}$ and $\{K_n\}$ be two gibonacci sequences with characteristics μ and v, respectively. Let $C_n = \sum\limits_{i=1}^{n} H_i K_{n-i}$. Show that $C_{n+2} = C_{n+1} + C_n + A_n$, where $\{A_n\}$ is a gibonacci sequence with characteristic μv (Hoggatt, 1972 [291]).

48. Let p, q, r, and s be any four consecutive gibonacci numbers. Prove that $(pq - rs)^2 = (ps)^2 + (2qr)^2$ (Umansky and Tallman, 1968 [567]).

Deduce each from Exercise 7.48.

49. $(L_n L_{n+1} - L_{n+2} L_{n+3})^2 = (L_n L_{n+3})^2 + (2L_{n+1} L_{n+2})^2$ (Umansky and Tallman, 1968 [567]).

50. $(F_n F_{n+1} - F_{n+2} F_{n+3})^2 = (F_n F_{n+3})^2 + (2F_{n+1} F_{n+2})^2$.

51. Using the secant method in Exercise 4.9 with $f(x) = x^2, x_1 = a^{1/a}$, and $x_2 = b^{1/b}$, prove that $x_n = 1/G_n$.

52. Prove Theorem 7.6.

53. $G_{m+n} + G_{m-n} = \begin{cases} (G_{m+1} + G_{m-1})F_n & \text{if } n \text{ is odd} \\ G_m L_n & \text{otherwise.} \end{cases}$

54. $G_{m+n} - G_{m-n} = \begin{cases} G_m L_n & \text{if } n \text{ is odd} \\ (G_{m+1} + G_{m-1})F_n & \text{otherwise.} \end{cases}$

55. $G_{m+n}^2 - G_{m-n}^2 = (G_{m+1} + G_{m-1})G_m F_{2n}$.

Consider the sequence $\{a_n\}$ defined by $a_{2n+1} = a_{2n} + a_{2n-1}$ and $a_{2n} = a_n$, where $a_1 = a, a_2 = b$, and $n \geq 1$. Verify each (Lind, 1968 [409]).

56. $\sum\limits_{k=1}^{n} a_k = a_{2n+1} - a$.

57. $\sum\limits_{k=1}^{n} a_{2k-1} = a_{4n+1} - a_{2n+1}$.

Prove each.

58. $G_{4m} + b = (G_{2m} + G_{2m-1})F_{2m-1}$.

59. $G_{4m+1} + a = G_{2m+1} L_{2m}$.

60. $G_{4m+2} + b = G_{2m+2} L_{2m}$.

61. $G_{4m+3} + a = (G_{2m+3} + G_{2m+1})F_{2m+1}$.

62. $G_{4m} - b = G_{2m+1} L_{2m-1}$.

63. $G_{4m+1} - a = (G_{2m+2} + G_{2m})F_{2m}$.

64. $G_{4m+2} - b = (G_{2m+3} + G_{2m+1})F_{2m}$.

65. $G_{4m+3} - a = G_{2m+2} L_{2m+1}$.

66. $(G_{4m+1} + a, G_{4m+2} + b) = L_{2m}$.

67. $(G_{4m+1} - a, G_{4m+2} - b) = F_{2m}$.

8

ADDITIONAL FIBONACCI AND LUCAS FORMULAS

In Chapter 5, we found explicit formulas for both F_n and L_n, namely, Binet's formulas. In this chapter we will derive additional explicit formulas for both [280].

8.1 NEW EXPLICIT FORMULAS

To begin with, we will conjecture an explicit formula for F_n. To this end, recall that $|\beta| < 1$, so, as n gets larger and larger, $\beta^n \to 0$; and hence $F_n \approx \alpha^n/\sqrt{5}$. So we compute the value of $\alpha^n/\sqrt{5}$ for the first ten values of n and then look for a pattern:

$$\alpha/\sqrt{5} \approx 0.72 \qquad \alpha^2/\sqrt{5} \approx 1.17 \qquad \alpha^3/\sqrt{5} \approx 1.89 \qquad \alpha^4/\sqrt{5} \approx 3.07$$
$$\alpha^5/\sqrt{5} \approx 4.96 \qquad \alpha^6/\sqrt{5} \approx 8.02 \qquad \alpha^7/\sqrt{5} \approx 12.98 \qquad \alpha^8/\sqrt{5} \approx 21.01$$
$$\alpha^9/\sqrt{5} \approx 33.99 \qquad \alpha^{10}/\sqrt{5} \approx 55.00.$$

The pattern might not be obvious; so we will go one step further. Add 1/2 to each, and see if a pattern emerges:

$$\alpha/\sqrt{5} + 1/2 \approx 1.22 \qquad \alpha^2/\sqrt{5} + 1/2 \approx 1.67 \qquad \alpha^3/\sqrt{5} + 1/2 \approx 2.39$$
$$\alpha^4/\sqrt{5} + 1/2 \approx 3.57 \qquad \alpha^5/\sqrt{5} + 1/2 \approx 5.46 \qquad \alpha^6/\sqrt{5} + 1/2 \approx 8.52$$
$$\alpha^7/\sqrt{5} + 1/2 \approx 13.48 \qquad \alpha^8/\sqrt{5} + 1/2 \approx 21.51 \qquad \alpha^9/\sqrt{5} + 1/2 \approx 34.49$$
$$\alpha^{10}/\sqrt{5} + 1/2 \approx 55.50.$$

Fibonacci and Lucas Numbers with Applications, Volume One, Second Edition. Thomas Koshy.
© 2018 John Wiley & Sons, Inc. Published 2018 by John Wiley & Sons, Inc.

A pattern, surprisingly enough, does emerge. So we conjecture that $\lfloor \alpha^n/\sqrt{5} + 1/2 \rfloor = F_n$. Fortunately, the next theorem confirms our observation. The proof requires the following lemma.

Lemma 8.1.

$$0 < \frac{\beta^n}{\sqrt{5}} + \frac{1}{2} < 1.$$

Proof. Since $\beta < 0$, $|\beta| = -\beta$. Also, since $0 < |\beta| < 1$, $0 < |\beta|^n < 1$. So $0 < |\beta|^n < \frac{\sqrt{5}}{2}$; that is, $0 < \frac{|\beta|^n}{\sqrt{5}} < \frac{1}{2}$.

Case 1. Let n be even. Then $|\beta|^n = \beta^n$; so $0 < \beta^n/\sqrt{5} < 1/2$, and hence $0 < \beta^n/\sqrt{5} + 1/2 < 1$.

Case 2. Let n be odd. Then $|\beta|^n = -\beta^n$; so $0 < -\beta^n/\sqrt{5} < 1/2$, and hence $-1/2 < \beta^n/\sqrt{5} < 0$. Therefore, $0 < \beta^n/\sqrt{5} + 1/2 < 1/2$.

Thus, in both cases, $0 < \beta^n/\sqrt{5} + 1/2 < 1$, as desired.   ∎

We are now ready to establish the conjecture. We will do so using Binet's formula and Lemma 8.1.

Theorem 8.1.

$$F_n = \left\lfloor \frac{\alpha^n}{\sqrt{5}} + \frac{1}{2} \right\rfloor.$$

Proof. Using Binet's formula, we have

$$F_n = \left(\frac{\alpha^n}{\sqrt{5}} + \frac{1}{2} \right) - \left(\frac{\beta^n}{\sqrt{5}} + \frac{1}{2} \right)$$

$$\frac{\alpha^n}{\sqrt{5}} + \frac{1}{2} = F_n + \left(\frac{\beta^n}{\sqrt{5}} + \frac{1}{2} \right) \tag{8.1}$$

$$< F_n + 1.$$

Since $(\beta^n/\sqrt{5}) + 1/2 > 0$, it follows from equation (8.1) that $F_n < \alpha^n/\sqrt{5} + 1/2$.

Thus $F_n < \alpha^n/\sqrt{5} + 1/2 < F_n + 1$. Consequently, $F_n = \left\lfloor \dfrac{\alpha^n}{\sqrt{5}} + \dfrac{1}{2} \right\rfloor$.   ∎

For example, $\alpha^{20}/\sqrt{5} + 1/2 \approx 6765.5$; so $\left\lfloor \dfrac{\alpha^{20}}{\sqrt{5}} + \dfrac{1}{2} \right\rfloor = 6765 = F_{20}$, as expected. Since $\lfloor x \rfloor = \lceil x \rceil - 1$ for nonintegral real numbers x, it follows that

$$F_n = \left\lceil \frac{\alpha^n}{\sqrt{5}} + \frac{1}{2} \right\rceil - 1.$$

But $\lceil x + n \rceil = \lceil x \rceil + n$ for any integer n. So

$$F_n = \left\lceil \frac{\alpha^n}{\sqrt{5}} + \frac{1}{2} - 1 \right\rceil = \left\lceil \frac{\alpha^n}{\sqrt{5}} - \frac{1}{2} \right\rceil.$$

Thus we have the following result.

Corollary 8.1.

$$F_n = \left\lceil \frac{\alpha^n}{\sqrt{5}} - \frac{1}{2} \right\rceil.$$
∎

For example,

$$\frac{\alpha^{15}}{\sqrt{5}} - \frac{1}{2} \approx 609.4997$$

$$\left\lceil \frac{\alpha^{15}}{\sqrt{5}} - \frac{1}{2} \right\rceil = 610$$

$$= F_{15}.$$

Here are two interesting observations:

$$\left\lfloor \frac{\alpha^{2n}}{\sqrt{5}} \right\rfloor = F_{2n} \quad \text{and} \quad \left\lceil \frac{\alpha^{2n-1}}{\sqrt{5}} \right\rceil = F_{2n+1},$$

where $1 \le n \le 5$. The next corollary confirms them.

Corollary 8.2.

$$\left\lfloor \frac{\alpha^{2n}}{\sqrt{5}} \right\rfloor = F_{2n} \quad \text{and} \quad \left\lceil \frac{\alpha^{2n+1}}{\sqrt{5}} \right\rceil = F_{2n+1}.$$

Proof. Let n be even. By the proof of Lemma 8.1, we have $1/2 < \beta^n/\sqrt{5} < 1$; so $-1/2 > -\beta^n/\sqrt{5} > -1$. Then

$$\frac{\alpha^n}{\sqrt{5}} - \frac{1}{2} > F_n > \frac{\alpha^n}{\sqrt{5}} - 1.$$

That is,

$$\frac{\alpha^n}{\sqrt{5}} - 1 < F_n < \frac{\alpha^n}{\sqrt{5}} - \frac{1}{2}.$$

Since $\lfloor x \rfloor \leq x$ and $\lfloor x + n \rfloor = \lfloor x \rfloor + n$, this implies

$$\left\lfloor \frac{\alpha^n}{\sqrt{5}} \right\rfloor - 1 < F_n < \frac{\alpha^n}{\sqrt{5}} - \frac{1}{2}$$

$$\left\lfloor \frac{\alpha^n}{\sqrt{5}} \right\rfloor - 1 < F_n < \frac{\alpha^n}{\sqrt{5}}.$$

Thus $F_n = \left\lfloor \dfrac{\alpha^n}{\sqrt{5}} \right\rfloor$.

The case n odd follows similarly.   ∎

Theorem 8.1 has an analogous result for Lucas numbers; see Hoggatt, 1969 [280]. We leave its proof as an exercise.

Theorem 8.2 (Hoggatt).

$$L_n = \left\lfloor \alpha^n + \frac{1}{2} \right\rfloor.$$ ∎

For example, $\alpha^{13} + 1/2 \approx 521.5019$; so $\left\lfloor \alpha^{13} + \frac{1}{2} \right\rfloor = 521 = L_{13}$.

Corollaries 8.1 and 8.2 also have their counterparts for Lucas numbers, as the next corollary reveals.

Corollary 8.3.

1) $L_n = \left\lceil \alpha^n - \frac{1}{2} \right\rceil$

2) $L_{2n} = \left\lceil \alpha^{2n} \right\rceil$ and $L_{2n+1} = \left\lfloor \alpha^{2n+1} \right\rfloor$.   ∎

For example, $\left\lceil \alpha^7 - 1/2 \right\rceil = \lceil 28.5344... \rceil = 29 = L_7$; $\left\lceil \alpha^8 \right\rceil = \lceil 46.9787... \rceil = 47 = L_8$; and $\left\lfloor \alpha^{11} \right\rfloor = \lfloor 199.0050... \rfloor = 199 = L_{11}$.

8.2   ADDITIONAL FORMULAS

In every explicit formula we have developed thus far, we needed to know the value of n to compute F_n or L_n. Surprisingly, this is no longer the case. Knowing a Fibonacci (or Lucas) number, we can easily compute its immediate successor recursively. The next theorem provides such a formula, but first we need to lay some groundwork in the form of a lemma, similar to Lemma 8.1.

Lemma 8.2. *If $n \geq 2$, then $0 < 1/2 - \beta^n < 1$.*

Proof. We have $|\beta| < 0.62$, $|\beta|^2 < 1/2$; so $|\beta|^n < 1/2$, when $n \geq 2$. Since $|\beta|^n = |\beta^n|$, this implies $-1/2 < \beta^n < 1/2$. Then $-1 < \beta^n - 1/2 < 0$; that is, $0 < 1/2 - \beta^n < 1$, as desired. ∎

We are now ready to state and prove the recursive formula for F_{n+1}.

Theorem 8.3. *Let $n \geq 2$. Then $F_{n+1} = \lfloor \alpha F_n + 1/2 \rfloor$.*

Proof. Using Binet's formula, we have

$$(\alpha - \beta)\alpha F_n = \alpha^{n+1} - \alpha \beta^n$$
$$= \alpha^{n+1} - (\alpha\beta)\beta^{n-1} + \beta^{n+1} - \beta^{n+1}$$
$$= (\alpha^{n+1} - \beta^{n+1}) + \beta^{n-1} + \beta^{n+1}$$
$$= (\alpha - \beta)F_{n+1} + \beta^{n-1}(\beta^2 + 1)$$
$$= (\alpha - \beta)F_{n+1} + \beta^{n-1}(-\sqrt{5}\beta)$$
$$\alpha F_n = F_{n+1} - \beta^n$$
$$\alpha F_n + 1/2 = F_{n+1} + (1/2 - \beta^n). \tag{8.2}$$

Since $1/2 - \beta^n > 0$, this implies $F_{n+1} < \alpha F_n + 1/2$. Besides, since $1/2 - \beta^n < 1$, equation (8.2) yields $\alpha F_n + 1/2 < F_{n+1} + 1$. Thus $F_{n+1} < \alpha F_n + 1/2 < F_{n+1} + 1$; so $F_{n+1} = \lfloor \alpha F_n + 1/2 \rfloor$, as desired. ∎

For example, let $F_n = 4181$. Then $F_{n+1} = \lfloor 4181\alpha + 1/2 \rfloor = \lfloor 6765.500... \rfloor = 6765$, as expected.

The next result [280] is a direct consequence of this theorem.

Corollary 8.4 (Hoggatt). *Let $n \geq 2$. Then $F_{n+1} = \left\lfloor \dfrac{F_n + \sqrt{5}F_n + 1}{2} \right\rfloor$.* ∎

Since $\lfloor x \rfloor = \lceil x \rceil - 1$ for any nonintegral real number x, we can express the recursive formulas in Theorem 8.3 and Corollary 8.4 in terms of the ceiling function, as the next corollary shows.

Corollary 8.5. *Let $n \geq 2$. Then*

1) $F_{n+1} = \lceil \alpha F_n - 1/2 \rceil$

2) $F_{n+1} = \left\lceil \dfrac{F_n + \sqrt{5}F_n - 1}{2} \right\rceil$. ∎

For example, the immediate successor of the Fibonacci number 1597 is given by $\lceil 1597\alpha - 1/2 \rceil = \lceil 2583.5002...\rceil = 2584$.

Using the recursive formula in Theorem 8.3 (or Corollary 8.4), we can compute the ratio F_{n+1}/F_n as $n \to \infty$, as the next corollary demonstrates. Its proof employs the following fact: If $\lfloor x \rfloor = k$, then $x = k + \theta$, where $0 \le \theta < 1$.

Corollary 8.6. $\displaystyle\lim_{n\to\infty} \frac{F_{n+1}}{F_n} = \alpha.$

Proof. By Theorem 8.3, we have

$$F_{n+1} = \alpha F_n + \frac{1}{2} + \theta$$

$$\frac{F_{n+1}}{F_n} = \alpha + \frac{1}{2F_n} + \frac{\theta}{F_n}$$

$$\lim_{n\to\infty} \frac{F_{n+1}}{F_n} = \alpha + 0 + 0,$$

where $0 \le \theta < 1$. This gives the desired limit. ∎

Using Corollary 8.6, we can evaluate $\displaystyle\lim_{n\to\infty} \frac{\tan \theta_n}{\tan \theta_{n+1}}$, where θ_n denotes the acute angle between the adjacent sides of the parallelogram in Figure 6.28. To this end, let $s_n = F_{2n-3} + 2F_{n-3}F_{n-2} = F_{n-1}F_n + F_{n-3}F_{n-2}$. Then

$$\frac{s_{n+1}}{s_n} = \frac{F_n F_{n+1} + F_{n-2}F_{n-1}}{F_{n-1}F_n + F_{n-3}F_{n-2}}$$

$$= \frac{(F_{n+1}/F_{n-1}) + (F_{n-2}/F_n)}{1 + (F_{n-3}F_{n-2})/(F_{n-1}F_n)}$$

$$= \frac{(F_{n+1}/F_n)\cdot(F_n/F_{n-1}) + (F_{n-2}/F_{n-1})\cdot(F_{n-1}/F_n)}{1 + (F_{n-3}/F_{n-2})\cdot(F_{n-2}/F_{n-1})\cdot(F_{n-2}/F_{n-1})\cdot(F_{n-1}/F_n)}$$

$$\lim_{n\to\infty} \frac{s_{n+1}}{s_n} = \frac{\alpha^2 + (1/\alpha)(1/\alpha)}{1 + (1/\alpha)(1/\alpha)(1/\alpha)(1/\alpha)}$$

$$= \frac{\alpha^2 + (1/\alpha^2)}{1 + (1/\alpha^4)}$$

$$= \alpha^2.$$

That is, $\displaystyle\lim_{n\to\infty} \frac{\tan \theta_n}{\tan \theta_{n+1}} = \alpha^2.$

Let u_n and v_n denote the lengths of the sides of the parallelogram in Figure 6.28 (or 6.29), where $u_n > v_n$. Then $u_n = \sqrt{F_n^2 + F_{n-2}^2}$ and $v_n = \sqrt{F_{n-1}^2 + F_{n-3}^2}$; so $\displaystyle\lim_{n\to\infty} u_n/v_n = \alpha.$

LUCAS COUNTERPARTS

Theorem 8.3 and its corollaries have analogous recursive results for Lucas numbers [280]; see Theorem 8.4, and Corollaries 8.7 and 8.8. We leave their proofs as routine exercises.

Theorem 8.4 (Hoggatt).   *Let $n \geq 2$. Then $L_{n+1} = \lfloor \alpha L_n + 1/2 \rfloor$.* ∎

For example, the immediate successor of the Lucas number $L_{15} = 1364$ is $\lfloor 1364\alpha + 1/2 \rfloor = \lfloor 2207.4983... \rfloor = 2207 = L_{16}$.

Corollary 8.7 (Hoggatt).   *Let $n \geq 2$. Then*

1) $L_{n+1} = \left\lfloor \dfrac{L_n + \sqrt{5}L_n + 1}{2} \right\rfloor$

2) $L_{n+1} = \lceil \alpha L_n - 1/2 \rceil$

3) $L_{n+1} = \left\lceil \dfrac{L_n + \sqrt{5}L_n - 1}{2} \right\rceil$. ∎

For example, the immediate successor of $L_{13} = 521$ is given by $\lceil 521\alpha - 1/2 \rceil = \lceil 842.4957... \rceil = 843 = L_{14}$.

Using Theorem 8.4, we can also compute the limit of L_{n+1}/L_n as $n \to \infty$. Again we omit the proof in the interest of brevity; see Exercise 8.25.

Corollary 8.8. $\displaystyle \lim_{n \to \infty} \frac{L_{n+1}}{L_n} = \alpha$. ∎

There is another recursive formula that expresses a Fibonacci number in terms of its immediate predecessor, and also another for Lucas numbers [25]. Both are given in the following theorem.

Theorem 8.5 (Basin, 1964 [25]).

1) $F_{n+1} = \dfrac{F_n + \sqrt{5F_n^2 + 4(-1)^n}}{2}$

2) $L_{n+1} = \dfrac{L_n + \sqrt{5L_n^2 + 4(-1)^n}}{2}$. ∎

S.L. Basin of Sylvania Electronic Systems, Mountain View, California, discovered these two formulas. They can be derived using the identities $2F_{n+1} = F_n + L_n, 2L_{n+1} = 5F_n + L_n$, and $L_n^2 - 5F_n^2 = 4(-1)^n$ [25].

There is yet another recursive formula that expresses a Fibonacci number in terms of its immediate predecessor. Hoggatt and Lind discovered it in 1967 [314].

Theorem 8.6 (Hoggatt and Lind, 1967 [314]).   *Let $n \geq 2$. Then*

$$F_{n+1} = \left\lceil \frac{F_n + 1 + \sqrt{5F_n^2 - 2F_n + 1}}{2} \right\rceil.$$

Proof. First, notice that $L_n - F_n = (F_{n+1} + F_{n-1}) - F_n = 2F_{n-1}$. By Exercise 5.37, $L_n^2 - 5F_n^2 = 4(-1)^n$, where $n \geq 1$. Also, when $n \geq 2$, $4(-1)^n \leq 4F_{n-1}$. Consequently, when $n \geq 2$, we have

$$L_n^2 - 5F_n^2 \leq 4F_{n-1}$$

$$\leq 2(L_n - F_n)$$

$$(L_n - 1)^2 \leq 5F_n^2 - 2F_n + 1.$$

But $L_n = 2F_{n+1} - F_n$; so

$$(2F_{n+1} - F_n - 1)^2 \leq 5F_n^2 - 2F_n + 1$$

$$2F_{n+1} - F_n - 1 \leq \sqrt{5F_n^2 - 2F_n + 1}$$

$$F_{n+1} \leq \frac{F_n + 1 + \sqrt{5F_n^2 - 2F_n + 1}}{2}. \tag{8.3}$$

Since $L_n + F_n = 2F_{n+1}$, and $4(-1)^n > -4F_{n+1}$ when $n \geq 2$, we also have

$$L_n^2 - 5F_n^2 > -2(L_n + F_n)$$

$$L_n^2 + 2L_n > 5F_n^2 - 2F_n$$

$$(L_n + 1)^2 > 5F_n^2 - 2F_n + 1$$

$$(2F_{n+1} - F_n + 1)^2 > 5F_n^2 - 2F_n + 1$$

$$2F_{n+1} - F_n + 1 > \sqrt{5F_n^2 - 2F_n + 1}$$

$$F_{n+1} > \frac{F_n - 1 + \sqrt{5F_n^2 - 2F_n + 1}}{2}$$

$$> \left\lceil \frac{F_n - 1 + \sqrt{5F_n^2 - 2F_n + 1}}{2} \right\rceil. \tag{8.4}$$

By the inequalities (8.3) and (8.4),

$$\left\lceil \frac{F_n - 1 + \sqrt{5F_n^2 - 2F_n + 1}}{2} \right\rceil < F_{n+1} \leq \frac{F_n + 1 + \sqrt{5F_n^2 - 2F_n + 1}}{2}.$$

Since F_{n+1} is an integer, it follows that

$$F_{n+1} = \left\lfloor \frac{F_n + 1 + \sqrt{5F_n^2 - 2F_n + 1}}{2} \right\rfloor,$$

where $n \geq 2$, as desired. ∎

For example, the immediate successor of $F_{16} = 987$ is given by

$$\left\lfloor \frac{987 + 1 + \sqrt{5 \cdot 987^2 - 2 \cdot 987 + 1}}{2} \right\rfloor = \lfloor 1597.2760... \rfloor = 1597 = F_{17}.$$

Analogously, we have the following result for Lucas numbers, also developed by Hoggatt and Lind in 1967 [314]. Its proof is quite similar, so we leave it as an exercise.

Theorem 8.7 (Hoggatt and Lind, 1967 [314]). *Let $n \geq 4$. Then*

$$L_{n+1} = \left\lfloor \frac{L_n + 1 + \sqrt{5L_n^2 - 2L_n + 1}}{2} \right\rfloor.$$

∎

For instance, the immediate successor of the Lucas number 1364 is given by

$$\left\lfloor \frac{1364 + 1 + \sqrt{5 \cdot 1364^2 - 2 \cdot 13647 + 1}}{2} \right\rfloor = \lfloor 2207.2748... \rfloor = 2207.$$

Interestingly, we can use Theorem 8.3 in the reverse direction also. It can be employed to compute the immediate predecessor of a given Fibonacci number, as the next theorem shows.

Theorem 8.8. *Let $n \geq 2$. Then $F_n = \left\lfloor \frac{1}{\alpha}(F_{n+1} + 1/2) \right\rfloor$.*

Proof. Since $x - 1 < \lfloor x \rfloor \leq x$, Theorem 8.3 yields the double inequality

$$\alpha F_n - \frac{1}{2} < F_{n+1} \leq \alpha F_n + \frac{1}{2}$$

$$F_n - \frac{1}{2\alpha} < \frac{F_{n+1}}{\alpha} \leq F_n + \frac{1}{2\alpha}.$$

Then $F_n < \frac{1}{\alpha}(F_{n+1} + 1/2)$ and $F_n \geq \frac{1}{\alpha}(F_{n+1} - 1/2)$. Consequently,

$$\frac{1}{\alpha}(F_{n+1} - 1/2) < F_n \leq \frac{1}{\alpha}(F_{n+1} + 1/2).$$

Since $\dfrac{1}{\alpha}(F_{n+1} + 1/2) - \dfrac{1}{\alpha}(F_{n+1} - 1/2) = \dfrac{1}{\alpha} \approx 0.618$ and F_n is an integer, it follows that $F_n = \left\lfloor \dfrac{1}{\alpha}(F_n + 1/2) \right\rfloor$, where $n \geq 2$.   ∎

For example, the immediate predecessor of $F_{19} = 4181$ is given by $\lfloor 4181.5/\alpha \rfloor = \lfloor 2584.3091... \rfloor = 2584 = F_{18}$.

We have an analogous result for Lucas numbers. Again, we leave its proof as an exercise.

Theorem 8.9. *Let $n \geq 2$. Then $L_n = \left\lfloor \dfrac{1}{\alpha}(L_n + 1/2) \right\rfloor$.*   ∎

For example, the immediate predecessor of $L_{20} = 15{,}127$ is given by $\lfloor 15127.5/\alpha \rfloor = \lfloor 9349.3091... \rfloor = 9349 = L_{19}$.

In 1972, R. Anaya and J. Crump of then San Jose State College, California, established the following generalization of Theorem 8.3 [13].

Theorem 8.10 (Anaya and Crump, 1972 [13]).   *Let $n \geq k \geq 1$. Then $F_{n+k} = \lfloor \alpha^k F_n + 1/2 \rfloor$.*

Proof. The theorem is clearly true when $k = 1$. So we let $n \geq k \geq 2$. By Binet's formula, we have

$$(\alpha - \beta)\alpha^k F_n = \alpha^{n+k} - \alpha^k \beta^n$$

$$= (\alpha^{n+k} - \beta^{n+k}) + (\beta^{n+k} - \alpha^k \beta^n)$$

$$\alpha^k F_n = F_{n+k} - \beta^n F_k$$

$$\alpha^k F_n + 1/2 = F_{n+k} + (1/2 - \beta^n F_k). \tag{8.5}$$

Next we will prove that $0 < 1/2 - \beta^n F_k < 1$. When $n = k$, $|\beta^n F_k|$ has its largest value. Notice that $|\beta^n| \to 0$ as $n \to \infty$. Besides,

$$|\beta^k F_k| = \left| \frac{\beta^k(\alpha^k - \beta^k)}{\sqrt{5}} \right| = \left| \frac{(-1)^k - \beta^{2k}}{\sqrt{5}} \right|.$$

Case 1. Let k be even. Then

$$|\beta^k F_k| = \left| \frac{1 - \beta^{2k}}{\sqrt{5}} \right|$$

$$\lim_{k \to \infty} |\beta^k F_k| = \left| \frac{1 - 0}{\sqrt{5}} \right|$$

$$= 1/\sqrt{5}$$

$$< 1/2.$$

Since $|\beta^n| < |\beta^k|$, it follows that $0 < |\beta^n F_k| < 1/2$.

Case 2. Let k be odd. Then

$$|\beta^k F_k| = \left|\frac{-1 - \beta^{2k}}{\sqrt{5}}\right|$$

$$= \left|\frac{1 + \beta^{2k}}{\sqrt{5}}\right|.$$

When $k = 3$, $\beta^{2k} \approx 0.055726$, so

$$|\beta^k F_k| \approx \frac{1.055726}{\sqrt{5}} \approx 0.472135 < 1/2.$$

As k increases, β^{2k} gets smaller and smaller. So $|\beta^k F_k| < 1/2$ for $k > 3$ also. Thus $0 < |\beta^n F_k| < 1/2$, since $|\beta^n| < |\beta^k|$.

Consequently, $0 < |\beta^n F_k| < 1/2$ for all $n \geq k \geq 2$; that is $-1/2 < \beta^n F_k < 1/2$. Hence $0 < 1/2 - \beta^n F_k < 1$. Therefore, by equation (8.5), $F_{n+k} < \alpha^k F_n + 1/2 < F_{n+k} + 1$. Thus $\lfloor \alpha^k F_n + 1/2 \rfloor = F_{n+k}$. ∎

For example, $\lfloor \alpha^7 F_8 + 1/2 \rfloor = \lfloor 21\alpha^7 + 1/2 \rfloor = \lfloor 610.223... \rfloor = 610 = F_{15} = F_{8+7}$. Notice that $\lfloor \alpha^8 F_7 + 1/2 \rfloor = 611 \neq F_{15}$.

Using the ceiling function, the formula in Theorem 8.10 can be rewritten as in the following corollary.

Corollary 8.9. *Let $n \geq k \geq 1$. Then $F_{n+k} = \lceil \alpha^k F_n - 1/2 \rceil$.* ∎

For example, $\lceil \alpha^9 F_{11} - 1/2 \rceil = \lceil 89\alpha^9 - 1/2 \rceil = \lceil 6764.6708... \rceil = 6765 = F_{20} = F_{11+9}$.

In 1972, Anaya and Crump conjectured a similar formula for L_{n+k} [13]. Carlitz proved it in the same year [116].

Theorem 8.11. *Let $n \geq 4$ and $k \geq 1$. Then $L_{n+k} = \lfloor \alpha^k L_n + 1/2 \rfloor$.*

Proof. First, notice that

$$\alpha L_n - L_{n+1} = \alpha(\alpha^n + \beta^n) - (\alpha^{n+1} + \beta^{n+1})$$

$$= \beta^n(\alpha - \beta)$$

$$= \sqrt{5}\beta^n.$$

When $n \geq 4$,

$$|\sqrt{5}\beta^n| \leq \sqrt{5}\beta^4$$

$$= \sqrt{5}(7 - 3\sqrt{5})/2$$

$$< 1/2.$$

So $|\alpha L_n - L_{n+1}| < 1/2$. Then $0 < \alpha L_n - L_{n+1} + 1/2 < 1$. Consequently, $L_{n+1} = \lfloor \alpha L_n + 1/2 \rfloor$. Thus the theorem is true when $k = 1$.

Now assume $n \geq k + 2$, where $k \geq 2$. Notice that

$$\alpha^{-2} + \alpha^{-6} = \beta^2 + \beta^6$$
$$= (3 - \sqrt{5})/2 + 9 - 4\sqrt{5}$$
$$= (21 - 9\sqrt{5})/2.$$

Since $k \geq 2$, this implies $\alpha^{-2} + \alpha^{-2k-2} < 1/2$; that is, $\alpha^{-k-2}(\alpha^k + \alpha^{-k}) < 1/2$. Since $n \geq k + 2$, this means $\alpha^{-n}(\alpha^k + \alpha^{-k}) < 1/2$; so $|\beta^n(\alpha^k - \beta^k)| < 1/2$. That is,

$$|\alpha^k(\alpha^n + \beta^n) - (\alpha^{n+k} + \beta^{n+k})| < 1/2$$
$$|\alpha^k L_n - L_{n+k}| < 1/2.$$

As before, this implies $\lfloor \alpha^k L_n + 1/2 \rfloor = L_{n+k}$.   ∎

For example, let $n = 11$ and $k = 3$. Then $\lfloor \alpha^3 L_{11} + 1/2 \rfloor = \lfloor 199\alpha^3 + 1/2 \rfloor = \lfloor 843.4774... \rfloor = 843 = L_{14} = L_{11+3}$.

The explicit formula in Theorem 8.11 can be rewritten in terms of the ceiling function, as the next corollary shows.

Corollary 8.10.   *Let $n \geq 4$ and $k \geq 1$. Then $L_{n+k} = \lceil \alpha^k L_n - 1/2 \rceil$.*   ∎

For example, $\lceil \alpha^4 L_{10} - 1/2 \rceil = \lceil 123\alpha^4 - 1/2 \rceil = \lceil 842.5545... \rceil = 843 = L_{14} = L_{10+4}$.

EXERCISES 8

Using Theorem 8.1, compute F_n for each value of n.
   1. 23
   2. 25

   3–4. Using Corollary 8.1, compute F_n for each value of n in Exercises 8.1 and 8.2.

Verify that $\lfloor \alpha^n/\sqrt{5} \rfloor = F_n$ for each value of n.
   5. 12
   6. 20

Verify that $\lceil \alpha^n/\sqrt{5} \rceil = F_n$ for each value of n.
   7. 15
   8. 23

Using Theorem 8.2, compute L_n for each value of n.

9. 15

10. 20

11–12. Using the formula $L_n = \lceil \alpha^n - 1/2 \rceil$, compute L_n for each value of n in Exercises 8.9 and 8.10.

Verify that $L_n = \lceil \alpha^n \rceil$ for each value of n.

13. 10

14. 16

Verify that $L_n = \lfloor \alpha^n \rfloor$ for each value of n.

15. 13

16. 19

Using Theorem 8.3, compute the immediate successor of each Fibonacci number.

17. 2584

18. 6765

19–20. Redo Exercises 8.17 and 8.18 using Corollary 8.5.

Using Theorem 8.4, compute the immediate successor of each Lucas number.

21. 843

22. 9349

23–24. Redo Exercises 8.21 and 8.22 using Corollary 8.6.

25. Using Theorem 8.4, evaluate $\lim\limits_{n \to \infty} L_{n+1}/L_n$.

26. Using Theorem 8.5, evaluate $\lim\limits_{n \to \infty} F_{n+1}/F_n$.

27. Using Theorem 8.7, evaluate $\lim\limits_{n \to \infty} L_{n+1}/L_n$.

Compute the immediate predecessor of each Fibonacci number.

28. 610

29. 17,711

Compute the immediate predecessor of each Lucas number.

30. 1364

31. 39,603

Let u_n and v_n denote the lengths of the sides of the parallelogram in Figure 6.28, where $u_n > v_n$. Prove each (Horadam, 1962 [323]).

32. $\lim\limits_{n \to \infty} u_n/v_n = \alpha$.

33. $\lim\limits_{n \to \infty} u_n/F_{n+1} = -\sqrt{3}\beta$.

34. $\lim\limits_{n \to \infty} v_n/F_n = -\sqrt{3}\beta$.

Suppose every F_n in Figure 6.28 is replaced by the corresponding gibonacci number G_n. Let θ_n denote the acute angle between the adjacent sides of the parallelogram, and let $t_n = (3a - b)G_{2n-1} - \mu F_{2n-1} + 2G_{n-2}G_{n-1}$. Prove each (Horadam, 1962 [323]).

35. The lengths of the sides of the parallelogram are $x_n = \sqrt{G_n^2 + G_{n-2}^2}$ and $y_n = \sqrt{G_{n-1}^2 + G_{n-3}^2}$, where $x_n > y_n$.

36. $\displaystyle\lim_{n\to\infty} t_{n+1}/t_n = \alpha^2$.

37. $\displaystyle\lim_{n\to\infty} x_n/y_n = \alpha$.

38. $\tan \theta_n = \mu/t_n$.

9

THE EUCLIDEAN ALGORITHM

This chapter continues our investigation of the properties of Fibonacci numbers. We will re-confirm, using the Euclidean algorithm, that any two consecutive Fibonacci numbers are relatively prime. To this end, we first lay the necessary foundation for justifying the algorithm.

Among the several procedures for finding the greatest common divisor (gcd) of two positive integers, one efficient algorithm is the celebrated *Euclidean algorithm*. Although it seems to have been known before Euclid, it is named after him. Euclid published it in Book VII of his extraordinary work, *The Elements*.

The next theorem paves the way for the Euclidean algorithm.

Theorem 9.1. *Let a and b be any positive integers, where $a \geq b$. Let r be the remainder when a is divided by b. Then $(a, b) = (b, r)$.*

Proof. Let $d = (a, b)$ and $d' = (b, r)$. To prove that $d = d'$, it suffices to show that $d | d'$ and $d' | d$. By the division algorithm, there is a unique quotient q such that

$$a = bq + r. \tag{9.1}$$

To show that $d | d'$:
Since $d = (a, b)$, $d | a$ and $d | b$; so $d | bq$ by Theorem A.10. Then $d | (a - bq)$, again by Theorem A.10; that is, $d | r$, by equation (9.1). Thus $d | b$ and $d | r$, so $d | (b, r)$; that is, $d | d'$.

Fibonacci and Lucas Numbers with Applications, Volume One, Second Edition. Thomas Koshy.
© 2018 John Wiley & Sons, Inc. Published 2018 by John Wiley & Sons, Inc.

Little is known about Euclid's life. He was on the faculty at the University of Alexandria and founded the Alexandrian School of Mathematics. When the Egyptian ruler, King Ptolemy I, asked Euclid if there were an easier way to learn geometry than by studying *The Elements*, Euclid replied, "There is no royal road to geometry." Euclid* is known as the father of geometry.

Similarly, $d'|d$; see Exercise 9.17. Thus, by Theorem A.9, $d = d'$; that is, $(a, b) = (b, r)$.   ∎

The following example illustrates this theorem.

Example 9.1. *Illustrate Theorem 9.1 with $a = 120$ and $b = 28$.*

Solution. First, we can verify that $(120, 28) = 4$. Now, by the division algorithm, $120 = 4 \cdot 28 + 8$; so by Theorem 9.1, $(120, 28) = (28, 8)$. But $(28, 8) = 4$. So $(120, 28) = 4$.   ∎

Before formally presenting the Euclidean algorithm, we will illustrate it in the next example.

Example 9.2. *Illustrate the Euclidean algorithm by evaluating $(2076, 1776)$.*

Solution. Apply the division algorithm with 2076 (larger of the two) as the dividend, and 1776 as the divisor: $2076 = 1 \cdot 1776 + 300$. Apply the division

Figure source: https://commons.wikimedia.org/wiki/File:Euklid-von-Alexandria_1.jpg.

algorithm again with 1776 as the dividend, and 300 as the divisor: $1776 = 5 \cdot 300 + 276$. Continue this procedure until a zero remainder is reached:

$$2076 = 1 \cdot 1776 + 300$$
$$1776 = 5 \cdot \ 300 + 276$$
$$300 = 1 \cdot \ 276 + \ 24$$
$$276 = 11 \cdot \ 24 + \boxed{12} \quad \leftarrow \text{last nonzero remainder}$$
$$24 = 2 \cdot \ 12 + \ 0.$$

The last nonzero remainder in this procedure is the gcd, so $(2076, 1776) = 12$. ∎

Take a close look at the preceding steps to see why the gcd is 12. By the repeated application of Theorem 9.1, we have

$$(2076, 1776) = (1776, 300) = (300, 276)$$
$$= (276, 24) \quad = (24, 12)$$
$$= 12.$$

We will justify this algorithm, although it should be obvious by now.

9.1 THE EUCLIDEAN ALGORITHM

Let a and b be any two positive integers, with $a \geq b$. If $a = b$, then $(a, b) = a$, so assume $a > b$. Let $r_0 = b$. Then by successive application of the division algorithm, we get a finite sequence of equations:

$$a = q_0 r_0 + r_1, \qquad 0 \leq r_1 < r_0$$
$$r_0 = q_1 r_1 + r_2, \qquad 0 \leq r_2 < r_1$$
$$r_1 = q_2 r_2 + r_3, \qquad 0 \leq r_3 < r_2$$
$$\vdots$$

Continuing like this, we get a sequence of remainders:

$$b = r_0 > r_1 > r_2 > r_3 > \cdots \geq 0.$$

Since the remainders are nonnegative, and getting progressively smaller, this sequence should eventually terminate with remainder $r_n = 0$. Thus the last two equations in the preceding procedure are:

$$r_{n-2} = q_{n-1}r_{n-1} + r_n, \qquad 0 \le r_n < r_{n-1}$$
$$r_{n-1} = q_n r_n.$$

It follows by PMI that $(a, b) = (a, r_0) = (r_0, r_1) = (r_1, r_2) = \cdots = (r_{n-1}, r_n) = r_n$, the last nonzero remainder; see Exercise 9.18.

Example 9.3. *Using the Euclidean algorithm, find $(4076, 1024)$.*

Solution. By the successive application of the division algorithm, we have

$$4076 = 3 \cdot 1024 + 1004$$
$$1024 = 1 \cdot 1004 + 20$$
$$1004 = 50 \cdot 20 + \boxed{4} \leftarrow \text{last nonzero remainder}$$
$$20 = 5 \cdot 4 + 0.$$

Since the last nonzero remainder is 4, $(4076, 1024) = 4$. ∎

The Euclidean algorithm is purely mechanical. All we need is to make our divisor the new dividend and the remainder the new divisor. That is, just follow the southwest arrows in the solution.

The Euclidean algorithm provides a mechanism for expressing the (a, b) as a linear combination of a and b, as the next example demonstrates.

Example 9.4. *Using the Euclidean algorithm, express $(4076, 1024)$ as a linear combination of 4076 and 1024.*

Solution. We use the equations in Example 9.3 in the reverse order, each time substituting for the remainder from the previous equation:

$$
\begin{aligned}
(4076, 1024) &= 4 \\
&= 1004 - 50 \cdot 20 \\
&= 1004 - 50(1024 - 1 \cdot 1004) \\
&= 51 \cdot 1004 - 50 \cdot 1024 \\
&= 51(3076 - 3 \cdot 1024) - 50 \cdot 1024 \\
&= 51 \cdot \boxed{4076} + (-203) \cdot \boxed{1024}. \quad \leftarrow \text{linear combination}
\end{aligned}
$$

(You can confirm this by direct computation.) ∎

Recall from Corollary 5.2 that any two consecutive Fibonacci numbers are relatively prime. We will now re-confirm it using the Euclidean algorithm.

Example 9.5. *Prove that any two consecutive Fibonacci numbers are relatively prime.*

Proof. Using the Euclidean algorithm with F_n as the original dividend and F_{n-1} as the original divisor, we get the following system of linear equations:

$$F_n = 1 \cdot F_{n-1} + F_{n-2}$$
$$F_{n-1} = 1 \cdot F_{n-2} + F_{n-3}$$
$$F_{n-2} = 1 \cdot F_{n-3} + F_{n-4}$$
$$\vdots$$
$$F_4 = 1 \cdot F_3 + \boxed{F_2}$$
$$F_3 = 2 \cdot F_2 + 0.$$

Thus, by the Euclidean algorithm, $(F_n, F_{n-1}) = F_2 = 1$. ∎

9.2 FORMULA (5.5) REVISITED

In 1990, Ian Cook of the University of Essex, United Kingdom, developed the summation formula (5.5) as a nice application of the Euclidean algorithm [128]. To see this, first consider the algorithm with $a = 1976$ and $b = 1776$:

$$1976 = 1 \cdot 1776 + 200$$
$$1776 = 8 \cdot 200 + 176$$
$$200 = 1 \cdot 176 + 24$$
$$176 = 7 \cdot 24 + \boxed{8} \leftarrow \text{gcd}$$
$$24 = 3 \cdot 8 + 0.$$
$$\uparrow$$
$$\text{quotients}$$

It follows from these equations by successive substitutions that

$$1976 \cdot 1776 = 1 \cdot 1776^2 + 8 \cdot 200^2 + 1 \cdot 176^2 + 7 \cdot 24^2 + 3 \cdot 8^2.$$

Notice that the coefficients on the RHS are the various quotients in the algorithm, and the corresponding factors are the squares of the corresponding divisors.

More generally, the equations

$$a = q_0 r_0 + r_1$$
$$r_0 = q_1 r_1 + r_2$$
$$r_1 = q_2 r_2 + r_3$$
$$\vdots$$
$$r_{i-1} = q_i r_i + r_{i+1}$$
$$\vdots$$
$$r_{n-2} = q_{n-1} r_{n-1} + \boxed{r_n} \quad \leftarrow \quad \text{gcd}$$
$$r_{n-1} = q_n r_n + 0$$

imply that

$$ab = \sum_{i=0}^{n} q_i r_i^2. \tag{9.2}$$

We can confirm this using PMI.

In particular, let $a = F_{n+1}$ and $b = F_n$. Then we have

$$F_{n+1} = 1 \cdot F_n + F_{n-1}$$
$$F_n = 1 \cdot F_{n-1} + F_{n-2}$$
$$\vdots$$
$$3 = 1 \cdot 2 + \boxed{1} \quad \leftarrow \quad (F_{n+1}, F_n)$$
$$2 = 2 \cdot 1 + 0.$$

With $q_n = 2$ and $q_i = 1$ for $0 \le i < n$, formula (9.2) yields

$$F_{n+1} F_n = \sum_{i=3}^{n} 1 \cdot F_i^2 + 2 \cdot 1^2,$$

that is,

$$\sum_{i=1}^{n} F_i^2 = F_{n+1} F_n,$$

which is formula (5.5).

We will now estimate the number of divisions in the Euclidean algorithm, for which we need the following result. We will establish it with the strong version of PMI.

Lemma 9.1. *Let $n \geq 3$. Then $\alpha^{n-2} < F_n < \alpha^{n-1}$.*

Proof. (We will prove by PMI that $\alpha^{n-2} < F_n$, and leave the other half as an exercise.) Let $P(n) : \alpha^{n-2} < F_n$, where $n \geq 3$.

Since $\alpha = \dfrac{1 + \sqrt{5}}{2} < \dfrac{1 + 3}{2} = 2 = F_3$, and $\alpha^2 = \left(\dfrac{1 + \sqrt{5}}{2}\right)^2 = \dfrac{3 + \sqrt{5}}{2} <$

$\dfrac{3 + 3}{2} = 3 = F_4$, both $P(3)$ and $P(4)$ are true.

Assume $P(i)$ is true for all integers $3 \leq i \leq k$, where k is an arbitrary integer ≥ 3. We will show that $P(k + 1)$ is also true.

Since $\alpha^2 = \alpha + 1$, we have

$$\alpha^{k-1} = \alpha^{k-2} + \alpha^{k-3}$$
$$< F_k + F_{k-1}$$
$$= F_{k+1}.$$

So $P(k + 1)$ is also true.

Thus, by the strong version of PMI, $P(n)$ is true for all integers $n \geq 3$. ∎

9.3 LAMÉ'S THEOREM

With this lemma at hand, we have the tools needed to estimate the number of divisions required to compute (a, b) by the Euclidean algorithm. It was established in 1844 by Lamé.

Theorem 9.2 (Lamé's Theorem). *The number of divisions needed to compute (a, b) by the Euclidean algorithm is no more than five times the number of decimal digits in b, where $a \geq b \geq 2$.*

Proof. Let $a = r_0$ and $b = r_1$. By the repeated application of the division algorithm, we have

$$r_0 = q_1 r_1 + r_2, \qquad 0 \leq r_2 < r_1$$
$$r_1 = q_2 r_2 + r_3, \qquad 0 \leq r_3 < r_2$$
$$\vdots$$
$$r_{n-2} = q_{n-1} r_{n-1} + r_n, \quad 0 \leq r_n < r_{n-1}$$
$$r_{n-1} = q_n r_n.$$

Clearly, it takes n divisions to compute $(a, b) = r_n$. Since $r_i < r_{i-1}$, $q_i \geq 1$ for $1 \leq i \leq n$. In particular, since $r_n < r_{n-1}$, $q_n \geq 2$; so $r_n \geq 1$ and $r_{n-1} \geq 2 = F_3$.

Consequently, we have

$$r_{n-2} = q_{n-1} r_{n-1} + r_n$$
$$\geq r_{n-1} + r_n$$
$$\geq F_3 + 1$$
$$= F_4$$

and

$$r_{n-3} = q_{n-2} r_{n-2} + r_{n-1}$$
$$\geq r_{n-2} + r_{n-1}$$
$$\geq F_4 + F_3$$
$$= F_5.$$

Continuing like this,

$$r_1 = q_2 r_2 + r_3$$
$$\geq r_2 + r_3$$
$$\geq F_n + F_{n-1}$$
$$= F_{n+1}.$$

That is, $b \geq F_{n+1}$.

By Lemma 9.1, $F_{n+1} > \alpha^{n-1}$, where $n \geq 3$. Therefore $b > \alpha^{n-1}$ and hence $\log b > (n-1) \log \alpha$. Since $\alpha \approx 1.618033989$, $\log \alpha \approx 0.2089876403 > 1/5$. So $\log b > (n-1)/5$.

Suppose b contains k decimal digits. Then $b < 10^k$. So $\log b < k$ and hence $k > (n-1)/5$. Thus $n < 5k + 1$ or $n \leq 5k$. That is, the number of divisions is no more than five times the number of decimal digits in n.   ■

Let us pursue this theorem a bit further. Since $\log b > (n-1)/5$, $n < 1 + 5 \log b$. Also, since $b \geq 2$, $5 \log b \geq 5 \log 2$; so $5 \log b > 1$. Thus

$$n < 1 + 5 \log b$$
$$< 5 \log b + 5 \log b$$
$$= 10 \log b$$
$$= O(\log b).$$

Thus it takes $O(\log b)$[†] divisions to compute (a, b) by the Euclidean algorithm.

[†] Let $f, g : \mathbb{N} \to \mathbb{R}$. Then $f(n)$ is said to be of *order at most* $g(n)$ if there exist a positive constant C and a positive integer n_0 such that $|f(n)| \leq C|g(n)|$ for every $n \geq n_0$. In symbols, we write $f(n) = O(g(n))$. (Read this as $f(n)$ is *big-oh of* $g(n)$.)

EXERCISES 9

Using the Euclidean algorithm, find the gcd of the given integers.
1. 1024, 1000
2. 2024, 1024
3. 2076, 1076
4. 2076, 1776
5. 1976, 1776
6. 3076, 1776
7. 3076, 1976
8. 4076, 2076

9–16. Using the Euclidean algorithm, express the gcd of each pair of integers in Exercises 9.1–9.8 as a linear combination of the given numbers.

17. Let a and b be any two positive integers, and r the remainder when a is divided by b. Let $d = (a, b)$ and $d' = (b, r)$. Prove that $d'|d$.

18. Let a and b be any two positive integers, where $a \geq b$. Using the sequence of equations in the Euclidean algorithm, prove that $(a, b) = (r_{n-1}, r_n)$, where $n \geq 1$.

19. Prove summation formula (5.10) using the Euclidean algorithm.

10

DIVISIBILITY PROPERTIES

In this chapter, we will explore a few divisibility properties of Fibonacci and Lucas numbers.

10.1 FIBONACCI DIVISIBILITY

In Chapter 5, we learned that $F_{2n} = F_n L_n$, so $F_n | F_{2n}$. Can we generalize this? In other words, under what conditions does $F_m | F_n$? The next theorem gives us a partial answer. We will establish it using PMI.

Theorem 10.1. *Let $m \geq 1$. Then $F_m | F_{mn}$.*

Proof. Clearly the given statement is true when $n = 1$.

Now assume it is true for all positive integers $n \leq k$, where $k \geq 1$. We will now show that $F_m | F_{m(k+1)}$. To this end, we invoke the *addition formula*

$$F_{r+s} = F_{r-1}F_s + F_r F_{s+1}. \tag{10.1}$$

(It can be confirmed using Binet's formula; see Chapter 20 for an alternate method.) Then

$$F_{m(k+1)} = F_{mk+m} = F_{mk-1}F_m + F_{mk}F_{m+1}.$$

Since $F_m | F_{mk}$ by the inductive hypothesis, it follows that $F_m | F_{m(k+1)}$.

Thus, by the strong version of PMI, the result is true for all integers $n \geq 1$. ∎

Fibonacci and Lucas Numbers with Applications, Volume One, Second Edition. Thomas Koshy.
© 2018 John Wiley & Sons, Inc. Published 2018 by John Wiley & Sons, Inc.

For example, $F_6 = 8$ and $F_{24} = 46{,}368$. Since $6|24$, it follows by the theorem that $F_6|F_{24}$; this can easily be verified.

The following corollary is a re-statement of the theorem.

Corollary 10.1. *Every mth Fibonacci number is divisible by F_m, where $m \geq 1$.* ∎

For example, every third Fibonacci number is divisible by F_3. Every fifth is divisible by F_5; that is, $F_5, F_{10}, F_{15}, \ldots$ are all divisible by F_5. Likewise, $F_6, F_{12}, F_{18}, F_{24}, \ldots$ are divisible by F_6.

What are the chances that F_n is divisible by a Fibonacci number F_k, selected at random, where $k \geq 3$? In 1964, F.D. Parker of the University of Alaska investigated this interesting problem, given in the next example [471].

Example 10.1 (Parker, 1964 [471]).   *Find the probability that a Fibonacci number F_n is divisible by another Fibonacci number F_k, selected at random, where $k \geq 3$.*

Solution. By Corollary 10.1, $F_3|F_{3m}$; that is, every third Fibonacci number is divisible by 3. So the probability that F_n is divisible by 2 is 1/3. The probability that F_n is divisible by $F_4 = 3$ is 1/4; so the probability that F_n is divisible by 3, but not by 2, is $1/4 \cdot 2/3 = 2/(3 \cdot 4)$. Likewise, the probability that F_n is divisible by $F_5 = 5$, but not by 2 or 3, is $2/(4 \cdot 5)$. In general, the probability that F_n is divisible by F_k, but *not* by F_j, is $2/(k-1)k$, where $3 \leq j \leq k$.

Thus, by the addition principle, the probability that F_n is divisible by F_k is

$$\sum_{i=2}^{k-1} \frac{2}{i(i+1)} = 2 \sum_{i=2}^{k-1} \left(\frac{1}{i} - \frac{1}{i+1} \right)$$
$$= 2 \left(\frac{1}{2} - \frac{1}{k} \right),$$

where $k \geq 3$. As $k \to \infty$, this probability approaches unity. ∎

In 1964, Carlitz established the converse of Theorem 10.1 using the identity

$$F_r = F_{r-s+1} F_s + F_{r-s} F_{s-1}, \tag{10.2}$$

where $r \geq s \geq 1$. This is a slight variation of the addition formula. The proof of the converse invokes the division algorithm [101].

Theorem 10.2 (Carlitz, 1964 [101]).   *Let $m \geq 1$. If $F_m|F_n$, then $m|n$.*

Proof. By the division algorithm, $n = qm + r$, where $0 \leq r < m$. Suppose $F_m|F_n$. Then, by Theorem A.10 and identity (10.2), $F_m|F_{n-m}F_{m-1}$. But $(F_m, F_{m-1}) = 1$, so $F_m|F_{n-m}$.

Similarly, $F_m|F_{n-2m}$. Continuing like this, $F_m|F_{n-qm}$; that is, $F_m|F_r$. This is impossible unless $r = 0$. Therefore $n = qm$. Thus $F_m|F_n$ implies $m|n$. ∎

Combining Theorems 10.1 and 10.2, we get the next corollary.

Corollary 10.2. *Let $m, n \geq 1$. Then $F_m | F_n$ if and only if $m | n$.* ∎

Recall from Corollary 5.2 and Example 9.5 that $(F_{n-1}, F_n) = 1$ for every $n \geq 1$. The next lemma generalizes this result in light of Theorem 10.1.

Lemma 10.1. *Let $q \geq 1$. Then $(F_{qn-1}, F_n) = 1$.*

Proof. Let $d = (F_{qn-1}, F_n)$. Then $d | F_{qn-1}$ and $d | F_n$. Since $F_n | F_{qn}$ by Theorem 10.1, $d | F_{qn}$. Thus $d | F_{qn-1}$ and $d | F_{qn}$. But $(F_{qn-1}, F_{qn}) = 1$ by Corollary 5.2. Therefore $d | 1$ and hence $d = 1$. Thus $(F_{qn-1}, F_n) = 1$, as desired. ∎

This lemma helps us derive another property, which bears a close resemblance to Theorem 10.1. The proof employs the addition formula (10.1) and Theorem 10.1.

Lemma 10.2. *Let $m = qn + r$, where $0 \leq r < n$. Then $(F_m, F_n) = (F_n, F_r)$.*

Proof. By the addition formula, Theorem 10.1, and Lemma 10.1, we have

$$\begin{aligned}
(F_m, F_n) &= (F_{qn+r}, F_n) \\
&= (F_{qn-1}F_r + F_{qn}F_{r+1}, F_n) \\
&= (F_{qn-1}F_r, F_n) \\
&= (F_r, F_n) \\
&= (F_n, F_r).
\end{aligned}$$
∎

The next theorem is really a gem. It shows that the gcd of two Fibonacci numbers is also a Fibonacci number. Its proof uses the Euclidean algorithm in Theorem 10.1 and Lemma 10.2.

Theorem 10.3. $(F_m, F_n) = F_{(m,n)}$.

Proof. Suppose $m \geq n$. Using the Euclidean algorithm with m as the dividend and n as the divisor, we get the following sequence of equations:

$$\begin{aligned}
m &= q_0 n + r_1, && 0 \leq r_1 < n \\
n &= q_1 r_1 + r_2, && 0 \leq r_2 < r_1 \\
r_1 &= q_2 r_2 + r_3, && 0 \leq r_3 < r_2 \\
&\ \ \vdots \\
r_{s-2} &= q_{s-1} r_{s-1} + r_s, && 0 \leq r_s < r_{s-1} \\
r_{s-1} &= q_s r_s.
\end{aligned}$$

By Lemma 10.2, $(F_m, F_n) = (F_n, F_{r_1}) = (F_{r_1}, F_{r_2}) = \cdots = (F_{r_{s-1}}, F_{r_s})$. But $r_s | r_{s-1}$; so by Theorem 10.1, $F_{r_s} | F_{r_{s-1}}$. So $(F_{r_{s-1}}, F_{r_s}) = F_{r_s}$. Thus $(F_m, F_n) = F_{r_s}$. But, by the Euclidean algorithm, $r_s = (m, n)$. Consequently, $(F_m, F_n) = F_{(m,n)}$. ∎

For example, $(144, 2584) = (F_{12}, F_{18}) = F_{(12,18)} = F_6 = 8$.

We now present an alternate proof of Theorem 10.3, given by G. Michael of Washington State University in 1964 [439].

An alternate proof. Let $d = (m, n)$ and $d' = (F_m, F_n)$. By Theorem 10.1, $F_d | F_m$ and $F_d | F_n$; so $F_d | d'$. Since $d = (m, n)$, there exist integers a and b such that $d = am + bn$. Since $d, m, n > 0$, either $a \le 0$ or $b \le 0$. Suppose $a \le 0$. Let $a = -k$, where $k \ge 0$. Then $bn = d + km$.

By identity (10.1),

$$F_{bn} = F_{d+km} = F_{d-1}F_{km} + F_d F_{km+1}. \tag{10.3}$$

Since $d' | F_m$, $d' | F_{km}$ by Theorem 10.1. Now $d' | F_n$ and $F_n | F_{bn}$, so $d' | F_{bn}$. Thus $d' | F_{km}$ and $d' | F_{bn}$. Therefore, by equation (10.3), $d' | F_d F_{km+1}$. But $(d', F_{km+1}) = 1$, since $d' | F_{km}$ and $(F_{km}, F_{km+1}) = 1$; so $d' | F_d$.

Thus $F_d | d'$ and $d' | F_d$, so $d' = F_d$. In other words, $(F_m, F_n) = F_{(m,n)}$. ∎

The next corollary also follows by Theorems 10.1 and 10.3.

Corollary 10.3. *If* $(m, n) = 1$ *then* $F_m F_n | F_{mn}$.

Proof. By Theorem 10.1, $F_m | F_{mn}$ and $F_n | F_{mn}$. Therefore, $[F_m, F_n] | F_{mn}$. But $(F_m, F_n) = F_{(m,n)} = F_1 = 1$, so $[F_m, F_n] = F_m F_n$. Thus $F_m F_n | F_{mn}$, as desired. ∎

For example, $(4, 7) = 1$, $F_4 = 3$, $F_7 = 13$, and $F_{28} = 317,811$. We can verify that $3 \cdot 13 | 317811$; that is, $F_4 F_7 | F_{28}$.

The next two corollaries follow from Theorem 10.3.

Corollary 10.4. *If* $(m, n) = 1$ *then* $(F_m, F_n) = 1$. ∎

For example, $(12, 25) = 1$, so $(F_{12}, F_{25}) = (144, 75025) = 1$.

Corollary 10.5. *If* $F_m | F_n$, *then* $m | n$.

Proof. Suppose $F_m | F_n$. Then $(F_m, F_n) = F_m = F_{(m,n)}$, by Theorem 10.3. So $m = (m, n)$, and hence $m | n$. ∎

Corollary 10.5, coupled with Theorem 10.1, provides an alternate proof of Corollary 10.2.

Theorem 10.3 has an intriguing byproduct. In 1965, M. Wunderlich of the University of Colorado employed the theorem to provide a beautiful proof that there are infinitely many primes, a fact that is universally known. The following corollary takes us to its proof [612].

Corollary 10.6. *There are infinitely many primes.*

Proof. Suppose there are only a finite number of primes, $p_1, p_2, \ldots$, and p_k. Consider the corresponding Fibonacci numbers $F_{p_1}, F_{p_2}, \ldots$, and F_{p_k}. Clearly, they are pairwise relatively prime. Since there are only k primes, each of these Fibonacci numbers has exactly one prime factor; that is, each is a prime. This is a contradiction, since $F_{19} = 4181 = 37 \cdot 113$. Thus our assumption is false. Thus there are infinitely many primes. ∎

In 1966, L. Weinstein of the Massachusetts Institute of Technology established the following divisibility property [587], which is a direct consequence of Erdös' theorem in the Appendix and Theorem 10.3.

Theorem 10.4 (Weinstein, 1966 [587]). *Every set S of $n + 1$ Fibonacci numbers, selected from $F_1, F_2, \ldots, F_{2n}$, contains two elements such that one divides the other.*

Proof. Let $S = \{F_{a_1}, F_{a_2}, \ldots, F_{a_{n+1}}\}$, where $1 \le a_i \le 2n$ and $1 \le i \le n + 1$. Since $A = \{a_1, a_2, \ldots, a_n, a_{n+1}\} \subseteq \{1, 2, \ldots, 2n\}$, by Erdös' theorem, A contains two elements a_i and a_j such that $a_i | a_j$. Then $(a_i, a_j) = a_i$, so $(F_{a_i}, F_{a_j}) = F_{(a_i, a_j)} = F_{a_i}$, by Theorem 10.3. Thus $F_{a_i} | F_{a_j}$. This yields the desired result. ∎

The following example features an interesting divisibility problem studied in 2011 by M. Bataille of Rouen, France [29]. The solution is based on the Catalan identity $F_{m+n} F_{m-n} - F_m^2 = (-1)^{m+n+1} F_n^2$ in Theorem 5.11. As you will see, it has an interesting byproduct.

Example 10.2. *Find all positive integers n such that $F_{m+n} | \left[F_m^2 + (-1)^{m+1} F_n^2 \right]$, where $m \ge 0$.*

Solution. Suppose n is even. Then, by the Catalan identity, $F_{m+n} F_{m-n} = F_m^2 + (-1)^{m+1} F_n^2$. So $F_{m+n} | \left[F_m^2 + (-1)^{m+1} F_n^2 \right]$.

On the other hand, suppose n is odd. We will now show that $F_{m+n} \nmid \left[F_m^2 + (-1)^{m+1} F_n^2 \right]$.

Suppose $m = 3$ and $n = 1$. Then $F_{m+n} = F_4 = 3$, and $F_m^2 + (-1)^{m+1} F_n^2 = F_3^2 + F_1^2 = 5$; clearly, $F_4 \nmid (F_3^2 + F_1^2)$. Suppose $n > 1$ and $n = m$. Then

$$\frac{F_m^2 + (-1)^{m+1} F_n^2}{F_{m+n}} = \frac{F_m^2 + F_m^2}{F_{2m}}$$

$$= \frac{2F_m}{L_m}$$

$$= \frac{2F_m}{F_{m+1} + F_{m-1}}$$

$$< 1.$$

So $F_{m+n} \nmid \left[F_m^2 + (-1)^{m+1} F_n^2 \right]$, if $m = n$ and $n > 1$.

Thus $F_{m+n} | \left[F_m^2 + (-1)^{m+1} F_n^2 \right]$ if and only if n is even. ∎

AN INTERESTING BYPRODUCT

This example has an interesting consequence. We will now show that the sum

$$S_n = F_{n-1} + \sum_{m=0}^{2n} \frac{F_m^2 + (-1)^{m+1} F_n^2}{F_{m+n}}$$

is a Fibonacci number when n is an even positive integer $2k$.

Proof. By Catalan's identity, we have

$$S_n = F_{n-1} + \sum_{m=0}^{4k} F_{m-2k}$$

$$= F_{n-1} + \sum_{m=0}^{2k-1} F_{m-2k} + \sum_{m=2k+1}^{4k} F_{m-2k}$$

$$= F_{n-1} + \sum_{m=0}^{2k-1} (-1)^{m+1} F_{2k-m} + \sum_{m=2k+1}^{4k} F_{m-2k}$$

$$= F_{n-1} + \sum_{i=1}^{2k} [1 - (-1)^i] F_i$$

$$= F_{n-1} + 2 \sum_{i=1}^{k} F_{2i-1}$$

$$= F_{n-1} + 2 F_{2k}$$

$$= F_{n-1} + 2 F_n$$

$$= F_{n+2},$$

where n is even. (Notice that the formula works if $n = 0$ also.)   ∎

For example,

$$S_4 = F_3 + \sum_{m=0}^{8} \frac{F_m^2 + (-1)^{m+1} F_4^2}{F_{m+4}}$$

$$= 2 + \sum_{m=0}^{8} F_{m-4}$$

$$= 2 + 6$$

$$= F_6, \quad \text{as expected.}$$

We will now turn to some divisibility properties of Lucas numbers.

10.2 LUCAS DIVISIBILITY

A quick look at Lucas numbers shows that every third Lucas number is even; that is, $2|L_{3n}$. This is, in fact, always the case; see Exercise 10.39.

Carlitz discovered the next two divisibility properties in 1964 [101].

Theorem 10.5 (Carlitz, 1964 [101]). *Let $m \geq 2$. Then $L_m|F_n$ if and only if $2m|n$.*
∎

For example, $10|20$, so $L_5|F_{20}$; so $11|6765$.

Theorem 10.6 (Carlitz, 1964 [101]). *Let $m \geq 2$ and $k \geq 1$. Then $L_m|L_n$ if and only if $n = (2k - 1)m$.*
∎

For example, let $m = 4$, and $n = 3 \cdot 4 = 12$. We have $L_4 = 7$ and $L_{12} = 322$. Clearly, $L_4|L_{12}$.

10.3 FIBONACCI AND LUCAS RATIOS

In 1965, George C. Cross and Helen G. Renzi of Williamstown Public Schools in Massachusetts proved [133] that if the ratio $a : b = 2 : 3$, then $[a, b] - (a, b) = a + b$. For example, let $a = 12$ and $b = 18$. Then $[a, b] - (a, b) = 36 - 6 = 30 = 12 + 18$. They also proved that if $a : b = 3 : 5$, then $[a, b] + (a, b) = 2(a + b)$. For instance let $a = 45$ and $b = 75$. Then $[a, b] + (a, b) = 225 + 15 = 240 = 2(45 + 75)$.

More generally, suppose $a : b = F_n : F_{n+1}$ or $a : b = L_n : L_{n+1}$. How are $[a, b]$, (a, b), and $a + b$ related? These two questions were investigated two years later by G.F. Freeman of Williams College, Williamstown, Massachusetts. The next two theorems provide their answers [185].

Theorem 10.7 (Freeman, 1967 [185]).

1) *Let $n \geq 2$ and $a : b = F_n : F_{n+1}$. Then $(a + b)F_{n-1} = [a, b] + (-1)^n(a, b)$.*
2) *Let $n \geq 3$ and $a : b = c : d$, where $(c, d) = 1$; and $(a + b)F_{n-1} = [a, b] + (-1)^n(a, b)$. Then the number of solutions of the ratio $c : d$ equals one-half the number of positive factors of F_nF_{n-2}, one of them being $F_n : F_{n+1}$.*

Proof.

1) Let $a : b = F_n : F_{n+1}$. Since $(F_n, F_{n+1}) = 1$, then $a = F_nk$, $b = F_{n+1}k$, $(a, b) = k$, and $[a, b] = F_nF_{n+1}k$ for some positive integer k. By Cassini's formula, we then have

$$(a+b)F_{n-1} = F_{n-1}(F_n + F_{n+1})k$$
$$= F_{n-1}F_{n+2}k$$
$$= (F_{n+1} - F_n)F_{n+2}k$$
$$= F_{n+1}(F_n + F_{n+1})k - F_n F_{n+2}k$$
$$= F_n F_{n+1}k + (F_{n+1}^2 - F_n F_{n+2})k$$
$$= [a,b] + (-1)^n(a,b).$$

2) Let $a : b = c : d$, where $(c,d) = 1$. Then $a = ck$, $B = dk$, $(a,b) = k$, and $[a,b] = cdk$ for some positive integer k. Since $(a+b)F_{n-1} = [a,b] + (-1)^n(a,b)$, $c + d = cd + (-1)^n$. This yields

$$c = \frac{dF_{n-1} - (-1)^n}{d - F_{n-1}}$$

$$= F_{n-1} + \frac{F_{n-1}^2 - (-1)^n}{d - F_{n-1}}$$

$$= F_{n-1} + \frac{F_n F_{n-2}}{d - F_{n-1}}. \tag{10.4}$$

If $0 < d < F_{n-1}$, then $c < 0$; so $d > F_{n-1}$. Since c is an integer, $(d - F_{n-1})|F_n F_{n-2}$. Thus equation (10.4) yields a value of c for every positive factor of $F_n F_{n-2}$. But, if $c = A$, $d = B$ is a solution of the ratio $c : d$, so is $c = B$, $d = A$. Thus the number of distinct solutions of the ratio $c : d$ equals one-half the number of positive factors of $F_n F_{n-2}$.

In particular, let $d = F_{n+1}$. Then

$$c = F_{n-1} + \frac{F_n F_{n-2}}{F_n}$$

$$= F_{n-1} + F_{n-2}$$

$$= F_n.$$

Thus $c : d = F_n : F_{n+1}$ is a solution of the ratio.   ■

The following example demonstrates this theorem.

Example 10.3.

1) Let $a : b = F_9 : F_{10} = 34 : 55$, so $n = 9$. Let $a = 238$ and $b = 385$, so $a : b = 34 : 55$. Then

$$[a,b] + (-1)^n(a,b) = 13{,}090 - 7 = 13{,}083$$

$$= (238 + 385) \cdot 21$$

$$= (a + b)F_8.$$

2) Since $(a+b)F_8 = [a,b] + (-1)^n(a,b)$, it follows that

$$c = F_8 + \frac{F_9 F_7}{d - F_8}$$
$$= 21 + \frac{34 \cdot 13}{d - 21}$$
$$= 21 + \frac{442}{d - 21}.$$

$442 = 2 \cdot 13 \cdot 17$, 442 has eight positive factors: 1, 2, 13, 17, 26, 34, 221, and 442. So d has eight possible choices: 22, 23, 34, 38, 47, 55, 242, and 463. Consequently, the possible values of the ratio $c : d$ are $463:22$, $242:23$, $55:34$, $47:38$, $38:47$, $34:55$, $23:242$, and $463:22$. Since one-half of them are duplicates, the four distinct ones are $38:47$, $34:55$, $23:242$, and $22:463$, keeping the numerators smaller. Notice that one of the ratios is $34 : 55 = F_9 : F_{10}$, as expected.   ∎

Interestingly, Theorem 10.7 has a Lucas counterpart. It requires the following lemma; we leave its proof as an exercise.

Lemma 10.3. $F_{2n-1} = F_{n+1}L_{n+2} - L_n L_{n+1}$, *where* $n \geq 2$.

We are now ready for the theorem.

Theorem 10.8 (Freeman, 1967 [185]).

1) *Let* $n \geq 2$ *and* $a : b = L_n : L_{n+1}$. *Then* $(a+b)F_{n+1} = [a,b] + (a,b)F_{2n-1}$.
2) *Let* $n \geq 3$ *and* $a : b = F_{n-2} : F_{n-1}$. *Then* $(a+b)F_{n+1} = [a,b] + (a,b)F_{2n-1}$.
3) *Let* $n \geq 2$, $a : b = c : d$, *where* $(c,d) = 1$, *and* $(a+b)F_{n+1} = [a,b] + (a,b)F_{2n-1}$. *Then the ratios* $c : d$ *are determined by the positive factors of* $F_{n+1}^2 - F_{2n-1}$, *one of them being* $L_n : L_{n+1}$. *When* $n \geq 3$, $F_{n-2} : F_{n-1}$ *is also a solution.*

Proof.

1) Let $a : b = L_n : L_{n+1}$. Since $(L_n, L_{n+1}) = 1$, $a = kL_n$, $b = kL_{n+1}$, $(a,b) = k$, and $[a,b] = L_n L_{n+1}k$ for some positive integer k. Then

$$(a+b)F_{n+1} = (L_n + L_{n+1})kF_{n+1}$$
$$= F_{n+1}L_{n+2}k$$
$$= (F_{2n-1} + L_n L_{n+1})k$$
$$= [a,b] + (a,b)F_{2n-1}.$$

2) Suppose $a : b = F_{n-2} : F_{n-1}$. Then $a = kF_{n-2}$, $b = kF_{n-1}$, $(a, b) = k$, and $[a, b] = F_{n-1}F_{n-2}k$ for some positive integer k. Then

$$\begin{aligned}
(a + b)F_{n+1} &= (F_{n-2} + F_{n-1})kF_{n+1} \\
&= F_n F_{n+1} k \\
&= (F_{2n-1} + F_{n-1}F_{n-2})k \\
&= [a, b] + (a, b)F_{2n-1}.
\end{aligned}$$

3) Let $a : b = c : d$, where $(c, d) = 1$. As before, $a = ck$, $b = dk$, $(a, b) = k$, and $[a, b] = cdk$ for some positive integer k. Since $(a + b)F_{n+1} = [a, b] + (a, b)F_{2n-1}$, we have

$$\begin{aligned}
(c + d)F_{n+1} &= cd + F_{2n-1} \\
c &= \frac{dF_{n+1} - F_{2n-1}}{d - F_{n+1}} \\
&= F_{n+1} + \frac{F_{n+1}^2 - F_{2n-1}}{d - F_{n-1}}.
\end{aligned}$$ (10.5)

Since c and d are positive integers, it follows that the ratio $c : d$ is determined by the positive factors of $F_{n+1}^2 - F_{2n-1}$.

In particular, let $d = F_{n+1}$. Then, by Lemma 10.3,

$$\begin{aligned}
c &= F_{n+1} + \frac{F_{n+1}^2 - F_{n+1}L_{n+2} + L_n L_{n+1}}{L_{n+1} - F_{n+1}} \\
&= \frac{F_{n+1}(L_{n+1} - L_{n+2}) + L_n L_{n+1}}{L_{n+1} - F_{n+1}} \\
&= \frac{L_n L_{n+1} - L_n F_{n+1}}{L_{n+1} - F_{n+1}} \\
&= L_n.
\end{aligned}$$

Thus $L_n : L_{n+1}$ is a solution of the ratio $c : d$. (By symmetry, $L_{n+1} : L_n$ is also a solution.)

Unlike Theorem 10.7, *not* all solutions are obtained by considering the case $d > F_{n+1}$. For instance, let $d = F_{n-1}$. Then, by equation (10.5),

$$\begin{aligned}
c &= \frac{F_{n-1}F_{n+1} - F_{2n-1}}{F_{n-1} - F_{n+1}} \\
&= \frac{F_{n-1}F_{n+1} - (F_n F_{n+1} - F_{n-2}F_{n-1})}{-F_n}
\end{aligned}$$

$$= \frac{F_{n+1}(F_n - F_{n-1}) - F_{n-2}F_{n-1}}{F_n}$$

$$= \frac{F_{n-2}(F_{n+1} - F_{n-1})}{F_n}$$

$$= F_{n-2}.$$

Thus $F_{n-2} : F_{n-1}$ is also a solution. ∎

The following example illustrates this theorem.

Example 10.4. *Illustrate Theorem 10.8 with $n = 8$, $F_{n+1} = F_9 = 34$, and $F_{2n-1} = F_{15} = 610$.*

Solution.

1) Let $a : b = L_n : L_{n+1} = L_8 : L_9 = 47 : 76$. Let $a = 235$ and $b = 380$. Then

$$[a, b] + (a, b)F_{2n-1} = [235, 380] + (235, 380) \cdot 610$$
$$= 17{,}860 + 5 \cdot 610 = 20{,}910$$
$$= (235 + 380) \cdot 34$$
$$= (a + b)F_{n+1}.$$

2) Let $a : b = F_{n-2} : F_{n-1} = F_6 : F_7 = 8 : 13$. Let $a = 96$ and $b = 156$. Then

$$[a, b] + (a, b)F_{2n-1} = [96, 156] + 12 \cdot 610$$
$$= 8568$$
$$= (96 + 156) \cdot 34$$
$$= (a + b)F_{n+1}.$$

3) Let $a : b = 180 : 204 = 15 : 17$, where $c : d = 15 : 17$ and $(15, 17) = 1$. Then, by part 3) of Theorem 10.8,

$$c = F_{n+1} + \frac{F_{n+1}^2 - F_{2n-1}}{d - F_{n+1}}$$
$$= 34 + \frac{34^2 - 610}{d - 34}$$
$$= 34 + \frac{546}{d - 34}.$$

$546 = 2 \cdot 3 \cdot 7 \cdot 13$, 546 has 16 positive factors: 1, 2, 3, 6, 7, 13, 14, 21, 26, 39, 42, 78, 91, 182, 273, and 546. The corresponding ratios are $35 : 580$, $36 : 307$,

37 : 216, 40 : 125, 41 : 112, 47 : 76, 48 : 73, 55 : 60, 60 : 55, 73 : 48, 76 : 47, 112 : 41, 125 : 40, 216 : 37, 307 : 36, and 580 : 35. They yield eight distinct ratios $c : d$ with $(c, d) = 1$, namely, 7 : 116, 8 : 25, 11 : 12, 36 : 307, 37 : 216, 41 : 112, 47 : 76, and 48 : 73. Notice that 8 : 13 is also a solution. Among these ratios, we find $L_8 : L_9 = 47 : 76$ and $F_6 : F_7 = 8 : 13$, as expected.   ■

10.4   AN ALTERED FIBONACCI SEQUENCE

In 1971, Underwood Dudley and Bessie Tucker of DePauw University, Green-castle, Indiana, investigated a slightly altered Fibonacci sequence [152], defined by $S_n = F_n + (-1)^n$, where $n \geq 1$. They made two interesting observations; see Table 10.1:

- The 1st, 3rd, 5th entries (circled numbers) in row 3 are the 2nd, 4th, 6th, ... Fibonacci numbers.
- The 2nd, 4th, 6th entries in row 3 are the 3rd, 5th, 7th, ... Lucas numbers.

TABLE 10.1.

n	1	2	3	4	5	6	7	8	9	10	11	12	13	14	15
S_n	0	2	1	4	4	9	12	22	33	56	88	145	232	378	609
(S_n, S_{n+1})	①		4		③		11		⑧		29		㉑		

To establish these observations, we need the following theorem.

Theorem 10.9 (Dudley and Tucker, 1971 [152]).

1) $F_{4n} + 1 = F_{2n-1} L_{2n+1}$ 2) $F_{4n} - 1 = F_{2n+1} L_{2n-1}$
3) $F_{4n+1} + 1 = F_{2n+1} L_{2n}$ 4) $F_{4n+1} - 1 = F_{2n} L_{2n+1}$
5) $F_{4n+2} + 1 = F_{2n+2} L_{2n}$ 6) $F_{4n+2} - 1 = F_{2n} L_{2n+2}$
7) $F_{4n+3} + 1 = F_{2n+1} L_{2n+2}$ 8) $F_{4n+3} - 1 = F_{2n+2} L_{2n+1}$.

Proof. The proof requires the following identities from Chapter 5:

$$F_{m+n} + F_{m-n} = \begin{cases} F_n L_m & \text{if } n \text{ is odd} \\ F_m L_n & \text{otherwise;} \end{cases}$$

$$F_{m+n} - F_{m-n} = \begin{cases} F_m L_n & \text{if } n \text{ is odd} \\ F_n L_m & \text{otherwise.} \end{cases}$$

Then

$$F_{4n} + 1 = F_{4n} + F_2$$
$$= F_{(2n+1)+(2n-1)} + F_{(2n+1)-(2n-1)}$$
$$= F_{2n-1} L_{2n+1}$$

and

$$F_{4n+1} + 1 = F_{4n+1} + F_1$$
$$= F_{(2n+1)+2n} + F_{(2n+1)-2n}$$
$$= F_{2n+1} L_{2n}.$$

The other formulas can be established similarly; see Exercises 10.59–10.64. ∎

The next corollary follows easily from this theorem.

Corollary 10.7 (Dudley and Tucker, 1971 [152]).

1) $(F_{4n+1} + 1, F_{4n+2} + 1) = L_{2n}$ 2) $(F_{4n+1} + 1, F_{4n+3} + 1) = F_{2n+1}$
3) $(F_{4n+1} - 1, F_{4n+2} - 1) = F_{2n}$ 4) $(F_{4n+1} - 1, F_{4n+3} - 1) = L_{2n+1}$
5) $(F_{4n-1} - 1, F_{4n+1} - 1) = F_{2n}$ 6) $(F_{4n-1} + 1, F_{4n+1} + 1) = L_{2n}$
7) $(F_{4n+3} + 1, F_{4n} - 1)\ \ \ = F_{2n+1}$ 8) $(F_{4n+3} + 1, F_{4n+2} - 1) = F_{2n}$
9) $(F_{4n+4} - 1, F_{4n+3} - 1) = L_{2n+1}.$

∎

Although it is not yet known whether there are infinitely many Fibonacci primes, Theorem 10.9 establishes their finiteness in the sequences $\{F_n \pm 1\}$, as the following corollary shows.

Corollary 10.8. $F_n + 1$ *is composite if* $n \geq 4$; *and* $F_n - 1$ *is composite if* $n \geq 7$.

Proof. When $n = 1$, $F_{4n} + 1 = 4$ is composite. When $n \geq 2$, it follows by Theorem 10.9 that $F_{4n+1} + 1$, $F_{4n+2} + 1$, and $F_{4n+3} + 1$ have nontrivial factors. Thus $F_n + 1$ is composite if $n \geq 4$.

Likewise, $F_n - 1$ is composite if $n \geq 7$. ∎

Notice that $F_n + 1$ is prime if $n < 4$; and $F_n - 1$ is prime if $n < 7$.
The next corollary confirms the observation made by Dudley and Tucker.

Corollary 10.9. Let $n \geq 1$ and $S_n = F_n + (-1)^n$. Then $(S_{4n}, S_{4n+1}) = L_{2n+1}$; $(S_{4n+1}, S_{4n+3}) = L_{2n+1}$; and $(S_{4n+2}, S_{4n+3}) = F_{2n+2}$.
Proof. By Theorem 10.9, we have

$$(S_{4n}, S_{4n+1}) = (F_{4n+1} + 1, F_{4n+1} - 1)$$
$$= (F_{2n-1} L_{2n+1}, F_{2n} L_{2n+1})$$
$$= (F_{2n-1}, F_{2n}) L_{2n+1}$$
$$= L_{2n+1}.$$

We can similarly establish the other two cases; see Exercises 10.65 and 10.66. ∎

The next result also follows by Theorem 10.9; see Exercises 10.67–10.69.

Corollary 10.10. Let $n \geq 1$ and $H_n = F_n - (-1)^n$. Then $(H_{4n}, H_{4n+1}) = F_{2n+1}$; $(H_{4n+1}, H_{4n+3}) = F_{2n+1}$; and $(H_{4n+2}, H_{4n+3}) = L_{2n+2}$.

∎

ADDITIONAL DIVISIBILITY PROPERTIES

The following divisibility properties were discovered in 1974 by Hoggatt and G.E. Bergum (some independently), where p and q are odd primes, and $k, m, n, r, s, t \geq 1$ [306]. We omit their proofs in the interest of brevity.

- If $p|L_n$, then $p^k|L_{np}^{k-1}$ (Carlitz and Bergum, independently).
- Let $p|L_{2 \cdot 3^k}$ and $n = 2 \cdot 3^k p^t$. Then $n|L_n$.
- Let $p \neq q$, $p|L_n$, and $q|L_m$, where m and n are odd. Then $(pq)^k|L_{mn(pq)^{k-1}}$.
- Let $p, q > 3$, $p \neq q$, $p|L_{2 \cdot 3^k}$, $q|L_{2 \cdot 3^k}$, and $n = 2 \cdot 3^k p^t q^r$, where $r, t \geq 0$. Then $n|L_n$.
- If $p|L_n$, then $p^k|F_{2np^{k-1}}$ (Carlitz and Bergum, independently).
- Let $p \neq q$, $p|L_n$, and $q|L_m$, where m and n are odd. Then $(pq)^k|F_{2mn(pq)^{k-1}}$ (Carlitz and Bergum, independently).
- If $p|F_n$, then $p^k|F_{np^{k-1}}$ (Carlitz and Bergum, independently).
- Let $p \neq q$, $p|F_n$, and $q|F_m$. Then $(pq)^k|F_{mn(pq)^{k-1}}$.
- If $n = 3^m 2^{r+1}$, then $n|F_n$.
- Let $n = 2^{r+1} 3^m 5^k$. Then $n|F_n$.
- Let $p > 3$ such that $p|F_{2^{r+1} 3^m}$, and $n = 2^{r+1} 3^m p^k$. Then $n|F_n$.
- Let $s = 2^{r+1} 3^m$, $p \neq q$, $p|F_s$, $q|F_s$, and $n = sp^k q^t$, where $k, t \geq 0$. Then $n|F_n$.
- $2^{k+2}|F_{3 \cdot 2^k}$.
- If n is odd, then $L_n = 4^t M$, where M is odd, and $t = 0$ or 1.
- If n is odd, then $L_n = 4^t M$, and the prime factors of M are of the form $10m \pm 1$, where $t = 0$ or 1 (Hoggatt).

EXERCISES 10

Verify each.

1. $F_7|F_{21}$.
2. $F_6|F_{24}$.
3. $(F_{12}, F_{18}) = F_{(12,18)}$.
4. $(F_{10}, F_{21}) = F_{(10,21)}$.
5. $(F_{144}, F_{1925}) = 1$.
6. $L_5|F_{10}$.
7. $L_6|F_{24}$.
8. $L_4|L_{12}$.

Find each.

9. (F_{144}, F_{440}).
10. (F_{80}, F_{100}).

11. Prove Theorem 10.1 using Binet's formula.

Prove that (F_n, F_{n+1}) using each method.

   12. PMI.

   13. The well-ordering principle (WOP).

   14. Prove that $(L_n, L_{n-1}) = 1$.

Disprove each.

   15. $m|n$ implies $L_m|L_n$.

   16. $(L_m, L_n) = L_{(m,n)}$.

   17. Let $m, n \geq 3$. Then $F_m F_n | F_{mn}$.

   18. Let $m, n \geq 2$. Then $L_m L_n | L_{mn}$.

   19. $[F_m, F_n] = F_{[m,n]}$.

   20. $[L_m, L_n] = L_{[m,n]}$.

   21. Compute (F_n, L_n) for $1 \leq n \leq 10$, and make a conjecture about (F_n, L_n).

   22. Identify the integers n for which $(F_n, L_n) = 2$.

Compute each.

   23. $F_{(F_5, F_{10})}$.

   24. $F_{(F_6, F_{18})}$.

   25. $L_{(F_5, F_{15})}$.

   26. $L_{(F_6, L_6)}$.

   27. $F_{(F_5, F_{10}, F_{15})}$.

   28. $F_{(F_6, F_{18}, F_{21})}$.

   29. $L_{(F_5, F_{10}, F_{15})}$.

   30. $L_{(F_6, L_6, L_9)}$.

   31. Disprove: If n is a prime, then F_n is a prime.

Prove each.

   32. If F_n is a prime, then n is a prime, where $n \geq 5$.

   33. $2|F_{3n}$.

   34. $3|n$ if and only if $2|F_n$.

   35. $4|n$ if and only if $3|F_n$.

   36. $6|n$ if and only if $4|F_n$.

   37. $5|n$ if and only if $5|F_n$.

   38. $(F_a, F_b, F_c) = F_{(a,b,c)}$.

   39. $2|L_{3n}$.

   40. F_{60} ends in 0.

   41. $(F_n, L_n) = 2$ if and only if $3|n$.

   42. $L_n|L_{3n}$.

43. $L_n | L_{(2k-1)n}$, where $k \geq 1$.

44. $(F_n, L_n) = 1$ or 2, where $n \geq 1$.

45. Using the identity $F_{m+n} = F_m F_{n+1} + F_{m-1} F_n$, prove that $F_m | F_{mn}$.

46. There are n consecutive composite Fibonacci numbers, where $n \geq 1$ (Litvack, 1964 [414]).

Verify that $(a + b)F_{n-1} = [a, b] + (-1)^n(a, b)$ for each ratio $a : b$.

47. $21 : 34$.

48. $89 : 144$.

Prove Lemma 10.3 using:

49. PMI.

50. Binet's formula.

Verify that $(a + b)F_{n+1} = [a, b] + (a, b)F_{2n-1}$ for each ratio $a : b$.

51. $11 : 18$.

52. $21 : 34$.

53. $72 : 116$.

54. $65 : 105$.

Prove each.

55. $(F_m, F_n) = (F_m, F_{m+n}) = (F_n, F_{m+n})$ (Brown, 1964 [74]).

56. Let $a_n = \sum_{d|n} F_d$, where $n \geq 1$. Then $\{a_n\}$ is an increasing sequence (Lind, 1967 [406]).

57. If $k > 4$, then $F_k \nmid L_n$ (Brousseau, 1969 [68]).

58. Let $F_m | L_n$, where $0 < m < n$. Then $m = 1, 2, 3$, or 4 (Lang, 1973 [380]).

59–64. Establish the identities 2) and 4)–8) in Theorem 10.9 (Dudley and Tucker, 1971 [152]).

Prove each, where $S_n = F_n + (-1)^n$, $H_n = F_n - (-1)^n$, and $n \geq 1$ (Dudley and Tucker, 1971 [152]).

65. $(S_{4n+1}, S_{4n+3}) = L_{2n+1}$.

66. $(S_{4n+2}, S_{4n+3}) = F_{2n+2}$.

67. $(H_{4n}, H_{4n+1}) = F_{2n+1}$.

68. $(H_{4n+1}, H_{4n+3}) = F_{2n+1}$.

69. $(H_{4n+2}, H_{4n+3}) = L_{2n+2}$.

Use the function $g_n = F_{4n-2} + F_{4n} + F_{4n+2}$ for Exercises 10.70–10.72 (Grassl, 1971 [242]).

70. Define g_n recursively.

71. Prove that $12 | g_n$ for every $n \geq 0$.

72. $168 | (F_{8n-4} + F_{8n} + F_{8n+4})$ (Grassl, 1971 [243]).

73. $(L_{2r} + 1)F_k|(F_{kn-2r} + F_{kn} + F_{kn+2r})$ (Hillman, 1971 [264]).
74. There are no even perfect Fibonacci numbers (Whitney, 1971 [594]).
75. Let $h = 5^k$, where $k \geq 1$. Then $h|F_h$ (Hoggatt, 1973 [297]).
76. Let $g = 2 \cdot 3^k$, where $k \geq 1$. Then $g|L_g$ (Hoggatt, 1973 [297]).

PASCAL'S TRIANGLE

We will see in this chapter how Fibonacci numbers can be extracted from Pascal's triangle. In addition, we will derive a host of Fibonacci and Lucas identities involving binomial coefficients.

We will begin with a brief discussion of binomial coefficients, which are coefficients occurring in the binomial expansion of $(x + y)^n$. The term *binomial coefficient* was introduced by the greatest German algebraist Michel Stifel (1486–1567).

The earliest known occurrence of binomial coefficients can be traced back to a tenth-century commentary by the Indian mathematician Halayudha on Pingala's *Chandas Shastra*; interestingly, it also contains a discourse on what is now called Pascal's triangle (called *Meru Prastara*). Bhaskara (1114–1185?) gives a concise discussion of binomial coefficients in his 1150 classic work *Leelavati*.

11.1 BINOMIAL COEFFICIENTS

Let n and k be nonnegative integers. The *binomial coefficient* $\binom{n}{k}$ is defined by

$$\binom{n}{k} = \begin{cases} \dfrac{n!}{k!(n-k)!} & \text{if } 0 \leq k \leq n \\ 0 & \text{otherwise.} \end{cases}$$

Fibonacci and Lucas Numbers with Applications, Volume One, Second Edition. Thomas Koshy.
© 2018 John Wiley & Sons, Inc. Published 2018 by John Wiley & Sons, Inc.

Pascal*

It is also denoted by $C(n, k)$ and $_nC_k$. Read $\binom{n}{k}$ as "n choose k" to be consistent with the typesetting language *LaTeX*.

The parenthesized bi-level notation for binomial coefficient was introduced by the German mathematician and physicist Andreas von Ettinghausen (1796–1878) in his 1826 book *Die Combinatorische Analysis*.

There are many instances when we need to compute the binomial coefficients $\binom{n}{k}$ and $\binom{n}{n-k}$. The next theorem shows that they are indeed equal; this can greatly reduce our workload. Its proof is straightforward, so we omit it in the interest of brevity.

Theorem 11.1. *Let $0 \le k \le n$. Then* $\binom{n}{k} = \binom{n}{n-k}$.   ∎

For example, $\binom{25}{20} = \binom{25}{25-20} = \binom{20}{5} = 53,130.$

The next theorem gives a recurrence satisfied by binomial coefficients. It is called *Pascal's identity*, after the outstanding French mathematician and physicist, Blaise Pascal (1623–1662). Again the proof is fairly simple, so we omit it.

Theorem 11.2 (Pascal's identity). *Let n and k be positive integers, where $k \leq n$. Then*

$$\binom{n}{k} = \binom{n-1}{k-1} + \binom{n-1}{k}.$$
■

11.2 PASCAL'S TRIANGLE

The various binomial coefficients $\binom{n}{k}$, where $0 \leq k \leq n$, can be arranged as a triangular array, called *Pascal's triangle*[†]; see Figures 11.1 and 11.2.

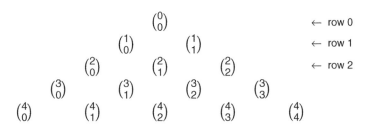

← row 0

← row 1

← row 2

Figure 11.1. Pascal's triangle.

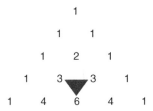

Figure 11.2. Pascal's triangle.

Pascal's triangle has many intriguing properties. Some of them are:

- Every row begins with and ends in 1.
- Pascal's triangle is symmetric about a vertical line through the middle. This is so by Theorem 11.1.

[†]Although Pascal's triangle is named after Pascal in the West, the array appeared in its present form in a work by the Chinese mathematician, Chu Shi-Kie in 1303.

- Any interior number in each row is the sum of the numbers immediately to its left and right in the preceding row. This is true by virtue of Pascal's identity.
- The sum of the numbers in row n is 2^n.

The next theorem shows how the binomial coefficients can be used to find the *binomial expansion* of $(x + y)^n$. It can be proved using PMI or combinatorics; the former is a bit long, while the latter is short[†].

Theorem 11.3 (The Binomial Theorem). [‡] *Let x and y be any real numbers, and n any nonnegative integer. Then* $(x + y)^n = \sum_{r=0}^{n} \binom{n}{r} x^{n-r} y^r$. ∎

The binomial theorem has some interesting and useful byproducts. They are given in the next two corollaries. Their proofs are also straightforward, so we omit them.

Corollary 11.1.

$$(1 + x)^n = \sum_{r=0}^{n} \binom{n}{r} x^r$$

$$(1 - x)^n = \sum_{r=0}^{n} (-1)^r \binom{n}{r} x^r.$$ ∎

Corollary 11.2.

$$\sum_{r=0}^{n} \binom{n}{r} = 2^n$$

$$\sum_{r=0}^{n} (-1)^r \binom{n}{r} = 0$$

$$\sum_{r \text{ odd}} \binom{n}{r} = \sum_{r \text{ even}} \binom{n}{r}.$$ ∎

11.3 FIBONACCI NUMBERS AND PASCAL'S TRIANGLE

How can Fibonacci numbers be extracted from Pascal's triangle? To see this, consider the array in Figure 11.2. Now add the numbers along the northeast diagonals; see Figure 11.3. The sums are $1, 1, 2, 3, 5, 8, \ldots$; and they seem to be Fibonacci numbers. In fact, they are, as the next theorem, discovered by Lucas in 1876, confirms.

[†] For a combinatorial proof, see the author's *Triangular Arrays with Applications*.
[‡] The binomial theorem for $n = 2$ can be found in Euclid's work (ca. 300 B.C.).

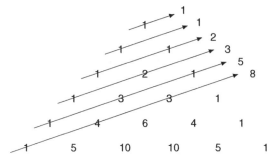

Figure 11.3. Pascal's triangle.

Theorem 11.4 (Lucas, 1876). *Let $n \geq 1$. Then*

$$F_n = \sum_{i=0}^{\lfloor (n-1)/2 \rfloor} \binom{n-i-1}{i}. \tag{11.1}$$

Proof. We will prove the result using the strong version of PMI. Since $\binom{0}{0} = 1 = F_1$, the statement is true when $n = 1$.

Now assume it is true for all positive integers $\leq k$, where $k \geq 1$. By Pascal's identity, we then have

$$\sum_{i=0}^{\lfloor k/2 \rfloor} \binom{k-i}{i} = \sum_{i=0}^{\lfloor k/2 \rfloor} \binom{k-i-1}{i-1} + \sum_{i=0}^{\lfloor k/2 \rfloor} \binom{k-i-1}{i}.$$

Suppose k is even. Then

$$\sum_{i=0}^{\lfloor k/2 \rfloor} \binom{k-i}{i} = \sum_{i=0}^{k/2} \binom{k-i-1}{i-1} + \sum_{i=0}^{k/2} \binom{k-i-1}{i}$$

$$= \sum_{i=0}^{(k-1)/2} \binom{k-i-2}{i} + \sum_{i=0}^{k/2-1} \binom{k-i-1}{i} + \binom{k/2-1}{k/2}$$

$$= \sum_{i=0}^{\lfloor (k-2)/2 \rfloor} \binom{k-i-2}{i} + \sum_{i=0}^{\lfloor (k-1)/2 \rfloor} \binom{k-i-1}{i} - 0$$

$$= F_{k-1} + F_k$$

$$= F_{k+1}.$$

So the formula works when k is even. It can similarly be shown that it works when n is odd. Consequently, it is true when $n = k + 1$.

Thus, by the strong version of PMI, the formula is true for all positive integers n. ∎

For example,

$$F_6 = \sum_{i=0}^{2} \binom{5-i}{i} = 1 + 4 + 3 = 8;$$

$$F_7 = \sum_{i=0}^{3} \binom{6-i}{i} = 1 + 5 + 6 + 1 = 13.$$

It follows by the Lucas formula in Theorem 11.4 that

$$K_n = \sum_{i=0}^{\lfloor n/2 \rfloor} \binom{n-i}{i}$$

satisfies Fibonacci recurrence, where $K_1 = 1 = F_2$ and $K_2 = 2 = F_3$.

The Lucas formula is a special case of an interesting result derived in 1950 by Steven Vajda (1901–1995) [570]. To see this, we use a bit of operator theory. Let

$$S_n = S_n(x) = \sum_{i \geq 0} \binom{n-i}{i} x^i$$

and $\Delta S_n = S_{n+1} - S_n$. Then

$$\Delta S_n = \sum_{i \geq 0} \left[\binom{n+1-i}{i} - \binom{n-i}{i} \right] x^i$$

$$= \sum_{i \geq 0} \binom{n-i}{i-1} x^i$$

$$= \sum_{j \geq 0} \binom{n-j-1}{j} x^{j+1};$$

$$\Delta^2 S_n = \Delta(\Delta S_n)$$

$$= \Delta S_{n+1} - \Delta S_n$$

$$= \sum_{i \geq 0} \left[\binom{n+1-i}{i-1} - \binom{n-i}{i-1} \right] x^i$$

$$= \sum_{i \geq 0} \binom{n-i}{i-2} x^i$$

$$= \sum_{j \geq 0} \binom{n-j-1}{j-1} x^{j+1}.$$

Then

$$\Delta^2 S_n + \Delta S_n = \sum_{j \geq 0} \left[\binom{n-j-1}{j} + \binom{n-j-1}{j-1} \right] x^{j+1}$$

$$S_{n+2} - S_{n+1} = \sum_{j \geq 0} \binom{n-j}{j} x^{j+1}$$

$$= xS_n.$$

Thus S_n satisfies the recurrence $S_{n+2} - S_{n+1} - xS_n = 0$, where $S_0 = 1 = S_1$. Its characteristic equation is $t^2 - t - x = 0$ with roots

$$r = \frac{1 + \sqrt{1 + 4x}}{2} \quad \text{and} \quad s = \frac{1 - \sqrt{1 + 4x}}{2}.$$

So the general solution of the recurrence is $S_n = Ar^n + Bs^n$, where A and B are constants. It follows by the initial conditions $S_0 = 1 = S_1$ that $A = \dfrac{r}{r-s}$ and $B = \dfrac{s}{s-r}$.

Thus

$$S_n(x) = \frac{r^{n+1} - s^{n+1}}{r - s}$$

$$= \frac{1}{\sqrt{1+4x}} \left[\left(\frac{1 + \sqrt{1+4x}}{2} \right)^{n+1} - \left(\frac{1 - \sqrt{1+4x}}{2} \right)^{n+1} \right]. \qquad (11.2)$$

In particular, $S_n(1)$ gives Binet's formula for F_{n+1}.

Formula (11.2) has other interesting byproducts. Suppose, for example, $x = 2$. Then

$$\sum_{i \geq 0} \binom{n-i}{i} 2^i = \frac{2^{n+1} - (-1)^{n+1}}{3}.$$

Numbers of the form $J_{n+1} = \dfrac{2^{n+1} - (-1)^{n+1}}{3}$ are the well-known *Jacobsthal numbers*, named after the German mathematician Ernst Jacobsthal (1882–1965), where $n \geq 0$. Formula (11.2) is Binet's formula for the *Jacobsthal polynomial* $J_{n+1}(x) = S_n(x)$, where $n \geq 0$:

$$J_{n+1}(x) = \sum_{i=0}^{\lfloor n/2 \rfloor} \binom{n-i}{i} x^i,$$

so Jacobsthal numbers and polynomials can be computed from Pascal's triangle with appropriate weights.

When $x = -1$, formula (11.2) yields another interesting case. Then
$r = \dfrac{1 + \sqrt{3}i}{2} = e^{i\pi/3}$ and $s = \dfrac{1 - \sqrt{3}i}{2} = e^{-i\pi/3}$, so

$$\sum_{i=0}^{\lfloor n/2 \rfloor} (-1)^i \binom{n-i}{i} = \frac{1}{\sqrt{3}i} \left[e^{(n+1)i\pi/3} - e^{-(n+1)i\pi/3} \right]$$

$$= \frac{2 \sin(n+1)\pi/3}{\sqrt{3}}$$

$$= \begin{cases} 0 & \text{if } n \equiv 2 \pmod 3 \\ (-1)^{\lfloor n/3 \rfloor} & \text{otherwise,} \end{cases}$$

where $i = \sqrt{-1}$ and $\lfloor t \rfloor$ denotes the floor of the real number t.

The case $x = -1/4$ is also an interesting one. We leave it for the curious-minded to pursue.

It follows by the identity $L_n = F_{n+1} + F_{n-1}$ and Theorem 11.4 that Lucas numbers also can be extracted from Pascal's triangle. Each L_n is the sum of the diagonal sums on rising diagonals $n + 1$ and $n - 1$.

11.4   ANOTHER EXPLICIT FORMULA FOR L_n

Using Theorem 11.4 and the identity $L_n = F_{n+1} + F_{n-1}$, we can develop another explicit formula for L_n:

$$L_n = F_{n+1} + F_{n-1}$$

$$= \sum_{k=0}^{\lfloor n/2 \rfloor} \binom{n-k}{k} + \sum_{k=0}^{\lfloor (n-2)/2 \rfloor} \binom{n-k-2}{k}$$

$$= \sum_{k=0}^{\lfloor n/2 \rfloor} \binom{n-k}{k} + \sum_{k=1}^{\lfloor n/2 \rfloor} \binom{n-k-1}{k-1}$$

$$= \sum_{k=0}^{\lfloor n/2 \rfloor} \binom{n-k}{k} + \sum_{k=0}^{\lfloor n/2 \rfloor} \binom{n-k-1}{k-1} + 0$$

$$= \sum_{k=0}^{\lfloor n/2 \rfloor} \left[\binom{n-k}{k} + \binom{n-k-1}{k-1} \right]$$

$$= \sum_{k=0}^{\lfloor n/2 \rfloor} \frac{n}{n-k} \binom{n-k}{k}. \tag{11.3}$$

For example,

$$L_5 = \sum_{k=0}^{2} \frac{5}{5-k} \binom{5-k}{k}$$

$$= \frac{5}{5}\binom{5}{0} + \frac{5}{4}\binom{4}{1} + \frac{5}{3}\binom{3}{2}$$

$$= 1 + 5 + 5 = 11.$$

11.5 CATALAN'S FORMULA

In lieu of using the rising diagonals of Pascal's triangle, we can use its rows to compute Fibonacci numbers. To see this, we expand Binet's formula using the binomial theorem:

$$F_n = \frac{1}{\sqrt{5}} \left[\left(\frac{1+\sqrt{5}}{2} \right)^n - \left(\frac{1+\sqrt{5}}{2} \right)^n \right]$$

$$= \frac{1}{2^{n-1}} \sum_{k=0}^{\lfloor (n-1)/2 \rfloor} \binom{n}{2k+1} 5^k;$$

see Exercise 11.14. Catalan discovered this formula in 1846.

But, by Corollary 11.2, $2^{n-1} = \sum_{k=0}^{\lfloor n/2 \rfloor} \binom{n}{2k}$. Consequently, we can rewrite Catalan's formula with a more aesthetic appeal:

$$F_n = \frac{\binom{n}{1} + \binom{n}{3}5 + \binom{n}{5}5^2 + \binom{n}{7}5^3 + \cdots}{\binom{n}{0} + \binom{n}{2} + \binom{n}{4} + \binom{n}{6} + \cdots}.$$

We can similarly show that

$$L_n = \frac{\binom{n}{0} + \binom{n}{2}5 + \binom{n}{4}5^2 + \binom{n}{6}5^3 + \cdots}{\binom{n}{1} + \binom{n}{3} + \binom{n}{5} + \binom{n}{7} + \cdots};$$

see Exercise 11.15.

For example,

$$F_7 = \frac{\binom{7}{1} + \binom{7}{3}5 + \binom{7}{5}5^2 + \binom{7}{7}5^3}{\binom{n}{0} + \binom{n}{2} + \binom{n}{4} + \binom{n}{6}} = \frac{832}{64} = 13;$$

$$L_7 = \frac{\binom{7}{0} + \binom{7}{2}5 + \binom{7}{4}5^2 + \binom{7}{6}5^3}{\binom{7}{1} + \binom{7}{3} + \binom{7}{5} + \binom{n}{7}} = \frac{1856}{64} = 29.$$

11.6   ADDITIONAL IDENTITIES

We can use Binet's and Lucas' formulas with the binomial theorem in tandem to derive an array of Fibonacci and Lucas identities.

To begin with, notice that

$$\sum_{i=0}^{5} \binom{5}{i} F_i = \binom{5}{0} F_0 + \binom{5}{1} F_1 + \binom{5}{2} F_2 + \binom{5}{3} F_3 + \binom{5}{4} F_4 + \binom{5}{5} F_5$$

$$= 0 + 5 + 10 + 20 + 15 + 5 = 55$$

$$= F_{10}.$$

More generally, we have the following identity.

Theorem 11.5 (Lucas).   *Let $n \geq 0$. Then*

$$\sum_{i=0}^{n} \binom{n}{i} F_i = F_{2n}. \tag{11.4}$$

Proof. Since $\alpha^2 = \alpha + 1$ and $\beta^2 = \beta + 1$, by Binet's formula, we have

$$(\alpha - \beta) \sum_{i=0}^{n} \binom{n}{i} F_i = \sum_{i=0}^{n} \binom{n}{i}(\alpha^i - \beta^i)$$

$$= \sum_{i=0}^{n} \binom{n}{i}\alpha^i - \sum_{i=0}^{n} \binom{n}{i}\beta^i$$

$$= (1 + \alpha)^n - (1 + \beta)^n$$

$$= \alpha^{2n} - \beta^{2n}$$

$$\sum_{i=0}^{n} \binom{n}{i} F_i = F_{2n}. \qquad\blacksquare$$

A similar argument yields yet another identity by Lucas, where $n \geq 0$:

$$\sum_{i=0}^{n} \binom{n}{i} L_i = L_{2n}. \tag{11.5}$$

For example,

$$\sum_{i=0}^{4} \binom{4}{i} L_i = \binom{4}{0} L_0 + \binom{4}{1} L_1 + \binom{4}{2} L_2 + \binom{4}{3} L_3 + \binom{4}{4} L_4$$

$$= 2 + 4 + 18 + 16 + 7 = 47$$

$$= L_8.$$

Theorem 11.6. *Let $n \geq 0$. Then*

$$\sum_{i=0}^{n} (-1)^{i+1} \binom{n}{i} F_i = F_n. \tag{11.6}$$

Proof. By Binet's formula and Corollary 11.1, we have

$$(\alpha - \beta)\text{LHS} = -\sum_{i=0}^{n} \binom{n}{i} \left[(-\alpha)^i - (-\beta)^i\right]$$

$$= -\left[(1-\alpha)^n - (1-\beta)^n\right]$$

$$= \alpha^n - \beta^n$$

$$\text{LHS} = F_n,$$

as desired. ∎

For example,

$$\sum_{i=0}^{5} (-1)^{i+1} \binom{n}{i} F_i = -\binom{5}{0} F_0 + \binom{5}{1} F_1 - \binom{5}{2} F_2 + \binom{5}{3} F_3 - \binom{5}{4} F_4 + \binom{5}{5} F_5$$

$$= 0 + 5 - 10 + 20 - 15 + 5 = 5$$

$$= F_5.$$

We can show similarly that

$$\sum_{i=0}^{n} (-1)^{i} \binom{n}{i} L_i = L_n, \tag{11.7}$$

where $n \geq 0$.

For example,

$$\sum_{i=0}^{4} (-1)^{i} \binom{n}{i} L_i = \binom{4}{0} L_0 - \binom{4}{1} L_1 + \binom{4}{2} L_2 - \binom{4}{3} L_3 + \binom{4}{4} L_4$$

$$= 2 - 4 + 18 - 16 + 7 = 7$$

$$= L_4.$$

Next we will study the number of paths of a rook on a chessboard.

11.7 FIBONACCI PATHS OF A ROOK ON A CHESSBOARD

In 1970, Edward T. Frankel of New York showed that Fibonacci numbers can be derived by enumerating the number of different paths open to a rook on an empty chessboard from one corner to the opposite corner, where its moves are restricted by a pattern of horizontal and vertical fences [183].

To see this, consider rows 0 through 7. This time, left-justify the elements in every row and then move up each column j by j elements, where $j \geq 0$. In other words, rotate Pascal's triangle to its left by $45°$. Figure 11.4 shows the resulting 8×8 array. Every rising diagonal of this array is a row of Pascal's triangle.

Row 3 in Pascal's triangle

j	0	1	2	3	4	5	6	7
i								
0	1	1	1	1	1	1	1	1
1	1	2	3	4	5	6	7	8
2	1	3	6	10	15	21	28	36
3	1	4	10	20	35	56	84	120
4	1	5	15	35	70	126	210	330
5	1	6	21	56	126	252	462	792
6	1	7	28	84	210	462	924	1716
7	1	8	36	120	330	792	1716	3432

Figure 11.4.

Every element $A(i,j)$ of this array can be realized by adding the element immediately to its left and the element immediately above it:

$$A(i,j) = A(i,j-1) + A(i-1,j),$$

where $i,j \geq 1$; and $A(0,j) = 1 = A(i,0)$ for all i,j.

A rook on a chessboard moves any number of cells either horizontally or vertically, but not in both directions in the same move. Suppose it moves horizontally to the right (R) or vertically down (D). It is well known that each entry in Figure 11.4 indicates the number of moves a rook can make from the upper left-hand corner to that cell. For example, $A(2,1) = 3$ implies that there are three different ways the rook can move from position (0,0) to position (2,1); they are 1R, 2D; 1D, 1R, 1D; and 2D, 1R. The rook has 3432 possible moves from the upper left corner to the lower right-hand corner (7,7). Using the combinatorial notation, we can rewrite the array in Figure 11.4, as Figure 11.5 shows.

i \ j	0	1	2	3	4	5	6	7
0	$\binom{0}{0}$	$\binom{1}{0}$	$\binom{2}{0}$	$\binom{3}{0}$	$\binom{4}{0}$	$\binom{5}{0}$	$\binom{6}{0}$	$\binom{7}{0}$
1	$\binom{1}{1}$	$\binom{2}{1}$	$\binom{3}{1}$	$\binom{4}{1}$	$\binom{5}{1}$	$\binom{6}{1}$	$\binom{7}{1}$	$\binom{8}{1}$
2	$\binom{2}{2}$	$\binom{3}{2}$	$\binom{4}{2}$	$\binom{5}{2}$	$\binom{6}{2}$	$\binom{7}{2}$	$\binom{8}{2}$	$\binom{9}{2}$
3	$\binom{3}{3}$	$\binom{4}{3}$	$\binom{5}{3}$	$\binom{6}{3}$	$\binom{7}{3}$	$\binom{8}{3}$	$\binom{9}{3}$	$\binom{10}{3}$
4	$\binom{4}{4}$	$\binom{5}{4}$	$\binom{6}{4}$	$\binom{7}{4}$	$\binom{8}{4}$	$\binom{9}{4}$	$\binom{10}{4}$	$\binom{11}{4}$
5	$\binom{5}{5}$	$\binom{6}{5}$	$\binom{7}{5}$	$\binom{8}{5}$	$\binom{9}{5}$	$\binom{10}{5}$	$\binom{11}{5}$	$\binom{12}{5}$
6	$\binom{6}{6}$	$\binom{7}{6}$	$\binom{8}{6}$	$\binom{9}{6}$	$\binom{10}{6}$	$\binom{11}{6}$	$\binom{12}{6}$	$\binom{13}{6}$
7	$\binom{7}{7}$	$\binom{8}{7}$	$\binom{9}{7}$	$\binom{10}{7}$	$\binom{11}{7}$	$\binom{12}{7}$	$\binom{13}{7}$	$\binom{14}{7}$

Figure 11.5.

Figure 11.6 shows the number of moves of the rook from the upper left-hand corner, where the moves are restricted by staggered horizontal and vertical fences. Oddly enough, all entries in this band array are Fibonacci numbers. The band is made up of four strands of Fibonacci numbers. The array begins with a 1 at the top left corner. Every other entry is the sum of the entries immediately to its left and immediately above it, assuming the entries outside the squares are 0s.

1	1	1					
1	2	3	3				
	2	5	8	8			
		5	13	21	21		
			13	34	55	55	
				34	89	144	144
					89	233	377
						233	610

Figure 11.6.

Figure 11.7 displays the same chessboard array using the Fibonacci labels. Notice that the subscripts of any two adjacent Fibonacci numbers on each strand

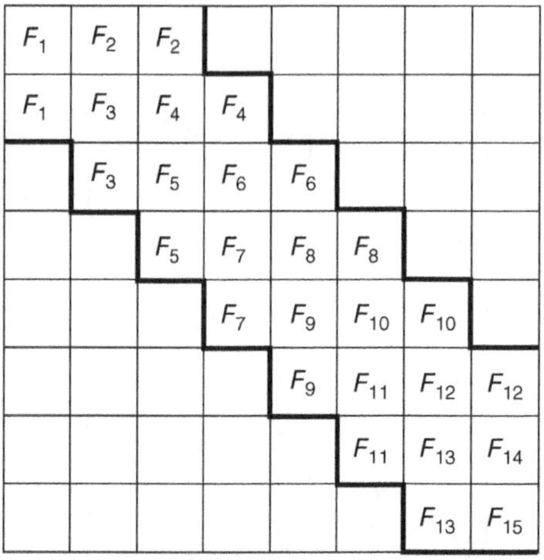

F_1	F_2	F_2					
F_1	F_3	F_4	F_4				
	F_3	F_5	F_6	F_6			
		F_5	F_7	F_8	F_8		
			F_7	F_9	F_{10}	F_{10}	
				F_9	F_{11}	F_{12}	F_{12}
					F_{11}	F_{13}	F_{14}
						F_{13}	F_{15}

Figure 11.7.

differ by two. Also, the Fibonacci numbers on the two upper strands have even
subscripts, whereas those on the two lower strands have odd subscripts.

It follows from Figures 11.4–11.7 that the rook has $F_{15} = 610$ possible moves
from position (0,0) to position (7,7). More generally, on an $n \times n$ chessboard, the
rook has F_{2n-1} restricted moves from the upper left-hand corner; see Exercise
11.30.

EXERCISES 11

Using Theorem 11.4, compute each.
1. F_{11}
2. F_{12}

Prove each.

3. $\sum_{i=0}^{n} \binom{n}{i} L_i = L_{2n}.$

4. $\sum_{i=0}^{n} (-1)^i \binom{n}{i} L_i = L_n.$

5. $\sum_{i=0}^{n} \binom{n}{i} F_{i+j} = F_{2n+j}.$

6. $\sum_{i=0}^{n} \binom{n}{i} L_{i+j} = L_{2n+j}.$

7. $\sum_{i=0}^{n} (-1)^i \binom{n}{i} F_{i+j} = (-1)^{j+1} F_{n-j}.$

8. $\displaystyle\sum_{i=0}^{n}(-1)^i\binom{n}{i}L_{i+j} = (-1)^j L_{n-j}$.

9. Verify that $5\displaystyle\sum_{i=0}^{n}\binom{n}{i}F_i^2 = \sum_{i=0}^{n}\binom{n}{i}L_{2i}$ for $n = 4$ and $n = 5$.

Establish the formula in Exercise 11.9 using:

10. The binomial theorem.

11. Exercise 5.40.

12. Verify that $\displaystyle\sum_{i=0}^{n}\binom{n}{i}L_i^2 = \sum_{i=0}^{n}\binom{n}{i}L_{2i}$ for $n = 4$ and $n = 5$.

13. Establish the formula in Exercise 11.12.

Prove each.

14. $2^{n-1}F_n = \displaystyle\sum_{i=0}^{\lfloor (n-1)/2 \rfloor}\binom{n}{2i+1}5^i$ (Catalan, 1846).

15. $2^{n-1}L_n = \displaystyle\sum_{i=0}^{\lfloor n/2 \rfloor}\binom{n}{2i}5^i$ (Catalan, 1846).

16. $\displaystyle\sum_{i=0}^{n-1}(-2)^i\binom{n}{i}F_i = \begin{cases} 2^i F_i - 2 \cdot 5^{(n-1)/2} & \text{if } n \text{ is odd} \\ -2^i F_i & \text{otherwise (Ferns, 1964 [166]).} \end{cases}$

17. $\displaystyle\sum_{i=0}^{n-1}(-2)^i\binom{n}{i}G_i = 5^{(n-1)/2}[c(-1)^n - d]$.

18. $\displaystyle\sum_{i=0}^{n}(-1)^i\binom{n}{i}L_{2i} = (-1)^n L_n$ (Gould, 1963 [232]).

19. $\displaystyle\sum_{i=0}^{n}(-1)^i\binom{n}{i}F_{2i} = (-1)^n F_n$ (Gould, 1963 [232]).

20. $\displaystyle\sum_{i=0}^{n}\binom{n}{i}G_i = G_{2n}$ (Ruggles, 1963 [504]).

21. $\displaystyle\sum_{i=0}^{n}(-1)^i\binom{n}{i}G_i = (-1)^n G_{-n}$.

22. $\displaystyle\sum_{i=0}^{n}\binom{n}{i}G_{i+j} = G_{2n+j}$.

23. $\displaystyle\sum_{i=0}^{n}(-1)^i\binom{n}{i}G_{i+j} = (-1)^n G_{j-n}$.

24. $\displaystyle\sum_{i=0}^{n}\binom{n}{i}F_m^i F_{m-1}^{n-i} F_{r+i} = F_{mn+r}$ (Vinson, 1963 [572]).

25. $\displaystyle\sum_{i=0}^{n}\binom{n}{i}F_{k+2i} = \begin{cases} 5^{(n-1)/2}L_{n+k} & \text{if } n \text{ is odd} \\ 5^{n/2}F_{n+k} & \text{otherwise.} \end{cases}$

26. $\sum_{i=0}^{n} \binom{n}{i} L_{k+2i} = \begin{cases} 5^{(n+1)/2} F_{n+k} & \text{if } n \text{ is odd} \\ 5^{n/2} L_{n+k} & \text{otherwise.} \end{cases}$

27. $\sum_{i=0}^{2n} (-1)^i \binom{2n}{i} 2^{i-1} L_i = 5^n$ (Brown, 1967 [77]).

28. $\sum_{i=0}^{2n} (-1)^i \binom{2n}{i} 2^{i-1} F_i = 0$ (Brown, 1967 [77]).

29. $\sum_{i=0}^{n} \binom{n}{k} F_{4mk} = L_{2m}^n F_{2mn}$ (Hoggatt, 1968 [278]).

30. A rook on an $n \times n$ chessboard has F_{2n-1} restricted moves from the upper left-hand corner to the lower right-hand corner.

12

PASCAL-LIKE TRIANGLES

In the preceding chapter, we learned how Fibonacci numbers can be generated from Pascal's triangle. We now turn our attention to how Fibonacci and Lucas numbers can be computed from similar triangular arrays that have Pascal-like features.

In 1966, N.A. Draim of Ventura, California, and M. Bicknell of A.C. Wilcox High School, Santa Clara, California, studied the sums and differences of like-powers of the solutions of the quadratic equation $x^2 - px - q = 0$ [149]. They were also studied in 1977 by J.E. Woko of Abia State Polytechnic, Aba, Nigeria [600]. As we will see shortly, an intriguing relationship exists between these expressions, and Fibonacci and Lucas numbers.

12.1 SUMS OF LIKE-POWERS

Let r and s be the solutions of the quadratic equation $x^2 - px - q = 0$. Then

$$r = \frac{p + \sqrt{p^2 + 4q}}{2} \quad \text{and} \quad s = \frac{p - \sqrt{p^2 + 4q}}{2};$$

so $r + s = p$ and $rs = -q$. Consequently, $r^2 + s^2 = (r + s)^2 - 2rs = p^2 + 2q$ and $r^3 + s^3 = (r + s)^3 - 3rs(r + s) = p^3 + 3pq$. Continuing like this, we can compute

Fibonacci and Lucas Numbers with Applications, Volume One, Second Edition. Thomas Koshy.
© 2018 John Wiley & Sons, Inc. Published 2018 by John Wiley & Sons, Inc.

the various sums $r^n + s^n$:

$$r + s = p$$
$$r^2 + s^2 = p^2 + 2q$$
$$r^3 + s^3 = p^3 + 3pq$$
$$r^4 + s^4 = p^4 + 4p^2q + 2q^2$$
$$r^5 + s^5 = p^5 + 5p^3q + 5pq^2$$
$$r^6 + s^6 = p^6 + 6p^4q + 9p^2q^2 + 2q^3$$
$$r^7 + s^7 = p^7 + 7p^5q + 14p^3q^2 + 7pq^3.$$

More generally, using PMI, Drain and Bicknell showed that

$$r^n + s^n = \sum_{i=0}^{\lfloor n/2 \rfloor} A(n,i)p^{n-2i}q^i, \tag{12.1}$$

where $A(n,i) = 2\binom{n-i}{i} - \binom{n-i-1}{i}$.

Using Pascal's identity, we can simplify the formula for $A(n,i)$:

$$A(n,i) = \binom{n-i}{i} + \left[\binom{n-i}{i} - \binom{n-i-1}{i}\right]$$

$$= \binom{n-i}{i} + \binom{n-i-1}{i-1}$$

$$= \binom{n-i}{i} + \frac{i}{n-i}\binom{n-i}{i}$$

$$= \frac{n}{n-i}\binom{n-i}{i}. \tag{12.2}$$

Consequently, we can rewrite formula (12.1) as

$$r^n + s^n = \sum_{i=0}^{\lfloor n/2 \rfloor} A(n,i)p^{n-2i}q^i, \tag{12.3}$$

where $A(n,i) = \frac{n}{n-i}\binom{n-i}{i}$ and $0 \leq i \leq \lfloor n/2 \rfloor$.

We can arrange the values of $A(n,i)$ in a Pascal-like triangle; see Table 12.1.

TABLE 12.1. Array A

n \ i	0	1	2	3	4	5	Row sum
1	1						1
2	1	2					3
3	1	3					4
4	1	4	2				7
5	1	5	5				11
6	1	6	9	2			18
7	1	7	14	7			29
8	1	8	20	16	2		47
9	1	9	27	30	9		76
10	1	10	35	50	25	2	123

$$\uparrow$$
$$L_n$$

In particular, let r and s be the solutions of the equation $x^2 = x + 1$. Then $r = \alpha$ and $s = \beta$. Equation (12.3) yields the formula

$$L_n = \sum_{i=0}^{\lfloor n/2 \rfloor} A(n, i). \tag{12.4}$$

This should not come as a surprise, since Table 12.1 is exactly the same triangular array we obtained by computing the topological indices of cycloparaffins $C_n H_{2n}$ in Chapter 3; see Table 3.9.

Using equation (12.2), this yields formula (11.3):

$$L_n = \sum_{i=0}^{\lfloor n/2 \rfloor} \frac{n}{n-i} \binom{n-i}{i}. \tag{12.5}$$

The array in Table 12.1 can be defined recursively:

$$A(1,0) = 1, \quad A(2,1) = 1$$
$$A(n, i) = A(n-1, i) + A(n-2, i-1), \tag{12.6}$$

where $0 \leq i \leq n$ and $n \geq 3$.

Every row begins with a 1. When $n \geq 3$, $A(n, i)$ can be obtained by adding the entry just above it in the previous row, and the entry to the left of that in the row above it; see the arrows in the table.

The array satisfies the following additional properties:

- If $i > \lfloor n/2 \rfloor$, then $A(n, i) = 0$.
- If n is odd, row n ends in n; and row $n + 1$ contains one more entry than row n.
- If n is even, then row n ends in 2; and row n and row $n + 1$ contain the same number of entries.

These results can be proved fairly easily.
  For example, let n be odd. Then

$$
\begin{aligned}
A(n, \lfloor n/2 \rfloor) &= A(n, (n - 1)/2) \\
&= \frac{2n}{n + 1} \binom{(n + 1)/2}{(n - 1)/2} \\
&= \left(\frac{2n}{n + 1} \right) \left(\frac{n + 1}{2} \right) \\
&= n.
\end{aligned}
$$

When n is odd, $\lfloor (n + 1)/2 \rfloor = (n + 1)/2 = (n - 1)/2 + 1 = \lfloor n/2 \rfloor + 1$; so row $n + 1$ contains one element more than row n.

12.2 AN ALTERNATE FORMULA FOR L_n

The preceding discussion yields a wonderful dividend in the form of an alternate formula for L_n. To see this, in the interest of brevity and convenience, we let $\Delta = \sqrt{p^2 + 4q}$. Then $2r = p + \Delta$ and $2s = p - \Delta$; so

$$
(2r)^n = \sum_{i=0}^{n} \binom{n}{i} p^{n-i} \Delta^i
$$

$$
(2s)^n = \sum_{i=0}^{n} \binom{n}{i} p^{n-i} (-\Delta)^i
$$

$$
(2r)^n + (2s)^n = 2 \sum_{i \text{ even}} \binom{n}{i} p^{n-i} \Delta^i
$$

$$
2^{n-1}(r^n + s^n) = \sum_{i=0}^{\lfloor n/2 \rfloor} \binom{n}{2i} p^{n-2i} \Delta^{2i}.
$$

In particular, when $p = 1 = q$, $\Delta = \sqrt{5}$. Then this yields the Catalan formula for L_n we found in Section 11.5 and Exercise 11.15:

$$L_n = \frac{1}{2^{n-1}} \sum_{i=0}^{\lfloor n/2 \rfloor} \binom{n}{2i} 5^i. \tag{12.7}$$

For example,

$$L_5 = \frac{1}{2^4} \sum_{i=0}^{2} \binom{5}{2i} 5^i$$

$$= \frac{1}{2^4} \left[\binom{5}{0} + \binom{5}{2} 5 + \binom{5}{4} 5^2 \right]$$

$$= (1 + 50 + 125)/16$$

$$= 11.$$

12.3   DIFFERENCES OF LIKE-POWERS

We now turn our attention to differences of like-powers of r and s:

$$r - s = \Delta$$
$$r^2 - s^2 = p\Delta$$
$$r^3 - s^3 = (p^2 + q)\Delta$$
$$r^4 - s^4 = (p^3 + 2pq)\Delta$$
$$r^5 - s^5 = (p^4 + 3p^2q + q^2)\Delta$$
$$r^6 - s^6 = (p^5 + 4p^3q + 3pq^2)\Delta$$
$$r^7 - s^7 = (p^6 + 5p^4q + 6p^2q^2 + q^3)\Delta$$
$$\vdots$$

More generally,

$$r^n - s^n = \Delta \sum_{i=0}^{\lfloor (n-1)/2 \rfloor} \binom{n-i-1}{i} p^{n-2i-1} q^i. \tag{12.8}$$

We can establish this also by PMI; see Exercise 12.11.

As before, we can arrange the coefficients $B(n, i) = \binom{n-i-1}{i}$ in a triangular array B, where $0 \leq i \leq \lfloor (n-1)/2 \rfloor$; see Table 12.2. We encountered this array in Table 3.8 when we computed the topological indices of paraffins C_nH_{2n+2}.

TABLE 12.2.   Array B

i n	0	1	2	3	4	5	Row sum
1	1						1
2	1						1
3	1	1					2
4	1	2					3
5	1	3	1				5
6	1	4	3				8
7	1	5	6	1			13
8	1	6	10	4			21
9	1	7	15	10	1		34
10	1	8	21	20	5		55
		$\uparrow$ t_n	$\uparrow$ T_n				$\uparrow$ F_n

Array B also satisfies several interesting properties:

- $B(n, i)$ satisfies the recurrence $B(n, i) = B(n-1, i) + B(n-2, i-1)$. This is so since

$$B(n-1, i) + B(n-2, i-1) = \binom{n-i-2}{i} + \binom{n-i-2}{i-1}$$
$$= \binom{n-i-1}{i}$$
$$= B(n, i).$$

- Since $B(n, 0) = 1$, every row begins with a 1.
- If n is odd, row n ends in 1. This is so since

$$B(n, \lfloor (n-1)/2 \rfloor) = B(n, (n-1)/2) = \binom{(n-1)/2}{(n-1)/2} = 1.$$

- Suppose n is odd. Since $\lfloor (n-1)/2 \rfloor = (n-1)/2 = \lfloor n/2 \rfloor$, row n and row $n+1$ contain the same number of elements.
- Suppose n is even. Then row n ends in $n/2$, and row $n+1$ contains one more element than row n.
- Since $B(n,2) = \binom{n-3}{2}$, column 2 consists of *triangular numbers* t_n, where $n \geq 3$.
- $B(n,3) = \binom{n-4}{3}$, so column 3 consists of *tetrahedral numbers* T_n. The remaining columns give higher dimensional figurate numbers.

Suppose we let $p = 1 = q$. Then $r = \alpha$, $s = \beta$, and $\Delta = \sqrt{5}$. Then equation (12.8) gives Lucas' combinatorial formula for F_n we established in Theorem 11.4:

$$F_n = \sum_{i=0}^{\lfloor (n-1)/2 \rfloor} \binom{n-i-1}{i}.$$

12.4 CATALAN'S FORMULA REVISITED

Using equations (12.1) and (12.8), we can derive Catalan's formula for F_n:

$$(2r)^n - (2s)^n = 2 \sum_{i \text{ odd}} \binom{n}{i} p^{n-i} \Delta^i$$

$$2^{n-1}(r^n - s^n) = \sum_{i=0}^{\lfloor (n-1)/2 \rfloor} \binom{n}{2i+1} p^{n-2i-1} \Delta^{2i+1}$$

$$r^n - s^n = \frac{1}{2^{n-1}} \sum_{i=0}^{\lfloor (n-1)/2 \rfloor} \binom{n}{2i+1} p^{n-2i-1} \Delta^{2i+1}.$$

In particular, let $p = 1 = q$. Since $\Delta = \sqrt{5}$, this yields Catalan's combinatorial formula for F_n we derived in Chapter 11:

$$F_n = \frac{1}{2^{n-1}} \sum_{i=0}^{\lfloor (n-1)/2 \rfloor} \binom{n}{2i+1} 5^i.$$

For example,

$$F_6 = \frac{1}{2^5} \sum_{i=0}^{2} \binom{6}{2i+1} 5^i$$

$$= (6 + 100 + 150)/32$$

$$= 8.$$

Next we construct a triangular array with Lucas implications.

12.5 A LUCAS TRIANGLE

In 1967, M. Feinberg, then a student at the University of Pennsylvania, studied the coefficients in the expansion of $f_n(x, y) = (x + y)^{n-1}(x + 2y)$, and discovered an invaluable treasure, where $n \geq 1$ [165].

The first six expansions are:

$$f_1(x, y) = x + 2y$$
$$f_2(x, y) = x^2 + 3xy + 2y^2$$
$$f_3(x, y) = x^3 + 4x^2y + 5xy^2 + 2y^3$$
$$f_4(x, y) = x^4 + 5x^3y + 9x^2y^2 + 7xy^3 + 2y^4$$
$$f_5(x, y) = x^5 + 6x^4y + 14x^3y^2 + 16x^2y^3 + 9xy^4 + 2y^5$$
$$f_6(x, y) = x^6 + 7x^5y + 20x^4y^2 + 30x^3y^3 + 25x^2y^4 + 11xy^5 + 2y^6.$$

TABLE 12.3. Array C: a Lucas Triangle

i \ j	0	1	2	3	4	5	6
1	1	2					
2	1	3	2				
3	1	4 →	5	2			
			↓				
4	1	5	9	7	2		
5	1	6	14	16	9	2	
6	1	7	20	(30)	25	11	2

Arranging the coefficients in these polynomials, we get a truncated triangular array C; see Table 12.3. Feinberg called it a *Lucas triangle*. Every row begins with a 1 and ends in a 2; this is so because the coefficient of x^n in $f_n(x, y)$ is 1; and that of y^n is 2. Since $f_n(1, 1) = 3 \cdot 2^{n-1}$, the sum of the numbers in row n is $3 \cdot 2^{n-1}$, where $n \geq 1$.

Let $C(n, j)$ denote the element in row n column j, where $n \geq 1$ and $j \geq 0$. We can easily find an explicit formula for $C(n, j)$:

$$(x + y)^{i-1} = \sum_{j=0}^{i-1} \binom{i-1}{j} x^{i-j-1} y^j$$

$$f_i(x, y) = (x + y)^{i-1}(x + 2y)$$

$$= \left[\sum_{j=0}^{i-1} \binom{i-1}{j} x^{i-j-1} y^j \right] (x + 2y)$$

$$C(n, j) = \text{Coefficient of } x^{n-j} y^j$$

$$= \binom{n-1}{j} + 2\binom{n-1}{j-1}$$

$$= \binom{n}{j} + \binom{n-1}{j-1}. \tag{12.9}$$

For example, $C(6, 3) = \binom{6}{3} + \binom{5}{2} = 20 + 10 = \boxed{30}$; see Table 12.3.

The triangular array C in Table 12.3 satisfies three additional properties:

- $\displaystyle\sum_{k=1}^{n} C(k, j) = C(n + 1, j + 1)$

- $\displaystyle\sum_{k=1}^{n} C(k, 1) = C(n + 1, 2)$

- $C(n, n - 2) = (n - 1)^2;$

see Exercises 12.15–12.17.

A RECURSIVE DEFINITION FOR $C(n, j)$

Using formula (12.9), we can easily verify that $C(n, j)$ satisfies the following recursive definition:

$$C(1, 0) = 1, \quad C(1, 1) = 2$$
$$C(n, j) = C(n - 1, j - 1) + C(n - 1, j),$$

where $n, j \geq 2$. This recurrence is the same as Pascal's identity; for example, $C(4, 2) = 14 = 5 + 9 = C(3, 1) + C(3, 2)$. Also $C(n, n) = 1$.

Formula (12.9) contains a hidden secret: We can obtain $C(n, j)$ from rows n and $n - 1$ of Pascal's triangle. Shift row $n - 1$ by one place to the right, and place the resulting row just above row n; then add the corresponding elements; this yields the elements $C(n, j)$ in row n of Table 12.1. This algorithm is illustrated below for $n = 4$:

	1	3	3	1		← Row 3 in Pascal's triangle
+	1	4	6	4	1	← Row 4 in Pascal's triangle
	1	5	9	7	2	← Row 4 in Table 12.3

Suppose we add the elements on the rising diagonals in Table 12.3. It appears from Figure 12.1 that the sums are Lucas numbers. This result is not a fluke. Let

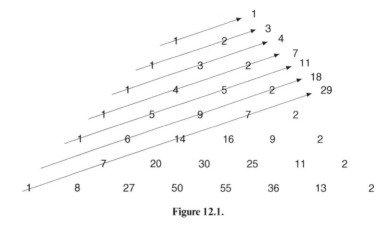

Figure 12.1.

us see why this is true. By formula (12.9), the sum of the elements on the *n*th rising diagonal is given by

$$\sum_{j=0}^{\lfloor n/2 \rfloor} C(n-j, j) = \sum_{j=0}^{\lfloor n/2 \rfloor} \binom{n-j}{j} + \sum_{j=0}^{\lfloor n/2 \rfloor} \binom{n-j-1}{j-1}$$

$$= \sum_{j=0}^{\lfloor n/2 \rfloor} \binom{n-j}{j} + \sum_{j=0}^{\lfloor (n-2)/2 \rfloor} \binom{n-j-2}{j}$$

$$= F_{n+1} + F_{n-1}$$

$$= L_n.$$

For example, the sum of the elements on the sixth rising diagonal is $18 = L_6$.

TABLE 12.4.   Array *D*: a Reflection of the Lucas Triangle

j \ *i*	0	1	2	3	4	5	6
1	2	1					
2	2	3	1				
3	2	5 → 4	1				
4	2	7	⑨	5	1		
5	2	9	16	14	6	1	
6	2	11	25	30	20	7	1

Suppose we flip the Lucas triangle in Figure 12.1 about a vertical line on the left and then left-justify the elements. Table 12.4 shows the resulting triangular array D.

Let $D(n,j)$ denote the element in row n and column j of this array. Then

$$D(1,0) = 2, \quad D(1,1) = 1$$

$$D(n,j) = D(n-1,j-1) + D(n-1,j),$$

where $n,j \geq 2$. This is the same recurrence satisfied by $C(n,j)$.

Since the array in Table 12.4 is a reflection of the Lucas triangle, it follows that

$$D(n,j) = C(n, n-j)$$

$$= \binom{n}{n-j} + \binom{n-1}{n-j-1}$$

$$= \binom{n}{j} + \binom{n-1}{j}.$$

Consequently, we can obtain every row of the array in Table 12.4 by adding rows $n-1$ and n (both left-justified) of Pascal's triangle; see the following demonstration of the algorithm:

	1 3 3 1	← Row 3 in Pascal's triangle
+	1 4 6 4 1	← Row 4 in Pascal's triangle
	2 7 9 5 1	← Row 4 in Table 12.4

The array in Table 12.4 also provides a fascinating bonus. Add the elements on each rising diagonal; every sum is a Fibonacci number; see Figure 12.2.

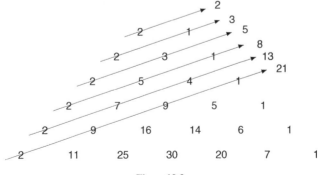

Figure 12.2.

This is so, because

$$\sum_{j=0}^{\lfloor n/2 \rfloor} D(n-j,j) = \sum_{j=0}^{\lfloor n/2 \rfloor} \binom{n-j}{j} + \sum_{j=0}^{\lfloor n/2 \rfloor} \binom{n-j-1}{j-1}$$

$$= F_{n+1} + F_n$$

$$= F_{n+2}.$$

The array in Table 12.4 satisfies three additional properties:

- $\sum_{k=1}^{n} D(k,j) = D(n+1,j+1)$

- $\sum_{k=1}^{n} D(k,1) = n^2$

- $D(n,2) = (n-1)^2$;

see Exercises 12.20–12.22.

In 1970, Hoggatt described an interesting relationship between the Lucas triangle in Table 12.3 and the triangular array of coefficients in the expansion of L_n^m, as we will see shortly [283].

To establish such a link, shift down column j (≥ 0) of the Lucas triangle by j elements. Table 12.5 shows the resulting array E. Let $E(n,j)$ denote the element in row n and column j of the array. Then $E(n,j) = A(n-j,j)$, where $0 \leq j \leq \lfloor n/2 \rfloor$. $E(n,j)$ satisfies the recurrence $E(n,j) = E(n-1,j) + E(n-2,j-1)$, where $E(n,0) = 1$, $E(2j,j) = 2$, $E(2j+1,j) = 2j+1$, $E(2j+1,j+1) = 0$ and $1 \leq j \leq \lfloor n/2 \rfloor$.

TABLE 12.5.   Array E

i \ n	0	1	2	3	4	5
1	1					
2	1	2				
3	1	3				
4	1	4	2			
5	1	5	5			
6	1	6	9	2		
7	1	7	(14)	7		
8	1	8	20	16	2	

12.6 POWERS OF LUCAS NUMBERS

In 1970, Harlan L. Umansky of Emerson High School, Union City, New Jersey, developed the following list of formulas of powers of Lucas numbers [565]:

$$L_n^1 = L_n$$
$$L_n^2 = L_{2n} + 2(-1)^n$$
$$L_n^3 = L_{3n} + 3(-1)^n L_n$$
$$L_n^4 = L_{4n} + 4(-1)^n L_n^2 - 2$$
$$L_n^5 = L_{5n} + 5(-1)^n L_n^3 - 5L_n$$
$$L_n^6 = L_{6n} + 6(-1)^n L_n^4 - 9L_n^2 + 2(-1)^n$$
$$L_n^7 = L_{7n} + 7(-1)^n L_n^5 - 14L_n^3 + 7(-1)^n L_n$$
$$L_n^8 = L_{8n} + 8(-1)^n L_n^6 - 20L_n^4 + 16(-1)^n L_n^2 - 2.$$

A few months later, Hoggatt observed that the absolute values of the coefficients of the Lucas numbers and their powers on the RHS are the same as the elements in array E in Table 12.5. As a result, he established the following result [283]. We will confirm it using PMI.

Theorem 12.1 (Hoggatt, 1970 [283]).

$$L_n^m = L_{mn} + \sum_{j=1}^{\lfloor m/2 \rfloor} E(m,j)(-1)^{nj+j-1} L_n^{m-2j}.$$

Proof. The formula is clearly true when $m = 1$. Now assume it is true for all positive integers $\leq k$, where $k \leq m$:

$$L_n^k = L_{kn} + \sum_{j=1}^{\lfloor k/2 \rfloor} E(k,j)(-1)^{nj+j-1} L_n^{k-2j}.$$

Then

$$L_n^{m+1} = L_n L_{mn} + \sum_{j=1}^{\lfloor m/2 \rfloor} E(m,j)(-1)^{nj+j-1} L_n^{m-2j+1}.$$

But $L_n L_{mn} = L_{(m+1)n} + (-1)^n L_{(m-1)n}$, so

$$L_n^{m+1} = L_{(m+1)n} + (-1)^n L_{(m-1)n} + \sum_{j=1}^{\lfloor m/2 \rfloor} E(m,j)(-1)^{nj+j-1} L_n^{m-2j+1}. \qquad (12.10)$$

By the inductive hypothesis,

$$L_{(m-1)n} = L_n^{m-1} - \sum_{j=1}^{\lfloor (m-1)/2 \rfloor} E(m-1,j)(-1)^{nj+j-1} L_n^{m-2j-1}$$

$$(-1)^n L_{(m-1)n} = (-1)^n L_n^{m-1} + \sum_{j=1}^{\lfloor (m-1)/2 \rfloor} E(m-1,j)(-1)^{n(j+1)+(j+1)-1} L_n^{m-2j-1}.$$

Letting $j + 1 = r$, this becomes

$$(-1)^n L_{(m-1)n} = (-1)^n L_n^{m-1} + \sum_{r=2}^{\lfloor (m+1)/2 \rfloor} E(m-1,r-1)(-1)^{nr+r-1} L_n^{m-2r+1}.$$

Then equation (12.10) becomes

$$L_n^{m+1} = L_{(m+1)n} + (-1)^n L_n^{m-1} + \sum_{r=2}^{\lfloor (m+1)/2 \rfloor} E(m-1,r-1)(-1)^{nr+r-1} L_n^{m-2r+1}$$

$$+ \sum_{r=1}^{\lfloor m/2 \rfloor} E(m,r)(-1)^{nr+r-1} L_n^{m-2r+1}. \tag{12.11}$$

Suppose $m = 2t$. Then $\lfloor m/2 \rfloor = \lfloor (m+1)/2 \rfloor$, $E(2t,t) = 2$, $E(2t-1,t-1) = 2t-1$; so $E(2t+1,t) = 2t+1$. On the other hand, let $m = 2t+1$. Then $\lfloor m/2 \rfloor + 1 = \lfloor (m+1)/2 \rfloor + 1 = t+1$, $E(2t+1,t+1) = 0$ and $E(2t,t) = 2$; thus $E(2t+2,t+1) = 2$. Consequently, $E(m+1,r) = E(m,r) + E(m-1,r-1)$, where $1 \le r \le \lfloor (m+1)/2 \rfloor$.

Thus equation (12.11) becomes

$$L_n^{m+1} = L_{(m+1)n} + \sum_{r=1}^{\lfloor (m+1)/2 \rfloor} E(m+1,r)(-1)^{nr+r-1} L_n^{m-2r+1}.$$

Consequently, by the strong version of PMI, the formula works for every $n \ge 1$.   ∎

We will now present additional Pascal-like triangles that contain Fibonacci and Lucas numbers as hidden treasures.

12.7 VARIANTS OF PASCAL'S TRIANGLE

The following three variants of Pascal's triangle were studied extensively by Henry W. Gould of West Virginia University [234].

Gould*

At the 1963 Joint Automatic Control Conference held at the University of Minnesota, P.C. Parks presented the variant of Pascal's triangle in Figure 12.3. The first few row sums are Fibonacci numbers, so we conjecture that the nth row sum is F_{n+1}, where $n \geq 0$.

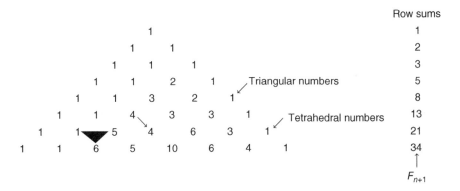

Figure 12.3.

To establish this, let $f(i,j)$ denote the element in row i and column j, where $i \geq j \geq 0$; $f(i,j) = 0$ if $j > i$; $f(i,0) = 1$; and $f(i,i) = 1$ for every i. The inner elements are defined by the recurrences $f(i+1, 2j+1) = f(i, 2j)$ and

Figure source: https://en.wikipedia.org/wiki/Henry_W._Gould#/media/File:Prof._Henry_W._Gould.jpg.

$f(i + 1, 2j) = f(i, 2j - 1) + f(i, 2j)$; see the arrow and black triangle in Figure 12.3. These two recurrences can in fact be combined into a single recurrence:

$$f(i + 1, j) = f(i, j - 1) + \frac{1 + (-1)^j}{2} f(i, j).$$

Using PMI, we can show that

$$f(n, 2k) = \binom{n - k}{k} \quad \text{and} \quad f(n, 2k + 1) = \binom{n - k - 1}{k}.$$

These two cases can be combined into a single formula for $f(n, r)$:

$$f(n, r) = \binom{n - \lfloor (r + 1)/2 \rfloor}{\lfloor r/2 \rfloor}.$$

We will now prove that every row sum in Figure 12.3 is a Fibonacci number. We will employ PMI to accomplish this.

Theorem 12.2. *Let $n \geq 0$. Then*

$$\sum_{r=0}^{n} f(n, r) = F_{n+2}.$$

Proof. Since $\sum_{r=0}^{0} f(n, r) = f(0, 0) = 1 = F_2$, the statement is true when $n = 0$.

Assume it works for all nonnegative integers $i \leq k$, where k is arbitrary. Then

$$\sum_{r=0}^{k+1} f(k + 1, r) = \sum_{r \text{ odd}} f(k + 1, r) + \sum_{r \text{ even}} f(k + 1, r)$$

$$= \sum_{r=0}^{\lfloor k/2 \rfloor} f(k + 1, r) + \sum_{r=0}^{\lfloor (k+1)/2 \rfloor} f(k + 1, r)$$

$$= F_{k+1} + F_{k+2}$$

$$= F_{k+3}.$$

Thus, by the strong version of PMI, the formula works for all $n \geq 0$. ∎

The next theorem provides another fascinating property of the triangular array in Figure 12.3.

Theorem 12.3. *Let $n \geq 0$. Then*

$$\sum_{r=0}^{n} (-1)^r f(n, r) = F_{n-1}.$$ ∎

For example,

$$\sum_{r=0}^{8}(-1)^r f(8,r) = 1 - 1 + 7 - 6 + 15 - 10 + 10 - 4 + 1$$

$$= 13 = F_7.$$

Using the rules of the previous array, we can construct a new variant of Pascal's triangle simply by changing $f(1,1)$ to 2; see Figure 12.4. This time, the row sums yield Lucas numbers.

									Row sums
				1					1
			1	2					3
		1	1	2					4
	1	1	3	2					7
1	1	4	3	2					11
1	1	5	4	5	2				18
1	1	6	5	9	5	2			29
1	1	7	6	14	9	7	2		47
									↑
									L_n

Figure 12.4.

Let $g(i,j)$ denote the element in row i and column j; where $i \geq j \geq 0$. Then $g(i,j) = 0$ if $j > i$; $g(i,0) = 1$; $g(1,1) = 2$; $g(i+1,2j+1) = g(i,2j)$; and $g(i+1,2j) = g(i,2j-1) + g(i,2j)$. We can combine the last two recurrences into a single recurrence:

$$g(i+1,j) = g(i,j-1) + \frac{1+(-1)^j}{2}g(i,j).$$

Using PMI, we can show that

$$g(n,2r) = \frac{n}{n-r}\binom{n-r}{r} \quad \text{and} \quad g(n,2r+1) = \frac{n-1}{n-r-1}\binom{n-r-1}{r},$$

where $g(1,1) = 2$. Consequently,

$$g(n,r) = \frac{n}{n-r}\binom{n-r}{r}.$$

Since $L_n = \sum_{r=0}^{\lfloor n/2 \rfloor} \frac{n}{n-r} \binom{n-r}{r}$, this array yields two properties corresponding to Theorems 12.2 and 12.3, as the following theorem shows.

Theorem 12.4. *Let $n \geq 0$. Then*

1) $\sum_{r=0}^{n} g(n,r) = L_{n+1}$

2) $\sum_{r=0}^{n} (-1)^r g(n,r) = L_{n-2}.$ ■

For example,

$$\sum_{r=0}^{5} g(5,r) = 1 + 1 + 5 + 4 + 5 + 2 = 18 = L_6;$$

$$\sum_{r=0}^{7} (-1)^r g(7,r) = 1 - 1 + 7 - 6 + 14 - 9 + 7 - 2 = 11 = L_5.$$

Interestingly, the arrays in Figures 12.3 and 12.4 are special cases of the generalized array in Figure 12.5. Let $h(i,j)$ denote the element in row i and column j, where $i \geq j \geq 0$. Then $h(i,j) = 0$ if $j > i$; $h(i,0) = a$; $h(1,1) = b$; and

$$h(i+1,j) = h(i,j-1) + \frac{1 + (-1)^j}{2} h(i,j),$$

where $i \geq 1$.

```
                                    a
                         a                    b
                 a               a                 b
          a           a               a + b             b
      a         a           2a + b          a + b            b
   a       a         3a + b         2a + b          a + 2b          b
 a      a       4a + b         3a + b         3a + 3b         a + 2b         b
a    a    5a + b        4a + b          6a + 4b         3a + 3b         a + 3b      b
```

Figure 12.5.

In this recurrence, we have imposed $i \geq 1$ to avoid an awkward situation. To see this, if we let $i = 0$ and $j = 1$, then we get $h(1,1) = h(0,0) + 0 = a$; but $h(1,1) = b$.

What can we say about the nth row sum $S_n(a,b)$? To predict it, first notice that:

$$\begin{aligned}
S_0(a,b) &= a & S_3(a,b) &= 3a + 2b \\
S_1(a,b) &= a + b & S_4(a,b) &= 5a + 3b \\
S_2(a,b) &= 2a + b & &\vdots
\end{aligned}$$

Clearly, a pattern emerges: $S_n(a,b) = aF_{n+1} + bF_n$, where $n \geq 0$.

Likewise, the alternating row sums also display an interesting pattern:

$$
\begin{aligned}
T_0(a,b) &= a & T_4(a,b) &= a+b \\
T_1(a,b) &= a-b & T_5(a,b) &= 2a+b \\
T_2(a,b) &= b & T_6(a,b) &= 3a+2b \\
T_3(a,b) &= a & &\vdots
\end{aligned}
$$

These patterns lead us to the following theorem. We will omit its proof; see Exercises 12.33 and 12.35.

Theorem 12.5. *Let $S_n(a,b)$ denote the sum of the elements in row n of the array in Figure 12.5, and $T_n(a,b)$ their alternating row sum. Then $S_n(a,b) = aF_{n+1} + bF_n$, where $n \geq 0$; and $T_n(a,b) = aF_{n-2} + bF_{n-3}$, where $n \geq 1$.* ∎

In particular, $S_n(1,1) = F_{n+1} + F_n = F_{n+2}$; and $T_n(1,1) = F_{n-2} + F_{n-3} = F_{n-1}$. These are consistent with Theorems 12.2 and 12.3. Likewise, $S_n(1,2) = F_{n+1} + 2F_n = L_{n+1}$ and $T_n(1,2) = F_{n-2} + 2F_{n-3} = L_{n-2}$; these are consistent with Theorem 12.4.

Figure 12.6 shows yet another triangular array developed by Hoggatt in 1971 [290]. This array F also possesses several interesting properties, in addition to the obvious ones:

- The first element in row n is F_{2n-1}, where $n \geq 0$.
- The nth row sum is F_{2n+1}, where $n \geq 0$; see Exercise 12.24.

							Row sums
1							1
1	1						2
2	2	1					5
5	4	3	1				13
13	9 → 7	4	1				34
	↓						
34	22	16	11	5	1		89
89	56	38	27	16	6	1	233
↑							↑
F_{2n-1}							F_{2n+1}

Figure 12.6. Array F.

EXERCISES 12

Compute L_8 using each formula.

1. Formula (12.5).
2. Formula (12.7).

Verify each, where r and s are the roots of the equation $x^2 - px - q = 0$.

3. $r^5 + s^5 = p^5 + 5p^3q + 5pq^2$.
4. $r^6 + s^6 = p^6 + 6p^4q + 9p^2q^2 + 2q^3$.
5. Establish formula (12.3) using PMI.

6. Prove that $\displaystyle\sum_{i=0}^{n}(-1)^i A(n,i) = F_{n-1}$.

Compute F_{10} using each formula.

7. Lucas' formula for F_n.
8. Catalan's formula for F_n.

Verify each, where r and s are the roots of the equation $x^2 - px - q = 0$.

9. $r^5 - s^5 = (p^4 + 3p^2q + q^2)\Delta$.
10. $r^6 - s^6 = (p^5 + 4p^3q + 3pq^2)\Delta$.
11. Establish formula (12.8) using PMI.
12. Verify that $C(n,j) = C(n-1,j-1) + C(n-1,j)$, where $n,j \geq 1$.

Compute the sum of the elements on the given rising diagonal in Figure 12.1.

13. Diagonal 6.
14. Diagonal 7.

Prove each.

15. $\displaystyle\sum_{k=1}^{n} C(k,j) = C(n+1,j+1)$.

16. $\displaystyle\sum_{k=1}^{n} C(k,1) = C(n+1,2)$.

17. $C(n, n-2) = (n-1)^2$.

Compute the sum of the elements on the given rising diagonal in Figure 12.2.

18. Diagonal 5.
19. Diagonal 6.

Prove each.

20. $\displaystyle\sum_{k=1}^{n} D(k,j) = D(n+1,j+1)$.

21. $D(n,2) = (n-1)^2$.

22. $\sum_{k=1}^{n} D(k, 1) = n^2.$

23. Let $A(n,j)$ denote the element in row n and column j of the array in Figure 12.6, where $n, j \geq 0$. Define $A(n,j)$ recursively.

24. Prove that the nth row sum of the elements of the array in Figure 12.6 is F_{2n+1}, where $n \geq 0$.

Using the array in Figure 12.7, developed by Hoggatt in 1972 [294], prove the results in Exercises 12.25–12.27, where $n \geq 0$.

Figure 12.7.

25. The nth row sum is F_{2n+1}.

26. If the columns are multiplied by $1, 2, 3, \ldots$, to the right, then the nth row sum is F_{2n+1}.

27. The sum of the elements on the nth rising diagonal is F_{n+1}^2.

Using the array in Figure 12.8, developed by Hoggatt in 1977 [301], prove the results in Exercises 12.28–12.30, where $n \geq 0$.

Figure 12.8.

28. The nth row sum is F_{2n+2}, where $n \geq 0$.

29. The sum of the elements on the nth rising diagonal is $F_{n+1}F_{n+2}$, where $n \geq 0$.

30. Multiply the columns by $1, 2, 3, \ldots$, to the right. Then the nth row sum is $F_{2n+3} - 1$.

Let $S_n(a, b)$ denote the sum of the elements in row n of the array in Figure 12.5.

31. Define $S_n(a, b)$ recursively.

32. Prove that $S_n(a, b) = S_{n-1}(a, b) + S_{n-2}(a, b)$, where $n \geq 2$.

33. Prove that $S_n(a, b) = aF_{n+1} + bF_n$, where $n \geq 0$.

Let $T_n(a, b)$ denote the alternating sum of the elements in row n of the array in Figure 12.5.

34. Define $T_n(a, b)$ recursively.

35. Prove that $T_n(a, b) = aF_{n-2} + bF_{n-3}$, where $n \geq 0$.

13

RECURRENCES AND GENERATING FUNCTIONS

In preceding chapters, we came across several recurrences. We will now develop a method for solving a large and important class of recurrences. *Solving* a recurrence for a_n means finding an explicit formula for a_n using initial conditions. We will employ this powerful method to confirm Binet's formulas for F_n and L_n.

13.1   LHRWCCs

A kth-order *linear homogeneous recurrence with constant coefficients* (LHRWCCs) is a recurrence of the form

$$a_n = c_1 a_{n-1} + c_2 a_{n-2} + \cdots + c_k a_{n-k}, \tag{13.1}$$

where $c_1, c_2, \ldots, c_k \in \mathbb{R}$ and $c_k \neq 0$.

We now add a few words of explanation about the definitional terms. The term *linear* means that every term on the RHS of equation (13.1) contains at most the first power of any predecessor a_i. A recurrence is *homogeneous* if every term on the RHS is a multiple of some a_i; in other words, the recurrence is satisfied by the sequence $\{0\}$, that is, $a_n = 0$ for every n. All coefficients c_i are constants. Since a_n depends on its k immediate predecessors, the *order* of the recurrence is k.

Fibonacci and Lucas Numbers with Applications, Volume One, Second Edition. Thomas Koshy.
© 2018 John Wiley & Sons, Inc. Published 2018 by John Wiley & Sons, Inc.

Consequently, to solve a kth-order recurrence, we will need k initial conditions, say, $a_0 = C_0, a_1 = C_1, \ldots, a_{k-1} = C_{k-1}$.

The following example illustrates in detail the various terms in this definition.

Example 13.1.

1) *The recurrence* $s_n = 2s_{n-1}$ *is a LHRWCCs. Its order is one.*
2) *The recurrence* $a_n = na_{n-1}$ *is linear and homogeneous. But the coefficient on the RHS is not a constant; so it is not a LHRWCCs.*
3) $h_n = h_{n-1} + (n - 1)$ *is a linear recurrence. But it is not homogeneous because of the term* $n - 1$.
4) *The recurrence* $a_n = a_{n-1}^2 + 3a_{n-2}$ *is homogeneous. But it is not linear, since the power of* a_{n-1} *is 2.*
5) $a_n = a_{n-1} + 2a_{n-2} + 3a_{n-6}$ *is a LHRWCCs with order six.*   ∎

To begin with, consider the first-order LHRWCCs $s_n = 2s_{n-1}$, where $s_0 = 1$. Its solution is $s_n = 2^n$, where $n \geq 0$. More generally, consider the recurrence $a_n = \gamma a_{n-1}$, where $a_0 = c$. Its solution is $a_n = c\gamma^n$, where $n \geq 0$.

Let us now turn our attention to the second-order LHRWCCs

$$a_n = aa_{n-1} + ba_{n-2}, \tag{13.2}$$

where a and b are nonzero constants. If it has a nonzero solution of the form $c\gamma^n$, then $c\gamma^n = ac\gamma^{n-1} + bc\gamma^{n-2}$. Since $c\gamma \neq 0$, this implies $\gamma^2 = a\gamma + b$; that is, $\gamma^2 = a\gamma - b = 0$; so γ must be a solution of the *characteristic equation*

$$x^2 - ax - b = 0. \tag{13.3}$$

The roots of equation (13.3) are the *characteristic roots* of recurrence (13.2).

The next theorem shows the power of characteristic roots in solving LHRWCCs.

Theorem 13.1. *Let* γ *and* δ *be the distinct solutions of the equation* $x^2 - ax - b = 0$, *where* $a, b \in \mathbb{R}$ *and* $b \neq 0$. *Then every solution of the LHRWCCs* $a_n = aa_{n-1} + ba_{n-2}$ *is of the form* $a_n = A\gamma^n + B\delta^n$ *for some constants* A *and* B, *where* $a_0 = C_0$ *and* $a_1 = C_1$.

Proof. The proof consists of two parts:

1) We will show that $a_n = A\gamma^n + B\delta^n$ satisfies the recurrence for any constants A and B.
2) We will find the values of A and B satisfying the given initial conditions.

Since γ and δ are solutions of equation (13.3), $\gamma^2 = a\gamma + b$ and $\delta^2 = a\delta + b$.

1) To show that $a_n = A\gamma^n + B\delta^n$ is a solution of the recurrence:

$$aa_{n-1} + ba_{n-2} = a(A\gamma^{n-1} + B\delta^{n-1}) + b(A\gamma^{n-2} + B\delta^{n-2})$$
$$= A\gamma^{n-2}(a\gamma + b) + B\delta^{n-2}(a\delta + b)$$
$$= A\gamma^{n-2} \cdot \gamma^2 + B\delta^{n-2} \cdot \delta^2$$
$$= A\gamma^n + B\delta^n$$
$$= a_n.$$

Thus $a_n = A\gamma^n + B\delta^n$ is a solution of the recurrence.

2) Next we will find the values of A and B. Using the initial conditions $a_0 = C_0$ and $a_1 = C_1$, we get a 2×2 linear system:

$$C_0 = A + B$$
$$C_1 = A\gamma + b\delta.$$

Solving this system, we get $A = \dfrac{C_1 - C_0\delta}{\gamma - \delta}$ and $B = \dfrac{C_0\gamma - C_1}{\gamma - \delta}$, where $\gamma \neq \delta$.

With these values for A and B, a_n satisfies the initial conditions and the recurrence. Since the recurrence together with the initial conditions determine a unique sequence $\{a_n\}$, $a_n = A\gamma^n + B\delta^n$ is indeed the unique solution of the recurrence. ∎

A FEW NOTES

1. The solutions γ and δ are nonzero. If $\gamma = 0$, for instance, that would imply that $b = 0$.

2. Theorem 13.1 *cannot* be applied if $\gamma = \delta$. However, it works even if γ and δ are complex numbers.

3. The solutions γ^n and δ^n are the *basic solutions* of the recurrence. In general, the number of basic solutions equals the order of the recurrence. The *general solution* $a_n = A\gamma^n + B\delta^n$ is a *linear combination* of the basic solutions. The particular solution is obtained by selecting A and B in such a way that the initial conditions are satisfied, as in the theorem.

The next two examples illustrate Theorem 13.1.

Example 13.2. *Solve the recurrence* $a_n = 5a_{n-1} - 6a_{n-2}$, *where* $a_0 = 4$ *and* $a_1 = 7$.

Solution.

1) *To find the general solution of the recurrence*:
   The characteristic equation of the recurrence is $x^2 - 5x - 6 = 0$; the characteristic roots are 2 and 3. Therefore, by Theorem 13.1, the general solution of the recurrence is $a_n = A \cdot 2^n + B \cdot 3^n$.

2) *To find the values of A and B*:
   Using the initial conditions, we get the linear system

$$A + B = 4$$
$$2A + 3B = 7.$$

Solving this system, we get $A = 5$ and $B = -1$.
Thus the desired solution of the recurrence is $a_n = 5 \cdot 2^n - 3^n$, where $n \geq 0$. ∎

We are now ready to derive Binet's formula for F_n using Theorem 13.1.

Example 13.3.   *Solve the Fibonacci recurrence $F_n = F_{n-1} + F_{n-2}$, where $F_1 = 1 = F_2$.*

Solution. The characteristic equation of the recurrence is $x^2 - x - 1$, and its solutions are α and β. Recall that $\alpha + \beta = 1$ and $\alpha\beta = -1$.

The general solution is $F_n = A\alpha^n + B\beta^n$. Using the initial conditions, we get

$$A\alpha + B\beta = 1$$
$$A\alpha^2 + B\beta^2 = 1.$$

Solving this linear system, we get $A = \dfrac{\alpha}{1 + \alpha^2} = \dfrac{1}{\alpha - \beta}$ and similarly $B = -\dfrac{1}{\alpha - \beta}$.

Thus the solution of the Fibonacci recurrence satisfying the given initial conditions is

$$F_n = \frac{\alpha^n - \beta^n}{\alpha - \beta},$$

which is Binet's formula for F_n.   ∎

The same technique can be employed to derive Binet's formula for L_n; see Exercise 13.15.

EXAMPLE 5.3 REVISITED

In Example 5.3, we learned that the general solution of the Pell's equation $x^2 - 5y^2 = 1$ is given by $(x_n, y_n) = (\frac{1}{2}L_{6n}, \frac{1}{2}F_{6n})$, where $n \geq 1$. Clearly, α^6 and

β^6 are solutions of the equation $(z - \alpha^6)(z - \beta^6) = 0$; that is, they are solutions of the equation $z^2 - 18z + 1 = 0$. This is the characteristic equation of the recurrence $z_{n+2} - 18z_{n+1} + z_n = 0$. In other words, both x_n and y_n satisfy the recurrence $z_{n+2} = 18z_{n+1} - z_n$.

For example, $x_3 = 18x_2 - x_1 = 18 \cdot 161 - 9 = 2889$, and $y_3 = 18y_2 - y_1 = 18 \cdot 72 - 4 = 1292$. Notice that $(2889, 1292)$ is indeed a solution of $x^2 - 5y^2 = 1$.

Since $z_{n+2} = 18z_{n+1} - z_n$, it follows that $L_{6n+12} = 18L_{6n+6} - L_{6n}$ and $F_{6n+12} = 18F_{6n+6} - F_{6n}$. For example,

$$L_{24} = 103{,}682$$
$$= 18 \cdot 5778 - 322$$
$$= 18L_{18} - L_{12}.$$

Similarly, $F_{30} = 832{,}040 = 18F_{24} - F_{18}$.

Next we turn to generating functions.

13.2 GENERATING FUNCTIONS

Generating functions are a powerful tool for solving LHRWCCs, as will be seen shortly. In 1718, the French mathematician Abraham De Moivre (1667–1754) invented generating functions in order to solve the Fibonacci recurrence.

First, notice that the polynomial $1 + x + x^2 + x^3 + x^4 + x^5$ can be written as $(x^6 - 1)/(x - 1)$. We can confirm this by either cross-multiplication or the familiar long-division method. Accordingly, $f(x) = (x^6 - 1)/(x - 1)$ is called the *generating function* of the sequence of coefficients 1, 1, 1, 1, 1, and 1 in the polynomial.

More generally, we make the following definition.

Let $a_0, a_1, a_2, \ldots$, be a sequence of real numbers. Then the function

$$g(x) = a_0 + a_1 x + a_2 x^2 + \cdots + a_n x^n + \cdots \tag{13.4}$$

is the *generating function* of the sequence $\{a_n\}$. We can also define a generating function for the finite sequence $a_0, a_1, a_2, \ldots, a_n$ by letting $a_0 = 0$ for $i > n$; thus $g(x) = a_0 + a_1 x + a_2 x^2 + \cdots + a_n x^n$ is the generating function of the finite sequence $a_0, a_1, a_2, \ldots, a_n$.

For example, $g(x) = 1 + 2x + 3x^2 + \cdots + (n + 1)x^n + \cdots$ is the generating function of the sequence of positive integers; and

$$f(x) = 1 + 3x + 6x^2 + \cdots + \frac{n(n + 1)}{2} x^n + \cdots$$

is the generating function of the sequence of triangular numbers $1, 3, 6, 10, \ldots$.
Since

$$\frac{x^n - 1}{x - 1} = 1 + x + x^2 + \cdots + x^{n-1},$$

$g(x) = \dfrac{x^n - 1}{x - 1}$ is the generating function of the sequence of n ones.

A word of caution. The RHS of equation (13.4) is a *formal power series* in x. The letter x does not represent anything. We use the powers x^n simply to keep track of the corresponding terms a_n of the sequence. In other words, think of the powers x^n as place-holders. Consequently, we are *not* interested in the convergence of the series.

EQUALITY OF GENERATING FUNCTIONS

Two generating functions $f(x) = \sum\limits_{n=0}^{\infty} a_n x^n$ and $g(x) = \sum\limits_{n=0}^{\infty} b_n x^n$ are *equal if* $a_n = b_n$ for every $n \geq 0$.

For example, let $f(x) = 1 + 3x + 6x^2 + 10x^3 + \cdots$ and $g(x) = 1 + \dfrac{2 \cdot 3}{2} x + \dfrac{3 \cdot 4}{2} x^2 + \dfrac{4 \cdot 5}{2} x^3 + \cdots$. Then $f(x) = g(x)$.

A generating function we will use frequently is

$$\frac{1}{1 - ax} = 1 + ax + a^2 x^2 + \cdots + a^n x^n + \cdots. \tag{13.5}$$

In particular,

$$\frac{1}{1 - x} = 1 + x + x^2 + \cdots + x^n + \cdots. \tag{13.6}$$

Can we add and multiply generating functions? Yes. Such operations are performed exactly the same way as polynomials are combined.

ADDITION AND MULTIPLICATION OF GENERATING FUNCTIONS

Let $f(x) = \sum\limits_{n=0}^{\infty} a_n x^n$ and $g(x) = \sum\limits_{n=0}^{\infty} b_n x^n$ be two generating functions. Then

$$f(x) + g(x) = \sum_{n=0}^{\infty} (a_n + b_n) x^n \quad \text{and} \quad f(x) g(x) = \sum_{n=0}^{\infty} \left(\sum_{i=0}^{n} a_i b_{n-i} \right) x^n.$$

For example,

$$\frac{1}{(1-x)^2} = \frac{1}{1-x} \cdot \frac{1}{1-x}$$

$$= \left(\sum_{i=0}^{\infty} x^i\right)\left(\sum_{i=0}^{\infty} x^i\right)$$

$$= \sum_{n=0}^{\infty} \left(\sum_{i=0}^{n} 1 \cdot 1\right) x^n$$

$$= \sum_{n=0}^{\infty} (n+1)x^n$$

$$= 1 + 2x + 3x^2 + \cdots + (n+1)x^n + \cdots; \qquad (13.7)$$

$$\frac{1}{(1-x)^3} = \frac{1}{1-x} \cdot \frac{1}{(1-x)^2}$$

$$= \left(\sum_{i=0}^{\infty} x^i\right)\left[\sum_{i=0}^{\infty} (i+1)x^i\right]$$

$$= \sum_{n=0}^{\infty} \left[\sum_{i=0}^{n} 1 \cdot (n+1-i)\right] x^n$$

$$= \sum_{n=0}^{\infty} \left(\sum_{i=1}^{n+1} i\right) x^n$$

$$= \sum_{n=0}^{\infty} \frac{(n+1)(n+2)}{2} x^n$$

$$= 1 + 3x + 6x^2 + 10x^3 + \cdots. \qquad (13.8)$$

PARTIAL FRACTION DECOMPOSITION

Before exploring how valuable generating functions are in solving LHRWCCs, we will study how the technique of *partial fraction decomposition*, used in integral calculus, enables us to express the quotient $p(x)/q(x)$ of two polynomials $p(x)$ and $q(x)$ as a sum of proper fractions, where $\deg p(x) < \deg q(x)$[†].

For example, $\dfrac{6x+1}{(2x-1)(2x+3)} = \dfrac{1}{2x-1} + \dfrac{2}{2x+3}$. (You may confirm this.)

[†]$\deg f(x)$ denotes the degree of the polynomial $f(x)$.

PARTIAL FRACTION DECOMPOSITION RULE

Suppose $q(x)$ has a factor of the form $(ax + b)^m$. Then the decomposition contains a sum of the form

$$\frac{A_1}{ax + b} + \frac{A_2}{(ax + b)^2} + \cdots + \frac{A_m}{(ax + b)^m},$$

where $A_i \in \mathbb{R}$.

Examples 13.4–13.6 illustrate the partial fraction decomposition technique. We will use their partial fraction decompositions to solve the recurrences in Examples 13.7–13.9.

Example 13.4. *Express* $\dfrac{x}{(1 - x)(1 - 2x)}$ *as a sum of partial fractions.*

Solution. Since the denominator contains two linear factors, we let

$$\frac{x}{(1 - x)(1 - 2x)} = \frac{A}{1 - x} + \frac{B}{1 - 2x}.$$

To determine the constants A and B, multiply both sides by $(1 - x)(1 - 2x)$:

$$x = A(1 - 2x) + B(1 - x).$$

Now give convenient values to x. Setting $x = 1$ yields $A = -1$; and setting $x = 1/2$ yields $B = 1$. (We can also find the values of A and B by equating the coefficients of like terms from either side of the equation, and then solving the resulting linear system.) Thus

$$\frac{x}{(1 - x)(1 - 2x)} = \frac{-1}{1 - x} + \frac{1}{1 - 2x}.$$

(You may confirm this by combining the sum on the RHS into a single fraction.) ∎

Example 13.5. *Express* $\dfrac{x}{1 - x - x^2}$ *as a sum of partial fractions.*

Solution. First, factor $1 - x - x^2$: $1 - x - x^2 = (1 - \alpha x)(1 - \beta x)$. We then let

$$\frac{x}{1 - x - x^2} = \frac{A}{1 - \alpha x} + \frac{B}{1 - \beta x}$$

$$x = A(1 - \beta x) + B(1 - \alpha x).$$

Equating the coefficients of like terms, we get

$$A + B = 0$$

$$-\beta A - \alpha B = 1.$$

Solving this linear system, we get $A = 1/(\alpha - \beta) = -B$. (Verify this.) Thus

$$\frac{x}{1 - x - x^2} = \frac{1}{\alpha - \beta} \left(\frac{1}{1 - \alpha x} - \frac{1}{1 - \beta x} \right).$$

∎

We will use this decomposition in Example 13.8.

Example 13.6. *Express* $\dfrac{2 - 9x}{1 - 6x + 9x^2}$ *as a sum of partial fractions.*

Solution. Again, factor the denominator: $1 - 6x + 9x^2 = (1 - 3x)^2$. Using the decomposition rule, we let

$$\frac{2 - 9x}{1 - 6x + 9x^2} = \frac{A}{1 - 3x} + \frac{B}{(1 - 3x)^2}.$$

Then

$$2 - 9x = A(1 - 3x) + B.$$

This yields $A = 3$ and $B = -1$. (Verify this.) Thus

$$\frac{2 - 9x}{1 - 6x + 9x^2} = \frac{3}{1 - 3x} - \frac{1}{(1 - 3x)^2}.$$

∎

We will use this result in Example 13.9.

We are now ready to use the partial fraction decompositions and generating functions to solve recurrences in the next three examples.

Example 13.7. *Using generating functions, solve the recurrence* $b_n = 2b_{n-1} + 1$, *where* $b_1 = 1$.

Solution. First, notice that $b_1 = 1$ implies $b_0 = 0$. To find the sequence $\{b_n\}$, we consider the corresponding generating function $g(x) = b_0 + b_1 x + b_2 x^2 + b_3 x^3 + \cdots + b_n x^n + \cdots$. Then

$$2xg(x) = 2b_1 x^2 + 2b_2 x^3 + 2b_3 x^4 + \cdots + 2b_n x^{n+1} + \cdots.$$

Since

$$\frac{1}{1 - x} = 1 + x + x^2 + x^3 + \cdots + x^n + \cdots,$$

we have

$$g(x) - 2xg(x) - \frac{1}{1 - x} = -1 + (b_1 - 1)x + (b_2 - 2b_1 - 1)x^2 + \cdots$$

$$+ (b_n - 2b_{n-1} - 1)x^n + \cdots$$

$$= -1.$$

Using Example 13.4, we then have

$$(1 - 2x)g(x) = \frac{1}{1-x} - 1 = \frac{x}{1-x}$$

$$g(x) = \frac{x}{(1-x)(1-2x)}$$

$$= -\frac{1}{1-x} + \frac{1}{1-2x}$$

$$= -\sum_{n=0}^{\infty} x^n + \sum_{n=0}^{\infty} 2^n x^n$$

$$= \sum_{n=0}^{\infty} (2^n - 1)x^n.$$

But $g(x) = \sum_{n=0}^{\infty} b_n x^n$; so $b_n = 2^n - 1$, where $n \geq 1$. ∎

Example 13.8. *Using generating functions, solve the Fibonacci recurrence* $F_n = F_{n-1} + F_{n-2}$, *where* $F_1 = 1 = F_2$, *and* $n \geq 3$.

Solution. Recall that $F_0 = 0$. Let

$$g(x) = F_0 + F_1 x + F_2 x^2 + F_3 x^3 + \cdots + F_n x^n + \cdots$$

be the generating function of the Fibonacci sequence. Since the orders of F_{n-1} and F_{n-2} are 1 and 2 less than the order of F_n, respectively, we find $xg(x)$ and $x^2 g(x)$, and then invoke the Fibonacci recurrence and the initial conditions:

$$g(x) = F_0 + F_1 x + F_2 x^2 + \sum_{n=3}^{\infty} F_n x^n$$

$$xg(x) = \qquad\qquad F_1 x^2 + \sum_{n=3}^{\infty} F_{n-1} x^n$$

$$x^2 g(x) = \qquad\qquad\qquad \sum_{n=3}^{\infty} F_{n-2} x^n$$

$$g(x) - xg(x) - x^2 g(x) = F_1 x + (F_2 - F_1)x^2 + \sum_{n=3}^{\infty} (F_n - F_{n-1} - F_{n-2})x^n + \cdots$$

$$= x.$$

By Example 13.5, we then have

$$(1 - x - x^2)g(x) = x$$

$$g(x) = \frac{x}{1 - x - x^2}$$

$$(\alpha - \beta)g(x) = \frac{1}{1 - \alpha x} - \frac{1}{1 - \beta x}$$

$$= \sum_{n=0}^{\infty} \alpha^n x^n - \sum_{n=0}^{\infty} \beta^n x^n$$

$$= \sum_{n=0}^{\infty} (\alpha^n - \beta^n) x^n$$

$$g(x) = \sum_{n=0}^{\infty} \frac{\alpha^n - \beta^n}{\alpha - \beta} x^n.$$

Thus, by the equality of generating functions, we get Binet's formula for F_n:

$$F_n = \frac{\alpha^n - \beta^n}{\alpha - \beta}.$$ ∎

Since $\dfrac{x}{1 - x - x^2} = \sum\limits_{n=0}^{\infty} F_n x^n$, it follows that

$$\frac{1}{n!} \frac{d^n}{dx^n} \left[\frac{x}{1 - x - x^2} \right]_{x=0} = F_n.$$

Example 13.9. *Using generating functions, solve the recurrence* $a_n = 6a_{n-1} - 9a_{n-2}$, *where* $a_0 = 2$ *and* $a_1 = 3$, *and* $n \geq 2$.

Solution. Let $g(x) = a_0 + a_1 x + a_2 x^2 + \cdots + a_n x^n + \cdots$. Then, by the given recurrence, initial conditions, and Example 13.6, we have

$$g(x) = a_0 + a_1 x + \sum_{n=2}^{\infty} a_n x^n$$

$$6xg(x) = \qquad 6a_0 x + \sum_{n=2}^{\infty} 6a_{n-1} x^n$$

$$9x^2 g(x) = \qquad \sum_{n=2}^{\infty} 9a_{n-2} x^n$$

$$g(x) - 6xg(x) + 9x^2 g(x) = a_0 + (a_1 - 6a_0)x$$

$$(1 - 6x - 9x^2)g(x) = 2 - 9x$$

$$g(x) = \frac{2 - 9x}{1 - 6x - 9x^2}$$

$$= \frac{3}{1 - 3x} - \frac{1}{(1 - 3x)^2}$$

$$= 3 \sum_{n=0}^{\infty} 3^n x^n - \sum_{n=0}^{\infty} (n+1)3^n x^n$$

$$= \sum_{n=0}^{\infty} [3^{n+1} - (n+1)3^n] x^n$$

$$= \sum_{n=0}^{\infty} (2 - n)3^n x^n.$$

Thus $a_n = (2 - n)3^n$, where $n \geq 0$.   ∎

The next example is a generalization of an interesting problem that appeared in the 1999 William Lowell Putnam Mathematical Competition; see [321]. As you will see, it has Fibonacci implications.

Example 13.10.   *Prove that the power series expansion*

$$\frac{1}{1 - cx - x^2} = a_0 + a_1 x + a_2 x^2 + \cdots + a_n x^n + \cdots$$

has the property that the sum of the squares of any two consecutive coefficients is also the coefficient of a later term in the series, where c is a nonzero constant.

Proof. Let $1 - cx - x^2 = (1 - rx)(1 - sx)$, where $rs = -1$. Then

$$\frac{1}{1 - cx - x^2} = \frac{1}{r - s} \left(\frac{r}{1 - rx} - \frac{s}{1 - sx} \right)$$

$$= \sum_{n=0}^{\infty} \left(\frac{r^{n+1} - s^{n+1}}{r - s} \right) x^n.$$

Consequently, $a_n = \dfrac{r^{n+1} - s^{n+1}}{r - s}$, where $n \geq 0$.

Then

$$(r - s)^2 (a_n^2 + a_{n+1}^2) = (r^{n+1} - s^{n+1})^2 + (r^{n+2} - s^{n+2})^2$$

$$= r^{2n+3} \left(r + \frac{1}{r} \right) + s^{2n+3} \left(r + \frac{1}{r} \right)$$

$$= (r - s) \left(r^{2n+3} - s^{2n+3} \right)$$

$$a_n^2 + a_{n+1}^2 = \frac{r^{2n+3} - s^{2n+3}}{r - s} \tag{13.9}$$

$$= a_{n+2},$$

as desired.   ∎

As byproducts, it is easy to show that

$$a_n + ca_{n+1} = a_{n+2} \tag{13.10}$$

$$a_{n+1}^2 - a_{n-1}^2 = ca_{2n+1}; \tag{13.11}$$

see Exercises 13.55 and 13.56. Formula (13.10) gives a recurrence for the coefficients a_n in the expansion. It follows from formula (13.11) that the difference of the squares of all other coefficients is c times a later coefficient.

Suppose we let $c = 1$ in the example. Then $a_n = F_{n+1}$. Formula (13.10) then yields the Fibonacci recurrence; and formulas (13.9) and (13.11) give the familiar Lucas formulas $F_{n+1}^2 + F_{n+2}^2 = F_{2n+3}$ and $F_{n+2}^2 - F_n^2 = F_{2n+2}$.

When $c = 2$, formula (13.10) gives the well known *Pell recurrence* $P_{n+2} = 2P_{n+1} + P_n$, where $P_0 = 0, P_1 = 1$, and $n \geq 0$ [370].

The next example presents an interesting identity linking binomial coefficients and Fibonacci numbers. The identity, developed by Hoggatt in 1968 [279], is an application of Example 13.8. The neat proof featured here is due to Carlitz [110].

Example 13.11. *Prove that* $\displaystyle\sum_{2j \leq n} j\binom{n-j}{j} = \sum_{j=0}^{n} F_{n-j}F_j$.

Proof. Let $C_n = \displaystyle\sum_{2j \leq n} j\binom{n-j}{j}$. Then

$$\sum_{n=0}^{\infty} C_n x^n = \sum_{n=0}^{\infty} \left[\sum_{2j \leq n} j\binom{n-j}{j} \right] x^n$$

$$= \sum_{j=0}^{\infty} jx^{2j} \left[\sum_{n=0}^{\infty} \binom{n+j}{j} x^n \right]$$

$$= \sum_{j=0}^{\infty} jx^{2j}(1-x)^{-j-1}$$

$$= \frac{1}{1-x} \sum_{j=1}^{\infty} j \left(\frac{x^2}{1-x} \right)^j$$

$$= \frac{x^2}{(1-x)^2} \sum_{j=0}^{\infty} (j+1) \left(\frac{x^2}{1-x} \right)^j$$

$$= \frac{x^2}{(1-x)^2} \left(1 - \frac{x^2}{1-x} \right)^{-2}$$

$$= \frac{x^2}{(1-x-x^2)^2}$$

$$= \left(\sum_{n=0}^{\infty} F_n x^n \right) \left(\sum_{j=0}^{\infty} F_j x^j \right)$$

$$= \sum_{n=0}^{\infty} \left(\sum_{j=0}^{n} F_j F_{n-j} \right) x^n.$$

Thus $C_n = \sum_{j=0}^{n} F_j F_{n-j}$. This yields the desired result. ∎

A PASCAL LINK

The identity in this example has a direct link with the well-known Pascal's triangle. To see this, consider the left-justified Pascal's triangle. Multiply each column by j, and then add the rising diagonals, where $j \geq 0$. The resulting sum on the nth diagonal is C_n; see Figure 13.1.

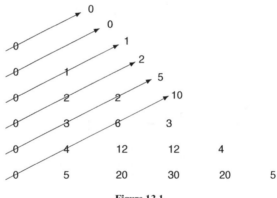

Figure 13.1.

The next example was proposed as a problem in 1970 by R.C. Drake of North Carolina A & T State University, Greensboro, North Carolina [150]. It deals with paths connecting lattice points. A *lattice point* on the Cartesian plane is a point (x, y), where both x and y are integers. The solution given here is based on the one by Carlitz [112].

Example 13.12.   *Let $f(n)$ denote the number of paths on the Cartesian plane from the origin to the lattice point $(n, 0)$. Each path is made up of directed line segments of one or more of the types in Table 13.1. For example, the next point on the path from $(k, 0)$ can be $(k, 1)$ or $(k + 1, 0)$; and that from $(k, 1)$ can be $(k + 1, 1)$ or $(k + 1, 0)$. Find a formula for $f(n)$.*

TABLE 13.1. Possible Paths

Type	1	2	3	4
Initial point	$(k,0)$	$(k,0)$	$(k,1)$	$(k,1)$
End point	$(k,1)$	$(k+1,0)$	$(k+1,1)$	$(k+1,0)$

Solution. There are $f(1) = 2 = F_3$ paths from $(0,0)$ to $(1,0)$; $f(2) = 5 = F_5$ paths from $(0,0)$ to $(2,0)$; and $f(3) = 13 = F_7$ paths from $(0,0)$ to $(3,0)$; see Figure 13.2.

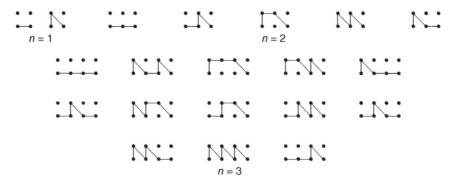

Figure 13.2. Paths from $(0, 0)$ to $(n, 0)$, where $0 \le n \le 3$.

Let $f_2(n)$ denote the number of paths ending with a line segment of type 2, and $f_4(n)$ the number of paths ending with a line segment of type 4. Then

$$f_2(n + 1) = f_2(n) + f_4(n)$$
$$= f(n);$$
$$f_4(n + 1) = f(0) + f(1) + \cdots + f(n)$$
$$= \sum_{k=0}^{n} f(k).$$

Consequently,

$$f(n + 1) = f_2(n + 1) + f_4(n + 1)$$
$$= f(n) + \sum_{k=0}^{n} f(k).$$

So $f(1) = f(0) + f(0) = 2f(0)$. But $f(1) = 2$; so $f(0) = 1$.

Now consider the power series

$$F(x) = \sum_{n=0}^{\infty} f(n)x^n$$

$$= f(0) + \sum_{n=1}^{\infty} f(n)x^n$$

$$= 1 + \sum_{n=0}^{\infty} f(n+1)x^{n+1}$$

$$= 1 + \sum_{n=0}^{\infty} \left[f(n) + \sum_{k=0}^{n} f(k) \right] x^{n+1}$$

$$= 1 + xF(x) + \frac{x}{1-x}F(x)$$

$$= \frac{1-x}{1-3x+x^2}.$$

But $\dfrac{1-x}{1-3x+x^2} = \sum_{n=0}^{\infty} F_{2n+1}x^n$; see the list of generating functions in Section 13.6.
Thus $f(n) = F_{2n+1}$, where $n \geq 0$.   ∎

Next we will demonstrate the power of generating functions in deriving
Fibonacci and Lucas identities. The ones we developed in earlier chapters will
come in handy in this endeavor.

To begin with, we will develop a generating function for F_{3n}. To this end, we
will need the second-order identity $F_{3n} = 4F_{3n-3} + F_{3n-6}$.

13.3   A GENERATING FUNCTION FOR F_{3n}

Let $g(x)$ be a generating function for F_{3n}. Then

$$g(x) = F_0 + F_3x + \sum_{n=2}^{\infty} F_{3n}x^n$$

$$4xg(x) = \qquad 4F_0x + 4\sum_{n=2}^{\infty} F_{3n-3}x^n + \cdots$$

$$x^2 g(x) = \qquad \sum_{n=2}^{\infty} F_{3n-6}x^n + \cdots$$

$$(1 - 4x - x^2)g(x) = 2x.$$

Thus $g(x) = \dfrac{2x}{1 - 4x - x^2}.$

Next we will derive a generating function for F_n^3.

13.4   A GENERATING FUNCTION FOR F_n^3

Let $g(x)$ be a generating function for F_n^3. Using the identity $F_{n+4}^3 = 3F_{n+3}^3 + 6F_{n+2}^3 - 3F_{n+1}^3 - F_n^3$ (see Section 5.8), we then have

$$g(x) = F_0^3 + F_1^3 x + F_2^3 x^2 + F_3^3 x^3 + \sum_{n=4}^{\infty} F_n^3 x^n$$

$$3xg(x) = \quad 3F_0^3 x + 3F_1^3 x^2 + 3F_2^3 x^3 + 3 \sum_{n=4}^{\infty} F_{n-1}^3 x^n$$

$$6x^2 g(x) = \quad 6F_0^3 x^2 + 6F_1^3 x^3 + 6 \sum_{n=4}^{\infty} F_{n-2}^3 x^n$$

$$3x^3 g(x) = \quad 3F_0^3 x^3 + 3 \sum_{n=4}^{\infty} F_{n-3}^3 x^n$$

$$x^4 g(x) = \quad \sum_{n=4}^{\infty} F_{n-4}^3 x^n$$

$$(1 - 3x - 6x^2 + 3x^3 + x^4)g(x) = x + x^2 - 3x^2 + 2x^3 - 3x^3 - 6x^3$$
$$= x - 2x^2 - 7x^3.$$

Thus $g(x) = \dfrac{x - 2x^2 - 7x^3}{1 - 3x - 6x^2 + 3x^3 + x^4}.$

13.5   SUMMATION FORMULA (5.1) REVISITED

In 1948, J. Ginzburg employed generating functions to establish the summation formula (5.1) that $\sum_{i=1}^{n} F_i = F_{n+2} - 1$ [220]. To see how this was done, first we derive a generating function for the sum $s_n = \sum_{i=0}^{n} F_i$, where $s_0 = 0$. Let $g(x)$ be a generating function of the sequence $\{s_n\}$. Since s_n satisfies the recurrence $s_n = 2s_{n-1} - s_{n-3}$ (you may confirm this), we then have

$$g(x) = \sum_{n=0}^{\infty} s_n x^n$$

$$2xg(x) = \sum_{n=2}^{\infty} 2s_{n-1} x^n$$

$$x^3 g(x) = \sum_{n=3}^{\infty} s_{n-3} x^n$$

$$(1 - 2x + x^3)g(x) = x.$$

Thus $g(x) = \dfrac{x}{1 - 2x + x^3}$ is the desired generating function; s_n is the coefficient of x^n in the power series expansion of the function g.

Since $1 - 2x + x^3 = (1 - x - x^2)(1 - x)$, we can convert the rational expression into partial fractions:

$$g(x) = \frac{1 + x}{1 - x - x^2} - \frac{1}{1 - x}.$$

Since $\dfrac{x}{1 - x - x^2} = \displaystyle\sum_{n=0}^{\infty} F_n x^n$, this yields

$$\sum_{n=0}^{\infty} s_n x^n = (1 + x) \sum_{n=1}^{\infty} F_n x^{n-1} - \sum_{n=0}^{\infty} x^n$$

$$= \sum_{n=1}^{\infty} (F_n x^{n-1} + F_n x^n) - \sum_{n=0}^{\infty} x^n$$

$$= \sum_{n=0}^{\infty} (F_{n+1} + F_n - 1) x^n$$

$$= \sum_{n=0}^{\infty} (F_{n+2} - 1) x^n.$$

Thus $s_n = F_{n+2} - 1$, as desired.

13.6   A LIST OF GENERATING FUNCTIONS

In 1967, Hoggatt and Lind compiled the following list of 18 generating functions [315]; they generate powers and products of Fibonacci and Lucas numbers.

1) $\dfrac{x}{1 - x - x^2} = \displaystyle\sum_{n=0}^{\infty} F_n x^n.$

2) $\dfrac{1}{1 - x - x^2} = \displaystyle\sum_{n=1}^{\infty} F_{n+1} x^n.$

3) $\dfrac{2 - x}{1 - x - x^2} = \displaystyle\sum_{n=0}^{\infty} L_n x^n.$

4) $\dfrac{1 + 2x}{1 - x - x^2} = \displaystyle\sum_{n=0}^{\infty} L_{n+1} x^n.$

5) $\dfrac{x - x^2}{1 - 2x - 2x^2 + x^3} = \displaystyle\sum_{n=1}^{\infty} F_n^2 x^n.$

6) $\dfrac{1 - x}{1 - 2x - 2x^2 + x^3} = \displaystyle\sum_{n=0}^{\infty} F_{n+1}^2 x^n.$

7) $\dfrac{1+2x-x^2}{1-2x-2x^2+x^3} = \sum\limits_{n=0}^{\infty} F_{n+2}^2 x^n.$

8) $\dfrac{x}{1-2x-2x^2+x^3} = \sum\limits_{n=0}^{\infty} F_n F_{n+1} x^n.$

9) $\dfrac{4-7x-x^2}{1-2x-2x^2+x^3} = \sum\limits_{n=0}^{\infty} L_n^2 x^n.$

10) $\dfrac{1+7x-4x^2}{1-2x-2x^2+x^3} = \sum\limits_{n=0}^{\infty} L_{n+1}^2 x^n.$

11) $\dfrac{9-2x-x^2}{1-2x-2x^2+x^3} = \sum\limits_{n=0}^{\infty} L_{n+2}^2 x^n.$

12) $\dfrac{x-2x^2-x^3}{1-3x-6x^2+3x^3+x^4} = \sum\limits_{n=0}^{\infty} F_n^3 x^n.$

13) $\dfrac{1-2x-x^2}{1-3x-6x^2+3x^3+x^4} = \sum\limits_{n=0}^{\infty} F_{n+1}^3 x^n.$

14) $\dfrac{1+5x-3x^2-x^3}{1-3x-6x^2+3x^3+x^4} = \sum\limits_{n=0}^{\infty} F_{n+2}^3 x^n.$

15) $\dfrac{8+3x-4x^2-x^3}{1-3x-6x^2+3x^3+x^4} = \sum\limits_{n=0}^{\infty} F_{n+3}^3 x^n.$

16) $\dfrac{2x}{1-3x-6x^2+3x^3+x^4} = \sum\limits_{n=0}^{\infty} F_n F_{n+1} F_{n+2} x^n.$

17) $\dfrac{F_k x}{1-L_k x+(-1)^k x^2} = \sum\limits_{n=0}^{\infty} F_{kn} x^n.$

18) $\dfrac{F_r+(-1)^r F_{k-r} x}{1-L_k x+(-1)^k x^2} = \sum\limits_{n=0}^{\infty} F_{kn+r} x^n.$

Hoggatt derived the following four generating functions in 1971 [285].

19) $\dfrac{x}{1-3x+x^2} = \sum\limits_{n=0}^{\infty} F_{2n} x^n.$

20) $\dfrac{1-x}{1-3x+x^2} = \sum\limits_{n=0}^{\infty} F_{2n+1} x^n.$

21) $\dfrac{3-2x}{1-3x+x^2} = \sum\limits_{n=0}^{\infty} L_{2n+2} x^n.$

22) $\dfrac{x+x^2}{1-3x+x^2} = \sum\limits_{n=0}^{\infty} L_{2n+1} x^{n+1}.$

Next we will derive generating functions of $\{F_{m+n}\}$ and $\{L_{m+n}\}$.

GENERATING FUNCTIONS OF $\{F_{m+n}\}$ AND $\{L_{m+n}\}$

In 1972, R.T. Hansen of Montana State University also employed generating functions in his investigations of Fibonacci and Lucas numbers [256]. For example, the generating function of $\{F_{m+n}\}$ is given by

$$(\alpha - \beta) \sum_{n=0}^{\infty} F_{m+n} x^n = \sum_{n=0}^{\infty} (\alpha^{m+n} - \beta^{m+n}) x^n$$

$$= \alpha^m \sum_{n=0}^{\infty} \alpha^n x^n - \beta^m \sum_{n=0}^{\infty} \beta^n x^n$$

$$= \frac{\alpha^m}{1 - \alpha x} - \frac{\beta^m}{1 - \beta x}$$

$$\sum_{n=0}^{\infty} F_{m+n} x^n = \frac{(\alpha^m - \beta^m) + (\alpha^{m-1} - \beta^{m-1})x}{(\alpha - \beta)(1 - \alpha x)(1 - \beta x)}$$

$$= \frac{F_m + F_{m-1} x}{1 - x - x^2}. \tag{13.12}$$

We can show similarly that

$$\sum_{n=0}^{\infty} L_{m+n} x^n = \frac{L_m + L_{m-1} x}{1 - x - x^2}; \tag{13.13}$$

see Exercise 13.71.

IDENTITIES USING GENERATING FUNCTIONS

We can employ generating functions (13.12) and (13.13) to derive a host of identities. To this end, let $D = 1 - x - x^2$. Notice that

$$\sum_{n=0}^{\infty} F_{n+1} x^n = \frac{1}{D}, \qquad \sum_{n=0}^{\infty} F_{n-1} x^n = \frac{1-x}{D}, \qquad \text{and} \qquad \sum_{n=0}^{\infty} L_n x^n = \frac{2-x}{D}.$$

Since

$$\frac{2-x}{D} = \frac{1}{D} + \frac{1-x}{D},$$

it follows that

$$\sum_{n=0}^{\infty} L_n x^n = \sum_{n=0}^{\infty} F_{n+1} x^n + \sum_{n=0}^{\infty} F_{n-1} x^n$$

$$= \sum_{n=0}^{\infty} (F_{n+1} + F_{n-1}) x^n.$$

Consequently, $F_{n+1} + F_{n-1} = L_n$, a fact we learned in Chapter 5.

Next we will prove that $F_m L_n + F_{m-1} L_{n-1} = L_{m+n-1}$:

$$\sum_{m=0}^{\infty} (F_m L_n + F_{m-1} L_{n-1}) x^m = L_n \sum_{m=0}^{\infty} F_m x^m + L_{n-1} \sum_{m=0}^{\infty} F_{m-1} x^m$$

$$= L_n \frac{x}{D} + L_{n-1} \frac{1-x}{D}$$

$$= \frac{L_{n-1} + L_{n-2} x}{D}$$

$$= \sum_{m=0}^{\infty} L_{m+n-1} x^m.$$

Thus $F_m L_n + F_{m-1} L_{n-1} = L_{m+n-1}$, as desired.

We can show similarly that $F_m F_n + F_{m-1} F_{n-1} = F_{m+n-1}$ and $L_m L_n + L_{m-1} L_{n-1} = 5 F_{m+n-1}$; see Exercises 13.62 and 13.63.

13.7   COMPOSITIONS REVISITED

Recall from Chapter 4 that the number of compositions C_n of a positive integer n using 1s and 2s satisfies the recurrence $C_n = C_{n-1} + C_{n-2}$, where $C_1 = 1$, $C_2 = 2$, and $n \geq 3$. We also found that $C_n = F_{n+1}$, where $n \geq 1$. Using generating functions, we will now re-confirm this fact. Let $C(x) = C_1 x + C_2 x^2 + C_3 X^3 + \cdots + C_n x^n + \cdots$. Then

$$C(x) = C_1 x + C_2 x^2 + \sum_{n=3}^{\infty} C_n x^n$$

$$x C(x) = \qquad C_1 x^2 + \sum_{n=3}^{\infty} C_{n-1} x^n$$

$$x^2 C(x) = \qquad\qquad \sum_{n=3}^{\infty} C_{n-2} x^n$$

$$(1 - x - x^2) C(x) = x + x^2$$

$$C(x) = \frac{x + x^2}{1 - x - x^2}$$

$$= \sum_{n=0}^{\infty} F_n x^n + \sum_{n=0}^{\infty} F_n x^{n+1}$$

$$= \sum_{n=1}^{\infty} (F_n + F_{n-1}) x^n$$

$$= \sum_{n=1}^{\infty} F_{n+1} x^n.$$

Thus $C_n = F_{n+1}$, as expected.

Next we will develop generating functions for $F_n/n!$ and $L_n/n!$.

13.8   EXPONENTIAL GENERATING FUNCTIONS

Since $e^t = \sum_{n=0}^{\infty} (t^n/n!)$, it follows that

$$e^{\alpha x} = \sum_{n=0}^{\infty} \frac{\alpha^n x^n}{n!} \qquad \text{and} \qquad e^{\beta x} = \sum_{n=0}^{\infty} \frac{\beta^n x^n}{n!}.$$

Then

$$\frac{e^{\alpha x} - e^{\beta x}}{\alpha - \beta} = \sum_{n=0}^{\infty} \left(\frac{\alpha^n - \beta^n}{\alpha - \beta} \right) \frac{x^n}{n!}$$

$$= \sum_{n=0}^{\infty} \frac{F_n}{n!} x^n. \qquad (13.14)$$

Thus the exponential function $\dfrac{e^{\alpha x} - e^{\beta x}}{\alpha - \beta}$ generates the numbers $\dfrac{F_n}{n!}$.

More generally, we can show that

$$\frac{e^{\alpha^k x} - e^{\beta^k x}}{\alpha - \beta} = \sum_{n=0}^{\infty} \frac{F_{nk}}{n!} x^n.$$

Likewise,

$$e^{\alpha^k x} + e^{\beta^k x} = \sum_{n=0}^{\infty} \frac{L_{nk}}{n!} x^n.$$

We can employ the generating functions for $F_n/n!$ and $L_n/n!$ to derive two identities; they link F_n and L_n with the hyperbolic functions sinh and cosh, respectively.

To see this, it follows by equation (13.14) that

$$e^{x/2}(e^{\sqrt{5}x/2} - e^{-\sqrt{5}x/2} = \sqrt{5} \sum_{n=0}^{\infty} \frac{F_n}{n!} x^n$$

$$2e^{x/2} \sinh(\sqrt{5}x/2) = \sqrt{5} \sum_{n=0}^{\infty} \frac{F_n}{n!} x^n.$$

Thus

$$2e^x \sinh(\sqrt{5}x) = \sqrt{5} \sum_{n=0}^{\infty} \frac{2^n F_n}{n!} x^n. \tag{13.15}$$

The generating function $e^{\alpha x} + e^{\beta x} = \sum_{n=0}^{\infty} \frac{L_n}{n!} x^n$ can similarly be used to derive the formula

$$2e^x \cosh(\sqrt{5}x) = \sqrt{5} \sum_{n=0}^{\infty} \frac{2^n L_n}{n!} x^n. \tag{13.16}$$

Using the exponential generating functions for $F_n/n!$ and $L_n/n!$, we can also develop a host of combinatorial identities, as C.A. Church and Bicknell did in 1973 [124]. To see this, we let

$$A(t) = \sum_{n=0}^{\infty} a_n \frac{t^n}{n!} \quad \text{and} \quad B(t) = \sum_{n=0}^{\infty} b_n \frac{t^n}{n!}.$$

Then

$$A(t)B(t) = \sum_{n=0}^{\infty} \left[\sum_{k=0}^{n} \binom{n}{k} a_k b_{n-k} \right] \frac{t^n}{n!}; \tag{13.17}$$

$$A(t)B(-t) = \sum_{n=0}^{\infty} \left[\sum_{k=0}^{n} (-1)^{n-k} \binom{n}{k} a_k b_{n-k} \right] \frac{t^n}{n!}. \tag{13.18}$$

In particular, let $A(t) = (e^{\alpha t} - e^{\beta t})/(\alpha - \beta)$ and $B(t) = e^t$. Then, by equation (13.17),

$$\frac{e^t(e^{\alpha t} - e^{\beta t})}{\alpha - \beta} = \sum_{n=0}^{\infty} \left[\sum_{k=0}^{n} \binom{n}{k} F_k \right] \frac{t^n}{n!}$$

$$\frac{e^{(\alpha+1)t} - e^{(\beta+1)t}}{\alpha - \beta} = \sum_{n=0}^{\infty} \left[\sum_{k=0}^{n} \binom{n}{k} F_k \right] \frac{t^n}{n!}$$

$$\frac{e^{\alpha^2 t} - e^{\beta^2 t}}{\alpha - \beta} = \sum_{n=0}^{\infty} \left[\sum_{k=0}^{n} \binom{n}{k} F_k \right] \frac{t^n}{n!}$$

$$\sum_{n=0}^{\infty} F_{2n} \frac{t^n}{n!} = \sum_{n=0}^{\infty} \left[\sum_{k=0}^{n} \binom{n}{k} F_k \right] \frac{t^n}{n!}.$$

$$\tag{13.19}$$

Equating the coefficients of $\dfrac{t^n}{n!}$ yields the combinatorial identity

$$\sum_{k=0}^{n} \binom{n}{k} F_k = F_{2n}.$$

Using $B(t) = e^{-t}$ and equation (13.18), it follows similarly that

$$\sum_{n=0}^{\infty} \left[\sum_{k=0}^{n} (-1)^{n-k} \binom{n}{k} F_k \right] \frac{t^n}{n!} = \frac{e^{(\alpha-1)t} - e^{(\beta-1)t}}{\alpha - \beta}$$

$$= \frac{e^{-\beta t} - e^{-\alpha t}}{\alpha - \beta}$$

$$= \sum_{n=0}^{\infty} (-1)^{n-1} F_n \frac{t^n}{n!}.$$

This yields the identity

$$\sum_{k=0}^{n} (-1)^{k+1} \binom{n}{k} F_k = F_n. \tag{13.20}$$

Obviously, by selecting $A(t)$ and $B(t)$ as suitable exponential functions, we can apply this technique to derive an array of Fibonacci and Lucas identities.

For example, by choosing

$$A(t) = \frac{e^{\alpha^2 t} - e^{\beta^2 t}}{\alpha - \beta} \qquad \text{and} \qquad B(t) = e^{-t},$$

we can show that

$$\sum_{k=0}^{n} (-1)^{n-k} \binom{n}{k} F_{2k} = F_n;$$

see Exercise 13.64.

13.9   HYBRID IDENTITIES

By choosing $A(t)$ and $B(t)$ strategically, we can develop a family of identities containing both Fibonacci and Lucas numbers.

To this end, we let

$$A(t) = \frac{e^{\alpha t} - e^{\beta t}}{\alpha - \beta} \qquad \text{and} \qquad B(t) = e^{\alpha t} + e^{\beta t}.$$

By equation (13.17), we have

$$\sum_{n=0}^{\infty} \left[\sum_{k=0}^{n} \binom{n}{k} F_k L_{n-k} \right] \frac{t^n}{n!} = \frac{e^{2\alpha t} - e^{2\beta t}}{\alpha - \beta}$$

$$= \sum_{n=0}^{\infty} 2^n F_n \frac{t^n}{n!}.$$

This yields the combinatorial identity

$$\sum_{k=0}^{n} \binom{n}{k} F_k L_{n-k} = 2^n F_n. \tag{13.21}$$

Likewise, we can show that

$$\sum_{k=0}^{n} \binom{n}{k} F_k F_{n-k} = \frac{2^n L_n - 2}{5} \tag{13.22}$$

$$\sum_{k=0}^{n} \binom{n}{k} L_k L_{n-k} = 2^n L_n + 2; \tag{13.23}$$

see Exercises 13.65 and 13.66.

In fact, identities (13.21)–(13.23) can be generalized:

$$\sum_{k=0}^{n} \binom{n}{k} F_{mk} L_{mn-mk} = 2^n F_{mn} \tag{13.24}$$

$$\sum_{k=0}^{n} \binom{n}{k} F_{mk} F_{mn-mk} = \frac{2^n L_{mn} - 2L_m^n}{5} \tag{13.25}$$

$$\sum_{k=0}^{n} \binom{n}{k} L_{mk} L_{mn-mk} = 2^n L_{mn} + 2L_m^n; \tag{13.26}$$

see Exercises 13.79–13.81.

13.10   IDENTITIES USING THE DIFFERENTIAL OPERATOR

We can realize more generalized families of identities using the differential operator d/dt. Since $A(t) = \sum_{n=0}^{\infty} a_n \frac{t^n}{n!}$, it follows that

$$\frac{d^r}{dt^r} A(t) = \sum_{n=0}^{\infty} a_{n+r} \frac{t^n}{n!}.$$

This fact will come in handy in our pursuit.

Let $A(t) = \dfrac{e^{\alpha t} - e^{\beta t}}{\alpha - \beta}$ and $B(t) = e^t$. By equation (13.17), we then have

$$\sum_{n=0}^{\infty} \left[\sum_{k=0}^{n} \binom{n}{k} F_{k+r} \right] \frac{t^n}{n!} = e^t \frac{d^r}{dt^r} \left(\frac{e^{\alpha t} - e^{\beta t}}{\alpha - \beta} \right)$$

$$= \frac{\alpha^r e^{(\alpha+1)t} - \beta^r e^{(\beta+1)t}}{\alpha - \beta}$$

$$= \frac{\alpha^r e^{\alpha^2 t} - \beta^r e^{\beta^2 t}}{\alpha - \beta}$$

$$= \sum_{n=0}^{\infty} F_{2n+r} \frac{t^n}{n!}.$$

This gives the identity

$$\sum_{k=0}^{n} \binom{n}{k} F_{k+r} = F_{2n+r}. \tag{13.27}$$

We can show similarly that

$$\sum_{k=0}^{n} \binom{n}{k} F_{4mk+r} = L_{2m}^{n} F_{2mn+4mr} \tag{13.28}$$

$$\sum_{k=0}^{n} \binom{n}{k} F_{m-1}^{n-k} F_{m}^{k} F_{k} = F_{mn} \tag{13.29}$$

$$\sum_{k=0}^{n} \binom{n}{k} F_{m-1}^{n-k} F_{m}^{k} F_{k+rm} = F_{mn+rm}; \tag{13.30}$$

see Exercises 13.82–13.84.

EXERCISES 13

Determine if each is a LHRWCCs.
1. $L_n = L_{n-1} + L_{n-2}$.
2. $D_n = n D_{n-1} + (-1)^n$.
3. $a_n = 1.08 a_{n-1}$.
4. $b_n = 2b_{n-1} + 1$.
5. $a_n = a_{n-1} + n$.
6. $a_n = 2a_{n-1} + 2^n - 1$.

7. $a_n = a_{n-1} + 2a_{n-2} + 3a_{n-5}$.

8. $a_n = a_{n-1} + 2a_{n-3} + n^2$.

Solve each LHRWCCs.

9. $a_n = a_{n-1} + 2a_{n-2}; a_0 = 3, a_1 = 0$.

10. $a_n = 5a_{n-1} - 6a_{n-2}; a_0 = 4, a_1 = 7$.

11. $a_n = a_{n-1} + 6a_{n-2}; a_0 = 5, a_1 = 0$.

12. $a_n = 4a_{n-2}; a_0 = 2, a_2 = -8$.

13. $a_n = a_{n-1} + a_{n-2}; a_0 = 1, a_1 = 2$.

14. $a_n = a_{n-1} + a_{n-2}; a_0 = 2, a_1 = 3$.

15. $L_n = L_{n-1} + L_{n-2}; L_1 = 1, L_2 = 3$.

16. $D_n = 2D_{n-1} + F_{2n-1}; D_1 = 1$ (Hoggatt, 1971 [289]).

17. $a_{n+2} = a_{n+1} + a_n - 1; a_1 = 1, a_2 = 3$ (Hunter, 1969 [341]).

18. $a_{n+2} = a_{n+1} + a_n - 1; a_1 = 5, a_2 = 3$.

19. $a_{n+2} = a_{n+1} + a_n + m^n$, where a_0 and a_1 are arbitrary (Hunter, 1966 [338]).

20. $a_{n+3} - 17a_{n+2} - 17a_{n+1} + a_n = 0; a_0 = 0, a_1 = 4, a_2 = 64$.

21. $a_{n+3} = 2a_{n+2} + 2a_{n+1} - a_n$, where $a_0 = 4, a_1 = 2, a_2 = 10$, and $n \geq 0$ (Beasley, 2000 [31]).

22. $b_{n+3} = 2b_{n+2} + 2b_{n+1} - b_n$, where $b_1 = 0, b_2 = 8, b_3 = 12$, and $n \geq 1$.

Find a recurrence satisfied by each.

23. F_n^2.

24. L_n^2.

25. $L_n^2 + F_n^2$.

26. $L_n^2 - F_n^2$.

Express each quotient as a sum of partial fractions.

27. $\dfrac{x+7}{(x-1)(x+3)}$.

28. $\dfrac{4x^2 - 3x - 25}{(x+1)(x-2)(x+3)}$.

29. $\dfrac{5}{1 - x - 6x^2}$.

30. $\dfrac{2 + 4x}{1 + 8x + 15x^2}$.

31. $\dfrac{x(x+2)}{(2+3x)(x^2+1)}$.

32. $\dfrac{-2x^2 - 2x + 2}{(x-1)(x^2+2x)}$.

33. $\dfrac{x^3 + x^2 + x + 3}{x^4 + 5x^2 + 6}$.

34. $\dfrac{-x^3 + 2x^2 + x}{x^4 + x^3 + x + 1}$.

35. $\dfrac{3x^3 - x^2 + 4x}{x^4 - x^3 + 2x^2 - x + 1}$.

36. $\dfrac{x^3 + x^2 + 5x - 2}{x^4 - x^2 + x - 1}$.

Using generating functions, solve each LHRWCCs.

37. $a_n = 2a_{n-1}, a_0 = 1$.

38. $a_n = a_{n-1} + 1, a_1 = 1$.

39. $a_n = a_{n-1} + 2, a_1 = 1$.

40. $a_n = a_{n-1} + 2a_{n-2}, a_0 = 3, a_1 = 0$.

41. $a_n = 4a_{n-2}, a_0 = 2, a_1 = -8$.

42. $a_n = a_{n-1} + 6a_{n-2}, a_0 = 5, a_1 = 0$.

43. $a_n = 5a_{n-1} - 6a_{n-2}, a_0 = 4, a_1 = 7$.

44. $a_n = a_{n-1} + a_{n-2}, a_0 = 1, a_1 = 2$.

45. $a_n = a_{n-1} + a_{n-2}, a_0 = 2, a_1 = 3$.

46. $L_n = L_{n-1} + L_{n-2}, L_1 = 1, L_2 = 3$.

47. $a_n = 4a_{n-1} - 4a_{n-2}, a_0 = 3, a_1 = 10$.

48. $a_n = 6a_{n-1} - 9a_{n-2}, a_0 = 2, a_1 = 3$.

49. $a_n = 3a_{n-1} + 4a_{n-2} - 12a_{n-3}, a_0 = 3, a_1 = -7, a_2 = 7$.

50. $a_n = 8a_{n-1} - 21a_{n-2} + 18a_{n-3}, a_0 = 0, a_1 = 2, a_2 = 13$.

51. $a_n = 7a_{n-1} + 16a_{n-2} - 12a_{n-3}, a_0 = 0, a_1 = 5, a_2 = 19$.

52. $a_n = 6a_{n-1} - 12a_{n-2} - 8a_{n-3}, a_0 = 0, a_1 = 2, a_2 = -2$.

53. $a_n = 13a_{n-2} - 36a_{n-4}, a_0 = 7, a_1 = -6, a_2 = 38, a_3 = -212/3$.

54. $a_n = -a_{n-1} + 3a_{n-2} + 5a_{n-3} + 2a_{n-4}, a_0 = 0, a_1 = -8, a_2 = 4, a_3 = -42$.

Using Example 13.10, prove Exercises 13.55 and 13.56.

55. $a_n + ca_{n+1} = a_{n+2}$.

56. $a_{n+1}^2 - a_{n-1}^2 = ca_{2n+1}$.

57. Let $f(x) = \displaystyle\sum_{n=0}^{\infty} \dfrac{F_n}{n!} x^n$. Show that $f(x) = -e^x f(-x)$ (Lehmer, 1936 [385]).

58. Show that $e^{\alpha x} + e^{\beta x} = \displaystyle\sum_{n=0}^{\infty} \dfrac{L_n}{n!} x^n$.

59. Let $g(x) = \displaystyle\sum_{n=0}^{\infty} \dfrac{L_n}{n!} x^n$. Show that $g(x) = -e^x g(-x)$.

Verify each.

60. $\dfrac{x}{1 + x - x^2} = \sum_{n=0}^{\infty} (-1)^{n+1} F_n x^n.$

61. $\dfrac{x + 2x^2}{(1 - x - x^2)^2} = \sum_{n=0}^{\infty} n F_{n+1} x^n$ (Grassl, 1974 [245]).

Using generating functions, prove each.

62. $F_m F_n + F_{m-1} F_{n-1} = F_{m+n-1}$ (Hansen, 1972 [256]).

63. $L_m L_n + L_{m-1} L_{n-1} = 5 F_{m+n-1}$ (Hansen, 1972 [256]).

64. $\displaystyle\sum_{k=0}^{n} (-1)^{n-k} \binom{n}{k} F_{2k} = F_n.$

65. $\displaystyle\sum_{k=0}^{n} \binom{n}{k} F_k F_{n-k} = \dfrac{2^n L_n - 2}{5}.$

66. $\displaystyle\sum_{k=0}^{n} \binom{n}{k} L_k L_{n-k} = 2^n L_n + 2.$

For Exercises 13.67–13.70, use the function $A_n(x) = \sum_{i=2}^{n} F_i x^i$ (Lind, 1967 [404]).

67. Show that $A_n(x) = \dfrac{F_n x^{n+2} + F_{n+1} x^{n+1} - x}{1 - x - x^2}.$

68. Deduce the value of $\displaystyle\sum_{i=1}^{n} F_i.$

69. Derive a formula for $B(x) = \displaystyle\sum_{i=1}^{\infty} \dfrac{A_n(x)}{n!}.$

70. Deduce the value of $B(1)$.

71. Show that $\displaystyle\sum_{m=0}^{\infty} L_{m+n} x^m = \dfrac{L_n + L_{n-1} x}{1 - x - x^2}$ (Hansen, 1972 [256]).

Let $C_{n+2} = C_{n+1} + C_n + F_{n+2}$, where $C_1 = 1$ and $C_2 = 2$. Verify each (Hoggatt, 1964 [270]).

72. $C_n = \displaystyle\sum_{i=1}^{n-1} F_i F_{n-i}.$

73. $C_{n+1} = \displaystyle\sum_{i=1}^{\lfloor n/2 \rfloor} (n - i + 1) \binom{n - i}{i}.$

74. $C_n = \dfrac{n L_{n+1} + 2 F_n}{5}$, where $n \geq 1$.

Using exponential generating functions, prove each (Church and Bicknell, 1973 [124]).

75. $\displaystyle\sum_{k=0}^{n}\binom{n}{k}L_k = L_{2n}.$

76. $\displaystyle\sum_{k=0}^{n}(-1)^{k+1}\binom{n}{k}L_k = L_n.$

77. $\displaystyle\sum_{k=0}^{n}(-1)^{n-k}\binom{n}{k}F_{2k} = F_n.$

78. $\displaystyle\sum_{k=0}^{n}(-1)^{n-k}\binom{n}{k}L_{2k} = L_n.$

79. $\displaystyle\sum_{k=0}^{n}\binom{n}{k}F_{mk}L_{mn-mk} = 2^n F_{mn}.$

80. $\displaystyle\sum_{k=0}^{n}\binom{n}{k}F_{mk}F_{mn-mk} = \frac{2^n L_{mn} - 2L_m^n}{5}.$

81. $\displaystyle\sum_{k=0}^{n}\binom{n}{k}L_{mk}L_{mn-mk} = 2^n L_{mn} + 2L_m^n.$

82. $\displaystyle\sum_{k=0}^{n}\binom{n}{k}F_{4mk+r} = L_{2m}^n F_{2mn+4mr}.$

83. $\displaystyle\sum_{k=0}^{n}\binom{n}{k}F_{m-1}^{n-k}F_m^k F_k = F_{mn}.$

84. $\displaystyle\sum_{k=0}^{n}\binom{n}{k}F_{m-1}^{n-k}F_m^k F_{k+rm} = F_{mn+rm}.$

85. Show that $\displaystyle\sum_{k=0}^{\infty}\frac{F_k}{2^k} = 2$ (Lind, 1968 [407]).

COMBINATORIAL MODELS I

> Nothing ever becomes real till it is experienced; even a
> proverb is no proverb to you till your life has illustrated it.
> —John Keats

In Chapters 4 and 5, we studied several interesting applications of the Fibonacci
and Lucas family to combinatorics, including the theory of partitioning. In this
chapter, we will present additional applications to combinatorics.

In Section 4.2, we briefly studied compositions of positive integers n with summands 1 and 2. We found that the number of distinct compositions C_n of n is F_{n+1};
see Table 14.1.

TABLE 14.1.

n	Compositions	C_n
1	1	1
2	$1 + 1, 2$	2
3	$1 + 1 + 1, 1 + 2, 2 + 1$	3
4	$1 + 1 + 1 + 1, 1 + 1 + 2, 1 + 2 + 1, 2 + 1 + 1, 2 + 2$	5
5	$1 + 1 + 1 + 1 + 1, 1 + 1 + 1 + 2, 1 + 1 + 2 + 1, 1 + 2 + 1 + 1,$ $2 + 1 + 1 + 1, 1 + 2 + 2, 2 + 1 + 2, 2 + 2 + 1$	8

$$\uparrow$$
$$F_{n+1}$$

Fibonacci and Lucas Numbers with Applications, Volume One, Second Edition. Thomas Koshy.
© 2018 John Wiley & Sons, Inc. Published 2018 by John Wiley & Sons, Inc.

14.1   A FIBONACCI TILING MODEL

Theorem 4.1 has a spectacular combinatorial interpretation. To see this, suppose we would like to tile a $1 \times n$ board (an array of n unit squares) with 1×1 tiles (unit squares) and 1×2 tiles (dominoes). A tiling of a $1 \times n$ board is a tiling of *length n* or an *n-tiling*.

Figure 14.1 shows the possible tilings of a $1 \times n$ board, where $1 \leq n \leq 5$. It appears from the figure that the number of n-tilings T_n is F_{n+1}, where $n \geq 1$.

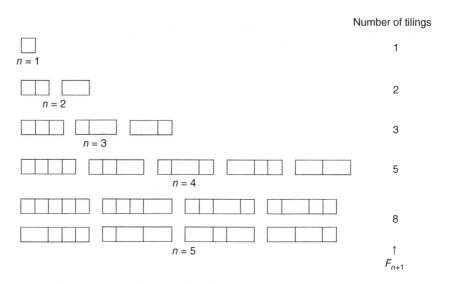

Figure 14.1.   Tilings of a $1 \times n$ board.

Let T be an arbitrary n-tiling. Suppose it ends in a domino: subtiling $\underbrace{\qquad}_{\text{length } n-2}$.

By definition, there are T_{n-2} such n-tilings.

On the other hand, suppose T ends in a square: subtiling $\underbrace{\square}_{\text{length } n-1}$. There are T_{n-1} such n-tilings.

These two cases are mutually exclusive; so, by the addition principle, $T_n = T_{n-1} + T_{n-2}$, where $T_1 = 1$ and $T_2 = 2$. Thus $T_n = F_{n+1}$, as conjectured.

Thus we have the following theorem.

Theorem 14.1.   *The number of tilings of a $1 \times n$ board with square tiles and dominoes is F_{n+1}, where $n \geq 1$.*   ∎

The tilings in Figure 14.1 are pictorial representations of the compositions in Table 14.1. Clearly, the process is reversible. Thus there is a bijection between the set of compositions of n and the set of n-tilings.

We can use the Fibonacci tiling model to construct combinatorial proofs of a number of Fibonacci identities. We will establish a few to illustrate the beauty and power of this delightful technique [35].

The essence of the combinatorial approach lies in the *Fubini principle* from Section 5.3. So, in our proofs, we will count the same objects (tilings, in our case) in two different ways, and then equate the two counts to get the desired result.

We begin with the summation formula (5.1).

Example 14.1. *Prove combinatorially that* $\sum_{k=1}^{n} F_k = F_{n+2} - 1$.

Proof. The RHS of the formula indirectly indicates the size of the board we must choose. By virtue of Theorem 14.1, it must be a $1 \times (n+1)$ board. By the theorem, we can tile it with squares and dominoes in F_{n+2} different ways. But exactly one of them consists of squares. So there are $F_{n+2} - 1$ tilings containing at least one domino.

We will now count, in a different way, the number of tilings of the board containing at least one domino. Suppose the last domino appears in cells k and $k+1$:

subtiling ⬚ all squares.

length $k-1$ $k\ k+1$ length $n-k$

There are F_k tilings of length $k-1$; so there are F_k tilings of length $n+1$ with the last domino covering cells k and $k+1$. Since $1 \leq k \leq n$, it follows by the addition principle that the total number of $(n+1)$-tilings containing at least one domino equals $\sum_{k=1}^{n} F_k$.

Equating the two counts, we get the given summation formula. ∎

For a specific example, let $n = 4$. There are $8 = F_6$ tilings of a 1×5 board; see Figure 14.2.

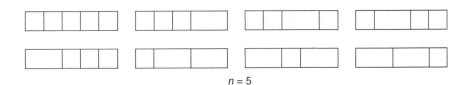

$n = 5$

Figure 14.2. 5-Tilings.

Seven of them contain at least one domino. The last domino in such tilings occupies cells k and $k+1$, where $1 \leq k \leq 4$; see Figure 14.3. There is one 5-tiling with $k = 1$, one with $k = 2$, two with $k = 3$, and three with $k = 4$. So there is total of $1 + 1 + 2 + 3 = 7$ 5-tilings with at least one domino each.

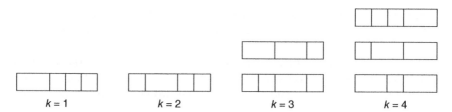

Figure 14.3. 5-Tilings with one or more dominoes.

Similarly, by considering the location of the last square of a tiling of a $1 \times 2n$ board, we can prove that

$$\sum_{k=1}^{n} F_{2k} = F_{2n+1} - 1;$$

see Exercise 14.1.

Next we confirm Lucas's formula for F_{n+1}.

Example 14.2.   *Prove combinatorially that* $F_{n+1} = \displaystyle\sum_{k=0}^{\lfloor n/2 \rfloor} \binom{n-k}{k}.$

Proof. Recall that there are F_{n+1} tilings of a $1 \times n$ board.

To count them in a different way, we will focus on the number of dominoes in a tiling. Suppose there are exactly k dominoes in an arbitrary tiling. They cover $2k$ cells and there are $n - 2k$ square tiles remaining. So there is a total of $n - k$ tiles in the tiling. Consequently, the k dominoes can be placed in the $n - k$ tiling positions in $\binom{n-k}{k}$ different ways. Thus there are $\binom{n-k}{k}$ tilings, each with exactly k dominoes. Since $0 \le k \le n/2$, the total number of tilings equals $\displaystyle\sum_{k=0}^{\lfloor n/2 \rfloor} \binom{n-k}{k}$.

Equating the two counts yields the desired result. ∎

Again, we will study a specific case. Let $n = 5$; see Figure 14.1. There is one tiling with $k = 0$ dominoes; four with $k = 1$; and three with $k = 2$. So the total number of 5-tilings equals

$$1 + 4 + 3 = \sum_{k=0}^{\lfloor 5/2 \rfloor} \binom{5-k}{k} = F_6.$$

Next we will establish the *addition formula* $F_{m+n} = F_{m+1}F_n + F_m F_{n-1}$; see Corollary 20.2 for an algebraic proof.

To accomplish this, we introduce the concept of *breakability*. A tiling is *unbreakable* at cell k if a domino occupies cells k and $k + 1$; see Figure 14.4. If a tiling is not unbreakable at cell k, then it is *breakable* at cell k; see Figure 14.5. Thus a tiling is breakable at cell k if and only if it can be partitioned into two subtilings, one of length k and the other of length $n - k$.

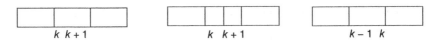

Figure 14.4.   Tiling unbreakable at cell k.   **Figure 14.5.**   Tilings breakable at cell k.

With this tool at our disposal, we can now prove the addition formula.

Example 14.3.   *Prove combinatorially the addition formula $F_{m+n} = F_{m+1}F_n + F_m F_{n-1}$.*

Proof.   Consider a board of length $m + n - 1$. There are F_{m+n} tilings of the board.

To count them differently, we will pivot on the breakability of an arbitrary tiling T at cell m.

Case 1.   Suppose tiling T is breakable at cell m: subtiling subtiling. There are $F_{m+1}F_n$

such tilings.

$$\underbrace{}_{\text{length } m} \underbrace{}_{\text{length } n-1}$$

Case 2.   Suppose tiling T is not breakable at cell m: subtiling ☐ subtiling.

There are $F_m F_{n-1}$ such tilings.

$$\underbrace{}_{\text{length } m-1} \overset{m \; m+1}{} \underbrace{}_{\text{length } n-2}$$

Combining the two cases, there are $F_{m+1}F_n + F_m F_{n-1}$ tilings. Thus $F_{m+n} = F_{m+1}F_n + F_m F_{n-1}$, as desired.   ∎

Although the Lucas identity $F_{2n+1} = F_{n+1}^2 + F_n^2$ follows from the addition formula, we will find it instructive to confirm it independently; see Exercise 14.3.

The next example features another Lucas formula.

Example 14.4.   *Prove combinatorially that $\sum_{k=0}^{n} \binom{n}{k} F_k = F_{2n}$.*

Proof.   Consider a $1 \times (2n - 1)$ board. It has F_{2n} tilings.

Since the length of the board is odd, a $(2n - 1)$-tiling must contain an odd number of squares. Clearly, a tiling must contain at least n tiles. We now focus on the number of square tiles among the first n tiles.

Suppose there are k squares and hence $n - k$ dominoes among the first n tiles. They account for a total of $k + 2(n - k) = 2n - k$ cells; they are then followed by a subtiling of length $k - 1$:

$$\underbrace{k \text{ squares and } n - k \text{ dominoes}}_{n \text{ tiles}} \underbrace{\text{subtiling}}_{\text{length } k-1}.$$

The k square tiles can be placed in the n tile positions in $\binom{n}{k}$ different ways. Since there are F_k different tilings of length $k - 1$, there are $\binom{n}{k} F_k$ tilings of the board with k squares among the first n tiles. Since $0 \le k \le n$, it follows that there are $\sum_{k=0}^{n} \binom{n}{k} F_k$ tilings of the board.

By equating the two counts, we get the desired formula.   ∎

To illuminate the beauty of this proof, consider a 1×5 board, so $n = 3$. Recall that there are $8 = F_6$ 5-tilings. Although $0 \le k \le 3$, there are *no* tilings with no squares among the first three tiles; so $1 \le k \le 3$; see Figure 14.1. There are three tilings with $k = 1$; see Figure 14.6. There are three tilings with $k = 2$; see Figure 14.7. There are two tilings with $k = 3$; see Figure 14.8.

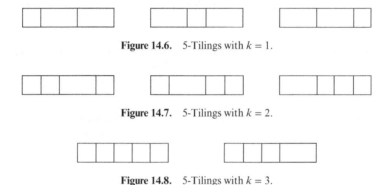

Figure 14.6.   5-Tilings with $k = 1$.

Figure 14.7.   5-Tilings with $k = 2$.

Figure 14.8.   5-Tilings with $k = 3$.

So the total number of 5-tilings equals

$$F_6 = 8 = 3 + 3 + 2 = \sum_{k=1}^{n} \binom{3}{k} F_k = \sum_{k=0}^{n} \binom{3}{k} F_k.$$

Next we will establish a beautiful combinatorial identity we have not yet discussed. Its charming proof is based on the concept of the *median square* [35] in a tiling of a board of odd length $2n + 1$. Since the length is odd, we are guaranteed that a $(2n + 1)$-tiling contains an odd number of squares. The square M that has an equal number of squares on either side is the *median square*. A median square exists in every $(2n + 1)$-tiling.

For example, the up arrows in Figure 14.9 indicate the median squares in the eight 5-tilings.

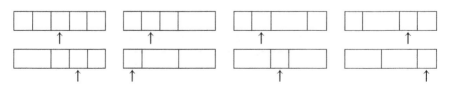

Figure 14.9.   5-Tilings.

Example 14.5. *Prove combinatorially that* $\displaystyle\sum_{\substack{i,j \geq 0 \\ i+j \leq n}} \binom{n-i}{j}\binom{n-j}{i} = F_{2n+2}.$

Proof. Consider a $1 \times (2n+1)$ board. It can be tiled in F_{2n+2} different ways.

This time, we focus on the median square M of an arbitrary tiling. We then pivot on the number of dominoes i to the left of M and the number of dominoes j to its right: $\underbrace{i \text{ dominoes}}_{\text{here}} \underset{\underset{M}{\uparrow}}{\square} \underbrace{j \text{ dominoes}}_{\text{here}}$.

Such a tiling has $i+j$ dominoes, so it contains $(2n+1) - 2(i+j)$ squares. Hence there are $n-i-j$ squares on either side of M. Since $i,j \geq 0$ and $n-i-j \geq 0$, $0 \leq i+j \leq n$. Consequently, there are $(n-i-j)+i = n-j$ tiles to the left of M; similarly, there are $n-i$ tiles to the right of M.

Since there are $n-j$ tiles to the left of M and i of them are dominoes, the i dominoes can be placed among the $n-j$ tile positions in $\binom{n-j}{i}$ different ways. Similarly, the j dominoes to the right of M can be placed among the $n-i$ tile positions in $\binom{n-i}{j}$ different ways. Consequently, there are $\binom{n-i}{j}\binom{n-j}{i}$ such $(2n+1)$-tilings.

Thus the total number of $(2n+1)$-tilings equals $\displaystyle\sum_{\substack{i,j \geq 0 \\ i+j \leq n}} \binom{n-i}{j}\binom{n-j}{i}$. This, coupled with the initial count, yields the given formula. ∎

Again, for a specific example, consider the 5-tilings in Figure 14.9 with $n = 2$. Then

$$\sum_{\substack{i,j \geq 0 \\ i+j \leq 2}} \binom{2-i}{j}\binom{2-j}{i} = 1+2+1+2+1+1 = 8 = F_6.$$

Next we present a circular tiling model for Lucas numbers.

14.2 A CIRCULAR TILING MODEL

In lieu of a linear board, consider a circular board with n cells; we number them 1 through n in a counterclockwise direction. Suppose we would like to tile the board with circular squares and dominoes. (Although such tiles are not available, we use the same terminology for consistency.) A tiling of a circular board is an *n-bracelet*. A bracelet is *out-of-phase* if a domino occupies cells n and 1; otherwise, it is *in-phase*.

Figure 14.10 shows the tilings of a circular board with n cells, where $1 \leq n \leq 4$. Notice that there are two 2-bracelets covered by a single domino, one out-of-phase and the other in-phase.

It appears from Figure 14.10 that the number of n-bracelets is L_n. The next theorem confirms this observation.

Number of tilings

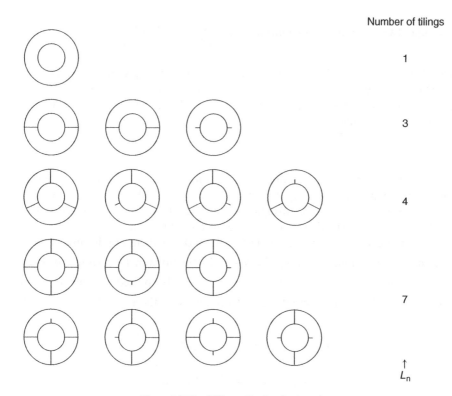

1

3

4

7

↑
L_n

Figure 14.10. Tilings of a circular board.

Theorem 14.2. *The number of n-bracelets is L_n, where $n \geq 1$.*

Proof. Let B_n denote the number of n-bracelets. It follows from Figure 14.10 that $B_1 = 1$ and $B_2 = 3$.

We will now show that B_n satisfies the Fibonacci recurrence, when $n \geq 3$. Consider an arbitrary n-bracelet, where $n \geq 3$; see Figure 14.11.

Figure 14.11.

Case 1. Suppose the bracelet begins with a square. Deleting the square yields an $(n - 1)$-bracelet. Since there are B_{n-1} $(n - 1)$-bracelets, it follows that there are B_{n-1} such n-bracelets.

Case 2. Suppose the bracelet begins with a domino. It occupies cells 1 and 2. We can use the remaining $n - 2$ cells to form an $(n - 2)$-bracelet. There are B_{n-2} $(n - 2)$-bracelets, so there are B_{n-2} such n-bracelets.

So the total number of n-bracelets equals $B_{n-1} + B_{n-2}$. Thus $B_n = B_{n-1} + B_{n-2}$. This, together with the initial conditions, implies that $B_n = L_n$, as conjectured. ∎

Using circular tilings, we now will establish the summation formula (12.5) for L_n.

Example 14.6. *Prove combinatorially that* $L_n = \sum_{k=0}^{\lfloor n/2 \rfloor} \dfrac{n}{n-k} \binom{n-k}{k}$.

Proof. Consider a circular board with n cells. By Theorem 14.2, there are L_n n-bracelets.

Consider an arbitrary n-bracelet. Suppose it contains exactly k dominoes. Then it contains $n - 2k$ square tiles.

Case 1. Suppose a domino occupies cells n and 1 of the bracelet; see Figure 14.12.

Then there are $(n - 2k) + k - 1 = n - k - 1$ tiles covering cells 2 through $n - 1$, of which $k - 1$ are tiles. So the remaining $k - 1$ dominoes can be placed in the $n - k - 1$ slots in $\binom{n-k-1}{k-1}$ different ways. Thus there are $\binom{n-k-1}{k-1}$ bracelets with a domino in cells n and 1.

Figure 14.12.

Case 2. Suppose a domino does *not* occupy cells n and 1; that is, the bracelet is breakable at cell n; see Figure 14.13. So we can straighten it and form an n-tiling with exactly k dominoes; see Figure 14.14.

Figure 14.13.   **Figure 14.14.**

Since the bracelet contains $n - 2k$ squares and k dominoes, it employs $(n - 2k) + k = n - k$ tiles. So the k dominoes can be placed in the $n - k$ tile positions in $\binom{n-k}{k}$ different ways. In other words, there are $\binom{n-k}{k}$ bracelets without a domino in cells n and 1.

By Cases 1 and 2, the total number of n-bracelets with k dominoes each equals

$$\binom{n-k-1}{k-1} + \binom{n-k}{k} = \frac{n}{n-k}\binom{n-k}{k}.$$

Since $0 \le k \le \lfloor n/2 \rfloor$, the total number of n-bracelets equals $\displaystyle\sum_{k=0}^{\lfloor n/2 \rfloor} \frac{n}{n-k}\binom{n-k}{k}$.

This, coupled with the earlier count, gives the desired formula. ∎

Again, for a specific example, consider the $L_3 = 4$ 3-bracelets in Figure 14.15. Here $0 \le k \le 1$. There is one bracelet with no dominoes, and three with one domino each. They account for all four 3-bracelets.

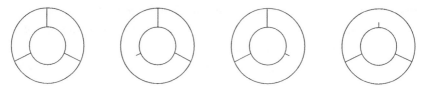

Figure 14.15.

Using the linear and circular models, we can now establish hybrid formulas, as the next two examples demonstrate.

Example 14.7.   *Confirm combinatorially that $L_n = F_{n+1} + F_{n-1}$.*

Proof. By Theorem 14.2, there are L_n ways of tiling a circular board with squares and dominoes.

To count those tilings in a different way, we will now consider two cases.

Case 1. Suppose a domino does *not* occupy cells n and 1. So the bracelet is breakable at cell n. We can then straighten it into an n-tiling; see Figure 14.16. There are F_{n+1} such n-bracelets.

Figure 14.16.

Case 2. Suppose a domino occupies cells n and 1. Now remove those two cells. We can then use the remaining $n - 2$ cells to form an $(n - 2)$-tiling; see Figure 14.17. Since there are F_{n-1} $(n - 2)$-tilings, it follows that there are F_{n-1} n-bracelets with a domino in cells n and 1.

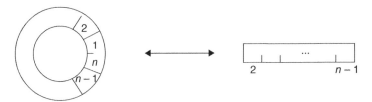

Figure 14.17.

Thus the total number of n-bracelets equals $F_{n+1} + F_{n-1}$. Combining the two counts yields the desired identity.   ■

Finally, we will establish the charming identity $F_{2n} = F_n L_n$. Since this identity also links Fibonacci and Lucas numbers, we need to design a clever technique. The LHS gives us a clue to the size of the linear board we should choose: a $1 \times (2n - 1)$ board. Recall that F_n denotes the number of $(n - 1)$-tilings T, and L_n the number of n-bracelets B. Since the RHS is the product of F_n and L_n, we will form pairs of tilings (T, B); and then exhibit a bijection between the set of tiling-pairs (T, B) and the set of $(2n - 1)$-tilings.

Example 14.8. *Establish the identity $F_{2n} = F_n L_n$.*

Proof. Consider a $1 \times (2n - 1)$ board. It can be tiled in F_{2n} different ways.

Let X be an arbitrary $(2n - 1)$-tiling. There are two cases we need to consider.

Case 1. Suppose X is breakable at cell $n - 1$: $X = \underbrace{\text{subtiling } A}_{\text{length } n-1} \underbrace{\text{subtiling } B}_{\text{length } n}$. Then

subtiling A is an $(n - 1)$-tiling. We can now glue the ends of subtiling B to form an n-bracelet; see Figure 14.18. Thus X determines a unique pair (A, B).

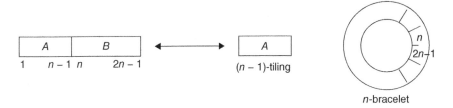

Figure 14.18.

Case 2. Suppose X is unbreakable at cell $n-1$: $X = \underbrace{\text{subtiling } C}_{\text{length } n-2} \;\square\; \underbrace{\text{subtiling } D}_{\text{length } n-1}$.

Then subtiling D is an $(n-1)$-tiling. We can use subtiling C, coupled with the domino in cells $n-1$ and n, to form an n-bracelet; see Figure 14.19. This way also X generates a unique pair of an $(n-1)$-tiling and an n-bracelet.

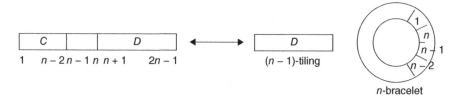

Figure 14.19.

Thus we can employ a $(2n-1)$-tiling X to construct a pair of tilings (T, B), where T is an $(n-1)$-tiling and B an n-bracelet. This constructive algorithm is clearly reversible. Consequently, the algorithm establishes a bijection between the set of $(2n-1)$-tilings and the set of tiling-pairs (T, B). Since there are $F_n L_n$ tiling-pairs, it follows that $F_{2n} = F_n L_n$, as desired.   ■

Next we show how path graphs can be used to extract interesting Fibonacci identities. Our technique continues to hinge on Fubini's principle.

14.3   PATH GRAPHS REVISITED

Recall that P_n denotes the path graph with n vertices and V_n its vertex set. Let $I(G)$ denote the family of independent subsets of the vertex set of graph G, and $i(G) = |I(G)|$. Clearly, $i(P_n) = F_{n+2}$.

In the following example, we will establish combinatorially a simple Fibonacci identity; there is beauty and elegance in the combinatorial proof, studied by J. DeMaio and J. Jacobson of Kennesaw State University, Atlanta, Georgia [140]. Its algebraic proof takes only a few seconds; see Exercise 14.12.

Example 14.9.   *Let $n \geq 3$. Then $F_{2n} = 2F_{n-1}F_n + F_n^2$.*

Proof.   Consider the path graph P_{2n-2} with vertices 1 through $2n-2$. Clearly, $i(P_{2n-2}) = F_{2n}$.

We will now compute this count in a different way. We will partition the set $I(P_{2n-2})$ into three pairwise disjoint independent subsets, X, Y, and Z, where

1) X consists of subsets that do *not* contain the vertex $n-1$ or n;
2) Y consists of subsets that contain the vertex n; and
3) Z consists of subsets that contain the vertex $n-1$;

see Figures 14.20, 14.21, and 14.22, respectively.

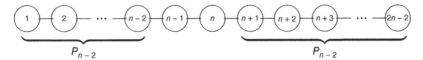

Figure 14.20.

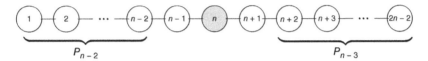

Figure 14.21.

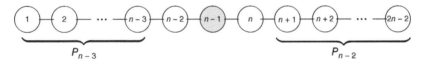

Figure 14.22.

Consider the independent subsets A in X. Since $n-1, n \notin A$, X consists of the independent subsets of the path from 1 through $n-2$ and the path from $n+1$ through $2n-2$. By the multiplication principle, there are $F_n F_n = F_n^2$ such independent subsets; that is, $|X| = F_n^2$.

Now consider the subsets B in Y. Since $n \in B$, $n-1$, $n+1 \notin B$. Consequently, Y consists of the elements of independent subsets from the path 1 through $n-2$ and the path from $n+2$ through $2n-2$. Again, by the multiplication principle, $|Y| = F_n F_{n-1}$. Similarly, $|Z| = F_n F_{n-1}$.

Since these are mutually exclusive cases, it follows by the addition principle that $i(P_{2n-2}) = |X| + |Y| + |Z| = 2F_n F_{n-1} + F_n^2$. Thus $F_{2n} = 2F_n F_{n-1} + F_n^2$, as desired. ∎

Next we will establish combinatorially a charming cubic Fibonacci identity; we will leave its algebraic proof as an exercise; see Exercise 14.13.

Theorem 14.3 (DeMaio and Jacobson, 2013 [140]).   *Let $n \geq 4$. Then*

$$F_{3n+2} = F_{n+2}^3 - 2F_n^2 F_{n+2} + F_n^2 F_{n-2}.$$

Proof.   Consider the path graph P_{3n} with vertices 1 through $3n$. Clearly, $i(P_{3n}) = F_{3n+2}$. We will now compute this count in a different way.

Consider the paths from 1 through n, $n + 1$ through $2n$, and $2n + 1$ through $3n$; see Figure 14.23. Each contains exactly n vertices. Since $i(P_n) = F_{n+2}$, the number of *possible* independent subsets of V_{3n} is $F_{n+2}F_{n+2}F_{n+2} = F_{n+2}^3$.

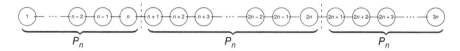

Figure 14.23.

But this count includes *unacceptable* subsets. Two cases lead to such unacceptable subsets, namely, those that contain vertex pair n and $n + 1$, or the pair $2n$ and $2n + 1$; so we must discount them.

Consider the subsets of V_{3n} that include vertices n and $n + 1$; see Figure 14.24. There are $F_nF_nF_{n+2} = F_n^2F_{n+2}$ such subsets.

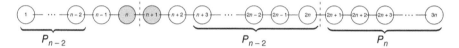

Figure 14.24.

There are $F_nF_nF_{n+2} = F_n^2F_{n+2}$ subsets of V_{3n} that include vertices $2n$ and $2n + 1$; see Figure 14.25.

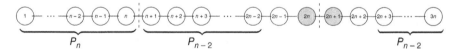

Figure 14.25.

Thus there are $2F_n^2F_{n+2}$ subsets of V_{3n} that contain the vertex pair n and $n + 1$, or $2n$ and $2n + 1$.

This sum counts twice the subsets that contain the vertices $n, n + 1, 2n$, and $2n + 1$; see Figure 14.26. There are $F_nF_{n-2}F_n = F_n^2F_{n-2}$ such subsets.

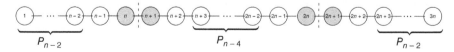

Figure 14.26.

Thus, by the inclusion–exclusion principle, $i(P_{3n}) = F_{n+2}^3 - 2F_n^2 F_{n+2} + F_n^2 F_{n-2}$. This, combined with the original count F_{3n+2}, yields the given result. ∎

Using the same technique, we can establish that [140]

$$F_{4n+2} = F_{n+2}^4 - 3F_n^2 F_{n+2}^2 + 2F_n^2 F_{n-2} F_{n+2} + F_n^4 - F_n^2 F_{n-2}^2;$$

see Exercise 14.14.

14.4 CYCLE GRAPHS REVISITED

Recall that there are $L_n = i(C_n)$ independent subsets of the vertex set V_n of the cycle graph C_n, where $n \geq 3$. The technique employed in Theorem 14.3 can be successfully applied to extract interesting identities linking Fibonacci and Lucas numbers, as the next theorem demonstrates [141].

Theorem 14.4 (DeMaio and Jacobson, 2014 [141]).   *Let $n \geq 4$. Then*

$$L_{3n} = F_{n+2}^3 - 3F_n^2 F_{n+2} + 3F_n^2 F_{n-2} - F_{n-2}^3.$$

Proof.   Consider the cycle graph C_{3n} in Figure 14.27. Clearly, $i(C_{3n}) = L_{3n}$. We will now compute this in a different way.

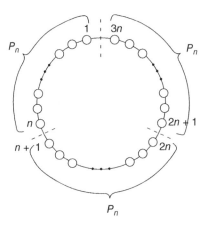

Figure 14.27.

Vertices 1 through n, $n + 1$ through $2n$, and $2n + 1$ through $3n$ form three paths, each with exactly n vertices. Since $i(P_n) = F_{n+2}$, the paths account for F_{n+2}^3 potential independent subsets of the vertex set of C_{3n}.

But *not* all of them are independent. So we must discount the unacceptable ones; such subsets contain the vertex pair $3n$ and 1, n and $n + 1$, or $2n$ and $2n + 1$.

It follows from Figure 14.28 that are $F_n F_{n+2} F_n = F_n^2 F_{n+2}$ subsets that contain the vertex pair $3n$ and 1. There are two such additional ones; they contribute $2F_n^2 F_{n+2}$ unacceptable subsets. So there are $3F_n^2 F_{n+2}$ such unacceptable subsets.

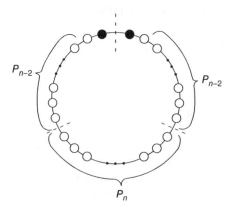

Figure 14.28.

There are other occurrences of unacceptable subsets. For example, they can contain vertex pairs $3n$ and 1, and $2n$ and $2n+1$; see Figure 14.29. There are $F_{n-2} F_n F_n = F_n^2 F_{n-2}$ such unacceptable subsets. Here also there are two other such cases, and they additionally produce $2F_n^2 F_{n-2}$ such subsets. So there are $3F_n^2 F_{n-2}$ such unacceptable subsets.

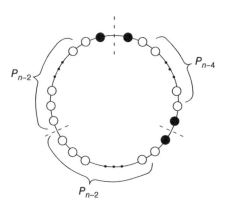

Figure 14.29.

We must now count the number of unacceptable subsets that contain the vertex pairs $3n$ and 1, $2n$ and $2n+1$, and n and $n+1$; see Figure 14.30. There are $F_{n-2} F_{n-2} F_{n-2} = F_{n-2}^3$ such unacceptable subsets.

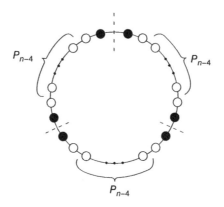

Figure 14.30.

Thus, by the inclusion–exclusion principle, $i(C_{3n}) = F_{n+2}^3 - 3F_n^2 F_{n+2} + 3F_n^2 F_{n-2} - F_{n-2}^3$. This, together with the initial count, gives the desired result. ∎

Using a similar technique, we can show that

$$L_{4n} = F_{n+2}^4 - 4F_n^2 F_{n+2}^2 + 4F_n^2 F_{n-2} F_{n+2} + 2F_n^4 - 4F_n^2 F_{n-2}^2 + F_{n-2}^4,$$

where $n \geq 4$ [140].

Next we study a new family of graphs, called tadpole graphs. They are obtained by combining cycle and path graphs, and were investigated by DeMaio and Jacobson [141].

14.5 TADPOLE GRAPHS

The *tadpole graph* $T_{n,k}$ is obtained by merging the cycle graph C_n and the path graph P_k in such a way that a vertex of C_n is adjacent to a terminal vertex of P_k, where $n \geq 3$ and $k \geq 0$. In the interest of clarity and convenience, we will let vertex c_1 be adjacent to vertex p_1, where $\{c_1, c_2, \ldots, c_n\}$ denotes the vertex set of C_n and $\{p_1, p_2, \ldots, p_k\}$ denotes the vertex set of P_k. Let $V_{n,k}$ denote the vertex set of the tadpole graph $T_{n,k}$.

Figure 14.31 shows the tadpole graph $T_{n,k}$.

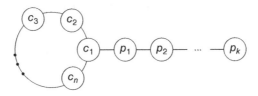

Figure 14.31. Tadpole graph $T_{n,k}$.

Knowing that $i(P_n) = F_{n+2}$, and that $i(C_n) = i(T_{n,0}) = L_n$ from Example 5.7, can we compute $i(T_{n,k})$? To answer this, first consider $T_{3,1}, T_{3,2}$, and $T_{4,1}$; see Figures 14.32 and 14.33.

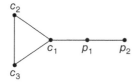

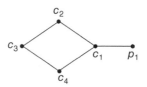

Figure 14.32. Tadpole graph $T_{3,2}$.   **Figure 14.33.** Tadpole graph $T_{4,1}$.

$V_{3,1}$ has 7 independent subsets: $\emptyset, \{c_1\}, \{c_2\}, \{c_3\}, \{p_1\}, \{c_2,p_1\}, \{c_3,p_1\}$; $V_{3,2}$ has 11 independent subsets: $\emptyset, \{c_1\}, \{c_2\}, \{c_3\}, \{p_1\}, \{p_2\}, \{c_1,p_2\}, \{c_2,p_1\}$, $\{c_2,p_2\}, \{c_3,p_1\}, \{c_3,p_2\}$; and $V_{4,1}$ has 12 independent subsets: $\emptyset, \{c_1\}, \{c_2\}, \{c_3\}$, $\{c_4\}, \{p_1\}, \{c_1,c_3\}, \{c_2,p_1\}, \{c_2,p_4\}, \{c_3,c_1\}, \{c_4,p_1\}, \{c_2,c_4,p_1\}$.

In our search for an explicit formula for $i(T_{n,k})$, the next theorem gives a second-order recurrence satisfied by $i(T_{n,k})$.

Theorem 14.5 (DeMaio and Jacobson, 2014 [141]).   *Let $n \geq 3$ and $k \geq 2$. Then* $i(T_{n,k}) = i(T_{n,k-1}) + i(T_{n,k-2})$.

Proof.   There are $i(T_{n,k})$ independent subsets of $V_{n,k}$. Let W be an arbitrary independent subset of $V_{n,k}$. Suppose $p_k \notin W$. Such subsets are the same as those of $V_{n,k-1}$; there are $i(T_{n,k-1})$ such independent subsets.

On the other hand, suppose $p_k \in W$. Then $p_{k-1} \notin W$. There are $i(T_{n,k-2})$ independent subsets of $V_{n,k-2}$. Now insert p_k in each of them. This results in $i(T_{n,k-2})$ independent subsets of $V_{n,k}$.

Thus the total number of independent subsets of $V_{n,k}$ is $i(T_{n,k-1}) + i(T_{n,k-2}) = i(T_{n,k})$.   ∎

The next theorem gives an explicit formula for $i(T_{n,k})$. We will establish it using PMI and Theorem 14.5.

Theorem 14.6 (DeMaio and Jacobson, 2014 [141]).   *Let $n \geq 3$ and $k \geq 0$. Then* $i(T_{n,k}) = L_{n+k} + F_{n-3}F_k$.

Proof.   Clearly, $i(T_n, 0) = L_n = L_{n+0} + F_{n-3}F_0$ and $i(T_n, 1) = i(C_n) + i(P_{n-1}) = L_n + F_{n+1} = L_{n+1} + F_{n-3}F_1$. So the formula works for $k = 0$ and $k = 1$.

Suppose it works for k and $k - 1$, where $k \geq 1$. Then, by Theorem 14.5,

$$i(T_{n,k+1}) = i(T_{n,k}) + i(T_{n,k-1})$$
$$= (L_{n+k} + F_{n-3}F_k) + (L_{n+k-1} + F_{n-3}F_{k-1})$$
$$= L_{n+k+1} + F_{n-3}F_{k+1}.$$

So the formula works for $k + 1$ also.

Thus, by the strong version of PMI, the formula is true for all $k \geq 0$.   ∎

The next theorem expresses L_{n+k} in three different ways. They can be confirmed algebraically. But we will do one of them combinatorially and leave the other two as exercises; see Exercises 14.18 and 14.19.

Theorem 14.7 (DeMaio and Jacobson, 2014 [141]).   *Let $n \geq 3$ and $k \geq 0$. Then*

1) $L_{n+k} = F_{n-1}F_{k+1} + F_{n+1}F_{k+2} - F_{n-3}F_k$

2) $L_{n+k} = F_{n+1}F_k + L_nF_{k+1} - F_{n-3}F_k$

3) $L_{n+k} = F_{n-1}F_{k+2} + F_{n+k+1} - F_{n-3}F_k.$

Proof.   We will prove part 1). By Theorem 14.6, $L_{n+k} + F_{n-3}F_k = i(T_{n,k})$. So it suffices to show that $i(T_{n,k}) = F_{n-1}F_{k+1} + F_{n+1}F_{k+2}$.

Let W be an arbitrary independent subset of $V_{n,k}$. Suppose $c_1 \in W$. Then $c_2, c_n, p_1 \notin W$. There are $i(P_{n-3})i(P_{k-1}) = F_{n-1}F_{k+1}$ such independent subsets. On the other hand, suppose $c_1 \notin W$. There are $i(P_{n-1})i(P_k) = F_{n+1}F_{k+2}$ such subsets. So the total number of independent subsets equals $F_{n-1}F_{k+1} + F_{n+1}F_{k+2}$; that is, $i(T_{n,k}) = F_{n-1}F_{k+1} + F_{n+1}F_{k+2}$, as desired. ∎

The next theorem gives a second recurrence for $i(T_{n,k})$. We will confirm it algebraically using Theorem 14.6.

Theorem 14.8 (DeMaio and Jacobson, 2014 [141]).   *Let $n \geq 3$ and $k \geq 0$. Then* $i(T_{n,k}) = i(T_{n-1,k}) + i(T_{n-2,k})$.

Proof.   By Theorem 14.6, we have

$$i(T_{n-1,k}) + i(T_{n-2,k}) = (L_{n-1+k} + F_{n-4}F_k) + (L_{n-2+k} + F_{n-5}F_k)$$
$$= (L_{n+k-1} + L_{n+k-2}) + (F_{n-4} + F_{n-5})F_k$$
$$= L_{n+k} + F_{n-3}F_k$$
$$= i(T_{n,k}). \quad ∎$$

Using the recurrence in Theorem 14.5, and the initial conditions $i(T_{3,0}) = 4$, $i(T_{3,1}) = 7 = i(T_{4,0})$, and $i(T_{4,1}) = 12$, we can now compute $i(T_{n,k})$ for every $n \geq 3$ and $k \geq 0$; see Table 14.2.

TABLE 14.2.   Numbers of Independent Subsets of $V_{n,k}$

n \ k	0	1	2	3	4	5	6
3	4	7	11	18	29	47	76
4	7	12	19	31	50	81	131
5	11	19 → 30	49	79	128	207	
6	18	31	49	80	129	209	338
7	29	50	79	129	208	337	545
8	47	81	128	209	337	546	883

We will now investigate some interesting properties of the rectangular array in Table 14.2.

TADPOLE TRIANGLE

Suppose we rotate the array through 45° in the clockwise direction; then the nth northeast diagonal becomes the nth row of a triangular array; see Figure 14.34. It is called the *tadpole triangle*.

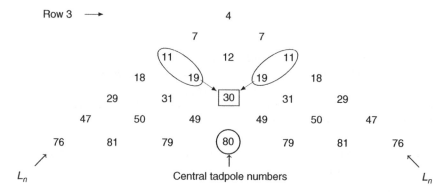

Figure 14.34.   Tadpole triangle.

Let $t_{n,k}$ denote the kth element in row n of the tadpole triangle, where $n \geq 3$ and $k \geq 0$. It follows from Table 14.2 that, as the row index increases by k, the column index decreases by k. So $i(T_{n-k}, k) = t_{n,k}$. Thus $t_{n,k} = L_n + F_{n-k-3}F_k$.

For example, $t_{9,4} = L_9 + F_2F_4 = 76 + 3 = 79$; see Figure 14.34.

Since $t_{n,k} = L_n + F_{n-k-3}F_k$, it follows that the *central tadpole number* is given by $t_{n,(n-3)/2} = L_n + F^2_{(n-3)/2}$, where $n \geq 3$ and is odd. For example, $t_{9,3} = L_9 + F^2_3 = \boxed{80}$; see Figure 14.34.

It follows by Theorems 14.5 and 14.8 that

$$t_{n,k} = \begin{cases} t_{n-1,k} + t_{n-2,k} & \text{if } 0 \leq k \leq \lfloor (n-2)/2 \rfloor \\ t_{n-1,k-1} + t_{n-2,k-2} & \text{otherwise,} \end{cases}$$

where $n \geq 3$.

For example, $t_{7,2} = t_{6,1} + t_{5,0} = 19 + 11 = \boxed{30} = t_{6,2} + t_{5,2}$; see Figure 14.34.

The tadpole triangle is symmetric about the vertical line through the middle; that is, $t_{n,k} = t_{n,n-k-3}$. The following theorem confirms this combinatorially.

Theorem 14.9 (DeMaio and Jacobson, 2014 [141]).   *Let $n \geq 3$ and $k \geq 0$. Then* $t_{n,k} = t_{n,n-k-3}$.

Proof. Since $t_{n,k} = i(T_{n-k,k})$, it follows that $t_{n,n-k-3} = i(T_{k+3,n-k-3})$. Consequently, $t_{n,k} = t_{n,n-k-3}$ if and only if $i(T_{n-k,k}) = i(T_{k+3,n-k-3})$. Clearly, both tadpole graphs $T_{n-k,k}$ and $T_{k+3,n-k-3}$ contain n vertices.

Let W be an arbitrary independent subset of $V_{n-k,k}$, and W' that of $V_{k+3,n-k-3}$. Suppose $c_2 \notin W$ and $c_2' \notin W'$. Then the numbers of independent subsets of $V_{n-k,k}$ and $V_{k+3,n-k-3}$ are each equal to the number of independent subsets of a path with $(n-k-1)+k = n-1 = (k+2)+(n-k-3)$ vertices.

On the other hand, suppose $c_2 \in W$ and $c_2' \in W'$. Then $c_1, c_3 \notin W$ and $c_1', c_3' \notin W'$. This decomposes $T_{n-k,k}$ into two disjoint paths: $c_4-\cdots-c_{n-k}$ and P_k, with $n-k-3$ and k vertices, respectively; likewise, $T_{k+3,n-k-3}$ is split into two disjoint paths: $c_4'-\cdots-c_{n-k}'$ and P_{n-k-3}, with k and $n-k-3$ vertices, respectively. Both decompositions contribute the same number of independent subsets of $V_{n-k,k}$ and $V_{k+3,n-k-3}$.

Combining the two cases yields the desired result. ∎

For example, $t_{10,4} = L_{10} + F_3 F_4 = 123 + 6 = 129 = t_{10,10-4-3}$.

The next tadpole property is an elegant application of the *d'Ocagne identity* $F_a F_{b+1} - F_{a+1} F_b = (-1)^b F_{a-b}$ (see Chapter 20) and Theorem 14.6. In the interest of brevity, we omit the proof; see Exercise 14.22. See [141] for a combinatorial proof.

Theorem 14.10 (DeMaio and Jacobson, 2014 [141]). *Let* $n \geq 3$ *and* $0 \leq k \leq \lfloor (n-3)/2 \rfloor$. *Then* $t_{n,k+1} - t_{n,k} = (-1)^k F_{n-2k-4}$. ∎

For example, $t_{9,4} - t_{9,3} = 79 - 80 = -1 = (-1)^3 F_{9-2 \cdot 3-4}$.

ROW SUMS

Since $t_{n,k} = L_n + F_{n-k-3} F_k$, we can compute the row sums of the tadpole triangle, as the next theorem shows. The proof is a straightforward application of the summation formula

$$5 \sum_{k=0}^{n} F_k F_{n-k} = n L_n - F_n;$$

so we omit it; see Exercise 14.23.

Theorem 14.11 (DeMaio and Jacobson, 2014 [141]). *Let* $n \geq 3$. *Then*

$$\sum_{k=0}^{n-3} t_{n,k} = (n-2) L_n + \frac{(n-3) L_{n-3} - F_{n-3}}{5}.$$ ∎

This theorem has an interesting byproduct, as the next corollary shows. We omit its proof also; see Exercise 14.24.

Corollary 14.1 (DeMaio and Jacobson, 2014 [141]).   *Let S_n denote the nth row sum of the tadpole triangle, where $n \geq 3$. Then $\lim\limits_{n \to \infty} \dfrac{S_{n+1}}{S_n} = \alpha$, the golden ratio.*   ■

Since the tadpole array is symmetric, it follows that $\sum\limits_{k=0}^{n-3}(-1)^k t_{n,k} = 0$, where n is even. But when n is odd, this sum has an interesting result, as the next theorem shows. The proof follows by PMI; see Exercise 14.25.

Theorem 14.12 (DeMaio and Jacobson, 2014 [141]).   *Let $n \geq 3$. Then*

$$\sum_{k=0}^{n-3}(-1)^k t_{n,k} = \begin{cases} 0 & \text{if } n \text{ is even} \\ 2F_n & \text{otherwise.} \end{cases}$$ ■

For example, $\sum\limits_{k=0}^{4}(-1)^k t_{7,k} = 29 - 31 + 30 - 31 + 29 = 2 \cdot 13 = 2F_7$.

EXERCISES 14

Prove each combinatorially.

1. $\sum\limits_{k=1}^{n} F_{2k} = F_{2n+1} - 1$ (Lucas, 1876).

2. $\sum\limits_{k=1}^{n} F_{2k-1} = F_{2n}$ (Lucas, 1876).

3. $F_{2n+1} = F_{n+1}^2 + F_n^2$ (Lucas, 1876).

4. $F_n = F_{m+1}F_{n-m} + F_m F_{n-m-1}$, where $0 \leq m < n$.

Compute each.

5. $\sum\limits_{\substack{i,j \geq 0 \\ i+j \leq 3}} \binom{3-i}{j}\binom{3-j}{i}$.

6. $\sum\limits_{\substack{i,j \geq 0 \\ i+j \leq 4}} \binom{4-i}{j}\binom{4-j}{i}$.

Establish each combinatorially.

7. $L_{m+n} = F_{m+1}L_n + F_m L_{n-1}$.

8. $L_n = F_{m+1}L_{n-m} + F_m L_{n-m-1}$, where $0 \leq m < n$.

9. $F_{n+1}F_{n-1} - F_n^2 = (-1)^n$.

10. $\sum_{k=1}^{n} F_k^2 = F_n F_{n+1}$ (Lucas, 1876).

11. $L_{n+1} + L_{n-1} = 5F_n$.

12. $F_{2n} = 2F_{n-1}F_n + F_n^2$.

13. $F_{3n+2} = F_{n+2}^3 - 2F_n^2 F_{n+2} + F_n^2 F_{n-2}$ (DeMaio and Jacobson, 2013 [140]).

14. $F_{4n+2} = F_{n+2}^4 - 3F_n^2 F_{n+2}^2 + 2F_n^2 F_{n-2}F_{n+2} + F_n^4 - F_n^2 F_{n-2}^2$ (DeMaio and Jacobson, 2013 [140]).

15. $L_{3n} = F_{n+2}^3 - 3F_n^2 F_{n+2} + 3F_n^2 F_{n-2} - F_{n-2}^3$ (DeMaio and Jacobson, 2013 [140]).

16. $L_{4n} = F_{n+2}^4 - 4F_n^2 F_{n+2}^2 + 4F_n^2 F_{n-2}F_{n+2} + 2F_n^4 - 4F_n^2 F_{n-2}^2 + F_{n-2}^4$ (DeMaio and Jacobson, 2013 [140]).

17. $L_n + F_{n+1} = L_{n+1} + F_{n-3}F_1$.

18. Prove combinatorially that $L_{n+k} = F_{n+1}F_k + L_n F_{k+1} - F_{n-3}F_k$ (DeMaio and Jacobson, 2014 [141]).

19. Prove combinatorially that $L_{n+k} = F_{n-1}F_{k+2} + F_{n+k+1} - F_{n-3}F_k$ (DeMaio and Jacobson, 2014 [141]).

20. Define the tadpole array $(t_{n,k})$ recursively.

Prove each algebraically, where $n \geq 3$.

21. $t_{n,k} = t_{n,n-k-3}$ (DeMaio and Jacobson, 2014 [141]).

22. Let $0 \leq k \lfloor (n-3)/2 \rfloor$. Then $t_{n,k+1} - t_{n,k} = (-1)^k F_{n-2k-4}$ (DeMaio and Jacobson, 2014 [141]).

23. $\sum_{k=0}^{n-3} t_{n,k} = (n-2)L_n + \dfrac{(n-3)L_{n-3} - F_{n-3}}{5}$ (DeMaio and Jacobson, 2014 [141]).

24. Let S_n denote the nth row sum of the tadpole triangle. Then $\lim_{n \to \infty} \dfrac{S_{n+1}}{S_n} = \alpha$ (DeMaio and Jacobson, 2014 [141]).

25. Let n be an odd integer ≥ 3. Then $\sum_{k=0}^{n-3} (-1)^k t_{n,k} = 2F_n$ (DeMaio and Jacobson, 2014 [141]).

15

HOSOYA'S TRIANGLE

In 1976, H. Hosoya of Ochanomizu University, Tokyo, introduced an interesting triangular array H; see Figure 15.1 [332]. It is closely linked to Fibonacci numbers. We call it *Hosoya's triangle*. The array is symmetric about the vertical line through the middle, and the top two northeast and southeast diagonals consist of Fibonacci numbers. We can obtain every interior element by adding the two immediate neighbors along the northeast and northwest diagonals; for example, $15 = 5 + 10 = 6 + 9$.

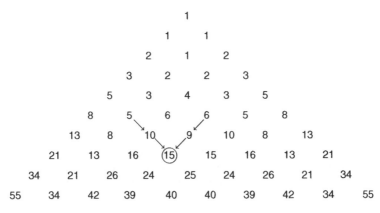

Figure 15.1. Hosoya's triangle H.

Fibonacci and Lucas Numbers with Applications, Volume One, Second Edition. Thomas Koshy.
© 2018 John Wiley & Sons, Inc. Published 2018 by John Wiley & Sons, Inc.

15.1   RECURSIVE DEFINITION

We can define the array H recursively, where $H(n,j)$ denotes the element in row n and column j:

$$H(0,0) = H(1,0) = H(1,1) = H(2,1) = 1$$

$$H(n,j) = H(n-1,j) + H(n-2,j) \tag{15.1}$$

$$= H(n-1,j-1) + H(n-2,j-2), \tag{15.2}$$

where $n \geq j \geq 0$ and $n \geq 2$.

Since $H(n,0) = H(n-1,0) + H(n-2,0)$, where $H(0,0) = 1 = F_1$ and $H(1,0) = 1 = F_2$, it follows that $H(n,0) = F_{n+1}$; likewise, since $H(n,n) = H(n-1,n) + H(n-2,n)$, it follows that $H(n,n) = F_{n+1}$. Similarly, we can show that $H(n,1) = H(n,n-1) = F_n$; see Exercises 15.1–15.3.

Successive application of recurrence (15.1) gives an interesting pattern (watch the coefficients):

$$H(n,j) = H(n-1,j) + H(n-2,j)$$

$$= 2H(n-2,j) + H(n-3,j)$$

$$= 3H(n-3,j) + 2H(n-4,j)$$

$$\vdots$$

Continuing like this, we find a close link between $H(n,j)$ and Fibonacci numbers:

$$H(n,j) = F_{k+1}H(n-k,j) + F_k H(n-k-1,j), \tag{15.3}$$

where $1 \leq k \leq n-j-1$; see Exercise 15.4. In particular, let $k = n-j-1$. Then, by Exercise 15.3, we have

$$H(n,j) = F_{n-j}H(j+1,j) + F_{n-j-1}H(j,j)$$

$$= F_{n-j}F_{j+1} + F_{n-j-1}F_{j+1}$$

$$= F_{j+1}(F_{n-j} + F_{n-j-1})$$

$$= F_{j+1}F_{n-j+1}. \tag{15.4}$$

Thus every element in the array is the product of two Fibonacci numbers.

For example, $H(7,3) = 15 = 3 \cdot 5 = F_4 F_5$; and $H(9,6) = 39 = 3 \cdot 13 = F_4 F_7$.

Since $H(n,j) = H(n,n-j)$, it follows from equation (15.4) that $H(n,j) = H(n,n-j) = F_{j+1}F_{n-j+1}$.

Let $n = 2m$ and $j = m$. Then equation (15.4) yields $H(2m,m) = F_{m+1}F_{m+1} = F_{m+1}^2$. Thus $H(2m,m)$ is the square of a Fibonacci number. In other words, the elements along the vertical line through the middle are Fibonacci squares.

For example, $H(8,4) = 25 = F_5^2$ and $H(10,5) = 64 = F_6^2$.

A LINK BETWEEN $H(n,j)$ AND L_m

Using equation (15.4), we can compute $H(n,j)$ using Lucas numbers:

$$5H(n,j) = (\alpha^{j+1} - \beta^{j+1})(\alpha^{n-j+1} - \beta^{n-j+1})$$
$$= (\alpha^{n+2} + \beta^{n+2}) + (\alpha\beta)^j(\alpha^{n-2j} + \beta^{n-2j})$$
$$H(n,j) = \frac{L_{n+2} + (-1)^j L_{n-2j}}{5}. \tag{15.5}$$

For example, let $n = 10$ and $j = 3$. Then

$$\frac{L_{12} + (-1)^3 L_4}{5} = \frac{322 - 7}{5} = 63 = H(10,3).$$

As a bonus, it follows from formula (15.5) that $L_{n+2} \equiv (-1)^{j-1}L_{n-2j}$ (mod 5). In particular, $L_{2m} \equiv 2(-1)^m$ (mod 5) and $L_{2m+1} \equiv (-1)^m$ (mod 5).

For example, $L_{12} = 322 \equiv 2 \equiv 2(-1)^6$ (mod 5); and $L_{15} = 1364 \equiv -1 \equiv (-1)^7$ (mod 5).

15.2   A MAGIC RHOMBUS

Recall that Hosoya's triangle was constructed using four initial conditions, that is, four 1s, and they form a rhombus. In fact, we can employ any rhombus with vertices $H(i,j)$, $H(i-1,j-1)$, $H(i-2,j-1)$, and $H(i-1,j)$ to generate their neighbors.

For example, consider the rhombus in Figure 15.2, where the letters A through H represent the numbers 4, 6, 9, 6, 5, 25, 5, and 1, respectively. Then $F = A + B + C + D$, $H = A + D - B - C$, $E = C + D - A - B$,   and $G = B + D - A - C$; see Exercise 15.7. We can represent these facts pictorially; see Figure 15.3.

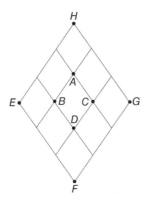

Figure 15.2.   A magic rhombus.

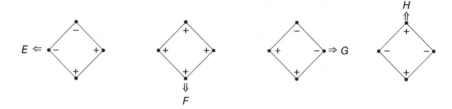

Figure 15.3.

ADDITIONAL FORMULAS

More generally, we have the following additional formulas:

$$H(n-1, j-2) = H(n,j) + H(n-1,j) - H(n-1,j-1) - H(n-2,j-1)$$
$$H(n+2, j+1) = H(n,j) + H(n-1,j) - H(n-1,j-1) + H(n-2,j-1)$$
$$H(n-1, j+1) = H(n-1,j-1) + H(n-1,j) - H(n-2,j-1) - H(n,j)$$
$$H(n-4, j-2) = H(n-2,j-1) + H(n-1,j) - H(n-1,j-1) - H(n,j);$$

see Figure 15.4.

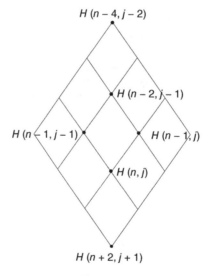

Figure 15.4.

Since $(F_{j+1}F_{n-j+1})(F_jF_{n-j}) = (F_jF_{n-j+1})(F_{j+1}F_{n-j})$, it follows by equation (15.4) that

$$H(n,j) \cdot H(n-2, j-1) = H(n-1, j-1) \cdot H(n-1, j); \qquad (15.6)$$

that is, the product of the opposite vertices A and D in the rhombus $ABCD$ equals that of the remaining two opposite vertices B and C.

For example, consider the rhombus formed by 15, 24, 40, and 25. Clearly, $15 \cdot 40 = 24 \cdot 25$. Likewise, $8 \cdot 26 = 13 \cdot 16$.

We can write equation (15.6) as

$$\{[H(n,j) \div H(n-1,j)] * H(n-2,j-1)\} \div H(n-1,j-1) = 1; \qquad (15.7)$$

see Figure 15.5.

Figure 15.5.   Figure 15.6.

Interestingly enough, we can extend equation (15.6) and hence equation (15.7) to the corners of any parallelogram:

$$H(n,j) \cdot H(n-k-l,j-k) = H(n-k,j-k) \cdot H(n-l,j);$$

see Figure 15.6.

For example, consider the array of parallelograms in Figure 15.7. Notice that $40 \cdot 4 = 16 \cdot 10$ and $15 \cdot 10 = 25 \cdot 6$. The other products can be verified similarly.

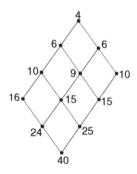

Figure 15.7.

Consider the downward-pointing triangles with vertices belonging to two adjacent rows. For example, consider the adjacent rows in Figure 15.8. The sum of their vertices is 34, a constant.

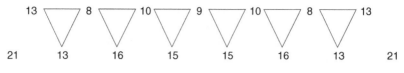

Figure 15.8.

More generally, $H(n,j) + H(n-1,j) + H(n-1,j-1)$ is a constant for every n; see Figures 15.9 and 15.10, and Exercise 15.8.

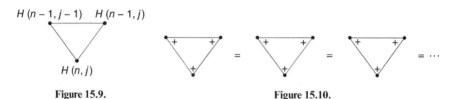

Figure 15.9.   **Figure 15.10.**

In particular,

$$H(n,0) + H(n-1,0) + H(n-1,-1) = F_{n+1} + F_n + 0 = F_{n+2}.$$

Thus

$$H(n,j) + H(n-1,j) + H(n-1,j-1) = F_{n+2}. \tag{15.8}$$

In words, the magic constant for the downward-pointing triangle with the lowest vertex in row n is F_{n+2}.

For instance, the constant for the triangles in Figure 15.8 is $34 = F_9$, as observed earlier.

Using recurrence (15.1), we can rewrite equation (15.8) as

$$H(n,j) + H(n-2,j-1) = F_{n+1}. \tag{15.9}$$

Thus the sum of any two vertical neighbors is a constant for a horizontal slide. That is, the sum of the north and south vertices in a magic rhombus is a Fibonacci number; see Figure 15.11.

Figure 15.11.   **Figure 15.12.**

For instance, the sum of the north and south vertices in the rhombus in Figure 15.12 is $25 + 64 = 89 = F_{11}$.

It follows from equations (15.8) and (15.9) that

$$H(n,j) + H(n,j-1) - H(n-1,j-1) = F_{n+1}; \qquad (15.10)$$

see Exercise 15.9. That is, the sum of the two lower vertices of an upward-pointing triangle minus the vertex $n-1$ is F_{n+1}; see Figures 15.13–15.15.

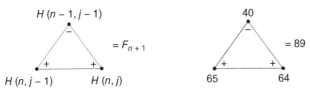

| Figure 15.13. | Figure 15.14. |

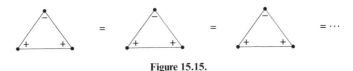

Figure 15.15.

Using equation (15.9), we can show that

$$H(n,j) + H(n-6,j-3) = 2F_{n-1}; \qquad (15.11)$$

see Exercise 15.10 and Figure 15.16.
For example, $H(10,4) + H(4,1) = 65 + 3 = 68 = 2F_9$.

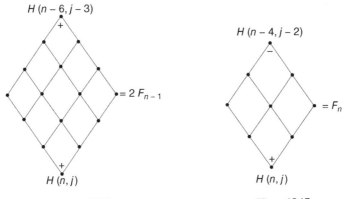

| Figure 15.16. | Figure 15.17. |

Equation (15.10) yields yet another treasure:

$$H(n,j) - H(n-4, j-2) = F_n; \qquad\qquad (15.12)$$

see Exercise 15.11 and Figure 15.17.

For instance, $H(10,4) - H(6,2) = 65 - 10 = 55 = F_{10}$.

EXERCISES 15

H denotes Hosoya triangle in Figure 15.1. Prove each, where $n \geq 1$.

1. $H(n,1) = F_n$.
2. $H(n,j) = H(n, n-j)$
3. $H(n, n-1) = F_n$.
4. $H(n,j) = F_{k+1} H(n-k, j) + F_k H(n-k-1, j)$, where $1 \leq k \leq n-j-1$.
5. $L_{2m} \equiv 2(-1)^m \pmod{5}$.
6. $L_{2m+1} \equiv (-1)^m \pmod{5}$.

7. Using Figure 15.2, show that $F = A + B + C + D$, $H = A + D - B - C$, $E = C + D - A - B$, and $G = B + D - A - C$.

Prove each.

8. $H(n,j) + H(n-1, j) + H(n-1, j-1) = F_{n+2}$.
9. $H(n,j) + H(n, j-1) - H(n-1, j-1) = F_{n+1}$.
10. $H(n,j) + H(n-6, j-3) = 2F_{n-1}$.
11. $H(n,j) - H(n-4, j-2) = F_n$.

16

THE GOLDEN RATIO

> He that holds fast the golden mean,
> And lives contentedly between
> The little and the great,
> Feels not the wants that pinch the poor
> Nor plagues that haunt the rich man's door,
> Embittering all his state.
> –William Cowper, English poet (1731–1800)

What can we say about the sequence of ratios F_{n+1}/F_n of consecutive Fibonacci numbers? Does it converge? If it does, what is its limit? If the limit exists, does it have any geometric significance? We will pursue these interesting questions, along with their Lucas counterparts, in this chapter.

16.1 RATIOS OF CONSECUTIVE FIBONACCI NUMBERS

To begin with, let us compute the ratios F_{n+1}/F_n of the first 20 Fibonacci and Lucas numbers, and then examine them for a possible pattern; see Table 16.1. As n gets larger and larger, it appears that F_{n+1}/F_n approaches a limit, namely, 1.618033....

This phenomenon was observed by the German astronomer and mathematician Johannes Kepler (1571–1630). It also appears that L_{n+1}/L_n approaches the same magic number as $n \to \infty$.

Fibonacci and Lucas Numbers with Applications, Volume One, Second Edition. Thomas Koshy.
© 2018 John Wiley & Sons, Inc. Published 2018 by John Wiley & Sons, Inc.

TABLE 16.1. Ratios of Consecutive Fibonacci and Lucas Numbers

n	F_{n+1}/F_n	L_{n+1}/L_n
1	$\frac{1}{1} \approx 1.0000000000$	$\frac{3}{1} \approx 3.0000000000$
2	$\frac{2}{1} \approx 2.0000000000$	$\frac{4}{3} \approx 1.3333333333$
3	$\frac{3}{2} \approx 1.5000000000$	$\frac{7}{4} \approx 1.7500000000$
4	$\frac{5}{3} \approx 1.6666666667$	$\frac{11}{7} \approx 1.5714285714$
5	$\frac{8}{5} \approx 1.6000000000$	$\frac{18}{11} \approx 1.6363636364$
6	$\frac{13}{8} \approx 1.6250000000$	$\frac{29}{18} \approx 1.6111111111$
7	$\frac{21}{13} \approx 1.6153846154$	$\frac{47}{29} \approx 1.6206896551$
8	$\frac{34}{21} \approx 1.6190476191$	$\frac{76}{47} \approx 1.6170212766$
9	$\frac{55}{34} \approx 1.6176470588$	$\frac{123}{76} \approx 1.6184210526$
10	$\frac{89}{55} \approx 1.6181818182$	$\frac{199}{123} \approx 1.6178861788$
11	$\frac{144}{89} \approx 1.6179775281$	$\frac{322}{199} \approx 1.6180904523$
12	$\frac{233}{144} \approx 1.6180555556$	$\frac{521}{322} \approx 1.6180124224$
13	$\frac{377}{233} \approx 1.6180257511$	$\frac{843}{521} \approx 1.6180422265$
14	$\frac{610}{377} \approx 1.6180371353$	$\frac{1364}{843} \approx 1.6180308422$
15	$\frac{987}{610} \approx 1.6180327869$	$\frac{2207}{1364} \approx 1.6180351906$
16	$\frac{1597}{987} \approx 1.6180344478$	$\frac{3571}{2207} \approx 1.6180335297$
17	$\frac{2584}{1597} \approx 1.6180338134$	$\frac{5778}{3571} \approx 1.6180341641$
18	$\frac{4181}{2584} \approx 1.6180340557$	$\frac{9349}{5578} \approx 1.6180339218$
19	$\frac{6765}{4181} \approx 1.6180339632$	$\frac{15127}{9349} \approx 1.6180340143$
20	$\frac{10946}{6765} \approx 1.6180339985$	$\frac{24476}{15127} \approx 1.6180339789$

Interestingly enough, $\alpha = (1 + \sqrt{5})/2 = 1.61803398875\ldots$. So it is reasonable to predict that both ratios converge to the very same limit α, the positive root of the quadratic equation $x^2 - x - 1 = 0$.

We will now confirm this observation. Let $x = \lim_{n\to\infty} F_{n+1}/F_n$. By the Fibonacci recurrence, we then have

$$\frac{F_{n+1}}{F_n} = 1 + \frac{F_{n-1}}{F_n}$$

$$= 1 + \frac{1}{F_n/F_{n-1}}$$

$$\lim_{n\to\infty} \frac{F_{n+1}}{F_n} = 1 + \lim_{n\to\infty} \frac{1}{F_n/F_{n-1}}$$

$$x = 1 + \frac{1}{x}.$$

Thus x satisfies the quadratic equation $x^2 - x - 1 = 0$; so $x = (1 \pm \sqrt{5})/2$. Since the limit is positive, it follows that

$$\lim_{n \to \infty} \frac{F_{n+1}}{F_n} = \frac{1 + \sqrt{5}}{2} = \alpha.$$

This confirms our observation.

The solutions α and β of the equation $x^2 - x - 1 = 0$ are the only numbers such that the reciprocal of each is obtained by subtracting 1 from it; that is, $x - 1 = 1/x$, where $x = \alpha$ or β. Thus α is the only positive number that has this property.

Since $\alpha - 1 = 1/\alpha$, it follows that α and $1/\alpha$ have the same infinite decimal portion:

$$\alpha = 1.61803398875\ldots$$

$$\frac{1}{\alpha} = 0.61803398875\ldots.$$

An interesting observation: It appears from Table 16.1 that when n is even, $F_{n+1}/F_n > \alpha$; and when it is odd, $F_{n+1}/F_n < \alpha$. The same behavioral pattern holds for L_{n+1}/L_n also. We will revisit this pattern in Chapter 19.

It follows from the preceding discussion that

$$\lim_{n \to \infty} \frac{L_{n+1}}{L_n} = \frac{1 + \sqrt{5}}{2} = \alpha.$$

16.2   THE GOLDEN RATIO

The number α is so intriguing a number that it was known to the ancient Greeks at least sixteen centuries before Fibonacci. They called it the *Golden Section*, for reasons that will be clear shortly.

Before the Greeks, the ancient Egyptians used it in the construction of their great pyramids. The *Papyrus of Ahmes*, written hundreds of years before ancient Greek civilization existed, and now in the British Museum, contains a detailed account of how the number was used in the building of the Great Pyramid of Giza around 3070 B.C. Ahmes refers to this number as a "sacred ratio."

The height of the Great Pyramid (see Figure 16.1a) is 414.4 feet, which is about 5813 inches; notice that three consecutive Fibonacci numbers appear in the height: 5, 8, and 13.

Herodotus, a Greek historian of the fifth century B.C., wrote that he was told by Egyptian priests that the proportions of the Great Pyramid were chosen in such a way that "the area of a square with a side of length equaling the height of the Pyramid is the same as the area of a slanted [triangular] face."

We will now confirm this. To this end, let $2b$ denote the length of the base of the Pyramid, a the altitude of a slanted [triangular] face, and h the height of the Pyramid. According to Herodotus' formula, $h^2 = (2b \cdot a)/2 = ab$. But, by

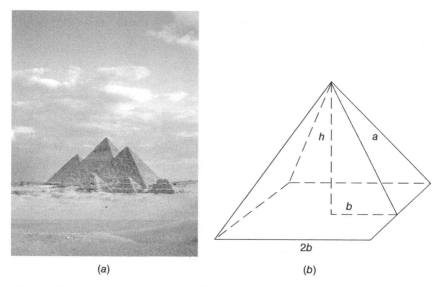

(a)   (b)

Figure 16.1.   (a) The pyramids at Giza.* (b) Diagram of pyramid showing the golden section.

the Pythagorean theorem, $h^2 = a^2 - b^2$, so $a^2 - b^2 = ab$, and hence $(a/b)^2 = 1 + (a/b)$. Thus a/b satisfies the quadratic equation $x^2 = x + 1$, so $a/b = \alpha$.

Interestingly, the actual measurements are $a = 188.4$ meters, $b = 116.4$ meters, and $h = 148.2$ meters. So $a/b = 188.4/116.4 \approx 1.618$.

In 1938, L. Hogben observed that the ratio of the base perimeter $8b$ of the Pyramid to its vertical height equals that of the circumference of a circle to its radius, that is, $8b = 2\pi h$ [266]. Thus

$$\pi = \frac{4b}{h} = \frac{4b}{\sqrt{ab}} = 4\sqrt{\frac{b}{a}} = \frac{4}{\sqrt{\alpha}}$$

$$\approx 3.1446.$$

Clearly, this estimate is accurate for two decimal places.

From Herodotus' statement, it follows that the ratio of the sum of the areas of the lateral faces of the Great Pyramid to the base area is also α; see Exercise 16.18.

Like the Egyptian pyramids, Mexican pyramids also exemplify the basic principles of aesthetics and perfect proportion. Both appear to have been built by people of common ancestry, and both have incorporated the magic ratio in their construction.

The golden ratio is often denoted by φ, the Greek letter *phi*. Theodore Andrea Cook relates that in about 1909 the American mathematician Mark Barr gave this name to the ratio, because it is the first Greek letter in the name of Phidias

Figure source: Stephen Studd. The Image Bank. Reproduced with permission of Getty Images.

(490?–430? B.C.), the great Greek sculptor, who constantly employed the Golden Section in his work. However, we will continue to denote the Golden Section by α for convenience and consistency.

Interestingly, *phi* appears 34 times and its numeric approximation 1.618 five times in *The Da Vinci Code*, both Fibonacci numbers. According to Robert Langdon in the novel, "phi is generally considered the most beautiful number in the universe."

The golden ratio is also often denoted by another name, τ, the Greek letter *tau*. According to the British-born Canadian geometer H.S.M. Coxeter (1907–2003) of the University of Toronto, this usage comes from the fact that τ is the first letter of the Greek word "$\tau\varnothing\mu\eta$," which means "the section."

Kepler referred to α as "*sectio divina*" (divine section); and the great Italian artist Leonardo da Vinci (1452–1519) employed the magic number in many of his great works, calling it "*sectio aurea*" (the golden chapter), a term still in popular use.

Kepler in his *Mysterium Cosmographicum de Admirabili Proportione Orbium Coelestium* singles out α as one of the two "great treasures" of geometry, the other being the Pythagorean theorem:

> Geometry has two great treasures; one is the Theorem of Pythagoras; the other, the division of a line into extreme and mean ratio. The first we may compare to a measure of gold, the second we may name a precious jewel.

The mysterious number α was the principal character of the book *De Divina Proportione*, by the Italian mathematician and Franciscan friar Fra Pacioli de Borgo (1445–1517), published in 1509 in Venice. Pacioli describes the properties of α, stopping at thirteen "for the sake of our salvation."

Today, the magical number is also called the *golden mean*, the *golden ratio*, the *golden proportion*, the *divine section*, and the *divine proportion*. Interestingly, the term *divine proportion* appears 13 times in *The Da Vinci Code*, again a Fibonacci number.

The concept of a golden mean has its origin in plane geometry. It stems from locating a point on a line segment such that it divides the line segment into two in a certain ratio. To elucidate this more clearly, we first define the concept of *mean proportional*.

MEAN PROPORTIONAL

Let a, b, and c be any three positive integers such that $a^2 = bc$. Then a is the *mean proportional* of b and c. Notice that $a^2 = bc$ if and only if $a/b = c/a$; that is, $a^2 = bc$ if and only if $a : b = c : a$.

For example, $6^2 = 4 \cdot 9$; so 6 is the mean proportional of 4 and 9. Likewise, $\sqrt{6}$ is the mean proportional of 2 and 3.

We now give a geometric interpretation of mean proportional.

A GEOMETRIC INTERPRETATION

Geometrically, we would like to find a point C on a line segment $\overline{AB}$ such that the length of the greater segment $\overline{AC}$ is the mean proportional of the whole length AB and the length BC of the smaller segment; see Figure 16.2. Thus we would like to find C such that $AC/BC = AB/AC$; then C divides $\overline{AB}$ in the golden ratio.

Figure 16.2.

To locate the point C, let $AC = x$ and $BC = y$. Then the equation $AC/BC = AB/AC$ yields

$$\frac{x}{y} = \frac{x+y}{x}$$

$$= 1 + \frac{1}{x/y};$$

that is, $(x/y)^2 - (x/y) - 1 = 0$. So x/y satisfies the familiar quadratic equation $t^2 - t - 1 = 0$. Since $x/y > 0$, this implies that $x/y = \alpha$; that is, $x : y = \alpha : 1$. Thus we must choose the point C in such a way that $AC/BC = \alpha$. Then C divides $\overline{AB}$ in the golden ratio.

Although this process determines the point C algebraically, how do we locate it geometrically? In other words, how do we locate it with a ruler and compass?

RULER AND COMPASS CONSTRUCTION

To this end, draw $\overline{BX} \perp \overline{AB}$; see Figure 16.3. Select a point D on $\overline{BX}$ such that $AB = 2BD$. With D as center, draw an arc of radius DB to intersect $\overline{AD}$ at E. Now with A as center, draw an arc of radius AE to meet $\overline{AB}$ at C. We now claim that C divides $\overline{AB}$ in the desired ratio.

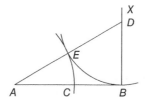

Figure 16.3.

We will now confirm this claim. We have $AC = AE$, $BD = ED$, and $AB = AC + BC = 2BD = 2ED$. Since $\triangle ABD$ is a right triangle, by the Pythagorean theorem, we have $AD = \sqrt{5}BD = \sqrt{5}ED$. Therefore, $AC = AE = AD - ED = (\sqrt{5} - 1)ED$. Thus

$$\frac{AB}{AC} = \frac{2ED}{(\sqrt{5} - 1)ED} = \frac{2}{\sqrt{5} - 1} = \frac{1 + \sqrt{5}}{2} = \alpha.$$

Then $AB/(AB - AC) = \alpha$. This yields $1 - (BC/AB) = 1/\alpha$; that is,

$$\frac{BC}{AB} = \frac{\alpha}{\alpha - 1} = -\frac{\alpha}{\beta} = \alpha^2$$

$$\frac{AC}{BC} = \frac{AC}{AB} \cdot \frac{AB}{BC} = \frac{1}{\alpha} \cdot \alpha^2 = \alpha.$$

Thus $AC/BC = AB/AC = \alpha$, so C is the desired point.

Next we present Euler's method for locating the point C, named after the great Swiss mathematician Leonhard Euler (1707–1783). This construction is in fact attributed to the Pythagoreans, since Euler included it among the theorems and constructions developed by them.

EULER'S CONSTRUCTION

To locate the point C that divides $\overline{AB}$ in the golden ratio, first we complete the square $ABDE$; see Figure 16.4. Let F bisect $\overline{AE}$. With F as center, draw an arc of radius FB to cut ray $\overrightarrow{FA}$ at G. Now, with A as center and AG as radius, draw an arc to intersect $\overline{AB}$ at C. Then C is the desired point.

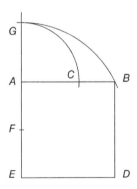

Figure 16.4.

To confirm this, we have $AB = AE = 2AF$. So, by the Pythagorean theorem, $FB = FG = \sqrt{5}AF$. Then $FA + AG = FG = \sqrt{5}AF$; so $AG = AC = (\sqrt{5} - 1)AF$. Consequently,

$$\frac{AB}{AC} = \frac{2AF}{(\sqrt{5}-1)AF} = \frac{2}{\sqrt{5}-1} = \alpha.$$

Moreover, since $BC = AB - AC$,

$$\frac{BC}{AC} = \frac{AB}{AC} - 1 = \alpha - 1 = -\beta,$$

so

$$\frac{AC}{BC} = -\frac{1}{\beta} = \alpha.$$

Thus

$$\frac{AB}{AC} = \frac{AC}{BC} = \alpha,$$

so C divides $\overline{AB}$ in the golden ratio, as desired.   ∎

Next we will express the golden ratio in terms of nested radicals.

16.3   GOLDEN RATIO AS NESTED RADICALS

Recall that α is the positive square root of the quadratic equation $x^2 = x + 1$. Now consider the cubic equation $x^3 = 1 + 2x$; that is, $(x + 1)(x^2 - x - 1) = 0$. This equation has exactly one positive solution, namely, α. Since $x = \sqrt[3]{1 + 2x}$, it follows by iteration that

$$\alpha = \sqrt[3]{1 + 2\sqrt[3]{1 + 2\sqrt[3]{1 + \cdots}}}.$$

Next consider the equation $x^4 = 2 + 3x$. Since $\alpha^4 = 3\alpha + 2$, it follows that α is a solution of the equation. Since $x^4 - 3x - 2 < 0$ when $0 \le x < \alpha$, and is monotonically increasing when $x > \alpha$, it is the only positive solution. Since $x = \sqrt[4]{2 + 3x}$, we also have

$$\alpha = \sqrt[4]{2 + 3\sqrt[4]{2 + 3\sqrt[4]{2 + \cdots}}}.$$

More generally, consider the equation $x^n = F_{n-1} + F_n x$, studied by C.R. Wall of Texas Christian University, Fort Worth, Texas, in 1964, where $n \ge 2$ [578]. Interestingly, α is its only positive solution.

To see this, let $a(x) = x^n - F_n x - F_{n-1}$, $b(x) = x^2 - x - 1$, and $c(x) = x^{n-2} + x^{n-3} + 2x^{n-4} + \cdots + F_k x^{n-k-1} + \cdots + F_{n-2}x + F_{n-1}$. Then $a(x) = b(x)c(x)$;

$b(x) < 0$ when $\beta < x < \alpha$; $b(x) > 0$ when $x > \alpha$; and $c(x) > 0$ when $x \geq 0$. It now follows that α is the only positive solution of $a(x) = 0$, as desired [1]. Thus

$$\alpha = \sqrt[n]{F_{n-1} + F_n \sqrt[n]{F_{n-1} + F_n \sqrt[n]{F_{n-1} + \cdots}}}.$$

Next we would like to find the value of x in the equation $n^n + (n+x)^n = (n + 2x)^n$ as $n \to \infty$ [6]. Then

$$\left(1 + \frac{x}{n}\right)^n + 1 = \left(1 + \frac{2x}{n}\right)^n.$$

Since

$$\lim_{n \to \infty} \left(1 + \frac{x}{n}\right)^n = e^x,$$

this implies that $e^x + 1 = e^{2x}$. Clearly, its positive solution is $x = \ln \alpha$.

Next we investigate how the golden ratio can be computed using Newton's method of approximation; for a detailed discussion of this method, see [491].

16.4 NEWTON'S APPROXIMATION METHOD

In 1999, J.W. Roche of LaSalle High School, Wyndmoor, Pennsylvania, studied how to estimate the golden ratio using Newton's method of approximation and the quadratic function $f(x) = x^2 - x - 1$ [501]. In the process he found a spectacular relationship between approximations and Fibonacci numbers.

With $x_1 = 2$ as the initial seed and the recursive formula $x_{n+1} = x_n - \dfrac{f(x_n)}{f'(x_n)}$, we can find the next three approximations: $x_2 = 5/3$, $x_3 = 34/21$, and $x_4 = 1597/987$. Interestingly, they are all ratios of consecutive Fibonacci numbers. Based on this observation, we conjecture that $x_n = \dfrac{F_{2^n+1}}{F_{2^n}}$, where $n \geq 1$.

We will now establish the validity of this conjecture using PMI. Since $F_3/F_2 = 2 = x_1$, the formula works when $n = 1$. Assume it is true for an arbitrary positive integer n: $x_n = F_{m+1}/F_m$, where $m = 2^n$. Since $f'(x) = 2x - 1$, we have

$$x_{n+1} = x_n - \frac{x_n^2 - x_n - 1}{2x_n - 1}$$

$$= \frac{x_n^2 + 1}{2x_n - 1}$$

$$= \frac{(F_{m+1}^2/F_m^2) + 1}{2(F_{m+1}/F_m) - 1}$$

$$= \frac{F_{m+1}^2 + F_m^2}{F_m(2F_{m+1} - F_m)}$$

$$= \frac{F_{2m+1}}{F_m(F_{m+1} + F_{m-1})}$$

$$= \frac{F_{2m+1}}{F_m L_m}$$

$$= \frac{F_{2m+1}}{F_{2m}}.$$

Thus, by PMI, the formula works for all $n \geq 1$.

An interesting byproduct: Since $\lim_{m \to \infty} F_{m+1}/F_m = \alpha$, it follows that the sequence of approximations $\{x_n\}$ approaches α as $n \to \infty$, as expected.

Newton's iterative formula can be also used to approximate square roots of positive real numbers a. This technique demonstrates an interesting confluence of algebra, analytic geometry, and trigonometry. The ubiquitous number α makes serendipitous appearances in the process [396].

To see this, we let $f(x) = x^2 - a$. Then Newton's formula yields the recurrence $x_n = \frac{1}{2}\left(x_{n-1} + \frac{a}{x_{n-1}}\right)$. So the approximations x_n are the iterates of the function $g(x) = \frac{1}{2}\left(x + \frac{a}{x}\right)$, where $x > 0$.

The graph of the function g is the branch of the hyperbola $y = \frac{1}{2}\left(x + \frac{a}{x}\right)$, that is, $x^2 - 2xy + a = 0$, in the first quadrant; see Figure 16.5 [413]. Its asymptotes are the lines $x = 0$ and $y = \frac{1}{2}x$.

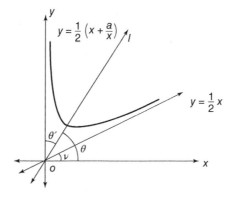

Figure 16.5.   Graph of $x^2 - 2xy + a = 0$, where $a, x > 0$.

Let v (Greek letter *nu*) be the angle between the x-axis and the asymptote $y = \frac{1}{2}x$, θ that between the x-axis and the axis l of the hyperbola, and θ' that between l and the y-axis. Then $2\theta' + v = \pi/2$, so $\theta = \frac{\pi}{4} + \frac{v}{2}$. Since $\tan v = 1/2$, it follows by the addition formula for the tangent function that

$$\tan\theta = \frac{1 + \tan v/2}{1 - \tan v/2}.$$

But $\tan v/2 = \sqrt{5} - 2 = 2\alpha - 3$, by the double-angle formula for tan. Thus

$$\tan\theta = \frac{1 + (2\alpha - 3)}{1 - (2\alpha - 3)}$$

$$= \frac{-\beta}{\beta + 1} = \alpha;$$

that is, the slope of the axis of the hyperbola is α.

By the addition formula, we also have

$$\tan(\theta + v) = \frac{\alpha + 1/2}{1 - \alpha/2}$$

$$= \frac{\alpha(\alpha + 1)}{1 + \beta} = \alpha^5.$$

Similarly, $\tan(\theta - v) = -\beta$.

16.5   THE UBIQUITOUS GOLDEN RATIO

On January 21, 1911, William Schooling wrote in the *Daily Telegraph* of London that there is a "very wonderful number which may be called by the Greek letter φ, of which nobody has heard as much as yet, but of which, perhaps, a great deal is likely to be heard in the course of time." It is intriguing that his prediction has come true.

According to R. Fisher, the great medieval philosopher and theologian Saint Thomas Aquinas (1225–1274) "described one of the basic rules of aesthetics – man's senses enjoy objects that are properly proportioned. He referred to the direct relationship between beauty and mathematics, which is often measurable and can be found in nature." Aquinas was of course referring to the golden ratio.

Just as Fibonacci and Lucas numbers occur in extremely unlikely and unrelated places, so does the golden ratio. Accordingly, we devote the rest of this chapter to demonstrate such occurrences, and the beauty and power of the golden ratio.

π AND α

Zerger observed two fascinating relationships between π and α:

- The first ten digits of α can be permuted to obtain the first ten digits of $1/\pi$: $\alpha = 1.618033988\ldots$ and $1/\pi = 0.3183098861\ldots$.
- The first nine digits of $1/\alpha$ can be permuted to form the first nine digits of $1/\pi$: $1/\alpha = 0.618033988\ldots$ and $1/\pi = 0.3183098861\ldots$.

ILLINOIS AND α

Zerger made another striking observation concerning the State of Illinois. Both the telephone area code 618 and the Zip Code prefix 618 are assigned to Illinois. Notice that 6, 1, and 8 are the first three digits in the decimal portion of α.

URANIUM AND THE GOLDEN RATIO

Uranium, an important source of nuclear energy, enjoys a unique place among the chemical elements. The ratio of the number of neutrons to protons is maximum for uranium; curiously enough, this ratio is approximately α:

$$\frac{\text{Number of neutrons}}{\text{Number of protons}} = \frac{146}{92} \approx 1.5869565 \approx \alpha.$$

16.6   HUMAN BODY AND THE GOLDEN RATIO

Studies have shown that several proportions of the human body exemplify the golden ratio. For instance, consider the drawing of a typical athlete in Figure 16.6. Then

$$\frac{AE}{CE} = \frac{\text{height}}{\text{navel height}} \approx \alpha,$$

and

$$\frac{CE}{AC} = \frac{\text{navel height}}{\text{height of the top of the head above the navel height}} \approx \alpha.$$

Thus height $\approx \alpha \times$ (navel height).

Moreover,

$$\frac{bc}{ab} = \frac{\text{arm length}}{\text{shoulder width}} \approx \alpha.$$

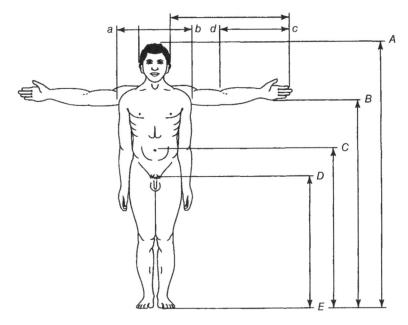

Figure 16.6.   Proportions of the human body.*

In fact, using Figure 16.6, we can find several other remarkable ratios that approximate the golden ratio.

Certain bones in the human body also show a relationship to the magical ratio. Figure 16.7, for instance, illustrates such a relationship between the hand and forearm. Because of the golden ratio's close association with human body proportions, α is often referred to as "the number of our physical body."

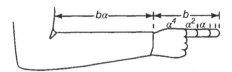

Figure 16.7.   *Source*: Holt, 1964 [319]. Reproduced with permission of Kappa Mu Epsilon.

According to S. Vajda of the University of Sussex, England, J. Gordon in 1938 detected the golden ratio in the English landscape painting *The Cornfield* by John Constable (1776–1837); in *Portrait of a Lady* by the Dutch artist Rembrandt Harmenszoon van Rijn (1606–1669); and in *Venus and Adonis* by the Venetian painter Titian (Tiziano Vecellio, 1497?–1576). Both *The Cornfield* and *Portrait of a Lady* are displayed in the National Gallery in London.

---

* Figure source: Smith, 1995 [532]. Reproduced with permission of Brooks/Cole, a division of Cengage Learning.

16.7   VIOLIN AND THE GOLDEN RATIO

The golden ratio plays a significant role in the design of the violin, one of the beautiful orchestral instruments; see Figure 16.8. The point B, where the two lines through the centers of the f holes intersect, divides the body in the golden ratio: $AB/BC = \alpha$. Besides, $AC/CD = \alpha$, so the body and the neck are in the golden proportion. It now follows that

$$\frac{AD}{AC} = \frac{AC}{AB} = \frac{CD}{BC} = \alpha.$$

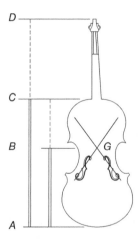

Figure 16.8.   The point B on the violin, where the two lines drawn through the centers of the f holes intersect, divides the instrument in the golden ratio; the body and the neck are likewise in the golden ratio.*

16.8   ANCIENT FLOOR MOSAICS AND THE GOLDEN RATIO

In 1970, R.E.M. Moore of Guy's Hospital Medical School, London, studied numerous two-thousand-year-old floor mosaics from Syria, Greece, and Rome [447]. He observed an interesting phenomenon: all mosaic patterns in these cultures showed exactly the same dimensions. Consequently, the mosaics in all these cultures must have used the same measuring technique and device. In fact, the calibrations on rulers employed by the mosaicists clearly and convincingly underscore the application of the golden proportions in the mosaic patterns; see Figure 16.9.

Figure source: Garland, 1987 [213]. Reproduced with permission of Pearson Education, Inc.

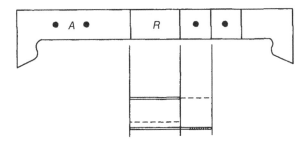

Figure 16.9. Calibrations on a ruler used in ancient mosaics.*

16.9 GOLDEN RATIO IN AN ELECTRICAL NETWORK

Figure 16.10 represents an infinite network consisting of resistors, each with resistance r. Suppose we would like to compute the resistance between the points A and B. (This problem appeared in the 1967 International Physics Olympiad, Poland.)

Figure 16.10. An infinite network consisting of resistors.

Let s denote the resistance between the points E and F of the infinite network to their right side. Then the resistance r_{CD} betwen the points C and D is given by

$$\frac{1}{r_{CD}} = \frac{1}{r} + \frac{1}{s}.$$

Now add the resistance r to this. The resistance r_{AB} between A and B of the given network is given by

$$r_{AB} = r + r_{CD}$$
$$= r + \frac{rs}{r+s}.$$

Since the resulting network is again infinite, $r_{AB} = s$. Thus

$$s = r + \frac{rs}{r+s}.$$

This implies $s = r\alpha$.

Suppose the network in Figure 16.10 consists of n resistors; see Figure 16.11. It follows from Chapter 3 that the resistance between A and B is given by

$$Z_i(n) = r + \frac{1}{(1/r) + [1/Z_i(n-1)]},$$

Figure source: Moore, 1970 [447]. Reproduced with permission of the Fibonacci Association.

where $Z_i(1) = 2r$. As $n \to \infty$, this recurrence yields

$$s = r + \frac{1}{(1/r) + (1/s)};$$

that is, $s = r + \dfrac{rs}{r+s}$, as we just found.

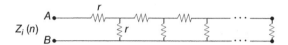

Figure 16.11.

Thus $\lim\limits_{n\to\infty} Z_i(n) = r\alpha$; so when $r = 1$, $\lim\limits_{n\to\infty} Z_i(n) = \alpha$. This we already knew, since, from Chapter 3, $Z_i(n) = \dfrac{F_{2n+1}}{F_{2n}}$; so $\lim\limits_{n\to\infty} Z_i(n) = \lim\limits_{n\to\infty} \dfrac{F_{2n+1}}{F_{2n}} = \alpha$.

Next we turn to an occurrence of α in electrostatics, the branch of physics that deals with properties and effects of static electricity.

16.10   GOLDEN RATIO IN ELECTROSTATICS

In 1972, B. Davis of the Indian Statistical Institute studied the following problem [137]:

A positive charge $+e$ and two negative charges $-e$ are to be placed on a line in such a way that the potential energy of the whole system is zero.

Suppose the charges are at points $A, B,$ and C; and let $AB = x$ and $BC = y$; see Figure 16.12. The potential energy of a system of static charges is the work done in bringing the charges from infinity to these points. The potential energy between two charges is the product of the charges divided by the distance between them.

A x B y C

$+e$ $-e$ $-e$

Figure 16.12.

The potential energy due to the charges at A and B is $\dfrac{(+e)(-e)}{x} = -\dfrac{e^2}{x}$; that due to the charges at A and C is $\dfrac{(+e)(-e)}{x+y} = -\dfrac{e^2}{x+y}$; and that due to the charges at B and C is $\dfrac{(-e)(-e)}{y} = \dfrac{e^2}{y}$.

For the potential energy of the system to be zero, we must have

$$-\frac{e^2}{x} - \frac{e^2}{x+y} + \frac{e^2}{y} = 0$$

$$-y(x+y) - xy + x(x+y) = 0$$

$$x^2 - xy - y^2 = 0$$

$$(x/y)^2 - (x/y) - 1 = 0.$$

So $x/y = \alpha$. Thus x/y must be the golden ratio for the potential energy to be zero.

Next we demonstrate a mysterious occurrence of our ubiquitous friend α in the Japanese art of *origami* (folding paper into decorative shapes).

16.11   GOLDEN RATIO BY ORIGAMI

In 1999, P. Glaister of Reading University, England, employed origami to illustrate yet another occurrence of α in a strange place [226].

To see this, take a 2×1 rectangular piece of paper and fold it in half both ways; see Figure 16.13. Make a crease along $\overline{AD}$; see Figure 16.14. Place $\overline{AD}$ along $\overline{AB}$, and form a crease along the fold $\overline{AQ}$ so that $\overline{AQ}$ bisects $\angle DAB$.

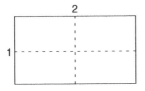

Figure 16.13.

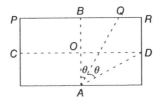

Figure 16.14.

Let $\angle DAB = 2\theta$. From $\triangle AOD$, $\tan 2\theta = 2$. Using the double-angle formula for the tangent function, this yields

$$\frac{2\tan\theta}{1 - \tan^2\theta} = 2$$

$$\tan^2\theta + \tan\theta - 1 = 0$$

$$\tan\theta = -\beta.$$

Therefore, $PQ = PB + BQ = 1 + AB\tan\theta = 1 - \beta = \alpha$, and hence $QR = 2 - \alpha = \beta$.

The next example is based on a calendar problem that appeared in the *Mathematics Teacher* in October, 1999 [95]

Example 16.1. *The points A and C on the axes are each one unit away from the origin. The point B lies one unit away from both axes in the first quadrant. Find the value of x such that the y-axis bisects the area ABCD, where D is the point $(-x, -x)$ and $x > 0$; see Figure 16.15.*

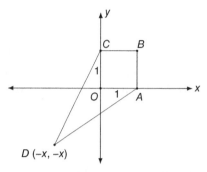

Figure 16.15.

Solution. $OABC$ is a square of unit area. The problem is to find x such that area $CDE = 1 +$ area OAE; see Figure 16.16. The slope of the line $\overline{AD}$ is $x/(x+1)$, so its equation is $y = \dfrac{x}{x+1}(x-1)$.

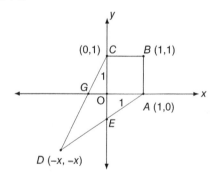

Figure 16.16.

Therefore, the point E is $(0, -x(x+1))$. Thus $OE = x/(x+1)$, and hence

$$CE = 1 + \frac{x}{x+1}$$

$$= \frac{2x+1}{x+1};$$

$$\text{area } CDE = \frac{1}{2} \cdot CE \cdot x$$

$$= \frac{x(2x+1)}{2(x+1)};$$

$$\text{area } OAE = \frac{1}{2} \cdot OA \cdot OE$$

$$= \frac{1}{2} \cdot 1 \cdot \frac{x}{x+1}$$

$$= \frac{x}{2(x+1)}.$$

Thus

$$\frac{x(2x+1)}{2(x+1)} = 1 + \frac{x}{2(x+1)}.$$

This yields $x^2 = x + 1$, so $x = \alpha$.

Then

$$CE = \frac{2\alpha + 1}{\alpha + 1}$$

$$= \frac{\alpha + (\alpha + 1)}{\alpha + 1}$$

$$= \frac{\alpha + \alpha^2}{\alpha + 1}$$

$$= \alpha.$$

Thus the point O divides $\overline{CE}$ in the golden ratio. Besides, $BD : OD = \sqrt{2}\alpha^2 : \sqrt{2}\alpha = \alpha : 1$, so O divides $\overline{BD}$ also in the golden ratio.

In addition, since E is the point $(0, \beta)$, $DA^2 = \alpha^2(1 + \alpha^2)$, and $DE^2 = \alpha^2 + (\alpha + \beta)^2 = 1 + \alpha^2$. Therefore, $DA^2 : DE^2 = \alpha^2 : 1$, so $DA : DE = \alpha : 1$. Thus E divides $\overline{DA}$ in the same magical ratio. By symmetry, it follows that G divides $\overline{DC}$ in the same ratio. ∎

The next geometric example illustrates another occurrence of the golden ratio.

Example 16.2. *Consider an equilateral triangle $\triangle ABC$ inscribed in a circle. Let Q and R be the midpoints of the sides $\overline{AB}$ and $\overline{BC}$. Let $\overleftrightarrow{QR}$ meet the circle at P and S, as in Figure 16.17. Find the length QR.*

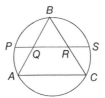

Figure 16.17.

Solution. Let $PQ = RS = 1$ and $QR = x$; see Figure 16.18. By the intersecting chord theorem, $PR \cdot RS = BR \cdot RC$; that is, $1 + x = x^2$. Thus $QR = x = \alpha$.

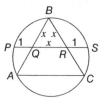

Figure 16.18.   ∎

G. Odom of Poughkeepsie, New York, proposed this example as a problem in 1983 [459]. It re-surfaced five years later in an article by J.F. Rigby of University College at Cardiff, with an interesting byproduct [502]: The ratio of the length of a side of one of the four large triangles in Figure 16.19 to that of a side of one of the three small (black) triangles is the golden ratio.

Figure 16.19.

The next geometric example gives still another occurrence of the golden ratio, including β. Hoggatt proposed it as a problem on the congruence and similarity of triangles in 1964. M.H. Holt gave an interesting solution in the following year [319].

Example 16.3. *Do there exist triangles $\triangle ABC$ and $\triangle PQR$ that have five of their six parts (three sides and three angles) congruent, but are still not congruent?*

Solution. This problem also appears in a high school geometry book by E. Moise and F. Downs, who gave two such triangles as a solution [443]; see Figure 16.20.

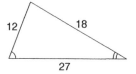
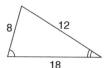

Figure 16.20.

Are there other solutions? If there are, how are they related?

To answer these questions, first notice that the five equal parts cannot include the three sides, since the triangles would then be congruent. Consequently, the five parts must consist of three angles and two sides, so the triangles are indeed similar. But the equal sides cannot be in the same order; otherwise, the triangles would be congruent by the side–angle–side (SAS) theorem or the angle–side–angle (ASA) theorem.

This yields two possibilities for $\triangle PQR$, as Figure 16.21 shows. Since $\triangle ABC \cong \triangle P'Q'R'$, $a/b = b/d = 1/k$ (say), where $k > 0$ and $k \neq 1$. Then $b = ak$ and $d = bk = ak^2$. Since $\triangle ABC \cong \triangle P''Q''R''$, $a/b = b/d = 1/k$, so $b = ak$ and $d = bk = ak^2$. In both cases, the lengths of the sides are in the same ratio $a : b : d = a : ak : ak^2 = 1 : k : k^2$. So, if there are triangles whose sides are in the same ratio $1 : k : k^2$, their parts would be congruent, but the triangles still would not be congruent.

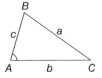 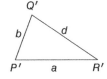 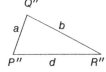

Figure 16.21.

To determine the values of k that yield such triangles, suppose such a triangle exists. Then, by the triangle inequality, $1 + k > k^2$, $1 + k^2 > k$, and $k + k^2 > 1$.

Case 1. Suppose $k > 1$. Then $k^2 > k$, so $1 + k^2 > 1 + k > k$. Also $k + k^2 > k > 1$. Thus, if $k > 1$, then $1 + k^2 > k$ and $k + k^2 > 1$. So it suffices to identify the values of k for which $1 + k > k^2$; that is, $k^2 - k - 1 < 0$.

Since $k^2 - k - 1 = (k - \alpha)(k - \beta)$, $k^2 - k - 1 < 0$ if and only if $\beta < k < \alpha$. But $k > 1$, so $1 < k < \alpha$.

Graphically, k is the value of x for which the line $1 + x = y$ lies above the parabola $y = x^2$, where $x > 1$; see Figure 16.22.

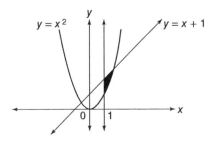

Figure 16.22.

Thus, if k is such that $1 < k < \alpha$, then every $\triangle PQR$ with sides a, ak, and ak^2 will meet the desired conditions.

Case 2. Suppose $k < 1$. Then $k > k^2$; so $1 + k > k^2$. Also, since $1 > k, 1 + k^2 > k$. Thus, $1 + k > k^2$ and $1 + k^2 > k$. So it suffices to look for values of k for which $k^2 + k > 1$; that is, $k^2 + k - 1 > 0$, where $k < 1$.

Since $k^2 + k - 1 = (k + \alpha)(k + \beta), k^2 + k - 1 > 0$ if and only if either $k < -\alpha$ or $k > \beta$. Since $k < 1$, this implies $\beta < k < 1$.

Graphically, k is the value of x for which the parabola $y = x + x^2$ lies above the line $y = 1$, where $x < 1$; see Figure 16.23.

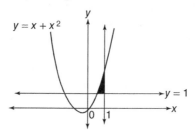

Figure 16.23.

Thus, if k is such that $\beta < k < 1$, then every $\triangle PQR$ with sides a, ak, and ak^2 will satisfy the given conditions.

To sum up, there are infinitely many triangles PQR whose five parts are congruent to those of $\triangle ABC$, but are still *not* congruent to $\triangle ABC$.   ∎

Strange as it may seem, the golden ratio occurs in the solution of differential equations.

16.12   DIFFERENTIAL EQUATIONS

Consider the second-order differential equation $y'' - y' - y = 0$, where $y' = dy/dx$. Its characteristic equation is $t^2 - t - 1 = 0$, so the characteristic roots are α and β. Thus the basic solutions of the differential equation are $e^{\alpha x}$ and $e^{\beta x}$, so the general solution is $y = Ae^{\alpha x} + Be^{\beta x}$, where A and B are arbitrary constants.

Using the initial conditions $y(0) = 0$ and $y'(0) = 1$, we get $y = \dfrac{e^{\alpha x} - e^{\beta x}}{\alpha - \beta}$. Its *Taylor expansion* is $y = \sum\limits_{n=0}^{\infty} \dfrac{F_n}{n!} x^n$.

Conversely, let $y = \sum\limits_{n=0}^{\infty} a_n x^n$. Then the differential equation yields

$$(n + 1)(n + 2)a_{n+2} - (n + 1)a_{n+1} - a_n = 0,$$

where $a_0 = 0$ and $a_1 = 1$. Letting $a_n = \dfrac{b_n}{n!}$, this gives the Fibonacci recurrence $b_{n+2} = b_{n+1} + b_n$, where $b_0 = 0$ and $b_1 = 1$. Consequently, $b_n = F_n$, and hence $a_n = \dfrac{F_n}{n!}$, as expected.

On the other hand, suppose $y(0) = 2$ and $y'(0) = 1$. Then the solution is

$$y = e^{\alpha x} + e^{\beta x} = \sum_{n=0}^{\infty} \frac{L_n}{n!} x^n.$$

GATTEI'S DISCOVERY OF THE GOLDEN RATIO

When P. Gattei was a young sixth former at Queen Elizabeth's Grammar School in Blackburn, England, he stumbled upon a problem involving the inverse f^{-1} of a real-valued function f [214]. Accidentally, he dropped the minus sign and ended up taking the derivative f' of f. This led him to investigate if there were real functions f such that

$$f'(x) = f^{-1}(x), \tag{16.1}$$

where $x \geq 0$.

Gattei discovered an interesting solution: $f(x) = Ax^n$, where A is a constant. Then $f'(x) = Anx^{n-1}$ and $f^{-1}(x) = (x/A)^{1/n}$. By equation (16.1), this implies $Anx^{n-1} = (x/A)^{1/n}$; that is, $A^{n+1}n^n x^{n(n-1)-1} = 1$. So $n^2 - n - 1 = 0$ and $A^{n+1}n^n = 1$. Then $n = \alpha$ or β, and $A = n^{-n/(n+1)}$.

Suppose $n = \beta$. Then $n < 0$, but $n + 1 > 0$. Thus $A = $ (a negative number)$^{\text{(a positive irrational number)}}$, so A is not a real number. Consequently, f is not a real function. But when $n = \alpha$, f is a real function. Thus the only solution to equation (16.1) is $f(x) = (\alpha x)^{\alpha}$.

More generally, consider the function

$$f^{-1}(x) = f^{(m)}(x). \tag{16.2}$$

Let $f(x) = Ax^n$, so $f^{(m)}(x) = An(n-1)\cdots(n-m+1)x^{n-m}$. Then equation (16.2) yields $(x/A)^{1/n} = An(n-1)\cdots(n-m+1)x^{n-m}$. Then

$$A^{n+1}[n(n-1)\cdots(n-m+1)]^n x^{n(n-m)-1} = 1.$$

So $A^{n+1}[n(n-1)\cdots(n-m+1)]^n = 1$ and $n^2 - mn - 1 = 0$. Thus we obtain

$$n = \frac{m + \sqrt{m^2 + 4}}{2} \quad \text{and} \quad A = [n(n-1)\cdots(n-m+1)]^{-n/(n+1)}.$$

Suppose m is odd. If $n = \dfrac{m - \sqrt{m^2 + 4}}{2}$, then, as before, we can show that A is not a real number; but $n = \dfrac{m + \sqrt{m^2 + 4}}{2}$ leads to a valid solution: $f(x) = [n(n-1)\cdots(n-m+1)]^{-n/(n+1)}x^n$.

On the other hand, let m be even. Then $n(n-1)\cdots(n-m+1)$ is positive, whether n is positive or negative. We then have two solutions:

$$f(x) = [n(n-1)\cdots(n-m+1)]^{-n/(n+1)}x^n, \text{ where } n = \frac{m \pm \sqrt{m^2 + 4}}{2}.$$

Next we show an occurrence of the golden ratio in a snow plowing problem.

GOLDEN RATIO AND SNOW PLOWING

T. Ratliff of Wheaton College, Norton, Massachusetts, discussed an interesting problem at the 1996 Fall Meeting of the Northeast Section of the Mathematical Association of America, held at the University of Massachusetts, Boston [495]. A similar problem appeared in 1984 in *Mathematical Spectrum* (Problem 16.6). The solutions of both versions involve the golden ratio:

> One wintry morning, it started snowing at a constant and heavy rate. A snowplow started plowing at 8 a.m.; by 9 a.m., it plowed two miles; and by 10 a.m., it plowed another mile. Assuming that the snowplow removes a constant volume of snow per hour, what time did it start snowing?

Solution. Suppose it started snowing at time t (in hours) and the plowing began T hours before 8 o'clock. Let $x = x(t)$ denote the distance traveled by the snowplow in time t. Since the speed of the plow is inversely proportional to the depth of the snow, it follows that

$$\frac{dx}{dt} = \frac{k}{\text{depth at time } t}$$
$$= \frac{k}{ct}$$
$$= \frac{K}{t},$$

where k, c, and K are constants.

Solving this differential equation, we get $x = K \ln t + C$, where $x(T) = 0$, $x(T + 1) = 2$, and $x(T + 2) = 3$, and C is a constant. The condition $x(T) = 0$ implies $C = -k \ln T$, so $x = K \ln(t/T)$. The other two conditions yield $2 = K \ln \left(\frac{T + 1}{T} \right)$ and $3 = K \ln \left(\frac{T + 2}{T} \right)$. Then

$$\frac{\ln[(T + 1)/T]}{\ln[(T + 2)/T]} = \frac{2}{3}$$
$$2 \ln(T + 2) - 2 \ln T = 3 \ln(T + 1) - 3 \ln T$$
$$T(T + 2)^2 = (T + 1)^3$$
$$T^2 + T - 1 = 0.$$

So $T = -\beta \approx 0.61803398875$. Thus $T \approx 37$ minutes, 5 seconds. Consequently, it started snowing at 7:22:55 a.m.

Next we cite an occurrence of the golden ratio in algebra.

16.13 GOLDEN RATIO IN ALGEBRA

In 1936, the well-known Scottish-American mathematician Eric T. Bell (1883–1960) proved that the only polynomial, symmetric function $\varphi(s, t)$ that satisfies the associativity condition $\varphi(x, \varphi(x, y)) = \varphi(\varphi(x, y), z)$ is $\varphi(s, t) = s \star t = a + b(s + t) + cst$, where a, b, and c are arbitrary constants such that $b^2 - b - ac = 0$, and s and t are complex numbers. In particular, let $ac = 1$. Then $b^2 - b - 1 = 0$, so $b = \alpha$ or β.

Thus the binary operation $\star$, defined by $\varphi(s, t) = s \star t = a + b(s + t) + cst$, is associative only if $b = \alpha$ or β, where $ac = 1$. We can easily confirm this; see Exercise 16.47.

BILINEAR TRANSFORMATION

In 1964, Hoggatt discovered a close relationship between the bilinear transformation $w = (az + b)/(cz + d)$ and Fibonacci numbers [269]. This is the essence of the following theorem.

Theorem 16.1. *The bilinear transformation $w = \dfrac{az + b}{cz + d}$ has two distinct fixed points, α and β, if and only if $a - d = b = c \neq 0$, where $a, d > 0$; $ad - bc = 1$; and a, b, c, and d are integers.*

Proof. Suppose the bilinear transformation has a fixed point. It is a solution of the equation $z = (az + b)/(cz + d)$; that is, $cz^2 - (a - d)z - b = 0$. Since there are two fixed points α and β, and $c \neq 0$, we then have

$$z^2 - \frac{a - d}{c} z - \frac{b}{c} = (z - \alpha)(z - \beta)$$
$$= z^2 - z - 1.$$

Equating the coefficients of like terms, we get $a - d = b = c \neq 0$.

Conversely, let $a - d = b = c \neq 0$. Then $w = \dfrac{az + b}{bz + (a - b)}$. Its fixed points are given by $z = \dfrac{az + b}{bz + (a - b)}$. Then $z^2 - z - 1 = 0$; so the fixed points are α and β. This completes the proof. ∎

16.14 GOLDEN RATIO IN GEOMETRY

In 1966, J.A.H. Hunter of Toronto, Canada, investigated a triangle with the property that the square of the length of one side equals the product of the lengths of the other two sides [340]. Suppose the lengths are ka^2, kab, and kb^2, where $a > b$ and $k \geq 1$; see Figure 16.24. Then, by the triangle inequality, $kab + kb^2 > ka^2$;

so $(a/b)^2 - (a/b) - 1 < 0$. Consequently, $\beta < a/b < \alpha$. Thus if we choose a and b such that $\dfrac{F_{2n-1}}{F_{2n}} < \dfrac{a}{b} < \dfrac{F_{2n}}{F_{2n-1}}$, then such a triangle will have the desired property.

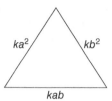

Figure 16.24.

Next we show how the golden ratio is related to an isosceles triangle.

ISOSCELES TRIANGLE AND THE GOLDEN RATIO

Let $\triangle ABC$ be an isosceles triangle with $AB = AC$. Let D be a point on $\overline{AB}$ such that $AD = CD = BC$. We will now show that $2\cos A = \dfrac{AB}{BC} = \alpha^{\dagger}$.

By the law of cosines, $CD^2 = AD^2 + AC^2 - 2AD \cdot AC \cos A$. Since $AD = CD$, this yields $2\cos A = \dfrac{AC}{AD} = \dfrac{AB}{BC}$, confirming one-half of the result.

Since $\triangle BCD$ is isosceles, $\angle B = \angle BDC$. But $\angle ADC + \angle BDC = \pi = \angle DAC + \angle ADC + \angle ACD$. Thus $\angle B = \angle BDC = \angle DAC + \angle ACD = 2\angle A$. This implies that $\angle A = \pi/5$. Since $\cos \pi/5 = \alpha/2$, it follows that $2\cos A = \alpha$, as desired. ∎

GOLDEN RATIO AND CENTROIDS OF CIRCLES

Consider two circles A and B, one inside the other, but tangential to each other at a point O; see Figure 16.25. Let their radii be a and $b\,(< a)$, respectively; so their areas are πa^2 and πb^2, respectively. Let C_A and C_B be the centroids of the circles, so the points O, C_A, and C_B are collinear. Then the centroid C of the remnant $A - B$ is the endpoint of the diameter of C_B through O.

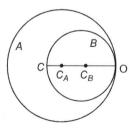

Figure 16.25.

---

†Source unknown.

Taking moments about O, we get

$$\pi b^2 \cdot OC_B + \pi(a^2 - b^2) = \pi a^2 \cdot OC_A$$

$$1 + \left(\frac{a^2}{b^2} - 1\right)\left(\frac{2b}{a}\right) = \left(\frac{a^2}{b^2}\right)\left(\frac{a}{b}\right)$$

$$2\left(\frac{a^2}{b^2} - 1\right) = \frac{a^3}{b^3} - 1.$$

Since $a \neq b$, this yields

$$2(a/b + 1) = a^2/b^2 + a/b + 1$$

$$(a/b)^2 - (a/b) - 1 = 0.$$

Since $a > b$, it follows that $a/b = \alpha$.

We can extend this discussion to any planar figure, as H.E. Huntley did in 1974 [346].

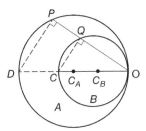

Figure 16.26.

As an additional exercise, let a chord $\overline{OP}$ of circle C_A intersect circle C_B at Q; see Figure 16.26. Since the angle in a semicircle is a right angle, it follows that $\triangle OPD \approx \triangle OQC$. Therefore,

$$\frac{OP}{OQ} = \frac{OD}{OC} = \frac{2a}{2b} = \alpha.$$

Thus Q divides the chord $\overline{OP}$ in the golden ratio.

MUTUALLY TOUCHING CIRCLES

The next example presents another occurrence of the divine proportion, studied by Colin Singleton of Sheffield, England [531].

Example 16.4. *Suppose the radii of four mutually touching circles are in geometric progression, say, $1, r, r^2$, and r^3; see Figure 16.27. Find the value of r. (Obviously, the geometric progression can be continued indefinitely in both directions. In the figure, the circle with a black dot represents the circle with radius $1/r$, and the shaded area is a portion of the circle with radius r^4.)*

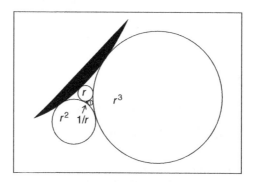

Figure 16.27.

Solution. We will employ the fact that when four circles mutually touch, as in the problem, "the sum of the squares of all four bends [that is, the reciprocals of the radii] is half of the square of their sum," as observed by F. Soddy in 1936 [533]. Consequently, it suffices to solve the equation

$$(1 + r + r^2 + r^3)^2 = 2(1 + r^2 + r^4 + r^6).$$

That is, $(r^2 + 1)(r^4 - 2r^3 - 2r^2 - 2r + 1) = 0$. Clearly, $r^2 + 1 \neq 0$; so we have $r^4 - 2r^3 - 2r^2 - 2r + 1 = 0$.

To solve this biquadratic equation, we make an important observation. If v (Greek letter *nu*) is a solution, then so is $1/v$. So $(r - v)(r - 1/v) = r^2 - (v + 1/v)r + 1 = r^2 - \kappa r + 1$ must be a factor of $r^4 - 2r^3 - 2r^2 - 2r + 1$, where κ (Greek letter *kappa*) equals $v + 1/v$. Now we can factor the biquadratic polynomial:

$$r^4 - 2r^3 - 2r^2 - 2r + 1 = (r^2 - \kappa r + 1)[r^2 - (\kappa - 2)r + 1].$$

Equating the coefficients of r^2 from both sides, we get the quadratic equation $\kappa^2 - 2\kappa - 4 = 0$. Solving this equation, we get $\kappa = 1 \pm \sqrt{5}$. But $\kappa > 0$, so $\kappa = 1 + \sqrt{5} = 2\alpha$.

Thus r is a solution of the equation $r^2 - 2\alpha r + 1 = 0$; and $r = \alpha \pm \sqrt{\alpha} \approx 2.89005, 0.34601$.   ∎

The next example, although simple in nature, features still another occurrence of the golden ratio. A.D. Rawlins of Brunel University, London, England, studied

it in 1995 [496]. It provides an interplay of two concentric circles, an ellipse, and
the divine ratio.

Example 16.5. *Consider an annulus (ring-shaped area) formed by two concentric
circles with radii a and b, where a > b. Inscribe an ellipse of major axis 2a and minor
axis 2b, so it touches the outer and inner circles; see Figure 16.28. Suppose the area
of the ellipse equals that of the annulus. Compute the ratio a/b.*

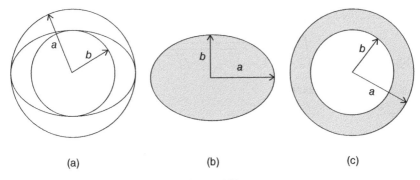

(a)   (b)   (c)

Figure 16.28.

Solution. Since the area of the ellipse equals that of the annulus, it follows that

$$\pi\, ab = \pi(a^2 - b^2)$$
$$a^2 - ab - b^2 = 0$$
$$(a/b)^2 - (a/b) - 1 = 0.$$

So $a/b = \alpha$, the divine ratio.   ∎

EXERCISES 16

1. Is $\alpha : 1 = 1 : \dfrac{1}{\alpha}$?

2. Is $\alpha : 1 = 1 : -\beta$?

3. Is $\alpha : 1 = 1 : \alpha - 1$?

4. Let C divide $\overline{AB}$ in the golden ratio, $\overline{AC}$ being the larger segment. Let $AC =
   1$. Show that $BC = 1/\alpha$ and $AB = \alpha$.

5. Let C divide $\overline{AB}$ of unit length in the golden ratio, $\overline{AC}$ being the larger segment. Show that $BC = 1/\alpha^2$ and $AC = 1/\alpha$.

6. Suppose $BD = 1$ in Figure 16.3. Find BC.

Let C divide the line segment $\overline{AB}$ in the golden ratio, where $AB = 1$ and $AC = t$.

7. Find a quadratic equation satisfied by t.

8. Solve the equation.

9. Find the value of t.

10. Show that $t = -\beta$.

11. Evaluate the sum $\sqrt{1 - \sqrt{1 - \sqrt{1 - \sqrt{1 - \cdots}}}}$, if possible.

12. Show that $-\beta = \sqrt{1 - \sqrt{1 - \sqrt{1 - \cdots \sqrt{1 - 1/\alpha}}}}$.

Let $a/b = c/d$. Prove each.

13. $\dfrac{b}{a} = \dfrac{d}{c}$.

14. $\dfrac{a+b}{b} = \dfrac{c+d}{d}$.

15. $\dfrac{a-b}{b} = \dfrac{c-d}{d}$.

16. $\dfrac{a+b}{a-b} = \dfrac{c+d}{c-d}$.

17. Suppose a side of the Great Pyramid is $2b$. Show that the altitude of a lateral face is $b\alpha$.

18. The base of the Great Pyramid is square. Show that the ratio of the sum of the areas of its lateral faces to the base area is α.

Prove each.

19. $\alpha = 1 + \dfrac{1}{\alpha}$.

20. $\alpha = \dfrac{1}{\alpha - 1}$.

21. $\alpha^n = \alpha^{n-1} + \alpha^{n-2}$, where $n \geq 2$.

22. $\alpha^{-n} = \alpha^{-(n+1)} + \alpha^{-(n+2)}$.

23. $\displaystyle\sum_{n=1}^{\infty} \alpha^{-n} = \alpha$.

24. $\displaystyle\sum_{n=0}^{\infty} \alpha^{-n} = \alpha^2$.

25. $\displaystyle\sum_{n=0}^{\infty} \alpha^{-2n} = -\beta$.

26. $\displaystyle\sum_{k=1}^{n} \lfloor \alpha^k \rfloor = L_{n+2} - \lfloor n/2 \rfloor - 3$ (Alfred, 1965 [8]).

27. $\displaystyle\sum_{n=1}^{\infty} \frac{1}{\alpha^{2n-1}} = 1.$

28. $\displaystyle\sum_{n=1}^{\infty} \frac{n}{\alpha^{n+1}} = \alpha^2.$

29. $\displaystyle\sum_{n=1}^{\infty} \frac{n}{\alpha^{2n+1}} = -\beta.$

30. $\displaystyle\sum_{n=1}^{\infty} \frac{n}{\alpha^{2n+2}} = \beta^2.$

Verify each.

31. $\alpha\sqrt{3 - \alpha} = \sqrt{\alpha + 2}.$

32. $\sqrt{3 - \alpha} = \dfrac{\sqrt{10 - 2\sqrt{5}}}{2}.$

33. $\sqrt{\alpha + 2} = \dfrac{\sqrt{10 + 2\sqrt{5}}}{2}.$

Using the fact that $\cos \pi/5 = \alpha/2$, express each in terms of α.

34. $\sin \pi/5.$

35. $\cos \pi/10.$

36. $\sin \pi/10.$

37. Let v (lowercase Greek letter *nu*) be a solution of the equation $x^2 = x + 1$. Show that $v + (1/v^2) = 2$.

Evaluate each, where G_n denotes the nth Gibonacci number.

38. $\displaystyle\lim_{n\to\infty} \frac{F_n}{F_{n+1}}.$

39. $\displaystyle\lim_{n\to\infty} \frac{L_n}{L_{n+1}}.$

40. $\displaystyle\lim_{n\to\infty} \frac{L_n}{F_n}.$

41. $\displaystyle\lim_{n\to\infty} \frac{G_{n+1}}{G_n}.$

42. $\displaystyle\lim_{n\to\infty} \frac{2n}{n + 1 + \sqrt{5n^2 - 2n + 1}}$ (Hoggatt and Lind, 1967 [314]).

The semi-vertical angle of a right-circular cone is $54°$, and its lateral side is one unit long. Compute each.

43. Base circumference.

44. Base area.

45. Volume of the cone.

46. Lateral surface area.

47. Show that the binary operation $\star$, defined by $s \star t = a + b(s + t) + cst$, is associative only if $b = \alpha$ or β, where $ac = 1$.

48. Show that the equation $x^n - xF_n - F_{n-1} = 0$ has no solution $> \alpha$, where $n \geq 2$ (Wall, 1964 [578]).

49. Show that the equation $x_{n+1} = \sqrt[n]{F_{n-1} + x_n F_n}$, where $x_0 \geq 0$ and $n \geq 2$. Find $\lim_{n \to \infty} x_n$, if it exists (Wall, 1964 [578]).

50. Find the value of x such that $n^n + (n + x)^n = (n + 2x)^n$, where $n \geq 1$ (Alfred, 1964 [7]).

51. Evaluate $\sum_{n=0}^{\infty} |\beta|^n$.

52. Let $I_n = \int_0^1 x^{n-1} dx$, where $n \geq 2$. Evaluate $\lim_{n \to \infty} I_n$.

53. Let t be a number such that $t = \int_0^1 x^t \, dx$. Find the value of t.

54. Derive a formula for $\sum_{i=0}^{n} \lfloor \alpha^i \rfloor$.

55. Evaluate $\lim_{n \to \infty} (F_{n+k}/L_n)$, where k is a positive integer (Dence, 1968 [142]).

56. Evaluate $\lim_{n \to \infty} (L_{n+k}/F_n)$, where k is a positive integer.

57. Let $a_n = a_{n-1} + a_{n-2} + k$, where $a_0 = 0$, $a_1 = 1$, and k is a constant. Find $\lim_{n \to \infty} (a_n/F_n)$ (Shallit, 1976 [519]).

Let $b_n = b_{n-1} + b_{n-2} + k$, where $b_0 = 2$, $b_1 = 1$, and k is a constant. Find each.

58. $\lim_{n \to \infty} (b_n/F_n)$.

59. $\lim_{n \to \infty} (b_n/L_n)$.

60. Let $c_n = c_{n-1} + c_{n-2} + k$, where $c_0 = a$, $c_2 = b$, and k is a constant. Find $\lim_{n \to \infty} (c_n/G_n)$, where G_n denotes the nth Gibonacci number.

61. Evaluate $\sum_{i=0}^{n} \binom{n}{i} \alpha^{3i-2n}$ (Freitag, 1975 [192]).

62. Evaluate the infinite product $\left(1 + \dfrac{1}{2}\right)\left(1 + \dfrac{1}{13}\right)\left(1 + \dfrac{1}{610}\right)\left(1 + \dfrac{1}{1346269}\right) \cdots$ (Shallit, 1980 [520]).

63. Consider the real sequence $\{x_n\}_0^{\infty}$, defined by $x_{n+1} = 1/(x_n + 1)$. Find x_0 such that $\lim_{n \to \infty} x_n$ exists, and then find the limit (Neumer, 1993 [455]).

Consider the vector space $V = \{(v_1, v_2, \ldots, v_n, \ldots) | v_n = v_{n-1} + v_{n-2}, v_i \in \mathbb{R}$, and $n \geq 3\}$ with the usual operations (Barbeau, 1993 [20]).

64. Do the vectors $u = (1, 0, 1, 1, 2, 3, 5, \ldots)$ and $v = (0, 1, 1, 2, 3, 5, 8, \ldots)$ belong to V?

65. Show that V is 2-dimensional.

66. Show that $(r, r^2, r^3, \ldots) \in V$ if and only if $r^2 = r + 1$.

67. Find r if $(r, r^2, r^3, \ldots) \in V$.

68. Let $F = (1, 1, 2, 3, 5, 8, \ldots) \in V$. Let $u = (\alpha, \alpha^2, \alpha^3, \ldots)$ and $v = (\beta, \beta^2, \beta^3, \ldots)$. Find the constants a and b such that $F = au + bv$.

69. With F, u, and v as in Exercise 16.68, deduce Binet's formula.

70. Let $f(x) = Ax^n$ and $f^{-1}(x) = [f^{(m)}(x)]^p$. Show that $n = \dfrac{mp \pm \sqrt{p^2 m^2 + 4p}}{2p}$ (Gattei, 1990 [214]).

In Exercises 16.71–16.73, G_n denotes the nth Gibonacci number (DiDomenico, 2007 [148]).

71. Prove that $G_{n+2}^4 - G_n G_{n+1} G_{n+3} G_{n+4}$ is a constant K.

72. Find a formula for K.

73. When $K = 0$, show that $\dfrac{b}{a} = \alpha$.

17

GOLDEN TRIANGLES AND RECTANGLES

> Beauty is truth, truth beauty, that is all
> Ye know on earth, and all ye need to know.
> –John Keats, *Ode on a Grecian Urn*

According to Martin Gardner (1914–2010), a popular *Scientific American* columnist, "Pi (π) is the best known of all irrational numbers. The irrational number α is not so well-known, but it expresses a fundamental ratio that is almost as ubiquitous as *pi*, and it has the same amusing habit of popping up where least expected." Gardner made this trenchant observation in 1959. In the preceding chapter, we found that the ubiquitous number α makes spectacular appearances in plane geometry. As can be predicted, it does also in solid geometry; see Chapter 18.

Some triangles, such as the golden triangle, are linked to the golden ratio in a mysterious way. We begin our pursuit with a simple definition.

17.1 GOLDEN TRIANGLE

An isosceles triangle is a *golden triangle* if the ratio of a lateral side to the base is α.

We now pursue some interesting properties of a golden triangle. The golden ratio acts as a chord intertwining them all.

Fibonacci and Lucas Numbers with Applications, Volume One, Second Edition. Thomas Koshy.
© 2018 John Wiley & Sons, Inc. Published 2018 by John Wiley & Sons, Inc.

Theorem 17.1. *Let $\triangle ABC$ be a golden triangle with base $\overline{AC}$. Let D divide $\overline{BC}$ in the golden ratio, $\overline{BD}$ being the larger segment. Then $\overline{AD}$ bisects $\angle A$.*

Proof. Since D divides $\overline{BC}$ in the golden ratio, $BD = \alpha CD$, $BD + CD = \alpha CD + CD = (\alpha + 1)CD = \alpha^2 CD$; that is, $BC = \alpha^2 CD$. Thus $BC = \alpha^2 CD = \alpha AC$, so $\alpha CD = AC$; see Figure 17.1.

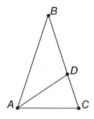

Figure 17.1.

Thus $\angle BCA = \angle ACD$ and $AB/AC = AC/CD$. So $\triangle BAC \sim \triangle ACD$. Consequently, $\angle ACD = \angle ADC$ and $\angle ABC = \angle CAD$. So $\triangle ACD$ is an isosceles triangle with $AD = AC = \alpha CD$. Thus $BD = AD$, so $\triangle ABD$ is also an isosceles triangle. Hence $\angle BAD = \angle ABD$. Thus $\angle BAD = \angle CAD$; that is, $\overline{AD}$ bisects $\angle BAC$.   ∎

The following corollary gives an interesting byproduct of Theorem 17.1.

Corollary 17.1. *Let $\triangle ABC$ be a golden triangle with base $\overline{AC}$. Let D divide $\overline{BC}$ in the golden ratio, $\overline{BD}$ being the larger segment. Then $\triangle CAD$ is also a golden triangle.*

Proof. Using Figure 17.1 and the preceding proof, $AD = AC = \alpha CD$; so $\triangle CAD$ is a golden triangle.   ∎

Thus $\overline{AD}$ cuts the golden triangle $\triangle ABC$ into two isosceles triangles, $\triangle ABD$ and $\triangle CAD$, the latter being similar to $\triangle ABC$.

The next result is an interesting property of a golden triangle.

Theorem 17.2. *The included angle between the equal sides of an isosceles golden triangle is 36°.*

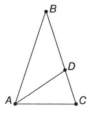

Figure 17.2.

Proof. Let $\triangle ABC$ be a golden triangle with $AB = BC = \alpha AC$. Let D divide $\overline{BC}$ in the golden ratio; see Figure 17.2. By Corollary 17.1, $\triangle CAD$ is a golden triangle similar to $\triangle ABC$.

Let $\angle ACD = 2x$. Then from $\triangle ACD$, $2x + 2x + x = 180°$, so $x = 36°$. Thus $\angle A = \angle C = 72°$, and $\angle B = 36°$. ∎

Is the converse true? Yes. If the non-repeating angle in an isosceles triangle is 36°, then the triangle is a golden triangle, as the next theorem shows.

Theorem 17.3. *If the non-repeating angle in an isosceles triangle is* 36°, *then it is a golden triangle.*

Proof. Let $\triangle ABC$ be an isosceles triangle with $AB = BC$, and $\angle A = 36°$. Then $\angle B = \angle C = 72°$; see Figure 17.2.

Let $\overline{AD}$ bisect $\angle A$. Then $\angle ADC = 72°$, so $AC = AD = BD = y$; and $\triangle ABC \sim \triangle ACD$. Then $AB/AC = BC/CD$; that is, $x/y = y(x - y)$. Thus $(x/y)^2 - (x/y) - 1 = 0$. Consequently, $x/y = \alpha$.

Thus $AB : AC = \alpha : 1$, so $\triangle ABC$ is a golden triangle. ∎

Combining Theorems 17.2 and 17.3, we get the following result.

Corollary 17.2. *An isosceles triangle is a golden triangle if and only if its angles are* 36°, 72°, *and* 72°. ∎

The next example is a delightful application of Corollary 17.2. It appeared as a Leningrad Mathematical Olympiad problem in 1989; see [321].

Example 17.1. *The perimeter of the star polygon ABCDE in Figure 17.3 is 1. Suppose the five angles θ at the vertices are equal. Find the perimeter p of the convex pentagon PQRST.*

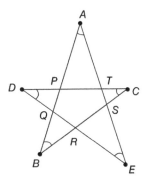

Figure 17.3.

Solution. Since $\triangle PBC$ is isosceles, it follows that $\angle DPQ = 2\theta$. Triangle QAE is also isosceles; so $\angle PQD = 2\theta$. Consequently, $\triangle PQD$ is isosceles. Since

$\theta + 2\theta + 2\theta = 180°$, it follows that $\theta = 36°$; so $\triangle PQD$ is an isosceles triangle with vertex angles $36°, 72°$, and $72°$. Thus, by Corollary 17.2, it is a golden triangle. Hence $DP = DQ = \alpha PQ$ and $DP + DQ = 2\alpha PQ$.

Similarly, $AP + AT = 2\alpha PT$, $BQ + BR = 2\alpha QR$, $ER + ES = 2\alpha RS$, and $CS + CT = 2\alpha ST$.

Since $(DP + DQ) + (AP + AT) + (BQ + BR) + (ER + ES) + (CS + CT) = 1$, it follows that

$$2\alpha(PQ + QR + RS + ST + TP) + p = 1$$

$$2\alpha p + p = 1$$

$$p = \frac{1}{2\alpha + 1}$$

$$= -\beta^3.$$ ∎

A golden triangle can also be characterized by areas, as the next theorem shows. We omit its proof for the sake of brevity; see Exercises 17.2 and 17.3.

Theorem 17.4. *Let D be a point on side $\overline{BC}$ of an isosceles triangle ABC such that $\triangle ABC \sim \triangle CAD$, where $AB = BC$. Then $\triangle ABC$ is a golden triangle if and only if area of $\triangle ABC$: area of $\triangle BDA = \alpha : 1$.* ∎

The next result is interesting in its own right.

Theorem 17.5. *Let the ratio a/b of two sides a and b of $\triangle ABC$ be greater than one. Remove a triangle with side b from $\triangle ABC$. The remaining triangle is similar to the original triangle if and only if $a/b = \alpha$.*

Proof. Remove $\triangle ABD$ from $\triangle ABC$; see Figure 17.4. Since $\triangle ADC \sim \triangle BAC$, $\dfrac{AC}{BC} = \dfrac{DC}{AC}$; that is,

$$\frac{b}{a} = \frac{a - b}{b}$$

$$= \frac{a}{b} - 1.$$

Since $a > b$, it follows that $a/b = \alpha$.

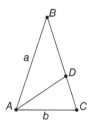

Figure 17.4.

Conversely, let $a/b = \alpha$. Then

$$\frac{DC}{AC} = \frac{a-b}{b}$$

$$= \frac{a}{b} - 1$$

$$= \alpha - 1$$

$$= \frac{1}{\alpha}$$

$$= \frac{b}{a}$$

$$= \frac{AC}{BC}.$$

Since $\angle C$ is common to triangles ADC and BAC, it follows that $\triangle ADC \sim \triangle BAC$. ∎

The next theorem is closely related to this.

Theorem 17.6. *Let the ratio of two sides of a triangle be $k > 1$. A triangle similar to the triangle can be removed from it in such a way that the ratio of the area of the original triangle and that of the remaining triangle is also k if and only if $k = \alpha$.*

Proof. Let $\triangle ADC \sim \triangle BAC$ such that $AC/BC = DC/AC = k$; see Figure 17.5. Let $\dfrac{\text{area} \triangle BAC}{\text{area} \triangle ABD} = k$.

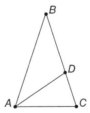

Figure 17.5.

Since $\triangle ABC$ and $\triangle ADC$ have the same altitude h from A, we have

$$\frac{\text{area} \triangle BAC}{\text{area} \triangle ABD} = \frac{1/2 \cdot BC \cdot h}{1/2 \cdot BD \cdot h}$$

$$= \frac{BC}{BD}$$

$$= \frac{BC}{BC - CD}$$

$$= \frac{BC/AC}{BC/AC - CD/AC}.$$

That is, $k = k/(k - 1/k)$; so $k = \alpha$.

Conversely, let $\dfrac{BC}{AC} = k = \alpha = \dfrac{\text{area } \triangle BAC}{\text{area } \triangle ADC}$. Then

$$\alpha = \frac{BC/AC}{BC/AC - DC/AC}$$

$$= \frac{\alpha}{\alpha - DC/AC}$$

$$\frac{DC}{AC} = \alpha - 1$$

$$= \frac{1}{\alpha};$$

that is, $AC/DC = \alpha$. Thus $\triangle BAC \sim \triangle ADC$.   ∎

REGULAR PENTAGRAM AND DECAGON

Regular pentagrams and decagons (10-gons) contain golden triangles. To see this, first consider a regular pentagram; see Figure 17.6. Since the angle at a vertex is 108°, it follows that $\angle BAC = 36°$, so $\triangle ABC$ is a golden triangle. The pentagon contains five golden triangles.

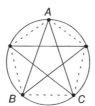

Figure 17.6.   **Figure 17.7.**

Since the central angle of a regular decagon is 360°, each side subtends an angle of 36° at the center; see Figure 17.7. It now follows by Corollary 17.2 that each of the triangles AOB is a golden triangle.

Next we study golden rectangles.

17.2 GOLDEN RECTANGLES

In *Der Goldene Schnitt* (1884), Adolf Zeising's classic work on the golden section, Zeising argued that "the golden ratio is the most artistically pleasing of all proportions and the key to the understanding of all morphology (including human anatomy), art, architecture, and even music."

Take a good look at the four picture frames of various proportions, represented in Figure 17.8. Which of them is aesthetically most pleasing? Most pleasing to the eyes? Frame (*a*) is too square; frame (*b*) looks narrow; and frame (*c*) appears too wide! So if you picked frame (*d*) as your top choice, you are right; it has aesthetically more pleasing proportions.

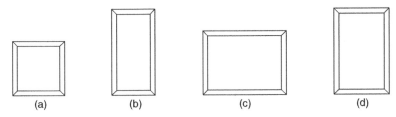

(a) (b) (c) (d)

Figure 17.8.

This choice puts you in good company. German psychologists Gustav Theodor Fechner (1801–1887) and Wilhelm Max Wundt (1832–1920) provide ample empirical support to Zeising's claims. They measured thousands of windows, picture frames, playing cards, books, mirrors, and other rectangular objects, and even checked the points where graveyard crosses were divided. They concluded that most people unconsciously select rectangular shapes in the golden ratio when selecting such objects. And, of course, such pleasing proportions were the basis of most ancient Greek art and architecture.

The American artist Jay Hambidge (1867–1924) of Yale University, in his extensive writings on *dynamic symmetry*, highlighted the prominent role the golden ratio has played in numerous Greek artworks, as well as modern art, architecture, and furniture design.

More recently, Frank A. Lone of New York confirmed one of Zeising's favorite theories. He measured the heights of 65 women and compared them to the heights of their navels. He found the ratio to be about 1.618, which he called the *lone relativity constant*. He also found a fascinating relationship between α and π:

$$\frac{6\alpha^2}{5} \approx \pi.$$

We now introduce a unique class of rectangles and present a number of occurrences of such rectangles.

Figure 17.8*d*, represented in Figure 17.9, has the fascinating property that the ratio of the length x of the longer side to the length y of the shorter side equals the ratio of their sum to the length of the longer side; that is, $x/y = \dfrac{x+y}{x}$. This yields the equation $x/y = 1 + y/x$; so $x/y = \alpha$, as we could have predicted. Such a rectangle is a *golden rectangle*.

Zerger devised a clever method for constructing a large rectangle that approximates a golden rectangle. Place 20 ordinary $8\frac{1}{2} \times 11$ sheets of paper in four rows of five each; see Figure 17.10. The resulting shape is a 34×55 rectangle, which is a pretty good approximation to a golden rectangle.

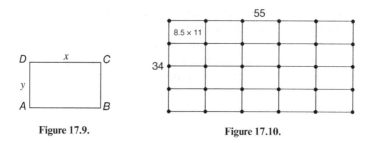

Figure 17.9.   Figure 17.10.

As another example, consider Figure 17.11, based on Land [379]. The lighthouse in the picture is drawn at a pivotal position. It divides the picture into two rectangular parts in such a way that if a denotes its distance from the left side and b that from the right side, then $a/b = (a + b)/a$. This is the golden ratio, so the rectangle in Figure 17.11 is indeed a golden rectangle.

Figure 17.11.   The lighthouse divides the picture in a way that creates a golden rectangle.*

Since the golden rectangle is the most pleasing rectangle, countless artists have used golden rectangles frequently in their work.

The *Holy Family* by Michelangelo Buonarroti (1475–1564), *Madonna of the Magnificat* by Sandro Botticelli (1444–1510), and, more recently, *Crucifixion* (*Corpus Hypercubus*) and *The Sacrament of the Last Supper* by the Spanish surrealist Salvador Dali (1904–1989) are wonderful illustrations of the visual power and beauty of the golden rectangle.

Figure source: Garland, 1987 [213]. Reproduced with permission of Pearson Education, Inc.

According to *Time* magazine, January 24, 1955, Dali originally titled his masterpiece *Corpus Hypercubus* (Hypercubic Body). His painting is based on "the harmonious division of a specific golden rectangle;" see Figure 17.12.

Figure 17.12.

Leonardo da Vinci(1452–1519) painted *St. Jerome* to fit very nicely into a golden rectangle; see Figure 17.13a. Art historians believe that da Vinci deliberately painted the figure according to the classical proportions he inherited from the Greeks. Michelangelo's *David* also illustrates a golden rectangle; see Figure 17.13b.

(a)　　　　　　　　　　　(b)

Figure 17.13. (a) *St. Jerome* by da Vinci fits into a golden rectangle.* (b) Michelangelo's *David* also illustrates a golden rectangle.*

Figure sources: Both images are from Scala/Art Resource, New York.

According to Sr. Marie Stephen of Rosary College, River Forest, Illinois, "da Vinci used [the golden ratio] in laying out canvases in such a manner that the points of interest would be at the intersections of the diagonals and perpendicular from the vertices;" see Figure 17.14.

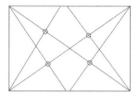

Figure 17.14. *Source*: Stephen, 1956 [543]. Reproduced with permission of The National Council of Teachers of Mathematics.

Golden rectangles are also evident in the work of Albrecht Dürer (1471–1528), a renowned German Renaissance painter, engraver, and designer. They also appear in modern abstract art such as *La Parade* by the French impressionist Georges Seurat (1859–1891). Seurat is said to have approached every canvas with the magical ratio in mind. The same can be said about many works by the Dutch abstractionist Pieter Cornelis Mondriaan (1872–1944). Juan Gris (1887–1927), the Spanish-born cubist greatly influenced by Pablo Picasso and Georges Braque, also lavishly applied the golden ratio in his work.

17.3   THE PARTHENON

The magnificent *Parthenon*, erected by the ancient Athenians in honor of Athena Parthenos, the patron goddess of Athens, stands on the Acropolis. It is a monument to the Greeks' worship of the golden rectangle; see Figure 17.15. The whole shape fits nicely into a golden rectangle. Even the Nashville, Tennessee,

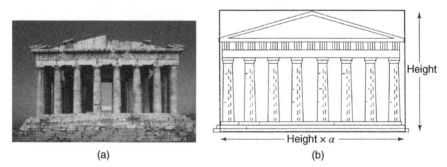

Figure 17.15. (a) A view of the Parthenon in Athens. (b) This magnificent building fits into a golden rectangle. *Figure source*: Photo Researchers. © Marcello Bertinetti, Photo Researchers, Inc., New York.

re-construction of the original Parthenon illustrates the aesthetic power of the golden rectangle; see Figure 17.16.

Figure 17.16.   The Parthenon in Nashville, Tennessee.*

According to R.F. Graesser of the University of Arizona, the golden ratio was used in the facade and floor plan of the Parthenon, as well as in facades and floor plans of other Greek temples. The occurrences of the golden rectangle in the architecture are depicted beautifully in Walt Disney's animated film *Donald Duck in Mathemagicland.*

Architect Le Corbusier (1887–1965) (Charles Edouard Jeanneret-Gris), one of the most influential designers of the twentieth century, developed a scale of proportions called the *modulator*. This unit was based on a human body whose height is divided by the navel into the golden ratio; see Figure 17.17.

The golden rectangle is also used extensively in the *Cathedral of Chartres* outside Paris, and the *Tower of Saint Jacques* in Paris. The royal doorway of the Cathedral clearly demonstrates a golden rectangle; see Figure 17.18.

According to Sr. Stephen, the *Tower of Saint Jacques* illustrates

the architecture leitmotif [α] in inverse progressions. At the corners, the buttresses rise in four superimposed layers, which diminish in size as they rise. The ratio thus established is exactly 1.618. The buttresses, like a human hand, whose proportions we shall see are the same, point toward the sky, while the three stories of windows which illumine the interior of the tower appear as a hand pointing down from the sky to the ground.

See Figure 17.19.

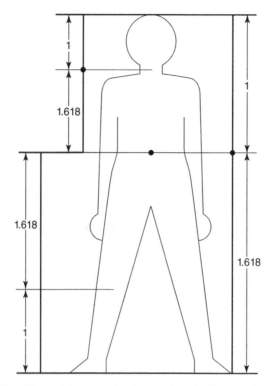

Figure 17.17.   The *modulator*, a scale of proportions developed by Le Corbusier.*

Figure 17.18.   The doorway of the *Cathedral of Chartres.*#

Figure source: 1977 *Yearbook of Science and the Future* by Britannica. Reproduced with permission of William Biderbost.

#*Figure source*: Stephen, 1956 [543]. Reproduced with permission of The National Council of Teachers of Mathematics.

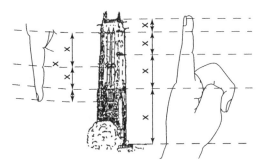

Figure 17.19. The Tower of Saint Jacques.*

Again, according to Sr. Stephen, "Dr. Christian Jacob of Buenos Aires has discovered the interesting proportion in the human brain" [543].

17.4 HUMAN BODY AND THE GOLDEN RECTANGLE

The ancient Greeks knew that the human body exemplifies the golden proportion. The head fits nicely into a golden rectangle, as Figure 17.20 demonstrates. In addition, the face provides visual examples of the golden ratio: $\dfrac{AC}{CD} = \dfrac{CD}{BC} = \dfrac{AD}{BD} = \alpha$. So do the fingers, as Figures 17.21 and 17.22 illustrate: $\dfrac{b}{a} = \dfrac{c}{d} = \dfrac{d}{c} = \alpha$.

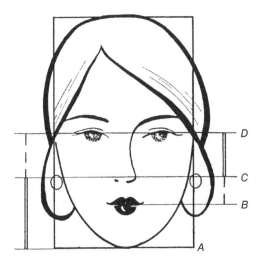

Figure 17.20. The golden proportions in a human head and face.[#]

Figure source: Stephen, 1956 [543]. Reproduced with permission of The National Council of Teachers of Mathematics.
[#]*Figure source*: Garland, 1987 [213]. Reproduced with permission of Pearson Education, Inc.

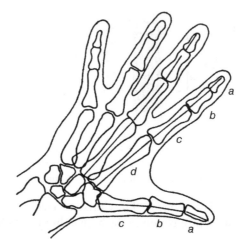

Figure 17.21.   The golden proportions in a human hand.*

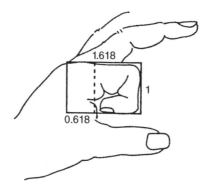

Figure 17.22.   A personal golden rectangle formed by a pointer finger.*

According to T.H. Garland, most of the ancient graveyard crosses in Europe exemplify the golden proportion: the point where the two arms meet divides the cross in the golden ratio; see Figure 17.23 [213]. Although many modern crosses do not display this magnificent characteristic, some still fit into a golden rectangle; see Figure 17.24.

Postage stamps often remind us of the golden rectangle. For example, consider the 1999 Prostate Cancer Awareness stamp in Figure 17.25. Its outer size is 4 cm × 2.5 cm and $4/2.5 \approx \alpha$; the size of the inner rectangle is 3.5 cm × 2.1 cm and $3.5/2.1 \approx \alpha$.

Figure sources: Garland, 1987 [213]. Reproduced with permission of Pearson Education, Inc.

Figure 17.23. An ancient graveyard cross.*　　　　**Figure 17.24.** A modern cross.*

The Greek urn in Figure 17.26 also displays the golden ratio; it fits nicely into a golden rectangle; so does the beautiful United Nations building in New York City, designed by Oscar Niemeyer (1907–2012) and Le Corbusier.

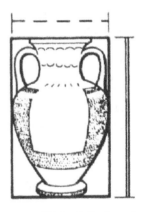

Figure 17.25. A Prostate Cancer Awareness stamp.　　　　**Figure 17.26.** A Greek urn.*

17.5　GOLDEN RECTANGLE AND THE CLOCK

It is common knowledge that the positions of the hour and minute hands on an analog wristwatch or clock in store displays and in advertisements tend to be approximately 10:09 or 8:18; see Figure 17.27.

Figure sources: Garland, 1987 [213]. Reproduced with permission of Pearson Education, Inc.

Figure 17.27. A wristwatch with hands set at 10:09.

One myth concerning the time 8:18 is that it was precisely the time Abraham Lincoln died by an assassin's bullet on April 15, 1865. Another misconception is that such a setting of the hands gives more space on the face of the clock to show the name of the manufacture clearly.

In 1983, M.G. Monzingo of Southern Methodist University, Dallas, Texas, made the case that such a setting is related to the golden rectangle, and hence is appealing aesthetically [445]. He showed that the angle θ in Figure 17.28 is about 58.3°. Suppose $OE = 1$. Then $EB \approx \tan 58.3° \approx \alpha$. So $AB : AD \approx 2\alpha : 2 = \alpha : 1$. In other words, such a setting pleases the human eye, since it creates an imaginary golden rectangle $OEBF$ on the face of the clock.

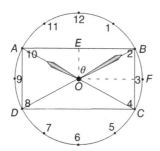

Figure 17.28. The setting on a watch is related to the golden ratio.

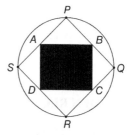

Figure 17.29.

Suppose the points A, B, C, and D in Figure 17.29 divide the respective sides of the square $PQRS$ in the golden ratio. Then $PA = PB$, $QB = QC$, and $PB/BQ = \alpha$. So

$$\frac{AB^2}{BC^2} = \frac{PA^2 + PB^2}{QB^2 + QC^2} = \frac{2PB^2}{2QB^2} = \alpha^2.$$

Thus $AB/BC = \alpha$, so $ABCD$ is indeed a golden rectangle.

We will now discuss how can we construct a golden rectangle with a straightedge and compass.

17.6 STRAIGHTEDGE AND COMPASS CONSTRUCTION

Consider a line segment $\overline{AB}$ with C dividing it in the golden ratio: $AC/CB = AB/AC = \alpha$.

Now with C as the center, draw an arc of radius CB; let the perpendicular $\overline{CH}$ to $\overline{CB}$ meet the arc at D. Complete the rectangle $ACDE$; see Figure 17.30. It is a golden rectangle, since $AC/CD = AC/CB = \alpha$.

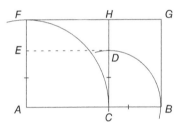

Figure 17.30.

Using the golden rectangle $ACDE$, we can draw another rectangle. With A as the center, draw an arc of radius AC so as to intersect the perpendicular to $\overline{AB}$ at F. Complete the rectangle $ABGF$, as Figure 17.30 shows. It is a golden rectangle, because $\dfrac{AB}{BG} = \dfrac{AB}{AC} = \alpha$.

We have in fact gained a third golden rectangle, namely, $BCHG$. This is so since $\dfrac{BG}{BC} = \dfrac{AC}{BC} = \alpha$.

As a byproduct, we can show that the ratio of the area of rectangle $ABGF$ to that of rectangle $BCHG$ is α^2; see Exercise 17.6.

Let us now continue this discussion a bit further. This will yield interesting dividends.

Suppose we remove the square $ACHF$ in Figure 17.30 from the golden rectangle $ABGF$; then the resulting rectangle $BCHG$ is also a golden rectangle. That is, if the ratio of the length to the width of rectangle $ABGF$ is the golden ratio, then that of rectangle $BCHG$ obtained by removing a square with one side equal to the width of the original rectangle is also a golden rectangle.

Conversely, suppose the ratio of the length to the width of a rectangle $ABCD$ is k; that is, $AB/BC = l/w = k$; see Figure 17.31. Let $BEFC$ be the rectangle obtained by deleting the square $AEFD$ from the rectangle $ABCD$. The ratio of the length to the width of the rectangle $BEFC$ is $BC/BE = w/(l - w)$. Suppose $AB/BC = BC/EB$. Then $l/w = w/(l - w)$, that is, $k = 1/(k - 1)$; so $k = \alpha$. Thus, if removing the square yields a rectangle similar to the original rectangle, then $k = \alpha$; that is, the original rectangle must be a golden rectangle.

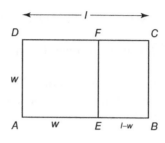

Figure 17.31.

On the other hand, suppose the ratio of the length to the width of a rectangle $ABCD$ is $k > 1$. Remove from rectangle $ABCD$ a rectangle $BEFC$ similar to it. Then the ratio of the area of the original rectangle to that of the remaining rectangle $AEFD$ is k if and only if $k = \alpha$. In fact, $AEFD$ is a square; see Exercise 17.7.

17.7   RECIPROCAL OF A RECTANGLE

We will now look at the golden rectangle $ABGF$ in Figure 17.30 from a slightly different perspective. Suppose we remove square $ACHF$ from the golden rectangle $ABGF$. The resulting rectangle $BCHG$ is also a golden rectangle; it is called the *reciprocal* of rectangle $ABGF$[†].

Thus the reciprocal of the rectangle is a smaller, similar rectangle such that one side of the original rectangle becomes a side of the new rectangle.

It now follows that the area of the reciprocal rectangle $BCHG$ in Figure 17.30 is (area $ABGF$)$/\alpha^2$.

In Figure 17.30, square $ACHF$ is the smallest figure that can be added to the golden rectangle $CBGH$ to yield a similar shape, another rectangle. Accordingly, square $ACHF$ is called the *gnomon* of the reciprocal rectangle $CBGH$, a term introduced by Sir D'Arcy Wentworth Thompson (1860–1948), a Scottish biologist, mathematician, and classics scholar.

[†]The term *reciprocal rectangle* was introduced by Hambidge.

Suppose the diagonals $\overline{BF}$ and $\overline{CG}$ of the reciprocal rectangles meet at P; see Figure 17.32. We can show that they are perpendicular; see Exercise 17.8. Suppose $\overline{BF}$ and $\overline{CH}$ intersect at Q. Let $\overline{QR} \perp \overline{BG}$. Then $BRQC$ is the reciprocal of $BGHC$ and is also a golden rectangle. This gives a systematic way of constructing the reciprocal of a (golden) rectangle.

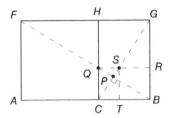

Figure 17.32.

Suppose we continue this procedure to draw the golden reciprocal of $BRQC$. Let $\overline{CG}$ meet $\overline{QR}$ at S, and draw $\overline{ST} \perp \overline{BC}$. Then $CTSQ$ is the (golden) reciprocal of $BRQC$.

17.8   LOGARITHMIC SPIRAL

Obviously, we can continue this algorithm indefinitely, producing a sequence of smaller and smaller rectangles; see Figure 17.33. The points that divide the sides of the various golden rectangles spiral inward to the point P, the intersection of the two original diagonals. They lie on the *logarithmic spiral*, as Figure 17.33 demonstrates. The spiral, with its pole at P, touches golden rectangles at the golden sections.

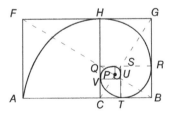

Figure 17.33.

The nautilus is one of the most gorgeous examples of the logarithmic spiral in nature. Figure 17.34 also shows shells that display this beautiful logarithmic spiral.

(a)

(b)

Figure 17.34. Shells displaying the logarithmic spiral.

The American writer and physician Oliver Wendell Holmes (1809–1894), in his poem "The Chambered Nautilus" (1858), describes the creation of the spiral:

Year after year beheld the silent toll
That spread his lustrous coil;
Still, as the spiral grew,
He left the past year's dwelling for the new,
Stole with soft step its shining archway through,
Built up its idle door,
Stretched in his last-found home, and knew the old no more.

The *Association for Women in Mathematics* (AWM) uses this spectacular spiral as their logo (see Figure 17.35); so do the hotel chain *Indigo*, their *Do not disturb signs,* and their *Phi* restaurants (see Figure 17.36). Interestingly, the nautilus design appears in the pediment of the Cathedral of the Episcopal Church in Boston, Massachusetts.

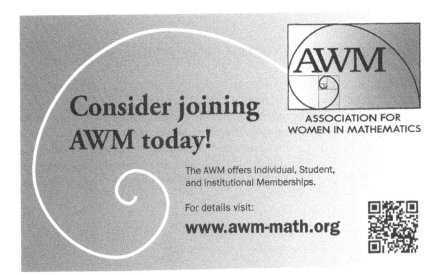

Figure 17.35. AWM logo. *Figure source*: Reproduced with permission of the Association for Women in Mathematics.

Figure 17.36. Indigo hotel.*

* Figure source: Courtesy of Ron Lancaster, Senior Lecturer in Mathematics Education, Ontario Institute for Studies in Education of The University of Toronto. ron2718nas.net.

Now consider the diagonals of the various squares in Figure 17.33 that are snipped off, namely, $\overline{AH}, \overline{HR}, \overline{RT}, \overline{TV}, \ldots$. Their lengths form a decreasing geometric sequence and their sum is $\sqrt{2}a\alpha^2$, where $AC = a$; see Exercise 17.12.

GOLDEN TRIANGLES REVISITED

Interestingly, we can employ golden triangles to generate the logarithmic spiral. Bisect a 72° angle in the golden triangle ABC in Figure 17.37. The point where the bisector meets the opposite side divides it in the golden ratio. The bisector produces a new similar golden triangle CBD. Now divide this triangle by a 72° angle bisector to yield another golden triangle. Continuing this algorithm indefinitely generates a sequence of whirling golden triangles, and hence the logarithmic spiral shown in the figure. Its pole P is the intersection of the two medians indicated by broken segments.

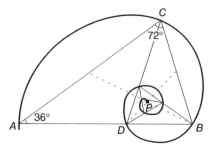

Figure 17.37.

Next we illustrate yet another surprising occurrence of the golden rectangle.

17.9   GOLDEN RECTANGLE REVISITED

Suppose we remove a $t \times t$ square from one of the corners of a unit square lamina; see Figure 17.38. We would like to find the value of t such that the center of gravity of the remaining gnomon is the corner G of the square removed. Taking moments about side $\overline{AD}$, we have

$$\begin{pmatrix} \text{Moment of the} \\ \text{removed square} \end{pmatrix} + \begin{pmatrix} \text{Moment of} \\ \text{the gnomon} \end{pmatrix} = \begin{pmatrix} \text{Moment of the} \\ \text{original square} \end{pmatrix}.$$

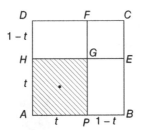

Figure 17.38.

That is,

$$\frac{1}{2}t \cdot t^2 + t(1 - t^2) = \frac{1}{2} \cdot 1$$

$$t^3 - 2t + 1 = 0$$

$$(t - 1)(t^2 + t - 1) = 0$$

$$(t - 1)(t + \alpha)(t + \beta) = 0.$$

Since $t > 0$ and $t \neq 1$, it follows that $t = -\beta = 1/\alpha$.

Then $1 - t = 1 - \dfrac{1}{\alpha} = \dfrac{\alpha - 1}{\alpha}$. So $t : 1 - t = 1 : \alpha - 1 = \alpha : \alpha^2 - \alpha = \alpha : 1$.

Thus P divides $\overline{AB}$ in the golden ratio. Consequently, both *PBEG* and *GFDH* are golden rectangles, as established in 1995 by Nick Lord of Tonbridge School, Kent, England [425].

17.10 SUPERGOLDEN RECTANGLE

In 1994, Tony Crilly of Middlesex University, London, England, introduced the concept of a *supergolden rectangle*, a variant of the classical golden rectangle [132]. To begin with, consider an $x \times 1$ rectangle; see Figure 17.39. Now divide it into a square *AEFD* and two rectangles *EBHG* and *GHCF* in such a way that *ABCD*, *EBHG*, and *GHCF* are similar.

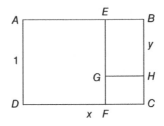

Figure 17.39. Supergolden rectangle.

Let $BH = y$. Since rectangles $ABCD$ and $GHCF$ are similar, $\dfrac{x}{1} = \dfrac{x-1}{1-y}$; so $xy = 1$. This implies that the shaded areas in Figure 17.40 are equal.

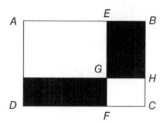

Figure 17.40.   Equal areas.

Now the rectangles $ABCD$ and $EBHG$ are similar; so $\dfrac{x}{1} = \dfrac{y}{x-1}$, where $xy = 1$. This yields the cubic equation $x^3 - x^2 - 1 = 0$. It has exactly one real root ω (Greek letter *omega*), where $1 < \omega < 2$; see Figure 17.41; ω is the *supergolden ratio*.

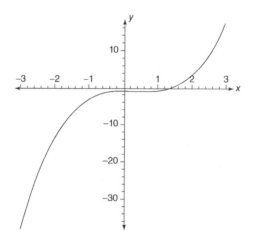

Figure 17.41.   Graph of the equation $y = x^3 - x^2 - 1$.

It is a bit easier to solve the corresponding equation in y: $y^3 + y - 1 = 0$; see Figure 17.42. Using *Cardano's formula*, named after the Italian mathematician Girolamo Cardano (1501–1576),

$$t = \sqrt[3]{-\frac{q}{2} + \sqrt{\frac{q^2}{4} + \frac{p^3}{27}}} + \sqrt[3]{-\frac{q}{2} - \sqrt{\frac{q^2}{4} + \frac{p^3}{27}}}$$

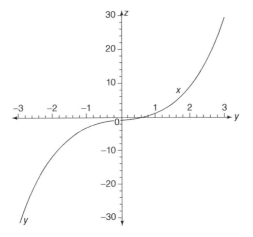

Figure 17.42. Graph of the equation $z = y^3 + y - 1$.

for the real solution of the equation $t^3 + pt + q = 0$, the real value θ of y is given by

$$\theta = \sqrt[3]{\frac{1}{2} + \sqrt{\frac{1}{4} + \frac{1}{27}}} + \sqrt[3]{\frac{1}{2} - \sqrt{\frac{1}{4} + \frac{1}{27}}}$$

$$= \sqrt[3]{\frac{1 + \sqrt{\frac{31}{27}}}{2}} + \sqrt[3]{\frac{1 - \sqrt{\frac{31}{27}}}{2}}$$

$$\approx 0.682327803828.$$

So $\omega \approx 1.46557123188$; see Figure 17.41.

Figure 17.43 shows some interesting properties of the supergolden rectangle; see Exercises 17.16–17.24.

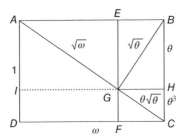

Figure 17.43. Geometric properties of the supergolden rectangle.

Recall that the golden ratio α satisfies the equation $x^2 = x + 1$, which is the characteristic equation of the Fibonacci recurrence. Likewise, the supergolden ratio ω satisfies the cubic equation $x^3 = x^2 + 1$, so it can be considered the characteristic equation of the third-order recurrence $a_{n+3} = a_{n+2} + a_n$, where $n \geq 1$. Using $a_1 = a_2 = a_3 = 1$ as the initial values, the sequence $\{a_n\}$ looks as follows:

$$1, 1, 1, 2, 3, 4, 6, 9, 13, 19, 28, 41, \ldots .$$

Suppose the ratios of consecutive elements of this sequence converge to a limit l. Since

$$\frac{a_{n+3}}{a_{n+2}} = 1 + \frac{a_n}{a_{n+1}} \cdot \frac{a_{n+1}}{a_{n+2}}$$

$$\lim_{n \to \infty} \frac{a_{n+3}}{a_{n+2}} = 1 + \lim_{n \to \infty} \left(\frac{a_n}{a_{n+1}} \cdot \frac{a_{n+1}}{a_{n+2}} \right),$$

we then have $l = 1 + \dfrac{1}{l^2}$, so $l = \omega$. Thus

$$\lim_{n \to \infty} \frac{a_{n+1}}{a_n} = \omega \quad \text{and hence} \quad \lim_{n \to \infty} \frac{a_n}{a_{n+1}} = \theta.$$

For example, $\dfrac{a_{17}}{a_{16}} = \dfrac{277}{189} \approx 1.46560846561$.

To our surprise, both α and ω enjoy similar properties. For example,

$$\alpha - 1 = \frac{1}{\alpha} \quad \text{and} \quad \omega - 1 = \frac{\theta}{\omega};$$

and

$$\sum_{k=0}^{\infty} \frac{1}{\alpha^k} = \alpha^2 \quad \text{and} \quad \sum_{k=0}^{\infty} \frac{1}{\omega^k} = \omega^3;$$

see Exercises 17.16–17.24.

EXERCISES 17

1. Let $\triangle ABC$ be an isosceles triangle, where the non-repeating angle $\angle B = 36°$. Let the bisector of $\angle A$ intersect $\overline{BC}$ at D. Prove that $\triangle CAD$ is a golden triangle.

Let D be a point on side $\overline{BC}$ of an isosceles triangle ABC such that $\triangle ABC \sim \triangle CAD$, where $AB = BC$. Prove each.

2. If $\triangle ABC$ is a golden triangle, then area $\triangle ABC$: area $\triangle BDA = \alpha : 1$.

3. If area $\triangle ABC$: area $\triangle BDA = \alpha$: 1, then $\triangle ABC$ is a golden triangle.

4. Let $\triangle ABC$ is a golden triangle, and D a point on $\overline{BC}$ such that area $\triangle ABC$: area $\triangle BDA = \alpha$: 1. Prove that area $\triangle ABC$: area $\triangle CAD = \alpha^2$: 1.

5. The lengths of the sides of a right triangle form a geometric sequence with common ratio r. Prove that $r = \sqrt{\alpha}$.

6. Using Figure 17.30, prove that the ratio of the area of rectangle $ABGF$ to that of rectangle $BCHG$ is α^2.

7. Let the ratio of the length to the width of a rectangle $ABCD$ be $k > 1$. From rectangle $ABCD$, remove a rectangle $BEFC$ similar to it. Prove that the ratio of rectangle $ABCD$ to the remaining area $AEFD$ is k if and only if $k = \alpha$. Besides, the remaining rectangle is a square.

8. Prove that the diagonals of two reciprocal rectangles are perpendicular.

9. Let $ABGF$ and $BGHC$ be two reciprocal golden rectangles. Let P be the point of intersection of the two diagonals $\overline{BF}$ and $\overline{CG}$. Prove that $FP/GP = BP/CP = \alpha$.

10. Let $ABGF$ be a golden rectangle and $BGHC$ its reciprocal, as in Figure 17.32. Prove that $ACHF$ is a square.

11. Let $BGHC$ be a golden rectangle. Complete the square $ACHF$ on its left. Prove that $ABGF$ is a golden rectangle.

12. Show that the sum of the lengths of the diagonals of the various "whirling squares" in Figure 17.33 is $\sqrt{2}a\alpha^2$, where $AC = a$.

13. Let P, Q, R, and S be points on the sides of a square $ABCD$, dividing each in the golden ratio. Prove that $PQRS$ is a golden rectangle.

Consider the sequence of decreasing smaller reciprocal golden rectangles in Figure 17.44, beginning with the golden rectangle $ABCD$. Let $DE = a$ and $EC = b$, where $a = \alpha b$ (Runion, 1972 [508]).

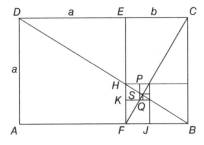

Figure 17.44.

14. Complete the following table.

Rectangle	ABCD	BCEF	BGHF	FJIH	HKLI	PQLI	QLRS
Shorter side	a						
Longer side	$a+b$						

15. Predict the size of the nth reciprocal rectangle in the sequence, where $n \geq 0$.

Using Figure 17.43, prove Exercises 17.16–17.24.

16. $BG = \sqrt{\theta}$.

17. $GC = \theta\sqrt{\theta}$.

18. $HC = \theta^3$.

19. $AG = \sqrt{\omega}$.

20. $\angle BGC$ is a right angle.

21. $AEGI$ is a supergolden rectangle.

22. $\dfrac{IG}{ID} = \omega^3$

23. $\omega - 1 = \dfrac{\theta}{\omega}$.

24. $\displaystyle\sum_{k=0}^{\infty} \dfrac{1}{\omega^k} = \omega^3$.

FIGEOMETRY

This chapter explores additional delightful properties of the golden ratio in Euclidean, solid, and analytic geometry. Always look for more occurrences of the sacred ratio. We begin our pursuit with an interesting problem studied by Hunter in 1963 [335].

18.1 THE GOLDEN RATIO AND PLANE GEOMETRY

Example 18.1. *Locate the points P and Q on two adjacent sides of a rectangle ABCD such that the areas of triangles APQ, BQC, and CDP are equal; see Figure 18.1.*

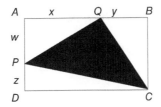

Figure 18.1.

Solution. Let $AQ = x$, $QB = y$, $AP = w$, and $PD = z$. Since the areas of $\triangle APQ$, $\triangle QBC$, and $\triangle CDP$ are equal, we have $xw/2 = y(w+z)/2 = z(x+y)/2$; that is,

$xw = y(w + z) = z(x + y)$. Equation $y(w + z) = z(x + y)$ yields $yw = zx$; that is, $x/y = w/z$. From the equation $xw = z(x + y)$, we have $w/z = (x + y)/x$; that is, $x/y = 1 + y/x$. Since $x/y > 0$, this implies $x/y = \alpha$. Thus

$$\frac{w}{z} = \frac{x}{y} = \alpha. \tag{18.1}$$

Consequently, we must choose the point P in such a way that P divides $\overline{AD}$ in the ratio $AP : PD = w : z = \alpha : 1$. Likewise, we must locate Q on $\overline{AB}$ such that $AQ : QB = x : y = \alpha : 1$. Thus $\overline{PQ}$ divides the two sides in the golden ratio. ∎

In 1964, H.E. Huntley of Somerset, England, pursued this problem further [343]. He proved that if $ABCD$ is a golden rectangle, then $\triangle PQC$ is an isosceles right triangle with right angle at Q, as the next example shows.

Example 18.2. *Suppose the rectangle ABCD in Figure 18.1 is a golden rectangle. Prove that $\triangle PQC$ is an isosceles right triangle and $\angle Q = 90°$.*

Proof. Since $ABCD$ is a golden rectangle, $\dfrac{AB}{BC} = \dfrac{x + y}{w + z} = \alpha$. Using equation (18.1), this implies $\dfrac{y(1 + \alpha)}{z(1 + \alpha)} = \alpha$. Thus $y = \alpha z$. But $\alpha z = w$, so $y = w$. Thus $AP = BQ$.

Then $x = \alpha y = \alpha(\alpha z) = \alpha^2 z = (\alpha + 1)z = \alpha z + z = w + z$; that is, $AQ = BC$. Therefore, by the side–angle–side (SAS) theorem, $\triangle APQ \cong \triangle BQC$. Consequently, $PQ = QC$, so $\triangle PQC$ is an isosceles triangle; see Figure 18.2.

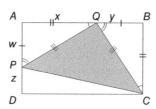

Figure 18.2.

Since $\triangle APQ \cong \triangle BQC$, $\angle AQP = \angle BCQ$. But $\angle BCQ + \angle BQC = 90°$; that is, $\angle AQP + \angle BQC = 90°$, so $\angle PQC = 90°$. Thus $\triangle PQC$ is an isosceles right triangle, as desired. ∎

INTERESTING BYPRODUCTS

The golden rectangle $ABCD$ in Figure 18.2 has some additional interesting properties.

1) Since $\angle CPQ = 45°$, $\angle PCQ = 45°$. So $\overleftrightarrow{PQ} \parallel \overleftrightarrow{BD}$.

2) We can derive a formula for the area of $\triangle PQC$:

$$\text{Area of } \triangle PQC = \frac{1}{2} PQ \cdot QC$$

$$= \frac{1}{2} PQ^2$$

$$= \frac{1}{2}(x^2 + w^2)$$

$$= \frac{1}{2}(\alpha^2 y^2 + y^2)$$

$$= \frac{1}{2}(\alpha^2 + 1)y^2$$

$$= \frac{1}{2}(\alpha + 2)y^2.$$

3) Since $PQ = \sqrt{\alpha + 2}\, y$, the length of the equal sides is $\sqrt{\alpha + 2}\, y$.
4) The area of the golden rectangle is given by

$$x(x + y) = \alpha y(\alpha y + y) = (\alpha^2 + \alpha)y^2 = (2\alpha + 1)y^2.$$

5) Area of $\triangle APQ$ + area of $\triangle QBC$ + area of $\triangle CDP = (2\alpha + 1)y^2 - \frac{1}{2}(\alpha + 2)y^2$

$$= \frac{3}{2}\alpha y^2.$$

In 1964, Hunter also proved that the ratios of the dimensions of a special rectangular prism are closely linked to the golden ratio [336], as the following example demonstrates.

Example 18.3. (Golden Cuboid) *Consider a rectangular prism with unit volume and a diagonal of two units long; see Figure 18.3. Find the ratios of the dimensions of the prism.*

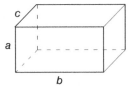

Figure 18.3.

Solution. Suppose the edges are a, b, and c units long. Then $abc = 1$ and $a^2 + b^2 + c^2 = 4$. Since our focus is on ratios, without loss of generality, we let $a = 1$.

Then $bc = 1$ and $b^2 + c^2 = 3$. Substituting for c, this equation yields

$$b^4 - 3b^2 + 1 = 0$$
$$(b^2 - \alpha^2)(b^2 - \beta^2) = 0.$$

This implies $b = \pm\alpha, \pm\beta$. Since $b > 0$, $b = \alpha$, and hence $c = 1/\alpha$.

Thus $a : b : c = 1 : \alpha : 1/\alpha$. [Notice that $a^2 + b^2 + c^2 = 1 + \alpha^2 + \alpha^{-2} = 1 + (\alpha^2 + \beta^2) = 1 + L_2 = 4$, as expected.]   ■

Example 18.3 leads to several properties of the cuboid:

- Ratios of the areas of the three different faces $= ab : bc : ca = \alpha : 1 : 1/\alpha$.
- Total surface area of the cuboid $= 2(ab + bc + ca) = 2(\alpha + 1 + 1/\alpha) = 2(2\alpha) = 4\alpha$.
- Since $a : b : c = 1 : \alpha : 1/\alpha$, it follows that the faces of the cuboid are indeed golden rectangles.

For example, consider the face $ABCD$ in Figure 18.4. We have $AB : BC = b : a = \alpha : 1$.

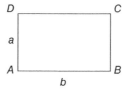

Figure 18.4.

- $\dfrac{\text{Surface area of the golden cuboid}}{\text{Surface area of the circumscribing sphere}} = \dfrac{4\alpha}{4\pi} = \dfrac{\alpha}{\pi}$.

The next example [344], although elementary in nature, is certainly interesting in its own right. It also illustrates the ubiquity of the marvelous number α.

Example 18.4. *Let P be a point on a chord $\overline{AB}$ of a circle, and $\overrightarrow{PT}$ a tangent to the circle at T such that $PT = AB$; see Figure 18.5. Compute the ratio $PB : AB$.*

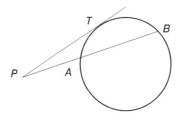

Figure 18.5.

Solution. It follows by elementary geometry that $PT^2 = PA \cdot PB$; so

$$AB^2 = PA \cdot PB$$
$$= PB(PB - AB)$$
$$(PB/AB)^2 = 1 + PB/AB.$$

It now follows that $PB : AB = \alpha : 1$; that is, A divides $\overline{PB}$ in the golden ratio. (As a bonus, it follows that $PA : AB = 1/\alpha : 1$. ∎

To continue this example a bit further, let C be a point on $\overline{AB}$ such that $PT = AB = PC$; see Figure 18.6. Since $PC = AB$, $PA + AC = AC + CB$; thus $PA = CB$. Since $PB/AB = \alpha$, $(PC + CB)/AB = \alpha$. That is,

$$\frac{AB + CB}{AB} = \alpha$$

$$1 + \frac{CB}{AB} = \alpha$$

$$\frac{AB}{CB} = \alpha.$$

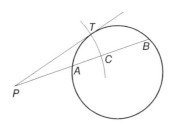

Figure 18.6.

Thus C divides $\overline{AB}$ in the golden ratio.
   Moreover,

$$\frac{PB}{PC} = \frac{PC + CB}{PC}$$

$$= 1 + \frac{CB}{AB}$$

$$= 1 + \frac{1}{\alpha}$$

$$= \alpha;$$

$$\frac{CB}{CA} = \frac{CB}{AB - CB}$$

$$= \frac{1}{AB/CB - 1}$$

$$= \frac{1}{\alpha - 1}$$

$$= \alpha;$$

$$\frac{PC}{PA} = \frac{PA + AC}{PA}$$

$$= 1 + \frac{AC}{CB}$$

$$= 1 + \frac{1}{\alpha}$$

$$= \alpha;$$

$$\frac{AB}{AC} = \frac{PC}{AC} = \frac{PC}{PA} \cdot \frac{PA}{AC}$$

$$= \alpha \cdot \alpha$$

$$= \alpha^2;$$

$$\frac{PB}{AC} = \frac{PA + AB}{AB - CB}$$

$$= \frac{PA/AB + 1}{1 - CB/AB}$$

$$= \frac{1/\alpha + 1}{1 - 1/\alpha}$$

$$= \alpha^3.$$

The next three examples also demonstrate occurrences of the golden ratio in geometry.

Example 18.5. *Suppose we inscribe a square BDEF in a semicircle such that one side of the square lies along its diameter; see Figure 18.7. Prove that D divides $\overline{AE}$ in the golden ratio.*

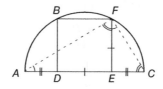

Figure 18.7.

Proof. Since the right triangles AFE and CFE are similar, $\dfrac{AE}{FE} = \dfrac{FE}{CE} = \dfrac{AF}{CF}$. But

$$\frac{AE}{FE} = \frac{AD + DE}{FE} = \frac{AD + FE}{FE} = \frac{AD}{FE} + 1.$$

Since $CE = AD$, this implies $AD/FE + 1 = FE/AD$. This implies $FE/AD = \alpha$. Thus $AE/DE = \alpha$; so D divides $\overline{AE}$ in the golden ratio, as desired. ∎

As a byproduct, it follows that $\dfrac{AE}{FE} = \dfrac{FE}{CE} = \dfrac{AF}{CF} = \alpha$; that is, the ratio of the corresponding sides of the similar triangles is the golden ratio.

Example 18.6. *Let A be the midpoint of the side $\overline{PQ}$ of the square $PQRS$ in Figure 18.8. Let $\overline{AR}$ be the tangent to the circle with center O. Since $AD^2 = AB \cdot AC$, it follows that $AD/AB = AC/AD$. In fact, we can show that $AD/AB = AC/AD = \alpha$.*

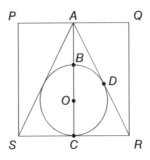

Figure 18.8.

∎

RIGHT TRIANGLES AND THE GOLDEN RATIO

The next example presents an occurrence of the golden ratio in the study of right triangles.

Example 18.7. *Suppose the lengths of the sides of a right triangle form a geometric sequence. Prove that the common ratio of the sequence is α.*

Proof. Suppose the lengths of the sides of the triangle are a, ar, and ar^2. Clearly, $r \neq 1$.

Case 1. Let $r < 1$. Then $ar^2 < ar < a$; see Figure 18.9. By the Pythagorean theorem, we then have

$$a^2 r^2 + a^2 r^4 = a^2$$
$$r^4 + r^2 = 1$$
$$r = 1/\sqrt{\alpha}.$$

Figure 18.9.   **Figure 18.10.**

Case 2. On the other hand, let $r > 1$. Then $a < ar < ar^2$; see Figure 18.10. Then $r = \sqrt{\alpha}$; see Exercise 18.19.

Thus the common ratio is either $1/\sqrt{\alpha}$ or $\sqrt{\alpha}$.   ∎

Next we study a very special cross, and see how it is related to the golden ratio.

18.2   THE CROSS OF LORRAINE

An interesting problem, related to the *Cross of Lorraine*[†] or the *Patriarchal Cross*, was studied by Martin Gardner. This ancient emblem, re-introduced in modern times by General Charles de Gaulle (1890–1970) of France, consists of three beams – two horizontal and one vertical – and covers an area of $13 = F_7$ square units; see Figure 18.11. We would like to cut the cross through C into two pieces of equal area, namely, 6.5 square units each.

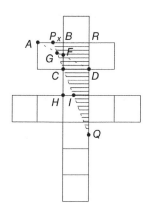

Figure 18.11.   The cross of Lorraine.

[†]Lothringen or Lorraine is a province on the border between France and Germany.

Suppose the line segment $\overline{PQ}$ has the desired property. Then shaded area of $\triangle PQR = 2.5$ square units. Let $BP = x$ and $DQ = y$. Since $\triangle BPC \sim \triangle DQC$, $x/1 = 1/y$; that is, $xy = 1$. Thus

$$\text{Area } \triangle PQR = \frac{1}{2} PR \cdot QR$$

$$= \frac{1}{2}(x+1)(y+1)$$

$$(x+1)(y+1) = 5$$

$$x^2 - 3x + 1 = 0$$

$$x = \alpha^2, \beta^2.$$

Since $0 < x < 1$, it follows that $x = \beta^2 = 1 + \beta$; so $y = 1/x = 1 + \alpha$. Thus $AP = 1 - x = -\beta$ and $EQ = y - 1 = \alpha$. So $\dfrac{AP}{PB} = \dfrac{-\beta}{1+\beta} = \alpha$ and $\dfrac{DE}{EQ} = \dfrac{1}{\alpha}$. Thus P and E divide $\overline{AB}$ and $\overline{DQ}$ in the golden ratio, respectively.

Can we locate the points P and Q geometrically? Yes.

LOCATING *P* AND *Q* GEOMETRICALLY

To locate the points P and Q geometrically, let $\overline{AD}$ meet $\overline{BC}$ at F, so $AF = FD$. With F as center and FB as radius, draw an arc to intersect $\overline{AF}$ at G. With G as center and AG as radius, draw an arc to intersect $\overline{AB}$ at P. Let $\overline{PC}$ meet $\overline{DE}$ at Q. Then P divides $\overline{AB}$ in the golden ratio, and E divides $\overline{DQ}$ in the same ratio; moreover, $\overline{PQ}$ divides the cross into two equal areas.

To confirm this, $\triangle PBC \cong \triangle CHI$ and $\triangle PBC \sim \triangle DQC$; so $xy = 1$. So shaded area $= 2 + $ area $\triangle EQI$. Since $\triangle PBC \sim \triangle EQI$, $x/1 = EI/(y-1)$, so

$$EI = x(y-1) = xy - x$$

$$= 1 - x;$$

$$\text{area } \triangle EQI = \frac{1}{2} EI \cdot EQ$$

$$= \frac{1}{2}(1-x)(y-1)$$

$$= \frac{1}{2}(1-x)(1/x - 1)$$

$$= \frac{(1-x)^2}{2x}.$$

But $\dfrac{1-x}{x} = \dfrac{-\beta}{1+\beta} = \alpha$; so area $\triangle EQI = \dfrac{\alpha(-\beta)}{2} = \dfrac{1}{2}$. Consequently, the shaded area equals 2.5 square units; so $\overline{PQ}$ partitions the cross equally, as desired. ∎

In 1959, Gardner, in his column in *Scientific American*, invited his readers to compute the length BC [207]. The same puzzle appeared five years later in an article by M.H. Holt in *The Pentagon* [319].

18.3 FIBONACCI MEETS APOLLONIUS

A mathematical giant of the third century B.C., Apollonius of Perga (262?–190?B.C) proposed the following problem: *Given three fixed circles, find a circle that touches each one of them.* In total, there are eight solutions. But if the given circles are mutually tangential, then there are exactly two solutions.

Assume, for convenience, that the given circles are not only tangential to each other, but their centers form a Pythagorean triangle; see Figure 18.12. (W.W. Horner of Pittsburgh studied this special case in 1973 [330].) Let r_1, r_2, and r_3 denote the radii of the given circles, and R and r those of the solutions. Assume $r_1 < r_2 < r_3$, and $r < R$.

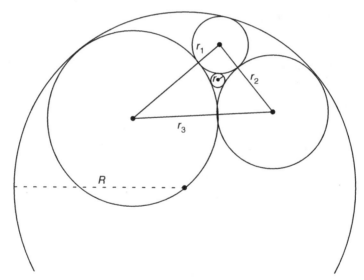

Figure 18.12.

Let $a = F_n, b = F_{n+1}, c = F_{n+2}$, and $d = F_{n+3}$. Since $(c^2 - b^2)^2 + (2bc)^2 = (c^2 + b^2)^2$, we can assume that the lengths of the sides of the Pythagorean triangle are $c^2 - b^2, 2bc$, and $c^2 + b^2$. Since the original circles are mutually tangential, it follows that $r_1 + r_2 = c^2 - b^2, r_2 + r_3 = c^2 + b^2$, and $r_3 + r_1 = 2bc$. Solving this linear system, we get $r_1 = b(c - b) = ab, r_2 = c(c - b) = ac$, and $r_3 = b(c + b) = bd$. Then $r_1r_2r_3 = a^2b^2cd, r_1r_2 = a^2bc, r_2r_3 = abcd$, and $r_3r_1 = ab^2d$.

In 1950, Col. R.S. Beard showed [30] that

$$R \text{ or } r = \frac{r_1 r_2 r_3}{r_1 r_2 + r_2 r_3 + r_3 r_1 \mp 2\sqrt{r_1 r_2 r_3(r_1 + r_2 + r_3)}},$$

where the negative root gives R and the positive root r.

Substituting for r_1, r_2, and r_3, we get

$$R \text{ or } r = \frac{a^2 b^2 cd}{a^2 bc + ab^2 d + abcd \mp 2\sqrt{a^2 b^2 c^2 d^2}}.$$

So

$$R = \frac{abcd}{ac + bd - cd}$$

$$= \frac{abcd}{ac - d(c - b)}$$

$$= \frac{abcd}{ac - ad}$$

$$= \frac{abcd}{ab}$$

$$= cd.$$

Similarly, $r = \dfrac{abcd}{4cd - ab}$.

Substituting for a, b, c, and d, we get $r_1 = F_n F_{n+1}$, $r_2 = F_n F_{n+2}$, $r_3 = F_{n+1} F_{n+3}$, $R = F_{n+2} F_{n+3}$, and $r = \dfrac{F_n F_{n+1} F_{n+2} F_{n+3}}{4 F_{n+2} F_{n+3} - F_n F_{n+1}}$.

Clearly, similar formulas exist for Lucas numbers also.

18.4 A FIBONACCI SPIRAL

We can arrange a series of $F_n \times F_n$ squares to construct a *Fibonacci spiral*, as in Figure 18.13, where $n \geq 1$. Moreover, their centers appear to lie on two lines and the two lines appear perpendicular. This is indeed the case.

We will now confirm this. To this end, suppose we choose the center of the first square as the origin, and the horizontal and vertical lines through it as the axes. Let n be odd. Then the change in the y-values in going from the $(n - 4)$th square to the nth square is $\pm(F_n + F_{n-4})/2$, and the change in the corresponding x-values is $\pm(F_{n-2} - 2F_{n-3} - F_{n-4})/2$. Therefore, the slope of the line passing through the centers is

$$\frac{F_n + F_{n-4}}{F_{n-2} - 2F_{n-3} - F_{n-4}} = \frac{3F_{n-2}}{F_{n-2}} = 3.$$

But this line passes through the origin. Thus the centers of all odd-numbered Fibonacci squares lie on the line $y = 3x$, as proved in 1983 by T. Gardiner of the

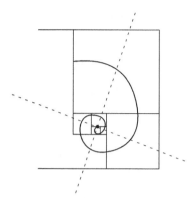

Figure 18.13.   A Fibonacci spiral.

University of Birmingham, England [205]. [In particular, the line containing the centers $(0, 0)$ and $(1/2, 3/2)$ is $y = 3x$.]

Similarly, the centers of all even-numbered Fibonacci squares lie on the line $x + 3y = 1$. Notice that the centers $(1, 0)$ and $(-2, 1)$ lie on it.

18.5   REGULAR PENTAGONS

Regular pentagons provide us with many examples of the golden ratio in everyday life. Some flowers have pentagonal shape; so do starfish (see Figure 18.14a) and the former Chrysler logo (see Figure 18.14b).

(a)   (b)

Figure 18.14.   (a) A starfish. (b) The former Chrysler logo.*

Figure source: Reproduced with the permission of Chrysler Group LLC.

In 1948, H.V. Baravalle of Adelphi University observed [19], "Outstanding among the mathematical facts connected with the [regular] pentagon are the manifold implications of the irrational ratio of the golden section." We will now investigate some of these implications.

Example 18.8. *The diagonals* $\overline{AC}$ *and* $\overline{BE}$ *of the regular pentagon* $ABCDE$ *in Figure 18.15 meet at* F. *Prove that* F *divides both diagonals in the golden ratio.*

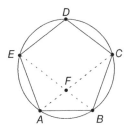

Figure 18.15.

Proof. [†]Let $AB = a$, $BF = b$, and $FE = c$; see Figure 18.16. By the side–angle–side theorem (SAS), $\triangle ABC \cong \triangle ABE$. Since $\angle ABC = 108°$, $\angle BAC = \angle ABE = 36°$. Therefore, $\angle CAE = 72° = \angle AFE$. Then $AF = BF = b$ and $AE = AF$, so $a = c$.

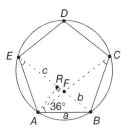

Figure 18.16.

Using the perpendicular $\overline{AR}$ to $\overline{BE}$, we can show that $ER = a\cos 36° = BR$; so $BE = BR + RE = b + c = 2a\cos 36°$. Likewise, $b = \dfrac{a}{2\cos 36°}$.

[†]Based on Hunter and Madachy, 1975 [342].

Then

$$c = 2a \cos 36° - \frac{a}{2\cos 36°}$$

$$= \frac{a(4\cos^2 36° - 1)}{2\cos 36°}.$$

Since $c = a$, this yields

$$4\cos^2 36° - 2\cos 36° - 1 = 0$$

$$\cos 36° = (1 + \sqrt{5})/4$$

$$= \alpha/2,$$

since $\cos 36° > 0$.

Therefore, $BE = b + c = \alpha a = AC$, and hence $BE : FE = \alpha a : a = \alpha : 1 = AC : FC$. Thus F divides both diagonals in the golden ratio. ∎

This is the ninth property of the divine proportion delineated by Pacioli in his book; see Section 16.2 and Baravalle [19].

Since $\cos \pi/5 = \alpha/2$, it follows that $\sin \pi/5 = \sqrt{3 - \alpha}/2$, $\cos \pi/10 = \sqrt{\alpha + 2}/2$, and $\sin \pi/10 = 1/(2\alpha)$.

We can derive the fact that $\cos \pi/5 = \alpha/2$ in a shorter and more elegant way.

AN ALTERNATE PROOF THAT COS $\pi/5 = \alpha/2$

Let $\theta = \pi/10$. Then $2\theta + 3\theta = \pi/2$, so 2θ and $\pi/2 - 3\theta$ are complementary angles. Since the values of cofunctions of complementary angles are equal, it follows that

$$\sin 2\theta = \cos 3\theta$$

$$2\sin \theta \cos \theta = 4\cos^3 \theta - 3\cos \theta$$

$$4\sin^2 \theta + 2\sin \theta - 1 = 0$$

$$\sin \theta = -\alpha/2, -\beta/2.$$

Since $\sin\theta > 0$, it follows that $\sin\theta = -\beta/2 = 1/(2\alpha)$. Therefore,

$$\cos\pi/5 = 1 - 2\sin^2\pi/10$$
$$= 1 - 1/(2\alpha^2)$$
$$= \frac{2-\beta^2}{2}$$
$$= \frac{2-(1+\beta)}{2}$$
$$= \alpha/2,$$

as desired.   ∎

Knowing the values of $\sin\pi/10$ and $\cos\pi/10$, we can compute the exact values of sines and cosines of several acute angles that are multiples of $\pi/20 = 9°$; see Table 18.1.

TABLE 18.1.   Exact Trigonometric Values

Angle θ	$\sin\theta$	$\cos\theta$
$\dfrac{\pi}{20}$	$\dfrac{\sqrt{2-\sqrt{2+\alpha}}}{2}$	$\dfrac{\sqrt{2+\sqrt{2+\alpha}}}{4}$
$\dfrac{\pi}{10}$	$\dfrac{1}{2\alpha}$	$\dfrac{\sqrt{2+\alpha}}{2}$
$\dfrac{3\pi}{20}$	$\dfrac{\sqrt{2-\sqrt{3-\alpha}}}{2}$	$\dfrac{\sqrt{2+\sqrt{3-\alpha}}}{2}$
$\dfrac{\pi}{5}$	$\dfrac{\sqrt{3-\alpha}}{2}$	$\dfrac{\alpha}{2}$
$\dfrac{\pi}{4}$	1	1
$\dfrac{3\pi}{10}$	$\dfrac{\alpha}{2}$	$\dfrac{\sqrt{3-\alpha}}{2}$
$\dfrac{7\pi}{20}$	$\dfrac{\sqrt{2+\sqrt{3-\alpha}}}{2}$	$\dfrac{\sqrt{2-\sqrt{3-\alpha}}}{2}$
$\dfrac{2\pi}{5}$	$\dfrac{\sqrt{2+\alpha}}{2}$	$\dfrac{1}{2\alpha}$
$\dfrac{9\pi}{20}$	$\dfrac{\sqrt{2+\sqrt{2+\alpha}}}{2}$	$\dfrac{\sqrt{2-\sqrt{2+\alpha}}}{2}$

We will now pursue Example 18.8 a bit further.

Example 18.9. *Suppose the perpendicular $\overleftrightarrow{AN}$ at A meets $\overline{ED}$ at N; see Figure 18.17. Show that N divides $\overline{ED}$ in the golden ratio.*

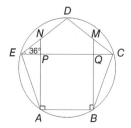

Figure 18.17.

Proof. Draw $\overline{BM} \perp \overline{AB}$. We have $CE = a\alpha$, $PQ = AB = a$, $EP = QC$, and $\angle DEC = \pi/5$. Then

$$EP = QC = \frac{a\alpha - a}{2} = \frac{a(\alpha - 1)}{2}$$

$$EN = \frac{EP}{\cos \pi/5} = \frac{2EP}{\alpha} = \frac{a(\alpha - 1)}{\alpha}$$

$$ND = DE - EN = a - \frac{a(\alpha - 1)}{\alpha} = \frac{a}{\alpha}.$$

Thus $DE : DN = a : a/\alpha = 1 : 1/\alpha = \alpha : 1$; so N divides $\overline{DE}$ in the golden ratio. ∎

Next we compute the area of a regular pentagon in terms of the length of a side and the golden ratio.

Example 18.10. *Compute the area of the regular pentagon ABCDE in Figure 18.18 with a side a units long.*

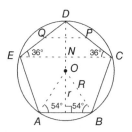

Figure 18.18.

Solution. Notice that $\triangle CDE$ is an isosceles triangle. (In fact, $\overleftrightarrow{CE} \parallel \overleftrightarrow{AB}$.) Let $\overline{DN} \perp \overline{CE}$. Then $CN = EN = a \cos \pi/5 = a\alpha/2$, so $CE = a\alpha$. Likewise, $PQ = a\alpha/2$, where P and Q at the midpoints of $\overline{CD}$ and $\overline{DE}$, respectively.

Let R be the circumradius, and r the inradius of the regular pentagon. Then $R = (a/2)\csc \pi/5$ and $r = (a/2)\cot \pi/5$. Consequently, since $\cos \pi/5 = \alpha/2$, we have

$$\text{area of } \triangle AOB = (1/2)a \cdot (a/2)\cot \pi/5$$
$$= (a^2/4)\cot \pi/5$$
$$= \frac{a^2}{4} \cdot \frac{\alpha}{\sqrt{3-\alpha}}$$
$$= \frac{a^2\alpha}{4\sqrt{3-\alpha}}.$$

Thus the area of the regular pentagon is $\dfrac{5a^2\alpha}{4\sqrt{3-\alpha}}$. ∎

Using the fact that $\cos \pi/5 = \alpha/2$, we will now derive trigonometric formulas for F_n.

18.6   TRIGONOMETRIC FORMULAS FOR F_n

To begin with, using the formula $\cos 3\theta = 4\cos^3 \theta - 3\cos \theta$ and Binet's formula, we have

$$\cos 3\pi/5 = 4\cos^3 \pi/5 - 3\cos \pi/5$$
$$= 4(\alpha/2)^3 - 3(\alpha/2)$$
$$= \alpha(\alpha^2 - 3)/2$$
$$= \alpha(-\beta^2)/2$$
$$= \beta/2;$$
$$F_n = \frac{\alpha^n - \beta^n}{\alpha - \beta}$$
$$= \frac{2^n(\cos^n \pi/5 - \cos^n 3\pi/5)}{\sqrt{5}}. \tag{18.2}$$

W. Hope-Jones discovered this formula in 1921 [322].

Since $\cos \pi/5 = \alpha/2$, $2\sin^2 \pi/10 = 1 - \cos \pi/5 = 1 - \alpha/2 = (2-\alpha)/2 = \beta^2/4$. So $\sin \pi/10 = |\beta|/2 = -\beta/2$; that is, $\beta = -2\sin \pi/10$. Thus $\alpha = 2\cos \pi/5$ and $\beta = -2\sin \pi/10$.

By Binet's formula, we then have

$$F_n = \frac{\alpha^n - \beta^n}{\alpha - \beta}$$

$$= \sum_{k=0}^{n-1} \alpha^{n-k-1} \beta^k$$

$$= \sum_{k=0}^{n-1} (2\cos \pi/5)^{n-k-1}(-2\sin \pi/10)^k$$

$$= 2^{n-1} \sum_{k=0}^{n-1} (-1)^k \cos^{n-k-1} \pi/5 \sin^k \pi/10. \tag{18.3}$$

For example,

$$F_3 = 4 \sum_{k=0}^{2} (-1)^k \cos^{2-k} \pi/5 \sin^k \pi/10$$

$$= 4 \left[(\alpha/2)^2(-\beta/2)^0 - (\alpha/2)(-\beta/2) + (\alpha/2)^0(-\beta/2)^2 \right]$$

$$= 4 \left(\frac{\alpha^2}{4} + \frac{\alpha\beta}{4} + \frac{\beta^2}{4} \right)$$

$$= \alpha^2 + \alpha\beta + \beta^2$$

$$= (\alpha + \beta)^2 - \alpha\beta$$

$$= 2,$$

as expected.

J.L. Brown, Jr., of Pennsylvania State University derived the summation formula (18.3) in 1964 [73].

We will now derive two additional trigonometric formulas for F_n.

Since $\cos \pi/5 = \alpha/2$, it follows that $\sin \pi/5 = \sqrt{3 - \alpha}/2$. Then

$$\sin 3\pi/5 = \sin(2\pi/5 + \pi/5)$$

$$= \sin 2\pi/5 \cos \pi/5 + \cos 2\pi/5 \sin \pi/5$$

$$= 2\sin \pi/5 \cos^2 \pi/5 + (2\cos^2 \pi/5 - 1)\sin \pi/5$$

$$= 4\sin \pi/5 \cos^2 \pi/5 - \sin \pi/5$$

$$= 4 \cdot \frac{\sqrt{3-\alpha}}{2} \cdot \frac{\alpha^2}{4} - \frac{\sqrt{3-\alpha}}{2}$$

$$= \frac{\alpha\sqrt{3-\alpha}}{2},$$

and hence

$$\sin \pi/5 \sin 3\pi/5 = \frac{\sqrt{3-\alpha}}{2} \cdot \frac{\alpha\sqrt{3-\alpha}}{2} = \frac{\alpha(3-\alpha)}{4} = \frac{\sqrt{5}}{4}.$$

Since $\sin 9\pi/5 = -\sin \pi/5$, it also follows that $\sin 3\pi/5 \sin 9\pi/5 = -\sqrt{5}/4$. Thus

$$\sqrt{5}F_n = \alpha^n - \beta^n$$

$$= (2\cos \pi/5)^n - (2\cos 3\pi/5)^n$$

$$5F_n = 2^{n+2}\left(\frac{\sqrt{5}}{4}\cos^n \pi/5 - \frac{\sqrt{5}}{4}\cos^n 3\pi/5\right)$$

$$F_n = \frac{2^{n+2}}{5}\left(\cos^n \pi/5 \sin \pi/5 \sin 3\pi/5 + \cos^n 3\pi/5 \sin 3\pi/5 \sin 9\pi/5\right). \tag{18.4}$$

It follows from this formula that

$$F_n = \frac{(-2)^{n+2}}{5}\left(\cos^n 2\pi/5 \sin 2\pi/5 \sin 6\pi/5 + \cos^n 4\pi/5 \sin 4\pi/5 \sin 12\pi/5\right); \tag{18.5}$$

see Exercise 18.36.

F. Stern of San Jose State University discovered formulas (18.4) and (18.5) in 1979 [544].

Since $\sqrt{5} = 4\sin \pi/5 \sin 3\pi/5$, we can rewrite the trigonometric formula (18.2) slightly differently:

$$F_n = \frac{(2\cos \pi/5)^n - (2\cos 3\pi/5)^n}{4\sin \pi/5 \sin 3\pi/5}$$

$$= 2^{n-2}\frac{\cos^n \pi/5 - \cos^n 3\pi/5}{\sin \pi/5 \sin 3\pi/5}. \tag{18.6}$$

Next we will study the pentagram. The terms *pentagram*, *pentacle*, *five-pointed star*, and *five-petal rose* appear a total of 21 times in *The Da Vinci Code*.

THE PENTAGRAM

Draw all diagonals of the regular pentagon in Figure 18.15. This produces the *star polygon APBQCRDSET*, called a *pentagram*; see Figure 18.19. It follows from Example 18.8 that the points P, Q, R, S, and T divide the diagonals in the golden ratio.

The pentagram was the logo of the Pythagorean School of the sixth century B.C. The flags of many countries of the world contain one or more five-pointed

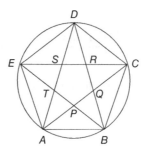

Figure 18.19.

stars. For example, the Australian flag contains six stars, the Chinese flag five, and the United States flag 50. The diameter of every star on the U.S. flag is $0.616 \approx 1/\alpha$. Even flags of several cities contain five-pointed stars: Chicago's flag contains four stars, and the flags of Dallas, Houston, and San Antonio each contain one.

Returning to Figure 18.19, the polygon $PQRST$ is also a regular polygon; see Exercise 18.5. Draw its diagonals to produce a new pentagram and a smaller regular pentagon; the points V, W, X, Y, and Z divide them in the golden ratio. Obviously, this process can be continued indefinitely; see Figure 18.20.

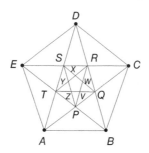

Figure 18.20.

Figure 18.20 contain many angles of various sizes, namely, $\pi/5, 2\pi/5, 3\pi/5,$ $4\pi/5, \pi, 6\pi/5, 7\pi/5, 8\pi/5, 9\pi/5$, and 2π; they form a finite arithmetic sequence with a common difference $\pi/5$.

We can partition the various line segments in Figure 18.20 into six different classes with representatives $\overline{BD}, \overline{AB}, \overline{BQ}, \overline{PQ}, \overline{QW}$, and $\overline{VW}$; the lengths of these line segments are different. Clearly, the longest of them is $\overline{BD}$.

There are five line segments of the same size as $\overline{BD}$; 15 of the same size as $\overline{AB}$; 15 of the same size as $\overline{BQ}$; 15 of the same size as $\overline{PQ}$; 10 of the same size as $\overline{QW}$; and five of the same size as $\overline{VW}$, a total of 65 line segments.

Let $BD = a$. Since R divides $\overline{BD}$ in the golden ratio, it follows that $AB = BR = BD/\alpha = a/\alpha$.

Notice that $\triangle BER \sim \triangle BCQ$, so $BR/BQ = BE/BC$; that is, $BR/BQ = a/(a\alpha) = \alpha$. Therefore, $BQ = BR/\alpha = a/\alpha^2$. Thus Q divides $\overline{BR}$ in the golden ratio. Similarly, R divides $\overline{DQ}$ in the golden ratio. Obviously, we can extend this property to other diagonals as well.

Consider the triangles SPQ and BPQ. Clearly, they are congruent, so $SQ = BQ$. Therefore,

$$PQ = \frac{SQ}{\alpha} = \frac{BQ}{\alpha} = \frac{BR}{\alpha^2} = \frac{a}{\alpha^3};$$

$$QW = \frac{PQ}{\alpha} = \frac{a}{\alpha^4};$$

$$VW = \frac{QW}{\alpha} = \frac{a}{\alpha^5}.$$

Thus $BD = a$, $AB = a/\alpha$, $BQ = a/\alpha^2$, $PQ = a/\alpha^3$, $QW = a/\alpha^4$, and $VW = a/\alpha^5$. They form a decreasing geometric sequence with the first term a and common ratio $1/\alpha$; so

$$BD : AB : BQ : PQ : QW : VW = 1 : \frac{1}{\alpha} : \frac{1}{\alpha^2} : \frac{1}{\alpha^3} : \frac{1}{\alpha^4} : \frac{1}{\alpha^5}.$$

Every element in the sequence, except the last, is α times its successor.

Suppose we continue this procedure indefinitely. Then the sum of the resulting geometric sequence of the different lengths of the various line segments is given by

$$\frac{a}{1 - 1/\alpha} = \frac{a\alpha}{\alpha - 1} = a\alpha^2$$

$$\approx 2.61803398875a.$$

Here is an interesting observation: $\triangle APT$ is a golden triangle. In fact, Figure 18.20 contains several golden triangles, which will become apparent if you search for them; see Exercise 18.18.

We now investigate a link between a regular decagon and α.

18.7   REGULAR DECAGON

Consider a regular decagon with a side of length l and circumradius R; see Figure 18.21. The central angle subtended by a side is $2\pi/10 = \pi/5$. Then, since $\beta < 0$, we have

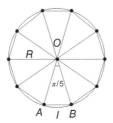

Figure 18.21.

$$R = \frac{l}{2} \csc \pi/5$$

$$= \frac{l}{2} \sqrt{\frac{2}{1 - \cos \pi/5}}$$

$$= \frac{l}{2} \sqrt{\frac{2}{1 - \alpha/2}}$$

$$= \frac{l}{\sqrt{2 - \alpha}}$$

$$= \frac{l}{\sqrt{\beta^2}}$$

$$= -\frac{l}{\beta}$$

$$= l\alpha.$$

In particular, let $l = 1$. Then $R = \alpha$; that is, the circumradius of a decagon of side with unit length is α. This is the seventh property of α described by Pacioli in his classic book.

Returning to Figure 18.21, we also have

$$\text{area of } \triangle AOB = 2lR \cos \pi/10$$

$$= 2l^2\alpha \left(\frac{\sqrt{\alpha + 2}}{2} \right)$$

$$= l^2\alpha\sqrt{\alpha + 2}.$$

So the area of the decagon equals $10l^2\alpha\sqrt{\alpha + 2}$.

Consequently, the area enclosed by the decagon of unit side is $10\alpha\sqrt{\alpha + 2}$.

Next we employ the fifth roots of unity to explore some properties of the divine proportion related to the regular pentagon.

18.8   FIFTH ROOTS OF UNITY

By De Moivre's theorem, the nth roots of unity are given by the complex numbers $z = \text{cis } 2k\pi/n$, where $0 \leq k \leq n, i = \sqrt{-1}$, and $\text{cis } \theta = \cos \theta + i \sin \theta$. They are equally spaced on the unit circle $|z| = 1$ on the complex plane.

In particular, the fifth roots of unity are given by $z = \text{cis } 2k\pi/5$, where $0 \leq k \leq 5$. They are $z_0 = \text{cis } 0 = 1$, $z_1 = \text{cis } 2\pi/5$, $z_2 = \text{cis } 4\pi/5$, $z_3 = \text{cis } 6\pi/5 = \text{cis }(-4\pi/5)$, and $z_4 = \text{cis } 8\pi/5 = \text{cis }(-2\pi/5)$.

Since $\cos \pi/5 = \alpha/2$, it follows that

$$\cos 2\pi/5 = 2\cos^2 \pi/5 - 1$$
$$= 2(\alpha^2/4) - 1$$
$$= \frac{\alpha^2 - 2}{2}$$
$$= \frac{\alpha - 1}{2};$$

$$\sin 2\pi/5 = 2\sin \pi/5 \cos \pi/5$$
$$= 2 \cdot \frac{\sqrt{3 - \alpha}}{2} \cdot \frac{\alpha}{2}$$
$$= \frac{\alpha\sqrt{3 - \alpha}}{2}.$$

Therefore, $z_1 = \dfrac{\alpha - 1}{2} + \dfrac{\alpha\sqrt{3 - \alpha}}{2}i$ and $z_4 = \dfrac{\alpha - 1}{2} - \dfrac{\alpha\sqrt{3 - \alpha}}{2}i$.

We also have

$$\cos 4\pi/5 = 2\cos^2 2\pi/5 - 1$$
$$= 2\left(\frac{\alpha - 1}{2}\right)^2 - 1$$
$$= -\frac{\alpha}{2};$$

$$\sin 4\pi/5 = 2\sin 2\pi/5 \cos 2\pi/5$$
$$= 2 \cdot \frac{\alpha\sqrt{3 - \alpha}}{2} \cdot \frac{\alpha - 1}{2};$$
$$= \frac{\sqrt{3 - \alpha}}{2}.$$

Consequently, $z_2 = -\dfrac{\alpha}{2} + \dfrac{\sqrt{3 - \alpha}}{2}i$ and $z_3 = -\dfrac{\alpha}{2} - \dfrac{\sqrt{3 - \alpha}}{2}i$.

Thus the fifth roots of unity are 1, $\dfrac{\alpha - 1}{2} \pm \dfrac{\alpha\sqrt{3 - \alpha}}{2}i$ and $-\dfrac{\alpha}{2} \pm \dfrac{\sqrt{3 - \alpha}}{2}i$. You may verify this without resorting to De Moivre's theorem; see Exercise 18.23. The roots are represented by the points C, D, B, E, and A, respectively, as in Figure 18.22.

We can extract many properties of the regular pentagon by using the coordinates of its vertices. For example, we can show that $BD = \sqrt{\alpha + 2}$ and $AB = \sqrt{\alpha + 2}/\alpha$, so $BD = \alpha AB$, as expected; see Exercises 18.24–18.27.

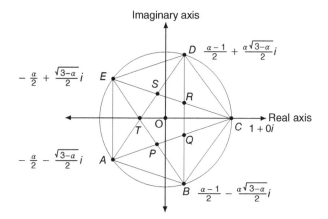

Figure 18.22. Fifth roots of unity.

Interestingly, our mysterious number α appears even in geometric figures not connected to pentagons or decagons. For example, consider a square $ABFE$ inscribed in a semicircle; see Figure 18.23. It can be shown that $\dfrac{AE}{AC} = \dfrac{AD}{AE} = \alpha$; see Exercise 18.33.

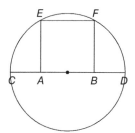

Figure 18.23.

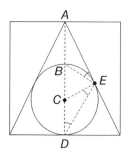

Figure 18.24.

The golden ratio also appears in a circle inscribed in an isosceles triangle which is in turn inscribed in a square; see Figure 18.24. Since the angle in a semicircle is a right angle, it follows that $\angle AEB = \angle CED$. But $\angle CED = \angle CDE$, since $\triangle CDE$ is isosceles. Therefore, $\angle AEB = \angle CDE$.

It now follows that $\triangle ABE \sim \triangle AED$, so $AE/AB = AD/AE$. Let $AE/AB = AD/AE = x$.

Now consider $\triangle ADF$ and $\triangle ACE$. They are similar, so $AD/DF = AE/CE$. But $AD = 2DF$, so $AE = 2CE = BD$. Therefore,

$$\frac{AD}{AE} = \frac{AB + BD}{AE}$$
$$= \frac{AB + AE}{AE}$$
$$= 1 + \frac{AB}{AE}$$
$$x = 1 + \frac{1}{x}$$
$$= \alpha.$$

Thus $\dfrac{AD}{AE} = \dfrac{AE}{AB} = \alpha$.

18.9   A PENTAGONAL ARCH

In 1974, D.W. DeTemple of Washington State University studied the pentagonal arch by rolling a regular pentagon along a line [146]; see Figure 18.25. As the leftmost pentagon is rolled to the right, vertex A moves toward B, then to C, D, and finally to E as the successive sides touch the base line. Connecting these five points, we generate the pentagonal arch $ABCDE$. Surprisingly, this arch is also related to the sacred number.

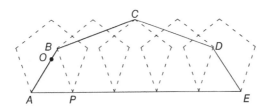

Figure 18.25.   Pentagonal arch. *Source*: DeTemple, 1974 [146]. Reproduced with permission of the Fibonacci Association.

To see this, let s denote the length of a side of the regular pentagon. Since a vertex angle of the pentagon is $108°$, it follows that $\angle APB = 72°$. Since $AP = BP = s$, by the law of cosines, we have

$$AB^2 = AP^2 + BP^2 - 2AP \cdot BP \cos 72°$$

$$= 2s^2(1 - \cos 72°)$$

$$= 4s^2 \sin^2 36°$$

$$= 4s^2 \left(\frac{3 - \alpha}{4} \right)$$

$$= (3 - \alpha)s^2$$

$$AB = s\sqrt{3 - \alpha}.$$

Likewise, we can show that $BC = s\sqrt{2 + \alpha}$. Consequently, $\dfrac{BC}{AB} = \sqrt{\dfrac{2 + \alpha}{3 - \alpha}}$. But $\alpha^2(3 - \alpha) = 3\alpha^2 - \alpha^3 = 3(\alpha + 1) - (2\alpha + 1) = 2 + \alpha$. So $\dfrac{BC}{AB} = \alpha$, the golden ratio.

We now turn to a few occurrences of golden rectangles and hence the golden ratio in solid geometry.

18.10 REGULAR ICOSAHEDRON AND DODECAHEDRON

A *regular icosahedron* is one of the five Platonic solids; see Figure 18.26. It has 12 vertices, 20 equilateral triangular faces, and 30 edges. Five faces meet at each vertex, and they form a pyramid with a regular pentagonal base. We can place three mutually perpendicular and symmetrically arranged golden rectangles (see Figure 18.27) inside the icosahedron in such a way that their 12 corners will coincide with those of the icosahedron; see Figure 18.28.

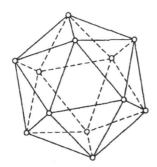

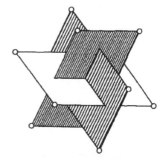

Figure 18.26. A regular icosahedron.*   **Figure 18.27.** Three golden rectangles.*

*Figure sources: Coxeter, 1969 [131], Figure 25.12. Reproduced with the permission of John Wiley & Sons, Inc.

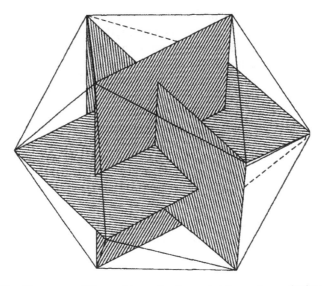

Figure 18.28. The corners of three golden rectangles meet at those of a regular icosahedron.*

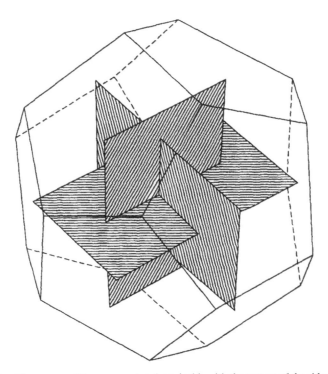

Figure 18.29. The corners of the same rectangles coincide with the centers of the sides of a regular dodecahedron.*

Figure sources: Coxeter, 1969 [131], Figure 25.14. Reproduced with the permission of John Wiley & Sons, Inc.

The length of a longer side of the golden rectangle equals the length of a diagonal of the pentagon. As we saw earlier, the length of a diagonal is α times that of a side of the pentagon. Thus the length of the golden rectangle is α times the length of an edge between any two adjacent vertices of the icosahedron. In particular, if the adjacent vertices are one unit away, then the length of a longer side of the golden rectangle is α. This is the essence of the "twelfth incomprehensible" property described by Pacioli in his celebrated work.

We can also arrange three mutually perpendicular and symmetrically placed golden rectangles in another Platonic solid, the *regular dodecahedron*. It has 12 pentagonal faces, 20 vertices, and 30 edges. The various corners of the rectangles meet the faces at their centers; see Figure 18.29.

Next we turn to golden ellipses and golden hyperbolas. They were introduced by H.E. Huntley of England in 1974 [344, 345]. He investigated the properties of such conic sections in detail. We will now pursue some of them. Again, we will encounter several occurrences of the golden ratio.

18.11   GOLDEN ELLIPSE

The ratio of the major axis to the minor axis of a *golden ellipse* is the magic ratio α.

Let $2a$ denote the length of the major axis, $2b$ that of the minor axis, and e the eccentricity of an ellipse. It is well known that $b^2 = a^2(1 - e^2)$. So for a golden ellipse, we must have $\dfrac{b^2}{a^2} = 1 - e^2 = \dfrac{1}{\alpha^2} = \beta^2$. So $e^2 = 1 - \beta^2 = -\beta$.

Thus the eccentricity of a golden ellipse is $e = \sqrt{-\beta}$; see Figure 18.30. Consequently, one-half of the length of the minor axis is given by $b^2 = a^2\beta^2$; so $b = a|\beta|$.

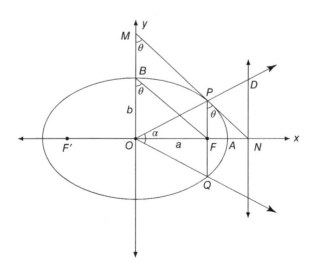

Figure 18.30.   The golden ellipse.

Suppose we inscribe a golden ellipse in a rectangle with its sides parallel to the axes. Then the rectangle would be a golden rectangle.

Let F and F' be the foci of the golden ellipse; see Figure 18.30. Then $OF = ae = a/\sqrt{\alpha} = a\sqrt{-\beta}$, and $BF = \sqrt{b^2 + a^2 e^2} = a$.

Let $\angle OBF = \theta$. Then $\sec\theta = BF/OB = a/b = \alpha$. Let $\overleftrightarrow{ON}$ be perpendicular to the directrix $\overleftrightarrow{ND}$. Then $ON = a/e = a\sqrt{\alpha}$; and

$$FN = ON - OF = a\sqrt{\alpha} - \frac{a}{\sqrt{\alpha}}$$

$$= \frac{a(\alpha - 1)}{\sqrt{\alpha}}$$

$$= -\frac{a\beta}{\sqrt{\alpha}}$$

$$= a|\beta|^{3/2}.$$

For any ellipse, the length of the minor axis is the geometric mean of that of the major axis and that of the latus rectum l; that is, $b^2 = al$. So for the golden ellipse,

$$PQ = l = \frac{b^2}{a} = b\left(\frac{b}{a}\right) = \frac{b}{\alpha}. \quad \text{Thus} \quad a : b : l = ba : b : b/\alpha = \alpha : 1 : 1/\alpha = \alpha^2 : \alpha : 1.$$

Notice that

$$\frac{ON}{FN} = \frac{a\sqrt{\alpha}}{a/\sqrt{\alpha}} = \alpha \quad \text{and} \quad \frac{OF}{FN} = \frac{a/\sqrt{\alpha}}{a(\alpha - 1)\sqrt{\alpha}} = \frac{1}{\alpha - 1} = \alpha;$$

so the focus F divides $\overline{ON}$ in the golden ratio.

Let $\overline{PQ}$ denote the latus rectum. Then

$$OP^2 = OF^2 + FP^2$$

$$= \frac{a^2}{\alpha} + \frac{b^2}{\alpha^2}$$

$$= b^2\alpha + \frac{b^2}{\alpha^2}$$

$$= \frac{b^2(\alpha^3 + 1)}{\alpha^2}$$

$$= \frac{b^2(2\alpha + 2)}{\alpha^2}$$

$$= 2b^2$$

$$OP = \sqrt{2}b.$$

An ellipse has the property that the tangent at P passes through N, and $\cot \angle FPN = e$. Since $\cot \theta = OB/OF = b/ae$, it follows in the case of a golden ellipse that

$$\cot \angle FPN = \frac{1}{\sqrt{\alpha}} = \sqrt{-\beta} \quad \text{and} \quad \cot \theta = \frac{1}{e\alpha} = \frac{1}{\sqrt{\alpha}} = \sqrt{-\beta};$$

so $\angle FPN = \theta$. Thus $MPFB$ is a parallelogram, and hence $MP = BF = a$.

In addition, since $\triangle OMN \sim \triangle FPN$, we have

$$\frac{MN}{PN} = \frac{ON}{FN}$$

$$\frac{MP}{PN} + 1 = \frac{OF}{FN} + 1.$$

Therefore, $\dfrac{MP}{PN} = \dfrac{OF}{FN} = \alpha$. Thus P divides $\overline{MN}$ in the golden ratio.

Finally, let $\overrightarrow{OP}$ intersect the directrix $\overleftrightarrow{ND}$ at D. Since $\triangle OND \sim \triangle OFP$, $OD/OP = ON/OF = \alpha$; so P divides $\overline{OD}$ also in the golden ratio.

18.12   GOLDEN HYPERBOLA

Next we investigate the *golden hyperbola*; see Figure 18.31. Huntley gives a fairly extensive account of its properties in his fascinating book, *The Divine Proportion* [344].

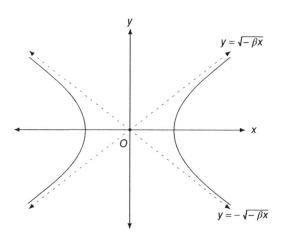

Figure 18.31.   The golden hyperbola.

The eccentricity e of the golden hyperbola is defined by $e^2 = \alpha$. Then $b^2 = a^2(e^2 - 1) = a^2(\alpha - 1) = -a^2\beta$. So $a/b = \sqrt{\alpha}$ and $b = a\sqrt{-\beta}$.

Consequently, the asymptotes of the golden hyperbola are given by $y = \pm\dfrac{b}{a}x$; that is, $y = \pm\sqrt{-\beta}x$.

Huntley also studied the parabola $y^2 = 4ax$ using $(a\alpha^2, 2a\alpha)$ as a point on it. Suppose we draw the parabola and the golden hyperbola to the same scale with the same origin. Then we can establish that the asymptotes of the hyperbola intersect the parabola at the origin and at the points $(4a\alpha, \pm 4a\sqrt{\alpha})$; see Exercise 18.39.

EXERCISES 18

1. Show that the triangle with sides $L_{n-1}L_{n+2}$, $2L_nL_{n+1}$, and $L_{2n}L_{2n+2}$ units long is Pythagorean (Freitag, 1975 [191]).
2. Compute the area of an equilateral trapezoid with bases F_{n-1} and F_{n+1}, and with lateral side F_n (Woodlum, 1968 [602]).
3. Consider a triangle ABC with sides $G_{n-1}G_{n+2}$, $2G_nG_{n+1}$, and $G_n^2 + G_{n+1}^2$ units long, where G_k denotes the kth gibonacci number. Prove that $\triangle ABC$ is a Pythagorean triangle.
4. In $\triangle ABC$, $AB = AC$. Let D be a point on side $\overline{AB}$ such that $AD = CD = BC$. Prove that $2\cos A = AB/BC = \alpha$. (Source unknown.)
5. Show that the polygon $PQRST$ in Figure 18.19 is a regular pentagon.

Using Figure 18.20, compute the area of each polygon, where $BD = a$.

6. $\triangle APB$.
7. $\triangle APT$.
8. $\triangle CDR$.
9. $\triangle CDS$.
10. Rhombus $CDEP$.
11. Rhombus $SPRD$.

Using Figure 18.20, compute each ratio.
12. $\triangle CDS : \triangle CDR$.
13. Rhombus $CDEP$: rhombus $SPRD$.

Compute the shaded area in each figure.

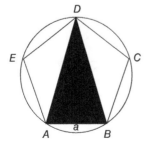

Figure 18.32.

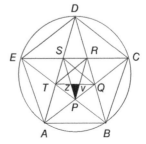

Figure 18.33.

14. Figure 18.32.
15. Figure 18.33.

Using Figure 18.20, compute the area of each polygon, where $BD = a$.

16. Pentagon $PQRST$.
17. Pentagram $APBQCRDSET$.
18. Find the number of golden triangles in Figure 18.20.
19. The lengths of the sides of a Pythagorean triangle form a geometric sequence with common ratio $r > 1$. Show that $r = \sqrt{\alpha}$.
20. Prove that the length of a side of a regular decagon with circumradius r is r/α.
21. The circumradius of the regular 10-pointed star in Figure 18.34 is r. Prove that $AD = r\alpha$.

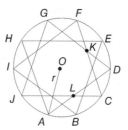

Figure 18.34.

22. Show that the shaded areas in Figure 18.35 form a geometric sequence with common ratio $1/\alpha$ (Baravalle, 1948 [19]).

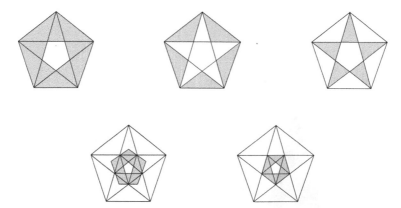

Figure 18.35.

23. Solve the equation $x^5 - 1 = 0$ algebraically.

Using Figure 18.22, answer Exercises 18.24–18.32.

24. Find BD.

25. Find AB.

26. Find $BD : AB$.

27. Compute area $ABDE$.

28. Using the fact that P and Q divide $\overline{AC}$ in the golden ratio, determine their coordinates.

29. Using Exercise 18.28, compute PQ, AP, and QC.

30. Find the ratio $\triangle ABQ : \triangle ABP$.

31. Find the ratio $BD : AB : BQ : PQ$.

32. Find the inradius of the circle inscribed in the pentagon.

33. Using Figure 18.23, show that $AE/AC = AD/AE = \alpha$.

34. A regular pentagon of side p, a regular hexagon of side h, and a regular decagon of side d are inscribed in the same circle. Prove that these lengths can be used to form the sides of a Pythagorean triangle (Bicknell, 1974 [43]).

35. Using formula (18.3), compute both F_2 and F_4.

36. Prove formula (18.5) (Stern, 1979 [544]).

37. Prove that $\dfrac{2^{n+1}}{5} \displaystyle\sum_{k=1}^{4} \cos^n k\pi/5 \sin k\pi/5 \sin 3k\pi/5 = \begin{cases} 0 & \text{if } n \text{ is odd} \\ F_n & \text{otherwise} \end{cases}$

(Hoggatt, 1979 [303]).

38. Compute F_5 using formula (18.6).

39. Show that the asymptotes of the golden hyperbola intersect the parabola $y^2 = 4ax$ at the origin and at the points $(4a\alpha, \pm 4a\sqrt{\alpha})$.

40. One endpoint P of the focal chord of the parabola $y^2 = 4ax$ is $(a\alpha^2, 2a\alpha)$. Find the other endpoint Q.

41. Compute the length of the focal chord $\overline{PQ}$ in Exercise 18.40.

42. Find the equations of the tangent and the normal to the parabola $y^2 = 4ax$ at the point $P(a\alpha^2, 2a\alpha)$.

43. Find the equations of the tangent and the normal to the parabola $y^2 = 4ax$ at the endpoint Q of the focal chord $\overline{PQ}$ in Exercise 18.40.

44. Find the point of intersection of the tangents at the ends of the focal chord $\overline{PQ}$ in Exercise 18.40.

45. Find the angle between the tangents in Exercises 18.42 and 18.43.

46. Find the point of intersection of the normals to the parabola $y^2 = 4ax$ at the ends of the focal chord $\overline{PQ}$ in Exercise 18.40.

47. Find the angle between the normals at P and Q.

48. Suppose the focal chord $\overline{PQ}$ in Exercise 18.40 intersects the y-axis at R. Show that the focus S divides $\overline{PR}$ in the golden ratio.

49. With S, Q, and R as in Exercise 18.48, show that Q divides $\overline{SR}$ in the golden ratio.

19

CONTINUED FRACTIONS

Recall from Chapter 16 that the sequence of ratios F_{n+1}/F_n of consecutive Fibonacci numbers approaches the golden ratio α as $n \to \infty$. Interestingly, we can employ these ratios, coupled with Fibonacci recurrence, to generate fractional numbers of a very special nature, called *continued fractions*[†]. The English mathematician John Wallis (1616–1703) coined the term *continued fractions*. Some continued fractions have finite decimal expansions, while others do not. We will now begin our pursuit with some basic vocabulary and a few characterizations of continued fractions.

19.1   FINITE CONTINUED FRACTIONS

A *finite continued fraction* is a multi-layered fraction of the form

$$x = a_0 + \cfrac{1}{a_1 + \cfrac{1}{a_2 + \cfrac{1}{a_3 + \cfrac{1}{\ddots + \cfrac{1}{a_{m-1} + \cfrac{1}{a_m}}}}}} \tag{19.1}$$

[†]The Italian mathematician Pietro Antonio Cataldi (1548–1626) is credited with laying the foundation for the theory of continued fractions.

Fibonacci and Lucas Numbers with Applications, Volume One, Second Edition. Thomas Koshy.
© 2018 John Wiley & Sons, Inc. Published 2018 by John Wiley & Sons, Inc.

where each a_i is a real number; $a_0 \geq 0$; $a_i > 0$ and $i \geq 1$. The numbers $a_1, a_2, \ldots, a_m$ are the *partial quotients* of the finite continued fraction. The fraction is *simple* if each a_i is an integer.

Since this notation is a bit cumbersome, this fraction is often written in a more compact form: $[a_0; a_1, a_2, a_3, \ldots, a_m]$, where $a_0 = \lfloor x \rfloor$ and the semicolon separates the fractional part from the integer part.

For example,

$$[1; 2, 3, 4, 5, 6] = 1 + \cfrac{1}{2 + \cfrac{1}{3 + \cfrac{1}{4 + \cfrac{1}{5 + \cfrac{1}{6}}}}}$$

$$= \frac{1393}{972}.$$

On the other hand, finding the continued fraction of the rational number $\dfrac{1393}{972}$ involves a repeated application of the Euclidean algorithm:

$$
\begin{aligned}
1393 &= 1 \cdot 972 + 421 \\
972 &= 2 \cdot 421 + 130 \\
421 &= 3 \cdot 130 + 31 \\
130 &= 4 \cdot \ 31 + \ 6 \\
31 &= 5 \cdot \ \ 6 + \ 1 \\
6 &= 6 \cdot \ \ 1 + \ 0.
\end{aligned}
$$

$$\uparrow$$
partial quotients

Now divide each dividend by the corresponding divisor, save the fractional remainder, and then apply substitution for the fractional remainder:

$$\frac{1393}{972} = 1 + \frac{421}{972} \qquad\qquad = 1 + \frac{1}{972/421}$$

$$= 1 + \cfrac{1}{2 + 130/421} \qquad\qquad = 1 + \cfrac{1}{2 + \cfrac{1}{421/130}}$$

$$= 1 + \cfrac{1}{2 + \cfrac{1}{3 + 31/130}} \qquad = 1 + \cfrac{1}{2 + \cfrac{1}{3 + \cfrac{1}{130/31}}}$$

$$= 1 + \cfrac{1}{2 + \cfrac{1}{3 + \cfrac{1}{4 + 6/31}}} = 1 + \cfrac{1}{2 + \cfrac{1}{3 + \cfrac{1}{4 + \cfrac{1}{31/6}}}}$$

$$= 1 + \cfrac{1}{2 + \cfrac{1}{3 + \cfrac{1}{4 + \cfrac{1}{5 + \cfrac{1}{6}}}}}$$

$$= [1; 2, 3, 4, 5, 6],$$

as expected. Notice that the partial quotients are the quotients in the Euclidean algorithm.

FLOOR TILINGS REVISITED

The Euclidean algorithm, floor tilings, and partial quotients are closely related. To see this, suppose we would like to tile a 31×13 floor with $n \times n$ tiles, where n is a positive integer. Assume that plenty of square tiles are available in every size. Always use the largest square tile(s) first.

By the Euclidean algorithm, we have

$$31 = 2 \cdot 13 + 5$$
$$13 = 2 \cdot 5 + 3$$
$$5 = 1 \cdot 3 + 2$$
$$3 = 1 \cdot 2 + 1$$
$$2 = 2 \cdot 1 + 0.$$
$$\uparrow$$

So the 31×13 floor can be tiled with two 13×13 tiles, two 5×5 tiles, one 3×3 tile, one 2×2 tile, and two 1×1 tiles; see Figure 19.1.

Clearly, $\dfrac{31}{13} = [2; 2, 1, 1, 2]$. Clearly, the partial quotients indicate the number of square tiles of various sizes required for the tiling.

In particular, for an $F_{n+1} \times F_n$ flooring, every $F_k \times F_k$ tile is used exactly once, where $1 \le k \le n$. Assuming $F_2 \times F_2$ and $F_1 \times F_1$ are of different sizes, there are no repetitions.

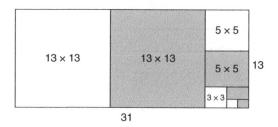

Figure 19.1.

19.2   CONVERGENTS OF A CONTINUED FRACTION

By chopping off the continued fraction for x in equation (19.1) at various plus signs, we get a sequence $\{C_k\}$ of approximations of x, where $0 \le k \le m$:

$$a_0, \quad a_0 + \frac{1}{a_1}, \quad a_0 + \cfrac{1}{a_1 + \cfrac{1}{a_2}}, \quad \dots \; .$$

Each $C_k = [a_0; a_1, a_2, a_3, \dots, a_k]$ is the kth *convergent* of x, where $k \ge 0$ and $C_0 = [a_0] = a_0$; Wallis introduced this concept in his 1695 book, *Opera Mathematica*.

For example, consider the Fibonacci ratio 21/13. As a simple finite continued fraction,

$$\frac{21}{13} = [1; 1, 1, 1, 1, 1, 1].$$

We can easily confirm this. The various convergents are:

$$
\begin{aligned}
C_0 &= [1]   &&= \frac{1}{1} = 1 \\[4pt]
C_1 &= [1; 1]   &&= \frac{2}{1} = 2 \\[4pt]
C_2 &= [1; 1, 1]   &&= \frac{3}{2} \approx 1.5 \\[4pt]
C_3 &= [1; 1, 1, 1]   &&= \frac{5}{3} \approx 1.6666666667 \\[4pt]
C_4 &= [1; 1, 1, 1, 1]   &&= \frac{8}{5} \approx 1.6 \\[4pt]
C_5 &= [1; 1, 1, 1, 1, 1]   &&= \frac{13}{8} \approx 1.625 \\[4pt]
C_6 &= [1; 1, 1, 1, 1, 1, 1] &&= \frac{21}{13} \approx 1.6153846154.
\end{aligned}
$$

We now make some interesting observations about these convergents.

SOME OBSERVATIONS ABOUT CONVERGENTS

- The convergents C_k approach the actual value $21/13$ as k increases, where $0 \le k \le 6$.
- The even-numbered convergents C_{2k} approach it from below, and the odd-numbered convergents C_{2k+1} from above.
- The convergents are alternately less than and greater than $21/13$, except the last one; that is, $C_0 < C_2 < C_4 < C_6 < C_5 < C_3 < C_1$; see Figure 19.2.

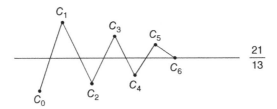

Figure 19.2.

- The convergents display a remarkable pattern: They are ratios of consecutive Fibonacci numbers; that is, $C_k = \dfrac{F_{k+2}}{F_{k+1}}$, where $0 \le k \le 6$. We can easily confirm this; see Exercise 19.9.

Evaluating the convergents may seem to be a tedious job, but recursion can speed up their computation.

RECURSIVE DEFINITION OF C_n

Let $C_k = \dfrac{p_k}{q_k}$ denote the kth convergent of the continued fraction (19.1). Then we can show [369] by PMI that

$$p_k = a_k p_{k-1} + p_{k-2}$$
$$q_k = a_k q_{k-1} + q_{k-2},$$

where $p_0 = a_0$, $q_0 = 1$, $p_1 = a_0 a_1 + 1$, and $q_1 = a_1$, and $2 \le k \le n$. Consequently, we can compute C_k using its predecessors C_{k-2} and C_{k-1}.

For example, consider the continued fraction $\dfrac{21}{13} = [1; 1, 1, 1, 1, 1, 1]$, where $a_i = 1$ for every $0 \le i \le 6$. Since $C_3 = \dfrac{p_3}{p_4} = \dfrac{5}{3}$ and

$C_4 = \dfrac{p_4}{p_5} = \dfrac{8}{5}$, we have

$$C_5 = \frac{p_5}{p_5} = \frac{a_5 p_4 + p_3}{a_5 q_4 + q_3} = \frac{1 \cdot 8 + 5}{1 \cdot 5 + 3} = \frac{13}{8},$$

as expected.

We can construct a table such as Table 19.1 to facilitate the computation of the convergents $C_k = \dfrac{p_k}{q_k}$, where $k \geq 2$. For example, consider the continued fraction $[2; 1, 3, 4, 2, 3, 5]$. Using the initial conditions $p_0 = a_0 = 2$, $q_0 = 1$, $p_1 = a_0 a_1 + 1 = 2 \cdot 1 + 1 = 3$, and $q_1 = a_1 = 1$, we can complete the remaining columns of the table.

TABLE 19.1.

k	0	1	2	3	4	5	6
a_k	2	1	3	4	2	3	5
p_k	2	3	(11)	47	105	362	1915
q_k	1	1	4	17	38	131	693

For example, $p_2 = a_2 p_1 + p_0 = 3 \cdot 3 + 2 = \boxed{11}$ and $q_3 = a_3 q_2 + q_1 + 4 \cdot 4 + 1 = \boxed{17}$; see Table 19.1.

It follows from the table that $C_6 = \dfrac{p_6}{q_6} = \dfrac{1915}{693}$. By direct computation, you may verify that $\dfrac{1915}{693} = [2; 1, 3, 4, 2, 3, 5]$.

We will now briefly study infinite continued fractions.

19.3 INFINITE CONTINUED FRACTIONS

Suppose there are infinitely many terms in the expression $[a_0; a_1, a_2, \ldots, a_n, \ldots]$, where $a_0 \geq 0$ and $a_i > 0$ when $i \geq 1$. Such a fraction is an *infinite continued fraction*. In particular, if each a_i is an integer, then it is an *infinite simple continued fraction*.

For example, $[1; 1, 1, 1, \ldots]$ is an infinite simple continued fraction, the simplest of them all. It is often written as $[1; \overline{1}]$ to indicate the periodic behavior.

It appears from our brief analysis of the finite simple continued fractions that the nth convergent C_n of the infinite simple continued fraction $[1; \overline{1}]$ is the Fibonacci ratio F_{n+1}/F_n. This is indeed the case, and can be established using PMI; see Exercise 19.9. Thus

$$C_n = \frac{p_n}{q_n} = \frac{F_{n+2}}{F_{n+1}},$$

where $n \geq 0$. This relationship was first observed by the Scottish mathematician Robert Simson (1687–1768).

Since $\lim\limits_{n \to \infty} C_n = \lim\limits_{n \to \infty} \dfrac{F_{n+2}}{F_{n+1}} = \alpha$, it follows that the infinite continued fraction $[1; \overline{1}]$ converges to the golden ratio. This yields a remarkably beautiful formula for α:

$$\alpha = [1; \overline{1}]$$

$$= 1 + \cfrac{1}{1 + \cfrac{1}{1 + \cfrac{1}{1 + \cfrac{1}{1 + \cdots}}}}.$$

This is consistent with the fact that the value of every infinite continued fraction is an irrational number.

We can establish the fact that $[1; \overline{1}] = \alpha$ by an alternate route, without employing convergents. To confirm this, we let $x = [1; \overline{1}]$. It is fairly obvious that the infinite continued fraction converges to a limit. So

$$[1; \overline{1}] = [1; [1; \overline{1}]]$$

$$x = [1; x]$$

$$= 1 + \frac{1}{x}.$$

Since $x > 0$, this implies $x = \alpha$. Thus

$$\lim\limits_{n \to \infty} C_n = \lim\limits_{n \to \infty} \frac{F_{n+2}}{F_{n+1}} = \alpha = [1; \overline{1}].$$

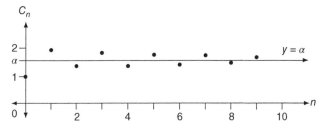

Figure 19.3.

It follows from the previous discussion that even-numbered convergents C_{2n} approach α from below, and odd-numbered convergents C_{2n+1} from above. Figure 19.3 exhibits this marvelous behavior for $0 \le n \le 9$.

Next we develop an infinite simple continued fraction for $-\beta = \dfrac{1}{\alpha}$.

AN INFINITE CONTINUED FRACTION FOR $-\beta$

In a letter to the editor of *The Scientific Monthly* in 1951 [460], F.C. Ogg of Bowling Green State University, Ohio, provided an elegant way of converting $\sqrt{5} - 1$ into an infinite simple continued fraction; it was a response to J.C. Pierce's article [484] on the Fibonacci series:

$$\sqrt{5} - 1 = 1 + (\sqrt{5} - 2) = 1 + \cfrac{1}{\sqrt{5} + 2}$$

$$= 1 + \cfrac{1}{4 + \sqrt{5} - 2}$$

$$= 1 + \cfrac{1}{4 + \cfrac{1}{\sqrt{5} + 2}}$$

$$= 1 + \cfrac{1}{4 + \cfrac{1}{4 + \sqrt{5} - 2}}$$

$$= 1 + \cfrac{1}{4 + \cfrac{1}{4 + \cfrac{1}{4 + \cdots}}}$$

$$= [1; 4, 4, 4, \ldots] = [1; \overline{4}].$$

As we might expect, this continued fraction has an interesting byproduct. To see this, notice that its convergents are $1, \dfrac{5}{4}, \dfrac{21}{17}, \dfrac{89}{72}, \ldots$. Now divide each by 2. The resulting numbers are $\dfrac{1}{2}, \dfrac{5}{8}, \dfrac{21}{34}, \dfrac{89}{144}, \ldots$; so the nth convergent of the infinite continued fraction for $(\sqrt{5} - 1)/2$ is $\dfrac{F_{3n+2}}{F_{3n+3}}$, where $n \ge 0$. Since $\lim\limits_{n \to \infty} \dfrac{F_{3n+2}}{F_{3n+3}} = \lim\limits_{n \to \infty} \dfrac{1}{F_{3n+3}/F_{3n+2}} = \dfrac{1}{\alpha} = -\beta$, it follows that

$$-\beta = \frac{\sqrt{5} - 1}{2} = \frac{1}{\alpha} = [0; \overline{1}].$$

We now add that $\alpha = [1; \overline{1}], \alpha^2 = [2; \overline{1}], \alpha^3 = [4; \overline{4}], \alpha^4 = [6; \overline{15}], \alpha^5 = [11; \overline{11}]$, and $\alpha^6 = [17; \overline{1, 16}]$. More generally, let $n \geq 0$; then

$$\alpha^n = \begin{cases} [L_n; \overline{L_n}] & \text{if } n \text{ is odd} \\ [L_n - 1; \overline{1, L_n - 2}] & \text{otherwise.} \end{cases}$$

Next we will introduce a Diophantine equation with Fibonacci and Lucas implications.

19.4   A NONLINEAR DIOPHANTINE EQUATION

In their 1967 study of continued fractions [419], C.T. Long and J.H. Jordan of Washington State University discovered a close relationship between F_n and L_n, and the *Pell's equation* $x^2 - dy^2 = k$ [370], where d is positive and nonsquare, and $k \neq 0$. (Although Pell's equation is named after the English mathematician John Pell (1611–1685), Pell added little to the study of such equations. Unfortunately, it is so called due to an innocent error by Euler.)

The following theorem [419], discovered by Long and Jordan, "provide[s] unusual characterizations of both Fibonacci and Lucas numbers." We omit its proof in the interest of brevity.

Theorem 19.1. *The positive solutions of the Pell's equation* $x^2 - 5y^2 = 4(-1)^n$ *are given by* $(x_n, y_n) = (L_n, F_n)$, *where* $n \geq 1$.   ■

For example, $L_5^2 - 5F_5^2 = 11^2 - 5 \cdot 5^2 = -4$, and $L_6^2 - 5F_6^2 = 18^2 - 5 \cdot 8^2 = 4$.

In Chapter 5, we found that $L_n^2 - 5F_n^2 = 4(-1)^n$. Consequently, (L_n, F_n) is a solution of the equation $x^2 - 5y^2 = 4(-1)^n$. So it remains to show that if (x_n, y_n) is a solution, then $(x_n, y_n) = (L_n, F_n)$.

EXERCISES 19

Represent each number as a continued fraction.

1. 51/35.
2. 68/89.

Rewrite each continued fraction as a rational number.

3. $[2; 3, 1, 5]$.
4. $[3; 1, 3, 2, 4, 7]$.

Compute the convergents of each continued fraction.

5. $[1; 2, 3, 4, 5]$.

6. $[1; 1, 1, 1, 1, 1, 1, 1]$.

7. The first and second convergents of the continued fraction $[1; 2, 3, 4, 5, 6]$ are 3/2 and 10/7. Compute its third and fourth convergents.

8. The seventh and eighth convergents of the continued fraction $[1; 1, 1, 1, 1, 1, 1, 1, 1]$ are 34/21 and 55/34. Compute its ninth convergent.

9. Let C_n denote the nth convergent of the finite simple continued fraction $[1; 1, 1, 1, \ldots, 1]$. Prove that $C_n = F_{n+2}/F_{n+1}$, where $n \geq 0$.

10. Let p_n/q_n denote the nth convergent of the finite simple continued fraction $[1; 1, 1, 1, \ldots, 1]$. Prove that $p_n q_{n-1} - q_n p_{n-1} = (-1)^{n-1}$, where $n \geq 1$.

11. Using Cassini's formula, prove that $\lim_{n \to \infty} (C_n - C_{n-1}) = 0$, where C_k denotes the kth convergent of the infinite simple continued fraction $[1; \overline{1}]$.

Evaluate each infinite simple continued fraction.

12. $[F_n; \overline{F_n}]$.

13. $[L_n; \overline{L_n}]$.

20

FIBONACCI MATRICES

In this chapter, we will take advantage of the power of matrices to extract new identities and results involving Fibonacci and Lucas numbers; see the Appendix for a short discussion of matrices.

20.1   THE Q-MATRIX

First we will demonstrate a close link between matrices and Fibonacci numbers. To this end, consider the matrix

$$Q = \begin{bmatrix} 1 & 1 \\ 1 & 0 \end{bmatrix}.$$

In 1960, Charles H. King studied this matrix, called the Q-matrix, for his Master's Thesis at then San Jose State College, California [360]. Notice that $|Q| = -1$, where $|M|$ denotes the determinant of the square matrix M. Then

$$Q^2 = \begin{bmatrix} 1 & 1 \\ 1 & 0 \end{bmatrix} \begin{bmatrix} 1 & 1 \\ 1 & 0 \end{bmatrix} = \begin{bmatrix} 2 & 1 \\ 1 & 1 \end{bmatrix}$$

$$Q^3 = \begin{bmatrix} 1 & 1 \\ 1 & 0 \end{bmatrix} \begin{bmatrix} 2 & 1 \\ 1 & 1 \end{bmatrix} = \begin{bmatrix} 3 & 2 \\ 2 & 1 \end{bmatrix}.$$

Similarly, $Q^4 = \begin{bmatrix} 5 & 3 \\ 3 & 2 \end{bmatrix}.$

Fibonacci and Lucas Numbers with Applications, Volume One, Second Edition. Thomas Koshy.
© 2018 John Wiley & Sons, Inc. Published 2018 by John Wiley & Sons, Inc.

Clearly, an interesting pattern is emerging. More generally, we have the following intriguing result. We will establish it using PMI.

Theorem 20.1. *Let $n \geq 1$. Then*

$$Q^n = \begin{bmatrix} F_{n+1} & F_n \\ F_n & F_{n-1} \end{bmatrix}.$$

Proof. Clearly, the result is true when $n = 1$. Now assume it is true for an arbitrary positive integer k:

$$Q^k = \begin{bmatrix} F_{k+1} & F_k \\ F_k & F_{k-1} \end{bmatrix}.$$

Then

$$\begin{aligned} Q^{k+1} &= Q^k Q^1 \\ &= \begin{bmatrix} F_{k+1} & F_k \\ F_k & F_{k-1} \end{bmatrix} \begin{bmatrix} 1 & 1 \\ 1 & 0 \end{bmatrix} \\ &= \begin{bmatrix} F_{k+2} & F_{k+1} \\ F_{k+1} & F_k \end{bmatrix}. \end{aligned}$$

So the result is true when $n = k + 1$ also.

Thus, by PMI, the result is true for all positive integers n.   ∎

It follows by Theorem 20.1 that the *trace* (sum of the diagonal elements) of the matrix Q^n is $F_{n+1} + F_{n-1} = L_n$.

CASSINI'S FORMULA REVISITED

Theorem 20.1 yields Cassini's formula as a delightful dividend; see the next corollary.

Corollary 20.1. *Let $n \geq 1$. Then $F_{n+1}F_{n-1} - F_n^2 = (-1)^n$.*

Proof. Since $|Q| = -1$, it follows by Theorem A.25 that $|Q^n| = (-1)^n$. But, by Theorem 20.1, $|Q^n| = F_{n+1}F_{n-1} - F_n^2$. Thus $F_{n+1}F_{n-1} - F_n^2 = (-1)^n$.   ∎

FORMULA (5.9) REVISITED

Interestingly, the Cassini-like formula $L_{n+1}L_{n-1} - L_n^2 = 5(-1)^{n-1}$ also follows by Theorem 20.1. To see this, first notice that $Q^2 + I = \begin{bmatrix} 2 & 1 \\ 1 & 1 \end{bmatrix} + \begin{bmatrix} 1 & 0 \\ 0 & 1 \end{bmatrix} = \begin{bmatrix} 3 & 1 \\ 1 & 2 \end{bmatrix}$.

So $|Q^2 + I| = 5$. Since $F_{n+1} + F_{n-1} = L_n$, we then have

$$Q^{n+1} + Q^{n-1} = \begin{bmatrix} F_{n+2} & F_{n+1} \\ F_{n+1} & F_n \end{bmatrix} + \begin{bmatrix} F_n & F_{n-1} \\ F_{n-1} & F_{n-2} \end{bmatrix}$$

$$Q^{n-1}(Q^2 + I) = \begin{bmatrix} L_{n+1} & L_n \\ L_n & L_{n-1} \end{bmatrix}$$

$$|Q^{n-1}(Q^2 + I)| = L_{n+1}L_{n-1} - L_n^2.$$

Since $|Q^{n-1}(Q^2 + I)| = |Q^{n-1}| \cdot |Q^2 + I| = 5(F_n F_{n-2} - F_{n-1}^2) = 5(-1)^{n-1}$, this implies $L_{n+1}L_{n-1} - L_n^2 = 5(-1)^{n-1}$, as desired.

FIBONACCI ADDITION FORMULA

Using Theorem 20.1, we can develop an addition formula for Fibonacci numbers, as the next corollary shows. Although the corollary lists four addition formulas, they are basically the same.

Corollary 20.2.

$$F_{m+n+1} = F_{m+1}F_{n+1} + F_m F_n \tag{20.1}$$

$$F_{m+n} = F_{m+1}F_n + F_m F_{n-1} \tag{20.2}$$

$$F_{m+n} = F_m F_{n+1} + F_{m-1}F_n \tag{20.3}$$

$$F_{m+n-1} = F_m F_n + F_{m-1}F_{n-1}. \tag{20.4}$$

Proof.

$$Q^{m+n} = Q^m Q^n$$

$$\begin{bmatrix} F_{m+n+1} & F_{m+n} \\ F_{m+n} & F_{m+n-1} \end{bmatrix} = \begin{bmatrix} F_{m+1} & F_m \\ F_m & F_{m-1} \end{bmatrix}\begin{bmatrix} F_{n+1} & F_n \\ F_n & F_{n-1} \end{bmatrix}$$

$$= \begin{bmatrix} F_{m+1}F_{n+1} + F_m F_n & F_{m+1}F_n + F_m F_{n-1} \\ F_m F_{n+1} + F_{m-1}F_n & F_m F_n + F_{m-1}F_{n-1} \end{bmatrix}.$$

Equating the corresponding elements yields the given identities. ∎

In particular, let $m = n$. Then identity (20.1) yields Lucas' formula $F_n^2 + F_{n+1}^2 = F_{2n+1}$; see identity (5.11). Likewise, identity (20.2) yields $F_{2n} = F_{n+1}F_n + F_n F_{n-1} = F_n(F_{n+1} + F_{n-1}) = F_n L_n$. F_{2n} also equals $(F_{n+1} - F_{n-1})(F_{n+1} + F_{n-1}) = F_{n+1}^2 - F_{n-1}^2$; see identity (5.12).

Addition formula (20.2), coupled with the charming identity $F_{2n} = F_n L_n$, can be used to evaluate an interesting infinite product, studied in 1980 by J. Shallit of Palo Alto, California [520].

Example 20.1. *Evaluate the infinite product* $P = \left(1 + \dfrac{1}{2}\right)\left(1 + \dfrac{1}{13}\right)\left(1 + \dfrac{1}{610}\right)\cdots$.

Solution. Notice that $P = \displaystyle\prod_{n=1}^{\infty}\left(1 + \dfrac{1}{F_{2^{n+1}-1}}\right)$. To evaluate this product, we will first prove two results.

1) *To prove that* $F_{2^n-1}L_{2^n} = F_{2^{n+1}-1} + 1$:
   Using the identity $F_{m+1} + F_{m-1} = L_m$, Cassini's formula, and the Fibonacci addition formula, we have

$$F_{2^n-1}L_{2^n} = F_{2^n-1}\left(F_{2^n+1} + F_{2^n-1}\right)$$

$$= F_{2^n-1}F_{2^n+1} + F_{2^n-1}^2$$

$$= F_{2^n-1}F_{2^n+1} + \left[F_{2^n}F_{2^n-2} - (-1)^{2^n-1}\right]$$

$$= \left(F_{2^n-1}F_{2^n+1} + F_{2^n}F_{2^n-2}\right) + 1$$

$$= F_{2^n+(2^n-1)} + 1$$

$$= F_{2^{n+1}-1} + 1.$$

2) *To prove that* $\displaystyle\prod_{i=1}^{n} L_{2^i} = F_{2^{n+1}}$:
   When $n = 1$, LHS $= L_2 = 3 = F_4 =$ RHS, so the result is true when $n = 1$.
   Now, assume it is true for an arbitrary positive integer k. Then using the identity $F_{2m} = F_m L_m$, we have

$$\prod_{i=1}^{k+1} L_{2^i} = \prod_{i=1}^{k} L_{2^i} \cdot L_{2^{k+1}}$$

$$= F_{2^{k+1}} L_{2^{k+1}}$$

$$= F_{2^{k+2}}.$$

So the result is true for all positive integers n by PMI.

With this machinery at our disposal, we are now ready to evaluate the given product. Let

$$P_m = \prod_{n=1}^{m}\left(1 + \dfrac{1}{F_{2^{n+1}-1}}\right).$$

Then

$$P_m = \prod_{n=1}^{m} \frac{1 + F_{2^{n+1}-1}}{F_{2^{n+1}-1}}$$

$$= \prod_{n=1}^{m} \frac{F_{2^n-1} L_{2^n}}{F_{2^{n+1}-1}}$$

$$= \prod_{n=1}^{m} \frac{F_{2^n-1}}{F_{2^{n+1}-1}} \cdot F_{2^{m+1}}$$

$$= \frac{F_{2^{m+1}}}{F_{2^{m+1}-1}};$$

$$P = \lim_{m \to \infty} \frac{F_{2^{m+1}}}{F_{2^{m+1}-1}}$$

$$= \alpha, \quad \text{the golden ratio.} \qquad \blacksquare$$

Another immediate consequence of the addition formula is the fact that $F_{mn} > F_m F_n$, where $m > n > 1$; see Exercise 20.9. Consequently, $F_{mn} > F_n^m$, where $m > n > 1$; see Exercise 20.10. It also follows by the addition formula that $F_{mn} | F_n$.

Next we will develop an addition formula for the Lucas family.

Corollary 20.3.

$$L_{m+n} = F_{m+1} L_n + F_m L_{n-1}. \qquad (20.5)$$

Proof. Using identities (20.1) and (20.4), we have

$$F_{m+n+1} = F_{m+1} F_{n+1} + F_m F_n$$

$$F_{m+n-1} = F_{m+1} F_{n-1} + F_m F_{n-2}.$$

Adding,

$$F_{m+n+1} + F_{m+n-1} = F_{m+1}(F_{n+1} + F_{n-1}) + F_m(F_n + F_{n-2})$$

$$L_{m+n} = F_{m+1} L_n + F_m L_{n-1},$$

as desired. $\qquad \blacksquare$

Corollary 20.2 can be used to derive two additional formulas linking the Fibonacci and Lucas families; see the next corollary. We omit their proofs; see Exercises 20.11 and 20.12.

Corollary 20.4.

$$2F_{m+n} = F_m L_n + F_n L_m \tag{20.6}$$

$$2L_{m+n} = L_m L_n + 5F_m F_n. \tag{20.7}$$

∎

Using the fact that $Q^{m-n} = Q^m Q^{-n}$, we can derive another Fibonacci identity:

$$F_m F_{n+1} - F_n F_{m+1} = (-1)^n F_{m-n}; \tag{20.8}$$

see Exercise 20.29. It is called *d'Ocagne's identity*, after the French mathematician Philbert Maurice d'Ocagne (1862–1938). Clearly, it is a generalization of Cassini's formula.

d'Ocagne's identity has an interesting Lucas counterpart:

$$L_m L_{n+1} - L_n L_{m+1} = 5(-1)^{n+1} F_{m-n}; \tag{20.9}$$

see Exercise 20.30.

For example, $L_7 L_5 - L_4 L_8 = 29 \cdot 11 - 7 \cdot 47 = -10 = 5(-1)^5 F_{7-4}$.

Identity (20.7) has an interesting byproduct. To see this, it follows from the identity that $2L_{m+n} \equiv L_m L_n \pmod 5$. Let $i+j = h+k$. Then $L_i L_j \equiv 2L_{i+j} \equiv 2L_{h+k} \equiv L_h L_k \pmod 5$. W.G. Brady of the University of Tennessee at Knoxville discovered this congruence in 1977 [56].

For example, $L_5 L_7 = 11 \cdot 29 \equiv 4 \equiv 47 \cdot 7 \equiv L_8 L_4 \pmod 5$.

THE *M*-MATRIX

Closely related to the Q-matrix is the M-matrix, studied by Sam Moore of the Community College of Allegheny County, Pennsylvania, in 1983 [448]:

$$M = \begin{bmatrix} 1 & 1 \\ 1 & 2 \end{bmatrix}.$$

We can show by PMI that

$$M^n = \begin{bmatrix} F_{2n-1} & F_{2n} \\ F_{2n} & F_{2n+1} \end{bmatrix},$$

where $n \geq 1$; see Exercise 20.31. Then

$$\frac{M^n}{F_{2n-1}} = \begin{bmatrix} 1 & F_{2n}/F_{2n-1} \\ F_{2n}/F_{2n-1} & F_{2n+1}/F_{2n-1} \end{bmatrix}.$$

Since $\lim\limits_{k\to\infty}(F_k/F_{k-1}) = \alpha$, it follows that

$$\lim\limits_{n\to\infty}\frac{M^n}{F_{2n-1}} = \begin{bmatrix} 1 & \alpha \\ \alpha & \alpha^2 \end{bmatrix}$$

$$= \begin{bmatrix} 1 & \alpha \\ \alpha & 1+\alpha \end{bmatrix}.$$

Thus the sequence of Fibonacci matrices $\{M_n/F_{2n-1}\}$ converges to the matrix $A = \begin{bmatrix} 1 & \alpha \\ \alpha & 1+\alpha \end{bmatrix}$, where $n \geq 1$. Likewise, sequence $\{Q^n/F_{2n-1}\}$ also converges to the matrix $\begin{bmatrix} 1+\alpha & \alpha \\ \alpha & 1 \end{bmatrix}$.

Next we will investigate a generalized version of the *M*-matrix.

A GENERALIZED *M*-MATRIX

Let $A = \begin{bmatrix} 1 & 1 \\ 1 & 1+x \end{bmatrix}$. We will compute its powers, scale them to make their leading entries 1, and then find the limit of the resulting sequence of scaled matrices [479].

The characteristic equation of A is given by $|A - \lambda I| = 0$; that is,

$$\begin{bmatrix} 1-\lambda & 1 \\ 1 & 1+x-\lambda \end{bmatrix} = 0$$

$$\lambda^2 - (x+2)\lambda + x = 0.$$

So the characteristic roots are $r(x) = 1 + \frac{1}{2}(x - \sqrt{x^2+4})$ and $s(x) = 1 + \frac{1}{2}(x + \sqrt{x^2+4})^\dagger$.

Next we will find a characteristic vector $\begin{pmatrix} u \\ v \end{pmatrix}$ associated with r. To this end, we solve the equation $A\begin{pmatrix} u \\ v \end{pmatrix} = r\begin{pmatrix} u \\ v \end{pmatrix}$. We can easily choose $\begin{pmatrix} u \\ v \end{pmatrix} = \begin{pmatrix} 1 \\ r-1 \end{pmatrix}$. Similarly, we can choose the characteristic vector associated with s to be $\begin{pmatrix} 1 \\ s-1 \end{pmatrix}$. Then

$$A = \begin{bmatrix} 1 & 1 \\ r-1 & s-1 \end{bmatrix} \begin{bmatrix} r & 0 \\ 0 & s \end{bmatrix} \begin{bmatrix} 1 & 1 \\ r-1 & s-1 \end{bmatrix}^{-1}.$$

---

†The expressions $\alpha(x) = \dfrac{x+\sqrt{x^2+4}}{2}$ and $\beta(x) = \dfrac{x-\sqrt{x^2+4}}{2}$ will play a significant role in the study of Fibonacci and Lucas polynomials.

Since $(r-1)(s-1) = -1$, we then have

$$A^n = \begin{bmatrix} 1 & 1 \\ r-1 & s-1 \end{bmatrix} \begin{bmatrix} r^n & 0 \\ 0 & s^n \end{bmatrix} \begin{bmatrix} 1 & 1 \\ r-1 & s-1 \end{bmatrix}^{-1}$$

$$= \frac{1}{s-r} \begin{bmatrix} 1 & 1 \\ r-1 & s-1 \end{bmatrix} \begin{bmatrix} r^n & 0 \\ 0 & s^n \end{bmatrix} \begin{bmatrix} s-1 & -1 \\ 1-r & 1 \end{bmatrix}$$

$$= \frac{1}{s-r} \begin{bmatrix} (s-1)r^n - (r-1)s^n & s^n - r^n \\ s^n - r^n & (s-1)s^n - (r-1)r^n \end{bmatrix}.$$

Scaling this matrix to make its leading entry 1 gives the matrix

$$B_n = \begin{bmatrix} 1 & \dfrac{s^n - r^n}{(s-1)r^n - (r-1)s^n} \\ \dfrac{s^n - r^n}{(s-1)r^n - (r-1)s^n} & \dfrac{(s-1)s^n - (r-1)r^n}{(s-1)r^n - (r-1)s^n} \end{bmatrix}$$

$$= \begin{bmatrix} 1 & \dfrac{1 - (r/s)^n}{(s-1)(r/s)^n - (r-1)} \\ \dfrac{1 - (r/s)^n}{(s-1)(r/s)^n - (r-1)} & \dfrac{(s-1) - (r-1)(r/s)^n}{(s-1)(r/s)^n - (r-1)} \end{bmatrix}.$$

Since $x > 0$, $(r/s)^n \to 0$ as $\to \infty$. So

$$\lim_{n \to \infty} B_n = \begin{bmatrix} 1 & \dfrac{1}{1-r} \\ \dfrac{1}{1-r} & \dfrac{s-1}{1-r} \end{bmatrix}$$

$$= \begin{bmatrix} 1 & s-1 \\ s-1 & (s-1)^2 \end{bmatrix}.$$

In particular, let $x = 1$. Then $s = \alpha + 1 = \alpha^2$. So

$$\lim_{n \to \infty} M^n = \begin{bmatrix} 1 & \alpha \\ \alpha & \alpha^2 \end{bmatrix},$$

as found earlier.

Next we will find the eigenvalues of the Q^n. In the process, we will employ the well-known formula $L_n^2 - 5F_n^2 = 4(-1)^n$.

20.2 EIGENVALUES OF Q^n

Let $A = (a_{ij})_{n \times n}$ and I the identity matrix of the same size. Then the equation $|A - \lambda I| = 0$ is the *characteristic equation* of matrix A. Its solutions are the *eigenvalues* of A.

To find the eigenvalues of Q^n, first we will find its characteristic equation. Using Cassini's formula, we have

$$|Q^n - \lambda I| = \begin{bmatrix} F_{n+1} - \lambda & F_n \\ F_n & F_{n-1} - \lambda \end{bmatrix}$$

$$= (F_{n+1} - \lambda)(F_{n-1} - \lambda) - F_n^2$$

$$= \lambda^2 - (F_{n+1} + F_{n-1})\lambda + F_{n+1}F_{n-1} - F_n^2$$

$$= \lambda^2 - L_n\lambda + (-1)^n.$$

So the characteristic equation of Q^n is

$$\lambda^2 - L_n\lambda + (-1)^n = 0. \tag{20.10}$$

Using the quadratic formula, the eigenvalues of Q^n are given by

$$\lambda = \frac{L_n \pm \sqrt{L_n^2 - 4(-1)^n}}{2}$$

$$= \frac{L_n \pm \sqrt{5}F_n}{2}$$

$$= \alpha^n, \beta^n.$$

Thus we have the following result.

Theorem 20.2. *The eigenvalues of Q^n are α^n and β^n.* ∎

In particular, we have the next corollary.

Corollary 20.5. *The eigenvalues of Q are α and β.* ∎

When $n = 1$, equation (20.10) becomes $\lambda^2 - \lambda - 1 = 0$, which is the characteristic equation of Q. But $Q^2 - Q - I = 0$; see Exercise 20.2. Thus Q satisfies its characteristic equation, illustrating the well-known *Cayley–Hamilton Theorem*: *Every square matrix satisfies its characteristic equation.*

Since $Q^2 = Q + I$, it follows by the binomial theorem that

$$Q^{2n} = (Q + I)^n$$

$$= \sum_{k=0}^{n} \binom{n}{k} Q^k.$$

Equating the corresponding elements from both sides, we get

$$F_{2n} = \binom{n}{1} F_1 + \binom{n}{2} F_2 + \cdots + \binom{n}{n} F_n$$

$$F_{2n+1} = \binom{n}{0} F_1 + \binom{n}{1} F_2 + \cdots + \binom{n}{n} F_{n+1},$$

as we found in Chapter 5.

Next we will see how I.D. Ruggles and Hoggatt in 1963 derived summation formula (5.1) using the Q-matrix [506].

SUMMATION FORMULA (5.1) REVISITED

Using PMI, we can establish that

$$(I + Q + Q^2 + \cdots + Q^n)(Q - I) = Q^{n+1} - I; \tag{20.11}$$

see Exercise 20.3.

Since $|Q - I| = -1 \neq 0$, $Q - I$ is invertible. Since $Q^2 = Q + I$, $Q^2 - Q = I$; that is, $Q(Q - I) = I$. Thus $(Q - I)^{-1} = Q$. Thus, by equation (20.11), we have

$$I + Q + Q^2 + \cdots + Q^n = (Q^{n+1} - I)Q$$

$$= Q^{n+2} - Q.$$

Equating the upper right-hand elements in this matrix equation, we get the desired summation formula: $\sum_{k=1}^{n} F_k = F_{n+2} - 1$.

Next we will briefly study four 2×2 matrices related to the Q-matrix. Joseph Ercolano of Baruch College, New York, investigated them in 1976 [161]. They, too, have interesting Fibonacci and Lucas implications.

We will begin with a definition. Let A and B be two $n \times n$ matrices. Then A is *similar* to B if there exists an invertible matrix M such that $A = MBM^{-1}$, that is, $AM = MB$.

1) The first matrix we will study is $A = \begin{bmatrix} \alpha & 0 \\ 0 & \beta \end{bmatrix}$. How are A and Q related? Both have the same characteristic polynomial $x^2 - x - 1$, and hence the same eigenvalues, α and β. Both have the same determinant: $|A| = -1 = |Q|$. Both have

the same trace $\alpha + \beta = 1$. Finally, Q is similar to A, since $Q = MAM^{-1}$, where $M = \begin{bmatrix} \alpha & 1 \\ 1 & -\alpha \end{bmatrix}$. This is true, since

$$QM = \begin{bmatrix} 1 & 1 \\ 1 & 0 \end{bmatrix}\begin{bmatrix} \alpha & 1 \\ 1 & -\alpha \end{bmatrix} = \begin{bmatrix} \alpha^2 & \beta \\ \alpha & 1 \end{bmatrix} = \begin{bmatrix} \alpha & 1 \\ 1 & -\alpha \end{bmatrix}\begin{bmatrix} \alpha & 0 \\ 0 & \beta \end{bmatrix} = MA.$$

Since $Q = MAM^{-1}$, it follows that $Q^n = MA^nM^{-1}$; see Exercise 20.39. So Q^n is similar to $A^n = \begin{bmatrix} \alpha^n & 0 \\ 0 & \beta^n \end{bmatrix}$. Since similar matrices have the same trace and determinant [58], it follows that trace$(A^n) = \alpha^n + \beta^n = L_n = F_{n+1} + F_{n-1} = $ trace(Q^n). Likewise, $|A^n| = |Q^n|$ yields Cassini's formula for Fibonacci numbers.

We can extract additional properties using the similarity of Q^n and A^n. Since $Q^n M = MA^n$, we have

$$\begin{bmatrix} F_{n+1} & F_n \\ F_n & F_{n-1} \end{bmatrix}\begin{bmatrix} \alpha & 1 \\ 1 & -\alpha \end{bmatrix} = \begin{bmatrix} \alpha & 1 \\ 1 & -\alpha \end{bmatrix}\begin{bmatrix} \alpha^n & 0 \\ 0 & \beta^n \end{bmatrix}$$

$$\begin{bmatrix} \alpha F_{n+1} + F_n & F_{n+1} - \alpha F_n \\ \alpha F_n + F_{n+1} & F_n - \alpha F_{n-1} \end{bmatrix} = \begin{bmatrix} \alpha^{n+1} & \beta^n \\ \alpha^n & \beta^{n-1} \end{bmatrix}.$$

This implies

$$\alpha F_{n+1} + F_n = \alpha^{n+1} \tag{20.12}$$

$$F_{n+1} - \alpha F_n = \beta^n \tag{20.13}$$

$$\alpha F_n + F_{n+1} = \alpha^n \tag{20.14}$$

$$F_n - \alpha F_{n-1} = \beta^{n-1}. \tag{20.15}$$

Notice that Binet's formula for F_n follows from equations (20.12) and (20.13); and also from equations (20.14) and (20.15).

2) The next matrix we will study is $B = \begin{bmatrix} \frac{1}{2} & 1 \\ \frac{5}{4} & \frac{1}{2} \end{bmatrix}$. Then

$$B = \begin{bmatrix} \frac{1}{2}L_1 & F_1 \\ \frac{5}{4}F_1 & \frac{1}{2}L_1 \end{bmatrix}$$

$$B^2 = \begin{bmatrix} \frac{1}{2}L_2 & F_2 \\ \frac{5}{4}F_2 & \frac{1}{2}L_2 \end{bmatrix}$$

$$B^3 = \begin{bmatrix} \frac{1}{2}L_3 & F_3 \\ \frac{5}{4}F_3 & \frac{1}{2}L_3 \end{bmatrix}$$

$$\vdots$$

More generally, we can confirm that

$$B^n = \begin{bmatrix} \frac{1}{2}L_n & F_n \\ \frac{5}{4}F_n & \frac{1}{2}L_n \end{bmatrix};$$

see Exercise 20.41.

The trace invariance between B^n and A^n yields Binet's formula for L_n. Since Q^n and B^n are similar, they have the same trace; this implies $L_n = F_{n+1} + F_{n-1}$. Since the determinant is also an invariant,

$$\frac{1}{4}L_n^2 - \frac{5}{4}F_n^2 = F_{n+1}F_{n-1} - F_n^2;$$

that is, $L_n^2 - 5F_n^2 = 4(-1)^n$; see Exercise 5.37. Similarity between Q^n and B^n yields the same result.

3) Next we will investigate the matrix $C = \begin{bmatrix} -1 & 1 \\ -1 & 2 \end{bmatrix}$. Then

$$C^n = \begin{bmatrix} -F_{n-2} & F_n \\ -F_n & F_{n+2} \end{bmatrix};$$

see Exercise 20.43.

Matrices C^n, Q^n, A^n, and B^n are all similar. Similarity with Q^n yields

$$F_{n+2} - F_{n-2} = L_n$$
$$F_{n+2}F_{n-2} - F_n^2 = (-1)^{n+1}.$$

Trace invariance between C^n and A^n gives $F_{n+2} - F_{n-2} = \alpha^n + \beta^n = L_n$; and the determinant invariance between C^n and B^n gives the identity

$$F_n^2 - F_{n+2}F_{n-2} = \frac{1}{4}L_n^2 - \frac{5}{4}F_n^2;$$

that is, $L_n^2 = 9F_n^2 - 4F_{n+2}F_{n-2}$.

4) Finally, consider the matrix $D = \begin{bmatrix} 3 & 1 \\ -5 & -2 \end{bmatrix}$. Then

$$D^n = \begin{bmatrix} L_{n+1} & F_n \\ -5F_n & -L_{n-1} \end{bmatrix};$$

see Exercise 20.44. Its similarity with Q^n, A^n, B^n, and C^n yields the following results. You may confirm them.

$$L_{n+1}L_{n-1} + F_{n+1}F_{n-1} = 6F_n^2$$
$$L_{n+1}L_{n-1} - 5F_n^2 = (-1)^{n+1}$$
$$L_n^2 + 4L_{n+1}L_{n-1} = 25F_n^2$$
$$L_{n+1} - L_{n-1} = F_{n+2} - F_{n-2}$$
$$L_{n+1}L_{n-1} - F_{n+2}F_{n-2} = 4F_n^2.$$

Consequently, both $L_n^2 + 4L_{n+1}L_{n-1}$ and $L_{n+1}L_{n-1} - F_{n+2}F_{n-2}$ are squares.

Next we introduce another 2×2 matrix R, introduced by Hoggatt and Ruggles in 1964 [316]. Coupled with the Q-matrix, it will give us Cassini's formula for Lucas numbers.

R-MATRIX

The R-matrix is given by

$$R = \begin{bmatrix} 1 & 2 \\ 2 & -1 \end{bmatrix}.$$

Using the identities $L_{n+1} = F_{n+1} + 2F_n$, $L_n = 2F_{n+1} - F_n$, $5F_{n+1} = L_{n+1} + 2L_n$, and $5F_n = 2L_{n+1} - L_n$, it follows that

$$RQ^n = \begin{bmatrix} 1 & 2 \\ 2 & -1 \end{bmatrix} \begin{bmatrix} F_{n+1} & F_n \\ F_n & F_{n-1} \end{bmatrix}$$

$$= \begin{bmatrix} L_{n+1} & L_n \\ L_n & L_{n-1} \end{bmatrix}.$$

Using Theorem A.25, this implies

$$\begin{bmatrix} 1 & 2 \\ 2 & -1 \end{bmatrix} \begin{bmatrix} F_{n+1} & F_n \\ F_n & F_{n-1} \end{bmatrix} = \begin{bmatrix} L_{n+1} & L_n \\ L_n & L_{n-1} \end{bmatrix};$$

that is, $L_{n+1}L_{n-1} - L_n^2 = (-5)(F_{n+1}F_{n-1} - F_n^2) = 5(-1)^{n-1}$. Thus

$$L_{n+1}L_{n-1} - L_n^2 = 5(-1)^{n-1}. \tag{20.16}$$

This is Cassini's formula for the Lucas family; see Exercise 5.36.

Next we will re-derive Cassini's formula for Fibonacci numbers, using Cramer's rule[†] for 2×2 linear systems.

[†]Named after the Swiss mathematician Gabriel Cramer (1704–1752).

CASSINI'S FORMULA REVISITED

We will first review Cramer's rule. The 2×2 linear system

$$ax + by = e$$
$$cx + dy = f$$

has a unique solution if and only if $ad - bc \neq 0$. It is given by

$$x = \frac{\begin{bmatrix} e & b \\ f & d \end{bmatrix}}{\begin{bmatrix} a & b \\ c & d \end{bmatrix}}, \quad y = \frac{\begin{bmatrix} a & e \\ c & f \end{bmatrix}}{\begin{bmatrix} a & b \\ c & d \end{bmatrix}}.$$

In particular, consider the linear system

$$F_n x + F_{n-1} y = F_{n+1}$$
$$F_{n+1} x + F_n y = F_{n+2}.$$

Since $(F_k, F_{k+1}) = 1$, it follows by the Fibonacci recurrence that $x = 1 = y$ is the unique solution to this system.

By Cramer's rule, we then have

$$y = \frac{\begin{bmatrix} F_n & F_{n+1} \\ F_{n+1} & F_n \end{bmatrix}}{\begin{bmatrix} F_n & F_{n-1} \\ F_{n+1} & F_n \end{bmatrix}} = 1.$$

Thus $F_n F_{n+2} - F_{n+1}^2 = F_n^2 - F_{n-1} F_{n+1}$; that is, $F_{n+2} F_n - F_{n+1}^2 = -(F_{n+1} F_{n-1} - F_n^2)$.

Let $p_n = F_{n+1} F_{n-1} - F_n^2$. Then this equation yields the recurrence $p_n = -p_{n-1}$, where $p_1 = F_2 F_0 - F_1^2 = -1$. Solving this recurrence, we get $p_n = (-1)^n$; see Exercise 20.34.

Thus $F_{n+1} F_{n-1} - F_n^2 = (-1)^n$, as desired.

Next we turn to vectors[†] formed by adjacent Fibonacci and Lucas numbers.

20.3 FIBONACCI AND LUCAS VECTORS

Consider the vectors $\mathbf{U}_n = (F_{n+1}, F_n)$ and $\mathbf{V}_n = (L_{n+1}, L_n)$. Their magnitudes are given by $|\mathbf{U}_n|^2 = F_{n+1}^2 + F_n^2 = F_{2n+1}$, and $|\mathbf{V}_n|^2 = L_{n+1}^2 + L_n^2 = 5F_{2n+1}$; see Exercise 5.35.

[†]In the rest of this chapter, the ordered pair (x, y) denotes a vector, and *not* the gcd of x and y.

Their directions are given by $\tan\theta = \dfrac{F_n}{F_{n+1}}$ and $\tan\theta' = \dfrac{L_n}{L_{n+1}}$, respectively.
Notice that

$$\mathbf{U}_0 Q^{n+1} = (1,0)\begin{bmatrix} F_{n+2} & F_{n+1} \\ F_{n+1} & F_n \end{bmatrix} = (F_{n+2}, F_{n+1}) = \mathbf{U}_{n+1} = \mathbf{U}_n Q.$$

Likewise, $\mathbf{V}_0 Q^{n+1} = \mathbf{V}_{n+1} = \mathbf{V}_n Q$. Besides, by the Fibonacci addition formula, we have

$$\begin{aligned}
\mathbf{U}_m Q^{n+1} &= (F_{m+1}, F_m)\begin{bmatrix} F_{n+2} & F_{n+1} \\ F_{n+1} & F_n \end{bmatrix} \\
&= \begin{bmatrix} F_{m+1}F_{n+2} + F_m F_{n+1} \\ F_{m+1}F_{n+1} + F_m F_n \end{bmatrix} \\
&= (F_{m+n+1}, F_{m+n}) \\
&= \mathbf{U}_{m+n+1}.
\end{aligned}$$

Likewise, $\mathbf{V}_m Q^{n+1} = \mathbf{V}_{m+n+1}$; see Exercise 20.51.

Next we investigate the impact of multiplying the vectors $\mathbf{V}_n$ and $\mathbf{U}_n$ by the R-matrix. To begin, notice that $|R| \neq 0$; so R is invertible and

$$R^{-1} = \frac{1}{5}\begin{bmatrix} 1 & 2 \\ 2 & -1 \end{bmatrix}.$$

We have

$$\begin{aligned}
\mathbf{V}_n R &= (L_{n+1}, L_n)\begin{bmatrix} 1 & 2 \\ 2 & -1 \end{bmatrix} \\
&= (L_{n+1} + 2L_n, 2L_{n+1} - L_n) \\
&= (5F_{n+1}, 5F_n) \\
&= 5\mathbf{U}_n; \\
\mathbf{V}_n &= (5\mathbf{U}_n)R^{-1} = 5(\mathbf{U}_n R^{-1}) \\
&= 5 \cdot \frac{1}{5}(F_{n+1}, F_n)\begin{bmatrix} 1 & 2 \\ 2 & -1 \end{bmatrix} \\
&= (F_{n+1} + 2F_n, 2F_{n+1} - F_n) \\
&= (L_{n+1}, L_n),
\end{aligned}$$

as expected.

Likewise, $\mathbf{U}_n R = \mathbf{V}_n$ and $\mathbf{U}_n = \mathbf{V}_n R^{-1}$.

What is the effect of multiplying by R on any nonzero vector $\mathbf{U} = (x, y)$? To see this, observe that

$$\mathbf{U}R = (x, y) \begin{bmatrix} 1 & 2 \\ 2 & -1 \end{bmatrix}$$

$$= (x + 2y, 2x - y);$$

$$|\mathbf{U}R|^2 = (x + 2y)^2 + (2x - y)^2$$

$$= 5(x^2 + y^2)$$

$$= 5|\mathbf{U}|^2.$$

Thus the Fibonacci matrix R magnifies every nonzero vector by a factor of $\sqrt{5}$.

To find the impact of R on the slope of $\mathbf{U}$, suppose the acute angles made with the x-axis by the directions of the vectors $\mathbf{U}$ and $\mathbf{U}R$ are θ and θ', respectively. Then $\tan \theta = y/x$ and $\tan \theta' = (2x - y)/(x + 2y)$. Consequently,

$$\tan(\theta + \theta') = \frac{\tan \theta + \tan \theta'}{1 - \tan \theta \tan \theta'}$$

$$= \frac{y/x + (2x - y)/(x + 2y)}{1 - (y/x) \cdot (2x - y)/(x + 2y)}$$

$$= \frac{y(x + 2y) + x(2x - y)}{x(x + 2y) - y(2x - y)}$$

$$= \frac{2(x^2 + y^2)}{x^2 + y^2}$$

$$= 2.$$

(Note that $\mathbf{U} \neq 0$.)

Let 2γ be the angle between the vectors $\mathbf{U}$ and $\mathbf{U}R$. Then $\gamma - \theta = \theta' - \gamma$, so $2\gamma = \theta + \theta'$. Since $\tan 2\gamma = \dfrac{\tan \gamma}{1 - \tan^2 \gamma}$, it follows that

$$\frac{\tan \gamma}{1 - \tan^2 \gamma} = 2;$$

that is, $\tan^2 \gamma + \tan \gamma - 1 = 0$. So $\tan \gamma = -\alpha, -\beta$. Since $2\gamma = \tan^{-1} 2 \approx 63.43°$, $\gamma \approx 31.7175°$; so we choose $\tan \gamma = -\beta$, which is the negative of an eigenvalue of Q. Thus the vector that bisects the angle between the vectors $\mathbf{U}$ and $\mathbf{U}R$ has slope $-\beta$; it is a vector of the form $\mathbf{W} = (\alpha x, x)$.

Thus we have the following results [316].

Theorem 20.3 (Hoggatt and Ruggles, 1964 [316]).   *The R-matrix transforms the nonzero vector $\mathbf{U} = (x, y)$ into a vector $\mathbf{U}R$ such that $|\mathbf{U}R| = \sqrt{5}|\mathbf{U}|$; and* *the bisector of the angle between them is the vector of the form* $(\alpha x, x)$ *with slope* $-\beta$. ∎

Corollary 20.6. *The R-matrix maps the vector* $\mathbf{U}_n$ *into* $\mathbf{V}_n$, *and* $\mathbf{V}_n$ *into* $\sqrt{5}\mathbf{U}_n$. ∎

Next we study a seemingly simple problem. But in the process, we will encounter an intriguing Fibonacci matrix.

20.4 AN INTRIGUING FIBONACCI MATRIX

In 1996, David M. Bloom of Brooklyn College, New York, proposed the following problem in *Math Horizons* [52]:

$$\text{Determine the sum } \sum_{\substack{i,j,k>0 \\ i+j+k=n}} F_i F_j F_k.$$

The neat solution, given by C. Libis of the University of West Alabama in the February, 1997 issue, involves an intriguing infinite-dimensional Fibonacci matrix [390]:

$$H = \begin{bmatrix} h_{0,n} \\ h_{1,n} \\ \vdots \\ h_{m,n} \\ \vdots \end{bmatrix},$$

where the element h_{ij} is defined recursively as follows:

$$
\begin{array}{ll}
h_{0,j} = 0 & \text{if } j \geq 0 \\
h_{j,j} = 1 & \text{if } j \geq 1 \\
h_{i,j} = 0 & \text{if } i > j \\
h_{i,j} = h_{i,j-2} + h_{i,j-1} + h_{i-1,j-1} & \text{if } i \geq 1 \text{ and } j \geq 2.
\end{array}
\tag{20.17}
$$

It follows by the recurrence (20.17) that $h_{1,j} = h_{1,j-2} + h_{1,j-1} + h_{0,j-1} = h_{1,j-2} + h_{1,j-1}$, where $h_{1,0} = 0$ and $h_{1,1} = 1$. Consequently, $h_{1,n} = F_n$. Thus

$$
\begin{aligned}
h_{2,5} &= h_{2,3} + h_{2,4} + h_{1,4} \\
&= (h_{2,1} + h_{2,2} + h_{1,2}) + (h_{2,2} + h_{2,3} + h_{1,3}) + F_4 \\
&= (0 + 1 + F_2) + [1 + (h_{2,1} + h_{2,2} + h_{1,1}) + F_3] + F_4 \\
&= 2 + [1 + (0 + 1 + F_2) + 2] + 3 \\
&= 10.
\end{aligned}
$$

The condition that $h_{0,j} = 0$ for every $j \geq 0$ implies that the top row of matrix H consists of zeros; $h_{i,i} = 1$ means that every element on the main diagonal is 1; and $h_{i,j} = 0$ for $i > j$ means the matrix H is upper triangular; that is, every element below the main diagonal is zero. Recurrence (20.17) implies that we can compute $h_{i,j}$ by adding the two previous elements $h_{i,j-2}$ and $h_{i,j-1}$ in the same row, and its northwest neighbor $h_{i-1,j-1}$, where $i \geq 1$ and $j \geq 2$. For example, $h_{3,6} = h_{3,4} + h_{3,5} + h_{2,5} = 3 + 9 + 10 = \boxed{22}$; see the arrows in Table 20.1.

Using these straightforward observations, we can determine the various elements of H; see Table 20.1.

TABLE 20.1.

$i \backslash j$	0	1	2	3	4	5	6	7	8	9	10	$\ldots$
0	0	0	0	0	0	0	0	0	0	0	0	
1	0	1	1	2	3	5	8	13	21	34	55	$\leftarrow F_n$
2	0	0	1	2	5	10	20	38	71	130	235	
3	0	0	0	1	3 $\rightarrow$ 9	(22)	51	111	233	474		
4	0	0	0	0	1	4	14	40	105	255	593	
5	0	0	0	0	0	1	15	56	176	487	918	

$H =$ (to the left of the table)

Notice that

$$h_{2,7} = 38$$
$$= 1 \cdot 8 + 1 \cdot 5 + 2 \cdot 3 + 3 \cdot 2 + 5 \cdot 1 + 8 \cdot 1$$
$$= F_1 h_{1,6} + F_2 h_{1,5} + F_3 h_{1,4} + F_4 h_{1,3} + F_5 h_{1,2} + F_6 h_{1,1}$$
$$= \sum_{j=1}^{7} F_j h_{1,7-j}$$
$$= \sum_{j=1}^{7} F_j F_{7-j}$$
$$= \sum_{\substack{j,k \geq 1 \\ j+k=7}}^{7} F_j F_k.$$

More generally, we have the following result. We will establish it using the strong version of PMI.

Theorem 20.4.

$$h_{2,n} = \sum_{\substack{j,k\geq 1 \\ j+k=n}} F_j F_k.$$

Proof. Since $h_{2,1} = 0 = \displaystyle\sum_{\substack{j,k\geq 1 \\ j+k=1}} F_j F_k$, the result is true when $n = 1$.

Now assume it is true for all positive integers $\leq m$, where $m \geq 2$:

$$h_{2,m} = \sum_{\substack{j,k\geq 1 \\ j+k=m}} F_j F_k.$$

Then, by the inductive hypothesis and recurrence (20.17), we have

$$\sum_{\substack{j,k\geq 1 \\ j+k=m+1}} F_j F_k = \sum_{j=1}^{m} F_j F_{m+1-j}$$

$$= \sum_{j=1}^{m} F_j (F_{m-j} + F_{m-j-1})$$

$$= \sum_{j=1}^{m} F_j F_{m-j} + \sum_{j=1}^{m} F_j F_{m-1-j}$$

$$= \sum_{j=1}^{m} F_j F_{m-j} + \sum_{j=1}^{m-1} F_j F_{m-1-j} + F_m F_{-1}$$

$$= h_{2,m} + h_{2,m-1} + F_m$$

$$= h_{2,m} + h_{2,m-1} + h_{1,m}$$

$$= h_{2,m+1}.$$

Thus by the strong version of PMI, the formula works for all positive integers n. ∎

Since $h_{1,k} = F_k$, this theorem implies the following result.

Corollary 20.7.

$$h_{2,n} = \sum_{i=1}^{n} F_i h_{1,n-i}. \qquad \blacksquare$$

Thus we can obtain every element $h_{2,n}$ by multiplying the elements $h_{1,n-1}, h_{1,n-2}, \ldots, h_{1,1}$ with weights $F_1, F_2, \ldots, F_{n-1}$, respectively, and then adding up the products, as we observed earlier in a numeric example. (Recall that $h_{1,0} = 0$.)

Corresponding to this corollary, there is a similar result for row 3 of matrix H. It also can be established by PMI, so we omit its proof.

Theorem 20.5.

$$h_{3,n} = \sum_{i=1}^{n} F_i h_{2,n-i}.$$

∎

It follows by this theorem that we can compute $h_{3,n}$ by multiplying the elements $h_{2,n-1}, h_{2,n-2}, \ldots, h_{2,1}$ with weights $F_1, F_2, \ldots, F_{n-1}$, respectively, and then by summing up the products.

For example,

$$h_{3,7} = \sum_{i=1}^{7} F_i h_{2,7-i}$$

$$= F_1 h_{2,6} + F_2 h_{2,5} + F_3 h_{2,4} + F_4 h_{2,3} + F_5 h_{2,2} + F_6 h_{2,1}$$

$$= 1 \cdot 20 + 1 \cdot 10 + 2 \cdot 5 + 3 \cdot 2 + 5 \cdot 1 + 8 \cdot 0$$

$$= \boxed{51};$$

see Table 20.1.

The next corollary provides the answer to the original problem.

Corollary 20.8.

$$h_{3,n} = \sum_{\substack{i,j,k \geq 1 \\ i+j+k=n}} F_i F_j F_k.$$

Proof. By Theorem 20.5, we have

$$h_{3,n} = \sum_{i=1}^{n} F_i h_{2,n-i}$$

$$= \sum_{i=1}^{n} F_i \left(\sum_{\substack{j,k \geq 1 \\ j+k=n-i}} F_j F_k \right)$$

$$= \sum_{\substack{i,j,k \geq 1 \\ i+j+k=n}} F_i F_j F_k.$$

∎

For example,

$$h_{3,5} = \sum_{\substack{i,j,k \geq 1 \\ i+j+k=5}} F_i F_j F_k$$

$$= F_1 \sum_{\substack{j,k \geq 1 \\ j+k=4}} F_j F_k + F_2 \sum_{\substack{j,k \geq 1 \\ j+k=3}} F_j F_k + F_3 \sum_{\substack{j,k \geq 1 \\ j+k=2}} F_j F_k$$

$$= F_1(F_1 F_3 + F_2 F_2 + F_3 F_1) + F_2(F_1 F_2 + F_2 F_1) + F_3(F_1 F_1)$$

$$= 1(1 \cdot 2 + 1 \cdot 1 + 2 \cdot 1) + 1(1 \cdot 1 + 1 \cdot 1) + (1 \cdot 1)$$

$$= 9,$$

as expected; see Table 20.1.

We can generalize Corollaries 20.7 and 20.8, as the following theorem shows.

Theorem 20.6 (Libis, 1997 [390]). *Let $m \geq 2$. Then*

$$h_{m,n} = \sum_{i=1}^{n} F_i h_{m-1,n-i}. \qquad \blacksquare$$

To pursue this problem a bit further, we will now find explicit formulas for both $h_{2,n}$ and $h_{3,n}$.

EXPLICIT FORMULAS FOR $h_{2,n}$ AND $h_{3,n}$

Murray Klamkin (1921–2004) of the University of Alberta, Canada, then editor of *Math Horizons*, presented explicit formulas for $h_{2,n}$ and $h_{3,n}$ in the same February issue, by introducing an operator E: $Eh_{m,n} = h_{m,n+1}$. Then, by the recurrence (20.17), we have

$$E^2 h_{m,n} = Eh_{m,n+1} = h_{m,n+2}$$
$$(E^2 - E - 1)h_{m,n} = h_{m,n+2} - h_{m,n+1} - h_{m,n}$$
$$= h_{m-1,n+1}.$$

Consequently, $(E^2 - E - 1)^m h_{m,n} = 0$.

It now follows that $h_{2,n}$ and $h_{3,n}$ must be of the forms

$$h_{2,n} = (an + b)F_n + (cn + d)F_{n-1} \qquad (20.18)$$
$$h_{3,n} = (an^2 + bn + c)F_n + (dn^2 + en + f)F_{n-1},$$

where a, b, c, d, e, and f are constants to be determined.

Since $h_{2,1} = 0, h_{2,2} = 1, h_{2,3} = 2$, and $h_{2,4} = 5$, equation (20.18) yields a 4×4 linear system in a, b, c, and d:

$$a + b \qquad\qquad = 0$$
$$2a + b + 2c + d = 1$$
$$6a + 2b + 3c + d = 2$$
$$12a + 3b + 8c + 2d = 5.$$

Solving this system, we get $a = 1/5 = -b$, $c = 2/5$, and $d = 0$. Thus

$$h_{2,n} = \frac{(n-1)F_n + 2nF_{n-1}}{5}. \tag{20.19}$$

For example,

$$h_{2,7} = \frac{6F_7 + 14F_6}{5}$$
$$= \frac{6 \cdot 13 + 14 \cdot 8}{5}$$
$$= 38;$$

see Table 20.1.

Likewise, it is a good exercise to verify that

$$h_{3,n} = \frac{(5n^2 - 3n - 2)F_n - 6nF_{n-1}}{50}. \tag{20.20}$$

For example,

$$h_{3,5} = \frac{(5 \cdot 5^2 - 3 \cdot 5 - 2)F_5 - 6 \cdot 5F_4}{50}$$
$$= \frac{108 \cdot 5 - 30 \cdot 3}{50}$$
$$= 9.$$

Since $h_{2,n}$ and $h_{3,n}$ are integers, it follows that $(n-1)F_n + 2nF_{n-1} \equiv 0 \pmod 5$ and $(5n^2 - 3n - 2)F_n \equiv 6nF_{n-1} \pmod{50}$.

20.5   AN INFINITE-DIMENSIONAL LUCAS MATRIX

We now introduce a similar infinite-dimensional Lucas matrix $K = (k_{i,j})$. This will also yield some interesting dividends.

We define K recursively, as follows, where $i, j \geq 0$:

1) $k_{0,j} = 0$.
2) $k_{1,1} = 1$.
3) $k_{j,j-1} = 2$, where $j \geq 1$.
4) $k_{i,j} = 0$, where $j < i - 1$.
5) $k_{i,j} = k_{i,j-2} + k_{i,j-1} + k_{i-1,j-1}$, where $i \geq 1$, and $j \geq 2$.

Condition 1) implies that row 0 consists of zeros; conditions 2) and 3) imply the first two elements in row 1 are 2 and 1; by condition 3), the diagonal right below the main diagonal consists of 2s; and by condition 4), every element below this diagonal is zero. We can now employ condition 5) to compute the remaining elements $k_{i,j}$ of matrix K: add the two previous elements $k_{i,j-2}$ and $k_{i,j-1}$ in the same row; add this sum to the element $k_{i-1,j-1}$ just above; the resulting sum is $\widehat{k_{i,j}}$; see Figure 20.1.

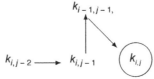

Figure 20.1.

Using recursion, we can construct matrix K; see Table 20.2.

TABLE 20.2.

	i \\ j	0	1	2	3	4	5	6	7	8	$\cdots$
	0	0	0	0	0	0	0	0	0	0	
	1	2	1	3	4	7	11	18	29	47	$\leftarrow L_n$
	2	0	2	3	8 → 15	30	56	104	189		
$K =$	3	0	0	2	5	15	35	80	171	355	
	4	0	0	0	2	7	24	66	170	407	
	5	0	0	0	0	2	9	35	110	315	
	6	0	0	0	0	0	2	11	48	169	
	$\vdots$								All 2s		

It follows by the recursive formula that

$$k_{1,n} = k_{1,n-2} + k_{1,n-1} + k_{0,n-2}$$
$$= k_{1,n-2} + k_{1,n-1} + 0$$
$$= k_{1,n-2} + k_{1,n-1},$$

where $k_{1,0} = 2$ and $k_{1,1} = 1$. Thus $k_{1,n} = L_n$, so row 1 consists of Lucas numbers.

Here is an interesting observation:

$$k_{2,7} = 104$$
$$= 1 \cdot 18 + 1 \cdot 11 + 2 \cdot 7 + 3 \cdot 4 + 5 \cdot 3 + 8 \cdot 1 + 13 \cdot 2$$
$$= \sum_{j=1}^{7} F_j k_{1,7-j}$$
$$= \sum_{\substack{j,k \geq 0 \\ j+k=7}} F_j L_k.$$

More generally, we have the following result. We will establish it using the strong version of PMI.

Theorem 20.7 (Koshy, 1999 [365]).

$$k_{2,n} = \sum_{\substack{j,k \geq 0 \\ j+k=n}} F_j L_k. \tag{20.21}$$

Proof. When $n = 0$, LHS = 0 = RHS; so the result is true.

Now assume it is true for every nonnegative integer $\leq m$. Since $2F_{m+1} - F_m = L_m = k_{1,m}$, by the recurrence, we then have

$$\sum_{\substack{j,k \geq 0 \\ j+k=m+1}} F_j L_k = \sum_{j=0}^{m+1} F_j L_{m+1-j}$$

$$= \sum_{j=0}^{m} F_j (L_{m-j} + L_{m-j-1}) + F_{m+1} L_0$$

$$= \sum_{j=0}^{m} F_j L_{m-j} + \sum_{j=0}^{m} F_j L_{m-j-1} + 2F_{m+1}$$

$$= \sum_{j=0}^{m} F_j L_{m-j} + \sum_{j=0}^{m-1} F_j L_{m-j-1} + F_m L_{-1} + 2F_{m+1}$$

$$= k_{2,m} + k_{2,m-1} + (2F_{m+1} - F_m)$$

$$= k_{2,m} + k_{2,m-1} + k_{1,m}$$

$$= k_{2,m+1}.$$

Thus, by the strong version of PMI, the formula holds for all $n \geq 0$.   ∎

Since $k_{1,n} = L_n$, we can rewrite formula (20.21) as

$$k_{2,n} = \sum_{\substack{j,t\geq0 \\ j+t=n}} F_j k_{1,t}. \tag{20.22}$$

As in Theorem 20.5, we can prove that

$$k_{3,n} = \sum_{i=0}^{n} F_i k_{2,n-i} = \sum_{\substack{i,j,k\geq0 \\ i+j+k=n}} F_i F_j L_k. \tag{20.23}$$

For example,

$$k_{3,5} = \sum_{i=0}^{5} F_i k_{2,5-i}$$

$$= F_0 k_{2,5} + F_1 k_{2,4} + F_2 k_{2,3} + F_3 k_{2,2} + F_4 k_{2,1} + F_5 k_{2,0}$$

$$= 0 \cdot 30 + 1 \cdot 15 + 1 \cdot 8 + 2 \cdot 3 + 3 \cdot 2 + 5 \cdot 0$$

$$= 35;$$

see Table 20.2.

Formulas (20.21) and (20.23) are special cases of the next result; we will establish that also using the strong version of PMI.

Theorem 20.8 (Koshy, 1999 [365]). *Let $m \geq 2$. Then*

$$k_{m,n} = \sum_{i=0}^{n} F_i k_{m-1,n-i}. \tag{20.24}$$

Proof. Assuming that the result is true for all m, we will prove that it is true for all $n \geq 0$. Since $k_{m,0} = 0 = \sum_{i=0}^{0} F_i k_{m-1,-i}$, the result is true when $n = 0$.

When $n = 1$, LHS $= k_{m,1}$; and RHS $= \sum_{0}^{1} F_i k_{m-1,1-i} = F_0 k_{m-1,1} + F_1 k_{m-1,0} =$ $0 + k_{m-1,0} = k_{m-1,0}$. Since $k_{2,1} = 2 = k_{1,0}$ and $k_{i,1} = 0 = k_{i-1,0}$ for $i > 2$, it follows that LHS $= k_{m,1} = k_{m-1,0} =$ RHS for $m \geq 2$.

Now assume the result is true for all integers $n = t$, where $t \geq 2$:

$$k_{m,t} = \sum_{i=0}^{t} F_i k_{m-1,t-i}.$$

Then

$$\sum_{i=0}^{t+1} F_i k_{m-1,t+1-i} = \sum_{i=0}^{t+1} F_i(k_{m-1,t-1-i} + k_{m-1,t-i} + k_{m-2,t-i})$$

$$= \sum_{i=0}^{t+1} F_i k_{m-1,t-1-i} + \sum_{i=0}^{t+1} F_i k_{m-1,t-i} + \sum_{i=0}^{t+1} F_i k_{m-2,t-i}$$

$$= \sum_{i=0}^{t-1} F_i k_{m-1,t-1-i} + \sum_{i=0}^{t} F_i k_{m-1,t-i} + \sum_{i=0}^{t} F_i k_{m-2,t-i}$$

$$= k_{m,t-1} + k_{m,t} + k_{m-1,t}$$

$$= k_{m,t+1}.$$

Thus the result is true for all $n \geq 0$.

On the other hand, assume formula (20.24) is true for all $n \geq 0$. We will now prove that it works for all $m \geq 2$. It is true when $m = 2$ by equation (20.21); and when $m = 3$ by equation (20.23).

Assume it is true for all integers $m \leq t$, where $t \geq 2$:

$$k_{t,n} = \sum_{i=0}^{n} F_i k_{t-1,n-i}.$$

Then

$$\sum_{i=0}^{n} F_i k_{t,n-i} = \sum_{i=0}^{n} F_i(k_{t,n-i-2} + k_{t,n-1-i} + k_{t-1,n-1-i})$$

$$= \sum_{i=0}^{n} F_i k_{t,n-2-i} + \sum_{i=0}^{n} F_i k_{t,n-1-i} + \sum_{i=0}^{n} F_i k_{t-1,n-1-i}$$

$$= \sum_{i=0}^{n-2} F_i k_{t,n-2-i} + \sum_{i=0}^{n-1} F_i k_{t,n-1-i} + \sum_{i=0}^{n-1} F_i k_{t-1,n-1-i}$$

$$= k_{t+1,n-2} + k_{t+1,n-1} + k_{t,n-1}$$

$$= k_{t+1,n}.$$

So the result is true for all $m \geq 2$. Thus it is true for all $m \geq 2$ and $n \geq 0$. ∎

For example,

$$k_{4,5} = \sum_{i=0}^{5} F_i k_{3,5-i}$$

$$= F_1 k_{3,4} + F_2 k_{3,3} + F_3 k_{3,2} + F_4 k_{3,1}$$

$$= 15 + 5 + 4 + 0$$

$$= 24;$$

see Table 20.2.

As in the case of $h_{2,n}$ and $h_{3,n}$, we can derive explicit formulas for $k_{2,n}$ and $k_{3,n}$ also.

EXPLICIT FORMULAS FOR $k_{2,n}$ AND $k_{3,n}$

First, notice that row 2 of matrix K contains an intriguing pattern:

$$
\begin{array}{rcccl}
k_{2,0} & = & 0 & = & 1 \cdot 0 \\
k_{2,1} & = & 2 & = & 2 \cdot 1 \\
k_{2,2} & = & 3 & = & 3 \cdot 1 \\
k_{2,3} & = & 8 & = & 4 \cdot 2 \\
k_{2,4} & = & 15 & = & 5 \cdot 3 \\
& & & & \uparrow \\
& \vdots & & & F_n
\end{array}
$$

So we conjecture that $k_{2,n} = (n+1)F_n$. The next theorem confirms this observation using the strong version of PMI.

Theorem 20.9 (Koshy, 1999 [365]).

$$k_{2,n} = (n+1)F_n. \tag{20.25}$$

Proof. Since $k_{2,0} = 0 = (0+1)F_0$, the result is true when $n = 0$.
Now assume it is true for all nonnegative integers $n \le t$. Then

$$
\begin{aligned}
(t+2)F_{t+1} &= (t+2)(F_t + F_{t-1}) \\
&= tF_{t-1} + (t+1)F_t + (F_t + 2F_{t-1}) \\
&= tF_{t-1} + (t+1)F_t + L_t \\
&= k_{2,t-1} + k_{2,t} + k_{1,t} \\
&= k_{2,t+1}.
\end{aligned}
$$

Thus, by the strong version of PMI, formula (20.25) works for all $n \ge 0$. ∎

We can also establish formula (20.25) by assuming that $k_{2,n}$ is of the form $(an+b)F_n + (cn+d)F_{n-1}$. Thus

$$k_{2,n} = (n+1)F_n = \sum_{j=0}^{n} F_j L_{n-j}. \tag{20.26}$$

We can also find an explicit formula for $k_{3,n}$. To this end, assume that it is of the form $(an^2 + bn + c)F_n + (dn^2 + en + f)F_{n-1}$. Using the values of $k_{3,0}$ through

$k_{3,5}$, we get $a = 1/10 = b$, $c = -1/5 = -d$, $e = 2/5$, and $f = 0$. This yields

$$k_{3,n} = \frac{(n^2 + n - 2)F_n + 2n(n + 2)F_{n-1}}{10}.$$

Since $F_n + 2F_{n-1} = L_n$, we can rewrite this as

$$k_{3,n} = \frac{(n + 2)(nL_n - F_n)}{10}. \qquad (20.27)$$

For example,

$$k_{3,7} = \frac{9(7L_7 - F_7)}{10}$$

$$= \frac{9(7 \cdot 29 - 13)}{10}$$

$$= 171,$$

as expected.

It follows from equation (20.27) that $(n + 2)(nL_n - F_n) \equiv 0 \pmod{10}$.

20.6   AN INFINITE-DIMENSIONAL GIBONACCI MATRIX

More generally, we can construct an infinite-dimensional gibonacci matrix G that satisfies conditions 1), 4), and 5) at the beginning of Section 20.5; and two additional conditions:

2') $g_{1,1} = a$
3') $g_{1,2} = b$,

where a and b are arbitrary integers, and $g_{1,0} = b - a$; see Table 20.3. Row 1 of matrix G consists of the gibonacci numbers G_n.

TABLE 20.3.

i \ j	0	1	2	3	4	5	6 $\cdots$	
0	0	0	0	0	0	0	0	
1	$b - a$	a	b	$a + b$	$a + 2b$	$2a + 3b$	$3a + 5b$	$\leftarrow G_n$
2	0	$b - a$	b	$3b - a$	$5b$	$10b$	$2a + 18b$	
3	0	0	$b - a$	$2b - a$	$6b - 3a \rightarrow 13b - 4a$	$29b - 7a$		

$G =$ (matrix label to the left of the table)

All equal

We can extend formula (20.24) to G, as the next theorem shows. Its proof follows along the same lines as in Theorem 20.8, so we omit it in the interest of brevity.

Theorem 20.10 (Koshy, 1999 [365]). *Let $m \geq 2$. Then*

$$g_{m,n} = \sum_{i=0}^{n} F_i g_{m-1,n-i}. \tag{20.28}$$

∎

For example,

$$g_{3,5} = \sum_{i=0}^{5} F_i g_{2,5-i}$$

$$= \sum_{i=1}^{4} F_i g_{2,5-i}$$

$$= F_1 g_{2,4} + F_2 g_{2,3} + F_3 g_{2,2} + F_4 g_{2,1}$$

$$= 5b + (3b - a) + 2b + 3(b - a)$$

$$= 13b - 4a;$$

see Table 20.3.

In particular, when $m = 2$ and $m = 3$, formula (20.28) yields

$$g_{2,n} = \sum_{i=0}^{n} F_i g_{1,n-i} = \sum_{i=0}^{n} F_i G_{n-i}$$

$$g_{3,n} = \sum_{i=0}^{n} F_i g_{2,n-i} = \sum_{\substack{i,j,k \geq 0 \\ i+j+k=n}} F_i F_j G_k.$$

Next we introduce the *lambda function* λ of a matrix, studied extensively in unpublished notes by Fenton S. Stancliff, a professional musician who had a life-long creative interest in mathematics, but no college training. It was reported and developed by Bicknell (now Bicknell-Johnson) [41, 46]. We can use it, coupled with Fibonacci matrices, to derive a host of Fibonacci identities.

20.7 THE LAMBDA FUNCTION

Let $A = (a_{ij})_{n \times n}$ and $A^* = (a_{ij} + 1)_{n \times n}$. Then $\lambda(A) = |A^*| - |A|$, the change in the value of the determinant.

For example, let $A = \begin{bmatrix} a & b \\ c & d \end{bmatrix}$. Then

$$A^* = \begin{bmatrix} a+1 & b+1 \\ c+1 & d+1 \end{bmatrix}$$

$$|A^*| = (a+1)(d+1) - (b+1)(c+1)$$

$$= (ad - bc) + (a + d - b - c)$$

$$\lambda(A) = a + d - b - c.$$

Suppose we add a constant k to each element in A. Then

$$|A^*| = \begin{bmatrix} a+k & b+k \\ c+k & d+k \end{bmatrix}$$

$$= (ad - bc) + k(a + d - b - c)$$

$$= |A| + k\lambda(A).$$

In particular, let $A = Q^n$. Then $|(Q^n)^*| = |Q^n| + k\lambda(Q^n)$. But $\lambda(Q^n) = F_{n+1} + F_{n-1} - 2F_n = F_{n-1} - F_{n-2} = F_{n-3}$. Therefore, by Cassini's formula, $|(Q^n)^*| = (-1)^n + kF_{n-3}$.

Now letting $k = F_n$, this yields

$$\begin{bmatrix} F_{n+1} + F_n & F_n + F_n \\ F_n + F_n & F_{n-1} + F_n \end{bmatrix} = (-1)^n + kF_{n-3}$$

$$\begin{bmatrix} F_{n+2} & 2F_n \\ 2F_n & F_{n+1} \end{bmatrix} = (-1)^n + kF_{n-3}.$$

This yields the identity

$$4F_n^2 = F_{n+2}F_{n+1} - F_nF_{n-3} + (-1)^{n-1}. \tag{20.29}$$

Next we study the lambda function of a special Fibonacci matrix, studied by Bicknell and Hoggatt [46], and Terry Brennan of Lockheed Missiles and Space Company, Sunnyvale, California.

THE P-MATRIX

Let

$$P = \begin{bmatrix} 0 & 0 & 1 \\ 0 & 1 & 2 \\ 1 & 1 & 1 \end{bmatrix}.$$

Then

$$P^2 = \begin{bmatrix} 1 & 1 & 1 \\ 2 & 3 & 4 \\ 1 & 2 & 4 \end{bmatrix}, \quad P^3 = \begin{bmatrix} 1 & 2 & 4 \\ 4 & 7 & 12 \\ 4 & 6 & 9 \end{bmatrix}, \quad \text{and} \quad P^4 = \begin{bmatrix} 4 & 6 & 9 \\ 12 & 19 & 30 \\ 9 & 15 & 25 \end{bmatrix}.$$

Do you see a pattern? If yes, can you conjecture P^n? The pattern is not that obvious, so try to find it before reading any further.

To see a clear pattern, we now rewrite each element in terms of Fibonacci numbers:

$$P = \begin{bmatrix} F_0^2 & F_0 F_1 & F_1^2 \\ 2F_0 F_1 & F_2^2 - F_0 F_1 & 2F_1 F_2 \\ F_1^2 & F_1 F_2 & F_2^2 \end{bmatrix}$$

$$P^2 = \begin{bmatrix} F_1^2 & F_1 F_2 & F_2^2 \\ 2F_1 F_2 & F_3^2 - F_1 F_2 & 2F_2 F_3 \\ F_2^2 & F_2 F_3 & F_3^2 \end{bmatrix}$$

$$P^3 = \begin{bmatrix} F_2^2 & F_2 F_3 & F_3^2 \\ 2F_2 F_3 & F_4^2 - F_2 F_3 & 2F_3 F_4 \\ F_3^2 & F_3 F_4 & F_4^2 \end{bmatrix},$$

and so on. Clearly, a pattern emerges. Can you predict P^n now?

We can now show by PMI that

$$P^n = \begin{bmatrix} F_{n-1}^2 & F_{n-1} F_n & F_n^2 \\ 2F_{n-1} F_n & F_{n+1}^2 - F_{n-1} F_n & 2F_n F_{n+1} \\ F_n^2 & F_n F_{n+1} & F_{n+1}^2 \end{bmatrix}.$$

Let

$$A = \begin{bmatrix} a & b & c \\ d & e & f \\ g & h & i \end{bmatrix}.$$

Then

$$\lambda(A) = \begin{vmatrix} a+e-b-d & b+f-c-e \\ d+h-g-e & e+i-h-f \end{vmatrix};$$

see Exercise 20.53.

In particular, let $A = P^n$. Then

$$\lambda(P^n) = \begin{bmatrix} F_{n-1}^2 + F_{n+1}^2 - 4F_{n-1} F_n & 2F_{n-1} F_n + 2F_n F_{n+1} - F_n^2 - F_{n+1}^2 \\ 3F_{n-1} F_n + F_n F_{n+1} - F_n^2 - F_{n+1}^2 & 2F_{n+1}^2 - 3F_{n-1} F_n - F_n F_{n-1} \end{bmatrix}.$$

The four expressions in this determinant can be simplified. For example,

$$2F_{n-1}F_n + 2F_nF_{n+1} - F_n^2 - F_{n+1}^2 = 2F_n(F_{n-1} + F_{n+1}) - (F_n^2 + F_{n+1}^2)$$
$$= 2F_nL_n - F_{2n+1}$$
$$= 2F_{2n} - F_{2n+1}$$
$$= F_{2n} - F_{2n-1}$$
$$= F_{2n-2}.$$

We can similarly simplify the other expressions also; see Exercises 20.62–20.64. The resultant determinant is

$$\lambda(P^n) = \begin{bmatrix} F_{2n-3} & F_{2n-2} \\ -F_{n-2}^2 & (-1)^n - F_{n-2}F_{n-1} \end{bmatrix}$$
$$= F_{2n-3}\left[(-1)^n - F_{n-2}F_{n-1}\right] + F_{n-2}^2 F_{2n-2}$$
$$= (-1)^n(F_{n-1}^2 - F_{n-3}F_{n-2})$$
$$= (-1)^n(\text{central element in } P^{n-2}).$$

EXERCISES 20

1. Let Q denote the Q-matrix. Prove that $Q^n = F_nQ + F_{n-1}I$, where I denotes the 2×2 identity matrix. (Notice the similarity between this result and the formula $\alpha^n = F_n\alpha + F_{n-1}$.)
2. Show that $Q^2 - Q - I = 0$.
3. Prove that $(I + Q + Q^2 + \cdots + Q^n)(Q - I) = Q^{n+1} - I$, where $n \geq 1$.
4. Using identity (20.6), prove identity (20.7).

Prove each.

5. $F_{m+n} = F_{m+1}F_n + F_mF_{n-1}$.
6. $L_{m+n} = F_{m+1}L_n + F_mL_{n-1}$.
7. $F_{m+n} = F_{m+1}F_{n+1} - F_{m-1}F_{n-1}$.
8. $L_{m+n} = F_{m+1}L_{n+1} - F_{m-1}L_{n-1}$.
9. $F_{m+n} > F_mF_n$, where $m, n \geq 2$.
10. $F_{nm} > F_n^m$, where $m > n > 1$ (Shapiro; see Cohen, 1978 [125]).
11. $2F_{m+n} = F_mL_n + F_nL_m$.
12. $2L_{m+n} = L_mL_n + 5F_mF_n$.

13. $2F_{m-n} = (-1)^n(F_mL_n - F_nL_m)$.

14. $2L_{m-n} = (-1)^n(L_mL_n - 5F_mF_n)$.

15. $5(L_mL_n + F_mF_n) = 6L_{m+n} + 4(-1)^nL_{m-n}$.

16. $5(L_mL_n - F_mF_n) = 4L_{m+n} + 6(-1)^nL_{m-n}$.

17. $F_{m-n} = (-1)^n(F_mF_{n-1} - F_{m-1}F_n)$.

18. $L_{m-n} = (-1)^n(F_{m+1}L_n - F_mL_{n+1})$.

19. $F_{m+n} - (-1)^nF_{m-n} = F_nL_m$.

20. $F_mL_{n+k} - F_kL_{n+m} = (-1)^kF_{m-k}L_n$ (Taylor, 1981 [556]).

21. $F_{m+n} + F_{m-n} = \begin{cases} L_mF_n & \text{if } n \text{ is odd} \\ F_mL_n & \text{otherwise.} \end{cases}$

22. $F_{m+n} - F_{m-n} = \begin{cases} F_mL_n & \text{if } n \text{ is odd} \\ L_mF_n & \text{otherwise.} \end{cases}$

23. $L_{m+n} + L_{m-n} = \begin{cases} 5F_mF_n & \text{if } n \text{ is odd} \\ L_mL_n & \text{otherwise.} \end{cases}$

24. $L_{m+n} - L_{m-n} = \begin{cases} L_mL_n & \text{if } n \text{ is odd} \\ 5F_mF_n & \text{otherwise.} \end{cases}$

25. $F_{m+n}^2 - F_{m-n}^2 = F_{2m}F_{2n}$.

26. $L_{m+n}^2 - L_{m-n}^2 = 5F_{2m}F_{2n}$.

27. $L_n + 2(-1)^mL_{n-2m-1}$ is divisible by 5 for all $m, n \geq 1$ (Deshpande, 2001 [144]).

28. Let $s(x, y) = \dfrac{xy + 1}{xy - 1}$. Evaluate $s(F_{m+1}/F_m, F_{n+1}/F_n)$ (Gill and Miller, 1981 [219]).

29. Prove that $F_mF_{n+1} - F_nF_{m+1} = (-1)^nF_{m-n}$.

30. Prove that $L_mL_{n+1} - L_nL_{m+1} = 5(-1)^{n+1}F_{m-n}$.

31. Let $M = \begin{bmatrix} 1 & 1 \\ 1 & 2 \end{bmatrix}$. Prove that $M^n = \begin{bmatrix} F_{2n-1} & F_{2n} \\ F_{2n} & F_{2n+1} \end{bmatrix}$ (Moore, 1983 [448]).

Let $A_n = \begin{bmatrix} F_n & L_n \\ L_n & F_n \end{bmatrix}$ (Rabinowitz, 1998 [488]).

32. Express A_{2n} in terms of A_n and A_{n+1}.

33. Express A_{2n} in terms of A_n and A_{n+1} only.

34. Let $p_{n+1} = -p_n$, where $p_1 = -1$. Prove that $p_n = (-1)^n$.

Use the matrix $M = \begin{bmatrix} 0 & 1 \\ \beta & \alpha \end{bmatrix}$ to answer Exercises 20.35 and 20.36.

35. Find the characteristic polynomial of M.

36. Find the eigenvalues of M.

Use the matrices $M = \begin{bmatrix} \alpha & 1 \\ 1 & -\alpha \end{bmatrix}$ and $A = \begin{bmatrix} \alpha & 0 \\ 0 & \beta \end{bmatrix}$ to answer Exercises 20.37 to 20.40.

37. Find M^{-1}.

38. Find MAM^{-1}.

39. Prove that $Q^n = MA^n M^{-1}$.

40. Find the trace of A^n.

Use the matrix $B = \begin{bmatrix} 1/2 & 1 \\ 5/4 & 1/2 \end{bmatrix}$ to answer Exercises 20.41 and 20.42.

41. Prove that $B^n = \begin{bmatrix} \frac{1}{2}L_n & F_n \\ \frac{5}{4}F_n & \frac{1}{2}L_n \end{bmatrix}$.

42. Prove that Q is similar to B.

43. Let $C = \begin{bmatrix} -1 & 1 \\ -1 & 2 \end{bmatrix}$. Prove that $C^n = \begin{bmatrix} -F_{n-2} & F_n \\ -F_n & F_{n+2} \end{bmatrix}$.

44. Let $D = \begin{bmatrix} 3 & 1 \\ -5 & -2 \end{bmatrix}$. Prove that $D^n = \begin{bmatrix} L_{n+1} & F_n \\ -5F_n & -L_{n-1} \end{bmatrix}$.

45. Consider the linear system

$$G_n x + G_{n-1} y = G_{n+1}$$
$$G_{n+1} x + G_n y = G_{n+2},$$

where $G_1 = a$ and $G_2 = b$. Using Cramer's rule, prove that $G_{n+1}G_{n-1} - G_n^2 = \mu(-1)^n$.

46. Using Exercise 20.45, deduce a formula for $L_{n+1}L_{n-1} - L_n^2$.

Let A be a 2×2 matrix and V_n a 2×1 matrix such that $V_{n+1} = AV_n$ (Thoro, 1963 [558]). Find V_n in each case.

47. $A = \begin{bmatrix} 1 & 1 \\ 1 & 0 \end{bmatrix}$, $V_1 = \begin{bmatrix} 1 \\ 1 \end{bmatrix}$.

48. $A = \begin{bmatrix} 2 & 1 \\ 1 & 1 \end{bmatrix}$, $V_1 = \begin{bmatrix} 2 \\ 1 \end{bmatrix}$.

49. $A = \begin{bmatrix} 0 & 1 \\ 1 & -1 \end{bmatrix}$, $\mathbf{V}_1 = \begin{bmatrix} 0 \\ 1 \end{bmatrix}$.

Let $\mathbf{U}_m = (F_{m+1}, F_m)$ and $\mathbf{V}_m = (L_{m+1}, L_m)$. Verify each (Ruggles and Hoggatt, 1963 [506]).

50. $\mathbf{U}_0 Q^{n+1} = \mathbf{U}_{n+1} = \mathbf{U}_n Q$.

51. $\mathbf{V}_m Q^{n+1} = \mathbf{V}_{m+n+1}$.

52. Find $\lambda(Q^n)$.

53. Let $M = \begin{bmatrix} a & b & c \\ d & e & f \\ g & h & i \end{bmatrix}$. Find $\lambda(M)$ (Bicknell and Hoggatt, 1963 [46]).

54. Compute $|P|$.

55. Find $\lambda(P)$.

Use the matrix $R = \begin{bmatrix} L_{n+1} & L_n \\ L_n & L_{n-1} \end{bmatrix}$ to answer Exercises 20.56 and 20.57.

56. Compute $|R|$.

57. Find $\lambda(R)$.

Let $A = \begin{bmatrix} G_{n+k} & G_n \\ G_n & G_{n-k} \end{bmatrix}$ and $B = \begin{bmatrix} G_{n+k}+i & G_n+i \\ G_n+i & G_{n-k}+i \end{bmatrix}$, where G_m denotes the mth gibonacci number. Compute each.

58. $|A|$.

59. $\lambda(A)$.

60. $|B|$.

61. Using PMI, establish the formula for P^n, where P denotes the P-matrix.

Prove each.

62. $F_{n-1}^2 + F_{n+1}^2 - 4F_{n-1}F_n = F_{2n-3}$.

63. $F_n^2 + F_{n+1}^2 - 3F_{n-1}F_n - F_nF_{n+1} = F_{n-2}^2$.

64. $3F_{n-1}F_n + F_nF_{n-1} - 2F_{n+1}^2 = F_{n-2}F_{n-1} - (-1)^n$.

65. The sum of any $2n$ consecutive Fibonacci numbers is divisible by F_n (Lind, 1964 [391]).

66. $L_{n+1}^3 + L_n^3 - L_{n-1}^3 = 5L_{3n}$ (Long, 1986 [418]).

67. $L_{n+2}^3 - 3L_n^3 + L_{n-2}^3 = 15L_{3n}$.

68. $L_{n+2}^3 - 3L_{n+1}^3 - 6L_n^3 + 3L_{n-1}^3 + L_{n-2}^3 = 0$.

69. $F_{m+2n}F_{m+n} - 2L_nF_{m+n}F_{m-n} - F_{m-n}F_{m-2n} = (L_{3n} - 2L_n)F_m^2$, where n is odd (Wulczyn, 1981 [610]).

70. $L_{m+2n}L_{m+n} - 2L_nL_{m+n}L_{m-n} - L_{m-n}L_{m-2n} = (L_{3n} - 2L_n)L_m^2$, where n is odd (Wulczyn, 1981 [609]).

71. $\displaystyle\sum_{k=1}^{\infty} \frac{F_{2^{k-1}}}{L_{2^k} + 1} = \frac{1}{\sqrt{5}}$ (Ohtsuka, 2014 [462]).

72. $\displaystyle\sum_{k=1}^{\infty} \frac{L_{2^{k+1}}}{F_{3\cdot 2^k}} = \frac{5}{4}$ (Ohtsuka, 2014 [463]).

73. $\displaystyle\sum_{k=1}^{\infty} \frac{F_{2^{k-1}}^2}{L_{2^k}^2 - 1} = \frac{3}{20}$ (Ohtsuka, 2014 [463]).

21

GRAPH-THEORETIC MODELS I

In Chapter 20, we used the Q-matrix to extract some interesting properties of both Fibonacci and Lucas numbers. We will now employ it to construct graph-theoretic models for them, and then explore some of the well-known identities. These models enable us to study Fibonacci and Lucas numbers in a different perspective.

To begin with, recall that

$$Q = \begin{bmatrix} 1 & 1 \\ 1 & 0 \end{bmatrix} \quad \text{and} \quad Q^n = \begin{bmatrix} F_{n+1} & F_n \\ F_n & F_{n-1} \end{bmatrix},$$

where $n \geq 1$.

We are now ready for the models.

21.1 A GRAPH-THEORETIC MODEL FOR FIBONACCI NUMBERS

The Q-matrix can geometrically be translated into a connected graph G with two vertices v_1 and v_2; see Figure 21.1. In fact, Q is the *adjacency matrix* $(q_{ij})_{2\times2}$ of the graph, where q_{ij} denotes the number of edges from v_i to v_j, and $1 \leq i \leq j \leq 2$.

For example, $q_{12} = 1$; so there is exactly one edge from v_1 to v_2. Likewise, $q_{22} = 0$; so there is no edge from v_2 to itself. The three edges of the graph can be denoted by 1–1, 1–2, and 2–1, or by the "words" 11, 12, and 21.

Fibonacci and Lucas Numbers with Applications, Volume One, Second Edition. Thomas Koshy.
© 2018 John Wiley & Sons, Inc. Published 2018 by John Wiley & Sons, Inc.

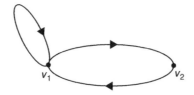

Figure 21.1.

Again recall that a *path* from vertex v_i to vertex v_j in a connected graph is a sequence $v_i - e_i - v_{i+1} - \cdots - v_{j-1} - e_j - v_j$ of vertices v_k and edges e_k, where each edge e_k is incident with vertices v_{k-1} and v_k. The path is *closed* if its endpoints are the same; otherwise, it is *open*. The *length* ℓ of a path is the number of edges in the path; that is, it takes ℓ steps to reach one endpoint of the path from the other.

The adjacency matrix of a connected graph can be employed to compute the number of paths of a given length n between any two vertices, as the next theorem shows. The proof follows by induction [368].

Theorem 21.1.   *Let A be the adjacency matrix of a connected graph with vertices $v_1, v_2, \ldots, v_k$, and n a positive integer. Then the ij-th entry of the matrix A^n records the number of paths of length n from v_i to v_j.*   ∎

The next result about the Q-matrix follows by this theorem.

Corollary 21.1.   *The ij-th entry of Q^n gives the number of paths of length n from v_i to v_j in G, where $1 \leq i \leq j \leq 2$.*   ∎

For example, we have $Q^4 = \begin{bmatrix} F_5 & F_4 \\ F_4 & F_3 \end{bmatrix}$. So there are $F_5 = 5$ paths of length 4 from v_1 to itself; $F_4 = 3$ such paths from v_1 to v_2, and from v_2 to v_1; and $F_3 = 2$ such paths from v_2 to itself. They are summarized in Table 21.1.

TABLE 21.1.   Paths of Length 4

From vertex	Paths				
1 to vertex 1	11111	11121	11211	12111	12121
1 to vertex 2	11112	11212	12112		
2 to vertex 1	21111	21121	21211		
2 to vertex 2	21112	21212			

The Fibonacci recurrence can now be interpreted in a new context: *The number of closed paths of length n beginning at v_1 equals the total number of paths of length n from v_1 to v_2 or v_2 to itself.*

A GRAPH-THEORETIC MODEL FOR LUCAS NUMBERS

Counting the number of closed paths of length n in G gives an interesting dividend. By Corollary 21.1, there are F_{n+1} closed paths beginning at v_1 and F_{n-1} closed paths beginning at v_2. So, by the addition principle, *the total number of closed paths of length n in G is $F_{n+1} + F_{n-1} = L_n$.*

For example, there are 5 closed paths of length 4 starting at v_1, and 2 such paths starting at v_2. So there is a total of $7 = L_4$ closed paths of length 4 in G:

$$11111 \quad 11121 \quad 11211 \quad 12111 \quad 12121$$
$$21112 \quad 21212$$

This leads us to two interesting observations; but first, two definitions from linear algebra [58]. The *eigenvalues λ* of a matrix $M = (m_{ij})_{n \times n}$ are the solutions of the equation $|M - \lambda I| = 0$, where I denotes the $n \times n$ *identity matrix*. The *trace* of M is $\sum_{i=1}^{n} m_{ii}$, the sum of the elements along its main diagonal.

INTERESTING OBSERVATIONS

The *eigenvalues* of the Q-matrix are α and β, and those of Q^n are α^n and β^n; see Exercises 21.1 and 21.2. Consequently, the number of closed paths of length n in G equals $F_{n+1} + F_{n-1}$, the trace of Q^n. Since $F_{n+1} + F_{n-1} = L_n = \alpha^n + \beta^n$, it also equals the sum of its eigenvalues.

21.2   BYPRODUCTS OF THE COMBINATORIAL MODELS

The graph-theoretic models can be used to illustrate some elegant Fibonacci and Lucas properties by providing combinatorial proofs. We will now establish a few using this approach.

Example 21.1.   *Confirm the beautiful property $F_{n+1}^2 + F_n^2 = F_{2n+1}$.*

Proof. To this end, we will count in two different ways the number of closed paths of length $2n$ starting at v_1. By Corollary 21.1, there are F_{2n+1} such paths.

We will now count them in a different way. The path will land at v_1 or v_2 after n steps.

Case 1.   Suppose the subpath ends at v_1 after n steps. Again, by Corollary 21.1, there are F_{n+1} paths of length n from v_1 to itself. The remaining subpath is also of length n and runs from v_1 to itself; there are F_{n+1} such paths. Thus, by the multiplication principle, there are $F_{n+1} \cdot F_{n+1} = F_{n+1}^2$ closed paths of length $2n$ that take us to v_1 after n steps.

Case 2. On the other hand, suppose the subpath stops at v_2 after n steps. There are F_n subpaths of length n from v_1 to v_2, and F_n subpaths of length n from v_2 to v_1. Consequently, there are $F_n \cdot F_n = F_n^2$ paths of length $2n$ from v_1 to itself that stop at v_2 after n steps.

Combining the two cases, there are exactly $F_{n+1}^2 + F_n^2$ paths of length $2n$ from v_1 to itself. The desired result follows by equating the two counts. ∎

For instance, let $n = 3$. There are $13 = F_7$ closed paths of length 6 starting at v_1. Exactly $9 = F_4^2$ of them land at v_1 after 3 steps, and $4 = F_3^2$ of them at v_2 after 3 steps; see Table 21.2.

TABLE 21.2. Closed Paths of Length 6 Starting at v_1

Closed Paths of Length 6 Starting at v_1 and Landing at v_1 in 3 Steps		Closed Paths of Length 6 Starting at v_1 and Landing at v_2 in 3 Steps
1111111	1121121	1112111
1111121	1211121	1112121
1111211	1121211	1212111
1121111	1211211	1212121
1211111		

Example 21.2. *Establish the addition formula* $F_{m+n} = F_{m+1}F_n + F_m F_{n-1}$.

Proof. This time we will count in two different ways the number of paths of length $m + n$ from v_1 to v_2. By Corollary 21.1, there are F_{m+n} such paths.

Case 1. Suppose the path takes us to v_1 after m steps. There are F_{m+1} such subpaths. The remaining subpath is of length n, and runs from v_1 to v_2. There are F_n such subpaths. So there is a total of $F_{m+1}F_n$ paths of length $m + n$ from v_1 to v_2 that run through v_1 after m steps.

Case 2. Suppose the path takes us to v_2 after m steps. As in Case 1, there are $F_m F_{n-1}$ such paths of length $m + n$.

Combining the two cases, there are $F_{m+1}F_n + F_m F_{n-1}$ paths of length $m + n$ from v_1 to v_2. The desired result follows by equating the two counts. ∎

Example 21.3. *Confirm the elegant formula* $F_{2n} = F_n L_n$.

Proof. We will count in two different ways the number of paths of length $2n$ from v_1 to v_2. Clearly, there are F_{2n} such paths in G.

There are F_{n+1} subpaths of length n ending at v_1 and F_n subpaths of length n ending at v_2; so there are $F_{n+1}F_n$ such paths from v_1 to v_2, which pass through v_1 after n steps.

On the other hand, there are F_n subpaths of length n from v_1 to v_2, and F_{n-1} subpaths of length n from v_2 to itself. So there are $F_n F_{n-1}$ such paths from v_1 to v_2, which pass through v_2 after n steps.

Thus there are $F_{n+1}F_n + F_nF_{n-1} = F_n(F_{n+1} + F_{n-1}) = F_nL_n$ paths of length $2n$ from v_1 to v_2. Consequently, $F_{2n} = F_nL_n$, as desired. ∎

For example, let $n = 3$. There are exactly $8 = F_3L_3$ paths of length 6 from v_1 to v_2:

$$111\boxed{1}112 \qquad\qquad 121\boxed{1}112$$
$$111\boxed{1}212 \qquad\qquad 121\boxed{2}112$$
$$111\boxed{2}112 \qquad\qquad 112\boxed{1}212$$
$$112\boxed{1}112 \qquad\qquad 121\boxed{1}212.$$

$F_4F_3 = 6$ of them land at v_1 after 3 steps, and $F_3F_2 = 2$ of them at v_2 after 3 steps.

Example 21.4. *Prove that $L_{n+1} + L_{n-1} = 5F_n$.*

Proof. We will accomplish this by constructing a *one-to-five correspondence* from the set of paths of length n from v_1 to v_2 to the set of closed paths of length $n + 1$ or $n - 1$.

There are L_{n+1} closed paths of length $n + 1$, and L_{n-1} closed paths of length $n - 1$. So there are $L_{n+1} + L_{n-1}$ closed paths of length $n + 1$ or $n - 1$.

There are F_n paths of length n from v_1 to v_2. Let $w = w_1w_2\cdots w_nw_{n+1}$ be such a path. Then $w_1 = 1$ and $w_{n+1} = 2$, and the immediate predecessor of every 2 is 1: $w = 1\underbrace{w_2\cdots w_{n-1}12}_{\text{length } n}$.

We will now describe an algorithm to recover all closed paths of length $n + 1$ or $n - 1$ in five steps, by constructing exactly F_n such closed paths in each step.

Step 1. Insert a 1 at the end of w: $1\underbrace{w_2\cdots w_{n-1}121}$. This produces a closed path $\underset{\text{length } n+1}{}$ of length $n + 1$ from v_1 to v_1. There are F_n such closed paths.

Step 2. Delete $w_{n+1} = 2$ from the end of w. This gives a closed path of length $n - 1$ from v_1 to v_1: $1\underbrace{w_2\cdots w_{n-1}1}$. There are F_n such closed paths. $\underset{\text{length } n-1}{}$

Step 3. Insert a 2 at the beginning of w: $21\underbrace{w_2\cdots w_{n-1}12}$. This is a closed path $\underset{\text{length } n+1}{}$ of length $n + 1$ from v_2 to itself. There are F_n such closed paths.

Step 4. Replace $w_{n+1} = 2$ with 11. This generates a closed path of length $n + 1$ from v_1 to v_1: $1\underbrace{w_2\cdots w_{n-1}111}$. There are F_n such closed paths. $\underset{\text{length } n+1}{}$

We are not quite done. Unfortunately, these four steps do not create all possible closed paths of length $n + 1$ or $n - 1$. The missing ones correspond to whether $w_1w_2 = 11$ or 12. Accordingly, we have one more case, with two parts.

Step 5A. Suppose $w_2 = 1$. Then delete w_1 and append 11 at the end. This gives a closed path of length $n + 1$ from v_1 to v_1: $\underbrace{w_2 \cdots w_n 211}_{\text{length } n+1}$.

Step 5B. Suppose $w_2 = 2$. Then delete w_1. This results in a closed path of length $n - 1$ from v_2 to itself: $\underbrace{w_2 \cdots w_{n+1}}_{\text{length } n-1}$.

Step 5 also generates F_n closed paths.

Steps 1–5 do not create duplicate closed paths. So they generate a total of $5F_n$ closed paths of length $n + 1$ or $n - 1$. Consequently, the two counts must be equal; that is, $L_{n+1} + L_{n-1} = 5F_n$. ∎

We will now illustrate this algorithm with a specific example. Let $n = 4$. There are $11 = L_5$ closed paths of length 5 and $4 = L_3$ of length 3, a total of $15 = L_5 + L_3$:

Length 5:   111111   111121   111211   112111   121111   112121   121121   121211
            211112   211212   212112
Length 3:   1111   1121   1211
            2112

There are $3 = F_4$ paths $w = w_1 w_2 w_3 w_4 w_5$ of length 4 from v_1 to v_2: 11112, 11212, 12112.

1) Inserting a 1 at the end of w yields three closed paths of length 5: 111121, 112121, 121121.
2) Deleting $w_5 = 2$ gives three closed paths of length 3: 1111, 1121, 1211.
3) Inserting a 2 at the beginning of w produces three closed paths of length 5: 211112, 211212, 212112.
4) Replace w_5 with 11. This generates three more closed paths of length 5: 111111, 112111, 121111.
5A) When $w_2 = 1$, delete w_1 and append 11 at the end. This yields two closed paths of length 5: 111211, 121211.
5B) When $w_2 = 2$, delete w_1. This creates exactly one closed path of length 3: 2112.

Interestingly, each step generates $3 = F_4$ closed paths of length 5 or 3. By now we have accounted for all $15 = 5 \cdot 3 = 5F_4$ of them.

Example 21.5. *Establish the Lucas formula* $F_{n+1} = \displaystyle\sum_{k=0}^{\lfloor n/2 \rfloor} \binom{n-k}{k}$.

Proof. Consider the closed paths of length n originating at v_1. Recall that there are F_{n+1} such paths.

Now consider an arbitrary closed path P. Suppose it contains exactly k closed paths 121 of two edges; for convenience, we call them *d-edges* ("d" for double). The k d-edges account for $2k$ edges, so there $n - 2k$ edges remaining in the path. Consequently, path P contains a total of $k + (n - 2k) = n - k$ *elements* (edges or d-edges). We can select the k d-edges from them in $\binom{n-k}{k}$ different ways, where $0 \le 2k \le n$. Thus the total number of closed paths originating at v_1 equals

$$\sum_{k=0}^{\lfloor n/2 \rfloor} \binom{n-k}{k}.$$

Lucas' formula now follows by equating the two counts. ∎

For example, consider the $8 = F_6$ closed paths of length 5 originating at v_1 in Table 21.3, where d-edges are parenthesized or boldfaced.

TABLE 21.3. Closed Paths of Length 5 from v_1 to Itself

111111	111**121**	111**211**	112**111**	**121**111	1(**121**)**21**	(**121**)**121**	(**121**)**211**

Exactly one of them contains zero d-edges, four contain exactly one d-edge, and three contain exactly two d-edges, a total of 8 edges.

The next example gives a combinatorial proof of another well-known summation formula.

Example 21.6. *Establish the identity* $F_{2n} = \sum\limits_{k=1}^{n} \binom{n}{k} F_k.$

Proof. Consider the closed paths of length $2n - 1$ from v_1 to v_1. There are F_{2n} such paths.

We will now count them in a different way. First, notice that edge 12 must be followed by edge 21. Thus every path must contain a d-edge corresponding to every visit to v_2. So we count every d-edge as one element of length 2.

Since $2n - 1$ is odd, every path must contain an odd number of edges (loops) 11. There can be a maximum of $n - 1$ d-edges; so every path must contain at least $(n - 1) + 1 = n$ elements.

Suppose there are k loops among the first n elements of the path. So there are $n - k$ d-edges among them; they account for a length of $k + 2(n - k) = 2n - k$. So the remaining subpath must be of length $(2n - 1) - (2n - k) = k - 1$. Consequently, such a path is of the form $\underbrace{\text{subpath}}_{\text{length } 2n-k}\ \underbrace{\text{subpath}}_{\text{length } k-1}$.

The k edges can be placed among the n elements in $\binom{n}{k}$ different ways, and the subpath of length $k - 1$ in F_k different ways. There are $\binom{n}{k} F_k$ such paths. So the total number of closed paths of length $2n - 1$ from v_1 to itself is $\sum\limits_{k=1}^{n} \binom{n}{k} F_k.$

Equating the two counts yields the desired result. ∎

We will now illustrate this proof with $n = 3$. There are $8 = F_6$ closed paths of length 5 originating at v_1; see Table 21.4.

TABLE 21.4.   Closed Paths of Length 5 Originating at v_1

111111	111121	111211	112111	121111
	112121	121211	121121	

There are $3 = \binom{3}{1}F_1$ such closed paths with $k = 1$: **112121**, **121211**, **121121**; the first three edges are boldfaced in Table 21.4. There are $3 = \binom{3}{2}F_2$ such closed paths with $k = 2$: **11121**1, **11211**1, **12111**1; the first three edges again are bold-faced in the table. There are $2 = \binom{3}{3}F_3$ such closed paths with $k = 3$: **111**111, **111**121; the first three edges are once again boldfaced in the table. Thus there are $3 + 3 + 2 = 8$ closed paths of length 5 from v_1 to v_1, as expected.

We will now re-confirm the combinatorial formula in Example 14.5. But before we do, we introduce the concept of a *median loop*, as in the case of the median square in Chapter 14. Consider a closed path of length $2n + 1$ beginning at v_1. Since every d-edge is of length 2, such a path must contain an odd number of 1–1 loops (loops at v_1). The middle 1–1 loop is the *median loop*.

For example, consider the $8 = F_6$ closed paths of length 5 originating at v_1; see Table 21.5. The bold edges indicate the median loops in the paths.

TABLE 21.5.   Closed Paths of Length 5 Originating at v_1

11**1**111	111**1**21	111**2**11	112**1**11
121**1**11	112**1**21	121**1**21	121**2**11

We are now ready for the proof.

Example 21.7.   *Using the graph-theoretic model, prove that*

$$\sum_{\substack{i,j \geq 0 \\ i+j \leq n}} \binom{n-i}{j}\binom{n-j}{i} = F_{2n+2}.$$

Proof. Consider the closed paths of length $2n + 1$ originating at v_1. By Corollary 21.1, there are F_{2n+2} such paths.

We will now count them in a different way. To this end, let P be such an arbitrary path. Let M denote the median loop in P. Suppose it has i d-edges to the left of M and j d-edges to the right: subpath A $\overset{\uparrow}{M}$ subpath B. Since path P con-

$$\underbrace{\qquad}_{i \text{ d-edges here}} \quad \underbrace{\qquad}_{j \text{ d-edges here}}$$

tains a total of $i + j$ d-edges, it must contain $(2n + 1) - 2(i + j)$ 1–1 loops. Hence there are $n - i - j$ 1–1 loops on each side of M.

Of the $n - j$ elements on the left side of M, i elements are d-edges. They can be placed in $\binom{n-j}{i}$ ways. So there are $\binom{n-j}{i}$ such paths A to the left of M.

Similarly, there are $\binom{n-i}{j}$ such paths A to the right of M. Thus there are

$\binom{n-i}{j}\binom{n-j}{i}$ paths P; each contains i d-edges to the left of M and j d-edges to the right.

It now follows that the total number of closed paths of length $2n + 1$ originating at v_1 equals $\sum_{i,j}\binom{n-i}{j}\binom{n-j}{i}$. Notice that $i, j \geq 0$ and $i + j \leq n$. Combining the two counts gives the desired formula. ∎

For example, consider the eight closed paths of length 5 in Table 21.5. To be consistent with the proof in Example 21.7, we have $n = 2$. Table 21.6 gives the closed path(s); indicates the corresponding median loops; lists the values of i and j; and computes the product $\binom{2-i}{j}\binom{2-j}{i}$. The total number of closed paths beginning at v_1 equals $\sum_{i,j}\binom{2-i}{j}\binom{2-j}{i} = 1 + 2 + 2 + 1 + 1 + 1 = 8$, as expected.

TABLE 21.6. Closed Paths of Length 5 Originating at v_1

Closed path(s)	i	j	$\binom{2-i}{j}\binom{2-j}{i}$
111111	0	0	1
111121	0	1	2
111211	.	.	.
112111	1	0	2
121111	.	.	.
112121	0	2	1
121121	1	1	1
121211	2	0	1

$\uparrow$

sum = 8

21.3 SUMMATION FORMULAS

The combinatorial models can be helpful in deriving and interpreting the following summation formulas:

1) $\displaystyle\sum_{i=1}^{n} F_i = F_{n+2} - 1$ 2) $\displaystyle\sum_{i=1}^{n} L_i = L_{n+2} - 3$

3) $\displaystyle\sum_{i=1}^{n} F_{2i-1} = F_{2n}$ 4) $\displaystyle\sum_{i=1}^{n} L_{2i-1} = L_{2n} - 2$

5) $\displaystyle\sum_{i=1}^{n} F_{2i} = F_{2n+1} - 1$ 6) $\displaystyle\sum_{i=1}^{n} L_{2i} = L_{2n+1} - 1.$

Lucas discovered the summation formulas in 1), 3), and 4) in 1876.

To work briefly with the combinatorial technique, we will establish formula 1).

Consider closed paths of length $n + 1$ beginning at v_1. There are F_{n+2} such paths. Exactly one of them does *not* visit v_2. So there are $F_{n+2} - 1$ closed paths that contain a d-edge.

We will now count such paths in a different way. Consider such an arbitrary closed path. Suppose the last d-edge occurs in edges i and $i + 1$ of the path. The remaining subpath must consist of the loops 11. So such a path is of the form

$$\underbrace{1 \cdots 1}_{\text{length } i-1} \quad \underbrace{121}_{\text{d-edge}} \quad \underbrace{\text{loops } 11.}$$

Since there are F_i closed paths of length $i - 1$ originating at v_1, it follows that there are F_i closed paths of length $n + 1$ originating at v_1, where $2 \le i + 1 \le n + 1$. So the total number of closed paths of length $n + 1$ starting at v_1 that contain at least one d-edge is $\displaystyle\sum_{i=1}^{n} F_i$.

Combining the two counts yields the desired formula. ∎

For example, consider the closed paths of length 5 from v_1 to itself in Table 21.3. Table 21.7 shows the paths with the last d-edge, its locations, and their counts.

TABLE 21.7. Closed Paths of Length 5 from v_1 to Itself with at Least One d-edge

i	Closed Paths of Length 5 with the Last d-edge Occupying Edges i and $i + 1$			Count
1	**121**111			1
2	1**121**11			1
3	111**21**1 121**21**1			2
4	1111**21** 112**121** 121**121**			3
				↑
				F_i

Finally, we can give graph-theoretic interpretations of the above summation formulas. To this end, we will need the following result [368].

Theorem 21.2. *Let A be the adjacency matrix of a connected graph with vertices $v_1, v_2, \ldots, v_k$. Then the ij-th entry of the matrix $A + A^2 + \cdots + A^n$ gives the number of paths of length $\le n$ from v_i to v_j, where $n \ge 1$.* ∎

For example, since

$$\sum_{i=1}^{n} Q^i = \begin{bmatrix} \displaystyle\sum_{i=1}^{n} F_{i+1} & \displaystyle\sum_{i=1}^{n} F_i \\ \displaystyle\sum_{i=1}^{n} F_i & \displaystyle\sum_{i=1}^{n} F_{i-1} \end{bmatrix}$$

$$= \begin{bmatrix} F_{n+3} - 2 & F_{n+2} - 1 \\ F_{n+2} - 1 & F_{n+1} - 1 \end{bmatrix},$$

$$\sum_{i=1}^{n} F_i = F_{n+2} - 1 \text{ gives the total number of paths of length } \leq n \text{ from } v_1 \text{ to } v_2.$$

Likewise, there are $\sum_{i=1}^{n} L_i = L_{n+2} - 3$ closed paths of length at most n.

EXERCISES 21

Prove each.
1. The eigenvalues of the Q-matrix Q are α and β.
2. The eigenvalues of Q^n are α^n and β^n.

List each.
3. Closed paths of length 4 originating at v_1.
4. Closed paths of length 4 originating at v_2.
5. Paths of length 4 from v_1 to v_2.
6. Paths of length 4 from v_2 to v_1.
7. Closed paths of length 3 starting at v_1 that contain at least one d-edge.
8. Closed paths of length 4 starting at v_1 that contain at least one d-edge.

Prove each combinatorially.
9. $F_n = F_{m+1}F_{n-m} + F_m F_{n-m-1}$.
10. $F_{n+1}F_{n-1} - F_n^2 = (-1)^n$.

Using the graph-theoretic model, interpret each summation formula.

11. $\sum_{i=1}^{n} F_{2i-1} = F_{2n}$.

12. $\sum_{i=1}^{n} L_{2i-1} = L_{2n} - 2$.

13. $\sum_{i=1}^{n} F_{2i} = F_{2n+1} - 1$.

14. $\sum_{i=1}^{n} L_{2i} = L_{2n+1} - 1$.

FIBONACCI DETERMINANTS

In Chapter 3, we found that Fibonacci and Lucas numbers occur in graph theory, specifically in the study of paraffins and cycloparaffins. We now turn our attention to an additional appearance of Lucas numbers in graph theory, and then to some Fibonacci and Lucas determinants.

22.1 AN APPLICATION TO GRAPH THEORY

In 1975, K.R. Rebman of California State University at Hayward established the occurrence of Lucas numbers in the study of spanning trees of wheel graphs [497]. Before we can present the main result, we need to lay some groundwork in the form of two lemmas and some basic vocabulary.

Lemma 22.1. *Let A_n denote the $n \times n$ matrix*

$$\begin{bmatrix} 3 & -1 & 0 & 0 & & & 0 \\ -1 & 3 & -1 & 0 & & & 0 \\ 0 & -1 & 3 & -1 & \cdots & & 0 \\ & & & & \vdots & & \\ & & & & \cdots & & \\ 0 & 0 & 0 & 0 & & 3 & -1 \\ 0 & 0 & 0 & 0 & & -1 & 3 \end{bmatrix}.$$

Then $|A_n| = F_{2n+2}$.

Fibonacci and Lucas Numbers with Applications, Volume One, Second Edition. Thomas Koshy.
© 2018 John Wiley & Sons, Inc. Published 2018 by John Wiley & Sons, Inc.

Proof. We will establish the lemma using PMI. Since $|A_1| = 3 = F_4$, and

$$|A_2| = \begin{vmatrix} 3 & -1 \\ -1 & 3 \end{vmatrix} = 8 = F_6,$$ the result is true when $n = 1$ and $n = 2$.

Now assume it is true for all positive integers $k < n$, where $n \geq 2$. Expanding $|A_n|$ by the first row, we get

$$|A_n| = 3|A_{n-1}| + \begin{vmatrix} -1 & -1 & 0 & & & 0 \\ 0 & 3 & -1 & \vdots & & 0 \\ & & & & & 0 \\ & & & \cdots & & \\ & & & & \vdots & -1 \\ 0 & 0 & 0 & & -1 & 3 \end{vmatrix}$$

$$= 3|A_{n-1}| + |A_{n-2}|$$

$$= 3F_{2(n-1)+2} - F_{2(n-2)+2}$$

$$= 3F_{2n} - F_{2n-2}$$

$$= F_{2n+2}.$$

Thus, by the strong version of PMI, the result is true for all positive integers n. ∎

Lemma 22.2. *Let B_n denote the $n \times n$ matrix*

$$\begin{bmatrix} 3 & -1 & 0 & 0 & & & -1 \\ -1 & 3 & -1 & 0 & \vdots & & 0 \\ 0 & -1 & 3 & -1 & \cdots & & 0 \\ & & & & \vdots & & 0 \\ & & & & & 3 & -1 \\ -1 & 0 & 0 & 0 & & -1 & 3 \end{bmatrix},$$

where $B_1 = [1]$ and $B_2 = \begin{bmatrix} 3 & -2 \\ -2 & 3 \end{bmatrix}$. Then $|B_n| = L_{2n} - 2$.

Proof. We will use $|A_n|$ in the proof. Since $|B_1| = 1 = L_2 - 2$ and $|B_2| = 5 = L_4 - 2$, the result is true when $n = 1$ and $n = 2$. So assume $n \geq 3$.

Expanding $|B_n|$ by the first row, $|B_n| = 3|A_{n-1}| + |R_{n-1}| + (-1)^{n+2}|S_{n-1}|$, where

$$|R_m| = \begin{vmatrix} -1 & -1 & 0 & 0 & & 0 & 0 \\ 0 & 3 & -1 & 0 & \vdots & 0 & 0 \\ 0 & -1 & 3 & -1 & \cdots & 0 & 0 \\ & & & & \vdots & & 0 \\ 0 & 0 & 0 & 0 & & 3 & -1 \\ 0 & 0 & 0 & 0 & & -1 & 3 \end{vmatrix},$$

and

$$|S_m| = \begin{vmatrix} -1 & 3 & -1 & 0 & & 0 & 0 \\ 0 & -1 & 3 & -1 & \vdots & 0 & 0 \\ 0 & 0 & -1 & 3 & \cdots & 0 & 0 \\ & & & & \vdots & & \\ 0 & 0 & 0 & 0 & & 3 & -1 \\ 0 & 0 & 0 & 0 & & -1 & 3 \\ -1 & 0 & 0 & 0 & & 0 & -1 \end{vmatrix}.$$

Expanding $|R_m|$ by the first column, we get

$$|R_m| = -|A_{m-1}| + (-1)^m \begin{vmatrix} -1 & 0 & 0 & & 0 & 0 \\ 3 & -1 & 0 & \vdots & 0 & 0 \\ -1 & 3 & -1 & \cdots & 0 & 0 \\ & & & \vdots & & \\ 0 & 0 & 0 & & 0 & -1 \end{vmatrix}$$

$$= -|A_{m-1}| + (-1)^m(-1)^{m-1}$$

$$= -|A_{m-1}| - 1$$

$$= -F_{2m} - 1.$$

Expanding $|S_m|$ by the first column, we get

$$|S_m| = (-1)^m + (-1)(-1)^{m-1}|A_{m-1}|$$

$$= (-1)^m(F_{2m} + 1).$$

Thus

$$|B_n| = 3F_{2n} + (-F_{2n-2} - 1) + (-1)^n(-1)^{n-1}(F_{2n-2} + 1)$$

$$= 3F_{2n} - 2F_{2n-2} - 2$$

$$= L_{2n} - 2,$$

as desired. ∎

Next we introduce a few additional basic terms and some fundamental results from graph theory for clarity and consistency.

ADDITIONAL BASIC GRAPH-THEORETIC FACTS

We can represent algebraically a graph G with n vertices by the *incidence matrix* $A(G) = (a_{ij})_{n \times n}$, where

$$a_{ij} = \begin{cases} 1 & \text{if there is an edge from vertex } i \text{ to vertex } j \\ 0 & \text{otherwise.} \end{cases}$$

For example, the incidence matrix of the graph in Figure 22.1 is

$$\begin{bmatrix} 0 & 1 & 0 & 0 \\ 1 & 0 & 1 & 1 \\ 0 & 1 & 0 & 1 \\ 0 & 1 & 1 & 0 \end{bmatrix}.$$

Figure 22.1.

The *degree* of a vertex v, denoted by $\deg(v)$, is the number of edges meeting at v. For instance, the degree of vertex 2 in Figure 22.1 is three.

Let $D(G) = (d_{ij})_{n \times n}$ denote the matrix defined by

$$d_{ij} = \begin{cases} \deg(i) & \text{if } i = j \\ 0 & \text{otherwise.} \end{cases}$$

For the graph in Figure 22.1,

$$D(G) = \begin{bmatrix} 1 & 0 & 0 & 0 \\ 0 & 3 & 0 & 0 \\ 0 & 0 & 2 & 0 \\ 0 & 0 & 0 & 2 \end{bmatrix}.$$

Recall from Chapter 5 that a spanning tree of a graph G is a subgraph of G that is a tree containing every vertex of G, and its complexity $k(G)$ is the number of distinct spanning trees of the graph. For any graph G, $k(G)$ equals the determinant of any one of the n principal $(n-1)$-rowed minors of the matrix $D(G) - A(G)$. The outstanding German physicist Gustav Robert Kirchhoff (1824–1887) discovered this remarkable result.

For example, using the graph in Figure 22.1,

$$D(G) - A(G) = \begin{bmatrix} 1 & -1 & 0 & 0 \\ -1 & 3 & -1 & -1 \\ 0 & -1 & 2 & -1 \\ 0 & -1 & -1 & 2 \end{bmatrix}.$$

Since

$$\begin{vmatrix} 3 & -1 & -1 \\ -1 & 2 & -1 \\ -1 & -1 & 2 \end{vmatrix} = 3,$$

it follows that the complexity of the graph is three, as we found in Chapter 5.

THE WHEEL GRAPH

Let $n \geq 3$. The *wheel graph* W_n is a graph with $n + 1$ vertices; n of them lie on a cycle (the rim), and the remaining vertex (the hub) is connected to every rim vertex. Figure 22.2 shows the wheel graphs W_3, W_4, and W_5.

W_3

W_4

W_5

Figure 22.2. Wheel graphs.

We are ready for the surprise.

Theorem 22.1. $k(W_n) = L_{2n} - 2$.

Proof. Denote the rim vertices of W_n by v_1 through v_n, and the hub vertex by v_{n+1}. Then

$$\deg(v_i) = \begin{cases} 3 & \text{if } i \neq n+1 \\ n & \text{otherwise.} \end{cases}$$

Consequently,

$$D(W_n) - A(W_n) = \begin{bmatrix} 3 & 0 & 0 & 0 & & & 0 \\ 0 & 3 & 0 & 0 & \vdots & & 0 \\ 0 & & 3 & 0 & \cdots & & 0 \\ 0 & & & & \vdots & & 0 \\ 0 & 0 & 0 & 0 & & 3 & 0 \\ 0 & 0 & 0 & 0 & & 0 & n \end{bmatrix} - \begin{bmatrix} 0 & 1 & 0 & 0 & & 0 & 1 \\ 1 & 0 & 1 & 0 & \vdots & 0 & 0 \\ 0 & 1 & 0 & 1 & \cdots & 0 & 0 \\ 0 & & & & & & 0 \\ 0 & & & & & & 0 & 1 \\ 0 & 0 & 0 & 0 & & 1 & 0 \end{bmatrix}$$

$$= \begin{bmatrix} & & & & & & \vdots & -1 \\ & & & & & & \vdots & -1 \\ & & B_n & & & & \vdots & \vdots \\ & & & & & & \vdots & \vdots \\ & & & & & & \vdots & -1 \\ \cdots\cdots\cdots\cdots\cdots\cdots\cdots\cdots\cdots\cdots\cdots \\ -1 & -1 & -1 & \cdots & \cdots & -1 & \vdots & n \end{bmatrix}$$

To compute $k(W_n)$, any principal $(n-1)$-rowed minor will suffice. Thus, by deleting row $n+1$ and column $n+1$, and invoking Lemma 22.2, we get $k(W_n) = |B_n| = L_{2n} - 2$. ∎

For example, W_3 has $L_6 - 2 = 18 - 2 = 16$ spanning trees. Figure 22.3 displays them.

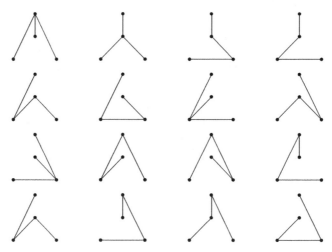

Figure 22.3.   Spanning trees.

In 1969, J. Sedlacek of The University of Calgary, Canada, discovered Theorem 22.1 [517]. B.R. Myers of the University of Notre Dame, Notre Dame, Indiana, independently discovered it two years later [452]. At the 1969 Calgary International Conference of Combinatorial Structures and their Applications, Sedlacek stated the formula as

$$k(W_n) = \left(\frac{3+\sqrt{5}}{2}\right)^n - \left(\frac{3-\sqrt{5}}{2}\right)^n - 2.$$

In contrast, Myers gave the result as $k(W_n) = F_{2n+2} - F_{2n-2} - 2$ in a problem he proposed in 1972 in *The American Mathematical Monthly*. Obviously, either formula can be rewritten in terms of L_{2n}, as in the theorem [453].

Next we study the singularity of Fibonacci matrices.

22.2   THE SINGULARITY OF FIBONACCI MATRICES

A square matrix A is *singular* if $|A| = 0$. For example, the matrix $A = \begin{bmatrix} 3 & 4 \\ 6 & 8 \end{bmatrix}$ is singular, since $|A| = 0$.

A square matrix $M_n = (a_{ij})_{n \times n}$ is a *Fibonacci matrix* if it contains the first n^2 Fibonacci numbers such that $a_{11} = F_1$, $a_{12} = F_2$, ..., $a_{1n} = F_n$, $a_{21} = F_{n+1}$, ..., $a_{nn} = F_{n^2}$. Thus $a_{ij} = F_{(i-1)n+j}$, where $1 \le i, j \le n$.

For example, $M_1 = [1]$, $M_2 = \begin{bmatrix} 1 & 1 \\ 2 & 3 \end{bmatrix}$,

$$M_3 = \begin{bmatrix} 1 & 1 & 2 \\ 3 & 5 & 8 \\ 13 & 21 & 34 \end{bmatrix}, \quad \text{and} \quad M_4 = \begin{bmatrix} 1 & 1 & 2 & 3 \\ 5 & 8 & 13 & 21 \\ 34 & 55 & 89 & 144 \\ 233 & 377 & 610 & 987 \end{bmatrix}.$$

Clearly, $|M_1| \ne 0$ and $|M_2| \ne 0$. But $|M_3| = \begin{vmatrix} 1 & 0 & 0 \\ 3 & 2 & 2 \\ 13 & 8 & 8 \end{vmatrix} = 0$. Using elementary column operations, it is easy to see that $|M_4| = 0$. Thus both M_3 and M_4 are singular matrices.

More generally, we claim that M_n is singular where $n \ge 3$. Graham Fisher observed this in 1995, when he was a student at Bournemouth School, England [178]. Before we give a formal proof of this fact, we need the following lemma; it was proposed as a problem by D.V. Jaiswal of Holkar Science College, Indore, India, in 1969 [349].

Lemma 22.3. *Let m, n, p, q, and r be positive integers, and G_n the n-th gibonacci number. Then*

$$\begin{vmatrix} G_p & G_{p+m} & G_{p+m+n} \\ G_q & G_{q+m} & G_{q+m+n} \\ G_r & G_{r+m} & G_{r+m+n} \end{vmatrix} = 0.$$

Proof. Since $G_{m+n} = G_m F_{n+1} + G_{m-1} F_n$ (see Exercise 7.23), it follows that

$$G_{k+m+n} = G_{k+m} F_{n+1} + G_{k+m-1} F_n. \tag{22.1}$$

Using equation (22.1), we can write the given determinant D as

$$D = F_{n+1} \begin{vmatrix} G_p & G_{p+m} & G_{p+m} \\ G_q & G_{q+m} & G_{q+m} \\ G_r & G_{r+m} & G_{r+m} \end{vmatrix} + F_n \begin{vmatrix} G_p & G_{p+m} & G_{p+m-1} \\ G_q & G_{q+m} & G_{q+m-1} \\ G_r & G_{r+m} & G_{r+m-1} \end{vmatrix}$$

$$= F_{n+1} \cdot 0 + F_n \begin{vmatrix} G_p & G_{p+m} & G_{p+m-1} \\ G_q & G_{q+m} & G_{q+m-1} \\ G_r & G_{r+m} & G_{r+m-1} \end{vmatrix}$$

$$= F_n \begin{vmatrix} G_p & G_{p+m} & G_{p+m-1} \\ G_q & G_{q+m} & G_{q+m-1} \\ G_r & G_{r+m} & G_{r+m-1} \end{vmatrix}$$

$$= F_n \begin{vmatrix} G_p & G_{p+m-2} & G_{p+m-1} \\ G_q & G_{q+m-2} & G_{q+m-1} \\ G_r & G_{r+m-2} & G_{r+m-1} \end{vmatrix}$$

$$= F_n \begin{vmatrix} G_p & G_{p+m-2} & G_{p+m-3} \\ G_q & G_{q+m-2} & G_{q+m-3} \\ G_r & G_{r+m-2} & G_{r+m-3} \end{vmatrix}.$$

Continuing in this way, we can reduce the subscripts of the elements in columns 2 and 3 further. At a certain stage, when m is even, columns 1 and 2 would be identical; when m is odd, columns 1 and 3 would be identical. In both cases, $D = 0$, as desired.   ∎

This lemma has several byproducts; see Exercises 22.2–22.5.

We are now ready to present the theorem. We will establish it using PMI.

Theorem 22.2. $|M_n| = 0$, *where* $n \geq 3$.

Proof. Since $|M_3| = 0$, assume the result is true when $n = k$, where $k \geq 3$.

Now consider M_{k+1}. Expanding $|M_{k+1}|$ by the first row and the resulting cofactors, we can express $|M_{k+1}|$ as a linear combination of a host of 3×3 matrices. Let

$$M = \begin{bmatrix} F_m & U_1 & U_2 \\ U_3 & U_4 & U_5 \\ U_6 & U_7 & U_8 \end{bmatrix},$$

where each U_i is a Fibonacci element belonging to M. Since we expanded $|M_{k+1}|$ and its successive cofactors by their first rows, M preserves the same order as in $|M_{k+1}|$; so $U_3 = F_{m+r}$ and $U_6 = F_{m+2r}$, where $r \geq 1$. Thus

$$M = \begin{bmatrix} F_m & F_{m+s} & F_{m+t} \\ F_{m+r} & F_{m+r+s} & F_{m+r+t} \\ F_{m+2r} & F_{m+2r+s} & F_{m+2r+t} \end{bmatrix},$$

where $s, t \geq 1$. It follows by Lemma 22.3 that $|M| = 0$. Since every 3×3 matrix M is singular, it follows that M_{k+1} is also singular.

Thus, by PMI, M_n is singular for every $n \geq 3$.   ∎

We will now illustrate the proof with $n = 4$. Using Lemma 22.3, we have

$$|M_4| = \begin{vmatrix} F_6 & F_7 & F_8 \\ F_{10} & F_{11} & F_{12} \\ F_{14} & F_{15} & F_{16} \end{vmatrix} - \begin{vmatrix} F_5 & F_7 & F_8 \\ F_9 & F_{11} & F_{12} \\ F_{13} & F_{15} & F_{16} \end{vmatrix} + 2 \begin{vmatrix} F_5 & F_6 & F_8 \\ F_9 & F_{10} & F_{12} \\ F_{13} & F_{14} & F_{16} \end{vmatrix} - 3 \begin{vmatrix} F_5 & F_6 & F_7 \\ F_9 & F_{10} & F_{11} \\ F_{13} & F_{14} & F_{15} \end{vmatrix}$$

$$= 0 - 0 + 2 \cdot 0 - 3 \cdot 0$$

$$= 0.$$

So M_4 is singular, as expected.

22.3   FIBONACCI AND ANALYTIC GEOMETRY

Next we study a few results in analytic geometry that involve Fibonacci numbers. Jaiswal developed them in 1974 [350].

Theorem 22.3 (Jaiswal, 1974 [350]).   *The area of the triangle with vertices* $(G_n, G_{n+r}), (G_{n+p}, G_{n+p+r})$, *and* (G_{n+q}, G_{n+q+r}) *is independent of n, where* G_k *denotes the kth gibonacci number.*

Proof. Twice the area of the triangle equals the absolute value of the determinant

$$D = \begin{vmatrix} G_n & G_{n+r} & 1 \\ G_{n+p} & G_{n+p+r} & 1 \\ G_{n+q} & G_{n+q+r} & 1 \end{vmatrix}.$$

Using the identity $G_{m+n} = G_m F_{n+1} + G_{m-1} F_n$ for the second column, we can write the determinant as

$$D = F_{r+1} \begin{vmatrix} G_n & G_n & 1 \\ G_{n+p} & G_{n+p} & 1 \\ G_{n+q} & G_{n+q} & 1 \end{vmatrix} + F_r \begin{vmatrix} G_n & G_{n-1} & 1 \\ G_{n+p} & G_{n+p-1} & 1 \\ G_{n+q} & G_{n+q-1} & 1 \end{vmatrix}$$

$$= 0 + F_r \begin{vmatrix} G_n & G_{n-1} & 1 \\ G_{n+p} & G_{n+p-1} & 1 \\ G_{n+q} & G_{n+q-1} & 1 \end{vmatrix}.$$

Add the negative of column 2 to column 1:

$$D = F_r \begin{vmatrix} G_{n-2} & G_{n-1} & 1 \\ G_{n+p-2} & G_{n+p-1} & 1 \\ G_{n+q-2} & G_{n+q-1} & 1 \end{vmatrix}.$$

Now add the negative of column 1 to column 2:

$$D = F_r \begin{vmatrix} G_{n-2} & G_{n-3} & 1 \\ G_{n+p-2} & G_{n+p-3} & 1 \\ G_{n+q-2} & G_{n+q-3} & 1 \end{vmatrix}.$$

Continuing like this, we get

$$D = \pm F_r \begin{vmatrix} G_1 & G_2 & 1 \\ G_{p+1} & G_{p+2} & 1 \\ G_{q+1} & G_{q+2} & 1 \end{vmatrix},$$

according as n is odd or even. Using the identity $G_n G_{m+k} - G_{n+k} G_m = (-1)^{n+1} F_k F_{m-n} \mu$ (see Exercise 7.25), now expand this determinant:

$$D = \pm F_r[(G_{p+1} G_{q+2} - G_{p+2} G_{q+1}) - (G_1 G_{q+2} - G_2 G_{q+1}) + (G_1 G_{p+2} - G_2 G_{p+1})]$$
$$= \pm F_r[(-1)^p F_{q-p} - F_q + F_p]\mu$$
$$= \pm F_r[(-1)^p F_{q-p} + F_p - F_q]\mu.$$

Since this is independent of n, the given area is independent of n. ∎

Corollary 22.1. *The area of the triangle with vertices* (F_n, F_{n+h}), (F_{n+2h}, F_{n+3h}), *and* (F_{n+4h}, F_{n+5h}) *is* $F_h(F_{4h} - 2F_{2h})/2$.

Proof. The proof follows from the theorem by letting $r = h, p = 2h, q = 4h$, and $a = b = \mu = 1$. ∎

Theorem 22.4 (Jaiswal, 1974 [350]). *The lines through the origin with direction ratios* G_n, G_{n+p}, G_{n+q} *are coplanar for every n, where p and q are constants.*

Proof. The direction ratios of three such lines can be taken as G_i, G_{i+p}, G_{i+q}; G_j, G_{j+p}, G_{j+q}; and G_k, G_{k+p}, G_{k+q}. The lines are coplanar if and only if

$$D = \begin{vmatrix} G_i & G_{i+p} & G_{i+q} \\ G_j & G_{j+p} & G_{j+q} \\ G_k & G_{k+p} & G_{k+q} \end{vmatrix} = 0.$$

Using the identity $G_{m+n} = G_m F_{n+1} + G_{m-1} F_n$, we can express D as the sum of four determinants, each of which is zero. Therefore, $D = 0$. So the three lines are coplanar, as desired. ∎

Theorem 22.5 (Jaiswal, 1974 [350]). *The plane containing the family of points* (G_n, G_{n+p}, G_{n+q}) *contains the origin, where p and q are constants.*

Proof. By Theorem 22.4, the given points are coplanar. An equation of the plane containing any three points of the family of points is

$$\begin{vmatrix} x & y & z & 1 \\ G_i & G_{i+p} & G_{i+q} & 1 \\ G_j & G_{j+p} & G_{j+q} & 1 \\ G_k & G_{k+p} & G_{k+q} & 1 \end{vmatrix} = 0.$$

Expanding this determinant with respect to row 1, we get the equation $Ax + By + Cz + D = 0$, where A, B, C, and D are constants.

But, by Theorem 22.4,

$$D = - \begin{vmatrix} G_i & G_{i+p} & G_{i+q} \\ G_j & G_{j+p} & G_{j+q} \\ G_k & G_{k+p} & G_{k+q} \end{vmatrix} = 0.$$

Thus the equation of the plane is $Ax + By + Cz = 0$, which clearly contains the origin. ∎

In fact, we can do better. We can show that the equation of the plane is $(-1)^p F_{q-p} x - F_q y + F_p z = 0$. In particular, the plane containing the points (G_k, G_{k+3}, G_{k+8}) is $(-1)^3 F_5 x - F_8 y + F_3 z = 0$; that is, $5x + 21y - 2z = 0$.

Theorem 22.6 (Jaiswal, 1974 [350]). *The family of planes $G_n x + G_{n+p} y + G_{n+q} z + G_{n+r} = 0$ intersect along a line whose equation is independent of n, where p, q, and r are constants.* ∎

The proof is a bit complicated, so we omit it in the interest of brevity. But we can show that the planes intersect along the line

$$\frac{(-1)^p F_p x - F_{r-p}}{F_{q-p}} = \frac{F_p y + F_r}{-F_q} = \frac{z}{F_p}.$$

For example, the planes $G_n x + G_{n+2} y + G_{n+3} z + G_{n+4} = 0$ intersect along the line

$$\frac{(-1)^2 F_2 x - F_3}{F_1} = \frac{F_2 y + F_4}{-F_3} = \frac{z}{F_2};$$

that is

$$\frac{x-2}{1} = \frac{y+3}{-2} = \frac{z}{1}.$$

In 1966, D.A. Klarner of the University of Alberta, Canada, developed a fascinating formula for computing a special class of determinants [362]. It has interesting Fibonacci and Lucas consequences. The following theorem gives the formula; but the proof is again a bit complicated, so we omit it for the sake of brevity.

Theorem 22.7 (Klarner, 1966 [362]). *Suppose a_n satisfies the second-order recurrence*

$$a_{n+2} = p a_{n+1} - q a_n, \tag{22.2}$$

where p and q are complex constants, and $n \geq 0$. Let

$$A_k(a_n) = \begin{vmatrix} a_n^k & a_{n+1}^k & & a_{n+k}^k \\ a_{n+1}^k & a_{n+2}^k & \cdots & a_{n+k+1}^k \\ & & \vdots & \\ a_{n+k}^k & a_{n+k+1}^k & \cdots & a_{n+2k}^k \end{vmatrix}.$$

Then $A_k(a_{mn+r}) = q^{nk(k+1)/2} A_k(a_r)$. ∎

When $p = 1 = -q$, recurrence (22.2) yields the Fibonacci recurrence, so we can employ this theorem to evaluate determinants of the form $A_k(F_n)$ (or $A_k(L_n)$), as the next four examples demonstrate.

Example 22.1. *Evaluate the determinant* $A = \begin{vmatrix} F_n & F_{n+1} \\ F_{n+1} & F_{n+2} \end{vmatrix}$.

Solution. Here $k = 1 = m$, $r = 0$, and $a_n = F_n$. By Theorem 22.7, we have

$$A = A_1(F_n)$$
$$= (-1)^{n \cdot 1 \cdot 2/2} A_1(F_0)$$
$$= (-1)^n \begin{vmatrix} F_0 & F_1 \\ F_1 & F_2 \end{vmatrix}$$
$$= (-1)^{n+1}.$$

That is, $F_{n+2}F_n - F_{n+1}^2 = (-1)^{n+1}$, which is Cassini's formula.   ∎

Example 22.2. *Evaluate the determinant* $B = \begin{vmatrix} F_n^2 & F_{n+1}^2 & F_{n+2}^2 \\ F_{n+1}^2 & F_{n+2}^2 & F_{n+3}^2 \\ F_{n+2}^2 & F_{n+3}^2 & F_{n+4}^2 \end{vmatrix}$.

Solution. Here $k = 2$, $m = 1$, $r = 0$, and $a_n = F_n$. By Theorem 22.7, we have

$$B = (-1)^{n \cdot 2 \cdot 3/2} A_2(F_0)$$
$$= (-1)^{3n} \begin{vmatrix} F_0^2 & F_1^2 & F_2^2 \\ F_1^2 & F_2^2 & F_3^2 \\ F_2^2 & F_3^2 & F_4^2 \end{vmatrix}$$
$$= (-1)^{3n} \begin{vmatrix} 0 & 1 & 1 \\ 1 & 1 & 4 \\ 1 & 4 & 9 \end{vmatrix}$$
$$= (-1)^n \begin{vmatrix} 0 & 1 & 1 \\ 1 & 1 & 4 \\ 0 & 3 & 5 \end{vmatrix}$$
$$= 2(-1)^{n+1}.$$

(Br. Alfred studied this problem in 1963 [3].)   ∎

Example 22.3. *Evaluate the determinant* $C = \begin{vmatrix} F_n^3 & F_{n+1}^3 & F_{n+2}^3 & F_{n+3}^3 \\ F_{n+1}^3 & F_{n+2}^3 & F_{n+3}^3 & F_{n+4}^3 \\ F_{n+2}^3 & F_{n+3}^3 & F_{n+4}^3 & F_{n+5}^3 \\ F_{n+3}^3 & F_{n+4}^3 & F_{n+5}^3 & F_{n+6}^3 \end{vmatrix}.$

Solution. Here $k = 3$, $m = 1$, $r = 0$, and $a_n = F_n$. By Theorem 22.7, we then have

$$C = (-1)^{n \cdot 3 \cdot 4/2} A_3(F_0)$$

$$= (-1)^{6n} \begin{vmatrix} F_0^3 & F_1^3 & F_2^3 & F_3^3 \\ F_1^3 & F_2^3 & F_3^3 & F_4^3 \\ F_2^3 & F_3^3 & F_4^3 & F_5^3 \\ F_3^3 & F_4^3 & F_5^3 & F_6^3 \end{vmatrix}$$

$$= \begin{vmatrix} 0 & 1 & 1 & 8 \\ 1 & 1 & 8 & 27 \\ 1 & 8 & 27 & 125 \\ 8 & 27 & 125 & 512 \end{vmatrix}$$

$$= \begin{vmatrix} 0 & 1 & 1 & 8 \\ 1 & 1 & 8 & 27 \\ 0 & 7 & 19 & 98 \\ 0 & 19 & 61 & 296 \end{vmatrix}$$

$$= - \begin{vmatrix} 1 & 1 & 8 \\ 7 & 19 & 98 \\ 19 & 61 & 296 \end{vmatrix}$$

$$= - \begin{vmatrix} 1 & 1 & 8 \\ 0 & 12 & 42 \\ 0 & 42 & 144 \end{vmatrix}$$

$$= 36.$$

(J. Erbacker et al. studied this problem in 1963 [160].)   ∎

Since Theorem 22.7 holds for any sequence that satisfies recurrence (22.1), we can invoke it with $a_n = G_n$, the nth gibonacci number, as the next example illustrates.

Example 22.4. *Evaluate the determinant* $D = \begin{vmatrix} G_n & G_{n+1} & G_{n+2} \\ G_{n+1} & G_{n+2} & G_{n+3} \\ G_{n+2} & G_{n+3} & G_{n+4} \end{vmatrix}.$

Solution. Here $k = 1 = m$, $r = 0$, and $a_n = G_n$. By Theorem 22.7, we then have

$$D = (-1)^{n \cdot 2 \cdot 3/2} A_2(G_0)$$

$$= (-1)^n \begin{vmatrix} G_0 & G_1 & G_2 \\ G_1 & G_2 & G_3 \\ G_2 & G_3 & G_4 \end{vmatrix}$$

$$= (-1)^n \begin{vmatrix} b - a & a & b \\ a & b & a + b \\ b & a + b & a + 2b \end{vmatrix}$$

$$= 0.$$

This follows trivially, since row 3 is the sum of rows 1 and 2.   ■

The following exercises offer additional opportunities to apply Theorem 22.7 in several cases.

EXERCISES 22

Evaluate each determinant.

1. $\begin{vmatrix} L_n & L_{n+1} \\ L_{n+1} & L_{n+2} \end{vmatrix}$.

2. $\begin{vmatrix} F_p & F_{p+m} & F_{p+m+n} \\ F_q & F_{q+m} & F_{q+m+n} \\ F_r & F_{r+m} & F_{r+m+n} \end{vmatrix}$ (Ivanoff, 1968 [347]).

3. $\begin{vmatrix} F_a & F_{a+d} & F_{a+2d} \\ F_{a+3d} & F_{a+4d} & F_{a+5d} \\ F_{a+6d} & F_{a+7d} & F_{a+8d} \end{vmatrix}$ (Finkelstein, 1969 [173]).

4. $\begin{vmatrix} L_a & L_{a+d} & L_{a+2d} \\ L_{a+3d} & L_{a+4d} & L_{a+5d} \\ L_{a+6d} & L_{a+7d} & L_{a+8d} \end{vmatrix}$ (Finkelstein, 1969 [173]).

5. $\begin{vmatrix} G_a & G_{a+d} & G_{a+2d} \\ G_{a+3d} & G_{a+4d} & G_{a+5d} \\ G_{a+6d} & G_{a+7d} & G_{a+8d} \end{vmatrix}$.

6. $\begin{vmatrix} F_n + k & F_{n+1} + k & F_{n+2} + k \\ F_{n+1} + k & F_{n+2} + k & F_{n+3} + k \\ F_{n+2} + k & F_{n+3} + k & F_{n+4} + k \end{vmatrix}$ (Alfred, 1963 [4]).

7. $\begin{vmatrix} G_p + k & G_{p+m} + k & G_{p+m+n} + k \\ G_q + k & G_{q+m} + k & G_{q+m+n} + k \\ G_r + k & G_{r+m} + k & G_{r+m+n} + k \end{vmatrix}$ (Jaiswal, 1969 [349]).

8. $\begin{vmatrix} F_{n+3} & F_{n+2} & F_{n+1} & F_n \\ F_{n+2} & F_{n+3} & F_n & F_{n+1} \\ F_{n+1} & F_n & F_{n+3} & F_{n+2} \\ F_n & F_{n+1} & F_{n+2} & F_{n+3} \end{vmatrix}$ (Ledin, 1967 [384]).

9. $\begin{vmatrix} G_{n+3} & G_{n+2} & G_{n+1} & G_n \\ G_{n+2} & G_{n+3} & G_n & G_{n+1} \\ G_{n+1} & G_n & G_{n+3} & G_{n+2} \\ G_n & G_{n+1} & G_{n+2} & G_{n+3} \end{vmatrix}$ (Jaiswal, 1969 [349]).

10. Show that $\begin{vmatrix} L_{n+3} & L_{n+2} & L_{n+1} & L_n \\ L_{n+2} & L_{n+3} & L_n & L_{n+1} \\ L_{n+1} & L_n & L_{n+3} & L_{n+2} \\ L_n & L_{n+1} & L_{n+2} & L_{n+3} \end{vmatrix} = 25 \begin{vmatrix} F_{n+3} & F_{n+2} & F_{n+1} & F_n \\ F_{n+2} & F_{n+3} & F_n & F_{n+1} \\ F_{n+1} & F_n & F_{n+3} & F_{n+2} \\ F_n & F_{n+1} & F_{n+2} & F_{n+3} \end{vmatrix}$

(Jaiswal, 1969 [349]).

Evaluate each determinant.

11. $\begin{vmatrix} L_n^2 & L_{n+1}^2 & L_{n+2}^2 \\ L_{n+1}^2 & L_{n+2}^2 & L_{n+3}^2 \\ L_{n+2}^2 & L_{n+3}^2 & L_{n+4}^2 \end{vmatrix}$.

12. $\begin{vmatrix} G_n^2 & G_{n+1}^2 & G_{n+2}^2 \\ G_{n+1}^2 & G_{n+2}^2 & G_{n+3}^2 \\ G_{n+2}^2 & G_{n+3}^2 & G_{n+4}^2 \end{vmatrix}$.

13. $\begin{vmatrix} L_n^3 & L_{n+1}^3 & L_{n+2}^3 & L_{n+3}^3 \\ L_{n+1}^3 & L_{n+2}^3 & L_{n+3}^3 & L_{n+4}^3 \\ L_{n+2}^3 & L_{n+3}^3 & L_{n+4}^3 & L_{n+5}^3 \\ L_{n+3}^3 & L_{n+4}^3 & L_{n+5}^3 & L_{n+6}^3 \end{vmatrix}$.

14. Let A be the $n \times n$ matrix
$$\begin{bmatrix} 3 & i & 0 & 0 & & 0 & 1 \\ i & 1 & i & 0 & & 0 & 0 \\ 0 & i & 1 & 0 & & 0 & 0 \\ 0 & 0 & i & 1 & \cdots & 0 & 0 \\ & & & \vdots & & & \\ & & & \cdots & & & \\ 0 & 0 & 0 & 0 & & 1 & i \\ 0 & 0 & 0 & 0 & & i & 1 \end{bmatrix},$$

where $i = \sqrt{-1}$. Prove that $|A| = L_{n+1}$ (Byrd, 1963 [93]).

Consider the $n \times n$ matrix $g_{n+1}(x) =$
$$\begin{bmatrix} 2x & i & 0 & 0 & & 0 & 1 \\ i & 2x & i & 0 & & 0 & 0 \\ 0 & i & 2x & i & & 0 & 0 \\ 0 & 0 & i & 2x & \cdots & 0 & 0 \\ & & & \vdots & & & \\ & & & \cdots & & & \\ 0 & 0 & 0 & 0 & & 2x & i \\ 0 & 0 & 0 & 0 & & i & 2x \end{bmatrix},$$

where $g_0(x) = 0$, $g_1(x) = 1$, $i = \sqrt{-1}$, and x is a real number (Byrd, 1963 [93]).

15. Find the recurrence satisfied by g_n.

16. Deduce the value of $g_{n+1}(1/2)$.

17. Show that $|e^{Q^n}| = e^{L_n}$, where e denotes the base of the natural logarithm, and Q the Q-matrix (Hoggatt and King, 1963 [313]).

18. Evaluate the $n \times n$ determinant
$$\begin{vmatrix} 1 & 1 & 0 & 0 & & 0 & 0 & 0 \\ -1 & 1 & 1 & 0 & & 0 & 0 & 0 \\ 0 & -1 & 1 & 1 & & 0 & 0 & 0 \\ 0 & 0 & -1 & 1 & \cdots & 0 & 0 & 0 \\ & & & \vdots & & & & \\ & & & \cdots & & & & \\ 0 & 0 & 0 & 0 & & -1 & 1 & 1 \\ 0 & 0 & 0 & 0 & & 0 & -1 & 1 \end{vmatrix}.$$ (Lucas)

19. Let g_n denote the number of nonzero terms in the expansion of the $n \times n$

$$\text{determinant}\begin{vmatrix} a_1 & b_1 & 0 & 0 & & 0 & 0 & 0 & 0 \\ -1 & a_2 & b_2 & 0 & & 0 & 0 & 0 & 0 \\ 0 & -1 & a_3 & b_3 & & 0 & 0 & 0 & 0 \\ 0 & 0 & -1 & a_4 & \cdots & 0 & 0 & 0 & 0 \\ & & & \vdots & & & & & \\ & & & \cdots & & & & & \\ 0 & 0 & 0 & 0 & & 0 & -1 & a_{n-1} & b_{n-1} \\ 0 & 0 & 0 & 0 & & 0 & 0 & -1 & a_n \end{vmatrix},$$

where $a_i b_i \neq 0$. Show that $g_n = F_{n+1}$ (Bridger, 1967 [59]).

20. Show that
$$\begin{vmatrix} 2 & 2 & -1 & 0 & 0 & & 0 & 0 & 0 & 0 \\ -1 & 2 & 2 & -1 & 0 & & 0 & 0 & 0 & 0 \\ 0 & -1 & 2 & 2 & -1 & & 0 & 0 & 0 & 0 \\ 0 & 0 & -1 & 2 & 2 & \cdots & 0 & 0 & 0 & 0 \\ & & & & \vdots & & & & & \\ & & & \cdots & & & & & & \\ 0 & 0 & 0 & 0 & 0 & & -1 & 1 & 1 & \\ 0 & 0 & 0 & 0 & 0 & & 0 & -1 & 1 & \end{vmatrix} = F_{n+1} F_{n+2}$$

(Lind, 1971 [411]).

21. Evaluate the $n \times n$ determinant
$$\begin{vmatrix} a+b & ab & 0 & 0 & & 0 & 1 \\ 1 & a+b & ab & 0 & & 0 & 0 \\ 0 & 1 & a+b & ab & & 0 & 0 \\ 0 & 0 & 1 & a+b & \cdots & 0 & 0 \\ & & & & \vdots & & \\ & & & \cdots & & & \\ 0 & 0 & 0 & 0 & & a+b & ab \\ 0 & 0 & 0 & 0 & & 1 & a+b \end{vmatrix}$$

(Church, 1964 [123]).

23

FIBONACCI AND LUCAS CONGRUENCES

We can employ the congruence relation to extract a host of Fibonacci and Lucas properties. To begin with, we will identify Fibonacci numbers ending in zero.

23.1 FIBONACCI NUMBERS ENDING IN ZERO

In Chapter 10, we found that $F_m | F_n$ if and only if $m|n$; that is, $F_n \equiv 0 \pmod{F_m}$ if and only if $n \equiv 0 \pmod{m}$. In particular, $F_n \equiv 0 \pmod{F_5}$ if and only if $n \equiv 0 \pmod 5$. Thus, *beginning with F_0, every fifth Fibonacci number is divisible by 5*, a fact we already knew.

Furthermore, $F_3 | F_n$ if and only if $3|n$; that is, $F_n \equiv 0 \pmod 2$ if and only if $n \equiv 0 \pmod 3$. Thus, *beginning with F_0, every third Fibonacci number is divisible by 2*, another fact we already knew.

Consequently, $F_n \equiv 0 \pmod 5$ if and only if $n \equiv 0 \pmod 5$; and $F_n \equiv 0 \pmod 2$ if and only if $n \equiv 0 \pmod 3$. Thus $F_n \equiv 0 \pmod{10}$ if and only if $n \equiv 0 \pmod{15}$. In other words, *beginning with F_0, every fifteenth Fibonacci number ends in zero*; and conversely, *if a Fibonacci number ends in zero, then n is divisible by 15*. For example, $F_{15} = 610, F_{30} = 832{,}040$, and $F_{45} = 1{,}134{,}903{,}170$ end in zero.

23.2   LUCAS NUMBERS ENDING IN ZERO

Are there Lucas numbers ending in zero? To answer this, first we will check if
there are Lucas numbers ending in 5. Suppose $L_n \equiv 0 \pmod 5$ for some integer
n. Then, by the binomial theorem, we have

$$2^n L_n \equiv (1 + \sqrt{5})^n + (1 - \sqrt{5})^n$$

$$= \sum_{i=0}^{n} \binom{n}{i} (\sqrt{5})^i + \sum_{i=0}^{n} \binom{n}{i} (-\sqrt{5})^i$$

$$= 2 \sum_{j=0}^{\lfloor n/2 \rfloor} \binom{n}{2j} 5^j.$$

This implies $2^{n-1} L_n = 1 + 5m$ for some integer m. Since $L_n \equiv 0 \pmod 5$, this
yields $1 \equiv 0 \pmod 5$, which is a contradiction. Thus *no Lucas number is divisible
by 5.* Consequently, *no Lucas numbers end in zero.*

23.3   ADDITIONAL CONGRUENCES

We now turn to some additional Fibonacci and Lucas congruences. We will prove
a few of them, and leave the others as routine exercises; see Exercises 23.20–23.22.

Theorem 23.1.

1) $L_n \equiv 0 \pmod 2$ if and only if $n \equiv 0 \pmod 3$. (23.1)

2) $L_n \equiv 0 \pmod 3$ if and only if $n \equiv 2 \pmod 4$. (23.2)

3) If $2 | k$ and $3 \nmid k$, then $L_n \equiv 3 \pmod 4$. (23.3)

4) $L_{n+2k} \equiv -L_n \pmod{L_k}$, where $2 | k$ and $3 \nmid k$. (23.4)

5) $F_{n+2k} \equiv -F_n \pmod{L_k}$, where $2 | k$ and $3 \nmid k$. (23.5)

6) $L_{n+12} \equiv L_n \pmod 8$. (23.6)

Proof. We will prove congruences (23.1), (23.5), and (23.6).

1) By Exercise 5.37, we have $5F_n^2 = L_n^2 - 4(-1)^n$. Suppose $L_n \equiv 0 \pmod 2$. Then
$5F_n^2 \equiv 0 \pmod 2$, so $F_n^2 \equiv 0 \pmod 2$. Then $F_n \equiv 0 \pmod 2$; that is, $F_n \equiv 0 \pmod{F_3}$. Therefore, $n \equiv 0 \pmod 3$.
   Conversely, let $n \equiv 0 \pmod 3$. Then $F_n \equiv 0 \pmod{F_3}$; that is, $F_n \equiv 0 \pmod 2$. This implies $L_n^2 - 4(-1)^n \equiv 0 \pmod 2$, so $L_n \equiv 0 \pmod 2$.
   Thus $L_n \equiv 0 \pmod 2$ if and only if $n \equiv 0 \pmod 3$.

5) Since $2F_{m+n} = F_m L_n + F_n L_m$ by Corollary 20.4, and k is even, we have

$$2F_{n+2k} = F_n L_{2k} + F_{2k} L_n$$
$$= F_n[L_k^2 + 2(-1)^{k-1}] + F_k L_k L_n$$
$$\equiv 2(-1)^{k-1} F_n + 0 \pmod{L_k}$$
$$\equiv -2F_n \pmod{L_k}.$$

Since $k \not\equiv 0 \pmod 3$, L_k is odd; so $(2, L_k) = 1$. Thus $F_{n+2k} \equiv -F_n \pmod{L_k}$.

6) Again, by Corollary 20.4 we have $2L_{m+n} = 5F_m F_n + L_m L_n$. Consequently,

$$2L_{n+12} = 5F_n F_{12} + L_n L_{12}$$
$$= 5 \cdot 144F_n + 322L_n$$
$$L_{n+12} = 360F_n + 161L_n$$
$$\equiv 0 + L_n \pmod 8$$
$$\equiv L_n \pmod 8. \qquad \blacksquare$$

For example, let $n = 15$ and $k = 8$. Clearly, k is even, and $3 \nmid k$. Then $n + 2k = 31$, and $L_k = L_8 = 47$. We have

$$L_{31} = 3,010,349 \equiv 46 \equiv -1364 = -L_{15} \pmod{47}$$
$$F_{31} = 1,346,269 \equiv 1 \equiv -610 = -F_{15} \pmod{47}.$$

Property (23.3) can be stated in words: *If n is of the form $6k \pm 2$, then $L_n \equiv 3$ (mod 4).* For example, let $n = 6 \cdot 3 + 2$. Then $L_{20} = 15,127 \equiv 3 \pmod 4$. Likewise, $L_{16} = 2207 \equiv 3 \pmod 4$.

Are there Lucas squares? We will now search for them.

23.4 LUCAS SQUARES

Clearly, $L_1 = 1$ and $L_3 = 4$ are squares. Are there others? In April 1964, J.H.E. Cohn of the University of London, England, established that there are *no* other Lucas squares [126]. His proof hinges on properties (23.3) and (23.4), a knowledge of quadratic residues [369], and the identity $L_{2n} = 2L_n^2 + 2(-1)^{n-1}$.

Theorem 23.2 (Cohn, 1964 [126]).   *The only Lucas numbers that are squares are 1 and 4.*

Proof. Suppose $L_n = x^2$ for some positive integer n.

Suppose n is even, say, $n = 2r$. Then $L_n = L_{2r} = L_r^2 \pm 2$. Since L_r^2 is a square, $L_r^2 \pm 2$ cannot be a square. This is a contradiction.

On the other hand, suppose n is odd. Let $n \equiv 1 \pmod 4$. Since L_1 is a square, assume $n > 1$. Then we can write n as $n = 1 + 2 \cdot 3^i k$, where $i \geq 0$, $2|k$, and $3 \nmid k$. Therefore, by property (23.4), $L_n \equiv -L_1 \equiv -1 \pmod{L_k}$. Since -1 is a quadratic nonresidue of L_k by property (23.3), it follows that L_n *cannot* be a square.

On the other hand, let $n \equiv 3 \pmod 4$. Since L_3 is a square, assume $n > 3$. Then $n = 3 + 2 \cdot 3^i k$, where $i \geq 0$, and $2|k$, and $3 \nmid k$. Then, by property (23.4), $L_n \equiv -L_3 \equiv -4 \pmod{L_k}$. Then also L_n *cannot* be a square.

Thus the only Lucas squares are 1 and 4.   ∎

The next theorem, also discovered by Cohn in 1964, identifies Lucas numbers of the form $2x^2$. Its proof employs properties (23.1), (23.2), and (23.6); and the basic facts $F_{-n} = (-1)^{n-1} F_n$ and $L_{-n} = (-1)^n L_n$ from Chapter 5.

Theorem 23.3 (Cohn, 1964 [126]).   *Let $L_n = 2x^2$ for some integer x. Then $n = 0$ or ± 6.*

Proof. Since $x^2 \equiv 0, 1$, or 4 $\pmod 8$, $L_n = 2x^2 \equiv 0$ or 2 $\pmod 8$. Since L_n is even, by property (23.1), $3|n$.

Suppose n is odd. Then n is of the form $12q + r$, where $0 \leq r < 12$. Since n is odd and $3|n$, $r = 3$ or 9. So n must be of the form $12q + 3$ or $12q + 9$. If $n = 12q + 3$, then by property (23.6), $L_n = L_{12q+3} \equiv L_3 \equiv 4 \pmod 8$. This is a contradiction, since $L_n \equiv 0$ or 2 $\pmod 8$.

If $n = 12q + 9$, $L_n = L_{12q+9} \equiv L_9 \equiv 4 \pmod 8$. Again, this is a contradiction. Thus n cannot be odd.

Suppose n is even. Then n is of the form $8t, 8t \pm 2$ or $8t + 4$. If $n = 8t$ or $8t + 4$, then $n \equiv 0 \pmod 4$. If $n = 0$, then $L_n = L_0 = 2 = 2 \cdot 1^2$ has the desired from. If $n \neq 0$, then $n = 2 \cdot 3^i \cdot k$. Then $2L_n \equiv -2L_0 \equiv -4 \pmod{L_k}$. Therefore, $2L_n$ cannot be a square y^2. Consequently, L_n cannot be of the form $2x^2$.

Suppose $n \equiv -2 \equiv 6 \pmod 8$. If $n = 6$, then $L_6 = 2 \cdot 3^2$ has the desired property. On the other hand, if $n \neq 6$, then n is of the form $n = 6 + 2 \cdot 3^i \cdot k$, where $4|k$ and $3 \nmid k$. Therefore, $2L_n \equiv -2L_6 \equiv 36 \pmod{L_k}$. By properties (23.2) and (23.3), this implies that -36 is a quadratic nonresidue of L_k. As before, this implies that L_n cannot be of the form $2x^2$.

Finally, let $n \equiv 2 \pmod 8$. Then $L_{-n} = L_n$. Therefore, $-n \equiv 6 \pmod 8$. This yields $-n = 6$ or $n = -6$.

Thus, if $L_n = 2x^2$, then $n = 0$ or ± 6, as desired.   ∎

23.5   FIBONACCI SQUARES

Historically, one of the oldest conjectures in the theory of Fibonacci numbers is that 0, 1, and 144 are the only Fibonacci numbers that are also squares. In 1963, M. Wunderlich conducted an extensive computer search among the first one million Fibonacci numbers, but did not find any new ones [611]. Surprisingly, Cohn confirmed the conjecture in the following year, as the next theorem shows [126].

Theorem 23.4 (Cohn, 1964 [126]). *If F_n is a square, then $n = 0, \pm 1, 2,$ or 12.*

Proof.
Case 1. Let n be odd. Then $n \equiv \pm 1$ (mod 4).

Suppose $n \equiv 1$ (mod 4). If $n = 1$, then $F_1 = 1 = 1^2$, a square. If $n \neq 1$, n must be of the form $n = 1 + 2 \cdot 3^i \cdot k$. By property (23.5), then $F_n \equiv -F_1 = -1$ (mod L_k). Consequently, F_n *cannot* be a square.

On the other hand, suppose $n \equiv -1$ (mod 4); that is, $-n \equiv 1$ (mod 4). Then $F_{-n} = F_n$. By the above subcase, $-n = 1$; that is, $n = -1$.

Case 2. Let n be even, say, $n = 2s$. Then $F_n = F_{2s} = F_s L_s = x^2$ for some integer x.

Suppose $3|n$. Then $F_3|F_n$; that is, $2|F_n$. So $F_s = 2y^2$ and $L_s = 2z^2$ for some integers y and z. Then, by Theorem 23.3, $n/2 = s = 0$ or ± 6; that is, $n = 0$ or ± 12.

When $n = 0$, $F_s = F_0 = 0 = 2 \cdot 0^2$; when $n = 12$, $F_s = F_6 = 8 = 2 \cdot 2^2$; but when $n = -12$, $F_s = F_{-6} = (-1)^5 F_6 = -8$, which is *not* of the form $2y^2$. Therefore, $n = 0$ or 12.

Suppose $3 \nmid n$; so F_n is not even. Then $F_s = y^2$ and $L_s = z^2$ for some integers y and z. By Theorem 23.2, $n/2 = s = 1$ or 3; so $n = 2$ or 6.

Finally, when $n = 2$, $F_s = F_1 = 1^2$; but when $n = 6$, $F_s = F_3 = 2$ is *not* a square. Thus, if F_n is a square, then $n = 0, \pm 1, 2,$ or 12, as claimed. ∎

It follows by this theorem that the only distinct positive Fibonacci squares are 1 and 144.

Cohn also established that 0, 2, and 8 are the only Fibonacci numbers that are of the form $2x^2$. This is the essence of the next theorem.

Theorem 23.5 (Cohn, 1964 [126]). *If F_n is of the form $2x^2$, then $n = 0, \pm 3,$ or 6.*

Proof.
Case 1. Let n be odd. Then $n \equiv \pm 1$ (mod 4).

Suppose $n \equiv -1 \equiv 3$ (mod 4). If $n = 3$, then $F_n = F_3 = 2 = 2 \cdot 1^2$ has the desired form. If $n \neq 3$, then n is of the form $n = 3 + 2 \cdot 3^i \cdot k$. By property (23.5), then $2F_n \equiv -2F_3 \equiv 4$ (mod L_k); that is, $x^2 \equiv -1$ (mod L_k). So F_n *cannot* be of the given form.

Suppose $n \equiv 1 \equiv -3$ (mod 4). Then $-n \equiv 3$ (mod 4) and $F_{-n} = F_n$, so $-n = 3$ by the preceding paragraph. This yields $n = -3$.

Case 2. Let n be even, say, $n = 2s$. Then $F_n = F_{2s} = F_s L_s = 2x^2$. So either $F_s = y^2$ and $L_s = 2z^2$; or $F_s = 2y^2$ and $L_s = z^2$ for some integers y and z.

By Theorems 23.3 and 23.4, the only value of s that satisfies the equations $F_s = y^2$ and $L_s = 2z^2$ is $s = 0$. Then $n = 0$.

Now consider the case $F_s = y^2$ and $L_s = 2z^2$. By Theorem 23.2, $s = 1$ or 3. But F_1 is *not* of the form $2y^2$, so $s \neq 1$. But $F_3 = 2 \cdot 1^2$, so $F_s = 2y^2$ is solvable when $s = 3$. Thus $n = 6$.

Thus, collecting all possible values of n, we get $n = 0, \pm 3,$ or 6. ∎

The next example deals with an interesting congruence, studied by C.R. Wall in 1964 [576]. The proof involves a fourth-order linear recurrence and PMI, and exemplifies a useful proof technique.

Example 23.1. *Prove that $nL_n \equiv F_n$ (mod 5).*

Proof. Let $a_n = nL_n - F_n$. It is easy to confirm that the congruence is true when $n = 0, 1, 2$, and 3; that is, $a_0 = a_1 = a_2 = a_3 \equiv 0$ (mod 5).

Suppose it works for four consecutive values of $n \geq 0$. We will now find a recurrence for a_n. To this end, we let

$$A = 2(n+3)L_{n+3} + (n+2)L_{n+2} - 2(n+1)L_{n+1} - nL_n.$$

Then

$$A = 2(n+3)L_{n+3} + (n+2)L_{n+2} - (n+2)L_{n+1} - n(L_{n+1} + L_n)$$

$$= 2(n+3)L_{n+3} + (n+2)L_{n+2} - (n+2)L_{n+1} - nL_{n+2}$$

$$= 2(n+3)L_{n+3} + (n+2)L_n - nL_{n+2}$$

$$= 2(n+3)L_{n+3} + (n+2)(L_{n+2} - L_{n+1}) - nL_{n+2}$$

$$= 2(n+3)L_{n+3} + 2L_{n+2} - (n+2)L_{n+1}$$

$$= 2(n+3)L_{n+3} + 2L_{n+2} - (n+2)(L_{n+3} - L_{n+2})$$

$$= 2(n+3)L_{n+3} - (n+2)L_{n+3} + (n+4)L_{n+2}$$

$$= (n+4)(L_{n+3} + L_{n+2})$$

$$= (n+4)L_{n+4}.$$

Let $B = 2F_{n+3} + F_{n+2} - 2F_{n+1} - F_n$. Then $B = 2F_{n+3} - F_{n+1} = F_{n+4}$.

Thus $A - B = (n+4)L_{n+4} - F_{n+4} = a_{n+4}$. In other words, a_n satisfies the fourth-order recurrence $a_{n+4} = 2a_{n+3} + a_{n+2} - 2a_{n+1} - a_n$. Then, by the initial conditions, $a_{n+4} \equiv 2 \cdot 0 + 0 - 2 \cdot 0 - 0 \equiv 0$ (mod 5).

Consequently, the congruence is true for all values of $n \geq 0$, as desired. ∎

For example, $5L_5 = 5 \cdot 11 = 55 \equiv 0 \equiv F_5$ (mod 5). Likewise, $13L_{13} = 6773 \equiv 3 \equiv 233 \equiv F_{13}$ (mod 5).

23.6 A GENERALIZED FIBONACCI CONGRUENCE

In 1963, J.A. Maxwell studied the following interesting problem [432]:

Generalize the congruences

$$F_{n+1}2^n + F_n 2^{n+1} \equiv 1 \quad (\text{mod } 5), \qquad F_{n+1}3^n + F_n 3^{n+1} \equiv 1 \quad (\text{mod } 11),$$

$$F_{n+1}5^n + F_n 5^{n+1} \equiv 1 \quad (\text{mod } 29),$$

where $n \geq 0$.

Their generalization, although not quite obvious, is

$$F_{n+1}p^n + F_n p^{n+1} \equiv 1 \pmod{p^2 + p - 1}, \tag{23.7}$$

where p is a prime.

Interestingly, we can generalize even this congruence to include gibonacci numbers G_n, as the next theorem demonstrates. We will prove it using the strong version of PMI.

Theorem 23.6 (Koshy, 1999 [365]). *Let $m \geq 2$ and $n \geq 0$. Then*

$$G_{n+1}m^n + G_n m^{n+1} \equiv a(1 - m) + bm \pmod{m^2 + m - 1}. \tag{23.8}$$

Proof. Since $G_1 m^0 + G_0 m = a + (b - a)m = a(1 - m) + bm \equiv a(1 - m) + bm$ (mod $m^2 + m - 1$) and $G_2 m + G_1 m^2 = bm + am^2 \equiv a(1 - m) + bm$ (mod $m^2 + m - 1$), the statement is true when $n = 0$ and $n = 1$.

Now assume it is true for all integers i, where $0 \leq i \leq k$, where $k \geq 1$. Then

$$G_k m^{k-1} + G_{k-1}m^k \equiv a(1 - m) + bm \pmod{m^2 + m - 1}$$

$$G_{k+1}m^k + G_k m^{k+1} \equiv a(1 - m) + bm \pmod{m^2 + m - 1}$$

$$\begin{aligned}
G_{k+2}m^k + G_{k+1}m^{k+1} &= (G_k + G_{k+1})m^{k+1} + (G_{k-1} + G_k)m^{k+2} \\
&= m(G_{k+1}m^k + G_k m^{k+1}) + m^2(G_k m^{k-1} + G_{k-1}m^k) \\
&\equiv m[a(1 - m) + bm] + m^2[a(1 - m) + bm] \pmod{m^2 + m - 1} \\
&\equiv (m + m^2)[a(1 - m) + bm] \pmod{m^2 + m - 1} \\
&\equiv 1 \cdot [a(1 - m) + bm] \pmod{m^2 + m - 1} \\
&\equiv a(1 - m) + bm \pmod{m^2 + m - 1}.
\end{aligned}$$

Thus congruence (23.8) works for all integers $n \geq 0$. ∎

The next corollary follows trivially from this theorem.

Corollary 23.1. *Let $m \geq 2$ and $n \geq 0$. Then*

$$F_{n+1}m^n + F_n m^{n+1} \equiv 1 \pmod{m^2 + m - 1} \tag{23.9}$$

$$L_{n+1}m^n + L_n m^{n+1} \equiv 1 + 2m \pmod{m^2 + m - 1}. \tag{23.10}$$

∎

For example, let $m = 15$ and $n = 12$; so $m^2 + m - 1 = 239$. Then

$$\begin{aligned}
F_{13}15^{12} + F_{12}15^{13} &= 233 \cdot 15^{12} + 144 \cdot 15^{13} \\
&\equiv 233 \cdot 80 + 144 \cdot 5 \pmod{239} \\
&\equiv 1 \pmod{239};
\end{aligned}$$

$$L_{17}7^{16} + L_{16}7^{17} = 3571 \cdot 7^{16} + 2207 \cdot 7^{17}$$

$$\equiv 3571 \cdot 26 + 2207 \cdot 17 \pmod{55}$$

$$\equiv 15 \pmod{55}$$

$$\equiv 1 + 2 \cdot 7 \pmod{55}.$$

Fortunately, congruence (23.8) can be extended to negative subscripts. To establish this generalization, we need the following lemma, which we will establish using strong induction.

Lemma 23.1 (Koshy, 1999 [365]). *Let $m \geq 2$ and $n \geq 0$. Then*

$$F_{n-1} - mF_n \equiv (-1)^n m^n \pmod{m^2 + m - 1}. \tag{23.11}$$

Proof. Since $F_{-1} - mF_0 = 1 - 0 = 1 \equiv (-1)^0 m^0 \pmod{m^2 + m - 1}$ and $F_0 - mF_1 = 0 - m \equiv (-1)^1 m^1 \pmod{m^2 + m - 1}$, the result is true when $n = 0$ and $n = 1$.

Now assume it is true for every nonnegative integer $\leq k$, where $k \geq 2$. Then

$$F_k - mF_{k+1} = (F_{k-2} + F_{k-1}) - m(F_{k-1} + F_k)$$

$$= (F_{k-2} - mF_{k-1}) + (F_{k-1} - mF_k)$$

$$\equiv (-1)^{k-1} m^{k-1} + (-1)^k m^k \pmod{m^2 + m - 1}$$

$$\equiv (-1)^{k-1} m^{k-1}(1 - m) \pmod{m^2 + m - 1}$$

$$\equiv (-1)^{k-1} m^{k-1} m^2 \pmod{m^2 + m - 1}$$

$$\equiv (-1)^{k+1} m^{k+1} \pmod{m^2 + m - 1}.$$

Thus, by the strong version of PMI, congruence (23.11) works for all $n \geq 0$. ∎

For instance, $F_{12} - 5F_{13} = 144 - 5 \cdot 233 \equiv 23 \equiv (-1)^{13} 5^{13} \pmod{29}$.

Curiously enough, congruence (23.11) does not hold for Lucas numbers; that is, $L_{n-1} - mL_n \not\equiv (-1)^n m^n \pmod{m^2 + m - 1}$. For example, $L_9 - 6L_{10} = 76 - 6 \cdot 123 \equiv 35 \pmod{41}$, whereas $(-1)10 6^{10} \equiv 32 \pmod{41}$.

This lemma has an interesting byproduct, as the next corollary reveals.

Corollary 23.2. $L_n \equiv (-1)^n 2^{n+1} \equiv 2 \cdot 3^n \pmod{5}$.

Proof. By Lemma 23.1, $F_n - 2F_{n+1} \equiv (-1)^{n+1} 2^{n+1} \pmod{5}$. But $2F_{n+1} - F_n = F_{n+1} + F_{n-1} = L_n$. Consequently, $-L_n \equiv (-1)^{n+1} 2^{n+1} \pmod{5}$. This yields the given result. ∎

For example, $L_{10} = 123 \equiv 3 \equiv (-1)^{11} 2^{11} \pmod{5}$.

The next corollary follows from Corollary 23.2, since $3^4 \equiv 1 \pmod{5}$.

Corollary 23.3. $L_{4n} \equiv 2 \pmod{5}$, $L_{4n+1} \equiv 1 \pmod{5}$, $L_{4n+2} \equiv 3 \pmod{5}$, *and* $L_{4n+3} \equiv 4 \pmod{5}$. ∎

For example, $L_{12} = 322 \equiv 2 \pmod{5}$, $L_{13} = 521 \equiv 1 \pmod{5}$, $L_{14} = 843 \equiv 3 \pmod{5}$, and $L_{15} = 1364 \equiv 4 \pmod{5}$.

Using Lemma 23.1 and the fact that $G_{-n} = (-1)^{n+1}(aF_{n+2} - bF_{n+1})$, we can now generalize Theorem 23.6, as the next theorem shows.

Theorem 23.7 (Koshy, 1999 [365]). *Let* $m \geq 2$ *and n any integer. Then*

$$G_{n+1}m^n + G_n m^{n+1} \equiv a(1 - m) + bm \pmod{m^2 + m - 1}.$$

Proof. In light of Theorem 23.6, it suffices to show that congruence (23.11) holds when n is negative. Let $M = m^2 + m - 1$. We then have

$$G_{-n+1}m^{-n} + G_{-n}m^{-n+1}$$

$$= m^{-n}G_{-(n-1)} + m^{-n+1}G_{-n}$$

$$= m^{-n}[(-1)^n(aF_{n+1} - bF_n)] + m^{-n+1}[(-1)^{n+1}(aF_{n+2} - bF_{n+1})]$$

$$= a(-1)^n m^{-n}(F_{n+1} - mF_{n+2}) - b(-1)^n m^{-n}(F_n - mF_{n+1})$$

$$\equiv a(-1)^n m^{-n}[(-1)^{n+2}m^{n+2}] - b(-1)^n m^{-n}[(-1)^{n+1}m^{n+1}] \pmod{M}$$

$$\equiv (am^2 + bm) \pmod{M}$$

$$\equiv a(1 - m) + bm \pmod{M},$$

as desired. ∎

For example, let $m = 6$ and $n = -8$. Then

$$L_{-7}6^{-8} + L_{-8}6^{-7} = (-29) \cdot 6^{-8} + 47 \cdot 6^{-7}$$

$$\equiv 12 \cdot 6^{-8} + 6 \cdot 6^{-7} \pmod{41}$$

$$\equiv 12 \cdot 6^{-8} + 6^{-6} \pmod{41}$$

$$\equiv (2 \cdot 6^{-1} + 1)6^{-6} \pmod{41}$$

$$\equiv (2 \cdot 7 + 1)7^6 \pmod{41}$$

$$\equiv 15 \cdot 20 \equiv 13 \pmod{41}$$

$$\equiv 1 + 2 \cdot 6 \pmod{41}.$$

Likewise, $F_{-7}6^{-8} + F_{-8}6^{-7} = 13 \cdot 5^8 + (-21) \cdot 5^7 \equiv 1 \pmod{29}$.

The next theorem shows that every prime p divides some Fibonacci number. Its proof employs the number-theoretic Legendre symbol, so we omit the proof [257].

Theorem 23.8. *Let p be a prime. Then $F_{p-1} \equiv 0$ (mod p) if $p \equiv \pm 1$ (mod 5); and $F_{p+1} \equiv 0$ (mod p) if $p \equiv \pm 2$ (mod 5).*   ∎

For example, let $p = 19 \equiv -1$ (mod 5). Then $F_{p-1} = F_{18} = 2584 \equiv 0$ (mod 19). Likewise, let $p = 23 \equiv -2$ (mod 5); then $F_{p+1} = F_{24} = 46{,}368 \equiv 0$ (mod 23).

Martin Pettet of Toronto discovered the following result in 1966 [482]. We will need it in the proof of the next theorem, so we label it a lemma. J.E. Desmond of Florida State University discovered the theorem in 1970 [145].

Lemma 23.2 (Pettet, 1966 [482]). *Let p be a prime. Then $L_p \equiv 1$ (mod p).*

Proof. Recall by the binomial theorem that

$$L_p = \frac{1}{2^{p-1}} \sum_{i=0}^{\lfloor p/2 \rfloor} \binom{p}{2i} 5^i.$$

Since p is a prime, $\binom{p}{j} \equiv 0$ (mod p), where $0 < j < p$ [369]. Therefore, $L_p \equiv \frac{1}{2^{p-1}}$ (mod p). But, by Fermat's little theorem, $2^{p-1} \equiv 1$ (mod p), so $L_p \equiv 1$ (mod p).   ∎

For example, $L_{43} = 969{,}323{,}029 \equiv 1$ (mod 43), as expected.

Unfortunately, the converse of the lemma fails: $L_n \equiv 1$ (mod n) does *not* imply the primality of n. Pettet found that 705, 2465, and 2737 are counterexamples [482].

We are now ready for the theorem. We will confirm it with PMI.

Theorem 23.9 (Desmond, 1970 [145]). *Let p be a prime and n any nonnegative integer. Then $F_{np} \equiv F_n F_p$ (mod p) and $L_{np} \equiv L_n L_p \equiv L_n$ (mod p).*

Proof. We will prove the first result; and leave the second as an exercise; see Exercise 23.34.

The statement is clearly true when $n = 0$ and $n = 1$. Now assume it is true for all nonnegative integers $n \leq k$, where $k \geq 1$.

Since $F_{r+s} = F_r L_s + (-1)^{s+1} F_{r-s}$, $F_{np+p} = F_{np} L_p + (-1)^{p+1} F_{np-p}$. Then, by Lemma 23.1, $F_{(n+1)p} \equiv F_{np} + F_{(n-1)p}$ (mod p). Consequently,

$$
\begin{aligned}
F_{(k+1)p} &\equiv F_{kp} + F_{(k-1)p} \quad (\text{mod } p) \\
&\equiv F_k F_p + F_{k-1} F_p \quad (\text{mod } p) \\
&\equiv (F_k + F_{k-1}) F_p \quad (\text{mod } p) \\
&\equiv F_{k+1} F_p \quad (\text{mod } p).
\end{aligned}
$$

Thus $F_{np} \equiv F_n F_p$ (mod p) for every prime and every integer $n \geq 0$.   ∎

For example, let $n = 4$ and $p = 7$. Then $F_n = F_4 = 3$, $F_p = F_7 = 13$, and $F_{np} = F_{28} = 317,811 \equiv 4 \equiv 3 \cdot 13 \equiv F_4 F_7 \pmod{7}$. Likewise, $L_{28} = 710,647 \equiv 7 \cdot 29 \equiv L_4 L_7 \pmod{7}$.

Carlitz discovered the next Lucas congruence in 1977 [117].

Theorem 23.10 (Carlitz, 1977 [117]). *Let p be a prime. Then $L_{p^2} \equiv 1 \pmod{p^2}$ if and only if $L_p \equiv 1 \pmod{p^2}$.*

Proof. Since $1 = (\alpha + \beta)^n = L_n + \sum_{k=1}^{n-1} \binom{n}{k} \alpha^k \beta^{n-k}$, we have

$$L_p = 1 - \sum_{k=1}^{p-1} \binom{p}{k} \alpha^k \beta^{p-k}$$

$$L_{p^2} = 1 - \sum_{k=1}^{p^2-1} \binom{p^2}{k} \alpha^k \beta^{p^2-k}.$$

Since $\binom{p}{k} = \frac{p}{k}\binom{p-1}{k-1}$ and $\binom{p-1}{k-1} \equiv (-1)^{k-1} \pmod{p}$, it follows that $L_{p^2} \equiv 1$ (mod p^2) if and only if

$$\sum_{k=1}^{p-1} \frac{(-1)^{k-1}}{k} \alpha^k \beta^{p-k} \equiv 0 \pmod{p}.$$

Since $p \nmid k$, $\binom{p^2}{k} \equiv 0 \pmod{p^2}$, and $\binom{p^2}{pk} = \frac{p}{k}\binom{p^2-1}{pk-1} \equiv (-1)^{k-1}\frac{p}{k} \pmod{p^2}$. Thus $L_{p^2} \equiv 1 \pmod{p^2}$ if and only if

$$\sum_{k=1}^{p-1} \frac{(-1)^{k-1}}{k} \alpha^{pk} \beta^{p^2-pk} \equiv 0 \pmod{p}.$$

Since

$$\left(\sum_{k=1}^{p-1} \frac{(-1)^{k-1}}{k} \alpha^k \beta^{p-k} \right)^p \equiv \sum_{k=1}^{p-1} \frac{(-1)^{k-1}}{k} \alpha^{pk} \beta^{p^2-k} \equiv 0 \pmod{p},$$

it follows that $L_{p^2} \equiv 1 \pmod{p^2}$ if and only if $\sum_{k=1}^{p-1} \frac{(-1)^{k-1}}{k} \alpha^k \beta^{p-k} \equiv 0 \pmod{p}$.

Thus $L_{p^2} \equiv 1 \pmod{p^2}$ if and only if $L_p \equiv 1 \pmod{p^2}$. ∎

More generally, we have the following result, also due to Carlitz [117].

Theorem 23.11 (Carlitz, 1977 [117]).   *Let $m \geq 2$ and p a prime. Then $L_{pm} \equiv 1$ (mod p^2) if and only if $L_p \equiv 1$ (mod p^2).*   ∎

The next example is interesting in its own right. G. Berzsenyi of Lamar University, Beaumont, Texas, studied it in 1976 [38].

Example 23.2.   *Prove that $L_{2mn}^2 \equiv 4$ (mod L_m^2), where $m, n \geq 1$.*

Proof. Since $L_{(2k+1)m} = (\alpha^m)^{2k+1} + (\beta^m)^{2k+1}$, it follows that $L_m | L_{(2k+1)m}$.

Case 1. Let $n = 2k + 1$ be odd. Then, by Exercise 5.39,

$$L_{2nm} = L_{nm}^2 - 2(-1)^{nm} = L_{(2k+1)m}^2 - 2(-1)^m.$$

So

$$L_{(2k+1)m}^2 = \begin{cases} L_{2nm} - 2 & \text{if } m \text{ is odd} \\ L_{2nm} + 2 & \text{otherwise.} \end{cases}$$

Consequently, $(L_{2nm} - 2)(L_{2nm} + 2) = L_{2mn}^2 - 4$ is divisible by $L_{(2k+1)m}^2$, and hence by L_m^2. Thus $L_{2mn}^2 \equiv 4$ (mod L_m^2), where m is odd.

Case 2. Let $n = 2k$ be even. We will prove this half by PMI on k. When $k = 1$,

$$L_{2nm} = L_{4m} = L_{2m}^2 - 2$$
$$= [L_m^2 - 2(-1)^m]^2 - 2$$
$$= L_m^4 - 4(-1)^m L_m^2 + 2.$$

Thus $L_{2nm} \equiv 2$ (mod L_m^2), so $L_{2mn}^2 \equiv 4$ (mod L_m^2).
Assume the result holds for all positive even integers less than $n = 2k$. Then

$$L_{2nm} = L_{4km} = L_{2km}^2 - 2$$
$$L_{2nm} - 2 = L_{2km}^2 - 4.$$

By Case 1 or the inductive hypothesis, $(L_{2nm} + 2)(L_{2nm} - 2)$ is divisible by L_m^2; that is, $L_{2km}^2 \equiv 4$ (mod L_m^2). Thus the congruence is true by PMI when m is even also.

Thus the result holds in both cases.   ∎

For example, let $m = 2$ and $n = 3$. Then $L_{2mn}^2 = L_{12}^2 = 322^2 \equiv 3$ (mod 4^2) $\equiv$ (mod L_3^2).

In 1965, D. Lind of Falls Church, Virginia, proposed an intriguing problem [392] in *The Fibonacci Quarterly*, that links Fibonacci numbers with *Euler's phi function φ*; $\varphi(n)$ denotes the number of positive integers $\leq n$ and relatively prime to it. For example, $\varphi(1) = 1, \varphi(5) = 4$, and $\varphi(6) = 2$. An incomplete proof by J.L.

Brown of Pennsylvania State University appeared in the *Quarterly* in the following year [76]. It resurfaced as a problem in 1976 by C. Kimberling of the University of Evansville, Indiana, in *The American Mathematical Monthly* [357]. In the following year, P.L. Montgomery of Huntsville, Alabama, provided an elegant solution using group theory [444]. Three years later, Hoggatt and H. Edgar of San Jose State University gave an alternate proof [312]. We now state it as a theorem.

Theorem 23.12. $\varphi(F_n) \equiv 0$ (mod 4), *where* $n \geq 5$. ∎

Notice that $\varphi(F_1) = \varphi(F_2) = \varphi(F_3) \equiv 1$ (mod 4), and $\varphi(F_4) \equiv 2$ (mod 4). But $\varphi(F_5) = \varphi(5) = 4 \equiv 0$ (mod 4); $\varphi(F_{10}) = \varphi(55) = \varphi(5 \cdot 11) = \varphi(5) \cdot \varphi(11) = 4 \cdot 10 \equiv 0$ (mod 4), and $\varphi(F_{17}) = \varphi(1597) = 1596 \equiv 0$ (mod 4).

We now turn our focus on the periodicities of Fibonacci and Lucas sequences.

23.7 FIBONACCI AND LUCAS PERIODICITIES

A cursory examination of the units digits of the Fibonacci numbers F_0 through F_{59} reveals no obvious or interesting pattern. But take a good look at the units digits in F_{60} and F_{61}; they are the same as those of F_0 and F_1, respectively: $F_{60} \equiv F_0$ (mod 10) and $F_{61} \equiv F_1$ (mod 10). Consequently, by virtue of the Fibonacci recurrence, the pattern continues: $F_{60+i} \equiv F_i$ (mod 10).

More generally, we have the following result. We will establish this using PMI.

Theorem 23.13. *Let* $n, i \geq 0$. *Then* $F_{60n+i} \equiv F_i$ (mod 10).

Proof. We just confirmed that the statement is true when $n = 0$ and $n = 1$.

Now assume it is true for an arbitrary integer $k \geq 0$: $F_{60k+i} \equiv F_i$ (mod 10). Then, by the Fibonacci addition formula, we have

$$
\begin{aligned}
F_{60(k+1)+i} &= F_{(60k+i)+60} \\
&= F_{60k+i+1}F_{60} + F_{60k+i}F_{59} \\
&= F_{60k+i+1} \cdot 0 + F_i \cdot 1 \quad \text{(mod 10)} \\
&= F_i \quad \text{(mod 10)}.
\end{aligned}
$$

Thus, by PMI, the statement is true for all nonnegative integers n. ∎

For example, $F_{14} = 377 \equiv 7$ (mod 10), so $F_{74} \equiv F_{14} \equiv 7$ (mod 10). To confirm this, notice that $F_{74} = 1,304,969,544,928,657$ ends in 7, as F_{14} does.

Let p be the smallest positive integer such that $F_{p+i} \equiv F_i$ (mod 10) for every integer i. Then p is the *period* of the Fibonacci sequence modulo 10. By virtue of Theorem 23.13, $p \leq 60$. But when we examine the units digits in F_0 through F_{59}, we see no repetitive pattern; so $p \geq 60$. Thus $p = 60$; that is, *the period of the Fibonacci sequence modulo 10 is 60.*

In 1963, using an extensive computer search, S.P. Geller of the University of Alaska established that the last two digits of F_n repeat every 300 Fibonacci numbers; the last three every 1500; the last four every 15,000; the last five every 150,000; and the last six digits every 1,500,000 Fibonacci numbers [216]:

$$F_{n+300} \equiv F_n \pmod{300} \qquad F_{n+1500} \equiv F_n \pmod{1500}$$

$$F_{n+15000} \equiv F_n \pmod{15,000} \qquad F_{n+150000} \equiv F_n \pmod{150,000}$$

$$F_{n+1500000} \equiv F_n \pmod{1,500,000}.$$

PERIODICITY OF THE LUCAS SEQUENCE

Is the Lucas sequence modulo 10 also periodic? If yes, what is its period? To answer these questions, consider the units digits in L_0 through L_{11}; they are 2, 1, 3, 4, 7, 1, 8, 9, 7, 6, 3, and 9. There is no pattern thus far, so the period is at least 12. But, by virtue of the Lucas recurrence, we can compute every residue modulo 10 by the sum of the two previous residues modulo 10. So the next two units digits are 2 and 1. Clearly, a pattern begins to emerge. Thus *the Lucas sequence modulo 10 is also periodic; its period is 12*; that is, $L_{12+i} \equiv L_i \pmod{10}$.

More generally, we have the following result. We leave its proof as a simple exercise; see Exercise 23.59.

Theorem 23.14. *Let $n, i \geq 0$. Then $L_{12n+i} \equiv L_i \pmod{10}$.*   ∎

For example, $L_{11} = 199 \equiv 9 \pmod{10}$; so $L_{47} \equiv 9 \pmod{10}$. This is true since $L_{47} = 6,643,838,879$ ends in 9.

23.8 LUCAS SQUARES REVISITED

In 1964, Br. Alfred published a neat and simple proof of the fact that 1 and 4 are the only Lucas numbers that are also squares [7]. His proof hinges on the periodicity of the Lucas sequence modulo 8.

Since $L_0 = 2 \pmod 8$ and $L_1 = 1 \pmod 8$, it follows that $L_i \equiv L_{i-1} + L_{i-2} \pmod 8$, where $i \geq 2$. For example, $L_8 = 7 \pmod 8$ and $L_9 = 4 \pmod 8$; so $L_{10} = 7 + 4 \equiv 3 \pmod 8$. Table 23.1 shows the residues of the Lucas numbers L_n, where $0 \leq n \leq 11$.

TABLE 23.1.   Lucas Numbers Modulo 8

n	0	1	2	3	4	5	6	7	8	9	10	11
$L_n \pmod 8$	2	1	3	4	7	3	2	5	7	4	3	7

It follows from the table that $L_{12} = 2$ (mod 8) and $L_{13} = 1$ (mod 8). Consequently, $L_0 = L_{12}$ (mod 8) and $L_1 = L_{13}$ (mod 8). It follows by the Lucas recurrence that the Lucas residues continue repeating. Thus there are exactly 12 distinct Lucas residues modulo 8, as the table shows; that is, *the period of the Lucas sequence modulo 8 is 12:* $L_{i+12} \equiv L_i$ (mod 8). More generally, $L_{12n+i} \equiv L_i$ (mod 8), where $n, i \geq 0$.

Since the least residue of a square modulo 8 is 0, 1, or 4, it follows from Table 23.1 that the only possible Lucas squares are of the form L_{12k+i}, where $i = 1, 3$, or 9. This observation narrows down our search for Lucas squares considerably.

In order to identify the Lucas squares, we need the results in the following lemma.

Lemma 23.3. *Let* $m, n \geq 0$, $r \geq 1$, *and* $t = 2^r$. *Then*

1) $L_{2t} = L_t^2 - 2$.
2) $(L_t, 2) = 1$.
3) $L_{2t+i} \equiv -L_i \pmod{L_t}$.
4) $L_{6n \pm 2} \equiv 3 \pmod 4$.

Proof.

1) Since $L_{2n} = L_n^2 - 2(-1)^n$, it follows that $L_{2t} = L_t^2 - 2$.
2) By congruence (23.1), $L_n \equiv 0$ (mod 2) if and only if $n \equiv 0$ (mod 3). Since $3 \nmid t$, $2 \nmid L_t$. Therefore, $(L_t, 2) = 1$.
3) Since $2L_{m+n} = 5F_m F_n + L_m L_n$, we have

$$2L_{2t+i} = 5F_{2t}F_i + L_{2t}L_i$$
$$= 5F_t L_t F_i + L_i(L_t^2 - 2)$$
$$\equiv -2L_i \pmod{L_t}.$$

But $(L_t, 2) = 1$, so $2L_{2t+i} \equiv -L_i \pmod{L_t}$.

We leave the proof of part 4) as an exercise; see Exercises 23.65 and 23.66. ∎

We are now in a position to establish the theorem. The featured proof is essentially the one given by Br. Alfred [7]; it uses a modicum of knowledge of quadratic residues.

Theorem 23.15. *The only Lucas squares are 1 and 4.*

Proof. Clearly, $L_1 = 1$ and $L_3 = 4$ are squares. We will now show that there are *no* more Lucas squares.

Let $i, k \geq 1$. Then $12k = 2mt$ for some odd integer m, and $t = 2^r$, where $r \geq 1$. By a repeated application of Lemma 23.3, we have

$$
\begin{aligned}
L_{12k+i} &= L_{2mt+i} \\
&\equiv -L_{2(m-1)t+i} \quad (\text{mod } L_t) \\
&\equiv (-1)^2 L_{2(m-2)t+i} \quad (\text{mod } L_t) \\
&\vdots \\
&\equiv (-1)^m L_i \quad (\text{mod } L_t) \\
&\equiv -L_i \quad (\text{mod } L_t),
\end{aligned}
$$

since m is odd.

Case 1. Let $i = 1$. Then $L_{12k+1} \equiv -L_1 \equiv -1 \pmod{L_t}$. Since $2|t$ and $3 \nmid t$, it follows by Lemma 23.3 that $L_t \equiv 3 \pmod 4$. Consequently, -1 is a quadratic nonresidue of L_t; that is, $x^2 \equiv -1 \pmod{L_t}$ has no solutions. Thus L_{12k+1} is *not* a square.

Case 2. Let $i = 3$. Then $L_{12k+3} \equiv -L_3 \equiv -4 \pmod{L_t}$. Since -4 is a quadratic nonresidue of L_t, L_{12k+3} *cannot* be a square.

Case 3. Let $i = 9$. Then L_{12k+9} can be factored: $L_{12k+9} = L_{4k+3}(L_{4k+3}^2 + 3)$; see Exercise 23.67.

Suppose $d | L_{4k+3}$ and $d | (L_{4k+3}^2 + 3)$. Then $d | 3$, so $d = 1$ or 3. But the only Lucas numbers divisible by 3 are of the form L_{4k+2}; see Exercises 23.69–23.71. So $d \nmid (4k + 3)$. Thus $d = 1$; so $(L_{4k+3}, L_{4k+3}^2 + 3) = 1$. Therefore, if L_{12k+9} is to be a square, then both factors must be squares.

Clearly, L_{4k+3} is not a square when $k = 1$ or 2. By the division algorithm, we have $k = 3s, 3s + 1$, or $3s + 2$, where $s \geq 1$. By Table 23.1, the corresponding Lucas numbers L_{12s+3}, L_{12s+7}, and L_{12s+11} are *not* squares.

Thus the only Lucas squares are 1 and 4, as desired.   ■

Next we investigate the periodicity of both Fibonacci and Lucas numbers modulo 10^n.

23.9   PERIODICITIES MODULO 10^n

In 1960, D.D. Wall of the IBM Corporation studied the periodicity of the Fibonacci sequence modulo a positive integer $m \geq 2$ [585]. He established that if $m = \prod_{i=1}^{k} p_i^{e_i}$ and h_i denotes the period of the sequence modulo $p_i^{e_i}$, then the period of the sequence modulo m is $[h_1, h_2, \ldots, h_k]$.

Twelve years later, J. Kramer and Hoggatt, both of San Jose State College, continued the investigation, and established the periodicity of both Fibonacci and Lucas numbers modulo 10^n [372]. To demonstrate this, we need the next three lemmas and four theorems.

Lemma 23.4. $L_{3n} \equiv 0$ (mod 2). ∎

This follows by Exercise 10.39.

Theorem 23.16 (Kramer and Hoggatt, 1972 [372]). *The period of the Fibonacci sequence modulo* 2^n *is* $3 \cdot 2^{n-1}$, *where* $n \geq 1$.

Proof. (We will establish this using PMI.) By virtue of Fibonacci recurrence, it suffices to prove that 1) $F_{3 \cdot 2^{n-1}} \equiv F_0$ (mod 2^n); and 2) $F_{3 \cdot 2^{n-1}+1} \equiv F_1$ (mod 2^n).

1) *To prove that* $F_{3 \cdot 2^{n-1}} \equiv F_0$ (mod 2^n):
   When $n = 1, F_{3 \cdot 2^{n-1}} = F_3 = 2 \equiv 0$ (mod 2); so the result is true when $n = 1$.
   Now assume it is true for an arbitrary positive integer k: $F_{3 \cdot 2^{k-1}} \equiv F_0$ (mod 2^k). Then, by Lemma 23.3 and the inductive hypothesis, $F_{3 \cdot 2^k} = F_{3 \cdot 2^{k-1}} L_{3 \cdot 2^{n-1}} \equiv 0$ (mod 2^{k+1}). Thus $F_{3 \cdot 2^{n-1}} \equiv 0$ (mod 2^n) for every $n \geq 1$.

2) *To prove that* $F_{3 \cdot 2^{n-1}+1} \equiv F_1$ (mod 2^n):
   Using the identity $F_{2m+1} = F_{m+1}^2 + F_m^2$,

$$F_{3 \cdot 2^{n-1}+1} = \left(F_{3 \cdot 2^{n-1}+1}\right)^2 + \left(F_{3 \cdot 2^{n-2}}\right)^2.$$

Since $F_{3 \cdot 2^{n-2}} \equiv 0$ (mod 2^{n-1}) by part 1), it follows that $\left(F_{3 \cdot 2^{n-2}}\right)^2 \equiv 0$ (mod 2^n). Then, by Cassini's formula, we have

$$\left(F_{3 \cdot 2^{n-1}+1}\right)^2 = F_{3 \cdot 2^{n-1}+2} F_{3 \cdot 2^{n-2}} - (-1)^{3 \cdot 2^{n-2}+1}$$

$$\equiv 0 + 1 \quad (\text{mod } 2^n)$$

$$\equiv 1 \quad (\text{mod } 2^n).$$

Thus $F_{3 \cdot 2^{n-1}+1} \equiv 1 \equiv F_1$ (mod 2^n). ∎

The next theorem generalizes this result to gibonacci numbers G_n.

Theorem 23.17 (Kramer and Hoggatt, 1972 [372]). *The period of the gibonacci sequence modulo* 2^n *is* $3 \cdot 2^{n-1}$, *where* $n \geq 1$. ∎

Its proof hinges on establishing the congruences $G_{3 \cdot 2^{n-1}+1} \equiv G_1$ (mod 2^n) and $G_{3 \cdot 2^{n-1}+2} \equiv G_2$ (mod 2^n), Cassini's formula, and the following identities:

$$G_{m+n+1} = G_{m+1} F_{n+1} + G_m F_n$$

$$G_{n+1} = a F_{n-1} + b F_n;$$

see Exercise 23.76.

We need two more lemmas. We can establish Lemma 23.5 using Binet's formula, and Lemma 23.6 from Lemma 23.4 using PMI; see Exercises 23.77 and 23.78.

Lemma 23.5. $F_{5n+1} = (L_{4 \cdot 5^n} - L_{2 \cdot 5^n} + 1)F_{5^n}$, where $n \geq 1$. ∎

Lemma 23.6. $F_{5^n} \equiv 0 \pmod{5^n}$, where $n \geq 1$. ∎

Theorem 23.18. *The period of the Fibonacci sequence modulo 5^n is $4 \cdot 5^n$.*

Proof. Again, in light of Fibonacci recurrence, it suffices to show that $F_{4 \cdot 5^n} \equiv F_0 \pmod{5^n}$ and $F_{4 \cdot 5^n + 1} \equiv F_1 \pmod{5^n}$.

1) *To prove that $F_{4 \cdot 5^n} \equiv F_0 \pmod{5^n}$:*
   Since $F_{5^n} | F_{4 \cdot 5^n}$, by Lemma 23.5, $F_{5^n} \equiv F_{4 \cdot 5^n} \equiv 0 \pmod{5^n}$.

2) *To prove that $F_{4 \cdot 5^n + 1} \equiv F_1 \pmod{5^n}$:*
   Using the identity $F_{2m+1} = F_{m+1}^2 + F_m^2$, we have

$$F_{4 \cdot 5^n + 1} = \left(F_{2 \cdot 5^n + 1}\right)^2 + \left(F_{2 \cdot 5^n}\right)^2$$

$$\equiv \left(F_{2 \cdot 5^n + 1}\right)^2 \pmod{5^n}.$$

By Cassini's formula, we have

$$\left(F_{2 \cdot 5^n + 1}\right)^2 = F_{2 \cdot 5^n + 2}F_{2 \cdot 5^n} - (-1)^{2 \cdot 5^n + 1}$$

$$\equiv 0 + 1 \pmod{5^n}$$

$$\equiv 1 \pmod{5^n}.$$

Thus $F_{4 \cdot 5^n + 1} \equiv 1 \pmod{5^n}$. ∎

The next theorem gives the Lucas counterpart of Theorem 23.18. We omit its proof in the interest of brevity.

Theorem 23.19. *The period of the Lucas sequence modulo 5^n is $4 \cdot 5^{n-1}$.* ∎

Combining Theorems 23.16–23.19, we get the periodicity of each sequence modulo 10^n.

Theorem 23.20.

1) *The period of the Fibonacci sequence modulo 10^n is* $\begin{cases} 60 & \text{if } n = 1 \\ 300 & \text{if } n = 2 \\ 15 \cdot 10^{n-1} & \text{otherwise.} \end{cases}$

2) *The period of the Lucas sequence modulo 10^n is* $\begin{cases} 12 & \text{if } n = 1 \\ 60 & \text{if } n = 2 \\ 3 \cdot 10^{n-1} & \text{otherwise.} \end{cases}$

Proof.

1) The period of the Fibonacci sequence modulo 10^n is given by

$$[3 \cdot 2^{n-1}, 4 \cdot 5^n] = \begin{cases} 60 & \text{if } n = 1 \\ 300 & \text{if } n = 2 \\ 15 \cdot 10^{n-1} & \text{otherwise.} \end{cases}$$

2) The period of the Lucas sequence modulo 10^n is given by

$$[3 \cdot 2^{n-1}, 4 \cdot 5^{n-1}] = \begin{cases} 12 & \text{if } n = 1 \\ 60 & \text{if } n = 2 \\ 3 \cdot 10^{n-1} & \text{otherwise.} \end{cases} \blacksquare$$

The next corollary follows immediately from this theorem.

Corollary 23.4.

1) *The units digits of Fibonacci numbers repeat in a cycle of period 60; the last two digits in a cycle of period 300; and the last $n \geq 3$ digits in a cycle of period $15 \cdot 10^{n-1}$.*

2) *The units digits of Lucas numbers repeat in a cycle of period 12; the last two digits in a cycle of period 60; and the last $n \geq 3$ digits in a cycle of period $3 \cdot 10^{n-1}$.* $\blacksquare$

Kramer and Hoggatt also established the following congruences in 1972 [372].

Theorem 23.21 (Kramer and Hoggatt, 1972 [372]).

1) $L_{2 \cdot 3^n} \equiv 0 \pmod{2 \cdot 3^n}$.
2) $F_{4 \cdot 3^n} \equiv 0 \pmod{4 \cdot 3^n}$. $\blacksquare$

For example, $L_{54} = 192,900,153,618 \equiv 0 \pmod{54}$, and $F_{36} = 14,930,352 \equiv 0 \pmod{36}$.

We can employ Theorem 23.20 to confirm the next result. H.T. Freitag of Virginia studied it in 1976 [193]. The proof presented is essentially the same as the one given by P.S. Bruckman, when he was at the University of Illinois [82].

Example 23.3. *Prove that $L_{2p^k} \equiv 3 \pmod{10}$, where p is a prime ≥ 5.*

Proof. Since $p \geq 5$, $p \equiv \pm 1 \pmod 6$; so $p^k \equiv \pm 1 \pmod 6$, and hence $2p^k \equiv \pm 2 \pmod{12}$. By Theorem 23.20, $L_{n+12} \equiv L_n \pmod{10}$. But $L_n \equiv 3 \pmod 3$ if and only if $n \equiv 2 \pmod{12}$. Therefore, $L_{2p^k} \equiv 3 \pmod{10}$ for primes $p \geq 5$. $\blacksquare$

For example, $L_{2 \cdot 5^2} = L_{50} = 28,143,753,123 \equiv 3 \pmod{10}$.

M.R. Turner of Regis University in Denver, Colorado, discovered the next result in 1974 [562]. It characterizes Fibonacci numbers that terminate in the same last two digits as their subscripts. The proof is fairly long, so we omit it.

Theorem 23.22. $F_n \equiv n$ (mod 100) *if and only if* $n = 1, 5, 25, 29, 41,$ *or* 49 (mod 60), *or* $n \equiv 0$ (mod 300). ∎

For example, $F_{41} = 165,580,141 \equiv 41$ (mod 100) and $n = 41 \equiv 41$ (mod 60). On the other hand, $n = 85 \equiv 25$ (mod 60) and $F_{85} = 259,695,496,911,122,585 \equiv 85$ (mod 100).

This theorem has an interesting byproduct.

Corollary 23.5. *Let p be a prime ≥ 5. Then $F_{p^2} \equiv p^2$ (mod 100).*

Proof. Clearly, $F_{5^2} \equiv 5^2$ (mod 100). So let $p > 5$. Then $p \equiv 1, 3, 7, 9, 11, 13, 17$ or 19 (mod 20). Then $p^2 \equiv 1$ or 9 (mod 20), so $p^2 \equiv 1$ or 49 (mod 20). Since $p^2 \equiv 1$ (mod 3) for every prime > 3, it follows that $p^2 \equiv 1$ or 49 (mod 60). Therefore, by Theorem 23.22, $F_{p^2} \equiv p^2$ (mod 100). ∎

For example, let $p = 7$. Clearly, $F_{49} = 7,778,742,049 \equiv 49$ (mod 100).

EXERCISES 23

Verify Corollary 23.1 for the given values of m and n.
1. $m = 13, n = 5$.
2. $m = 20, n = 11$.

Verify Corollary 23.2 for the given values of m and n.
3. $m = 11, n = 5$.
4. $m = 18, n = 7$.

Prove each, where $m, n, k \geq 1$ and p is a prime.
5. $F_n \equiv 0$ (mod 3) if and only if $n \equiv 0$ (mod 4).
6. $F_n \equiv 0$ (mod 4) if and only if $n \equiv 0$ (mod 6).
7. $F_n \equiv 0$ (mod 5) if and only if $n \equiv 0$ (mod 5).
8. $L_n^2 \equiv L_{2n}$ (mod 2).
9. $L_n^2 \equiv F_n^2$ (mod 4).
10. $L_n^2 \equiv L_{n-1}L_{n+1}$ (mod 5), where $n \geq 2$.
11. $2L_{m+n} \equiv L_m L_n$ (mod 5).
12. $F_{15n} \equiv 0$ (mod 10).
13. $L_{(2k-1)n} \equiv 0$ (mod L_n).
14. $F_{2n} \equiv n(-1)^n$ (mod 5).

15. $F_{n+24} \equiv F_n \pmod 9$ (Householder, 1963 [333]).
16. $F_{n+3} \equiv F_n \pmod 2$.
17. $F_{3n} \equiv 0 \pmod 2$.
18. $F_{n+5} \equiv 3F_n \pmod 5$.
19. $F_{5n} \equiv 0 \pmod 5$.
20. Congruence (23.2).
21. Congruence (23.3).
22. Congruence (23.4).
23. $L_{3n} \equiv 0 \pmod 2$.
24. $L_{3n} \equiv 0 \pmod{L_n}$.
25. $(F_n, L_n) = 2$ if and only if $n \equiv 0 \pmod 3$.
26. $L_n \equiv 3L_{n-1} \pmod 5$.
27. $3L_{2n-2} \equiv (-1)^{n-1} \pmod 5$.
28. $L_{2n} \equiv 2(-1)^n \pmod 5$ (Wessner, 1968 [590]).
29. $L_{2n+1} \equiv (-1)^n \pmod 5$.
30. $L_{2n}L_{2n+2} \equiv 1 \pmod 5$.
31. $L_{m+n}^2 \equiv L_{2m}L_{2n} \pmod 5$.
32. $2^n L_n \equiv 2 \pmod 5$ (Wall, 1965 [580]).
33. $2^n F_n \equiv 2n \pmod 5$ (Wall, 1967 [581]).
34. $L_{np} \equiv L_n \pmod p$ (Desmond, 1970 [145]).
35. $F_{5^n} \equiv 5^n \pmod{5^{n+3}}$ (Bruckman, 1982 [89]).
36. $L_{5^n} \equiv L_{5^{n+1}} \pmod{5^{n+3}}$ (Bruckman, 1982 [89]).
37. $25F_n^4 + 16 \equiv L_n^4 \pmod{5F_n^2}$ (Freitag, 1979 [194]).
38. Let $t_n = n(n+1)/2$ be the nth triangular number. Then $L_{n^2} \equiv (-1)^{t_n-1} L_n$ (mod 5). (Freitag, 1982 [199]).
39. $\displaystyle\sum_{i=1}^{20} F_{n+i} \equiv 0 \pmod{F_{10}}$ (Ruggles, 1963 [504]).
40. $L_{2^n} \equiv 0 \pmod{10}$.
41. $F_n \equiv L_n \pmod 2$.
42. $F_m L_n \equiv L_m L_n \pmod 2$.
43. Let t_n denote the nth triangular number. Then $L_n = (-1)^{t_n-1}[L_{t_{n-1}} L_{t_n} - L_{n^2}]$. (Freitag, 1981 [198]).
44. $L_i L_j \equiv L_h L_k \pmod 5$, where $i + j = h + k$ (Brady, 1977 [56]).
45. $5L_{4n} - L_{2n}^2 - 6(-1)^n L_{2n} + 6 \equiv 0 \pmod{10F_n^2}$ (Freitag, 1975 [192]).
46. Let a, h, k and d be positive integers with d odd. Then $L_{a+hd} + L_{a+hd+d} \equiv L_{a+kd} + L_{a+kd+d} \pmod{L_d}$ (Freitag, 1982 [200]).
47. Compute $F_{F_5}, F_{L_5}, L_{F_5}$, and L_{L_5}.

Prove each, where $n \geq 1$.

48. $F_{L_n} \not\equiv 0 \pmod 5$.

49. $F_{F_n} \equiv 0 \pmod 5$ if and only if $n \equiv 0 \pmod 5$.
50. $L_{5n} \equiv 0 \pmod{L_n}$.
51. The units digit in F_{31} is 9, and that in F_{32} is also 9. Compute the units digit in F_{33}.
52. F_{38} ends in 9 and F_{39} in 6. Compute the units digit in F_{42}.
53. F_{43} ends in 7. Determine the units digit in F_{703}.
54. L_{20} ends in 7 and L_{21} in 6. Compute the units digits in L_{22} and L_{25}.
55. L_{45} ends in 6. Find the units digit in L_{93}.
56. Complete the following table.

Modulus m	Period of the Sequence $\{F_n\}$ (mod m)	Period of the Sequence $\{L_n\}$ (mod m)
2		
3		
4		
5		
6		
7		
8		
9		
10		

57. Given that $L_{23} \equiv 7 \pmod 8$ and $L_{24} \equiv 2 \pmod 8$, compute $L_{25} \pmod 8$.
58. Let $L_i \equiv 5 \pmod 8$ and $L_{i+1} \equiv 7 \pmod 8$. Compute $L_{i+2} \pmod 8$.
59. Let $L_i \equiv 3 \pmod 8$ and $L_{i-1} \equiv 4 \pmod 8$. Compute $L_{i-2} \pmod 8$.

Prove each, where $n, i \geq 0$.

60. $L_{12n+i} \equiv L_i \pmod{10}$.
61. $F_{6n+1} \equiv 1 \pmod 4$.
62. $F_{6n-2} \equiv 3 \pmod 4$.
63. $F_{6n-1} \equiv 1 \pmod 4$.
64. $L_{6n-1} \equiv 3 \pmod 4$.
65. $L_{6n+2} \equiv 3 \pmod 4$.
66. $L_{6n+4} \equiv 3 \pmod 4$.
67. $L_{12n+9} = L_{4n+3}(L_{4n+3}^2 + 3)$.
68. $F_{4n} \equiv 0 \pmod 3$.
69. $L_{4n+2} \equiv 0 \pmod 3$.

70. $L_{4n} \equiv \pm 1 \pmod 3$.

71. $L_{4n+1} \equiv \pm 1 \pmod 3$.

72. $L_{12n+i} \equiv L_i \pmod 3$.

73. Let $2 \mid t$ and $3 \nmid t$. Then $L_t \equiv 3 \pmod 4$.

74. $F_{60n+i} \equiv F_i \pmod{60}$. (*Hint:* $2F_{m+n} = F_m L_n + F_n L_m$.)

75. $L_{4n+2} \equiv 0 \pmod 3$. [*Hint:* $L_{m+2k} \equiv -L_m \pmod{L_k}$.]

76. Theorem 23.17.

77. Lemma 23.5.

78. Lemma 23.6.

79. $F_{60k} \equiv 20k \pmod{100}$ (Turner, 1974 [562]).

80. If $n \equiv 1 \pmod{60}$, then $F_n \equiv n \pmod{100}$ (Turner, 1974 [562]).

81. $F_{60k+n} \equiv 20k F_{n-1} + (60k+1)F_n \pmod{100}$ (Turner, 1974 [562]).

82. The sum of n consecutive Lucas numbers is divisible by 5 if and only if $4 \mid n$ (Freitag, 1974 [188]).

83. $F_{(n+2)k} \equiv F_{nk} \pmod{L_k}$, where k is odd (Freitag, 1974 [189]).

84. $F_{(n+2)k} + F_{nk} \equiv 2F_{(n+1)k} \pmod{L_k - 2}$, where k is even (Freitag, 1974 [190]).

85. $L_{2m(2n+1)} \equiv L_{2m} \pmod{F_{2m}^2}$ (Bruckman, 1974 [84]).

86. $L_{(2m+1)(4n+1)} \equiv L_{2m+1} \pmod{F_{2m+1}}$ (Bruckman, 1979 [86]).

87. $L_{(2m+1)(4n+1)} \equiv L_{2m+1} \pmod{F_{2n}F_{2n+1}}$.

88. $L_{2p^k} \equiv 3 \pmod{10}$, where p is a prime ≥ 5 (Freitag, 1976 [193]).

89. $F_{3n+1} + F_{n+3} \equiv 0 \pmod 3$ (Berzsenyi, 1979 [40]).

90. $F_{2n} \equiv n(-1)^{n+1} \pmod 5$ (Freitag, 1979 [194]).

91. $L_{2^n} \equiv 7 \pmod{10}$, where $n \geq 2$ (Shannon, 1979 [522]).

92. $F_{3 \cdot 2^n} \equiv 2^{n+2} \pmod{2^{n+3}}$, where $n \geq 1$ (Bruckman, 1979 [87]).

93. $L_{3 \cdot 2^n} \equiv 2 + 2^{n+2} \pmod{2^{n+4}}$, where $n \geq 1$ (Bruckman, 1979 [87]).

FIBONACCI AND LUCAS SERIES

In this chapter, we study the convergence of some interesting Fibonacci and Lucas series, and evaluate them when convergent.

24.1 A FIBONACCI SERIES

To begin with, suppose we place successively every Fibonacci number F_n after a decimal point, so its units digit falls in the $(n + 1)$st decimal place. The resulting real number, to our great surprise, is the decimal expansion of the rational number $\dfrac{1}{89} = \dfrac{1}{F_{11}}$. Be sure to account for the carries:

$$
\begin{array}{r}
0.\ 0\ 1\ 1\ 2\ 3\ 5\ 8 \\
1\ 3 \\
2\ 1 \\
3\ 4 \\
5\ 5 \\
8\ 9 \\
1\ 4\ 4 \\
\vdots
\end{array}
$$

$$= 0.\ 0\ 1\ 1\ 2\ 3\ 5\ 9\ 5\ 5\ 0\ 5\ \ldots$$

$$= \frac{1}{89};$$

that is, $\displaystyle\sum_{i=0}^{\infty} \frac{F_i}{10^{i+1}} = \frac{1}{F_{11}}$. F. Stancliff discovered this result in 1953 [537].

Fibonacci and Lucas Numbers with Applications, Volume One, Second Edition. Thomas Koshy.
© 2018 John Wiley & Sons, Inc. Published 2018 by John Wiley & Sons, Inc.

To establish this fact, we need to study the convergence of the Fibonacci series

$$S = \sum_{i=0}^{\infty} \frac{F_i}{k^{i+1}}, \tag{24.1}$$

where k is a positive integer.

Suppose the series converges. Then, by Binet's formula,

$$\sum_{i=0}^{\infty} \frac{F_i}{k^{i+1}} = \frac{1}{\sqrt{5}k} \left[\sum_{i=0}^{\infty} \left(\frac{1+\sqrt{5}}{2k} \right)^i - \left(\frac{1-\sqrt{5}}{2k} \right)^i \right]$$

$$= \frac{1}{\sqrt{5}k} \left[\frac{1}{1-(1+\sqrt{5})/2k} - \frac{1}{1-(1-\sqrt{5})/2k} \right] \tag{24.2}$$

$$= \frac{2}{\sqrt{5}k} \left(\frac{1}{2k-1-\sqrt{5}} - \frac{1}{2k-1+\sqrt{5}} \right)$$

$$S = \frac{1}{k^2 - k - 1}. \tag{24.3}$$

Notice that the denominator of the RHS of equation (24.3) is the characteristic polynomial of Fibonacci recurrence. Also, S is an integer if and only if $k = 2$.

It is well known that the power series $\dfrac{1}{1-x} = \displaystyle\sum_{i=0}^{\infty} x^i$ converges if and only if $|x| < 1$. Consequently, it follows from equation (24.2) that Fibonacci power series (24.1) converges if and only if $|\alpha| < k$ and $|\beta| < k$; that is, if and only if $k > \max\{|\alpha|, |\beta|\}$. But $\alpha = |\alpha| > |\beta|$. Thus the series (24.1) converges if and only if $k > \alpha$; that is, if and only if $k \geq 2$, which was somewhat obvious.

Equation (24.3) yields the following results:

$$\sum_{i=0}^{\infty} \frac{F_i}{2^{i+1}} = 1 = \frac{1}{F_1}$$

$$\sum_{i=0}^{\infty} \frac{F_i}{3^{i+1}} = \frac{1}{5} = \frac{1}{F_5}$$

$$\sum_{i=0}^{\infty} \frac{F_i}{8^{i+1}} = \frac{1}{55} = \frac{1}{F_{10}}$$

$$\sum_{i=0}^{\infty} \frac{F_i}{10^{i+1}} = \frac{1}{89} = \frac{1}{F_{11}}.$$

These values of k yield the value of the infinite sum to be of the form $\dfrac{1}{F_t}$ for some Fibonacci number F_t; in other words, they are such that $k^2 - k - 1 = F_t$; that is, $k(k-1) = 1 + F_t$.

Conversely, suppose $1 + F_t$ is the product $b(b-1)$ of two consecutive positive integers b and $b - 1$. Solving the equation $k^2 - k - (1 + F_t) = 0$ for k, we get

$$k = \frac{1 \pm \sqrt{1 + 4(1 + F_t)}}{2}$$

$$= \frac{1 \pm \sqrt{4F_t + 5}}{2}.$$

Since $k > 0$, this implies $k = \dfrac{1 + \sqrt{4F_t + 5}}{2}$.

But $4F_t + 5 = 4(1 + F_t) + 1 = 4b(b-1) + 1 = (2b-1)^2$. So $k = \dfrac{1 + 2b - 1}{2} = b$, and $S = 1/F_t$. Thus $1/S$ is a Fibonacci number F_t if and only if $1 + F_t$ is the product of two consecutive positive integers.

It now follows that there are at least four such values of k:

$$1 \cdot 2 = 1 + F_1, \quad 2 \cdot 3 = 1 + F_5, \quad 7 \cdot 8 = 1 + F_{10}, \quad \text{and} \quad 9 \cdot 10 = 1 + F_{11}.$$

Next we investigate the convergence of the corresponding Lucas series

$$S^* = \sum_{i=0}^{\infty} \frac{L_i}{k^{i+1}}.$$

24.2   A LUCAS SERIES

As before, we can show that this series converges to a finite sum S^* if and only if $k > \alpha$, and

$$S^* = \frac{2k - 1}{k^2 - k - 1}.$$

When $k = 2$, $S^* = 3$; and when $k = 3$, $S^* = 1$. In both cases, S^* is an integer.

Suppose we wish to investigate the integral values of k for which S^* is an integer t. When $t = 0$, $k = 1/2$; this is a contradiction; so $t \geq 1$. Let $\dfrac{2k - 1}{k^2 - k - 1} = t$. Then $tk^2 - (t + 2)k - (t - 1) = 0$, so

$$k = \frac{(t + 2) \pm \sqrt{(t + 2)^2 + 4t(t - 1)}}{2t}$$

$$= \frac{(t + 2) \pm \sqrt{5t^2 + 4}}{2t}. \tag{24.4}$$

Since k is an integer, $\sqrt{5t^2 + 4}$ must be a square. When $t = 1$, $k = 3$ or 0. Since $k \geq 1$, 0 is not acceptable. So $t = 1$, and hence $k = 3$.

When $t > 1$, $\sqrt{5t^2 + 4} > t + 2$. Consequently, since $k > 0$, the negative root in equation (24.3) is not acceptable. Thus

$$k = \frac{(t + 2) + \sqrt{5t^2 + 4}}{2t}. \tag{24.5}$$

Suppose $k \geq 4$. Then

$$\frac{(t + 2) + \sqrt{5t^2 + 4}}{2t} \geq 4$$

$$\sqrt{5t^2 + 4} \geq 7t - 2$$

$$5t^2 + 4 \geq 49t^2 - 28t + 4$$

$$11t^2 \leq 7t$$

$$t \leq 7/11,$$

which is a contradiction.

Thus the only positive values of k that yield an integral value for S^* are $k = 2$ and $k = 3$. The corresponding values of S^* from equation (24.2) are $3 = L_2$ and $1 = L_1$.

A similar argument shows that the only positive value of k that produces an integral value for $S = 1/(k^2 - k - 1)$ is $k = 2$, in which case $S = 1 = F_1$ (or F_2).

24.3   FIBONACCI AND LUCAS SERIES REVISITED

In 1981, Calvin T. Long of Washington State University at Pullman showed that the following summation results can be derived from a bizarre summation formula established in the next theorem [417]:

$$\frac{1}{89} = \sum_{n=1}^{\infty} \frac{F_{n-1}}{10^n} \tag{24.6}$$

$$\frac{19}{89} = \sum_{n=1}^{\infty} \frac{L_{n-1}}{10^n} \tag{24.7}$$

$$\frac{1}{109} = \sum_{n=1}^{\infty} \frac{F_{n-1}}{(-10)^n} \tag{24.8}$$

$$-\frac{21}{109} = \sum_{n=1}^{\infty} \frac{L_{n-1}}{(-10)^n}. \tag{24.9}$$

Theorem 24.1 (Long, 1981 [417]). *Let a, b, c, d, and B be integers. Let $U_{n+2} = aU_{n+1} + bU_n$, where $U_0 = c$, $U_1 = d$, and $n \geq 2$. Let integers m and N be defined by $B^2 = m + aB + b$ and $N = cm + dB + bc$. Then*

$$B^n N = m \sum_{i=1}^{n+1} B^{n-i+1} U_{i-1} + BU_{n+1} + bU_n \qquad (24.10)$$

for all $n \geq 0$.

Proof. (We will establish the theorem using PMI.) When $n = 0$, equation (24.10) yields $N = cm + dB + bc$. So the result is true when $= 0$.

Assume it is true for an arbitrary nonnegative integer k:

$$B^k N = m \sum_{i=1}^{k+1} B^{k-i+1} U_{i-1} + BU_{k+1} + bU_k.$$

Then

$$B^{k+1} N = m \sum_{i=1}^{k+1} B^{k-i+2} U_{i-1} + B^2 U_{k+1} + bBU_k$$

$$= m \sum_{i=1}^{k+1} B^{k-i+2} U_{i-1} + (m + aB + b) U_{k+1} + bBU_k$$

$$= m \sum_{i=1}^{k+2} B^{k-i+2} U_{i-1} + B(aU_{k+1} + bU_k) + bU_{k+1}$$

$$= m \sum_{i=1}^{k+2} B^{k-i+2} U_{i-1} + BU_{k+2} + bU_{k+1}.$$

So the formula also works when $n = k + 1$.

Thus, by PMI, it is true for all $n \geq 0$. ∎

Formula (24.10) yields the following result.

Theorem 24.2 (Long, 1981 [417]). *Let a, b, c, d, m, B and N be integers as in Theorem 24.1. Let*

$$r = \frac{a + \sqrt{a^2 + 4b}}{2} \quad \text{and} \quad s = \frac{a - \sqrt{a^2 + 4b}}{2},$$

where $|r| < |B|$ and $|s| < |B|$. Then

$$\frac{N}{mB} = \sum_{i=1}^{\infty} \frac{U_{i-1}}{B^i}. \qquad (24.11)$$

Proof. Using the recurrence in Theorem 24.1, we can show that

$$U_n = Pr^n + Qs^n, \tag{24.12}$$

where $P = \dfrac{c}{2} + \dfrac{2d - ca}{2\sqrt{a^2 + 4b}}$ and $Q = \dfrac{c}{2} - \dfrac{2d - ca}{2\sqrt{a^2 + 4b}}$; see Exercise 24.17. Then, since $|r|, |s| < 1$, by equation (24.10), we have

$$\frac{N}{mB} = \sum_{i=1}^{n+1} \frac{U_{i-1}}{B^i} + \frac{BU_{n+1} + bU_n}{mB^{n+1}}$$

$$\lim_{n \to \infty} \frac{N}{mB} = \sum_{i=1}^{\infty} \frac{U_{i-1}}{B^i} + 0$$

$$\frac{N}{mB} = \sum_{i=1}^{\infty} \frac{U_{i-1}}{B^i},$$

as desired.   ■

In particular, let $a = 1 = b = d$, $c = 0$, and $B = 10$. Then $m = B^2 - aB - b = 100 - 10 - 1 = 89$, and $N = cm + dB + bc = 0 + 10 + 0 = 10$. Equation (24.11) then yields

$$\frac{1}{89} = \sum_{i=1}^{\infty} \frac{F_{i-1}}{10^i}.$$

We can similarly derive the summation formulas (24.7)–(24.9); see Exercises 24.18–24.20.

The next corollary follows from Theorem 24.2.

Corollary 24.1.   *Let $a = b = d = 1$, and $c = 0$.*

1) *If $B = 10^h$, then* $\dfrac{1}{10^{2h} - 10^h - 1} = \displaystyle\sum_{i=1}^{\infty} \dfrac{F_{i-1}}{10^{ih}}.$

2) *If $B = (-10)^h$, then* $\dfrac{1}{10^{2h} - (-10)^h - 1} = \displaystyle\sum_{i=1}^{\infty} \dfrac{F_{i-1}}{(-10)^{ih}}.$

Proof. If $B = 10^h$, then $m = 10^{2h} - 10^h - 1$, and $N = 10^h$. If $B = (-10)^h$, then $m = 10^{2h} - (-10)^h - 1$, and $N = (-10)^h$. Both formulas now follow by substitution.   ■

The following summation formulas follow from this corollary:

$$\frac{1}{89} = \sum_{i=1}^{\infty} \frac{F_{i-1}}{10^i}$$

$$= 0.0112359350557\ldots$$

$$= 0.0112358$$

$$13$$
$$21$$
$$34$$
$$55$$
$$89$$
$$144$$
$$233$$

$$\vdots$$

$$\frac{1}{9899} = \sum_{i=1}^{\infty} \frac{F_{i-1}}{10^{2i}}$$

$$= 0.000101020305081321\ldots$$

$$\frac{1}{998,999} = \sum_{i=1}^{\infty} \frac{F_{i-1}}{10^{3i}}$$

$$= 0.000001001002003005008013\ldots$$

$$\frac{1}{109} = \sum_{i=1}^{\infty} \frac{F_{i-1}}{(-10)^{i}}$$

$$\frac{1}{10,099} = \sum_{i=1}^{\infty} \frac{F_{i-1}}{(-100)^{i}}$$

$$\frac{1}{1,000,999} = \sum_{i=1}^{\infty} \frac{F_{i-1}}{(-1000)^{i}}.$$

24.4   A FIBONACCI POWER SERIES

We now turn our attention to the convergence of the Fibonacci power series

$$T = \sum_{i=0}^{\infty} F_i x^i. \tag{24.13}$$

We have

$$T = F_1 x + F_2 x^2 + \sum_{i=3}^{\infty} (F_{i-1} + F_{i-2}) x^i$$

$$= x + x^2 + x \sum_{r=2}^{\infty} F_r x^r + x^2 \sum_{s=1}^{\infty} F_s x^s$$

$$= x + x^2 + x(T - x) + x^2 T$$

$$= \frac{x}{1 - x - x^2} \tag{24.14}$$

$$= \frac{x}{(1 - \alpha x)(1 - \beta x)}$$

$$= \frac{A}{1 - \alpha x} - \frac{A}{1 - \beta x},$$

where $A = 1/\sqrt{5}$. Accordingly, the series converges if and only if both $|\alpha x| < 1$ and $|\beta x| < 1$. This is true if and only if $|x| < \min\{1/|\alpha|, 1/|\beta|\}$. Since $\alpha\beta = -1$, $1/\alpha = -\beta$; so $\min\{1/|\alpha|, 1/|\beta|\} = \min\{-\beta, \alpha\} = -\beta$. Thus series (24.13) converges if and only if $\beta < x < -\beta$.

Next we will identify the rational values of x for which T is an integer $k \geq 1$, as P. Glaister did in 1995 [223]:

$$\frac{x}{1 - x - x^2} = k$$

$$kx^2 + (k + 1)x - k = 0$$

$$x = \frac{-(k + 1) \pm \sqrt{(k + 1)^2 + (2k)^2}}{2k}.$$

This implies that there are two possible values of x.

Since we require x to be rational, $(k + 1)^2 + (2k)^2$ must be a square. This can be realized using Pythagorean triples; so we let

$$k + 1 = m^2 - n^2 \quad \text{and} \quad 2k = 2mn \tag{24.15}$$

for some integers m and n, where $m > n \geq 1$. Then

$$(k + 1)^2 + (2k)^2 = (m^2 - n^2) + (2mn)^2$$

$$= (m^2 + n^2)^2$$

$$x = \frac{-(m^2 - n^2) \pm (m^2 - n^2)}{2mn}$$

$$= -\frac{m}{n}, \frac{n}{m}.$$

Suppose both these values lie in the interval of convergence $(\beta, -\beta)$; that is, $\beta < -m/n < -\beta$ and $\beta < n/m < -\beta$. Form the second double inequality, $m/n > -1/\beta$, so $-m/n < -\alpha$. This is a contradiction, since $-m/n > \beta > -\alpha$. Thus both possible values of x cannot lie within the interval at the same time.

From equation (24.15), $m^2 - n^2 = mn + 1$. Then

$$(m - n/2)^2 = m^2 - mn + n^2/4$$
$$= n^2 + 1 + n^2/4$$
$$= 1 + 5n^2/4$$
$$(2m - n)^2 = 4 + 5n^2.$$

Thus $4 + 5n^2$ must be a square r^2 for some positive integer r. So $2m - n = r$.

Let us now look at the first three possible values of n, and compute the corresponding values of r, m, and $T = k$.

Case 1. Let $n = 1$. Then $4 + 5n^2 = 3^2$; so $2m - 1 = 3$ and hence $m = 2$. Therefore, $T = k = mn = 2$, $n/m = 1/2$, and $-m/n = -2$. But $-2 \notin (\beta, -\beta)$.

Thus $\sum_{i=1}^{\infty} \dfrac{F_i}{2^i} = 2$.

Case 2. Let $n = 3$. Then $4 + 5n^2 = 7^2$; so $2m - 3 = 7$ and hence $m = 5$. So $T = mn = 15$, $n/m = 3/5$, and $-m/n = -5/3$. But $-5/3 \notin (\beta, -\beta)$. Thus

$$\sum_{i=1}^{\infty} \left(\frac{3}{5}\right)^i F_i = 15.$$

Case 3. Let $n = 8$. Then $4 + 5n^2 = 18^2$; so $2m - 8 = 18$ and hence $m = 13$. Therefore, $T = mn = 104$, $n/m = 8/13$, and $-m/n = -13/8$. Again, $-13/8 \notin (\beta, -\beta)$.

Thus $\sum_{i=1}^{\infty} \left(\dfrac{8}{13}\right)^i F_i = 104$.

The next choice of n is 21. Surprisingly, an interesting pattern begins to emerge. These four values of n are even-numbered Fibonacci numbers: $F_2 = 1$, $F_4 = 3$, $F_6 = 8$, and $F_8 = 21$; the corresponding m-values are their immediate successors: $F_3 = 2$, $F_5 = 5$, $F_7 = 13$, and $F_9 = 34$; and

$$\sum_{i=0}^{\infty} \frac{F_i}{2^i} = 1 \cdot 2 \qquad\qquad \sum_{i=0}^{\infty} \left(\frac{3}{5}\right)^i F_i = 3 \cdot 5$$

$$\sum_{i=0}^{\infty} \left(\frac{8}{13}\right)^i F_i = 8 \cdot 13 \qquad\qquad \sum_{i=0}^{\infty} \left(\frac{21}{34}\right)^i F_i = 21 \cdot 34.$$

Fortunately, this fascinating pattern does indeed hold.

To confirm this, let $n = F_{2k}$, where $k \geq 1$. Then, by Binet's formula, we have

$$4 + 5n^2 = 4 + 5F_{2k}^2$$
$$= 4 + (\alpha^{4k} + \beta^{4k} - 2)$$
$$= \alpha^{4k} + \beta^{4k} + 2$$
$$= (\alpha^{2k} + \beta^{2k})^2.$$

Thus $4 + 5n^2$ is a square, as desired.

Then, again by Binet's formula,

$$2m - n = \alpha^{2k} + \beta^{2k}$$

$$m = \frac{\alpha^{2k} + \beta^{2k} + F_{2k}}{2}$$

$$= \frac{\alpha^{2k+1} - \beta^{2k+1}}{\alpha - \beta}$$

$$= F_{2k+1}.$$

Since $\beta < 0$, it follows that

$$\frac{n}{m} = \frac{F_{2k}}{F_{2k+1}}$$

$$= \frac{\alpha^{2k} - \beta^{2k}}{\alpha^{2k+1} - \beta^{2k+1}}$$

$$< \frac{\alpha^{2k}}{\alpha^{2k+1}}$$

$$= \frac{1}{\alpha}$$

$$= -\beta.$$

Thus $0 < n/m < -\beta$, and hence n/m lies within the interval of convergence. Consequently, $T = k = mn = F_{2k}F_{2k+1}$, as expected.

Thus we have the following theorem.

Theorem 24.3. *The Fibonacci power series (24.13) converges if $\beta < x < -\beta$ and*

$$\sum_{i=0}^{\infty} \left(\frac{F_{2k}}{F_{2k+1}} \right)^i F_i = F_{2k}F_{2k+1}, \text{ where } k \geq 1.$$ ∎

For example, $F_{12} = 144$ and $F_{13} = 233$. Then

$$\sum_{i=0}^{\infty} \left(\frac{144}{233} \right)^i F_i = 144 \cdot 233 = 33{,}552.$$

A similar study of the related Lucas series

$$T = \sum_{i=0}^{\infty} L_i x^i \tag{24.16}$$

yields fascinating dividends. To see this, suppose the series converges. Then

$$T = L_0 + L_1 x + \sum_{i=2}^{\infty} (L_{i-1} + L_{i-2}) x^i$$

$$= 2 + x + x \sum_{i=1}^{\infty} L_i x^i + x^2 \sum_{i=0}^{\infty} L_i x^i$$

$$= 2 + x + x(T - 2) + x^2 T$$

$$(1 - x - x^2)T = 2 - x$$

$$T = \frac{2 - x}{1 - x - x^2}. \tag{24.17}$$

Since $1 - x - x^2 = (1 - \alpha x)(1 - \beta x)$, we can convert this into partial fractions:

$$T = \frac{A}{1 - \alpha x} + \frac{B}{1 - \beta x},$$

where A and B are constants.

Expanding the RHS yields

$$T = A \sum_{i=0}^{\infty} (\alpha x)^i + B \sum_{i=0}^{\infty} (\beta x)^i.$$

The two series on the RHS converge if and only if $|\alpha x| < 1$ and $|\beta x| < 1$; that is, if and only if $|x| < 1/|\alpha|$ and $|x| < 1/|\beta|$. Thus the series (24.16) converges if and only if $|x| < \min\{1/\alpha, 1/(-\beta)\}$; that is, if and only if $|x| < \min\{-\beta, \alpha\}$. Since $-\beta < \alpha$, $\min\{-\beta, \alpha\} = -\beta$. Thus the Lucas power series (24.16) converges if and only if $|x| < -\beta$; that is, if and only if $\beta < x < -\beta$. When the series converges, formula (24.17) gives the value of the infinite sum.

Formula (24.17) raises an interesting question for the curious-minded: *Are there rational numbers x in the interval of convergence for which T is an integer?*

We will answer this question shortly. But before we do, we will study a few examples and look for a possible pattern:

when $x = 1/3$, $\qquad T = \dfrac{2 - 1/3}{1 - 1/3 - (1/3)^2} \qquad = 3 = 3 \cdot 1$

when $x = 4/7$, $\qquad T = \dfrac{2 - 4/7}{1 - 4/7 - (4/7)^2} \qquad = 14 = 7 \cdot 2$

when $x = 11/18$, $\qquad T = \dfrac{2 - 11/18}{1 - 11/18 - (11/18)^2} \qquad = 90 = 18 \cdot 5.$

Clearly, a nice pattern emerges. In each case, x lies within the interval of convergence; x is of the form L_{2k-1}/L_{2k}; and T is an integer of the form $L_{2k} F_{2k-1}$, where $k \geq 1$.

Fortunately, the patterns do hold. We will now establish it using the identities $L_{n+1} + L_{n-1} = 5F_n$ and $L_{2n}^2 - L_{2n}L_{2n-1} - L_{2n-1}^2 = 5$; see Exercises 5.32 and 24.6.

Theorem 24.4 (Koshy, 1999 [365]).   *Let k be a positive integer. Then*

$$\sum_{i=0}^{\infty} \left(\frac{L_{2k-1}}{L_{2k}} \right)^i L_i = L_{2k} F_{2k-1}.$$

Proof. First, we will show that $\beta < \dfrac{L_{2k-1}}{L_{2k}} < -\beta$. By Binet's formula, we have

$$\frac{L_{2k-1}}{L_{2k}} = \frac{\alpha^{2k-1} + \beta^{2k-1}}{\alpha^{2k} + \beta^{2k}}$$

$$< \frac{\alpha^{2k-1} + \beta^{2k-1}}{\alpha^{2k}}$$

$$< \frac{\alpha^{2k-1}}{\alpha^{2k}}$$

$$= \frac{1}{\alpha} = -\beta.$$

Thus $0 < \dfrac{L_{2k-1}}{L_{2k}} < -\beta$, so $\beta < \dfrac{L_{2k-1}}{L_{2k}} < -\beta$.

Since $x = L_{2k-1}/L_{2k} \in (\beta, -\beta)$, by formula (24.17), we have

$$\sum_{i=0}^{\infty} \left(\frac{L_{2k-1}}{L_{2k}} \right)^i L_i = \frac{2 - L_{2k-1}/L_{2k}}{1 - L_{2k-1}/L_{2k} - (L_{2k-1}/L_{2k})^2}$$

$$= \frac{L_{2k}(2L_{2k} - L_{2k-1})}{L_{2k}^2 - L_{2k}L_{2k-1} - L_{2k-1}^2}$$

$$= \frac{L_{2k}(2L_{2k} - L_{2k-1})}{5}$$

$$= \frac{L_{2k}(L_{2k} + L_{2k-2})}{5}$$

$$= \frac{L_{2k}(5F_{2k-1})}{5}$$

$$= L_{2k} F_{2k-1},$$

as desired.   ∎

For example, let $k = 5$, $x = L_9/L_{10} = 76/123$, and $F_9 = 34$. Then

$$\sum_{i=0}^{\infty} \left(\frac{76}{123} \right)^i L_i = 123 \cdot 34 = 4182.$$

Interestingly, although when $x = L_{2k-1}/L_{2k}$, the Fibonacci power series (24.13) converges to a finite sum, it is *not* an integer:

$$\sum_{i=0}^{\infty} \left(\frac{L_{2k-1}}{L_{2k}} \right)^i F_i = \frac{L_{2k} L_{2k-1}}{5}; \tag{24.18}$$

see Exercise 24.44. For example,

$$\sum_{i=0}^{\infty} \left(\frac{L_9}{L_{10}} \right)^i F_i = \frac{123 \cdot 76}{5} = 1869.6.$$

As in the proof of Theorem 24.4, we can show that $\beta < F_{2k}/F_{2k+1} < -\beta$; so the Lucas series (24.16) converges to a finite sum when $x = F_{2k}/F_{2k+1}$, where $k \geq 0$. Then

$$\sum_{i=0}^{\infty} \left(\frac{F_{2k}}{F_{2k+1}} \right)^i L_i = \frac{2 - F_{2k}/F_{2k+1}}{1 - F_{2k}/F_{2k+1} - (F_{2k}/F_{2k+1})^2}$$

$$= \frac{F_{2k+1}(2F_{2k+1} - F_{2k})}{F_{2k+1}^2 - F_{2k}F_{2k+1} - F_{2k}^2}$$

$$= \frac{F_{2k+1}(F_{2k+1} + F_{2k-1})}{1}$$

$$= F_{2k+1}L_{2k},$$

again an integer. For the sake of brevity, we have omitted some details.

Thus we have the following result.

Theorem 24.5 (Koshy, 1999 [365]). *Let $k \geq 1$. Then $\sum_{i=0}^{\infty} \left(\dfrac{F_{2k}}{F_{2k+1}} \right)^i L_i = F_{2k+1}L_{2k}$.* ∎

For example, $\sum_{i=0}^{\infty} \left(\dfrac{F_6}{F_7} \right)^i L_i = F_7 L_6 = 13 \cdot 18 = 234$; and $\sum_{i=0}^{\infty} \left(\dfrac{F_{10}}{F_{11}} \right)^i L_i = F_{11} L_{10} = 89 \cdot 123 = 10{,}947$.

24.5 GIBONACCI SERIES

The next example involves a telescoping gibonacci sum (see Chapter 7).

Example 24.1. *Show that*

$$\sum_{k=1}^{\infty} \frac{1}{G_k G_{k+1}^2 G_{k+3}} + \sum_{k=1}^{\infty} \frac{1}{G_k G_{k+2}^2 G_{k+3}} = \frac{1}{ab(a+b)}.$$

Proof. For convenience, we let $p = G_k$, $q = G_{k+1}$, $r = G_{k+2}$, and $s = G_{k+3}$. Then $p + q = r$ and $q + r = s$. Then

$$\sum_{k=1}^{n} \frac{1}{pq^2s} + \sum_{k=1}^{n} \frac{1}{pr^2s} = \sum_{k=1}^{n} \left(\frac{1}{pq^2s} + \frac{1}{pr^2s} \right)$$

$$= \sum_{k=1}^{n} \left(\frac{r}{pq^2rs} + \frac{q}{pqr^2s} \right)$$

$$= \sum_{k=1}^{n} \left(\frac{s-q}{pq^2rs} + \frac{r-p}{pqr^2s} \right)$$

$$= \sum_{k=1}^{n} \left(\frac{1}{pq^2r} - \frac{1}{pqrs} + \frac{1}{pqrs} - \frac{1}{qr^2s} \right)$$

$$= \sum_{k=1}^{n} \left(\frac{1}{pq^2r} - \frac{1}{qr^2s} \right)$$

$$= \sum_{k=1}^{n} \left(\frac{1}{G_k G_{k+1}^2 G_{k+2}} - \frac{1}{G_{k+1} G_{k+2}^2 G_{k+3}} \right)$$

$$= \frac{1}{G_1 G_2^2 G_3} - \frac{1}{G_{n+1} G_{n+2}^2 G_{n+3}}$$

$$= \frac{1}{ab(a+b)} - \frac{1}{G_{n+1} G_{n+2}^2 G_{n+3}}$$

$$\lim_{n \to \infty} \text{LHS} = \frac{1}{ab(a+b)} - 0$$

$$= \frac{1}{ab(a+b)},$$

as required.   ∎

In particular, let $G_k = F_k$. Then

$$\sum_{k=1}^{\infty} \frac{1}{F_k F_{k+1}^2 F_{k+3}} + \sum_{k=1}^{\infty} \frac{1}{F_k F_{k+2}^2 F_{k+3}} = \frac{1}{2}.$$

Carlitz studied this special case in 1963 [100].

Suppose we let $G_k = L_k$. Then

$$\sum_{k=1}^{\infty} \frac{1}{L_k L_{k+1}^2 L_{k+3}} + \sum_{k=1}^{\infty} \frac{1}{L_k L_{k+2}^2 L_{k+3}} = \frac{1}{12}.$$

We can apply the same telescoping technique to derive the following summation formulas, where G_k denotes the kth gibonacci number:

$$\sum_{k=2}^{\infty} \frac{1}{G_{k-1}G_{k+1}} = \frac{1}{ab}$$

$$\sum_{k=2}^{\infty} \frac{G_k}{G_{k-1}G_{k+1}} = \frac{1}{a} + \frac{1}{b};$$

see Exercises 24.28 and 24.29.

In particular, $\sum_{k=2}^{\infty} \frac{1}{F_{k-1}F_{k+1}} = 1$ and $\sum_{k=2}^{\infty} \frac{F_k}{F_{k-1}F_{k+1}} = 2$. R.L. Graham of Bell Telephone Laboratories, Murray Hill, New Jersey (now called Lucent Technologies), developed these two formulas in 1963 [238, 239].

It also follows from the two gibonacci formulas that $\sum_{k=2}^{\infty} \frac{1}{L_{k-1}L_{k+1}} = \frac{1}{3}$ and

$$\sum_{k=2}^{\infty} \frac{L_k}{L_{k-1}L_{k+1}} = \frac{4}{3}.$$

24.6 ADDITIONAL FIBONACCI SERIES

Letting $x = 1/2$ in the power series

$$\frac{1}{1 - x - x^2} = \sum_{i=1}^{\infty} F_i x^{i-1}, \tag{24.19}$$

we get $\sum_{i=1}^{\infty} \frac{F_i}{2^{i-1}} = 4$; that is, $\sum_{i=1}^{\infty} \frac{F_i}{2^i} = 2$.

Differentiating the power series (24.19) with respect to x, we get

$$\frac{1 + 2x}{(1 - x - x^2)^2} = \sum_{i=1}^{\infty} (i - 1)F_i x^{i-2}.$$

Letting $x = 1/2$, this yields $\sum_{i=1}^{\infty} \frac{iF_i}{2^i} = 10$. Similarly, $\sum_{i=1}^{\infty} \frac{iL_i}{2^i} = 22$; see Exercise 24.7.

It follows by adding the sums $\sum_{k=1}^{\infty} \frac{F_k}{2^k} = 2$ and $\sum_{k=1}^{\infty} \frac{kF_k}{2^k} = 10$ that $\sum_{k=1}^{\infty} \frac{kF_{k-1}}{2^k} = 6$.

Using integration also, we can derive interesting summation formulas [156]. To demonstrate this technique, we let

$$s = \frac{x}{1 - x - x^2} = \sum_{k=0}^{\infty} F_k x^{k+1} \quad \text{and} \quad t = \sum_{k=1}^{\infty} (k + 1)F_k x^k.$$

Then

$$\int t\, dx = F_1 x^2 + F_2 x^3 + F_3 x^4 + \cdots + C$$

$$= xs + C,$$

where C is an arbitrary constant. This implies

$$t = s + x\frac{ds}{dx}$$

$$= \frac{x}{1 - x - x^2} + \frac{x(1 + x^2)}{(1 - x - x^2)^2}$$

$$\sum_{k=1}^{\infty}(k+1)F_k x^k = \frac{x(2-x)}{(1 - x - x^2)^2}.$$

Letting $x = 1/2$, this yields

$$\sum_{k=1}^{\infty}\frac{k+1}{2^k}F_k = 12.$$

Similarly, we can show that

$$\sum_{k=1}^{\infty}(k+1)^2 F_k x^k = \frac{(4 - 3x + 5x^2 - x^3)x}{(1 - x - x^2)^3}.$$

In particular,

$$\sum_{k=1}^{\infty}\frac{(k+1)^2}{2^k}F_k = 116;$$

see Exercises 24.8 and 24.9.

I.J. Good of Virginia Polytechnic Institute and State University at Blacksburg, Virginia, discovered the next result in 1974 [228].

Example 24.2. *Prove that* $\displaystyle\sum_{i=0}^{\infty}\frac{1}{F_{2^i}} = 4 - \alpha = 3 + \beta.$

Proof. First, we will establish by PMI that

$$\sum_{i=0}^{n}\frac{1}{F_{2^i}} = 3 - \frac{F_{2^n-1}}{F_{2^n}}. \tag{24.20}$$

When $n = 1$, LHS $= 2 = 3 - 1 =$ RHS; so the result is true when $n = 1$.

Now assume it works for all positive integers $\leq n$. Then

$$\sum_{i=0}^{n+1} \frac{1}{F_{2^i}} = 3 - \frac{F_{2^n-1}}{F_{2^n}} + \frac{1}{F_{2^{n+1}}}$$

$$= 3 - \frac{L_{2^n}F_{2^n-1}}{L_{2^n}F_{2^n}} + \frac{1}{F_{2^{n+1}}}$$

$$= 3 - \frac{L_{2^n}F_{2^n-1} - 1}{F_{2^{n+1}}}. \tag{24.21}$$

By Exercise 5.58, $F_{2^{n+1}-1} + F_1 = L_{2^n}F_{2^n-1}$; that is, $L_{2^n}F_{2^n-1} - 1 = F_{2^{n+1}-1}$. Then equation (24.21) becomes

$$\sum_{i=0}^{n+1} \frac{1}{F_{2^i}} = 3 - \frac{F_{2^{n+1}-1}}{F_{2^{n+1}}}.$$

Thus, by PMI, formula (24.20) is true for all positive integers n.
　　Consequently,

$$\lim_{n\to\infty} \sum_{i=0}^{n} \frac{1}{F_{2^i}} = 3 - \lim_{n\to\infty} \frac{F_{2^n-1}}{F_{2^n}}.$$

That is,

$$\sum_{i=0}^{\infty} \frac{1}{F_{2^i}} = 3 - (-\beta) = 3 + \beta = 4 - \alpha. \qquad\blacksquare$$

Next we pursue an interesting problem proposed by H.W. Gould of West Virginia University in 1963 [231].

Example 24.3. *Show that* $\dfrac{x(1-x)}{1 - 2x - 2x^2 + x^3} = \displaystyle\sum_{n=0}^{\infty} F_n^2 x^n.$

Proof. Let

$$\frac{x(1-x)}{1 - 2x - 2x^2 + x^3} = \sum_{n=0}^{\infty} a_n x^n. \tag{24.22}$$

Then $x - x^2 = \displaystyle\sum_{n=0}^{\infty} a_n(1 - 2x - 2x^2 + x^3)x^n$. Equating the coefficients of like terms, $a_0 = 0$, $1 = -2a_0 + a_1$, $-1 = -2a_0 - 2a_1 + a_2$, and $0 = a_{n-3} - 2a_{n-2} - 2a_{n-1} + a_n$, where $n \geq 3$. Thus $a_0 = 0$, $a_1 = 1 = a_2$, and

$$a_n = 2a_{n-1} + 2a_{n-2} - a_{n-3}, \tag{24.23}$$

where $n \geq 3$.

Since $a_n = F_n^2$ for $0 \leq n \leq 2$, it remains to show that a_n satisfies the same recurrence as F_n^2 for $n \geq 3$. This is so, since

$$F_n^2 = (F_{n-1} + F_{n-2})^2$$
$$= 2F_{n-1}^2 + 2F_{n-2}^2 - (F_{n-1} - F_{n-2})^2$$
$$= 2F_{n-1}^2 + 2F_{n-2}^2 - F_{n-3}^2.$$

Thus $a_n = F_n^2$ for $n \geq 0$; so

$$\frac{x(1-x)}{1 - 2x - 2x^2 + x^3} = \sum_{n=0}^{\infty} F_n^2 x^n.$$

∎

Let us pursue this example a bit further. The roots of the cubic equation $1 - 2x - 2x^2 + x^3 = (1 + x)(1 - 3x + x^2) = 0$ are -1, α^2, and β^2, of which β^2 has the least absolute value. Therefore, the power series (24.23) converges if and only if $|x| < \beta^2$, where $\beta^2 \approx 0.38196601125$.

In particular, the series converges when $x = 1/4$; and $\sum_{n=0}^{\infty} \frac{F_n^2}{4^n} = \frac{12}{25}$.

Next we list some summation formulas; Br. Alfred Brousseau developed them in 1969 [67].

SUMMATION FORMULAS

1) $\displaystyle\sum_{n=1}^{\infty} \frac{(-1)^{n-1}}{F_n F_{n+3}} = \frac{6\alpha - 9}{4}$

2) $\displaystyle\sum_{n=1}^{\infty} \frac{(-1)^{n-1} L_{2n+2}}{F_n F_{n+1} L_{n+1} L_{n+2}} = \frac{1}{3}$

3) $\displaystyle\sum_{n=1}^{\infty} \frac{(-1)^{n-1} F_{2n+2}}{L_n^2 L_{n+2}^2} = \frac{8}{45}$

4) $\displaystyle\sum_{n=1}^{\infty} \frac{1}{F_n F_{n+2} F_{n+3}} = \frac{1}{4}$

5) $\displaystyle\sum_{n=1}^{\infty} \frac{F_{n+3}}{F_n F_{n+2} F_{n+4} F_{n+6}} = \frac{17}{480}$

6) $\displaystyle\sum_{n=1}^{\infty} \frac{F_{4n+3}}{F_{2n} F_{2n+1} F_{2n+2} F_{2n+3}} = \frac{1}{2}$

7) $\displaystyle\sum_{n=1}^{\infty} \frac{L_{n+2}}{F_n F_{n+4}} = \frac{17}{6}$

8) $\displaystyle\sum_{n=1}^{\infty} \frac{(-1)^{n-1} F_{2n+1}}{F_n^2 F_{n+1}^2} = 1$

9) $\displaystyle\sum_{n=1}^{\infty} \frac{(-1)^{n-1} L_{n+1}}{F_n F_{n+1} F_{n+2}} = 1$

10) $\displaystyle\sum_{n=1}^{\infty} \frac{(-1)^{n-1}}{L_{3n} L_{3n+3}} = \frac{3 - \alpha}{40(1 + \alpha)}$

11) $\displaystyle\sum_{n=1}^{\infty} \frac{(-1)^{n-1} F_{6n+3}}{F_{3n}^3 F_{3n+3}^3} = \frac{1}{8}$

12) $\displaystyle\sum_{n=1}^{\infty} \frac{(-1)^{n-1}}{F_n F_{n+5}} = \frac{150 - 83\alpha}{105\alpha}$

13) $\displaystyle\sum_{n=1}^{\infty} \frac{(-1)^{n-1} F_{6n+3}}{F_{6n} F_{6n+6}} = \frac{1}{16}$

14) $\displaystyle\sum_{n=1}^{\infty} \frac{F_{2n+5}}{F_n F_{n+1} F_{n+2} F_{n+3} F_{n+4} F_{n+5}} = \frac{1}{15}$

15) $\displaystyle\sum_{n=1}^{\infty} \frac{(-1)^{n-1}}{F_{2n-1} F_{2n+3}} = \frac{1}{6}$

16) $\displaystyle\sum_{n=1}^{\infty} \frac{F_{2n}}{F_{n+2}^2 F_{n-2}^2} = \frac{85}{108}.$

EXERCISES 24

Evaluate each sum.

1. $\displaystyle\sum_{i=1}^{\infty} \frac{F_i}{2^i}$.

2. $\displaystyle\sum_{i=0}^{\infty} \frac{L_i}{3^i}$.

Consider the recurrence $u_n - (\alpha + \beta)u_{n-1} + \alpha\beta u_{n-2} = 0$, where $u_0 = a$ and $u_1 = b$.

3. Solve the recurrence.

4. Evaluate the sum $\displaystyle\sum_{i=0}^{\infty} \frac{u_i}{k^{i+1}}$, where $k > \alpha$ (Cross, 1966 [134]).

5. Evaluate the sum $\displaystyle\sum_{i=0}^{\infty} \frac{F_i}{k^{i+1}}$, where $k > a$.

6. Prove that $L_{2n}^2 - L_{2n}L_{2n-1} - L_{2n-1}^2 = 5$.

7. Prove that $\displaystyle\sum_{i=1}^{\infty} \frac{iL_i}{2^i} = 22$.

8. Prove that $\displaystyle\sum_{k=1}^{\infty} (k+1)^2 F_k x^k = \frac{(4 - 3x + 5x^2 - x^3)x}{(1 - x - x^2)^3}$.

9. Prove that $\displaystyle\sum_{k=1}^{\infty} \frac{(k+1)^2}{2^k} F_k = 116$.

10. Suppose $k > \alpha$. Show that the Lucas series $\displaystyle\sum_{i=0}^{\infty} \frac{L_i}{k^{i+1}}$ converges to the limit

$$S^* = \frac{2k - 1}{k^2 - k - 1}.$$

11. Show that $-m/n = F_{2k-1}/F_{2k} \notin (\beta, -\beta)$ (Glaister, 1994 [222]).

12. Show that the value of m corresponding to $n = -F_{2k}$ is F_{2k-1}, where $k \geq 1$ (Glaister, 1994 [222]).

13. Show that $F_{2k}/F_{2k-1} \notin (\beta, -\beta)$ (Glaister, 1994 [222]).

14. Show that $\beta < F_k/L_k < -\beta$, where $k \geq 0$.

15. Does the Fibonacci power series (24.13) converge when $x = F_k/L_k$, where $k \geq 0$?

16. Does the Lucas power series (24.16) converge when $x = F_k/L_k$, where $k \geq 0$?

17. Show that $U_n = Pr^n + Qs^n$, where P, Q, r, s, and n are defined as in Theorem 24.2.

Derive each.

18. Formula (24.7).

19. Formula (24.8).

20. Formula (24.9).

21. Show that the series $\sum\limits_{n=1}^{\infty} \dfrac{1}{F_n}$ converges (Lind, 1967 [402]).

22. Show that the series $\sum\limits_{n=3}^{\infty} \dfrac{1}{\ln F_n}$ diverges (Lind, 1967 [402]).

23. Show that $\sum\limits_{n=1}^{\infty} \dfrac{1}{F_n} > \dfrac{803}{240}$ (Guillotte, 1971 [251]).

24. Let $M(x) = \sum\limits_{n=1}^{\infty} \dfrac{L_n}{n} x^n$. Show that the Maclaurin expansion of $e^{M(x)}$ is $\sum\limits_{n=1}^{\infty} F_n x^{n-1}$ (Hoggatt, 1975 [299]).

Evaluate each infinite sum.

25. $\sum\limits_{n=1}^{\infty} \dfrac{1}{\alpha F_{n+1} + F_n}$ (Guillotte, 1971 [249]).

26. $\sum\limits_{n=1}^{\infty} \dfrac{1}{F_n + \sqrt{5}F_{n+1} + F_{n+2}}$ (Guillotte, 1971 [250]).

Prove each, where G_n denotes the nth gibonacci number.

27. $\sum\limits_{n=1}^{\infty} \dfrac{1}{F_n} = 3 + \sum\limits_{n=1}^{\infty} \dfrac{(-1)^{n+1}}{F_n F_{n+1} F_{n+2}}$ (Graham, 1963 [238]).

28. $\sum\limits_{k=2}^{\infty} \dfrac{1}{G_{k-1} G_{k+1}} = \dfrac{1}{ab}$.

29. $\sum\limits_{k=2}^{\infty} \dfrac{G_k}{G_{k-1} G_{k+1}} = \dfrac{1}{a} + \dfrac{1}{b}$.

30. $\sum\limits_{n=1}^{\infty} \dfrac{1}{G_n} = \dfrac{2}{a} + \dfrac{1}{b} + \sum\limits_{n=1}^{\infty} \dfrac{(-1)^{n+1}\mu}{G_n G_{n+1} G_{n+2}}$.

31. $\dfrac{F_{n+1}}{F_n} = 1 + \sum\limits_{i=2}^{n} \dfrac{(-1)^i}{F_i F_{i-1}}$ (Basin, 1964 [24]).

32. $\sum\limits_{i=2}^{n} \dfrac{(-1)^i}{F_i F_{i-1}} = |\beta|$ (Basin, 1964 [24]).

33. $\sum\limits_{n=1}^{\infty} \dfrac{F_{2n+1}}{L_n L_{n+1} L_{n+2}} = \dfrac{1}{3}$ (Ferns, 1967 [168]).

34. $\sum\limits_{n=0}^{\infty} \dfrac{1}{F_{2n+1}} = \sqrt{5} \sum\limits_{n=0}^{\infty} \dfrac{(-1)^n}{L_{2n+1}}$ (Carlitz, 1967 [106]).

35. $\sum\limits_{n=0}^{\infty} \dfrac{(-1)^n}{F_{4n+2}} = \sqrt{5} \sum\limits_{n=0}^{\infty} \dfrac{1}{L_{4n+2}}$ (Carlitz, 1967 [107]).

36. $\displaystyle\sum_{n=0}^{\infty} \frac{F_{n+1}}{2^n} = 4$ (Butchart, 1968 [92]).

37. $\displaystyle\sum_{n=1}^{\infty} \frac{1}{F_n F_{n+2}} = 1$ (Brousseau, 1969 [67]).

38. $\displaystyle\sum_{n=1}^{\infty} \frac{1}{F_n F_{n+4}} = \frac{7}{18}$ (Brousseau, 1969 [67]).

39. $\displaystyle\sum_{n=1}^{\infty} \frac{1}{F_n F_{n+2} F_{n+3}} = \frac{1}{4}$ (Brousseau, 1969 [67]).

40. $\displaystyle\sum_{n=1}^{\infty} \frac{F_n}{F_{n+1} F_{n+2}} = 1$ (Brousseau, 1969 [67]).

41. $\displaystyle\sum_{n=1}^{\infty} \frac{F_{n+1}}{F_n F_{n+3}} = \frac{5}{4}$ (Brousseau, 1969 [67]).

42. $\displaystyle\sum_{n=1}^{\infty} \frac{(-1)^{n-1}}{L_{n-1} L_n} = \frac{\sqrt{5}}{10}$ (Brousseau, 1969 [67]).

43. $\displaystyle\sum_{n=1}^{\infty} F_{2n-1} x^n = \frac{x - x^2}{1 - 3x + x^2}$, where $|x| < \beta^2$ (Hoggatt, 1971 [286]).

44. $\displaystyle\sum_{i=0}^{\infty} \left(\frac{L_{2k-1}}{L_{2k}}\right)^i F_i = \frac{L_{2k} L_{2k-1}}{5}$, where $k \geq 1$.

45. $\displaystyle\sum_{i=0}^{\infty} \left(\frac{F_k}{L_k}\right)^i F_i = \frac{F_{2k}}{L_k^2 - L_k F_k - F_k^2}$, where $k \geq 0$.

46. $\displaystyle\sum_{i=0}^{\infty} \left(\frac{F_k}{L_k}\right)^i L_i = \frac{2L_k - F_{2k}}{L_k^2 - L_k F_k - F_k^2}$, where $k \geq 0$.

25

WEIGHTED FIBONACCI AND LUCAS SUMS

Recall from Chapter 5 that

$$\sum_{i=1}^{n} F_i = F_{n+2} - 1$$

$$\sum_{i=1}^{n} L_i = L_{n+2} - 3.$$

Can we generalize these formulas using *weights* w_i, where w_i is a positive integer. Yes, as we will soon see.

25.1 WEIGHTED SUMS

To begin with, we will find a formula for each, with $w_i = i$; that is, a formula for $\sum_{i=1}^{n} iF_i$ and $\sum_{i=1}^{n} iL_i$.

Fibonacci and Lucas Numbers with Applications, Volume One, Second Edition. Thomas Koshy.
© 2018 John Wiley & Sons, Inc. Published 2018 by John Wiley & Sons, Inc.

To derive a formula for the weighted Fibonacci sum $\sum\limits_{i=1}^{n} iF_i$, we let $A_n = \sum\limits_{i=1}^{n} F_i$ and $B_n = \sum\limits_{i=1}^{n} iF_i$. Then

$$B_n = F_1 + 2F_2 + 3F_3 + \cdots + nF_n$$

$$= \sum_{i=1}^{n} F_i + \sum_{i=2}^{n} F_i + \sum_{i=3}^{n} F_i + \cdots + \sum_{i=n}^{n} F_i$$

$$= A_n + (A_n - A_1) + (A_n - A_2) + \cdots + (A_n - A_{n-1})$$

$$= nA_n - \sum_{i=1}^{n-1} A_i$$

$$= n(F_{n+2} - 1) - \sum_{i=1}^{n-1}(F_{i+2} - 1)$$

$$= nF_{n+2} - n - (F_{n+3} - 3) + (n - 1)$$

$$= nF_{n+2} - F_{n+3} + 2. \tag{25.1}$$

We can show similarly that

$$\sum_{i=1}^{n} iL_i = nL_{n+2} - L_{n+3} + 4. \tag{25.2}$$

For example,

$$\sum_{i=1}^{7} iF_i = 1 \cdot 1 + 2 \cdot 1 + 3 \cdot 2 + 4 \cdot 3 + 5 \cdot 5 + 6 \cdot 8 + 7 \cdot 13$$

$$= 185 = 7 \cdot 34 - 55 + 2$$

$$= 7F_9 - F_{10} + 2.$$

Similarly, $\sum\limits_{i=1}^{6} iL_i = 210 = 6L_8 - L_9 + 4$.

Now that we have formulas for B_n and $C_n = \sum\limits_{i=1}^{n} iL_i$, we can ask if there are formulas for

$$\sum_{i=1}^{n}(n - i + 1)F_i \quad \text{and} \quad \sum_{i=1}^{n}(n - i + 1)L_i.$$

To this end, notice that $B_n^* = \sum\limits_{i=1}^{n}(n - i + 1)F_i$ is the sum in formula (25.1) with the coefficients in the reverse order, and similarly for C_n^*.

For example, $B_1^* = F_1 = 1$, and $B_2^* = 2F_1 + F_2 = 2 + 1 = 3$. Similarly, $B_3^* = 7$, $B_4^* = 14$, $B_5^* = 26$, and $B_6^* = 46$. Although the values do not appear to follow an obvious pattern, we can easily derive a formula for B_n^*. Using formula (25.1), we have

$$B_n + B_n^* = \sum_{i=1}^{n} iF_i + \sum_{i=1}^{n} (n - i + 1)F_i$$

$$= \sum_{i=1}^{n} [i + (n - i + 1)]F_i$$

$$= (n + 1) \sum_{i=1}^{n} F_i$$

$$= (n + 1)(F_{n+2} - 1)$$

so

$$B_n^* = (n + 1)(F_{n+2} - 1) - B_n$$

$$= (n + 1)(F_{n+2} - 1) - (nF_{n+2} - F_{n+3} + 2)$$

$$= F_{n+4} - n - 3. \tag{25.3}$$

For example, $B_6^* = F_{10} - 6 - 3 = 55 - 9 = 46$, as expected.
Using the same technique, we can show that

$$\sum_{i=1}^{n} (n - i + 1)L_i = L_{n+4} - 3n - 7. \tag{25.4}$$

For example,

$$\sum_{i=1}^{5} (6 - i)L_i = 5L_1 + 4L_2 + 3L_3 + 2L_4 + L_5$$

$$= 54$$

$$= L_9 - 15 - 7.$$

Formula (25.1) tempts us to investigate Fibonacci sums with integer coefficients and subscripts with the same parity; that is, the sums $\sum_{i=1}^{n}(2i - 1)F_{2i-1}$ and $\sum_{i=1}^{n}(2i)F_{2i}$, and then the same sums with the coefficients reversed.

To derive a formula for $D_n = \sum_{i=1}^{n}(2i - 1)F_{2i-1}$, we first let $E_n = \sum_{i=1}^{n} F_{2i-1} = F_{2n}$; see formula (5.2). Then

$$D_n = F_1 + 3F_3 + 5F_5 + \cdots + (2n - 1)F_{2n-1}$$

$$= \sum_{i=1}^{n} F_{2i-1} + 2\sum_{i=2}^{n} F_{2i-1} + 2\sum_{i=3}^{n} F_{2i-1} + \cdots + 2\sum_{i=n}^{n} F_{2i-1}$$

$$= E_n + 2(E_n - E_1) + 2(E_n - E_2) + \cdots + 2(E_n - E_{n-1})$$

$$= E_n + 2(n - 1)E_n - 2\sum_{i=1}^{n-1} E_i$$

$$= (2n - 1)E_n - 2\sum_{i=1}^{n-1} F_{2j}$$

$$= (2n - 1)E_n - 2(F_{2n-1} - 1)$$

$$= (2n - 1)F_{2n} - 2F_{2n-1} + 2. \tag{25.5}$$

For example,

$$D_4 = \sum_{i=1}^{4} (2i - 1)F_{2i-1}$$

$$= 7F_8 - 2F_7 + 2$$

$$= 123.$$

You may confirm this by direct computation.
   Using formula (25.5), we now have

$$D_n + D_n^* = \sum_{i=1}^{n} (2i - 1)F_{2i-1} + \sum_{i=1}^{n} (2n - 2i + 1)F_{2i-1}$$

$$= 2n \sum_{i=1}^{n} F_{2i-1}$$

$$= 2nF_{2n}$$

so

$$D_n^* = 2nF_{2n} - D_n$$

$$= 2nF_{2n} - [(2n - 1)F_{2n} - 2F_{2n-1} + 2]$$

$$= F_{2n} + 2F_{2n-1} - 2$$

$$= F_{2n+1} + F_{2n-1} - 2$$

$$= L_{2n} - 2. \tag{25.6}$$

For example,

$$D_5^* = 9F_1 + 7F_3 + 5F_5 + 3F_7 + F_9$$

$$= 123 - 2$$

$$= L_{10} - 2.$$

Using formulas (5.3), (25.5), and (25.6), we can show that

$$\sum_{i=1}^{n}(2i)F_{2i} = 2(nF_{2n+1} - F_{2n}) \tag{25.7}$$

$$\sum_{i=1}^{n}(2n - 2i + 2)F_{2i} = 2F_{2n+2} - 2n - 2; \tag{25.8}$$

see Exercises 25.1 and 25.2.

For example,

$$\sum_{i=1}^{5}(2i)F_{2i} = 2(5F_{11} - F_{10}) = 780$$

$$\sum_{i=1}^{4}(10 - 2i)F_{2i} = 2F_{10} - 8 - 2 = 100.$$

You may confirm both by direct computation.

LUCAS COUNTERPARTS

Formulas (25.5) through (25.8) have analogous results for the Lucas family; we can derive and verify them as illustrated previously.

$$\sum_{i=1}^{n}(2i - 1)L_{2i-1} = (2n - 1)L_{2n} - 2L_{2n-1} \tag{25.9}$$

$$\sum_{i=1}^{n}(2n - 2i + 1)L_{2i-1} = 5F_{2n} - 4n \tag{25.10}$$

$$\sum_{i=1}^{n}(2i)L_{2i} = 2nL_{2n+1} - 2L_{2n} + 4 \tag{25.11}$$

$$\sum_{i=1}^{n}(2n - 2i + 2)L_{2i} = 2L_{2n+2} - 2n - 6. \tag{25.12}$$

Their proofs employ formulas (5.7), (5.8), and (25.1); they can be established using PMI also; see Exercises 25.3–25.6.

Interestingly, we can extend formula (25.1) to any Fibonacci sum, where the coefficients form an arbitrary arithmetic sequence with the first term a and common difference d. For example, let $S_n = \sum_{i=1}^{n} [a + (i-1)d]F_i$. Then

$$S_n = a\sum_{i=1}^{n} F_i + d\left(\sum_{i=1}^{n} iF_i\right) - d\left(\sum_{i=1}^{n} F_i\right)$$

$$= a(F_{n+2} - 1) + d(nF_{n+2} - F_{n+3} + 2) - d(F_{n+2} - 1)$$

$$= (a + nd - d)F_{n+2} - d(F_{n+3} - 3) - a. \tag{25.13}$$

Analogously, we have

$$\sum_{i=1}^{n} [a + (i-1)d]L_i = (a + nd - d)L_{n+2} - d(L_{n+3} - 7) - 3a; \tag{25.14}$$

see Exercise 25.7.

In particular,

$$\sum_{i=1}^{n} L_i = L_{n+2} - 3 \quad \text{and} \quad \sum_{i=1}^{n} iL_i = nL_{n+2} - L_{n+3} + 4.$$

Let S_n^* denote the Fibonacci sum in formula (25.13) with the coefficients reversed: $S_n^* = \sum_{i=1}^{n} [a + (n-i)d]F_i$. Then

$$S_n + S_n^* = \sum_{i=1}^{n} \{[a + (i-1)d] + [a + (n-i)d]\}\, F_i$$

$$= [2a + (n-1)d]\sum_{i=1}^{n} F_i$$

$$= [2a + (n-1)d](F_{n+2} - 1);$$

$$S_n^* = [2a + (n-1)d](F_{n+2} - 1) - [(a + nd - d)F_{n+2} - d(F_{n+3} - 3) - a]$$

$$= [2a + (n-1)d - (a + nd - d)]F_{n+2} - [2a + (n-1)d] + d(F_{n+3} - 3) + a$$

$$= aF_{n+2} + d(F_{n+3} - 3) - a - (n-1)d. \tag{25.15}$$

When $a = 1 = d$, this reduces to formula (25.3).

Using the same technique, we can show that

$$\sum_{i=1}^{n} [a + (n-i)d]L_i = aL_{n+2} + d(L_{n+3} - 7) - 3[a + (n-1)d]; \tag{25.16}$$

see Exercise 25.8.

Using the facts that

$$\sum_{i=1}^{n} F_i^2 = F_n F_{n+1} \quad \text{and} \quad \sum_{i=1}^{n} F_i F_{i+1} = \begin{cases} F_n^2 - 1 & \text{if } n \text{ is odd} \\ F_n^2 & \text{otherwise,} \end{cases}$$

we can show that

$$\sum_{i=1}^{n} [a + (i-1)d]F_i^2 = (a + nd - d)F_n F_{n+1} - d(F_n^2 - \gamma), \tag{25.17}$$

where

$$\gamma = \begin{cases} 1 & \text{if } n \text{ is odd} \\ 0 & \text{otherwise;} \end{cases}$$

see Exercise 25.9.

In particular, this yields

$$\sum_{i=1}^{n} iF_i^2 = nF_n F_{n+1} - F_n^2 + \gamma. \tag{25.18}$$

For example,

$$\sum_{i=1}^{5} iF_i^2 = 5F_5 F_6 - F_5^2 + 1 = 176, \quad \text{and} \quad \sum_{i=1}^{6} iF_i^2 = 6F_6 F_7 - F_6^2 + 0 = 560.$$

Again, you may confirm these by direct computation.

Let $D_n^* = \sum_{i=1}^{n} [a + (n-i)d]F_i^2$, the same sum as in (25.17) with the coefficients in the reversed order. Then

$$D_n + D_n^* = [2a + (n-1)d] \sum_{i=1}^{n} F_i^2$$

$$= [2a + (n-1)d]F_n F_{n+1};$$

$$D_n^* = [2a + (n-1)d]F_n F_{n+1} - (a + nd - d)F_n F_{n+1} + d(F_n^2 - \gamma)$$

$$= aF_n F_{n+1} + d(F_n^2 - \gamma). \tag{25.19}$$

In particular,

$$\sum_{i=1}^{n} (n - i + 1)F_i^2 = F_n F_{n+1} + F_n^2 - \gamma. \tag{25.20}$$

For example, $\sum_{i=1}^{5} (6 - i)F_i^2 = F_5 F_6 + F_5^2 - 1 = 64.$

Fortunately, formulas (25.17) and (25.19) have analogous counterparts for the Lucas family:

$$\sum_{i=1}^{n}[a + (i-1)d]L_i^2 = (a + nd - d)(L_n L_{n+1} - 2) - d(L_n^2 - 2n - v) \qquad (25.21)$$

$$\sum_{i=1}^{n}[a + (n-i)d]L_i^2 = a(L_n L_{n+1} - 2) + d(L_n^2 - 2n - v), \qquad (25.22)$$

where

$$v = \begin{cases} -1 & \text{if } n \text{ is odd} \\ 4 & \text{otherwise.} \end{cases}$$

It follows from identities (25.21) and (25.22) that

$$\sum_{i=1}^{n} iL_i^2 = nL_n L_{n+1} - L_n^2 + v \qquad (25.23)$$

$$\sum_{i=1}^{n}(n - i + 1)L_i^2 = L_n L_{n+2} - 2n - v - 2; \qquad (25.24)$$

see Exercises 25.10–25.13.

For example,

$$\sum_{i=1}^{6} iL_i^2 = 6L_6 L_7 - L_6^2 + 4 = 2812$$

$$\sum_{i=1}^{5}(6 - i)L_i^2 = L_5 L_7 - 10 + 1 - 2 = 308.$$

You may verify both by direct computation.

25.2 GAUTHIER'S DIFFERENTIAL METHOD

Thus far we have found formulas for the sums $S(m) = \sum_{i=1}^{n} i^m F_i$ and $T(m) = \sum_{i=1}^{n} i^m L_i$, with $m = 0$ and 1.

New formulas corresponding to $m = 2$ and $m = 3$ were developed algebraically by P. Glaister of the University of Reading, England [224], and

N. Gauthier of The Royal Military College, Kingston, Canada [215]:

$$\sum_{i=1}^{n} i^2 F_i = (n+1)^2 F_{n+2} - (2n+3)F_{n+4} + 2F_{n+6} - 8 \qquad (25.25)$$

$$\sum_{i=1}^{n} i^2 L_i = (n+1)^2 L_{n+2} - (2n+3)L_{n+4} + 2L_{n+6} - 18. \qquad (25.26)$$

Gauthier re-discovered formulas for $S(1)$ and $S(2)$ using a fascinating method involving the differential operator $x\dfrac{d}{dx}$. In the interest of brevity and economy, we will denote it by ∇. In addition to giving a general method for computing $S(m)$ in 1995, Gauthier gave an explicit formula for S_3 also.

Interestingly, we can employ Gauthier's differential method to derive a formula for $T(m)$ also. To see this, we need to use Binet's formulas, and the facts $1/(1-\alpha)^i = (-\alpha)^i$ and $1/(1-\beta)^i = (-\beta)^i$. Suppose we have a formula for $T(m)$. Since we can obtain F_i from L_i by changing β^i to $-\beta^i$, and then dividing the difference by $\sqrt{5}$, we can extract a formula for $S(m)$ from that of $T(m)$.

To derive a formula for $T(m)$, notice that $L_i = \alpha^i + \beta^i = (x^i)_{x=\alpha} + (x^i)_{x=\beta}$; we will abbreviate this by $L_i = (x^i)_\alpha + (x^i)_\beta$.

Let

$$f(x) = \sum_{i=1}^{n} x^i = x + x^2 + \cdots + x^n = \frac{1-x^{n+1}}{1-x} - 1,$$

where $x \neq 1$. Then $x\dfrac{df}{dx} = \sum_{i=1}^{n} ix^i$; that is, $\nabla f = \sum_{i=1}^{n} ix^i$. Then $\nabla^2 f = \nabla(\nabla f) = \sum_{i=1}^{n} i^2 x^i$. More generally,

$$\nabla^m f = \sum_{i=1}^{n} i^m x^i, \qquad (25.27)$$

where $m \geq 0$ and $\nabla^0 f = f$.

By formula (25.27), we have

$$\sum_{i=1}^{n} L_i = \left(\frac{1-\alpha^{n+1}}{1-\alpha} - 1 \right) + \left(\frac{1-\beta^{n+1}}{1-\beta} - 1 \right)$$

$$= \left(\frac{1}{1-\alpha} + \frac{1}{1-\beta} \right) - \left(\frac{\alpha^{n+1}}{1-\alpha} + \frac{\beta^{n+1}}{1-\beta} \right) - 2$$

$$= -(\alpha+\beta) + (\alpha^{n+2} + \beta^{n+2}) - 2$$

$$= L_{n+2} - 3. \qquad (25.28)$$

This is formula (5.6).

We can rewrite this formula as $\sum_{i=1}^{n} L_i = L_{n+2} - L_2$. Now change β^i to $-\beta^i$, and divide the resulting equation by $\sqrt{5}$. This yields $\sum_{i=1}^{n} F_i = F_{n+2} - F_2 = F_{n+2} - 1$. In other words, we can obtain this by simply changing L_i to F_i in formula (25.28).

Let $m \geq 1$ and $g(x) = \dfrac{1 - x^{n+1}}{1 - x}$. Then $\nabla g = \nabla f$, and hence $\nabla^m g = \sum_{i=1}^{n} i^m x^i$.

Thus $(\nabla^m g)_\alpha = \sum_{i=1}^{n} i^m \alpha^i$; so by Binet's formula,

$$\sum_{i=1}^{n} i^m L_i = (\nabla^m g)_\alpha + (\nabla^m g)_\beta. \qquad (25.29)$$

This gives a general formula for computing $T(m)$.

Notice that

$$g(x) = \frac{1}{1 - x} - \frac{x^{n+1}}{1 - x} = g_0(x) - g_{n+1}(x),$$

where $g_t = \dfrac{x^t}{1 - x}$. Then $\nabla g = \nabla(g_0 - g_{n+1}) = \nabla g_0 - \nabla g_{n+1}$; and more generally, $\nabla^m g = \nabla^m g_0 - \nabla^m g_{n+1}$, which we can find from $\nabla^m g_t$. Thus we can refine formula (25.29):

$$\sum_{i=1}^{n} i^m L_i = [(\nabla^m g_0)_\alpha + (\nabla^m g_0)_\beta] - [(\nabla^m g_{n+1})_\alpha + (\nabla^m g_{n+1})_\beta]. \qquad (25.30)$$

Since

$$\nabla g_t = t \frac{x^t}{1 - x} + \frac{x^{t+1}}{(1 - x)^2},$$

we have

$$\nabla g_0 = \frac{x}{(1 - x)^2} \quad \text{and} \quad \nabla g_{n+1} = (n + 1) \frac{x^n}{1 - x} + \frac{x^{n+2}}{(1 - x)^2}.$$

Consequently, when $m = 1$, formula (25.30) gives

$$\sum_{i=1}^{n} i L_i = \left[\frac{\alpha}{(1 - \alpha)^3} + \frac{\beta}{(1 - \beta)^3} \right] - (n + 1)\left(\frac{\alpha^{n+1}}{1 - \alpha} + \frac{\beta^{n+1}}{1 - \beta} \right)$$

$$- \left[\frac{\alpha^{n+2}}{(1 - \alpha)^2} + \frac{\beta^{n+2}}{(1 - \beta)^2} \right]$$

$$= (\alpha^3 + \beta^3) + (n + 1)(\alpha^{n+2} + \beta^{n+2}) - (\alpha^{n+4} + \beta^{n+4})$$

$$= (n + 1)L_{n+2} - L_{n+4} + 4, \qquad (25.31)$$

which is formula (25.2).

Changing L_i to F_i in equation (25.31), we get

$$\sum_{i=1}^{n} iF_i = (n+1)F_{n+2} - F_{n+4} + 2,$$

which is formula (25.1).

Now consider the case $m = 2$. Since

$$\nabla^2 g_t = \nabla(\nabla g_t) = t^2 \frac{x^t}{1-x} - (2t+1)\frac{x^{t+1}}{(1-x)^2} + 2\frac{x^{t+2}}{(1-x)^3},$$

we have

$$\sum_{i=1}^{n} i^2 L_i = \left[\frac{\alpha}{(1-\alpha)^2} + \frac{\beta}{(1-\beta)^2}\right] + 2\left[\frac{\alpha^2}{(1-\alpha)^3} + \frac{\beta^2}{(1-\beta)^3}\right]$$

$$- (n+1)^2\left(\frac{\alpha^{n+1}}{1-\alpha} + \frac{\beta^{n+1}}{1-\beta}\right) - (2n+3)\left[\frac{\alpha^{n+2}}{(1-\alpha)^2} + \frac{\beta^{n+2}}{(1-\beta)^2}\right]$$

$$- 2\left[\frac{\alpha^{n+3}}{(1-\alpha)^3} + \frac{\beta^{n+3}}{(1-\beta)^3}\right]$$

$$= (\alpha^3 + \beta^3) - 2(\alpha^5 + \beta^5) + (n+1)^2(\alpha^{n+2} + \beta^{n+2})$$

$$- (2n+3)(\alpha^{n+4} + \beta^{n+4}) + 2(\alpha^{n+6} + \beta^{n+6})$$

$$= (n+1)^2 L_{n+2} - (2n+3)L_{n+4} + 2L_{n+6} + L_3 - 2L_5,$$

which is formula (25.26).

Replacing L_j with F_j, this yields

$$\sum_{i=1}^{n} i^2 F_i = (n+1)^2 F_{n+2} - (2n+3)F_{n+4} + 2F_{n+6} + F_3 - 2F_5.$$

This is formula (25.25).

We will now develop formulas for $T(3)$ and hence $S(3)$. We have

$$\nabla^3 g_t = \nabla(\nabla^2 g_t)$$

$$= t^3 \frac{x^t}{1-x} + (3t^2 + 3t + 1)\frac{x^{t+1}}{(1-x)^2} + (6t+6)\frac{x^{t+2}}{(1-x)^3} + \frac{x^{t+3}}{(1-x)^4}.$$

Then

$$(\nabla^3 g_0)_\alpha + (\nabla^3 g_0)_\beta = \left[\frac{\alpha}{(1-\alpha)^2} + \frac{\beta}{(1-\beta)^2}\right] + 6\left[\frac{\alpha^2}{(1-\alpha)^3} + \frac{\beta^2}{(1-\beta)^3}\right]$$

$$+ 6\left[\frac{\alpha^3}{(1-\alpha)^4} + \frac{\beta^3}{(1-\beta)^4}\right]$$

$$= (\alpha^3 + \beta^3) - 6(\alpha^5 + \beta^5) + 6(\alpha^7 + \beta^7)$$

$$= L_3 - 6L_5 + 6L_7.$$

Likewise,

$$(\nabla^3 g_{n+1})_\alpha + (\nabla^3 g_{n+1})_\beta = -(n+1)^3 L_{n+2} + (3n^2 + 9n + 7)L_{n+4}$$
$$- (6n+12)L_{n+6} + 6L_{n+8}.$$

Thus

$$\sum_{i=1}^{n} i^3 L_i = (n+1)^3 L_{n+2} - (3n^2 + 9n + 7)L_{n+4} + (6n+12)L_{n+6} - 6L_{n+8}$$

$$+ L_3 - 6L_5 + 6L_7$$

$$= (n+1)^3 L_{n+2} - (3n^2 + 9n + 7)L_{n+4} + (6n+12)L_{n+6} - 6L_{n+8} + 112.$$
$$(25.32)$$

Changing L_j to F_j, this yields the formula

$$\sum_{i=1}^{n} i^3 F_i = (n+1)^3 F_{n+2} - (3n^2 + 9n + 7)F_{n+4} + (6n+12)F_{n+6} - 6F_{n+8}$$

$$+ F_3 - 6F_5 + 6F_7 \qquad\qquad (25.33)$$

$$= (n+1)^3 F_{n+2} - (3n^2 + 9n + 7)F_{n+4} + (6n+12)F_{n+6} - 6F_{n+8} + 50.$$
$$(25.34)$$

Clearly, we can continue this procedure for an arbitrary positive integer m. For the curious-minded, we give the formulas for $T(4)$ and $S(4)$:

$$\sum_{i=1}^{n} i^4 L_i = (n+1)^4 L_{n+2} - (4n^3 + 18n^2 + 28n + 15)L_{n+4} + (12n^2 + 48n + 50)L_{n+6}$$

$$- (24n+60)L_{n+8} + 24L_{n+10} + L_3 - 14L_5 + 36L_7 - 24L_9$$

$$= (n+1)^4 L_{n+2} - (4n^3 + 18n^2 + 28n + 15)L_{n+4} + (12n^2 + 48n + 50)L_{n+6}$$

$$- (24n+60)L_{n+8} + 24L_{n+10} - 930. \qquad\qquad (25.35)$$

Consequently,

$$\sum_{i=1}^{n} i^4 F_i = (n+1)^4 F_{n+2} - (4n^3 + 18n^2 + 28n + 15)F_{n+4} + (12n^2 + 48n + 50)F_{n+6}$$

$$- (24n + 60)F_{n+8} + 24F_{n+10} + F_3 - 14F_5 + 36F_7 - 24F_9 \qquad (25.36)$$

$$= (n+1)^4 F_{n+2} - (4n^3 + 18n^2 + 28n + 15)F_{n+4} + (12n^2 + 48n + 50)F_{n+6}$$

$$- (24n + 60)F_{n+8} + 24F_{n+10} - 416. \qquad (25.37)$$

For example, by formula (25.35),

$$\sum_{i=1}^{5} i^4 L_i = 1296L_7 - 1105L_9 + 590L_{11} - 180L_{13} + 24L_{15} - 930 = 9040.$$

You may verify this by direct computation.

A few interesting observations about the formulas for $S(m)$ and $T(m)$:

- Both $S(m)$ and $T(m)$ contain $m + 2$ terms.
- The coefficients in $S(m)$ and $T(m)$ alternate their signs, and the corresponding coefficients are identical.
- The leading term in $S(m)$ is $(n+1)^m F_{n+2}$, and that in $T(m)$ is $(n+1)^m L_{n+2}$. The subscripts in the Fibonacci and Lucas sums increase by 2, while the exponent of n in each coefficient decreases by one.
- We can obtain the formula for $S(m)$ from that of $T(m)$ and vice versa, by switching F_j and L_j.
- Except for the trailing constant term, we can obtain the formula for $S(m-1)$ from that of $S(m)$. The same is true for $T(m)$.

For example, consider formula (25.36) for $S(4)$. The nonconstant coefficients on the RHS are $(n+1)^4$, $-(4n^3 + 18n^2 + 28n + 15)$, $12n^2 + 48n + 50$, $-(24n + 60)$, and 24. Their derivatives with respect to n are $4(n+1)^3$, $-4(3n^2 + 9n + 7)$, $4(6n + 12)$, $4(-6)$ and 0. The derivative of i^4 with respect to i is $4i^3$. Dividing them by 4, we get the nonconstant coefficients in $S(3)$:

$$\sum_{i=1}^{n} i^3 F_i = (n+1)^3 F_{n+2} - (3n^2 + 9n + 7)F_{n+4} - (6n + 12)F_{n+6} + 6F_{n+8} + k,$$

where k is a constant; this is clearly consistent with the formula (25.33).

In fact, $k = (\nabla^m g_0)_\alpha - (\nabla^m g_0)_\beta$ in the case of $S(m)$, and $k = (\nabla^m g_0)_\alpha + (\nabla^m g_0)_\beta$ in the case of $T(m)$. For example, when $m = 3$, $k = (\nabla^3 g_0)_\alpha - (\nabla^3 g_0)_\beta = F_3 - 6F_5 + 6F_7 = 50$, as obtained earlier.

On the other hand, if we could use the coefficients in $S(m-1)$ to determine those in $S(m)$, it would be a tremendous advantage in the study of weighted Fibonacci and Lucas sums. The same would hold for $T(m)$ as well.

EXERCISES 25

Prove each.

1. $\sum_{i=1}^{n}(2i)F_{2i} = 2(nF_{2n+1} - F_{2n})$.

2. $\sum_{i=1}^{n}(2n - 2i + 2)F_{2i} = 2F_{2n+2} - 2n - 2$.

3. $\sum_{i=1}^{n}(2i - 1)L_{2i-1} = (2n - 1)L_{2n} - 2L_{2n-1}$.

4. $\sum_{i=1}^{n}(2n - 2i + 1)L_{2i-1} = L_{2n+1} + L_{2n-1} - 4n$.

5. $\sum_{i=1}^{n}(2i)L_{2i} = 2nL_{2n+1} - 2L_{2n} + 4$.

6. $\sum_{i=1}^{n}(2n - 2i + 2)L_{2i} = 2L_{2n+2} - 2n - 6$.

7. $\sum_{i=1}^{n}[a + (i - 1)d]L_i = (a + nd - d)L_{n+2} - d(L_{n+3} - 7) - 3a$.

8. $\sum_{i=1}^{n}[a + (n - i)d]L_i = aL_{n+2} + d(L_{n+3} - 7) - 3[a + (n - 1)d]$.

9. $\sum_{i=1}^{n}[a + (i - 1)d]F_i^2 = (a + nd - d)F_nF_{n+1} - d(F_n^2 - \gamma)$.

10. $\sum_{i=1}^{n} iL_i^2 = nL_nL_{n+1} - L_n^2 + v$.

11. $\sum_{i=1}^{n}(n - i + 1)L_i^2 = L_nL_{n+2} - 2(n + 1) - v$.

12. $\sum_{i=1}^{n}[a + (i - 1)d]L_i^2 = (a + nd - d)(L_nL_{n+1} - 2) - d(L_n^2 - 2n - v)$.

13. $\sum_{i=1}^{n}[a + (n - i)d]L_i^2 = a(L_nL_{n+1} - 2) + d(L_n^2 - 2n - v)$.

Let G_i denote the ith gibonacci number. Derive a formula for each sum.

14. $\sum_{i=1}^{n} G_i$.

15. $\sum_{i=1}^{n} iG_i$.

16. $\sum_{i=1}^{n}(n - i + 1)G_i$.

17. $\sum_{i=1}^{n} G_{2i-1}$.

18. $\displaystyle\sum_{i=1}^{n} G_{2i}$.

19. $\displaystyle\sum_{i=1}^{n}(2i-1)G_{2i-1}$.

20. $\displaystyle\sum_{i=1}^{n}(2n-2i+1)G_{2i-1}$.

Compute each sum.

21. $\displaystyle\sum_{i=1}^{10} i^2 F_i$.

22. $\displaystyle\sum_{i=1}^{10} i^2 L_i$.

23. $\displaystyle\sum_{i=1}^{5} i^3 F_i$.

24. $\displaystyle\sum_{i=1}^{5} i^3 L_i$.

25. $\displaystyle\sum_{i=1}^{5} i^4 F_i$.

26. $\displaystyle\sum_{i=1}^{5} i^4 L_i$.

Verify each formula for $n = 6$.

27. Formula (25.25).
28. Formula (25.26).
29. Formula (25.32).
30. Formula (25.34).
31. Formula (25.35).

32. Establish formula (25.25) algebraically.
33. Establish formula (25.26) algebraically.

26

FIBONOMETRY I

Several well-known trigonometric formulas are intimately tied with Fibonacci and Lucas numbers via the trigonometric functions.

For example, a number of interesting relationships exist linking Fibonacci and Lucas numbers with the inverse tangent function $\tan^{-1}$, and the ubiquitous irrational number π:

$$
\begin{aligned}
\frac{\pi}{4} &= \tan^{-1}\frac{1}{1} \\
&= \tan^{-1}\frac{1}{2} + \tan^{-1}\frac{1}{3} \\
&= 2\tan^{-1}\frac{1}{3} + \tan^{-1}\frac{1}{7} \\
&= \tan^{-1}\frac{1}{2} + \tan^{-1}\frac{1}{5} + \tan^{-1}\frac{1}{8} \quad \text{(Dase, 1844)} \\
&= 2\tan^{-1}\frac{1}{5} + \tan^{-1}\frac{1}{7} + 2\tan^{-1}\frac{1}{8} \\
&= \tan^{-1}\frac{1}{3} + \tan^{-1}\frac{1}{5} + \tan^{-1}\frac{1}{7} + \tan^{-1}\frac{1}{8}.
\end{aligned}
$$

Fibonacci and Lucas Numbers with Applications, Volume One, Second Edition. Thomas Koshy.
© 2018 John Wiley & Sons, Inc. Published 2018 by John Wiley & Sons, Inc.

26.1   GOLDEN RATIO AND INVERSE TRIGONOMETRIC FUNCTIONS

Several interesting relationships link the golden ratio with inverse trigonometric functions.

Since $\cos^{-1} x = \sin^{-1} \sqrt{1 - x^2}$ for $0 \le x \le 1$, it follows that $\cos^{-1}(1/\alpha) = \cos^{-1}(-\beta) = \sin^{-1} \sqrt{1 - \beta^2} = \sin^{-1} \sqrt{-\beta} = \sin^{-1}(1/\sqrt{\alpha})$. Likewise, $\sin^{-1}(1/\alpha) = \cos^{-1} \sqrt{1 - \beta^2} = \cos^{-1}(1/\sqrt{\alpha})$.

Suppose $\tan x = \cos x$. Then $\sin x = \cos^2 x$, so $\sin^2 x + \sin x - 1 = 0$. Since $\sin x \ge 0$, it follows that $\sin x = -\beta = 1/\alpha$, and $x = \sin^{-1}(1/\alpha)$. Then $\tan(\sin^{-1} 1/\alpha) = \cos(\sin^{-1} 1/\alpha) = \cos(\cos^{-1}(1/\sqrt{\alpha})) = 1/\sqrt{\alpha}$. In addition, $\cot(\cos^{-1}(1/\alpha)) = 1/\sqrt{\alpha} = \sin(\cos^{-1} 1/\alpha)$. Br. L. Raphael of St. Mary's College, California, studied these results in 1970 [494]; they are pictorially represented in Figure 26.1.

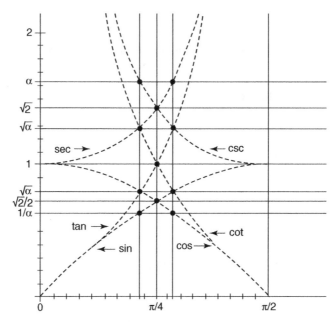

Figure 26.1. The golden ratio and trigonometric functions. *Source*: Raphael, 1970 [494]. Reproduced with permission of the Fibonacci Association.

We will now use the golden triangle with vertex angle 36° to extract a surprising trigonometric result.

26.2 GOLDEN TRIANGLE REVISITED

The exact trigonometric values of some acute angles are known. The smallest such integral angle is $\theta = 3°$. Oddly enough, we can use the golden triangle to compute the exact value of $\sin 3°$. Once we know it, we can easily compute the exact values of the remaining trigonometric functions of $3°$, and hence the trigonometric values of multiples of $3°$ using the sum formulas.

Recall from Chapter 17 that the ratio of a lateral side to the base of the golden triangle with vertex angle $36°$ is the golden ratio α: $AB/AC = x/y = \alpha$; see Figure 26.2. In particular, let $x = 1$. Then $1/y = \alpha$, so $y = -\beta = (\sqrt{5} - 1)/2$.

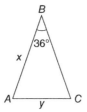

Figure 26.2.

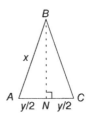

Figure 26.3.

Let $\overline{BN}$ be the perpendicular bisector of $\overline{AC}$; see Figure 26.3. Then

$$\sin 18° = \frac{AN}{BN} = \frac{y}{2} = \frac{1}{2\alpha} = \frac{\sqrt{5} - 1}{4}.$$

Since $\sin^2 u + \cos^2 u = 1$, this implies

$$\cos 18° = \frac{\sqrt{10 + 2\sqrt{5}}}{4} = \frac{\sqrt{\sqrt{5}\alpha}}{2}.$$

It is well known that

$$\sin 15° = \sin(45° - 30°)$$

$$= \frac{\sqrt{2}}{2} \cdot \frac{\sqrt{3}}{2} - \frac{\sqrt{2}}{2} \cdot \frac{1}{2} = \frac{\sqrt{6} - \sqrt{2}}{4}.$$

Likewise, $\cos 15° = \dfrac{\sqrt{6} + \sqrt{2}}{4}$.

We can now compute the exact value of $\sin 3°$:

$$\sin 3° = \sin(18° - 15°)$$

$$= \sin 18° \cos 15° - \cos 18° \sin 15°$$

$$= \frac{\sqrt{5}-1}{4} \cdot \frac{\sqrt{6}+\sqrt{2}}{4} - \frac{\sqrt{6}-\sqrt{2}}{4} \cdot \frac{\sqrt{10+2\sqrt{5}}}{4}$$

$$= \frac{(\sqrt{5}-1)(\sqrt{6}+\sqrt{2}) - (\sqrt{6}-\sqrt{2})(\sqrt{10+2\sqrt{5}})}{16}.$$

W.R. Ransom developed this formula in 1959 [492].

We can compute the exact value of sin 15° in a different way. Since sin 30° = 1/2, it follows by the double-angle formula for the sine function that sin 15° cos 15° = 1/4. Squaring both sides yields the biquadratic equation $16 \sin^4 15° - 16 \sin^2 15° + 1 = 0$. Solving this, we get $\sin 15° = \frac{\sqrt{6}-\sqrt{2}}{4}$, as before.

Knowing the exact values of trigonometric functions of 18° and 15°, we can compute them for 9°, 7.5°, and 1.5° [481].

Next we investigate some weaving patterns involving the golden ratio.

26.3 GOLDEN WEAVES

In 1978, W.E. Sharpe of the University of South Carolina at Columbia, while he was a member of the Norwegian Geological Survey, observed a set of remarkable weave patterns [524]. To see them, suppose the weave begins at the lower left-hand corner of a square loom of unit side. Suppose the first thread makes an angle θ with the base, where $\tan \theta = \frac{n+\alpha}{n+\alpha+1}$ and $n \geq 0$.

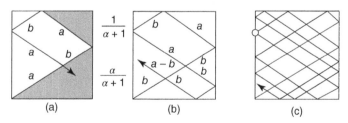

Figure 26.4. The development of a golden weave on a square loom of unit side for $n = 0$; (a) after 3 reflections; (b) after 5 reflections; (c) after 15 reflections. *Source*: Sharpe, 1978 [524]. Reproduced with permission of The Mathematical Association.

Initially, $n = 0$; see Figure 26.4a. Every time the thread (or line) meets a side of the square, it is reflected in the same way as a ray of light does and a new weave begins. After the third reflection, the thread crosses the original thread. Suppose the point of intersection divides the original thread into lengths a and b, as the figure shows. Using the properties of isosceles triangles and parallelograms, all the lengths marked a are equal, and so are those marked b.

Since $\tan \theta = \alpha/(\alpha + 1)$, it follows that the first thread meets the side of the square at the point that divides it into segments of lengths

$$\frac{\alpha}{\alpha+1} \quad \text{and} \quad 1 - \frac{\alpha}{\alpha+1} = \frac{1}{\alpha+1}.$$

The shaded triangles in Figure 26.4a are similar, so

$$\frac{a+b}{a} = \frac{\alpha/(\alpha+1)}{1/(\alpha+1)};$$

that is, $1 + b/a = \alpha$. So

$$\frac{a}{b} = \frac{1}{\alpha-1} = \frac{1}{-\beta} = \alpha.$$

Thus the first point of intersection divides the first thread in the golden ratio.

From Figure 26.4b, the second point of intersection divides the line segment of length a into two parts of length b and $a - b$. Since

$$\frac{b}{a-b} = \frac{1}{a/b-1} = \frac{1}{\alpha-1} = \alpha,$$

this line segment is also divided in the golden ratio at the point of intersection.

As additional crossovers occur, successive line segments are divided in the golden ratio at each intersection. For this reason, Sharpe called these weaves the *golden weaves*.

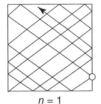 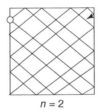 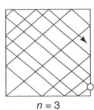

$n = 1$ $n = 2$ $n = 3$

Figure 26.5. The weaving patterns for the first 15 reflections. *Source*: Sharpe, 1978 [524]. Reproduced with permission of The Mathematical Association.

Figure 26.5 shows the weaving patterns for the first 15 reflections, for $n = 1, 2$, and 3. In fact, in all four cases, the first golden division occurs after $2n + 3$ reflections, marked with a tiny circle in these diagrams.

26.4 ADDITIONAL FIBONOMETRIC BRIDGES

In Chapter 18, we found an interesting trigonometric expansion of F_n:

$$F_n = 2^{n-1} \sum_{k=0}^{n-1} (-1)^k \cos^{n-k-1} \pi/5 \sin^k \pi/10.$$

We now explore a host of additional bridges linking them, beginning with the next result.

Theorem 26.1. *Let G_k denote the kth gibonacci number. Then*

$$\tan\left(\tan^{-1}\frac{G_n}{G_{n+1}} - \tan^{-1}\frac{G_{n+1}}{G_{n+2}}\right) = \frac{(-1)^{n+1}\mu}{G_{n+1}(G_n + G_{n+2})}. \tag{26.1}$$

Proof. Using the sum formula for the tangent function,

$$\begin{aligned} \text{LHS} &= \frac{G_n/G_{n+1} - G_{n+1}/G_{n+2}}{1 + (G_n/G_{n+1})(G_{n+1}/G_{n+2})} \\ &= \frac{G_n G_{n+2} - G_{n+1}^2}{G_n G_{n+1} + G_{n+1} G_{n+2}} \\ &= \frac{(-1)^{n+1}\mu}{G_{n+1}(G_n + G_{n+2})} \\ &= \text{RHS}. \end{aligned}$$ ∎

The following corollary follows immediately from this theorem.

Corollary 26.1.

$$\tan\left(\tan^{-1}\frac{F_n}{F_{n+1}} - \tan^{-1}\frac{F_{n+1}}{F_{n+2}}\right) = \frac{(-1)^{n+1}}{F_{2n+2}} \tag{26.2}$$

$$\tan\left(\tan^{-1}\frac{L_n}{L_{n+1}} - \tan^{-1}\frac{L_{n+1}}{L_{n+2}}\right) = \frac{(-1)^{n}}{F_{2n+2}}. \tag{26.3}$$

∎

We will establish the next result using PMI.

Theorem 26.2.

$$\sum_{i=1}^{n}(-1)^{i+1}\tan^{-1}\frac{1}{F_{2i}} = \tan^{-1}\frac{F_n}{F_{n+1}}.$$

Proof. When $n = 1$, LHS $= (-1)^2 \tan^{-1} 1/F_2 = \tan^{-1} 1 = $ RHS; so the result is true when $n = 1$.

Now assume it is true for an arbitrary positive integer k. Since $\tan^{-1}(-x) = -\tan^{-1}x$, by equation (26.2) we then have

$$\sum_{i=1}^{k+1}(-1)^{i+1}\tan^{-1}\frac{1}{F_{2i}} = \sum_{i=1}^{k}(-1)^{i+1}\tan^{-1}\frac{1}{F_{2i}} + (-1)^{k+2}\tan^{-1}\frac{1}{F_{2k+2}}$$

$$= \tan^{-1}\frac{F_k}{F_{k+1}} + (-1)^k\tan^{-1}\frac{1}{F_{2k+2}}$$

$$= \left(\tan^{-1}\frac{(-1)^{k+1}}{F_{2k+2}} + \tan^{-1}\frac{F_{k+1}}{F_{k+2}}\right) + (-1)^k\tan^{-1}\frac{1}{F_{2k+2}}$$

$$= \tan^{-1}\frac{F_{k+1}}{F_{k+2}}.$$

So the formula works when $n = k + 1$. Thus, by PMI, it works for all positive integers n. ∎

This summation formula has an interesting byproduct, as the next corollary shows.

Corollary 26.2.

$$\sum_{n=1}^{\infty}(-1)^{n+1}\tan^{-1}\frac{1}{F_{2n}} = \tan^{-1}(-\beta).$$

Proof. Since $\tan^{-1}$ is a continuous increasing function, $\tan^{-1}(1/F_{2n}) > \tan^{-1}(1/F_{2n+2})$. Also $\lim_{n\to\infty}\tan^{-1}(1/F_{2n}) = \tan^{-1}0 = 0$. Therefore, the series converges, and

$$\sum_{n=1}^{\infty}(-1)^{n+1}\tan^{-1}\frac{1}{F_{2n}} = \lim_{m\to\infty}\sum_{n=1}^{m}(-1)^{n+1}\tan^{-1}\frac{1}{F_{2n}}$$

$$= \lim_{m\to\infty}\tan^{-1}\frac{F_m}{F_{m+1}}$$

$$= \tan^{-1}\left(\lim_{m\to\infty}\frac{F_m}{F_{m+1}}\right)$$

$$= \tan^{-1}(-\beta),$$

as desired. ∎

This corollary has a companion result for odd-numbered Fibonacci numbers. Before we state and prove it, we need to lay some groundwork in the form of a lemma. The proof of Lemma 26.1 uses the sum formula for the tangent function and Cassini's formula.

Lemma 26.1.

$$\tan^{-1}\frac{1}{F_{2n+1}} = \tan^{-1}\frac{1}{F_{2n}} - \tan^{-1}\frac{1}{F_{2n+2}}.$$

Proof. Let $\theta_n = \tan^{-1}\dfrac{1}{F_{2n}} - \tan^{-1}\dfrac{1}{F_{2n+2}}$. Then

$$\tan\theta_n = \frac{1/F_{2n} - 1/F_{2n+2}}{1 + (1/F_{2n})(1/F_{2n+2})}$$

$$= \frac{F_{2n+2} - F_{2n}}{F_{2n+2}F_{2n} + 1}$$

$$= \frac{F_{2n+1}}{F_{2n+1}^2}$$

$$= \frac{1}{F_{2n+1}}.$$

This yields the desired result.   ∎

This lemma has an interesting byproduct. It follows by the lemma that

$$\tan^{-1}\frac{1}{F_{2n}} = \tan^{-1}\frac{1}{F_{2n+1}} + \tan^{-1}\frac{1}{F_{2n+2}}.$$

But $\tan^{-1} x + \tan^{-1}\dfrac{1}{x} = \dfrac{\pi}{2}$; see Exercise 26.7. Consequently,

$$\tan^{-1}\frac{1}{F_{2n}} + \tan^{-1} F_{2n+1} + \tan^{-1} F_{2n+2} = \pi;$$

see Exercise 26.9. W.W. Horner of Pittsburgh, Pennsylvania, found this result in 1968 [329].

The proof of the next lemma is quite similar to the one in Lemma 26.1, so we leave it as an exercise; see Exercise 26.3.

Lemma 26.2.

$$\tan^{-1}\frac{1}{F_{2n+1}} = \tan^{-1}\frac{1}{L_{2n}} + \tan^{-1}\frac{1}{L_{2n+2}}.$$ ∎

We are now ready to present the celebrated result for odd-numbered Fibonacci numbers promised above. The American mathematician Derrick H. Lehmer (1905–1991) discovered it in 1936 when he was at Lehigh University, Pennsylvania [387].

Theorem 26.3 (Lehmer, 1936 [387]).

$$\sum_{n=1}^{\infty} \tan^{-1}\frac{1}{F_{2n+1}} = \frac{\pi}{4}.$$

Proof. By Lemma 26.1, we have

$$\sum_{n=1}^{m} \tan^{-1} \frac{1}{F_{2n+1}} = \sum_{n=1}^{m} \left(\tan^{-1} \frac{1}{F_{2n}} - \tan^{-1} \frac{1}{F_{2n+2}} \right)$$

$$= \tan^{-1} \frac{1}{F_2} - \tan^{-1} \frac{1}{F_{2m+2}}$$

$$= \frac{\pi}{4} - \tan^{-1} \frac{1}{F_{2m+2}};$$

$$\sum_{n=1}^{\infty} \tan^{-1} \frac{1}{F_{2n+1}} = \lim_{m \to \infty} \sum_{n=1}^{m} \tan^{-1} \frac{1}{F_{2n+1}}$$

$$= \lim_{m \to \infty} \left(\frac{\pi}{4} - \tan^{-1} \frac{1}{F_{2m+2}} \right)$$

$$= \frac{\pi}{4} - \tan^{-1} 0$$

$$= \frac{\pi}{4} - 0$$

$$= \frac{\pi}{4}.$$ ∎

The next theorem gives the counterpart for even-numbered Lucas numbers. Its proof employs Lemma 26.2 and Theorem 26.3.

Theorem 26.4.

$$\sum_{n=1}^{\infty} \tan^{-1} \frac{1}{L_{2n}} = \tan^{-1}(-\beta).$$

Proof. By Lemma 26.2, we have

$$\sum_{n=1}^{m} \tan^{-1} \frac{1}{F_{2n+1}} = \sum_{n=1}^{m} \left(\tan^{-1} \frac{1}{L_{2n}} + \tan^{-1} \frac{1}{L_{2n+2}} \right)$$

$$= \tan^{-1} \frac{1}{3} + 2 \sum_{n=1}^{m} \tan^{-1} \frac{1}{L_{2n}} + \tan^{-1} \frac{1}{L_{2n+2}}$$

and

$$\sum_{n=1}^{\infty} \tan^{-1} \frac{1}{F_{2n+1}} = \tan^{-1} \frac{1}{3} + 2 \sum_{n=2}^{\infty} \tan^{-1} \frac{1}{L_{2n}} + 0$$

$$= 2 \sum_{n=1}^{\infty} \tan^{-1} \frac{1}{L_{2n}} - \tan^{-1} \frac{1}{3}.$$

Since $\pi/4 = \tan^{-1} 1$, by Theorem 26.3 and Exercises 26.4 and 26.5, this yields

$$2 \sum_{n=1}^{\infty} \tan^{-1} \frac{1}{L_{2n}} = \tan^{-1} \frac{1}{3} + \tan^{-1} 1$$

$$= \tan^{-1} 2;$$

$$\sum_{n=1}^{\infty} \tan^{-1} \frac{1}{L_{2n}} = \frac{1}{2} \tan^{-1} 2$$

$$= \tan^{-1} \frac{\sqrt{5} - 1}{2}$$

$$= \tan^{-1}(-\beta). \qquad \blacksquare$$

Lehmer proposed the next result as a problem in 1936 [386]. The featured proof is essentially the one derived two years later by M.A. Heaslet of then San Jose State College, California [260]; it uses the sum formula for the cotangent function, identity (5.2), and a telescoping sum. Hoggatt found alternate proofs in 1964 and 1965, and C.W. Trigg of California found one in 1973 [561].

Theorem 26.5 (Lehmer, 1936 [386]).

$$\sum_{n=1}^{\infty} \cot^{-1} F_{2n+1} = \cot^{-1} 1.$$

Proof.

$$\cot^{-1} F_{2k} - \cot^{-1} F_{2k+1} = \cot^{-1} \frac{F_{2k} F_{2k+1} + 1}{F_{2k+1} - F_{2k}}$$

$$= \cot^{-1} \frac{F_{2k} F_{2k+1} + 1}{F_{2k-1}}$$

$$= \cot^{-1} \frac{F_{2k-1} F_{2k+2}}{F_{2k-1}}$$

$$= \cot^{-1} F_{2k+2}$$

$$\cot^{-1} F_{2k+1} = \cot^{-1} F_{2k} - \cot^{-1} F_{2k+2}$$

$$\sum_{k=1}^{n} \cot^{-1} F_{2k+1} = \sum_{n=1}^{n} \left(\cot^{-1} F_{2k} - \cot^{-1} F_{2k+2} \right)$$

$$\sum_{k=1}^{n} \cot^{-1} F_{2k+1} = \cot^{-1} F_2 - \cot^{-1} F_{2n+2}.$$

As $m \to \infty$, $\cot^{-1} F_m \to 0$. Thus

$$\sum_{n=1}^{\infty} \cot^{-1} F_{2n+1} = \cot^{-1} 1 - 0$$

$$= \cot^{-1} 1. \qquad \blacksquare$$

J.R. Goggins of Glasgow, Scotland, developed the next formula in 1973 [227]. M. Harvey and P. Woodruff of St. Paul's School, London, re-discovered it 22 years later by an alternate method [259]. We will establish it using PMI.

Theorem 26.6 (Goggins, 1973 [227]).

$$\sum_{k=1}^{n} \tan^{-1} \frac{1}{F_{2k+1}} + \tan^{-1} \frac{1}{F_{2n+2}} = \frac{\pi}{4}.$$

Proof. When $n = 1$, LHS $= \tan^{-1} \frac{1}{2} + \tan^{-1} \frac{1}{3} = \frac{\pi}{4} =$ RHS. So the result is true when $n = 1$.

Now assume it works for an arbitrary positive integer n. Then

$$\sum_{k=1}^{n+1} \tan^{-1} \frac{1}{F_{2k+1}} + \tan^{-1} \frac{1}{F_{2n+4}} = \sum_{k=1}^{n} \tan^{-1} \frac{1}{F_{2k+1}}$$

$$+ \tan^{-1} \frac{1}{F_{2n+3}} + \tan^{-1} \frac{1}{F_{2n+4}}$$

$$= \left(\frac{\pi}{4} - \tan^{-1} \frac{1}{F_{2n+2}} \right) + \tan^{-1} \frac{1}{F_{2n+3}}$$

$$+ \tan^{-1} \frac{1}{F_{2n+4}}. \tag{26.4}$$

Let $\tan x = 1/F_{2n+2}$ and $\tan y = 1/F_{2n+4}$. Then, by Cassini's formula, we have

$$\tan(x - y) = \frac{1/F_{2n+2} - 1/F_{2n+4}}{1 + (1/F_{2n+2})(1/F_{2n+4})}$$

$$= \frac{F_{2n+4} - F_{2n+2}}{1 + F_{2n+4}F_{2n+2}}$$

$$= \frac{F_{2n+3}}{F_{2n+3}^2}$$

$$= \frac{1}{F_{2n+3}}$$

$$x - y = \tan^{-1} 1/F_{2n+3};$$

that is,

$$\tan^{-1} 1/F_{2n+2} - \tan^{-1} 1/F_{2n+4} = \tan^{-1} 1/F_{2n+3}.$$

It now follows from equation (26.4) that

$$\sum_{k=1}^{n+1} \tan^{-1} \frac{1}{F_{2k+1}} + \tan^{-1} \frac{1}{F_{2n+4}} = \frac{\pi}{4}.$$

So the result is true for $n + 1$ also.
Thus, by PMI, it is true for every $n \geq 1$.   ■

Clearly, Lehmer's formula in Theorem 26.3 follows as a byproduct of this theorem.

Next we investigate ways Fibonacci and Lucas numbers can be factored using trigonometric functions.

26.5 FIBONACCI AND LUCAS FACTORIZATIONS

In 1965, D. Lind of the University of Virginia at Charlottesville developed an interesting factorization of F_n [393]:

$$F_n = \prod_{k=1}^{n-1}(1 - 2i \cos k\pi/n), \qquad (26.5)$$

where $i = \sqrt{-1}$. Its proof requires a knowledge of Fibonacci polynomials[†] and their zeros, so will defer its proof to Volume Two of the book.
For example,

$$F_4 = \prod_{k=1}^{3}(1 - 2i \cos k\pi/4)$$

$$= (1 - 2i \cos \pi/4)(1 - 2i \cos 2\pi/4)(1 - 2i \cos 3\pi/4)$$

$$= (1 - \sqrt{2}i)(1 - 0)(1 + \sqrt{2}i)$$

$$= 3.$$

Two years after Lind's discovery, D. Zeitlin of Minnesota derived the Lucas counterpart of formula (26.5) using trigonometric factorizations of Chebyshev polynomials of the first and second kinds [615]:

$$L_n = \prod_{k=0}^{n-1} \left[1 - 2i \cos(2k + 1)\pi/2n \right]. \qquad (26.6)$$

[†]See Volume Two for a discussion of Fibonacci and Lucas polynomials.

For example,

$$L_3 = \prod_{k=1}^{2} \left[1 - 2i\cos(2k+1)\pi/6\right]$$

$$= (1 - 2i\cos\pi/6)(1 - 2i\cos 3\pi/6)(1 - 2i\cos 5\pi/6)$$

$$= (1 - 2i\cos\pi/6)(1 - 0)(1 + 2i\cos\pi/6)$$

$$= (1 + 4\cos^2\pi/6)$$

$$= 1 + 4 \cdot \frac{3}{4}$$

$$= 4,$$

as expected.

In 1967, Lind expressed both factorizations without using the complex number i [397]. They are given in the next theorem, which was posed as an advanced problem. The featured proof is based on the solution given by M.N.S. Swamy of the University of Saskatchewan, Canada [549].

Theorem 26.7 (Lind, 1967 [397]).

$$F_n = \prod_{k=1}^{\lfloor(n-1)/2\rfloor} (3 + 2\cos 2k\pi/n) \tag{26.7}$$

$$L_n = \prod_{k=0}^{\lfloor(n-2)/2\rfloor} \left[3 + 2\cos(2k+1)\pi/n\right]. \tag{26.8}$$

Proof. This is divided into odd-n and even-n cases.

Case 1. Let $n = 2m + 1$ be odd. Then formula (26.5) becomes

$$F_{2m+1} = \prod_{k=1}^{2m} \left(1 - 2i\cos\frac{k\pi}{2m+1}\right)$$

$$= \prod_{k=1}^{m} \left(1 - 2i\cos\frac{k\pi}{2m+1}\right) \prod_{k=m+1}^{2m} \left(1 - 2i\cos\frac{k\pi}{2m+1}\right)$$

$$= \prod_{k=1}^{m} \left(1 - 2i\cos\frac{k\pi}{2m+1}\right) \prod_{k=m+1}^{2m} \left[1 + 2i\cos\left(\pi - \frac{k\pi}{2m+1}\right)\right].$$

Letting $j = 2m + 1 - k$ in the second product, we get

$$F_{2m+1} = \prod_{k=1}^{m}\left(1 - 2i\cos\frac{k\pi}{2m+1}\right)\prod_{k=1}^{m}\left(1 + 2i\cos\frac{k\pi}{2m+1}\right)$$

$$= \prod_{k=1}^{m}\left(1 + 4\cos^2\frac{j\pi}{2m+1}\right)$$

$$= \prod_{k=1}^{m}\left(3 + 2\cos^2\frac{2j\pi}{2m+1}\right). \tag{26.9}$$

Case 2. Let $n = 2m$ be even. Then formula (26.5) becomes

$$F_{2m} = \prod_{k=1}^{2m-1}\left(1 - 2i\cos\frac{k\pi}{2m}\right).$$

As before, this yields

$$F_{2m} = \prod_{k=1}^{m-1}\left(1 - 2i\cos\frac{k\pi}{2m}\right)\prod_{j=1}^{m-1}\left(1 + 2i\cos\frac{j\pi}{2m}\right)\cdot(1 + 2i\cos\pi/2)$$

$$= \prod_{j=1}^{m-1}\left(1 + 4\cos^2\frac{j\pi}{2m}\right)$$

$$= \prod_{j=1}^{m-1}\left(3 + 2\cos^2\frac{2j\pi}{2m}\right). \tag{26.10}$$

Combining equations (26.9) and (26.10), we get formula (26.7).

To derive the formula for L_n, we have

$$F_{2n} = \prod_{k=1}^{\lfloor(2n-1)/2\rfloor}(3 + 2\cos k\pi/n)$$

$$= \prod_{\substack{\text{even integers}\\ i\leq n-1}}(3 + 2\cos i\pi/n)\prod_{\substack{\text{odd integers}\\ j\leq n-1}}(3 + 2\cos j\pi/n).$$

Letting $i = 2k$ and $j = 2k + 1$, this yields

$$F_{2n} = \prod_{k=1}^{\lfloor(n-1)/2\rfloor}(3 + 2\cos k\pi/n)\prod_{k=0}^{\lfloor(n-2)/2\rfloor}\left[3 + 2\cos(2k + 1)\pi/n\right]$$

$$= F_n\prod_{k=0}^{\lfloor(n-2)/2\rfloor}\left[3 + 2\cos(2k + 1)\pi/n\right].$$

Since $F_{2n} = F_n L_n$, this gives the desired formula for L_n.   ∎

For example,

$$F_6 = \prod_{k=1}^{2}(3 + 2\cos k\pi/3)$$

$$= (3 + 2\cos \pi/3)(3 + 2\cos 2\pi/3)$$

$$= (3 + 2\cos \pi/3)(3 - 2\cos \pi/3)$$

$$= 9 - 4\cos^2 \pi/3$$

$$= 9 - 1$$

$$= 8;$$

and

$$L_6 = \prod_{k=0}^{2} \big[3 + 2\cos(2k + 1)\pi/6\big]$$

$$= (3 + 2\cos \pi/6)(3 + 2\cos \pi/2)(3 + 2\cos 5\pi/6)$$

$$= (3 + 2\cos \pi/6)(3 - 2\cos \pi/6)(3 + 0)$$

$$= 3(9 - 4\cos^2 \pi/6)$$

$$= 3(9 - 4 \cdot 3/4)$$

$$= 18.$$

EXERCISES 26

1. Deduce formula (26.3) from formula (26.1).

Prove each.

2. Formula (26.1) using PMI.

3. Lemma 26.2.

4. $\tan^{-1} 1 + \tan^{-1} \dfrac{1}{3} = \tan^{-1} 2.$

5. $\tan^{-1} 2 = 2\tan^{-1} \beta.$

6. $\tan^{-1} \dfrac{1}{L_{2n}} + \tan^{-1} \dfrac{1}{L_{2n+2}} = \tan^{-1} \dfrac{1}{F_{2n+1}}.$

7. $\tan^{-1} x + \tan^{-1} \dfrac{1}{x} = \dfrac{\pi}{2}.$

8. Let θ be the angle between the vectors $u = (B_0, B_1, B_2, \ldots, B_n)$ and $v = (F_m, F_{m+1}, F_{m+2}, \ldots, F_{m+n})$ in the Euclidean $(n+1)$-space, where $B_i = \binom{n}{i}$. Find $\lim\limits_{n \to \infty} \theta$ (Gootherts, 1967 [229]).

9. $\pi = \tan^{-1}(1/F_{2n}) + \tan^{-1} F_{2n+1} + \tan^{-1} F_{2n+2}$ (Horner, 1968 [329]).

10. Prove that $\displaystyle\sum_{n=1}^{\infty} \frac{1}{\alpha F_{n+1} + F_n} = \sum_{n=1}^{\infty} \tan^{-1} \frac{1}{F_{2n+1}}$ (Guillotte, 1972 [253]).

27

COMPLETENESS THEOREMS

This chapter provides additional opportunities for studying patterns and making conjectures.

27.1 COMPLETENESS THEOREM

We begin with yet another interesting pattern:

$$1 = 1 \qquad\qquad 2 = 2$$
$$3 = 2 + 1 \qquad\qquad 4 = 3 + 1$$
$$5 = 3 + 2 \qquad\qquad 6 = 5 + 1$$
$$7 = 5 + 2 \qquad\qquad 8 = 5 + 3$$
$$9 = 8 + 1 \qquad\qquad 10 = 8 + 2.$$

Every integer on the LHS of each equation is a positive integer; and each number on the RHS is a Fibonacci number, and each occurs exactly once.

More generally, we have the following result.

Fibonacci and Lucas Numbers with Applications, Volume One, Second Edition. Thomas Koshy.
© 2018 John Wiley & Sons, Inc. Published 2018 by John Wiley & Sons, Inc.

Theorem 27.1 (Completeness Theorem). *Every positive integer n can be expressed as the sum of a finite number of distinct Fibonacci numbers.*

Proof. Let F_m be the largest Fibonacci number $\leq n$. Then $n = F_m + n_1$, where $n_1 \leq F_m$. Let F_{m_1} be the largest number $\leq n_1$. Then $n = F_m + F_{m_1} + n_2$, where $n \geq F_m > F_{m_1}$. Continuing like this, we get $n = F_m + F_{m_1} + F_{m_2} + \cdots$, where $n \geq F_m > F_{m_1} > F_{m_2} > \cdots$. Since this sequence of decreasing positive integers must terminate, the given result follows. ∎

We must emphasize that the representation of an integer n in terms of Fibonacci numbers is *not* unique. For example, $25 = 21 + 3 + 1 = 13 + 8 + 3 + 1$.

27.2 EGYPTIAN ALGORITHM FOR MULTIPLICATION

Every positive integer can be expressed as a unique sum of distinct powers of 2. This extremely useful fact is the basis of the well-known *Egyptian algorithm for multiplication*. For example, let $b = \sum\limits_{i=0}^{n} b_i 2^i$, where $b_i = 0$ or 1. Then $ab = \sum\limits_{i=0}^{n} b_i 2^i$. Thus, to compute ab, we need only keep doubling a until the product gets larger than 2^n, and then add the products corresponding to the 1s in the binary representation of b. The following example illustrates this algorithm.

Example 27.1. *Using the Egyptian algorithm for multiplication, compute $47 \cdot 73$.*

Solution. First, express 47 as a sum of powers of 2: $47 = 1 + 2 + 4 + 8 + 32$. Then

$$47 \cdot 73 = 1 \cdot 73 + 2 \cdot 73 + 4 \cdot 73 + 8 \cdot 73 + 32 \cdot 73.$$

Next, construct a table (see Table 27.1) consisting of two rows, one headed by 1 and the other by 73; each successive column is obtained by doubling the preceding column. Identify the numbers in the second row that correspond to the powers of 2 used in the representation of 47 by asterisks; they correspond to the terms in the binary expansion of 47.

TABLE 27.1.

1	2	4	8	16	32
73*	146*	292*	584*	1168	2336*

To find the desired sum, we now add the starred numbers:

$$47 \cdot 73 = 73 + 146 + 292 + 584 + 2336$$

$$= 3431.$$ ∎

By virtue of Theorem 27.1, we can use Fibonacci numbers also to effect multiplication of positive integers, as the next example demonstrates.

Example 27.2. *Using Fibonacci numbers, compute* $47 \cdot 73$.

Solution. First, we express 47 as a sum of Fibonacci numbers: $47 = 2 + 3 + 8 + 13 + 21$. Then

$$47 \cdot 73 = 2 \cdot 73 + 3 \cdot 73 + 8 \cdot 73 + 13 \cdot 73 + 21 \cdot 73.$$

Now construct a table as before (see Table 27.2). It follows from the table that

$$47 \cdot 73 = 146 + 219 + 584 + 949 + 1533$$
$$= 3431,$$

as expected.

TABLE 27.2.

1	2	3	5	8	13	21
73	146*	219*	365	584*	949*	1533*

∎

Although Fibonacci numbers F_n are complete, Lucas numbers L_n are *not*, where $n \geq 1$. For example, using the numbers $1, 3, 4, \ldots$, we cannot represent 2 as a Lucas sum. But all is not lost. If we add $2 = L_0$ to the list, the resulting set is complete; see the next theorem.

Theorem 27.2 (Completeness Theorem). *The set of Lucas numbers* L_n *is complete, where* $n \geq 0$. ∎

Again, we emphasize that the representation need not be unique. For instance, $43 = 29 + 11 + 3 = 29 + 7 + 4 + 3 = 29 + 11 + 2 + 1$.

As a byproduct of Theorem 27.2, we can employ Lucas numbers also to perform integer multiplication, as the next example illustrates.

Example 27.3. *Using Lucas numbers, compute* $47 \cdot 73$.

Solution. First, we express 47 as a sum of Lucas numbers: $47 = 3 + 4 + 11 + 29$. It follows by Table 27.3 that $47 \cdot 73 = 219 + 292 + 803 + 2117 = 3431$.

TABLE 27.3.

2	1	3	4	7	11	18	29
146	73	219*	292*	511	803*	1314	2117*

∎

EXERCISES 27

Express each number as a sum of distinct Fibonacci numbers.

    1. 43
    2. 99
    3. 137
    4. 343

  5–8. Express each number in Exercises 27.1–27.4 as a sum of distinct Lucas numbers, where $n \geq 0$.

Using Fibonacci numbers, compute each product.

    9. $43 \cdot 49$
    10. $99 \cdot 101$
    11. $111 \cdot 121$
    12. $243 \cdot 342$

13–16. Using Lucas numbers, compute each product in Exercises 27.9–27.12.

    17. Prove Theorem 12.2.

THE KNAPSACK PROBLEM

In this chapter, we investigate the well-known knapsack problem, with Fibonacci and Lucas numbers as weights.

Let G_i denote the ith gibonacci number. Recall from Chapter 7 that

$$\sum_{i=1}^{n} G_i = G_{n+2} - b$$

$$= \begin{cases} F_{n+2} - 1 & \text{if } G_i = F_i \\ L_{n+2} - 3 & \text{if } G_i = L_i. \end{cases}$$

Consequently,

$$\sum_{k=i}^{i+n-1} G_k = \sum_{j=1}^{i+n-1} G_j - \sum_{j=1}^{i-1} G_j$$

$$= G_{i+n+1} - G_{i+1}. \tag{28.1}$$

We will use this result in our discussion.

28.1 THE KNAPSACK PROBLEM

Given a knapsack of integer volume S, and n items of integer volumes $a_1, a_2, \ldots, a_n$, which of the items can fill the knapsack?

Fibonacci and Lucas Numbers with Applications, Volume One, Second Edition. Thomas Koshy.
© 2018 John Wiley & Sons, Inc. Published 2018 by John Wiley & Sons, Inc.

In other words, given the positive integers $a_1, a_2, \ldots, a_n$, called *weights*, and a positive integer S, solve the linear diophantine equation (LDE) $a_1 x_1 + a_2 x_2 + \cdots + a_n x_n = S$, where $x_i = 0$ or 1. This is the well-known *knapsack problem*.

In particular, consider the knapsack problem

$$G_i x_1 + G_{i+1} x_2 + \cdots + G_{i+n-1} x_n = G_{i+n}, \tag{28.2}$$

where $i \geq 2$. It follows by the Fibonacci recurrence that this LDE is solvable with $(0, 0, \ldots, 0, 1, 1)$ as a solution. In fact, since $G_i + G_{i+1} + \cdots + G_{i+n-2} = G_{i+n} - G_{i-1} < G_{i+n}$, no sum of $G_i, G_{i+1}, \ldots, G_{i+n-2}$ can add up to G_{i+n}. Consequently, $(0, 0, \ldots, 0, 1, 1)$ is the unique solution, with $x_{n-1} = 1$.

For example, consider the LDE $x_2 + 2x_3 + 3x_4 + 5x_5 + 8x_6 + 13x_7 = 21$. Its only solution with $x_6 = 1$ is $(0, 0, 0, 0, 1, 1)$.

This raises an interesting question: Is the LDE (28.2) solvable with $x_{n-1} = 0$? If yes, how many solutions does the problem have?

To answer this, first notice that

$$
\begin{aligned}
G_{i+n} &= G_{i+n-1} + G_{i+n-2} \\
&= G_{i+n-1} + G_{i+n-3} + G_{i+n-4} \\
&= G_{i+n-1} + G_{i+n-3} + G_{i+n-5} + G_{i+n-6} \\
&= G_{i+n-1} + G_{i+n-3} + G_{i+n-5} + G_{i+n-7} + G_{i+n-8} \\
&\;\;\vdots \\
&= G_{i+n-1} + G_{i+n-3} + G_{i+n-5} + G_{i+n-7} + \cdots + G_i.
\end{aligned} \tag{28.3}
$$

Since $i = (i + n) - n$, the RHS of equation (28.3) contains $\lfloor n/2 \rfloor$ additions. Thus there are $\lfloor n/2 \rfloor$ ways of expressing G_{i+n} as a sum of its predecessors through G_i. In other words, the knapsack problem (28.2) has $\lfloor n/2 \rfloor$ solutions, one of which corresponds to $x_{n-1} = 1$.

Accordingly, we have the following result.

Theorem 28.1. *The knapsack problem* $G_i x_1 + G_{i+1} x_2 + \cdots + G_{i+n-1} x_n = G_{i+n}$ *has* $\lfloor n/2 \rfloor$ *solutions, where* $i \geq 1$. ∎

For example, the LDE $F_5 x_1 + F_6 x_2 + \cdots + F_{10} x_6 = F_{11}$ has $\lfloor 6/2 \rfloor = 3$ solutions $(x_1, x_2, \ldots, x_6)$. They correspond to the internal nodes in the binary tree in Figure 28.1 and to the three different ways of expressing F_{11} in terms of its predecessors through F_5:

$$
\begin{aligned}
F_{11} &= F_{10} + F_9 \\
&= F_{10} + F_8 + F_7 \\
&= F_{10} + F_8 + F_6 + F_5.
\end{aligned}
$$

Correspondingly, the three solutions are $(0, 0, 0, 0, 1, 1)$, $(0, 0, 1, 1, 0, 1)$, and $(1, 1, 0, 1, 0, 1)$.

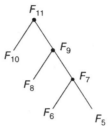

Figure 28.1.

The next result follows from Theorem 28.1.

Corollary 28.1. *The knapsack problem $G_1x_1 + G_2x_2 + \cdots + G_nx_n = G_{n+1}$ has $\lfloor n/2 \rfloor$ solutions.* ∎

For example, $F_1x_1 + F_2x_2 + \cdots + F_{10}x_{10} = F_{11}$ has $\lfloor 10/2 \rfloor = 5$ solutions $(x_1, x_2, \ldots, x_{10})$. They correspond to the five different ways of expressing F_{11} in terms of its predecessors:

$$
\begin{aligned}
F_{11} &= F_{10} + F_9 \\
&= F_{10} + F_8 + F_7 \\
&= F_{10} + F_8 + F_6 + F_5 \\
&= F_{10} + F_8 + F_6 + F_4 + F_3 \\
&= F_{10} + F_8 + F_6 + F_4 + F_2 + F_1.
\end{aligned}
$$

They are represented by the internal vertices in the binary tree in Figure 28.2. Correspondingly, the five solutions are $(0, 0, 0, 0, 0, 0, 0, 0, 1, 1)$, $(0, 0, 0, 0, 0, 0, 1, 1, 0, 1)$, $(0, 0, 0, 0, 1, 1, 0, 1, 0, 1)$, $(0, 0, 1, 1, 0, 1, 0, 1, 0, 1)$, and $(1, 1, 0, 1, 0, 1, 0, 1, 0, 1)$.

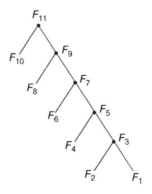

Figure 28.2.

Suppose the RHS of equation (28.2) is G_j:

$$G_i x_1 + G_{i+1} x_2 + \cdots + G_{i+n-1} x_n = G_j, \tag{28.4}$$

where $j \neq i + n$. If $i \leq j \leq i + n - 1$, the knapsack problem is solvable with a solution $(0, \ldots, 0, 1, 0, \ldots, 0)$, where the 1 occurs in position j; it need not be unique. If $j < i$, then no solution is possible. Suppose $j > i + n$. Since $G_i + G_{i+1} + \cdots + G_{i+n-1} = G_{i+n-1} - G_{i+1} < G_{i+n+1} \leq G_j$, the knapsack problem has no solutions.

Finally, let S be any positive integer $\leq G_{n+1}$. Then the knapsack problem $G_1 x_1 + G_2 x_2 + \cdots + G_n x_n = S$ is solvable. This follows by Theorem 28.1 and the following result.

Theorem 28.2. *The knapsack problem $G_1 x_1 + G_2 x_2 + \cdots + G_n x_n = S$ is solvable, where S is a positive integer $\leq G_{n+1}$.*   ■

For example, $x_1 + x_2 + 2x_3 + 3x_4 + 5x_5 + 8x_6 + 13x_7 = 7$ with Fibonacci weights is solvable; since $7 = 2 + 5$, $(0, 0, 1, 0, 1, 0, 0)$ is a solution. Likewise, the knapsack problem $x_1 + x_2 + 2x_3 + 3x_4 + 5x_5 + 8x_6 + 13x_7 + 21x_8 + 34x_9 = 48$ is solvable; since $48 = 1 + 5 + 8 + 34$, $(1, 0, 0, 0, 1, 1, 0, 0, 1)$ is a solution. Both problems have more than one solution.

The knapsack problem $2x_1 + x_2 + 3x_3 + 4x_4 + 7x_5 + 11x_6 = 15$ with Lucas weights is solvable; since $15 = 4 + 11$, $(0, 0, 0, 1, 0, 1)$ is a solution; so is $(0, 1, 1, 0, 0, 1)$. The problem $2x_1 + x_2 + 3x_3 + 4x_4 + 7x_5 + 11x_6 = 10$ is also solvable, $(0, 0, 1, 0, 1, 0)$ being a solution.

29

FIBONACCI AND LUCAS SUBSCRIPTS

In this chapter, we investigate Fibonacci and Lucas numbers with Fibonacci and Lucas subscripts; that is, numbers of the form $F_{F_n}, F_{L_n}, L_{F_n}$, and L_{L_n}. For example, $F_{F_6} = F_8 = 21$, $F_{L_6} = F_{18} = 2584$, $L_{F_6} = L_8 = 47$, and $L_{L_6} = L_{18} = 5778$. In the interest of brevity and clarity, we let $U_n = F_{F_n}$, $V_n = F_{L_n}$, $X_n = L_{F_n}$, and $W_n = L_{L_n}$.

29.1 FIBONACCI AND LUCAS SUBSCRIPTS

Although Binet's formulas for F_n and L_n are not very useful in our investigations, we can still employ them to find explicit formulas for U_n, V_n, X_n and W_n. For example,

$$U_n = \frac{\alpha^{F_n} - \beta^{F_n}}{\alpha - \beta} \quad \text{and} \quad V_n = \alpha^{F_n} + \beta^{F_n}.$$

In 1966, R.E. Whitney of then Lockhaven State College, Lockhaven, Pennsylvania, partially succeeded in defining U_n, V_n, X_n and W_n recursively in terms of hybrid relations [593].

Theorem 29.1.

$$5U_n U_{n+1} = X_{n+2} - (-1)^{F_n} X_{n-1}.$$

Fibonacci and Lucas Numbers with Applications, Volume One, Second Edition. Thomas Koshy.
© 2018 John Wiley & Sons, Inc. Published 2018 by John Wiley & Sons, Inc.

Proof. Recall that $\sqrt{5}F_n = \alpha^n - \beta^n$ and $L_n = \alpha^n + \beta^n$. Consequently,

$$\sqrt{5}U_n = \alpha^{F_n} - \beta^{F_n}$$
$$\sqrt{5}U_{n+1} = \alpha^{F_{n+1}} - \beta^{F_{n+1}}.$$

Multiplying these two equations, we get

$$\begin{aligned}
5U_nU_{n+1} &= \left(\alpha^{F_{n+2}} - \beta^{F_{n+2}}\right) - \left(\alpha^{F_n}\beta^{F_{n+1}} + \alpha^{F_{n+1}}\beta^{F_n}\right) \\
&= X_{n+2} - \left[\beta^{F_{n+1}}(-\alpha^{-1})^{F_n} + \alpha^{F_{n+1}}(-\beta^{-1})^{F_n}\right] \\
&= X_{n+2} - (-1)^{F_n}\left(\alpha^{F_{n+1}-F_n} + \beta^{F_{n+1}-F_n}\right) \\
&= X_{n+2} - (-1)^{F_n}\left(\alpha^{F_{n-1}} + \beta^{F_{n-1}}\right) \\
&= X_{n+2} - (-1)^{F_n}X_{n-1}.
\end{aligned}$$ ∎

For example, let $n = 7$. Then

$$\begin{aligned}
X_9 - (-1)^{13}X_6 &= L_{34} + L_8 \\
&= 12{,}752{,}043 + 47 \\
&= 12{,}752{,}090 \\
&= 5 \cdot 233 \cdot 10946 \\
&= 5F_{13}F_{21} \\
&= 5U_7U_8.
\end{aligned}$$

The next theorem gives a similar result, but involving only the numbers X_k.

Theorem 29.2.
$$X_nX_{n+1} = X_{n+2} + (-1)^{F_n}X_{n-1}.$$ ∎

Its proof follows along the same lines, so we leave it as an exercise; see Exercise 29.2.

For example,

$$\begin{aligned}
X_8 + (-1)^{F_6}X_5 &= L_{21} + (-1)^8 L_5 \\
&= 24476 + 11 \\
&= 24{,}487
\end{aligned}$$

$$= 47 \cdot 521$$
$$= L_8 L_{13}$$
$$= X_6 X_7.$$

Combining Theorems 29.1 and 29.2, we get the following result.

Corollary 29.1.

$$5U_n U_{n+1} + X_n X_{n+1} = 2X_{n+2}. \qquad \blacksquare$$

The next theorem follows by Binet's formulas for Fibonacci and Lucas numbers. We omit the proof in the interest of brevity; see Exercises 29.6 and 29.7.

Theorem 29.3.

1) $5V_n V_{n+1} = W_{n+2} - (-1)^{L_n} W_{n-1}$

2) $W_n W_{n+1} = W_{n+2} + (-1)^{L_n} W_{n-1}. \qquad \blacksquare$

The following result follows from this theorem.

Corollary 29.2.

$$5V_n V_{n+1} + W_n W_{n+1} = 2W_{n+2}. \qquad \blacksquare$$

The next theorem also follows by Binet's formulas for Fibonacci and Lucas numbers; see Exercises 29.10 and 29.11.

Theorem 29.4.

1) $X_{n-1} X_{n+1} = W_n + (-1)^{F_{n-1}} X_n$

2) $5U_{n-1} U_{n+1} = W_n - (-1)^{F_{n-1}} X_n. \qquad \blacksquare$

This theorem also has interesting byproducts, as the next corollary shows.

Corollary 29.3.

1) $X_{n-1} X_{n+1} + 5U_{n-1} U_{n+1} = 2W_n$

2) $X_{n-1} X_{n+1} - 5U_{n-1} U_{n+1} = 2(-1)^{F_{n-1}} X_n. \qquad \blacksquare$

29.2   GIBONACCI SUBSCRIPTS

In 1967, Lind succeeded in defining both $Y_n = F_{G_n}$ and $Z_n = L_{G_n}$ recursively, where G_n denotes the nth gibonacci number [400]. We will need the following identities from Chapter 5 to pursue such Fibonacci and Lucas numbers:

$$2F_{n+1} = F_n + L_n \tag{29.1}$$

$$2F_{n-1} = L_n - F_n \tag{29.2}$$

$$L_n^2 - 5F_n^2 = 4(-1)^n \tag{29.3}$$

$$2L_{n+1} = 5F_n + L_n. \tag{29.4}$$

It follows from identities (29.1) and (29.3) that

$$F_{n+1} = \frac{1}{2}\left(\sqrt{5F_n^2 + 4(-1)^n} + F_n\right); \tag{29.5}$$

and from identities (29.3) and (29.4) that

$$L_{n+1} = \frac{1}{2}\left(\sqrt{5L_n^2 - 20(-1)^n} + L_n\right). \tag{29.6}$$

For example,

$$\begin{aligned}
F_{10} &= \frac{1}{2}\left(\sqrt{5F_9^2 + 4(-1)^9} + F_9\right) \\
&= \frac{1}{2}\left(\sqrt{5 \cdot 1156 - 4} + 34\right) \\
&= 55.
\end{aligned}$$

Likewise, it follows by recurrence (29.6) that $L_{10} = 123$, as expected.

Identity (29.5) can be used to compute the immediate predecessor of a Fibonacci number:

$$F_{n-1} = \frac{1}{2}\left(\sqrt{5F_n^2 + 4(-1)^n} - F_n\right).$$

We will require two addition formulas from Chapter 20:

$$F_{m+n+1} = F_m F_n + F_{m+1} F_{n+1} \tag{29.7}$$

$$L_{m+n+1} = F_m L_n + F_{m+1} L_{n+1}. \tag{29.8}$$

Finally, let $s(n) = n^2 - 3\lfloor n^2/3 \rfloor$. Clearly,

$$s(n) = \begin{cases} 0 & \text{if } 3|n \\ 1 & \text{otherwise;} \end{cases}$$

see Exercise 29.16. So $(-1)^{s(n)} = (-1)^{F_n} = (-1)^{L_n} = (-1)^{G_n}$; see Exercise 29.18.

With these tools at hand, we are in a position to develop a recursive definition of Y_n.

29.3   A RECURSIVE DEFINITION OF Y_n

Using formula (29.7), we have

$$\begin{aligned} Y_{n+2} &= F_{G_{n+2}} \\ &= F_{G_{n+1}-1} F_{G_n} + F_{G_{n+1}} F_{G_n+1} \\ &= \frac{1}{2} \left[Y_n \sqrt{5 Y_{n+1}^2 + 4(-1)^{G_{n+1}}} + Y_{n+1} \sqrt{5 Y_n^2 + 4(-1)^{G_n}} \right]. \end{aligned}$$

Since $Y_1 = F_{G_1} = F_a$ and $Y_2 = F_{G_2} = F_b$, Y_n can now be defined recursively:

$$\begin{aligned} &Y_1 = F_a, \quad Y_2 = F_b \\ &Y_{n+2} = \frac{1}{2} \left[Y_n \sqrt{5 Y_{n+1}^2 + 4(-1)^{G_{n+1}}} + Y_{n+1} \sqrt{5 Y_n^2 + 4(-1)^{G_n}} \right], \end{aligned} \qquad (29.9)$$

where $n \geq 1$.

From this recursive definition, we can deduce one for both U_n and V_n.

RECURSIVE DEFINITIONS OF U_n AND V_n

When $a = 1 = b$, $Y_n = U_n$. Accordingly, we have the following recursive definition of U_n, where $n \geq 1$:

$$\begin{aligned} &U_1 = 1 = U_2 \\ &U_{n+2} = \frac{1}{2} \left[U_n \sqrt{5 U_{n+1}^2 + 4(-1)^{s(n+1)}} + U_{n+1} \sqrt{5 U_n^2 + 4(-1)^{s(n)}} \right]. \end{aligned}$$

For example, let $n = 6$. We have $U_6 = F_8 = 21$ and $U_7 = F_{13} = 233$. Then

$$U_8 = \frac{1}{2}\left[U_6\sqrt{5U_7^2 - 4} + U_7\sqrt{5U_6^2 + 4}\right]$$

$$= \frac{1}{2}\left[21\sqrt{5 \cdot 54289 - 4} + 233\sqrt{5 \cdot 441 + 4}\right]$$

$$= 10{,}946.$$

When $a = 1$ and $b = 3$, $Y_n = V_n$; consequently, we have the following recursive definition of V_n, where $n \geq 1$:

$$V_1 = 1, \quad V_2 = 4$$

$$V_{n+2} = \frac{1}{2}\left[V_n\sqrt{5V_{n+1}^2 + 4(-1)^{s(n+1)}} + V_{n+1}\sqrt{5V_n^2 + 4(-1)^{s(n)}}\right].$$

For example, $U_5 = F_{11} = 89$ and $V_6 = F_{18} = 2584$. Then

$$V_7 = \frac{1}{2}\left[V_5\sqrt{5V_6^2 + 4} + V_6\sqrt{5V_5^2 - 4}\right]$$

$$= \frac{1}{2}\left[89\sqrt{5 \cdot 6677056 + 4} + 2584\sqrt{5 \cdot 7921 - 4}\right]$$

$$= 514{,}229.$$

A RECURSIVE DEFINITION OF Z_n

Using formulas (29.2), (29.3), (29.6), and (29.8), we can now develop a recurrence for Z_n:

$$\begin{aligned}
Z_{n+2} &= L_{G_{n+2}} \\
&= L_{G_{n+1}+G_n} \\
&= F_{G_{n+1}-1}L_{G_n} + F_{G_{n+1}}L_{G_n+1} \\
&= \frac{1}{2}\left(L_{G_{n+1}} - F_{G_{n+1}}\right)L_{G_n} + F_{G_{n+1}}L_{G_n+1} \\
&= \frac{1}{2}L_{G_n}L_{G_{n+1}} - \frac{1}{2}L_{G_n}F_{G_{n+1}} + F_{G_{n+1}}L_{G_n+1} \\
&= \frac{1}{2}L_{G_n}L_{G_{n+1}} + \frac{1}{2}F_{G_{n+1}}\left(2L_{G_n+1} - L_{G_n}\right) \\
&= \frac{1}{2}\left\{Z_nZ_{n+1} + \sqrt{\left[Z_{n+1}^2 - 4(-1)^{G_{n+1}}\right]\left[Z_n^2 - 4(-1)^{G_n}\right]}\right\}.
\end{aligned}$$

We can now define Z_n recursively:

$$Z_1 = L_a, \quad Z_2 = L_b$$

$$Z_{n+2} = \frac{1}{2} \left\{ Z_n Z_{n+1} + \sqrt{\left[Z_{n+1}^2 - 4(-1)^{G_{n+1}} \right] \left[Z_n^2 - 4(-1)^{G_n} \right]} \right\}. \qquad (29.10)$$

In particular, this yields a recursive definition of X_n.

A RECURSIVE DEFINITION OF X_n

When $a = 1 = b$, it yields

$$X_1 = 1, \quad Z_2 = 3$$

$$X_{n+2} = \frac{1}{2} \left\{ X_n X_{n+1} + \sqrt{\left[X_{n+1}^2 - 4(-1)^{s(n+1)} \right] \left[X_n^2 - 4(-1)^{s(n)} \right]} \right\},$$

where $n \geq 1$.

For example, let $n = 5$. Since $X_5 = L_5 = 11$ and $X_6 = L_8 = 47$, we then have

$$X_7 = \frac{1}{2} \left[X_5 X_6 + \sqrt{(X_6^2 - 4)(X_5^2 + 4)} \right]$$

$$= \frac{1}{2} \left[11 \cdot 47 + \sqrt{(2209 - 4)(12 + 4)} \right]$$

$$= 521.$$

A recursive definition of W_n also follows from formula (29.10) by letting $G_n = L_n$.

A RECURSIVE DEFINITION OF W_n

$$W_1 = 1, \quad W_2 = 4$$

$$W_{n+2} = \frac{1}{2} \left\{ W_n W_{n+1} + \sqrt{\left[W_{n+1}^2 - 4(-1)^{s(n+1)} \right] \left[W_n^2 - 4(-1)^{s(n)} \right]} \right\},$$

where $n \geq 1$.

Using this definition, you can verify that $W_6 = 5778$.

EXERCISES 29

1. Verify Theorem 29.1 for $n = 5$.
2. Prove Theorem 29.2.
3. Verify Theorem 29.2 for $n = 5$.
4. Verify Corollary 29.1 for $n = 6$.
5. Verify Theorem 29.3 for $n = 5$.
6. Prove part 1) of Theorem 29.3.
7. Prove part 2) of Theorem 29.3.
8. Verify part 1) of Theorem 29.4 for $n = 5$.
9. Verify part 2) of Theorem 29.4 for $n = 5$.
10. Prove part 1) of Theorem 29.4.
11. Prove part 2) of Theorem 29.4.
12. Compute F_{13} using identity (29.5).
13. Compute L_{13} using identity (29.6).
14. Prove identity (29.5).
15. Prove identity (29.6).
16. Let $s(n) = n^2 - 3\lfloor n^2/3 \rfloor$, where n is an integer. Prove that
$$s(n) = \begin{cases} 0 & \text{if } 3|n \\ 1 & \text{otherwise.} \end{cases}$$
17. Prove that $2|L_n$ if and only if $3|n$.
18. Prove that $(-1)^{s(n)} = (-1)^{F_n} = (-1)^{L_n}$.

Compute each.
19. U_7.
20. V_6.
21. X_8.
22. W_5.

30

FIBONACCI AND THE COMPLEX PLANE

In 1963, A.F. Horadam of the University of New England, Armidale, Australia, examined Fibonacci numbers on the complex plane, and established some interesting properties about them [325]. Two years later, J.H. Jordan of Washington State University followed up with a study of his own [355]. We will now briefly study these numbers.

To begin with, we introduce the well-known Gaussian numbers. They were investigated in 1832 by the German mathematician Karl Friedrich Gauss (1777–1855), sometimes called the "prince of mathematics."

Gauss*

Figure source: https://en.wikipedia.org/wiki/Carl_Friedrich_Gauss#/media/File:Carl_Friedrich_Gauss.jpg.

Fibonacci and Lucas Numbers with Applications, Volume One, Second Edition. Thomas Koshy.
© 2018 John Wiley & Sons, Inc. Published 2018 by John Wiley & Sons, Inc.

30.1   GAUSSIAN NUMBERS

A *Gaussian number* is a complex number $z = a + bi$, where a and b are any integers, and $i = \sqrt{-1}$. Its *norm* $\|z\|$ is defined by $\|z\| = a^2 + b^2$; it is the square of the distance of z from the origin on the complex plane: $\|z\| = |z|^2$.

The norm function satisfies the following fundamental properties:

- $\|z\| \geq 0$.

- $\|z\| = 0$ if and only if $z = 0$.

- $\|wz\| = \|w\| \cdot \|z\|$.

We leave their proofs as routine exercises; see Exercises 30.1–30.3.

We now introduce Gaussian Fibonacci and Lucas numbers.

30.2   GAUSSIAN FIBONACCI AND LUCAS NUMBERS

Gaussian Fibonacci numbers f_n are defined recursively by $f_n = f_{n-1} + f_{n-2}$, where $f_0 = i$, $f_1 = 1$, and $n \geq 2$. The first six Gaussian Fibonacci numbers are 1, $1 + i$, $2 + i$, $3 + 2i$, $5 + 3i$, and $8 + 5i$. Clearly, $f_n = F_n + F_{n-1}i$. Consequently, $\|f_n\| = F_n^2 + F_{n-1}^2 = F_{2n-1}$.

Gaussian Lucas numbers l_n also can be defined recursively: $l_n = l_{n-1} + l_{n-2}$, where $l_0 = 2 - i$, $l_2 = 1 + 2i$, and $n \geq 2$. The first six Gaussian Lucas numbers are $1 + 2i$, $3 + i$, $4 + 3i$, $7 + 4i$, $11 + 7i$, and $18 + 11i$. It is easy to see that $l_n = L_n + L_{n-1}i$. Also $\|l_n\| = L_n^2 + L_{n-1}^2 = 5F_{2n-1}$.

As we can predict, the identities we established in Chapter 5 can be extended to Gaussian Fibonacci and Lucas numbers. We will prove a few and leave some as exercises; see Exercises 30.12–30.26. They can easily be validated using PMI or summation formula (5.1).

Theorem 30.1.

$$\sum_{j=0}^{n} f_j = f_{n+2} - 1.$$

Proof. Clearly, the formula works when $n = 0$; so assume it works for an arbitrary integer $k \geq 0$. Then

$$\sum_{j=0}^{k+1} f_j = \sum_{j=0}^{k} f_j + f_{k+1}$$

$$= (f_{k+2} - 1) + f_{k+1}$$

$$= f_{k+3} - 1.$$

Thus, by PMI, the formula is true for all nonnegative integers n.   ∎

The next two theorems also can be easily established using PMI, so we omit their proofs; see Exercises 30.12 and 30.13.

Theorem 30.2.

$$\sum_{k=0}^{n} l_k = l_{n+2} - (1 + 2i).$$ ∎

Theorem 30.3. *Let $n \geq 1$. Then*

$$f_{n+1}f_{n-1} - f_n^2 = (2 - i)(-1)^n.$$ ∎

This is Cassini's formula for Gaussian Fibonacci numbers. For example,

$$\begin{aligned}
f_5 f_3 - f_2^2 &= (5 + 3i)(2 + i) - (3 + 2i)^2 \\
&= 2 - i \\
&= (2 - i)(-1)^4.
\end{aligned}$$

We can extend the definitions of both Gaussian Fibonacci and Lucas numbers to negative subscripts:

$$f_{-n} = F_{-n} + F_{-n-1}i = (-1)^{n-1}F_n + (-1)^n F_{n+1}i = (-1)^{n-1}(F_n - F_{n+1}i);$$
$$l_{-n} = L_{-n} + L_{-n-1}i = (-1)^{n-1}L_n + (-1)^{n+1}L_{n+1}i = (-1)^n(L_n - L_{n+1}i).$$

For example, $f_{-3} = (-1)^2(F_3 - F_4 i) = 2 - 3i$; and $l_{-4} = (-1)^4(L_4 - L_5 i) = 7 - 11i$.

In Chapter 5, we learned that $L_n^2 - 5F_n^2 = 4(-1)^n$. The next theorem gives its counterpart for Gaussian Fibonacci and Lucas numbers.

Theorem 30.4.

$$l_n^2 - 5f_n^2 = 4(2 - i)(-1)^n.$$ ∎

This result has an interesting byproduct. Since $5 = (2 - i)(2 + i)$, it follows that $(2 - i)|l_n^2$; so $(2 - i)|l_n$, where $n \geq 2$. Accordingly, we have the following result.

Corollary 30.1 (Jordan, 1965 [355]). *l_n is composite for $n \geq 2$.* ∎

For example, $l_5 = 11 + 7i = (2 - i)(3 + 5i)$; so $(2 - i)|(11 + 7i)$.

Since $l_{-n} = (-1)^n(L_n - L_{n+1}i)$, it follows that $(2 - i)|l_{-n}$, where $n \geq 2$. This gives the next result.

Corollary 30.2. *l_n is composite for all integers $n \neq \pm 1$.*   ∎

For instance, $l_{-5} = -11 + 18i = (2 - i)(-8 + 5i)$; so $(2 - i)|l_{-5}$.

In Chapter 10, we found that $F_m|F_n$ if and only if $m|n$, where $m \geq 2$. This result also has a counterpart for Gaussian Fibonacci and Lucas numbers [355]. To establish it, we will need the next lemma.

Lemma 30.1 (Jordan, 1965 [355]). *If $(2m - 1)|(2n - 1)$, then $(2m - 1)|(m + n - 1)$.*

Proof. Suppose $(2m - 1)|(2n - 1)$. Then $(2m - 1)|[(2n - 1) - (2m - 1)]$; that is, $(2m - 1)|(2n - 2m)$. But $(2m - 1, 2) = 1$, so $(2m - 1)|(n - m)$. Therefore, $(2m - 1)|[(2n - 1) - (n - m)]$; that is, $(2m - 1)|(m + n - 1)$.   ∎

We are now ready for the counterpart.

Theorem 30.5 (Jordan, 1965 [355]). *Let $m \geq 3$. Then $f_m|f_n$ if and only if $(2m - 1)|(2n - 1)$.*

Proof. Suppose $f_m|f_n$. Then $\|f_m\| \mid \|f_n\|$; that is, $F_{2m-1}|F_{2n-1}$. So, by Corollary 10.2, $(2m - 1)|(2n - 1)$.

Conversely, suppose $(2m - 1)|(2n - 1)$. Then $F_{2m-1}|F_{2n-1}$. Consequently,

$$\left\| \frac{f_n}{f_m} \right\| = \frac{\|f_n\|}{\|f_m\|} = \frac{F_{2n-1}}{F_{2m-1}}$$

is a positive integer; see Exercise 30.4. But, by the Fibonacci addition formula,

$$\begin{aligned}
\frac{f_n}{f_m} &= \frac{F_n + F_{n-1}i}{F_m + F_{m-1}i} \\
&= \frac{F_m F_n + F_{m-1}F_{n-1} + (F_{m-1}F_n - F_m F_{n-1})i}{F_m^2 + F_{m-1}^2} \\
&= \frac{F_m F_n + F_{m-1}F_{n-1}}{F_{2m-1}} + \frac{F_{m-1}F_n - F_m F_{n-1}}{F_{2m-1}}i \\
&= \frac{F_{m+n-1}}{F_{2m-1}} + \frac{F_{m-1}F_n - F_m F_{n-1}}{F_{2m-1}}i.
\end{aligned}$$

By Lemma 30.1 and Corollary 10.2, $\dfrac{F_{m+n-1}}{F_{2m-1}}$ is an integer. Since $\dfrac{\|f_n\|}{\|f_m\|}$ is also an integer, it follows that $\dfrac{F_{m-1}F_n - F_m F_{n-1}}{F_{2m-1}}$ must also be an integer. So $\dfrac{f_n}{f_m}$ is a Gaussian integer. Thus $f_m|f_n$, as desired.   ∎

For example, let $m = 3$ and $n = 8$. Then $(2m - 1)|(2n - 1)$. Notice that $f_8 = 21 + 13i = (2 + i)(11 + i) = (11 + i)f_3$. So $f_3|f_8$, as expected.

Theorem 30.5 has an interesting byproduct, as the next corollary shows.

Corollary 30.3. *Let* $m \geq 2$. *Then* $F_{2m-1}|(F_{m-1}F_n - F_mF_{n-1})$ *if and only if* $(2m - 1)|(2n - 1)$.   ∎

For example, with $m = 3$ and $n = 8$, $F_{2m-1} = F_5 = 5$ and $F_{m-1}F_n - F_mF_{n-1} = F_2F_8 - F_3F_7 = 1 \cdot 21 - 2 \cdot 13 = -5$; clearly, $F_{2m-1}|(F_{m-1}F_n - F_mF_{n-1})$.

Before presenting the next result, we now extend the definition of gcd to Gaussian integers.

GCD OF GAUSSIAN INTEGERS

The Gaussian integer y is the *gcd of the Gaussian integers* w and z if:

- $y|w$ and $y|z$;

- if $x|w$ and $x|z$, then $\|x\| \leq \|y\|$.

The gcd is denoted by $y = (w, z)$.

For example, $(1 + 7i, 2 + 9i) = 1 + 2i$.

In Chapter 10, we learned that $(F_m, F_n) = F_{(m,n)}$. Correspondingly, we have the following beautiful result for Gaussian Fibonacci integers [355].

Theorem 30.6 (Jordan, 1965 [355]).   *Let* $(2m - 1, 2n - 1) = 2k - 1$.   *Then* $(f_m, f_n) = f_k$.

Proof. Since $(2k - 1)|(2m - 1)$ and $(2k - 1)|(2n - 1)$, it follows by Theorem 30.5 that $f_k|f_m$ and $f_k|f_n$; therefore, $f_k|(f_m, f_n)$.

Suppose $x|f_m$ and $x|f_n$. Then $\|x\| \mid \|f_m\|$ and $\|x\| \mid \|f_n\|$; that is, $\|x\| \mid F_{2m-1}$ and $\|x\| \mid F_{2n-1}$. Therefore, $\|x\| \mid (F_{2m-1}, F_{2n-1})$; that is, $\|x\| \mid F_{(2m-1,2n-1)}$. In other words, $\|x\| \mid F_{2k-1}$; that is, $\|x\| \mid f_k$. Consequently, $\|x\| \leq \|f_k\|$. Thus $(f_m, f_n) = f_k$.   ∎

For example, let $m = 5$ and $n = 8$. Then $(2m - 1, 2n - 1) = 3 = 2k - 1$, where $k = 2$. We have $f_5 = 5 + 3i$, $f_8 = 21 + 13i$, and $f_2 = 1 + i$. Notice that $f_5 = (1 + i)(4 - i)$ and $f_8 = (1 + i)(17 - 4i)$, so $(f_5, f_8) = 1 + i = f_2$, as expected.

A TELESCOPING FIBONACCI SUM

Next we study an infinite sum involving Gaussian Fibonacci numbers. The ubiquitous divine ratio makes a surprising visit in the process. Manuel Kauers of Johannes Kepler University, Linz, Austria, studied it in 2006 [356].

Example 30.1. *Prove that* $\displaystyle\sum_{k=0}^{\infty} \frac{F_{3^k} - 2F_{3^k+1}}{F_{3^k} + iF_{2 \cdot 3^k}} = \beta + i$, *where* $i = \sqrt{-1}$.

Proof. Let m be odd. Then, by Binet's formula, we have

$$\sqrt{5}(F_m + iF_{2m}) = (\alpha^m - \beta^m) + i(\alpha^{2m} - \beta^{2m})$$

$$= i\alpha^{2m} + \alpha^m - (-\alpha)^{-m} - i(-\alpha)^{-2m}$$

$$= i\alpha^{2m} + \alpha^m + \alpha^{-m} - i\alpha^{-2m}$$

$$= -i\alpha^{-2m}(i\alpha^m + 1)(i\alpha^{3m} + 1);$$

$$\sqrt{5}(F_m - 2F_{m+1}) = (\alpha^m - \beta^m) - 2(\alpha^{m+1} - \beta^{m+1})$$

$$= (1 - 2\alpha)\alpha^m + (1 + 2\alpha^{-1})\alpha^{-m}$$

$$= \sqrt{5}(-\alpha^m + \alpha^{-m})$$

$$F_m - 2F_{m+1} = -\alpha^m + \alpha^{-m}.$$

Consequently,

$$\frac{F_m - 2F_{m+1}}{F_m + iF_{2m}} = \sqrt{5} \cdot \frac{-\alpha^m + \alpha^{-m}}{-i\alpha^{-2m}(i\alpha^m + 1)(i\alpha^{3m} + 1)}$$

$$= \sqrt{5} \cdot \frac{-i\alpha^{3m} + i\alpha^m}{(i\alpha^m + 1)(i\alpha^{3m} + 1)}$$

$$= \frac{\sqrt{5}}{i\alpha^{3m} + 1} - \frac{\sqrt{5}}{i\alpha^m + 1}.$$

Letting $m = 3^k$, this yields a telescoping sum:

$$\sum_{k=0}^{\infty} \frac{F_{3^k} - 2F_{3^k+1}}{F_{3^k} + iF_{2\cdot 3^k}} = \sum_{k=0}^{\infty} \left(\frac{\sqrt{5}}{i\alpha^{3^{k+1}} + 1} - \frac{\sqrt{5}}{i\alpha^{3^k} + 1} \right)$$

$$= -\frac{\sqrt{5}}{i\alpha + 1}$$

$$= \frac{i\alpha - 1}{\alpha}$$

$$= \beta + i,$$

as desired. ∎

It follows by this example that

$$\sum_{k=0}^{\infty} \frac{F_{3^k} - 2F_{3^k+1}}{F_{3^k} - iF_{2\cdot 3^k}} = \beta - i.$$

30.3 ANALYTIC EXTENSIONS

In 1966, Whitney also investigated analytic generalizations of both Fibonacci and Lucas numbers, by extending Binet's formulas to the complex plane [592]:

$$f(z) = \frac{\alpha^z - \beta^z}{\alpha - \beta} \quad \text{and} \quad l(z) = \alpha^z + \beta^z,$$

where z is an arbitrary complex variable, $f(z)$ is the *complex Fibonacci function* and $l(z)$ the *Lucas Fibonacci function*. Clearly, $f(n) = F_n$ and $l(n) = L_n$, where n is an integer. Both functions f and l possess several interesting properties. We will examine a few of them here.

PERIODICITIES OF α^z AND β^z

Let p be the period of α^z. Then $\alpha^{z+p} = \alpha^z$, so $\alpha^p = 1 = e^{2\pi i}$. Thus $p = 2\pi i / \ln \alpha$.

On the other hand, let q be the period of β^z. Then $\beta^{z+q} = \beta^z$, so $\beta^q = 1 = e^{2\pi i}$. Since $\beta < 0$, we rewrite $\beta = e^{\pi i}(-\beta)$. Then $e^{q\pi i}(-\beta)^q = e^{2\pi i}$, so $(-\beta)^q = e^{\pi(2-q)i}$. That is, $\alpha^q = e^{\pi(2-q)i}$, so $q \ln \alpha = \pi(q-2)i$. Then

$$q = \frac{2\pi i}{\pi i - \ln \alpha} = \frac{2\pi(\pi - i \ln \alpha)}{\ln^2 \alpha + \pi^2}.$$

Thus α^z is periodic with period $\dfrac{2\pi i}{\ln \alpha}$, and β^z has period $\dfrac{2\pi(\pi - i \ln \alpha)}{\ln^2 \alpha + \pi^2}$.

Are $f(z)$ and $l(z)$ periodic? If yes, what are their periods? We will now answer these questions.

PERIODICITIES OF $f(z)$ AND $l(z)$

Suppose $f(z)$ is periodic with period w. Then $f(0) = 0 = f(w)$, which implies $\alpha^w = \beta^w$. So $f(z + w) = f(z)$ yields $\alpha^{z+w} - \beta^{z+w} = \alpha^z - \beta^z$; that is, $\alpha^w(\alpha^z - \beta^z) = \alpha^z - \beta^z$; so $\alpha^w = 1$. This implies the real part x of $w = x + yi$ must be zero. Then $\alpha^{yi} = 1$. But this is possible only if $y = 0$. Then $w = 0 + 0i = 0$, which is a contradiction. Thus $f(z)$ is *not* periodic. Likewise, we can show that $l(z)$ also is *not* periodic; see Exercise 30.33.

ZEROS OF $f(z)$ AND $l(z)$

We will now pursue the zeros of $f(z)$ and $l(z)$. First, notice that $f(z)$ has a real zero, namely 0; but $l(z)$ has no real zeros.

To find the complex zeros of $f(z)$, let $f(z) = 0$. This yields $(\alpha/\beta)^z = 1 = e^{2k\pi i}$, where k is an arbitrary integer. Then $z \ln(\alpha/\beta) = 2k\pi i$. But $\beta = e^{\pi i}(-\beta)$, so $\alpha/\beta = \alpha^2 e^{-\pi i}$ and $\ln(\alpha/\beta) = 2 \ln \alpha - \pi i$. Thus

$$z(2 \ln \alpha - \pi i) = 2k\pi i$$

$$
\begin{aligned}
z &= \frac{2k\pi i}{2 \ln \alpha - \pi i} \\
&= \frac{2k\pi i(2 \ln \alpha + \pi i)}{4 \ln^2 \alpha + \pi^2} \\
&= \frac{2k\pi(-\pi + 2i \ln \alpha)}{4 \ln^2 \alpha + \pi^2}.
\end{aligned}
$$

This equation gives the infinitely many zeros of $f(z)$.

Similarly, we can show that the infinitely many complex zeros of $l(z)$ are given by

$$z = \frac{(2k + 1)\pi(-\pi + 2i \ln \alpha)}{4 \ln^2 \alpha + \pi^2};$$

see Exercise 30.34.

Next we will study the behavior of the two functions on the real axis.

BEHAVIOR OF $f(z)$ AND $l(z)$ ON THE REAL AXIS

Let $z = x$, an arbitrary real number. Then $\alpha^z = \alpha^x$; and since $\ln \beta = -\ln \alpha$, $\beta^z = \beta^x = e^{x \ln \beta} = e^{x(\pi i - \ln \alpha)} = e^{-x \ln \alpha}(\cos \pi x + i \sin \pi x)$. Since $\mathrm{Im}\, f(z) = 0 = \mathrm{Im}\, l(z)$, this yields $e^{-x \ln \alpha} \sin \pi x = 0$, so $\sin \pi x = 0$. This implies that x must be an integer n. Thus $f(z)$ is an integer if and only if z is an integer. The same is true for $l(z)$.

IDENTITIES SATISFIED BY $f(z)$ AND $l(z)$

Many of the Fibonacci and Lucas properties we studied in Chapter 5 have their counterparts on the complex plane. A sampling of these are listed below:

1) $f(z + 2) = f(z + 1) + f(z)$.

2) $l(z + 2) = l(z + 1) + l(z)$.

3) $f(z + 1)f(z - 1) - f^2(z) = e^{\pi z i}$.

4) $l^2(z) - 5f^2(z) = 4e^{\pi z i}$.

5) $f(-z) = -f(z)e^{\pi z i}$.

6) $l(-z) = l(z)e^{\pi z i}$.

7) $f(2z) = f(z)l(z)$.

8) $f(z + w) = f(z)f(w + 1) + f(z - 1)f(w)$.

9) $f(3z) = f^3(z + 1) + f^3(z) - f^3(z - 1)$.

We can easily establish these identities using Binet's formulas. For example, we have

$$5[f(z + 1)f(z - 1) - f^2(z)] = (\alpha^{z+1} - \beta^{z+1})(\alpha^{z-1} - \beta^{z-1}) - (\alpha^z - \beta^z)^2$$

$$= \alpha^{2z} + \beta^{2z} - (\alpha\beta)^z(\alpha^2 + \beta^2) - [\alpha^{2z} + \beta^{2z} - 2(\alpha\beta)^z]$$

$$= 3(-1)^z + 2(-1)^z$$

$$= 5e^{\pi zi}.$$

Thus $f(z + 1)f(z - 1) - f^2(z) = e^{\pi zi}$, as expected.

We leave the proofs of the others as routine exercises; see Exercises 30.35–30.42.

Finally, we will study the Taylor expansions of $f(z)$ and $l(z)$. They have interesting byproducts to Fibonacci and Lucas numbers, respectively.

TAYLOR EXPANSIONS OF $f(z)$ AND $l(z)$

Since both $f(z)$ and $l(z)$ are entire functions, both have Taylor expansions. Since $\dfrac{d^k}{dw^k}(\alpha^w) = \alpha^w \ln^k \alpha$, it follows that

$$f(z) = \sum_{k=0}^{\infty} \frac{f^{(k)}(w)}{k!}(z - w)^k$$

$$= \frac{1}{\sqrt{5}} \sum_{k=0}^{\infty} \frac{\alpha^w \ln^k \alpha - \beta^w \ln^k \beta}{k!}(z - w)^k. \tag{30.1}$$

Likewise,

$$l(z) = \sum_{k=0}^{\infty} \frac{\alpha^w \ln^k \alpha + \beta^w \ln^k \beta}{k!}(z - w)^k. \tag{30.2}$$

In particular, let $z = n$ and $w = n - 1$. Then equations (30.1) and (30.2) yield interesting infinite series expansions of F_n and L_n:

$$F_n = \frac{1}{\sqrt{5}} \sum_{k=0}^{\infty} \frac{\alpha^{n-1} \ln^k \alpha - \beta^{n-1} \ln^k \beta}{k!}$$

$$L_n = \sum_{k=0}^{\infty} \frac{\alpha^{n-1} \ln^k \alpha + \beta^{n-1} \ln^k \beta}{k!}.$$

EXERCISES 30

Prove each, where w and z are Gaussian integers.

    1. $\|z\| \geq 0$.

    2. $\|z\| = 0$ if and only if $z = 0$.

    3. $\|wz\| = \|w\| \cdot \|z\|$.

    4. $\|w/z\| = \|w\| / \|z\|$.

    5. $f_n = F_n + F_{n-1}i$, where $n \geq 1$.

    6. $l_n = L_n + L_{n-1}i$, where $n \geq 1$.

    7. $\|\bar{z}\| = \|z\|$, where $\bar{z}$ denotes the complex conjugate of z.

    8. Compute f_{10} and l_{10}.

    9. Compute f_{-10} and l_{-10}.

  10. Verify Theorem 30.5 for $m = 4$ and $n = 7$.

  11. Verify Corollary 30.3 for $m = 4$ and $n = 11$.

Prove each (Jordan, 1965 [355]).

  12. $\displaystyle\sum_{k=0}^{n} l_k = l_{n+2} - (1 + 2i)$.

  13. $f_{n+1}f_{n-1} - f_n^2 = (2 - i)(-1)^n$.

  14. $f_{n+1} + f_{n-1} = l_n$.

  15. $l_{n+1} + l_{n-1} = 5f_n$.

  16. $f_{n+2} + f_{n-2} = 3f_n$.

  17. $f_{n+2} - f_{n-2} = l_n$.

  18. $l_{n+2} - l_{n-2} = 5f_n$.

  19. $f_n^2 + f_{n+1}^2 = (1 + 2i)F_{2n}$.

  20. $f_{n+1}^2 - f_{n-1}^2 = (1 + 2i)F_{2n-1}$.

  21. $f_n l_n = (1 + 2i)F_{2n-1}$.

  22. $f_{m+1}f_{n+1} + f_m f_n = (1 + 2i)f_{m+n}$.

  23. $l_{n+1}l_{n-1} - l_n^2 = 5(2 - i)(-1)^{n+1}$.

  24. $f_{n+2}^2 - f_{n-2}^2 = 3(1 + 2i)F_{2n-1}$.

  25. $l_n^2 - 5f_n^2 = 4(2 - i)(-1)^n$.

  26. $\displaystyle\sum_{k=1}^{n} f_k^2 = (1 + 2i)f_n^2 + i(-1)^n - i$.

  27. $\displaystyle\sum_{k=1}^{n} f_{2k-1} = f_{2n} - i$.

28. $\displaystyle\sum_{k=1}^{n} f_{2k} = f_{2n+1} - 1.$

29. $\displaystyle\sum_{k=1}^{2n} (-1)^k f_k = f_{2n-1} + i - 1.$

30. $\displaystyle\sum_{k=1}^{n} (-1)^k f_k = (-1)^{n+1} f_n + i - 1.$

Let $C_n = F_n + F_{n+1} i$. Prove each, where $\overline{C_n}$ denotes the complex conjugate of C_n.

31. $C_n \overline{C_n} = F_{2n+1}.$

32. $C_n \overline{C_{n+1}} = F_{2n+2} + i(-1)^n.$

33. Prove that $l(z)$ is not periodic.

34. Find the complex zeros of $l(z)$.

Prove each.

35–42. Identities 1), 2), and 4)–9) on pages 562–563.

43. $\displaystyle F_n = \frac{1}{\sqrt{5}} \sum_{k=0}^{\infty} \frac{(\ln^k \alpha - \ln^k \beta)n^k}{k!}.$

44. $\displaystyle L_n = \sum_{k=0}^{\infty} \frac{(\ln^k \alpha + \ln^k \beta)n^k}{k!}.$

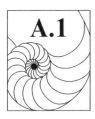

A.1

FUNDAMENTALS

This Appendix presents the fundamental symbols, definitions, and facts needed to pursue the essence of the theory of Fibonacci and Lucas numbers. For convenience, we have omitted examples and all proofs [369].

SEQUENCES

The sequence $s_1, s_2, s_3, \ldots, s_n, \ldots$ is denoted by $\{s_n\}_{n=1}^{\infty}$ or simply $\{s_n\}$ when there is no possibility of confusion. The nth term s_n is the *general term* of the sequence. Sequences can be classified as finite or infinite, as the next definition shows.

A sequence is *finite* if its domain is finite; otherwise, it is *infinite*.

SUMMATION AND PRODUCT NOTATIONS

$$\sum_{i=k}^{i=m} a_i = a_k + a_{k+1} + \cdots + a_m.$$

The variable i is the *summation index*. The values k and m are the *lower* and *upper limits* of the index i. The "$i =$" above the Σ is often omitted; in fact, the indices above and below the Σ are also omitted when there is no confusion. Thus

$$\sum_{i=k}^{i=m} a_i = \sum_{i=k}^{m} a_i = \sum_{k}^{m} a_i.$$

Fibonacci and Lucas Numbers with Applications, Volume One, Second Edition. Thomas Koshy.
© 2018 John Wiley & Sons, Inc. Published 2018 by John Wiley & Sons, Inc.

The index i is a *dummy variable*; we can use any variable as the index without affecting the value of the sum; so

$$\sum_{i=k}^{m} a_i = \sum_{j=k}^{m} a_j = \sum_{l=k}^{m} a_l.$$

The following results are extremely useful in evaluating finite sums. They can be established using the principle of mathematical induction (PMI).

Theorem A.1. *Let $n \in \mathbb{N}$ and $c \in \mathbb{R}$. Let $a_1, a_2, \ldots,$ and $b_1, b_2, \ldots$ be any two number sequences. Then*

$$\sum_{i=1}^{n} c = nc$$

$$\sum_{i=1}^{n} (ca_i) = c \left(\sum_{i=1}^{n} a_i \right)$$

$$\sum_{i=1}^{n} (a_i + b_i) = \left(\sum_{i=1}^{n} a_i \right) + \left(\sum_{i=1}^{n} b_i \right).$$

(These results can be extended to any integral lower limit.) ∎

INDEXED SUMMATION

The summation notation can be extended to sequences with *index sets I* as their domains. For instance, $\sum_{i \in I} a_i$ denotes the sum of the values a_i, as i runs over the values in I.

Often we need to evaluate sums of the form $\sum_{P} a_{ij}$, where the subscripts i and j satisfy certain properties P.

Multiple summations often arise in mathematics. They are evaluated in the right-to-left fashion. For example, the double summation $\sum_{i} \sum_{j} a_{ij}$ is evaluated as $\sum_{i} \left(\sum_{j} a_{ij} \right)$, and the triple summation $\sum_{i} \sum_{j} \sum_{k} a_{ijk}$ as $\sum_{i} \left[\sum_{j} \left(\sum_{k} a_{ijk} \right) \right]$.

THE PRODUCT NOTATION

The product $a_k a_{k+1} \cdots a_m$ is denoted by $\prod_{i=k}^{i=m} a_i$, where the *product symbol $\prod$* is the uppercase Greek letter *pi*. As in the case of the summation notation, the "$i =$"

above the product symbol can be dropped, if doing so leads to no confusion. Thus

$$\prod_{i=k}^{i=m} a_i = \prod_{i=k}^{m} a_i = \prod_{k}^{m} a_i = a_k a_{k+1} \cdots a_m.$$

Again, i is just a dummy variable.

THE FACTORIAL NOTATION

Let n be a nonnegative integer. The *factorial notation* $f(n) = n!$ (read n factorial) is defined by $n! = n(n-1)\cdots 2 \cdot 1$, where $0! = 1$. Thus $f(n) = n! = \prod_{i=1}^{n} i$.

FLOOR AND CEILING FUNCTIONS

The *floor* of a real number x, denoted by $\lfloor x \rfloor$, is the greatest integer $\leq x$. The *ceiling* of x, denoted by $\lceil x \rceil$, is the least integer $\geq x$.[†] The *floor function* $f(x) = \lfloor x \rfloor$ and the *ceiling function* $g(x) = \lceil x \rceil$ are also known as the *greatest integer function* and the *least integer function*, respectively.

Theorem A.2. *Let x be any real number, and n any integer. Then*

1) $\lfloor n \rfloor = n = \lceil n \rceil$ 2) $\lceil x \rceil = \lfloor x \rfloor + 1 \ (x \notin \mathbb{Z})$

3) $\lfloor x + n \rfloor = \lfloor x \rfloor + n$ 4) $\lceil x + n \rceil = \lceil x \rceil + n$

5) $\lfloor n/2 \rfloor = (n-1)/2$ if n is odd   6) $\lceil n/2 \rceil = (n+1)/2$ if n is odd. ∎

THE WELL-ORDERING PRINCIPLE (WOP)

The principle of mathematical induction (PMI) is a powerful proof technique we employ often. It is based on the following axiom.

Every nonempty set of positive integers has a least element.

[†]These two notations and the names, *floor* and *ceiling*, were introduced by Kenneth E. Iverson (1920–2004) in the early 1960s. Both notations are variations of the original greatest integer notation $[x]$.

MATHEMATICAL INDUCTION

We use the powerful proof technique of PMI[†] throughout the book.

The next result is the cornerstone of PMI. Its proof follows by the WOP.

Theorem A.3. *Let k be an arbitrary positive integer and S a set of positive integers satisfying the following properties:*

1) $1 \in S$.
2) *If* $k \in S$, *then* $k + 1 \in S$.

Then $S = \mathbb{N}$. ∎

This theorem can be generalized, as the following theorem shows.

Theorem A.4. *Let* n_0 *be a fixed integer, k an arbitrary positive integer, and S a set of positive integers satisfying the following properties:*

1) $n_0 \in S$.
2) *If* $k \geq n_0$ *and* $k \in S$, *then* $k + 1 \in S$.

Then S contains all integers $\geq n_0$. ∎

The next theorem gives the principle of mathematical induction.

Theorem A.5 (Simple Version of PMI). *Let* $n \in \mathbb{Z}$ *and* $P(n)$ *a statement satisfying the following conditions:*

1) $P(n_0)$ *is true for some integer* n_0.
2) *If* $P(k)$ *is true for an arbitrary integer* $k \geq n_0$, *then* $P(k + 1)$ *is also true.*

Then $P(n)$ *is true for every integer* $n \geq n_0$. ∎

Proving a result by PMI involves three key steps:

1) *Basis Step*. Verify that $P(n_0)$ is true.
2) *Induction Step*. Assume $P(k)$ is true for an arbitrary integer $k \geq n_0$ (*inductive hypothesis*). Then verify that $P(k + 1)$ is also true.
3) *Conclusion*. $P(n)$ is true for every integer $n \geq n_0$.

The next four summation formulas will come in handy in our discussions. They can be established easily using PMI.

[†]Although the Venetian scientist Francesco Maurocylus (1491–1575) applies mathematical induction in proofs in a book he wrote in 1575, the term "mathematical induction" was coined by the English mathematician Augustus De Morgan (1806–1871).

SUMMATION FORMULAS

1) $\displaystyle\sum_{i=1}^{n} i = \frac{n(n+1)}{2}$.

2) $\displaystyle\sum_{i=1}^{n} i^2 = \frac{n(n+1)(2n+1)}{6}$.

3) $\displaystyle\sum_{i=1}^{n} i^3 = \left[\frac{n(n+1)}{2}\right]^2$.

4) $\displaystyle\sum_{i=1}^{n} ar^{i-1} = \frac{a(r^n-1)}{r-1}$, where $\neq 1$.

Next we present a stronger version of PMI.

Theorem A.6 (Strong Version of PMI). *Let $n \in \mathbb{Z}$ and $P(n)$ a statement satisfying the following conditions:*

1) $P(n_0)$ *is true for some integer n_0.*
2) *Let k be an arbitrary integer $\geq n_0$ such $P(n_0), P(n_0 + 1), \ldots, P(k)$ are true. Then $P(k+1)$ is also true.*

Then $P(n)$ is true for every integer $n \geq n_0$. ∎

RECURSION

Recursion is one of the most elegant and powerful problem-solving techniques. It is the backbone of most programming languages. We will have numerous opportunities to apply recursion for solving problems throughout our discussion, so be prepared to use it.

Suppose you would like to solve a complex problem. The solution may not be obvious. It may be, however, that the problem can be defined in terms of a simpler version of itself. Such a definition is a *recursive definition*. Consequently, the given problem can be solved, provided the simpler version can be solved.

RECURSIVE DEFINITION OF A FUNCTION

The *recursive definition* of a function f consists of three parts, where $a \in \mathbb{W}$:

1) *Basis Clause.* A few initial values $f(a), f(a+1), \ldots, f(a+k-1)$ are specified. An equation that specifies such an initial value is an *initial condition*.
2) *Recursive Clause.* A formula to compute $f(n)$ from the k preceding functional values $f(n-1), f(n-2), \ldots, f(n-k)$ is made. Such a formula is a *recurrence* or *recursive formula*.
3) *Terminal Clause.* Only values thus obtained are valid functional values. (For convenience, we will omit stating this clause in the definition.)

Thus the recursive definition of f consists of one or more (a finite number of) initial conditions, and a recurrence.

The next theorem confirms that a recursive definition is indeed a valid definition.

Theorem A.7.   *Let $a \in \mathbb{W}$, $X = \{a, a + 1, a + 2, \ldots\}$, and $k \in \mathbb{N}$. Let $f : X \to \mathbb{R}$ such that $f(a), f(a + 1), \ldots, f(a + k - 1)$ are known. Let n be a positive integer $\geq a + k$ such that $f(n)$ is defined in terms of $f(n - 1), f(n - 2), \ldots, f(n - k)$. Then $f(n)$ is defined for every integer $n \geq a$.*   ∎

By virtue of this theorem, recursive definitions are also known as *inductive definitions*.

THE DIVISION ALGORITHM

The division algorithm is an application of the WOP, and is often used to check the correctness of a division problem.

Suppose an integer a is divided by a positive integer b. Then we get a unique *quotient q* and a unique *remainder r*, where $0 \leq r < b$; a is the *dividend* and b the *divisor*. This is formally stated in the following theorem.

Theorem A.8 (The Division Algorithm).   *Let a be any integer and b a positive integer. Then there exist unique integers q and r such that*

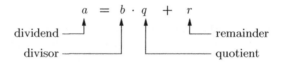

where $0 \leq r < b$.   ∎

Although this theorem does not present an algorithm for finding q and r, traditionally it has been called the *division algorithm*. Integers q and r can be found using the familiar long-division method.

Notice that the equation $a = bq + r$ can be rewritten as $\dfrac{a}{b} = q + \dfrac{r}{b}$, where $0 \leq r/b < 1$. Consequently, $q = \lfloor a/b \rfloor$ and $r = a - bq$.

DIV AND MOD OPERATORS

Two simple and useful operators, *div* and *mod*, are used often in discrete mathematics and computer science to find quotients and remainders:

$$a \text{ div } b = \text{quotient when } a \text{ is divided by } b$$

$$a \text{ mod } b = \text{remainder when } a \text{ is divided by } b.$$

It follows from these definitions that $q = a$ div $b = \lfloor a/b \rfloor$, and $r = a$ mod $b = a - bq = a - b \cdot \lfloor a/b \rfloor$.

DIVISIBILITY RELATION

Suppose $a = bq + 0 = bq$. We then say that b *divides* a, b is a *factor* of a, a is *divisible* by b, or a is a *multiple* of b, and write $b|a$. If b is not a factor of a, we write $b \nmid a$.

DIVISIBILITY PROPERTIES

Theorem A.9.   *Let a and b be positive integers such that $a|b$ and $b|a$. Then $a = b$.*   ∎

Theorem A.10.   *Let a, b, c, s, and t be any integers. Then the following hold.*

1) *If $a|b$ and $b|c$, then $a|c$ (transitive property).*
2) *If $a|b$ and $a|c$, then $a|(sb + tc)$.*
3) *If $a|b$, then $a|bc$.*   ∎

The expression $sb + tc$ is a *linear combination* of b and c. Thus, if a is a factor of both b and c, then a is also a factor of any linear combination of b and c. In particular, $a|(b + c)$ and $a|(b - c)$.

The floor function can be used to determine the number of positive integers less than or equal to a positive integer a and divisible by a positive integer b, as the next theorem shows.

Theorem A.11.   *Let a and b be any positive integers. Then the number of positive integers $\leq a$ and divisible by b is $\lfloor a/b \rfloor$.*   ∎

PIGEONHOLE PRINCIPLE

Suppose m pigeons fly into n pigeonholes to roost, where $m > n$. Since there are more pigeons than pigeonholes, at least two pigeons must roost in the same pigeonhole; in other words, there must be a pigeonhole containing two or more pigeons.

We now state the simple version of the pigeonhole principle.

Theorem A.12 (The Pigeonhole Principle).   *Suppose m pigeons are assigned to n pigeonholes, where $m > n$. Then at least two pigeons must occupy the same pigeonhole.*   ∎

The pigeonhole principle is also called the *Dirichlet box principle* after the German mathematician, Peter Gustav Lejeune Dirichlet (1805–1859), who used it extensively in his work on number theory.

We can generalize the pigeonhole principle, as the following theorem shows.

Theorem A.13 (The Generalized Pigeonhole Principle). *Suppose m pigeons are assigned to n pigeonholes. Then there must be a pigeonhole containing at least* $\lfloor (m-1)/n \rfloor + 1$ *pigeons.* ∎

ADDITION PRINCIPLE

Let A be a finite set; and $|A|$ its *cardinality*, the number of elements in A. (Often we use the vertical bars to denote the absolute value of a number, but here it denotes the number of elements in a set. The meaning of the notation should be clear from the context.)

UNION AND INTERSECTION

Let A and B be any two sets. Their *union* $A \cup B$ consists of elements belonging to A or B; their *intersection* $A \cap B$ consists of their common elements.

With this understanding, we can move on to the inclusion–exclusion principle.

Theorem A.14 (Inclusion–Exclusion Principle). *Let A, B, and C be any finite sets. Then*

$$|A \cup B| = |A| + |B| - |A \cap B|$$
$$|A \cup B \cup C| = |A| + |B| + |C| - |A \cap B| - |B \cap C| - |C \cap A| + |A \cap B \cap C|.$$

∎

This theorem can be extended to any finite number of finite sets.

Corollary A.1 (Addition Principle). *Let A, B, and C be any finite, pairwise disjoint sets. Then*

$$|A \cup B| = |A| + |B|$$
$$|A \cup B \cup C| = |A| + |B| + |C|.$$

∎

GCD AND LCM

A positive integer can be a factor of two positive integers, a and b. Such a factor is a *common divisor* or *common factor* of a and b. Often we are interested in their greatest common divisor.

GREATEST COMMON DIVISOR

The *greatest common divisor* (gcd) of two positive integers a and b is the greatest positive integer that divides both a and b; it is denoted by (a, b).

A SYMBOLIC DEFINITION OF GCD

A positive integer d is the gcd of two positive integers a and b:

1) if $d|a$ and $d|b$; and
2) if $d'|a$ and $d'|b$, then $d'|d$, where d' is a positive integer.

RELATIVELY PRIME INTEGERS

Two positive integers a and b are *relatively prime* if $(a, b) = 1$.

We now turn to a short discussion of some interesting and useful properties of gcds.

Theorem A.15. *Let $d = (a, b)$. Then $(a/d, b/d) = 1$ and $(a, a - b) = d$.* ∎

Theorem A.16. *The gcd of the positive integers a and b is a linear combination of a and b.* ∎

The gcd (a, b) is in fact the least positive linear combination of a and b.

Theorem A.17. *Let a, b, and c be any positive integers. Then $(ac, bc) = c(a, b)$.* ∎

Theorem A.18. *Two positive integers a and b are relatively prime if and only if there are integers s and t such that $sa + tb = 1$.* ∎

Corollary A.2. *If $a|c$ and $b|c$, and $(a, b) = 1$, then $ab|c$.* ∎

Remember that $a|bc$ does not mean $a|b$ or $a|c$, although under some conditions it does. The next corollary explains when it is true.

Corollary A.3 (Euclid). *If $a|bc$ and $(a, b) = 1$, then $a|c$.* ∎

The definition of gcd can be extended to any finite number of positive integers, as the next definition shows.

GCD OF *n* POSITIVE INTEGERS

The gcd of $n \geq 2$ positive integers $a_1, a_2, \ldots, a_n$ is the largest positive integer that divides each a_i. It is denoted by $(a_1, a_2, \ldots, a_n)$.

The next theorem shows how nicely recursion can be used to find the gcd of three or more integers.

Theorem A.19. *Let $a_1, a_2, \ldots, a_n$ be $n \geq 2$ positive integers. Then $(a_1, a_2, \ldots, a_n)$ $= ((a_1, a_2, \ldots, a_{n-1}), a_n)$.* ∎

The next result is an extension of Corollary A.3.

Corollary A.4. *If $d|a_1 a_2 \cdots a_n$, and $(d, a_i) = 1$ for $1 \leq i \leq n - 1$, then $d|a_n$.* ∎

FUNDAMENTAL THEOREM OF ARITHMETIC

Prime numbers are the building blocks of all positive integers, and hence of all integers. Positive integers, except 1, are made up of primes and every such integer can be decomposed into primes. This result, called the *fundamental theorem of arithmetic*, is the cornerstone of number theory. It appears in Euclid's *Elements*.

Before stating it formally, we need to lay some groundwork in the form of two lemmas and a corollary. Throughout, assume all letters denote positive integers.

Lemma A.1 (Euclid). *If p is a prime and $p|ab$, then $p|a$ or $p|b$.* ∎

The next lemma extends this result to any finite number of positive factors, using PMI.

Lemma A.2. *Let p be a prime and $p|a_1 a_2 \cdots a_n$, where $a_1, a_2, \ldots, a_n$ are positive integers. Then $p|a_i$ for some i, where $1 \leq i \leq n$.* ∎

The next result follows nicely from this result.

Corollary A.5. *If $p, q_1, q_2, \ldots, q_n$ are primes and $p|q_1 q_2 \cdots q_n$, then $p = q_i$ for some i, where $1 \leq i \leq n$.* ∎

We can now state the fundamental theorem of arithmetic, the most fundamental result in number theory.

Theorem A.20 (Fundamental Theorem of Arithmetic). *Every positive integer $n \geq 2$ is either a prime, or can be expressed as a product of primes. The factorization into primes is unique except for the order of the factors.* ∎

A factorization of a composite number n in terms of primes is a *prime factorization* of n. Using the exponential notation, this product can be rewritten in a compact way. Such a product is the *prime-power decomposition* of n; if the primes occur in ascending order, then it is the *canonical decomposition*.

CANONICAL DECOMPOSITION

The *canonical decomposition* of a positive integer n is of the form $n = p_1^{e_1} p_2^{e_2} \cdots p_k^{e_k}$, where $p_1, p_2, \ldots, p_k$ are distinct primes with $p_1 < p_2 < \cdots < p_k$ and each exponent e_i is a positive integer.

LEAST COMMON MULTIPLE

The least common multiple of two positive integers a and b is intimately related to their gcd. We will explore two methods for finding their least common multiple, but first its definition.

The *least common multiple* (lcm) of two positive integers a and b is the least positive integer divisible by both a and b; it is denoted by $[a, b]$.

We now rewrite this definition symbolically.

A SYMBOLIC DEFINITION OF LCM

The lcm of two positive integers a and b is the positive integer m such that:

1) if $a|m$ and $b|m$; and
2) if $a|m'$ and $b|m'$, then $m \leq m'$, where m' is a positive integer.

The following theorem shows the close relationship between the gcd and lcm of two positive integers.

Theorem A.21. *Let a and b be positive integers. Then $[a, b] = \dfrac{ab}{(a, b)}$.* ∎

Corollary A.6. *Two positive integers a and b are relatively prime if and only if $[a, b] = ab$.* ∎

As in the case of the gcd, we can invoke recursion to compute the lcm of a finite number of positive integers, as the next result shows.

Theorem A.22. *Let $a_1, a_2, \ldots, a_n$ be $n \geq 2$ positive integers. Then $[a_1, a_2, \ldots, a_n]$ $= [[a_1, a_2, \ldots, a_{n-1}], a_n]$.* ∎

This gives the next result.

Corollary A.7. *Let the positive integers $a_1, a_2, \ldots, a_n$ with $n \geq 2$ be pairwise relatively prime. Then $[a_1, a_2, \ldots, a_n] = a_1 a_2 \cdots a_{n-1} a_n$.* ∎

The converse of this result is also true.

Corollary A.8. *Let the positive integers $a, m_1, m_2, \ldots, m_k$ be such that $m_i | a$ for $1 \leq i \leq k$. Then $[m_1, m_2, \ldots, m_k] | a$.* ∎

MATRICES AND DETERMINANTS

Matrices contribute significantly to the study of Fibonacci and Lucas numbers. They were discovered jointly by two brilliant English mathematicians, Arthur Cayley (1821–1895) and James Joseph Sylvester (1814–1897). Matrix notation allows data to be summarized in a very compact form, and hence manipulated collectively in a convenient way.

MATRICES

A *matrix* is a rectangular arrangement of numbers enclosed by brackets. A matrix with m rows and n columns is an $m \times n$ (read m by n) matrix, its *size* being $m \times n$. If $m = 1$, it is a *row vector*; and if $n = 1$, it is a *column vector*. If $m = n$, it is a *square matrix* of *order n*. Each number in the array is an *element* of the matrix. Matrices are denoted by uppercase letters.

Let a_{ij} denote the element in row i and column j of matrix A, where $1 \leq i \leq m$ and $1 \leq j \leq n$. Then the matrix is abbreviated as $A = (a_{ij})_{m \times n}$, or simply (a_{ij}) if the size is clear from the context.

EQUALITY OF MATRICES

Two matrices $A = (a_{ij})$ and $B = (b_{ij})$ are *equal* if they have the same size and $a_{ij} = b_{ij}$ for every i and j.

The next definition presents two special matrices.

ZERO AND IDENTITY MATRICES

If every element of a matrix is zero, it is a *zero* matrix, denoted by O.

Let $A = (a_{ij})_{n \times n}$. The elements $a_{11}, a_{22}, \ldots, a_{nn}$ form its *main diagonal*. Suppose

$$a_{ij} = \begin{cases} 1 & \text{if } i = j \\ 0 & \text{otherwise.} \end{cases}$$

Then A is the *identity matrix* of order n; it is denoted by I_n, or I when there is no ambiguity.

Like numbers, matrices can be combined to construct new matrices. The various matrix operations are presented below.

MATRIX OPERATIONS

The *sum* of the matrices $A = (a_{ij})_{m \times n}$ and $B = (b_{ij})_{m \times n}$ is defined by $A + B = (a_{ij} + b_{ij})_{m \times n}$. (We can add only matrices of the same size.)

The *negative* or *additive inverse* of a matrix $A = (a_{ij})$, denoted by $-A$, is defined by $-A = (-a_{ij})$.

The *difference* of the matrices $A = (a_{ij})_{m \times n}$ and $B = (b_{ij})_{m \times n}$ is defined by $A - B = (a_{ij} - b_{ij})_{m \times n}$. (We can subtract only matrices of the same size.)

Let $A = (a_{ij})$, and k any real number, called a *scalar*. Then $kA = (ka_{ij})$.

The next theorem gives the fundamental properties of the matrix operations defined thus far.

Theorem A.23. *Let $A, B,$ and C be any $m \times n$ matrices; O the $m \times n$ zero matrix; and c and d any real numbers. Then*

1) $A + B = B + A$
3) $A + O = A$
5) $(-1)A = -A$
7) $(c + d)A = cA + dA$

2) $A + (B + C) = (A + B) + C$
4) $A + (-A) = O$
6) $c(A + B) = cA + cB$
8) $(cd)A = c(dA).$ ■

Next we define the product of two matrices.

MATRIX MULTIPLICATION

The *product* AB of the matrices $A = (a_{ij})_{m \times n}$ and $B = (b_{ij})_{n \times p}$ is the matrix $C = (c_{ij})_{m \times p}$, where $c_{ij} = a_{i1}b_{1j} + a_{i2}b_{2j} + \cdots + a_{in}b_{nj}$.

The product $C = AB$ is defined only if the number of columns in A equals the number of rows in B. The size of the product is $m \times p$.

The fundamental properties of matrix multiplication are stated in the following theorem.

Theorem A.24.   *Let A, B, and C be three matrices. Then*

1) $A(BC) = (AB)C$ 2) $AI = A = IA$
3) $(A + B)C = AC + BC$ 4) $(A + B)C = AC + BC,$

provided the indicated sums and products are defined.   ∎

Next we briefly discuss the concept of the determinant of a square matrix.

DETERMINANTS

With each square matrix $A = (a_{ij})_{n \times n}$, a unique real number can be associated. This number is the *determinant* of A, denoted by $|A|$. When A is $n \times n$, it is of order n.

The determinant of $A = \begin{bmatrix} a & b \\ c & d \end{bmatrix}$ is defined by $|A| = \begin{vmatrix} a & b \\ c & d \end{vmatrix} = ad - bc.$

If we know how to evaluate 2×2 determinants, we can evaluate higher-order determinants, but first, a few definitions.

MINORS AND COFACTORS

Let $A = (a_{ij})_{n \times n}$. The *minor* M_{ij} of the element a_{ij} is the determinant obtained by deleting row i and column j of $|A|$. The *cofactor* A_{ij} of the element a_{ij} is defined by $A_{ij} = (-1)^{i+j} M_{ij}$.

DETERMINANT OF A SQUARE MATRIX

The *determinant* of a matrix $A = (a_{ij})_{n \times n}$ is defined by $|A| = \sum_{j=1}^{n} a_{ij} A_{ij}$. This is the *Laplace expansion* of $|A|$ by the ith row, named after the French mathematician Pierre-Simon Laplace (1749–1827).

The following result will come in handy in our discussions.

Theorem A.25.   *Let A and B be two square matrices of the same size. Then $|AB| = |A| \cdot |B|$.*   ∎

CONGRUENCES

One of the most beautiful and powerful relations in number theory is the congruence relation, introduced and developed by the brilliant German mathematician Carl Friedrich Gauss (1777–1855). It shares many interesting properties with the equality relation, so it is denoted by the *congruence symbol* $\equiv$. The congruence relation facilitates the study of the divisibility theory and has numerous delightful applications.

We begin our short discussion with the following definition.

CONGRUENCE MODULO *m*

Let *m* be a positive integer. An integer *a* is *congruent* to an integer *b modulo m* if $m|(a - b)$. Symbolically, we then write $a \equiv b \pmod{m}$; *m* is the *modulus* of the *congruence relation*.

If *a* is not *congruent* to *b modulo m*, then *a* is *incongruent* to *b modulo m*; we then write $a \not\equiv b \pmod{m}$.

We now present a series of properties of congruence. Throughout we assume that all letters denote integers and all *moduli* (plural of modulus) are positive integers.

Theorem A.26. $a \equiv b \pmod{m}$ *if and only if* $a = b + km$ *for some integer k.* ■

A useful observation: It follows from the definition (also from this theorem) that $a \equiv 0 \pmod{m}$ if and only if $m|a$; that is, an integer is congruent to 0 modulo *m* if and only if it is divisible by *m*. Thus $a \equiv 0 \pmod{m}$ and $m|a$ mean exactly the same thing.

Theorem A.27.

1) $a \equiv a \pmod{m}$ (reflexive property).
2) If $a \equiv b \pmod{m}$, then $b \equiv a \pmod{m}$ (symmetric property).
3) If $a \equiv b \pmod{m}$ and $b \equiv c \pmod{m}$, then $a \equiv c \pmod{m}$ (transitive property). ■

The next theorem provides another useful characterization of congruence.

Theorem A.28. $a \equiv b \pmod{m}$ *if and only if a and b leave the same remainder when divided by m.* ■

The next result follows from this theorem.

Corollary A.9. *Let* $0 \leq r < m$. *If* $a \equiv r \pmod{m}$, *then r is the remainder when a is divided by m; and if r is the remainder when a is divided by m, then* $a \equiv r \pmod{m}$. ■

By this corollary, every integer a is congruent to its remainder r modulo m; r is called the *least residue* of a modulo m. Since r has exactly m choices $0, 1, 2, \ldots,$ $m - 1$, a is congruent to exactly one of them modulo m. Accordingly, we have the following result.

Corollary A.10. *Every integer is congruent to exactly one of the least residues* $0, 1, 2, \ldots, m - 1$ *modulo m.*   ∎

This corollary justifies the definition of the *mod* operator. If $a \equiv r \pmod{m}$ and $0 \leq r < m$, then $a \bmod m = r$; conversely, if $a \bmod m = r$, then $a \equiv r$ $\pmod{m}$ and $0 \leq r < m$.

The next theorem shows that congruences with the same modulus can be added and multiplied.

Theorem A.29. *Let* $a \equiv b \pmod{m}$ *and* $c \equiv d \pmod{m}$. *Then* $a + c \equiv b + d$ $\pmod{m}$ *and* $ac \equiv bd \pmod{m}$.   ∎

It follows by this theorem that one congruence can be subtracted from another, provided they have the same modulus.

Corollary A.11. *If* $a \equiv b \pmod{m}$ *and* $c \equiv d \pmod{m}$, *then* $a - c \equiv b - d$ $\pmod{m}$.   ∎

The next corollary also follows by Theorem A.29.

Corollary A.12. *Let* $a \equiv b \pmod{m}$ *and* c *any integer. Then* $a + c \equiv b + c$ $\pmod{m}$, $a - c \equiv b - c \pmod{m}$, $ac \equiv bc \pmod{m}$, *and* $a^2 \equiv b^2 \pmod{m}$.   ∎

The fourth part of this corollary can be extended to any positive integral exponent n, as the next theorem shows.

Theorem A.30. *If* $a \equiv b \pmod{m}$, *then* $a^n \equiv b^n \pmod{m}$ *for any positive integer* n.   ∎

The following theorem shows that the cancellation property of multiplication can be extended to congruences under special circumstances.

Theorem A.31. *If* $ac \equiv bc \pmod{m}$ *and* $(c, m) = 1$, *then* $a \equiv b \pmod{m}$.   ∎

Thus we can cancel the same number c from both sides of a congruence, provided c and m are relatively prime.

Theorem A.31 can be generalized as follows.

Theorem A.32. *If* $ac \equiv bc \pmod{m}$ *and* $(c, m) = d$, *then* $a \equiv b \pmod{m/d}$.   ∎

Congruences of two numbers with different moduli can be combined into a single congruence, as the next theorem shows.

Theorem A.33. *Suppose $a \equiv b \pmod{m_1}$, $a \equiv b \pmod{m_2}$, ..., $a \equiv b \pmod{m_k}$. Then $a \equiv b \pmod{[m_1, m_2, \ldots, m_k]}$.* ∎

The following corollary follows from this theorem.

Corollary A.13. *Suppose $a \equiv b \pmod{m_1}$, $a \equiv b \pmod{m_2}$, ..., $a \equiv b \pmod{m_k}$, where the moduli are pairwise relatively prime. Then $a \equiv b \pmod{m_1 m_2 \cdots m_k}$.* ∎

A.2

THE FIRST 100 FIBONACCI AND LUCAS NUMBERS

n	F_n	L_n
1	1	1
2	1	3
3	2	4
4	3	7
5	5	11
6	8	18
7	13	29
8	21	47
9	34	76
10	55	123
11	89	199
12	144	322
13	233	521
14	377	843
15	610	1,364
16	987	2,207

(*Continued*)

Fibonacci and Lucas Numbers with Applications, Volume One, Second Edition. Thomas Koshy.
© 2018 John Wiley & Sons, Inc. Published 2018 by John Wiley & Sons, Inc.

n	F_n	L_n
17	1,597	3,571
18	2,584	5,778
19	4,181	9,349
20	6,765	15,127
21	10,946	24,476
22	17,711	39,603
23	28,657	64,079
24	46,368	103,682
25	75,025	167,761
26	121,393	271,443
27	196,418	439,204
28	317,811	710,647
29	514,229	1,149,851
30	832,040	1,860,498
31	1,346,269	3,010,349
32	2,178,309	4,870,847
33	3,524,578	7,881,196
34	5,702,887	12,752,043
35	9,227,465	20,633,239
36	14,930,352	33,385,282
37	24,157,817	54,018,521
38	39,088,169	87,403,803
39	63,245,986	141,422,324
40	102,334,155	228,826,127
41	165,580,141	370,248,451
42	267,914,296	599,074,578
43	433,494,437	969,323,029
44	701,408,733	1,568,397,607
45	1,134,903,170	2,537,720,636
46	1,836,311,903	4,106,118,243
47	2,971,215,073	6,643,838,879
48	4,807,526,976	10,749,957,122
49	7,778,742,049	17,393,796,001
50	12,586,269,025	28,143,753,123

(*Continued*)

n	F_n	L_n
51	20,365,011,074	45,537,549,124
52	32,951,280,099	73,681,302,247
53	53,316,291,173	119,218,851,371
54	86,267,571,272	192,900,153,618
55	139,583,862,445	312,119,004,989
56	225,851,433,717	505,019,158,607
57	365,435,296,162	817,138,163,596
58	591,286,729,879	1,322,157,322,203
59	956,722,026,041	2,139,295,485,799
60	1,548,008,755,920	3,461,452,808,002
61	2,504,730,781,961	5,600,748,293,801
62	4,052,739,537,881	9,062,201,101,803
63	6,557,470,319,842	14,662,949,395,604
64	10,610,209,857,723	23,725,150,497,407
65	17,167,680,177,565	38,388,099,893,011
66	27,777,890,035,288	62,113,250,390,418
67	44,945,570,212,853	100,501,350,283,429
68	72,723,460,248,141	162,614,600,673,847
69	117,669,030,460,994	263,115,950,957,276
70	190,392,490,709,135	425,730,551,631,123
71	308,061,521,170,129	688,846,502,588,399
72	498,454,011,879,264	1,114,577,054,219,522
73	806,515,533,049,393	1,803,423,556,807,921
74	1,304,969,544,928,657	2,918,000,611,027,443
75	2,111,485,077,978,050	4,721,424,167,835,364
76	3,416,454,622,906,707	7,639,424,778,862,807
77	5,527,939,700,884,757	12,360,848,946,698,171
78	8,944,394,323,791,464	20,000,273,725,560,978
79	14,472,334,024,676,221	32,361,122,672,259,149
80	23,416,728,348,467,685	52,361,396,397,820,127
81	37,889,062,373,143,906	84,722,519,070,079,276
82	61,305,790,721,611,591	137,083,915,467,899,403
83	99,194,853,094,755,497	221,806,434,537,978,679
84	160,500,643,816,367,088	358,890,350,005,878,082

(*Continued*)

n	F_n	L_n
85	259,695,496,911,122,585	580,696,784,543,856,761
86	420,196,140,727,489,673	939,587,134,549,734,843
87	679,891,637,638,612,258	1,520,283,919,093,591,604
88	1,100,087,778,366,101,931	2,459,871,053,643,326,447
89	1,779,979,416,004,714,189	3,980,154,972,736,918,051
90	2,880,067,194,370,816,120	6,440,026,026,380,244,498
91	4,660,046,610,375,530,309	10,420,180,999,117,162,549
92	7,540,113,804,746,346,429	16,860,207,025,497,407,047
93	12,200,160,415,121,876,738	27,280,388,024,614,569,596
94	19,740,274,219,868,223,167	44,140,595,050,111,976,643
95	31,940,434,634,990,099,905	71,420,983,074,726,546,239
96	51,680,708,854,858,323,072	115,561,578,124,838,522,882
97	83,621,143,489,848,422,977	186,982,561,199,565,069,121
98	135,301,852,344,706,746,049	302,544,139,324,403,592,003
99	218,922,995,834,555,169,026	489,526,700,523,968,661,124
100	354,224,848,179,261,915,075	792,070,839,848,372,253,127

A.3

THE FIRST 100 FIBONACCI NUMBERS AND THEIR PRIME FACTORIZATIONS

n	F_n	Prime Factorization of F_n
1	1	1
2	1	1
3	2	2
4	3	3
5	5	5
6	8	2^3
7	13	13
8	21	$3 \cdot 7$
9	34	$2 \cdot 17$
10	55	$5 \cdot 11$
11	89	89
12	144	$2^4 \cdot 3^2$
13	233	233
14	377	$13 \cdot 29$
15	610	$2 \cdot 5 \cdot 61$
16	987	$3 \cdot 7 \cdot 47$

(*Continued*)

Fibonacci and Lucas Numbers with Applications, Volume One, Second Edition. Thomas Koshy.
© 2018 John Wiley & Sons, Inc. Published 2018 by John Wiley & Sons, Inc.

n	F_n	Prime Factorization of F_n
17	1,597	1597
18	2,584	$2^3 \cdot 17 \cdot 19$
19	4,181	$37 \cdot 113$
20	6,765	$3 \cdot 5 \cdot 11 \cdot 41$
21	10,946	$2 \cdot 13 \cdot 421$
22	17,711	$89 \cdot 199$
23	28,657	28657
24	46,368	$2^5 \cdot 3^2 \cdot 7 \cdot 23$
25	75,025	$5^2 \cdot 3001$
26	121,393	$233 \cdot 521$
27	196,418	$2 \cdot 17 \cdot 53 \cdot 109$
28	317,811	$2 \cdot 13 \cdot 29 \cdot 281$
29	514,229	514229
30	832,040	$2^3 \cdot 5 \cdot 11 \cdot 31 \cdot 61$
31	1,346,269	$577 \cdot 2417$
32	2,178,309	$3 \cdot 7 \cdot 47 \cdot 2207$
33	3,524,578	$2 \cdot 89 \cdot 19801$
34	5,702,887	$1597 \cdot 3571$
35	9,227,465	$5 \cdot 13 \cdot 141961$
36	14,930,352	$2^4 \cdot 3^3 \cdot 17 \cdot 19 \cdot 107$
37	24,157,817	$73 \cdot 149 \cdot 2221$
38	39,088,169	$37 \cdot 113 \cdot 9349$
39	63,245,986	$2 \cdot 233 \cdot 135721$
40	102,334,155	$3 \cdot 5 \cdot 7 \cdot 11 \cdot 41 \cdot 2161$
41	165,580,141	$2789 \cdot 59369$
42	267,914,296	$2^3 \cdot 13 \cdot 29 \cdot 211 \cdot 421$
43	433,494,437	433494437
44	701,408,733	$3 \cdot 43 \cdot 89 \cdot 199 \cdot 307$
45	1,134,903,170	$2 \cdot 5 \cdot 17 \cdot 61 \cdot 109441$
46	1,836,311,903	$139 \cdot 461 \cdot 28657$
47	2,971,215,073	2971215073
48	4,807,526,976	$2^6 \cdot 3^2 \cdot 7 \cdot 23 \cdot 47 \cdot 1103$
49	7,778,742,049	$13 \cdot 97 \cdot 6168709$
50	12,586,269,025	$5^2 \cdot 11 \cdot 101 \cdot 151 \cdot 3001$

(*Continued*)

n	F_n	Prime Factorization of F_n
51	20,365,011,074	$2 \cdot 1597 \cdot 6376021$
52	32,951,280,099	$3 \cdot 233 \cdot 521 \cdot 90481$
53	53,316,291,173	$953 \cdot 55945741$
54	86,267,571,272	$2^3 \cdot 17 \cdot 19 \cdot 53 \cdot 109 \cdot 5779$
55	139,583,862,445	$5 \cdot 89 \cdot 661 \cdot 474541$
56	225,851,433,717	$3 \cdot 7^2 \cdot 13 \cdot 29 \cdot 281 \cdot 14503$
57	365,435,296,162	$2 \cdot 37 \cdot 113 \cdot 797 \cdot 54833$
58	591,286,729,879	$59 \cdot 19489 \cdot 514229$
59	956,722,026,041	$353 \cdot 2710260697$
60	1,548,008,755,920	$2^4 \cdot 3^2 \cdot 5 \cdot 11 \cdot 31 \cdot 41 \cdot 61 \cdot 2521$
61	2,504,730,781,961	$4513 \cdot 555003497$
62	4,052,739,537,881	$557 \cdot 2417 \cdot 3010349$
63	6,557,470,319,842	$2 \cdot 13 \cdot 17 \cdot 421 \cdot 35239681$
64	10,610,209,857,723	$3 \cdot 7 \cdot 47 \cdot 1087 \cdot 2207 \cdot 4481$
65	17,167,680,177,565	$5 \cdot 233 \cdot 14736206161$
66	27,77,7890,035,288	$2^3 \cdot 89 \cdot 199 \cdot 9901 \cdot 19801$
67	44,915,570,212,853	$269 \cdot 116849 \cdot 1429913$
68	72,723,460,248,141	$3 \cdot 67 \cdot 1597 \cdot 3571 \cdot 63443$
69	117,669,030,460,994	$2 \cdot 137 \cdot 829 \cdot 18077 \cdot 28657$
70	190,392,490,709,135	$5 \cdot 11 \cdot 13 \cdot 29 \cdot 71 \cdot 911 \cdot 141961$
71	308,061,521,170,129	$6673 \cdot 46165371073$
72	498,454,011,879,264	$2^5 \cdot 3^3 \cdot 7 \cdot 17 \cdot 19 \cdot 23 \cdot 107 \cdot 103681$
73	806,515,533,049,393	$9375829 \cdot 86020717$
74	1,304,969,544,928,657	$73 \cdot 149 \cdot 2221 \cdot 54018521$
75	2,111,485,077,978,050	$2 \cdot 5^2 \cdot 61 \cdot 3001 \cdot 230686501$
76	3,416,454,622,906,707	$3 \cdot 37 \cdot 113 \cdot 9349 \cdot 29134601$
77	5,527,939,700,884,757	$13 \cdot 89 \cdot 988681 \cdot 4832521$
78	8,944,394,323,791,464	$2^3 \cdot 79 \cdot 233 \cdot 521 \cdot 859 \cdot 135721$
79	14,472,334,024,676,221	$157 \cdot 92180471494753$
80	23,416,728,348,467,685	$3 \cdot 5 \cdot 7 \cdot 11 \cdot 41 \cdot 47 \cdot 1601 \cdot 2161 \cdot 3041$
81	37,889,062,373,143,906	$2 \cdot 17 \cdot 53 \cdot 109 \cdot 2269 \cdot 4373 \cdot 19441$
82	61,305,790,721,611,591	$2789 \cdot 59369 \cdot 370248451$
83	99,194,853,094,755,497	99194853094755497
84	160,500,643,816,367,088	$2^4 \cdot 3^2 \cdot 13 \cdot 29 \cdot 83 \cdot 211 \cdot 281 \cdot 421 \cdot 1427$

(*Continued*)

n	F_n	Prime Factorization of F_n
85	259,695,496,911,122,585	$5 \cdot 1597 \cdot 9521 \cdot 3415914041$
86	420,196,140,727,489,673	$6709 \cdot 144481 \cdot 433494437$
87	679,891,637,638,612,258	$2 \cdot 173 \cdot 514229 \cdot 3821263937$
88	1,100,087,778,366,101,931	$3 \cdot 7 \cdot 43 \cdot 89 \cdot 199 \cdot 263 \cdot 307 \cdot 881 \cdot 967$
89	1,779,979,416,004,714,189	$1069 \cdot 1665088321800481$
90	2,880,067,194,370,816,120	$2^3 \cdot 5 \cdot 11 \cdot 17 \cdot 19 \cdot 31 \cdot 61 \cdot 181 \cdot 541 \cdot 109441$
91	4,660,046,610,375,530,309	$13^2 \cdot 233 \cdot 741469 \cdot 159607993$
92	7,540,113,804,746,346,429	$3 \cdot 139 \cdot 461 \cdot 4969 \cdot 28657 \cdot 275449$
93	12,200,160,415,121,876,738	$2 \cdot 557 \cdot 2417 \cdot 4531100550901$
94	19,740,274,219,868,223,167	$2971215073 \cdot 6643838879$
95	31,940,434,634,990,099,905	$5 \cdot 37 \cdot 113 \cdot 761 \cdot 9641 \cdot 67735001$
96	51,680,708,854,858,323,072	$2^7 \cdot 3^2 \cdot 7 \cdot 23 \cdot 47 \cdot 769 \cdot 1103 \cdot 2207 \cdot 3167$
97	83,621,143,489,848,422,977	$193 \cdot 389 \cdot 3084989 \cdot 3610402019$
98	135,301,852,344,706,746,049	$13 \cdot 29 \cdot 97 \cdot 6168709 \cdot 599786069$
99	218,922,995,834,555,169,026	$2 \cdot 17 \cdot 89 \cdot 197 \cdot 19801 \cdot 18546805133$
100	354,224,848,179,261,915,075	$3 \cdot 5^2 \cdot 11 \cdot 41 \cdot 101 \cdot 151 \cdot 401 \cdot 3001 \cdot 570601$

A.4

THE FIRST 100 LUCAS NUMBERS AND THEIR PRIME FACTORIZATIONS

n	L_n	Prime Factorization of L_n
1	1	1
2	3	3
3	4	2^2
4	7	7
5	11	11
6	18	$2 \cdot 3^2$
7	29	29
8	47	47
9	76	$2^2 \cdot 19$
10	123	$3 \cdot 41$
11	199	199
12	322	$2 \cdot 7 \cdot 23$
13	521	521
14	843	$3 \cdot 281$
15	1,364	$2^2 \cdot 11 \cdot 31$
16	2,207	2207

(*Continued*)

Fibonacci and Lucas Numbers with Applications, Volume One, Second Edition. Thomas Koshy.
© 2018 John Wiley & Sons, Inc. Published 2018 by John Wiley & Sons, Inc.

n	L_n	Prime Factorization of L_n
17	3,571	3571
18	5,778	$2 \cdot 3^3 \cdot 107$
19	9,349	9349
20	15,127	$7 \cdot 2161$
21	24,476	$2^2 \cdot 29 \cdot 211$
22	39,603	$3 \cdot 43 \cdot 307$
23	64,079	$139 \cdot 461$
24	103,682	$2 \cdot 47 \cdot 1103$
25	167,761	$11 \cdot 101 \cdot 151$
26	271,443	$3 \cdot 90481$
27	439,204	$2^2 \cdot 19 \cdot 5779$
28	710,647	$7^2 \cdot 4503$
29	1,149,851	$59 \cdot 19489$
30	1,860,498	$2 \cdot 3^2 \cdot 41 \cdot 2521$
31	3,010,349	3010349
32	4,870,847	$1087 \cdot 4481$
33	7,881,196	$2^2 \cdot 199 \cdot 9901$
34	12,752,043	$3 \cdot 67 \cdot 63443$
35	20,633,239	$11 \cdot 29 \cdot 71 \cdot 911$
36	33,385,282	$2 \cdot 7 \cdot 23 \cdot 103681$
37	54,018,521	54018521
38	87,403,803	$3 \cdot 29134601$
39	141,422,324	$2^2 \cdot 79 \cdot 521 \cdot 859$
40	228,826,127	$47 \cdot 1601 \cdot 3041$
41	370,248,451	370248451
42	599,074,578	$2 \cdot 3^2 \cdot 83 \cdot 281 \cdot 1427$
43	969,323,029	$6709 \cdot 144481$
44	1,568,397,607	$7 \cdot 263 \cdot 881 \cdot 967$
45	2,537,720,636	$2^2 \cdot 11 \cdot 19 \cdot 31 \cdot 181 \cdot 541$
46	4,106,118,243	$3 \cdot 4969 \cdot 275449$
47	6,643,838,879	6643838879
48	10,749,957,122	$2 \cdot 769 \cdot 2207 \cdot 3167$
49	17,393,796,001	$29 \cdot 599786069$
50	28,143,753,123	$3 \cdot 41 \cdot 401 \cdot 570601$

(Continued)

n	L_n	Prime Factorization of L_n
51	45,537,549,124	$2^2 \cdot 919 \cdot 3469 \cdot 3571$
52	73,681,302,247	$7 \cdot 103 \cdot 102193207$
53	119,218,851,371	119218851371
54	192,900,153,618	$2 \cdot 3^4 \cdot 107 \cdot 1128427$
55	312,119,004,989	$11^2 \cdot 199 \cdot 331 \cdot 39161$
56	505,019,158,607	$47 \cdot 10745088481$
57	817,138,163,596	$2^2 \cdot 229 \cdot 9349 \cdot 95419$
58	1,322,157,322,203	$3 \cdot 347 \cdot 1270083883$
59	2,139,295,485,799	$709 \cdot 8969 \cdot 336419$
60	3,461,452,808,002	$2 \cdot 7 \cdot 23 \cdot 241 \cdot 2161 \cdot 20641$
61	5,600,748,293,801	5600748293801
62	9,062,201,101,803	$3 \cdot 3020733700601$
63	14,662,949,395,604	$2^2 \cdot 19 \cdot 29 \cdot 211 \cdot 1009 \cdot 31249$
64	23,725,150,497,407	$127 \cdot 186812208641$
65	38,388,099,893,011	$11 \cdot 131 \cdot 521 \cdot 2081 \cdot 24571$
66	62,113,250,390,418	$2 \cdot 3^2 \cdot 43 \cdot 307 \cdot 261399601$
67	100,501,350,283,429	$4021 \cdot 24994118449$
68	162,614,600,673,847	$7 \cdot 23230657239121$
69	263,115,950,957,276	$2^2 \cdot 139 \cdot 461 \cdot 691 \cdot 1485571$
70	425,730,551,631,123	$3 \cdot 41 \cdot 281 \cdot 12317523121$
71	688,846,502,588,399	688846502588399
72	1,114,577,054,219,522	$2 \cdot 47 \cdot 1103 \cdot 10749957121$
73	1,803,423,556,807,921	$151549 \cdot 11899937029$
74	2,918,000,644,027,443	$3 \cdot 11987 \cdot 81143477963$
75	4,721,424,167,835,364	$2^2 \cdot 11 \cdot 31 \cdot 101 \cdot 151 \cdot 12301 \cdot 18451$
76	7,639,424,778,862,807	$7 \cdot 1091346396980401$
77	12,360,848,946,698,171	$29 \cdot 199 \cdot 229769 \cdot 9321929$
78	20,000,273,725,560,978	$2 \cdot 3^2 \cdot 90481 \cdot 12280217041$
79	32,361,122,672,259,149	32361122672259149
80	52,361,396,397,820,127	$2207 \cdot 23725145626561$
81	84,722,519,070,079,276	$2^2 \cdot 19 \cdot 3079 \cdot 5779 \cdot 62650261$
82	137,083,915,467,899,403	$3 \cdot 163 \cdot 800483 \cdot 350207569$
83	221,806,434,537,978,679	$35761381 \cdot 6202401259$
84	358,890,350,005,878,082	$2 \cdot 7^2 \cdot 23 \cdot 167 \cdot 14503 \cdot 65740583$

(Continued)

n	L_n	Prime Factorization of L_n
85	580,696,784,543,856,761	$11 \cdot 3571 \cdot 1158551 \cdot 12760031$
86	939,587,134,549,734,843	$3 \cdot 313195711516578281$
87	1,520,283,919,093,591,604	$2^2 \cdot 59 \cdot 349 \cdot 19489 \cdot 947104099$
88	2,459,871,053,643,326,447	$47 \cdot 93058241 \cdot 562418561$
89	3,980,154,972,736,918,051	$179 \cdot 22235502640988369$
90	6,440,026,026,380,244,498	$2 \cdot 3^3 \cdot 41 \cdot 107 \cdot 2521 \cdot 10783342081$
91	10,420,180,999,117,162,549	$29 \cdot 521 \cdot 689667151970161$
92	16,860,207,025,497,407,047	$7 \cdot 253367 \cdot 9506372193863$
93	27,280,388,024,614,569,596	$2^2 \cdot 63799 \cdot 3010349 \cdot 35510749$
94	44,140,595,050,111,976,643	$3 \cdot 563 \cdot 5641 \cdot 4632894751907$
95	71,420,983,074,726,546,239	$11 \cdot 191 \cdot 9349 \cdot 41611 \cdot 87382901$
96	115,561,578,124,838,522,882	$2 \cdot 1087 \cdot 4481 \cdot 11862575248703$
97	186,982,561,199,565,069,121	$3299 \cdot 56678557502141579$
98	302,544,139,324,403,592,003	$3 \cdot 281 \cdot 5881 \cdot 61025309469041$
99	489,526,700,523,968,661,124	$2^2 \cdot 19 \cdot 199 \cdot 991 \cdot 2179 \cdot 9901 \cdot 1513909$
100	792,070,839,848,372,253,127	$7 \cdot 2161 \cdot 9125201 \cdot 5738108801$

ABBREVIATIONS

Abbreviation	Meaning
LHRWCCs	linear homogeneous recurrence with constant coefficients
LNHRWCCs	linear nonhomogeneous recurrence with constant coefficients
RHS	right-hand side
LHS	left-hand side
PMI	principle of mathematical induction
rms	root-mean-square
AIME	American Invitational Mathematics Examinations
FSCF	finite simple continued fraction
LDE	linear diophantine equation
ISCF	infinite simple continued fraction

Fibonacci and Lucas Numbers with Applications, Volume One, Second Edition. Thomas Koshy.
© 2018 John Wiley & Sons, Inc. Published 2018 by John Wiley & Sons, Inc.

REFERENCES

[1] G.L. Alexanderson, Solution to Problem B-55, *Fibonacci Quarterly*, 3 (1965), 158.

[2] G.L. Alexanderson, Solution to Problem B-56, *Fibonacci Quarterly*, 3 (1965), 159.

[3] U. Alfred, Problem H-8, *Fibonacci Quarterly*, 1(1) (1963), 48.

[4] U. Alfred, Problem B-24, *Fibonacci Quarterly*, 1(4) (1963), 73.

[5] U. Alfred, On Square Lucas Numbers, *Fibonacci Quarterly*, 2 (1963), 11–12.

[6] U. Alfred, Problem H-29, *Fibonacci Quarterly*, 2 (1964), 49.

[7] U. Alfred, Solution to Problem B-22, *Fibonacci Quarterly*, 2 (1964), 78.

[8] U. Alfred, Problem B-79, *Fibonacci Quarterly*, 3 (1965), 324.

[9] U. Alfred, Solution to Problem B-79, *Fibonacci Quarterly*, 4 (1966), 287.

[10] K. Alladi and V.E. Hoggatt, Jr., Compositions with Ones and Twos, *Fibonacci Quarterly*, 12 (1974), 233–239.

[11] K. Alladi, P. Erdös, and V.E . Hoggatt, Jr., On Additive Partitions of Integers, *Discrete Mathematics*, 22 (1978), 201–211.

[12] C. Alsina and R.B. Nelsen, *Charming Proofs: A Journey into Elegant Mathematics*, Math. Association of America, Washington, D.C., 2010.

[13] R. Anaya and J. Crump, A Generalized Greatest Integer Function Theorem, *Fibonacci Quarterly*, 10 (1972), 207–211.

[14] R.H. Anglin, Problem B-160, *Fibonacci Quarterly*, 8 (1970), 107.

[15] R.C. Archibald, Golden Section, *American Mathematical Monthly*, 25 (1918), 232–238.

[16] R.A. Askey, Fibonacci and Related Sequences, *Mathematics Teacher*, 97 (2004), 116–119.

[17] J.H. Avial, Solution to Problem B-19, *Fibonacci Quarterly*, 2 (1964), 75–76.

[18] A. Baker and H. Davenport, The Equations $3x^2 - 2 = y^2$ and $8x^2 - 7 = z^2$, *Quarterly Journal of Mathematics*, 20 (1969), 129–137.

Fibonacci and Lucas Numbers with Applications, Volume One, Second Edition. Thomas Koshy.
© 2018 John Wiley & Sons, Inc. Published 2018 by John Wiley & Sons, Inc.

[19] H.V. Baravalle, The Geometry of the Pentagon and the Golden Section, *Mathematics Teacher*, 41 (1948), 22–31.

[20] R. Barbeau, Example 2, *College Mathematics Journal*, 24 (1993), 65–66.

[21] W.C. Barley, Problem B-234, *Fibonacci Quarterly*, 10 (1972), 330.

[22] W.C. Barley, Problem B-240, *Fibonacci Quarterly*, 11 (1973), 335.

[23] S.L. Basin, The Fibonacci Sequence as it Appears in Nature, *Fibonacci Quarterly*, 1(1) (1963), 53–56.

[24] S.L. Basin, Problem B-23, *Fibonacci Quarterly*, 2 (1964), 78–79.

[25] S.L. Basin, Problem B-42, *Fibonacci Quarterly*, 2 (1964), 329.

[26] S.L. Basin and V.E. Hoggatt, Jr., A Primer on the Fibonacci Sequence–Part I, *Fibonacci Quarterly*, 1(1) (1963), 65–72.

[27] S.L. Basin and V.E. Hoggatt, Jr., A Primer on the Fibonacci Sequence–Part II, *Fibonacci Quarterly*, 1(2) (1963), 61–68.

[28] S.L. Basin and V. Ivanoff, Problem B-4, *Fibonacci Quarterly*, 1(1) (1963), 74.

[29] M. Bataille, Problem 957, *College Mathematics Journal*, 42 (2011), 329.

[30] R.S. Beard, The Golden Section and Fibonacci Numbers, *Scripta Mathematica*, 16 (1950), 116–119.

[31] B.D. Beasley, Problem B-897, *Fibonacci Quarterly*, 38 (2000), 181.

[32] A.H. Beiler, *Recreations in the Theory of Numbers*, 2nd edition, Dover, New York, 1966.

[33] E.T. Bell, A Functional Equation in Arithmetic, *Trans. American Mathematical Society*, 39 (1936), 341–344.

[34] A.T. Benjamin and J.J. Quinn, Recounting Fibonacci and Lucas Identities, *College Mathematics Journal*, 30 (1999), 359–366.

[35] A.T. Benjamin and J.J. Quinn, *Proofs That Really Count*, Math. Association of America, Washington, D.C., 2003.

[36] A.A. Bennett, The "Most Pleasing Rectangle", *American Mathematical Monthly*, 30 (1923), 27–30.

[37] L. Beran, Schemes Generating the Fibonacci Sequence, *Mathematical Gazette*, 70 (1986), 38–40.

[38] G. Berzsenyi, Problem H-263, *Fibonacci Quarterly*, 14 (1976), 182.

[39] G. Berzsenyi, Solution to Problem H-263, *Fibonacci Quarterly*, 15 (1977), 373.

[40] G. Berzsenyi, Problem B-378, *Fibonacci Quarterly*, 17 (1979), 185–186.

[41] M. Bicknell, The Lambda Number of a Matrix: The Sum of its n^2 Cofactors, *American Mathematical Monthly*, 72 (1965), 260–264.

[42] M. Bicknell, A Primer for the Fibonacci Numbers: Part VII, *Fibonacci Quarterly*, 8 (1970), 407–420.

[43] M. Bicknell, Problem B-267, *Fibonacci Quarterly*, 12 (1974), 316.

[44] M. Bicknell, In Memoriam, *Fibonacci Quarterly*, 18 (1980), 289.

[45] M. Bicknell-Johnson and D. DeTemple, Visualizing Golden Ratio Sums with Tiling Patterns, *Fibonacci Quarterly*, 33 (1995), 289–303.

[46] M. Bicknell and V.E. Hoggatt, Jr., Fibonacci Matrices and Lambda Functions, *Fibonacci Quarterly*, 1(2) (1963), 47–52.

[47] M. Bicknell and V.E. Hoggatt, Jr., Golden Triangles, Rectangles, and Cuboids, *Fibonacci Quarterly*, 7 (1969), 73–91, 98.

[48] M.T. Bird, Solution to Problem 3802, *American Mathematical Monthly*, 45 (1938), 632–633.

[49] G. Blank, Another Fibonacci Curiosity, *Scripta Mathematica*, 21 (1956), 30.

[50] R. Blazej, Problem B-298, *Fibonacci Quarterly*, 13 (1975), 94.

[51] R. Blazej, Problem B-294, *Fibonacci Quarterly*, 13 (1975), 375.

[52] D.M. Bloom, Problem 55, *Math Horizons*, 4 (Sept., 1996), 32.

[53] A.P. Boblett, Problem B-29, *Fibonacci Quarterly*, 1(4) (1963), 75.

[54] W.G. Brady, The Lambert Function, *Fibonacci Quarterly*, 9 (1971), 199–200.

[55] W.G. Brady, Problem B-253, *Fibonacci Quarterly*, 12 (1974), 104–105.

[56] W.G. Brady, Problem B-366, *Fibonacci Quarterly*, 15 (1977), 375.

[57] W.G. Brady, Solution to Problem B-366, *Fibonacci Quarterly*, 16 (1978), 563.

[58] O. Bretscher, *Linear Algebra with Applications*, 5th edition, Pearson, Boston, Mass., 2013.

[59] C.A. Bridger, Problem B-94, *Fibonacci Quarterly*, 5 (1967), 203.

[60] R.C. Brigham, R.M. Caron, P.Z. Chinn, and R.P. Grimaldi, A Tiling Scheme for the Fibonacci Numbers, *Journal of Recreational Mathematics*, 28 (1996–1997), 10–16.

[61] M. Brooke, Fibonacci Fancy, *Mathematics Magazine*, 35 (1962), 218.

[62] M. Brooke, Solution to Problem H-14, *Fibonacci Quarterly*, 1(4) (1963), 51.

[63] M. Brooke, Fibonacci Formulas, *Fibonacci Quarterly*, 1(2) (1963), 60.

[64] M. Brooke, Fibonacci Numbers: Their History Through 1900, *Fibonacci Quarterly*, 2 (1964), 149–153.

[65] A. Brousseau, A Fibonacci Generalization, *Fibonacci Quarterly*, 5 (1967), 171–174.

[66] A. Brousseau, Problem H-92, *Fibonacci Quarterly*, 6 (1968), 94.

[67] A. Brousseau, Summation of Infinite Fibonacci Series, *Fibonacci Quarterly*, 7 (1969), 143–168.

[68] A. Brousseau, Fibonacci–Lucas Infinite Series – Research Topic, *Fibonacci Quarterly*, 7 (1969), 211–217.

[69] A. Brousseau, Fibonacci Numbers and Geometry, *Fibonacci Quarterly*, 10 (1972), 303–318, 323.

[70] A. Brousseau, Ye Olde Fibonacci Curiosity, *Fibonacci Quarterly*, 10 (1972), 441–443.

[71] D. Brown, *The Da Vinci Code*, Doubleday, New York, 2003.

[72] J.L. Brown, Jr., Solution to Problem B-6, *Fibonacci Quarterly*, 1(4) (1963), 75–76.

[73] J.L. Brown, Jr., A Trigonometric Sum, *Fibonacci Quarterly*, 2 (1964), 74–75.

[74] J.L. Brown, Jr., Solution to Problem B-41, *Fibonacci Quarterly*, 2 (1964), 328.

[75] J.L. Brown, Jr., Solution to Problem B-41, *Fibonacci Quarterly*, 2 (1964), 330.

[76] J.L. Brown, Jr., Solution to Problem H-54, *Fibonacci Quarterly*, 4 (1966), 334–335.

[77] J.L. Brown, Jr., Solution to Problem H-71, *Fibonacci Quarterly*, 5 (1967), 166–167.

[78] P.S. Bruckman, Solution to Problem B-231, *Fibonacci Quarterly*, 11 (1973), 111–112.

[79] P.S. Bruckman, Solution to Problem B-234, *Fibonacci Quarterly*, 11 (1973), 222.

[80] P.S. Bruckman, Solution to Problem B-236, *Fibonacci Quarterly*, 11 (1973), 223.

[81] P.S. Bruckman, Solution to Problem B-241, *Fibonacci Quarterly*, 11 (1973), 336.

[82] P.S. Bruckman, Solution to Problem B-314, *Fibonacci Quarterly*, 14 (1976), 288.

[83] P.S. Bruckman, Solution to Problem H-201, *Fibonacci Quarterly*, 12 (1974), 218–219.

[84] P.S. Bruckman, Problem B-277, *Fibonacci Quarterly*, 13 (1974), 96.

[85] P.S. Bruckman, Problem B-278, *Fibonacci Quarterly*, 13 (1975), 96.

[86] P.S. Bruckman, Problem B-391, *Fibonacci Quarterly*, 17 (1979), 372–373.

[87] P.S. Bruckman, Problem H-280, *Fibonacci Quarterly*, 17 (1979), 377.

[88] P.S. Bruckman, Problem H-286, *Fibonacci Quarterly*, 18 (1980), 281–282.

[89] P.S. Bruckman, Solution to Problem B-461, *Fibonacci Quarterly*, 20 (1982), 369.

[90] D.M. Burton, *Elementary Number Theory*, 3rd edition, W.C. Brown, Dubuque, Iowa, 1989.

[91] R.G. Bushman, Problem H-18, *Fibonacci Quarterly*, 2 (1964), 126–127.

[92] J.H. Butchart, Problem B-124, *Fibonacci Quarterly*, 6 (1968), 289–290.

[93] P.F. Byrd, Expansion of Analytic Functions in Polynomials Associated with Fibonacci Numbers, *Fibonacci Quarterly*, 1(1) (1963), 16–29.

[94] Calendar Problems, *Mathematics Teacher*, 86 (1993), 48–49.

[95] Calendar Problems, *Mathematics Teacher*, 92 (1999), 606–611.

[96] W.A. Camfield, Jaun Gris and the Golden Section, *Art Bulletin*, 47 (1965), 128–134.

[97] G. Candido, A Relationship Between the Fourth Powers of the Terms of the Fibonacci Series, *Scripta Mathematica*, 17 (1951), 230.

[98] G. Candido, Curiosa 272, *Scripta Mathematica*, 17 (1951), 230.

[99] W.A. Capstick, On the Fibonacci Sequence, *Mathematical Gazette*, 41 (1957), 120–121.

[100] L. Carlitz, Problem B-19, *Fibonacci Quarterly*, 1(3) (1963), 75.

[101] L. Carlitz, A Note on Fibonacci Numbers, *Fibonacci Quarterly*, 2 (1964), 15–28.

[102] L. Carlitz, Solution to Problem H-39, *Fibonacci Quarterly*, 3 (1965), 51–52.

[103] L. Carlitz, The Characteristic Polynomial of a Certain Matrix of Binomial Coefficients, *Fibonacci Quarterly*, 3 (1965), 81–89.

[104] L. Carlitz, A Telescoping Sum, *Fibonacci Quarterly*, 3 (1965), 75–76.

[105] L. Carlitz, Solution to Problem B-104, *Fibonacci Quarterly*, 5 (1967), 292.

[106] L. Carlitz, Problem B-110, *Fibonacci Quarterly*, 5 (1967), 469–470.

[107] L. Carlitz, Problem B-111, *Fibonacci Quarterly*, 5 (1967), 470–471.

[108] L. Carlitz, Solution to Problem H-92, *Fibonacci Quarterly*, 6 (1968), 145.

[109] L. Carlitz, Problem H-112, *Fibonacci Quarterly*, 7 (1969), 61–62.

[110] L. Carlitz, Solution to Problem H-131, *Fibonacci Quarterly*, 7 (1969), 285–286.

[111] L. Carlitz, Problem B-185, *Fibonacci Quarterly*, 8 (1970), 325.

[112] L. Carlitz, Solution to Problem B-180, *Fibonacci Quarterly*, 8 (1970), 547–548.

[113] L. Carlitz, Problem B-186, *Fibonacci Quarterly*, 9 (1971), 109–110.

[114] L. Carlitz, Problem B-213, *Fibonacci Quarterly*, 10 (1972), 224.

[115] L. Carlitz, Solution to Problem B-215, *Fibonacci Quarterly*, 10 (1972), 331–332.

[116] L. Carlitz, A Conjecture Concerning Lucas Numbers, *Fibonacci Quarterly*, 10 (1972), 526, 550.

[117] L. Carlitz, Problem H-262, *Fibonacci Quarterly*, 15 (1977), 372–373.

[118] L. Carlitz and J.A.H. Hunter, Some Powers of Fibonacci and Lucas Numbers, *Fibonacci Quarterly*, 7 (1969), 467–473.

[119] L. Carlitz and R. Scoville, Problem B-255, *Fibonacci Quarterly*, 12 (1971), 106.

[120] B.-Y. Chen, Solution to Problem 43.7, *Mathematical Spectrum*, 44 (2011–2012), 46.

[121] W. Cheves, Problem B-192, *Fibonacci Quarterly*, 8 (1970), 443.

[122] W. Cheves, Problem B-275, *Fibonacci Quarterly*, 13 (1975), 95.

[123] C.A. Church, Problem B-46, *Fibonacci Quarterly*, 2 (1964), 231.

[124] C.A. Church and M. Bicknell, Exponential Generating Functions for Fibonacci Identities, *Fibonacci Quarterly*, 11 (1973), 275–281.

[125] D.I.A. Cohen, *Basic Techniques of Combinatorial Theory*, John Wiley & Sons, Inc., New York, 1978.

[126] J.H.E Cohn, Square Fibonacci Numbers, etc., *Fibonacci Quarterly*, 2 (1964), 109–113.

[127] C.K. Cook, Solution to Problem B-962, *Fibonacci Quarterly*, 42 (2004), 183.

[128] I. Cook, The Euclidean Algorithm and Fibonacci, *Mathematical Gazette*, 74 (1990), 47–48.

[129] R. Cormier and J. Brinn, Solution to Problem 567, *Journal of Recreational Mathematics*, 10 (1977–1978), 314–315.

[130] H.S.M. Coxeter, The Golden Section, Phyllotaxis, and Wythoff's Game, *Scripta Mathematica*, 19 (1953), 135–143.

[131] H.S.M. Coxeter, *Introduction to Geometry*, 2nd edition, John Wiley & Sons, Inc., New York, 1969.

[132] T. Crilly, A Supergolden Rectangle, *Mathematical Gazette*, 78 (1994), 320–325.

[133] G. Cross and H. Renzi, Teachers Discover New Math Theorems, *Arithmetic Teacher*, 12 (1965), 625–626.

[134] T. Cross, A Fibonacci Fluke, *Mathematical Gazette*, 80 (1966), 398–400.

[135] H.M. Cundy and A.P. Rollett, *Mathematical Models*, 2nd edition, Oxford University Press, New York, 1961.

[136] Dali Makes Met, *Time*, January 27, 1965, 72.

[137] B. Davis, Fibonacci Numbers in Physics, *Fibonacci Quarterly*, 10 (1972), 659–660, 662.

[138] G.R. Deily, A Logarithmic Formula for Fibonacci Numbers, *Fibonacci Quarterly*, 4 (1966), 89.

[139] M.J. DeLeon, Solution to Problem B-319, *Fibonacci Quarterly*, 20 (1982), 96.

[140] J. DeMaio and J. Jacobson, Fibonacci and Lucas Identities via Graphs, *Topics from the 8th Annual UNCG Regional Mathematics Conference* (J. Rychtář et al. (eds.), Springer, New York, 2013.

[141] J. DeMaio and J. Jacobson, Fibonacci Number of the Tadpole graph, *Electronic Journal of Graph Theory and Applications*, 2 (2014), 129–138.

[142] T.P. Dence, Problem B-129, *Fibonacci Quarterly*, 6 (1968), 296.

[143] R.A. Deninger, Fibonacci Numbers and Water Pollution, *Fibonacci Quarterly*, 10 (1972), 299–300, 302.

[144] M.N. Deshpande, Problem B-911, *Fibonacci Quarterly*, 39 (2001), 85.

[145] J.E. Desmond, Problem B-182, *Fibonacci Quarterly*, 8 (1970), 549–550.

[146] D.W. DeTemple, A Pentagonal Arch, *Fibonacci Quarterly*, 12 (1974), 235–236.

[147] L.E. Dickson, *History of the Theory of Numbers*, Vol. 1, Chelsea, New York, 1952.

[148] A.S. DiDomenico, Problem 818, *College Mathematics Journal*, 38 (2007), 64.

[149] N.A. Draim and M. Bicknell, Sums of *n*th Powers of Roots of a Given Quadratic Function, *Fibonacci Quarterly*, 4 (1966), 170–178.

[150] R.C. Drake, Problem B-180, *Fibonacci Quarterly*, 8 (1970), 106.

[151] G.E. Duckworth, *Structural Patterns and Proportions in Vergil's Aeneid*, University of Michigan Press, Ann Arbor, Mich., 1962.

[152] U. Dudley and B. Tucker, Greatest Common Divisors in Altered Fibonacci Sequences, *Fibonacci Quarterly*, 9 (1971), 89–91.

[153] A. Dunn, *Mathematical Bafflers*, Dover, New York, 1980.

[154] A. Dunn, *Second Book of Mathematical Bafflers*, Dover, New York, 1983.

[155] Editor, Solution to Problem B-22, *Fibonacci Quarterly*, 2 (1964), 78.

[156] Editor, Consecutive Heads and Fibonacci, *Mathematical Gazette*, 69 (1985), 208–211.

[157] S. Edwards, Problem B-962, *Fibonacci Quarterly*, 41 (2003), 374.

[158] S. Edwards, Problem B-980, *Fibonacci Quarterly*, 42 (2004), 182.

[159] M.H. Eggar, Applications of Fibonacci Numbers, *Mathematical Gazette*, 63 (1979), 36–39.

[160] J. Erbacker, J.A Fuchs, and F.D. Parker, Problem H-25, *Fibonacci Quarterly*, 1(4) (1963), 47.

[161] J. Ercolano, Golden Sequences of Matrices with Applications to Fibonacci Algebra, *Fibonacci Quarterly*, 14 (1976), 419–426.

[162] D. Everman, A.E. Danese, K. Venkannayah, and E.M. Scheuer, Problem E1396, *American Mathematical Monthly*, 67 (1960), 694.

[163] H. Eves, *An Introduction to the History of Mathematics*, 3rd edition, Holt, Rinehart and Winston, New York, 1969.

[164] M. Feinberg, Fibonacci–Tribonacci, *Fibonacci Quarterly*, 1(3) (1963), 70–74.

[165] M. Feinberg, A Lucas Triangle, *Fibonacci Quarterly*, 5 (1967), 486–490.

[166] H.H. Ferns, Problem B-48, *Fibonacci Quarterly*, 2 (1964), 232.

[167] H.H. Ferns, Problem B-104, *Fibonacci Quarterly*, 4 (1966), 374.

[168] H.H. Ferns, Problem B-106, *Fibonacci Quarterly*, 5 (1967), 107.

[169] H.H. Ferns, Problem B-115, *Fibonacci Quarterly*, 6 (1968), 92–93.

[170] The Fibonacci Numbers, *Time*, April 4, 1969.

[171] Fibonacci or Forgery, *Scientific American*, 272 (May, 1995), 104–105.

[172] P. Filipponi and D. Singmaster, Average Age of Generalized Rabbits, *Fibonacci Quarterly*, 28 (1990), 281.

[173] R. Finkelstein, Problem B-143, *Fibonacci Quarterly*, 7 (1969), 220–221.

[174] R. Finkelstein, On Fibonacci Numbers Which Are One More Than a Square, *J. Reine Angewandte Mathematik*, 262/263 (1973), 171–182.

[175] R. Finkelstein, On Lucas Numbers Which Are One More Than a Square, *Fibonacci Quarterly*, 13 (1975), 340–342.

[176] R. Finkelstein, On Triangular Fibonacci Numbers, *Utilitas Mathematica*, 9 (1976), 319–327.

[177] K. Fischer, The Fibonacci Sequence Encountered in Nerve Physiology, *Fibonacci Quarterly*, 14 (1976), 377–379.

[178] G. Fisher, The Singularity of Fibonacci Matrices, *Mathematical Gazette*, 81 (1997), 295–298.

[179] R. Fisher, *Fibonacci Applications and Strategies for Traders*, John Wiley & Sons, Inc., New York, 1993.

[180] S. Fisk, Problem B-10, *Fibonacci Quarterly*, 1(2) (1963), 85.

[181] Forbidden Fivefold Symmetry May Indicate Quasicrystal Phase, *Physics Today*, February, 1985, 17–19.

[182] K. Ford, Soution to Problem 487, *College Mathematics Journal*, 24 (1993), 476.

[183] E.T. Frankel, Fibonacci Numbers as Paths of a Rook on a Chessboard, *Fibonacci Quarterly*, 8 (1970), 538–541.

[184] D.L. Freedman, Fibonacci and Lucas Connections, *Mathematics Teacher*, 89 (1996), 624–625.

[185] G.F. Freeman, On Ratios of Fibonacci and Lucas Numbers, *Fibonacci Quarterly*, 5 (1967), 99–106.

[186] H.T. Freitag, Problem B-256, *Fibonacci Quarterly*, 12 (1974), 221.

[187] H.T. Freitag, Problem B-257, *Fibonacci Quarterly*, 12 (1974), 222.

[188] H.T. Freitag, Problem B-262, *Fibonacci Quarterly*, 12 (1974), 314.

[189] H.T. Freitag, Problem B-270, *Fibonacci Quarterly*, 12 (1974), 405.

[190] H.T. Freitag, Problem B-271, *Fibonacci Quarterly*, 12 (1974), 405.

[191] H.T. Freitag, Problem B-282, *Fibonacci Quarterly*, 13 (1975), 192.

[192] H.T. Freitag, Problem B-286, *Fibonacci Quarterly*, 13 (1975), 286.

[193] H.T. Freitag, Problem B-314, *Fibonacci Quarterly*, 14 (1976), 288.

[194] H.T. Freitag, Problem B-379, *Fibonacci Quarterly*, 17 (1979), 186.

[195] H.T. Freitag, Problem B-455, *Fibonacci Quarterly*, 19 (1981), 377.

[196] H.T. Freitag, Problem B-457, *Fibonacci Quarterly*, 19 (1981), 377.

[197] H.T. Freitag, Problem B-462, *Fibonacci Quarterly*, 19 (1981), 466.

[198] H.T. Freitag, Problem B-463, *Fibonacci Quarterly*, 19 (1981), 466.

[199] H.T. Freitag, Problem B-487, *Fibonacci Quarterly*, 20 (1982), 367.

[200] H.T. Freitag, Problem B-488, *Fibonacci Quarterly*, 20 (1982), 367.

[201] H.T. Freitag, Problem B-489, *Fibonacci Quarterly*, 20 (1982), 367.

[202] J.A. Fuchs, Problem B-39, *Fibonacci Quarterly*, 2 (1964), 154.

[203] J.A. Fuchs, Solution to Problem B-32, *Fibonacci Quarterly*, 2 (1964), 235.

[204] S.E. Ganis, Notes on the Fibonacci Sequence, *American Mathematical Monthly*, 66 (1959), 129–130.

[205] T. Gardiner, The Fibonacci Spiral, *Mathematical Gazette*, 67 (1983), 120.

[206] M. Gardner, *Mathematics Magic and Mystery*, Dover, New York, 1956.

[207] M. Gardner, Mathematical Games, *Scientific American*, 201 (Aug., 1959), 128–134.

[208] M. Gardner, *Mathematical Puzzles and Diversions*, University of Chicago Press, Chicago, 1960.

[209] M. Gardner, The Multiple Fascination of the Fibonacci Sequence, *Scientific American*, 211 (March, 1969), 116–120.

[210] M. Gardner, Mathematical Games, *Scientific American*, 222 (Feb., 1970), 112–114.

[211] M. Gardner, Mathematical Games, *Scientific American*, 222 (March, 1970), 121–125.

[212] M. Gardner, The Cult of the Golden Ratio, *Skeptical Inquirer*, 18 (Spring, 1994), 243–247.

[213] T.H. Garland, *Fascinating Fibonaccis*, Dale Seymour Publications, Palo Alto, Calif., 1987.

[214] P. Gattei, The 'Inverse' Differential Equation, *Mathematical Spectrum*, 23 (1990–1991), 127–131.

[215] N. Gauthier, Fibonacci Sums of the Type $\sum r^m F_m$, *Mathematical Gazette*, 79 (1995), 364–367.

[216] S.P. Geller, A Computer Investigation of a Property of the Fibonacci Sequence, *Fibonacci Quarterly*, 1(2) (1963), 84.

[217] I. Gessel, Problem H-187, *Fibonacci Quarterly*, 10 (1972), 417–419.

[218] M. Ghyka, *The Geometry of Art and Life*, Dover, New York, 1977.

[219] J. Gill and G. Miller, Newton's Method and Ratios of Fibonacci Numbers, *Fibonacci Quarterly*, 19 (1981), 1–4.

[220] J. Ginsburg, Fibonacci Pleasantries, *Scripta Mathematica*, 14 (1948), 163–164.

[221] J. Ginsburg, A Relationship Between Cubes of Fibonacci Numbers, *Scripta Mathematica*, 19 (1953), 242.

[222] P. Glaister, Multiplicative Fibonacci Sequences, *Mathematical Gazette*, 78 (1994), 68.

[223] P. Glaister, Fibonacci Power Series, *Mathematical Gazette*, 79 (1995), 521–525.

[224] P. Glaister, Golden Earrings, *Mathematical Gazette*, 80 (1996), 224–225.

[225] P. Glaister, Two Fibonacci Sums – a Variation, *Mathematical Gazette*, 81 (1997), 85–88.

[226] P. Glaister, The Golden Ratio by Origami, *Mathematical Spectrum*, 3 (1998–1999), 52–53.

[227] J.R. Goggins, Formula for $\pi/4$, *Mathematical Gazette*, 57 (1973), 134.

[228] I.J. Good, A Reciprocal Series of Fibonacci Numbers, *Fibonacci Quarterly*, 12 (1974), 346.

[229] J.N. Gootherts, Problem H-67, *Fibonacci Quarterly*, 5 (1967), 78.

[230] J. Gordon, *A Step-ladder to Painting*, Faber & Faber, London, 1938.

[231] H.W. Gould, Problem B-7, *Fibonacci Quarterly*, 1(3) (1963), 80.

[232] H.W. Gould, General Functions for Products of Powers of Fibonacci Numbers, *Fibonacci Quarterly*, 1(2) (1963), 1–16.

[233] H.W. Gould, Associativity and the Golden Section, *Fibonacci Quarterly*, 2 (1964), 203.

[234] H.W. Gould, A Variant of Pascal's Triangle, *Fibonacci Quarterly*, 3 (1965), 257–271.

[235] H.W. Gould, The Case of the Strange Binomial Identities of Professor Moriarty, *Fibonacci Quarterly*, 10 (1972), 381–391, 402.

[236] S.W. Gould, Problem 4720, *American Mathematical Monthly*, 64 (1957), 749–750.

[237] R.F. Graesser, The Golden Section, *The Pentagon*, (Spring, 1964), 7–19.

[238] R.L. Graham, Problem H-10, *Fibonacci Quarterly*, 1(2) (1963), 53.

[239] R.L. Graham, Problem B-9, *Fibonacci Quarterly*, 1(4) (1963), 76.

[240] R.L. Graham, Problem H-45, *Fibonacci Quarterly*, 3 (1965), 127–128.

[241] R.L. Graham, D.E. Knuth, and O. Patashnik, *Concrete Mathematics*, Addison-Wesley, Reading, Mass., 1989.

[242] R.M. Grassl, Problem B-202, *Fibonacci Quarterly*, 9 (1971), 546–547.

[243] R.M. Grassl, Problem B-203, *Fibonacci Quarterly*, 9 (1971), 547–548.

[244] R.M. Grassl, Solution to Problem B-264, *Fibonacci Quarterly*, 11 (1973), 333.

[245] R.M. Grassl, Problem B-279, *Fibonacci Quarterly*, 12 (1974), 101.

[246] R.M. Grassl, Problem B-279, *Fibonacci Quarterly*, 13 (1975), 286.

[247] R.P. Grimaldi, MAA Minicourse No. 13, Joint Meetings, Orlando, Fl., 1996.

[248] S. Gudder, *A Mathematical Journey*, 2nd edition, McGraw-Hill, New York, 1964.

[249] G.A.R. Guillotte, Problem B-206, *Fibonacci Quarterly*, 9 (1971), 107.

[250] G.A.R. Guillotte, Problem B-207, *Fibonacci Quarterly*, 9 (1971), 107.

[251] G.A.R. Guillotte, Problem B-210, *Fibonacci Quarterly*, 9 (1971), 217–218.

[252] G.A.R. Guillotte, Problem B-218, *Fibonacci Quarterly*, 9 (1971), 439.

[253] G.A.R. Guillotte, Problem B-241, *Fibonacci Quarterly*, 10 (1972), 447–448.

[254] R.K. Guy, *Unsolved Problems in Number Theory*, 2nd edition, Springer, New York, 1994.

[255] J.H. Halton, On a General Fibonacci Identity, *Fibonacci Quarterly*, 3 (1965), 31–43.

[256] R.T. Hansen, Generating Identities for Fibonacci and Lucas Triples, *Fibonacci Quarterly*, 10 (1972), 571–578.

[257] G.H. Hardy and E.M. Wright, *An Introduction to the Theory of Numbers*, 5th edition, Oxford University Press, New York, 1995.

[258] V.C. Harris, On Identities Involving Fibonacci Numbers, *Fibonacci Quarterly*, 3 (1965), 214–218.

[259] M. Harvey and P. Woodruff, Fibonacci Numbers and Sums of Inverse Tangents, *Mathematical Gazette*, 79 (1995), 565–567.

[260] M.A. Heaslet, Solution to Problem 3801, *American Mathematical Monthly*, 45 (1938), 636–637.

[261] R.V. Heath, Another Fibonacci Curiosity, *Scripta Mathematica*, 16 (1950), 128.

[262] T. Herring et al., Solution to Problem B-1148, *Fibonacci Quarterly*, 53 (2015), 183–184.

[263] F. Higgins, Problem B-305, *Fibonacci Quarterly*, 14 (1976), 189.

[264] A.P. Hillman, Editorial Note, *Fibonacci Quarterly*, 9 (1971), 548.

[265] G. Hoare, Regenerative Powers of Rabbits, *Mathematical Gazette*, 73 (1989), 336–337.

[266] L. Hogben, *Science for the Citizen*, Allen & Unwin, London, 1938.

[267] V.E. Hoggatt, Jr., Problem H-3, *Fibonacci Quarterly*, 1(1) (1963), 46.

[268] V.E. Hoggatt, Jr., Problem B-2, *Fibonacci Quarterly*, 1(1) (1963), 73.

[269] V.E. Hoggatt, Jr., Problem H-31, *Fibonacci Quarterly*, 2 (1964), 49–50.

[270] V.E. Hoggatt, Jr., Problem H-39, *Fibonacci Quarterly*, 2 (1964), 124.

[271] V.E. Hoggatt, Jr., Fibonacci Numbers from a Differential Equation, *Fibonacci Quarterly*, 2 (1964), 176, 196.

[272] V.E. Hoggatt, Jr., Problem B-53, *Fibonacci Quarterly*, 2 (1964), 324.

[273] V.E. Hoggatt, Jr., Problem B-60, *Fibonacci Quarterly*, 3 (1965), 238.

[274] V.E. Hoggatt, Jr., Problem B-64, *Fibonacci Quarterly*, 3 (1965), 153.

[275] V.E. Hoggatt, Jr., Problem B-86, *Fibonacci Quarterly*, 4 (1966), 192.

[276] V.E. Hoggatt, Jr., Problem H-77, *Fibonacci Quarterly*, 5 (1967), 256–258.

[277] V.E. Hoggatt, Jr., Problem H-82, *Fibonacci Quarterly*, 6 (1968), 52–54.

[278] V.E. Hoggatt, Jr., Problem H-88, *Fibonacci Quarterly*, 6 (1968), 253–254.

[279] V.E. Hoggatt, Jr., Problem H-131, *Fibonacci Quarterly*, 6 (1968), 142.

[280] V.E. Hoggatt, Jr., *Fibonacci and Lucas Numbers*, The Fibonacci Association, Santa Clara, Calif., 1969.

[281] V.E. Hoggatt, Jr., Problem B-149, *Fibonacci Quarterly*, 7 (1969), 333–334.

[282] V.E. Hoggatt, Jr., Problem B-150, *Fibonacci Quarterly*, 7 (1969), 334.

[283] V.E. Hoggatt, Jr., An Application of the Lucas Triangle, *Fibonacci Quarterly*, 8 (1970), 360–364, 427.

[284] V.E. Hoggatt, Jr., Problem B-193, *Fibonacci Quarterly*, 8 (1970), 444.

[285] V.E. Hoggatt, Jr., Some Special Fibonacci and Lucas Generating Functions, *Fibonacci Quarterly*, 9 (1971), 121–133.

[286] V.E. Hoggatt, Jr., Problem B-204, *Fibonacci Quarterly*, 9 (1971), 106.

[287] V.E. Hoggatt, Jr., Problem B-205, *Fibonacci Quarterly*, 9 (1971), 107.

[288] V.E. Hoggatt, Jr., Problem B-208, *Fibonacci Quarterly*, 9 (1971), 217.

[289] V.E. Hoggatt, Jr., Problem B-211, *Fibonacci Quarterly*, 9 (1971), 218.

[290] V.E. Hoggatt, Jr., Problem H-183, *Fibonacci Quarterly*, 9 (1971), 389.

[291] V.E. Hoggatt, Jr., Problem B-231, *Fibonacci Quarterly*, 10 (1972), 219–220.

[292] V.E. Hoggatt, Jr., Generalized Fibonacci Numbers in Pascal's Pyramid, *Fibonacci Quarterly*, 10 (1972), 271–276, 293.

[293] V.E. Hoggatt, Jr., Problem B-216, *Fibonacci Quarterly*, 10 (1972), 332–333.

[294] V.E. Hoggatt, Jr., Problem H-201, *Fibonacci Quarterly*, 10 (1972), 630.

[295] V.E. Hoggatt, Jr., Problem B-222, *Fibonacci Quarterly*, 10 (1972), 665.

[296] V.E. Hoggatt, Jr., Problem B-302, *Fibonacci Quarterly*, 11 (1973), 95.

[297] V.E. Hoggatt, Jr., Problem B-248, *Fibonacci Quarterly*, 11 (1973), 553.

[298] V.E. Hoggatt, Jr., Problem B-249, *Fibonacci Quarterly*, 11 (1973), 553.

[299] V.E. Hoggatt, Jr., Problem B-313, *Fibonacci Quarterly*, 13 (1975), 285.

[300] V.E. Hoggatt, Jr., Number Theory: The Fibonacci Sequence, *1977 Yearbook of Science and the Future*, Encyclopedia Britannica, Chicago, Ill.

[301] V.E. Hoggatt, Jr., Problem H-257, *Fibonacci Quarterly*, 15 (1977), 283.

[302] V.E. Hoggatt, Jr., Problem B-313, *Fibonacci Quarterly*, 15 (1977), 288.

[303] V.E. Hoggatt, Jr., Problem B-375, *Fibonacci Quarterly*, 17 (1979), 93.

[304] V.E. Hoggatt, Jr., Problem H-319, *Fibonacci Quarterly*, 18 (1980), 96.

[305] V.E. Hoggatt, Jr. and K. Alladi, Generalized Fibonacci Tiling, *Fibonacci Quarterly*, 13 (1975), 137–145.

[306] V.E. Hoggatt, Jr. and G.E. Bergum, Divisibility and Congruence Relations, *Fibonacci Quarterly*, 12 (1974), 189–195.

[307] V.E. Hoggatt, Jr. and G.E. Bergum, A Problem of Fermat and the Fibonacci Sequence, *Fibonacci Quarterly*, 15 (1977), 323–330.

[308] V.E. Hoggatt, Jr. and M. Bicknell, Some New Fibonacci Identities, *Fibonacci Quarterly*, 2 (1964), 29–32.

[309] V.E. Hoggatt, Jr. and M. Bicknell, Convolution Triangles, *Fibonacci Quarterly*, 10 (1972), 599–608.

[310] V.E. Hoggatt, Jr. and M. Bicknell, Roots of Fibonacci Polynomials, *Fibonacci Quarterly*, 11 (1973), 271–274.

[311] V.E. Hoggatt, Jr. and M. Bicknell, A Primer for the Fibonacci Numbers: Part XIV, *Fibonacci Quarterly*, 12 (1974), 147–156.

[312] V.E. Hoggatt, Jr. and H. Edgar, Another Proof That $\varphi(F_n) \equiv 0 \pmod 4$ for $n > 4$, *Fibonacci Quarterly*, 18 (1980), 80–82.

[313] V.E. Hoggatt, Jr. and C.H. King, Problem H-20, *Fibonacci Quarterly*, 1(4) (1963), 52.

[314] V.E. Hoggatt, Jr. and D.A. Lind, The Heights of the Fibonacci Polynomials and an Associated Function, *Fibonacci Quarterly*, 5 (1967), 141–152.

[315] V.E. Hoggatt, Jr. and D.A. Lind, A Primer for the Fibonacci Numbers: Part VI, *Fibonacci Quarterly*, 5 (1967), 445–460.

[316] V.E. Hoggatt, Jr. and I.D. Ruggles, A Primer for the Fibonacci Numbers: Part V, *Fibonacci Quarterly*, 2 (1964), 65–71.

[317] V.E. Hoggatt, Jr., J.W. Phillips, and H.T. Leonard, Jr., Twenty-four Master Identities, *Fibonacci Quarterly*, 9 (1971), 1–17.

[318] V.E. Hoggatt, Jr., M. Bicknell, and E.L. King, Fibonacci and Lucas Triangles, *Fibonacci Quarterly*, 10 (1972), 555–560.

[319] M.H. Holt, The Golden Section, *The Pentagon*, (Spring, 1964), 80–104.

[320] M.H. Holt, Mystery Puzzler and Phi, *Fibonacci Quarterly*, 3 (1965), 135–138.

[321] R. Honsberger, *Mathematical Delights*, Math. Association of America, Washington, D.C., 2004.

[322] W. Hope-Jones, The Bee and the Regular Pentagon, *Mathematical Gazette*, 10 (1921); reprinted in 55 (1971), 220.

[323] A.F. Horadam, Fibonacci Sequences and a Geometrical Paradox, *Mathematics Magazine*, 35 (1962), 1–11.

[324] A.F. Horadam, Further Appearance of the Fibonacci Sequence, *Fibonacci Quarterly*, 1(4) (1963), 41–42, 46.

[325] A.F. Horadam, Complex Fibonacci Numbers and Fibonacci Quaternions, *American Mathematical Monthly*, 70 (1963), 289–291.

[326] A.F. Horadam, Pell Identities, *Fibonacci Quarterly*, 9 (1971), 245–252, 263.

[327] W.W. Horner, Fibonacci and Pascal, *Fibonacci Quarterly*, 2 (1964), 228.

[328] W.W. Horner, Fibonacci and Euclid, *Fibonacci Quarterly*, 4 (1966), 168–169.

[329] W.W. Horner, Problem B-146, *Fibonacci Quarterly*, 6 (1968), 289.

[330] W.W. Horner, Fibonacci and Apollonius, *Fibonacci Quarterly*, 11 (1973), 541–542.

[331] H. Hosoya, Topological Index and Fibonacci Numbers with Relation to Chemistry, *Fibonacci Quarterly*, 11 (1973), 255–265.

[332] H. Hosoya, Fibonacci Triangle, *Fibonacci Quarterly*, 14 (1976), 173–278.

[333] J.E. Householder, Problem B-3, *Fibonacci Quarterly*, 1(1) (1963), 73.

[334] D. Huang, Fibonacci Identities, Matrices, and Graphs, *Mathematics Teacher*, 98 (2005), 400–403.

[335] J.A.H. Hunter, Triangle Inscribed in a Rectangle, *Fibonacci Quarterly*, 1(3) (1963), 66.

[336] J.A.H. Hunter, The Golden Cuboid, *Fibonacci Quarterly*, 2 (1964), 303.

[337] J.A.H. Hunter, Problem H-48, *Fibonacci Quarterly*, 2 (1964), 303.

[338] J.A.H. Hunter, Problem H-79, *Fibonacci Quarterly*, 4 (1966), 57.

[339] J.A.H. Hunter, Fibonacci Yet Again, *Fibonacci Quarterly*, 4 (1966), 273.

[340] J.A.H. Hunter, Problem B-100, *Fibonacci Quarterly*, 4 (1966), 373.

[341] J.A.H. Hunter, Problem H-124, *Fibonacci Quarterly*, 7 (1969), 179.

[342] J.A.H. Hunter and J.S. Madachy, *Mathematical Diversions*, Dover, New York, 1975.

[343] H.E. Huntley, Fibonacci Geometry, *Fibonacci Quarterly*, 2 (1964), 104.

[344] H.E. Huntley, *The Divine Proportion*, Dover, New York, 1970.

[345] H.E. Huntley, The Golden Ellipse, *Fibonacci Quarterly*, 12 (1974), 38–40.

[346] H.E. Huntley, Phi: Another Hiding Place, *Fibonacci Quarterly*, 12 (1974), 65–66.

[347] V. Ivanoff, Problem H-107, *Fibonacci Quarterly*, 6 (1968), 358–359.

[348] W.D. Jackson, Problem B-142, *Fibonacci Quarterly*, 7 (1969), 220.

[349] D.V. Jaiswal, Determinants Involving Generalized Fibonacci Numbers, *Fibonacci Quarterly*, 7 (1969), 319–330.

[350] D.V. Jaiswal, Some Geometrical Properties of the Generalized Fibonacci Sequence, *Fibonacci Quarterly*, 12 (1974), 67–70.

[351] D. Jarden, Strengthened Inequalities for Fibonacci and Lucas Numbers, *Fibonacci Quarterly*, 2 (1964), 45–48.

[352] D. Jarden, A New Important Formula for Lucas Numbers, *Fibonacci Quarterly*, 5 (1967), 346.

[353] R.V. Jean, The Fibonacci Sequence, *The UMAP Journal*, 5 (1984), 23–47.

[354] R. André-Jeannin, Solution to Problem B-897, *Fibonacci Quarterly*, 39 (2001), 89.

[355] J.H. Jordan, Gaussian Fibonacci and Lucas Numbers, *Fibonacci Quarterly*, 3 (1965), 315–318.

[356] M. Kauers, Problem 11258, *American Mathematical Monthly*, 113 (2006), 939.

[357] C. Kimberling, Problem E2581, *American Mathematical Monthly*, 83 (1976), 197.

[358] B.W. King, Solution to Problem B-129, *Fibonacci Quarterly*, 6 (1968), 296.

[359] B.W. King, Problem B-184, *Fibonacci Quarterly*, 9 (1971), 107–108.

[360] C.H. King, Some Properties of the Fibonacci Numbers, Master's Thesis, San Jose State College, San Jose, Calif., 1960.

[361] C.H. King, Leonardo Fibonacci, *Fibonacci Quarterly*, 1(4) (1963), 15–19.

[362] D.A. Klarner, Determinants Involving kth Powers from Second Order Sequences, *Fibonacci Quarterly*, 4 (1966), 179–183.

[363] T. Koshy, Fibonacci Partial Sums, *Pi Mu Epsilon Journal*, 10 (1996), 267–268.

[364] T. Koshy, New Fibonacci and Lucas Identities, *Mathematical Gazette*, 82 (1998), 481–484.

[365] T. Koshy, The Convergence of a Lucas Series, *Mathematical Gazette*, 83 (1999), 272–274.

[366] T. Koshy, Weighted Fibonacci and Lucas Sums, *Mathematical Gazette*, 85 (2000), 93–96.

[367] T. Koshy, A Family of Polynomial Functions, *Pi Mu Epsilon Journal*, 12 (2001), 195–201.

[368] T. Koshy, *Discrete Mathematics with Applications*, Elsevier, Boston, Mass., 2004.

[369] T. Koshy, *Elementary Number Theory with Applications*, 2nd edition, Elsevier, Boston, Mass., 2007.

[370] T. Koshy, *Pell and Pell-Lucas Numbers with Applications*, Springer, New York, 2014.

[371] T. Koshy, Graph-theoretic Models for the Fibonacci Family, *Mathematical Gazette*, 98 (2014), 256–265.

[372] J. Kramer and V.E. Hoggatt, Jr., Special Cases of Fibonacci Periodicity, *Fibonacci Quarterly*, 10 (1972), 519–522.

[373] S. Kravitz, Problem B-58, *Fibonacci Quarterly*, 3 (1965), 74.

[374] S. Kravitz, Problem B-63, *Fibonacci Quarterly*, 3 (1965), 239–240.

[375] M. Krebs and N.C. Martinez, The Combinatorial Trace Method in Action, *College Mathematics Journal*, 44 (2013), 32–36.

[376] H.V. Krishna, Problem B-227, *Fibonacci Quarterly*, 10 (1972), 218.

[377] H. Kwong, Solution to Problem B-1155, *Fibonacci Quarterly*, 53 (2015), 277.

[378] R.A. Laird, Solution to Problem H-41, *Fibonacci Quarterly*, 3 (1965), 120–121.

[379] F. Land, *The Language of Mathematics*, John Murray, London, 1960.

[380] L. Lang, Problem B-247, *Fibonacci Quarterly*, 11 (1973), 552.

[381] A.G. Law, Solution to Problem B-157, *Fibonacci Quarterly*, 9 (1971), 65–67.

[382] G. Ledin, Solution to Problem H-29, *Fibonacci Quarterly*, 2 (1964), 305.

[383] G. Ledin, Problem H-57, *Fibonacci Quarterly*, 4 (1966), 336–338.

[384] G. Ledin, Problem H-117, *Fibonacci Quarterly*, 5 (1967), 162.

[385] D.H. Lehmer, Problem E621, *American Mathematical Monthly*, 43 (1936), 307.

[386] D.H. Lehmer, Problem 3801, *American Mathematical Monthly*, 43 (1936), 580.

[387] D.H. Lehmer, Problem 3802, *American Mathematical Monthly*, 43 (1936), 580.

[388] D.H. Lehmer, Solution to Problem 3802, *American Mathematical Monthly*, 45 (1938), 632–633.

[389] W.J. LeVeque, *Elementary Theory of Numbers*, Addison-Wesley, Reading, Mass., 1962.

[390] C. Libis, A Fibonacci Summation Formula, *Math Horizons*, (Feb., 1997), 33–34.

[391] D.A. Lind, Problem B-31, *Fibonacci Quarterly*, 2 (1964), 72.

[392] D.A. Lind, Problem H-54, *Fibonacci Quarterly*, 3 (1965), 44.

[393] D.A. Lind, Problem H-64, *Fibonacci Quarterly*, 3 (1965), 116.

[394] D.A. Lind, Problem B-70, *Fibonacci Quarterly*, 3 (1965), 235.

[395] D.A. Lind, Problem B-71, *Fibonacci Quarterly*, 3 (1965), 235.

[396] D.A. Lind, Solution to Problem B-58, *Fibonacci Quarterly*, 3 (1965), 236–237.

[397] D.A. Lind, Problem H-93, *Fibonacci Quarterly*, 4 (1966), 252.

[398] D.A. Lind, Solution to Problem H-57, *Fibonacci Quarterly*, 4 (1966), 337–338.

[399] D.A. Lind, The Q Matrix as a Counterexample in Group Theory, *Fibonacci Quarterly*, 5 (1967), 44.

[400] D.A. Lind, Iterated Fibonacci and Lucas Subscripts, *Fibonacci Quarterly*, 5 (1967), 80, 89–90.

[401] D.A. Lind, Solution to Problem B-89, *Fibonacci Quarterly*, 5 (1967), 108–109.

[402] D.A. Lind, Problem B-91, *Fibonacci Quarterly*, 5 (1967), 110.

[403] D.A. Lind, Solution to Problem B-93, *Fibonacci Quarterly*, 5 (1967), 111–112.

[404] D.A. Lind, Problem B-98, *Fibonacci Quarterly*, 5 (1967), 206.

[405] D.A. Lind, Problem B-99, *Fibonacci Quarterly*, 5 (1967), 206–207.

[406] D.A. Lind, Problem B-103, *Fibonacci Quarterly*, 5 (1967), 290–292.

[407] D.A. Lind, Solution to Problem B-118, *Fibonacci Quarterly*, 6 (1968), 186–187.

[408] D.A. Lind, Problem B-145, *Fibonacci Quarterly*, 6 (1968), 289.

[409] D.A. Lind, Problem H-134, *Fibonacci Quarterly*, 6 (1968), 404–405.

[410] D.A. Lind, Problem B-165, *Fibonacci Quarterly*, 8 (1970), 111–112.

[411] D.A. Lind, A Determinant Involving Generalized Binomial Coefficients, *Fibonacci Quarterly*, 9 (1971), 113–119, 162.

[412] P. Lindstrom, Problem B-341, *Fibonacci Quarterly*, 14 (1976), 470.

[413] G. Ling, Newton's Method and the Golden Ratio, *College Mathematics Journal*, 38 (2007), 355.

[414] B. Litvack, Problem B-47, *Fibonacci Quarterly*, 2 (1964), 232.

[415] E.K. Lloyd, Enumeration, *Handbook of Applicable Mathematics*, John Wiley & Sons, Inc., New York, 1985, 531–621.

[416] H. London and R. Finkelstein, On Fibonacci and Lucas Numbers Which Are Perfect Powers, *Fibonacci Quarterly*, 7 (1969), 476–481, 487.

[417] C.T. Long, The Decimal Expansion of 1/89 and Related Results, *Fibonacci Quarterly*, 19 (1981), 53–55.

[418] C.T. Long, Discovering Fibonacci Identities, *Fibonacci Quarterly*, 24 (1986), 160–167.

[419] C.T. Long and J.H. Jordan, A Limited Arithmetic on Simple Continued Fractions, *Fibonacci Quarterly*, 5 (1967), 113–128.

[420] G. Lord, Solution to Problem B-233, *Fibonacci Quarterly*, 11 (1973), 221–222.

[421] G. Lord, Solution to Problem B-248, *Fibonacci Quarterly*, 11 (1973), 553.

[422] G. Lord, Solution to Problem B-286, *Fibonacci Quarterly*, 13 (1975), 286.

[423] G. Lord, Solution to Problem B-313, *Fibonacci Quarterly*, 14 (1976), 288.

[424] G. Lord, Solution to Problem B-342, *Fibonacci Quarterly*, 15 (1977), 376.

[425] N. Lord, Balancing and Golden Rectangles, *Mathematical Gazette*, 79 (1995), 573–574.

[426] P. Mana, Problem B-152, *Fibonacci Quarterly*, 6 (1968), 401.

[427] P. Mana, Problem B-163, *Fibonacci Quarterly*, 7 (1969), 219.

[428] P. Mana, Solution to Problem B-167, *Fibonacci Quarterly*, 8 (1970), 328.

[429] P. Mana, Problem B-215, *Fibonacci Quarterly*, 10 (1972), 331–332.

[430] P. Mana, Problem B-354, *Fibonacci Quarterly*, 16 (1978), 185–186.

[431] Y.V. Matiyasevich and R.K. Guy, A New Formula for *pi*, *American Mathematical Monthly*, 93 (1986), 631–635.

[432] J.A. Maxwell, Problem B-8, *Fibonacci Quarterly*, 1(1) (1963), 75.

[433] J.R. McCowan and M.A. Sequeira, *Patterns in Mathematics*, Prindle, Weber, and Schmidt, Boston, Mass., 1994.

[434] M.McNabb, Phyllotaxis, *Fibonacci Quarterly*, 1(4) (1963), 57–60.

[435] D.G. Mead, Problem B-67, *Fibonacci Quarterly*, 3 (1965), 326–327.

[436] R.S. Melham, Families of Identities Involving Sums of Powers of the Fibonacci and Lucas Numbers, *Fibonacci Quarterly*, 37 (1999), 315–319.

[437] R.S. Melham, A Fibonacci Identity in the Spirit of Simson and Gelin–Cesàro, *Fibonacci Quarterly*, 41 (2003), 142–143.

[438] R.S. Melham, A Three-variable Identity Involving Cubes of Fibonacci Numbers, *Fibonacci Quarterly*, 41 (2003), 220–223.

[439] G. Michael, A New Proof of an Old Property, *Fibonacci Quarterly*, 2 (1964), 57–58.

[440] J.W. Milsom, Solution to Problem B-227, *Fibonacci Quarterly*, 11 (1973), 107–108.

[441] J.W. Milsom, Solution to Problem B-462, *Fibonacci Quarterly*, 20 (1982), 369–370.

[442] L. Ming, On Triangular Fibonacci Numbers, *Fibonacci Quarterly*, 27 (1989), 98–108.

[443] E.E. Moise and E. Downs, *Geometry*, Addison-Wesley, Reading, Mass., 1964.

[444] P. Montgomery, Solution to Problem E2581, *American Mathematical Monthly*, 84 (1977), 488.

[445] M.G. Monzingo, Why Are 8:18 and 10:09 Such Pleasant Times?, *Fibonacci Quarterly*, 21 (1983), 107–110.

[446] J.D. Mooney, Solution to Problem B-53, *Fibonacci Quarterly*, 3 (1965), 157.

[447] R.E.M. Moore, Mosaic Units: Patterns in Ancient Mosaics, *Fibonacci Quarterly*, 8 (1970), 281–310.

[448] S. Moore, Fibonacci Matrices, *Mathematical Gazette*, 67 (1983), 56–57.

[449] L. Moser, Problem B-5, *Fibonacci Quarterly*, 1(1) (1963), 74.

[450] L. Moser and J.L. Brown, Some Reflections, *Fibonacci Quarterly*, 1(4) (1963), 75.

[451] L. Moser and M. Wyman, Problem B-6, *Fibonacci Quarterly*, 1(1) (1963), 74.

[452] B.R. Myers, Number of Spanning Trees in a Wheel, *IEEE Transactions on Circuit Theory*, CT-18 (1971), 280–281.

[453] B.R. Myers, Advanced Problem 5795, *American Mathematical Monthly*, 79 (1972), 914–915.

[454] T. Nagel, *Introduction to Number Theory*, Chelsea, New York, 1964.

[455] G. Neumer, Problem 487, *College Mathematics Journal*, 24 (1993), 476.

[456] I. Niven and H.S. Zuckerman, *An Introduction to the Theory of Numbers*, 4th edition, John Wiley & Sons, Inc., New York, 1980.

[457] A. Nkwanta and L.W. Shapiro, Pell Walks and Riordan Matrices, *Fibonacci Quarterly*, 43 (2005), 170–180.

[458] M.K. O'Connell and D.T. O'Connell, Letter to the Editor, *Scientific Monthly*, 78 (1951), 333.

[459] G. Odom, Problem E3007, *American Mathematical Monthly*, 93 (1986), 572.

[460] F.C. Ogg, Letter to the Editor, *Scientific Monthly*, 78 (1951), 333.

[461] C.S. Ogilvy and J.T. Anderson, *Excursions in Number Theory*, Dover, New York, 1966.

[462] H. Ohtsuka, Problem B-1148, *Fibonacci Quarterly*, 52 (2014), 179.

[463] H. Ohtsuka, Problem B-1155, *Fibonacci Quarterly*, 52 (2014), 275.

[464] R.L. Ollerton, Proof Without Words: Fibonacci Tiles, *Mathematics Magazine*, 81 (2008), 302.

[465] T.J. Osler and A. Hilburn, An Unusual Proof That F_m Divides F_{mn} Using Hyperbolic Functions, *Mathematical Gazette*, 91 (2007), 510–512.

[466] J.C. Owings, Solution of the System $a^2 \equiv -1$ (mod b), $b^2 \equiv -1$ (mod a), *Fibonacci Quarterly*, 25 (1987), 245–249.

[467] S.L. Padwa, Solution to Problem H-188, *Fibonacci Quarterly*, 10 (1972), 631.

[468] G.C. Padilla, Solution to Problem B-98, *Fibonacci Quarterly*, 5 (1967), 206.

[469] G.C. Padilla, Problem B-172, *Fibonacci Quarterly*, 7 (1969), 545.

[470] G.C. Padilla, Problem B-173, *Fibonacci Quarterly*, 7 (1969), 545.

[471] F.D. Parker, Problem H-21, *Fibonacci Quarterly*, 2 (1964), 133.

[472] P.C. Parks, A New Proof of the Hurwitz Stability Criterion by the Second Method of Liapunov with Applications to "Optimum" Transfer Functions, *Proceedings of the Joint Automatic Control Conference*, University of Minnesota, 1963.

[473] C.B.A. Peck, Solution to Problem B-146, *Fibonacci Quarterly*, 7 (1969), 223.

[474] C.B.A. Peck, Editorial Note, *Fibonacci Quarterly*, 8 (1970), 392.

[475] C.B.A. Peck, Solution to Problem B-206, *Fibonacci Quarterly*, 9 (1971), 550.

[476] C.B.A. Peck, Solution to Problem B-253, *Fibonacci Quarterly*, 12 (1974), 105.

[477] C.B.A. Peck, Solution to Problem B-270, *Fibonacci Quarterly*, 12 (1974), 405.

[478] A. Pepperdine, Solution to Problem 2568, *Journal of Recreational Mathematics*, 35 (2006), 74.

[479] H. Perfect, A Matrix-Limit (1), *Mathematical Gazette*, 68 (1984), 42–44.

[480] J.M.H. Peters, An Approximate Relation Between π and the Golden Ratio, *Mathematical Gazette*, 62 (1978), 197–198.

[481] J.M. Peters, Getting Down to tan $1\frac{1}{2}^{\circ}$, *Mathematical Gazette*, 69 (1985), 211–212.

[482] M. Pettet, Problem B-93, *Fibonacci Quarterly*, 4 (1966), 191.

[483] R. Phillips, *Numbers: Facts, Figures, and Fiction*, Cambridge University Press, New York, 1994.

[484] J.C. Pierce, The Fibonacci Series, *Scientific Monthly*, 78 (1951), 224–228.

[485] J.C. Pond, Solution to Problem B-91, *Fibonacci Quarterly*, 5 (1967), 110.

[486] R.R. Prechter and A.J. Frost, *The Elliott Wave Principle*, New Classics Library, Gainesville, Georgia, 1978 (latest edition 2014).

[487] B. Prielipp, Solution to Problem B-463, *Fibonacci Quarterly*, 20 (1982), 370.

[488] S. Rabinowitz, An Arcane Formula for a Curious Matrix, *Fibonacci Quarterly*, 36 (1998), 376.

[489] S. Rabinowitz, Problem B-964, *Fibonacci Quarterly*, 41 (2003), 375.

[490] C.W. Raine, Pythagorean Triangles from the Fibonacci Series $1, 1, 2, 3, 5, 8, …$, *Scripta Mathematica*, 14 (1948), 164.

[491] A. Ralston, *A First Course in Numerical Analysis*, McGraw-Hill, New York, 1965.

[492] W.R. Ransom, *Trigonometric Novelties*, J. Weston Walch, Portland, Maine, 1959.

[493] K.S. Rao, Some Properties of Fibonacci Numbers, *American Mathematical Monthly*, 60 (1953), 680–684.

[494] L. Raphael, Some Results in Trigonometry, *Fibonacci Quarterly*, 8 (1970), 371, 392.

[495] T. Ratliff, A Few of My Favorite Calc III Things, MAA Fall Meeting, University of Massachusetts, Boston, 1996.

[496] A.D. Rawlins, A Note on the Golden Ratio, *Mathematical Gazette*, 79 (1995), 104.

[497] K.R. Rebman, The Sequence 1 5 16 45 121 320 … in Combinatorics, *Fibonacci Quarterly*, 13 (1975), 51–55.

[498] K.G. Recke, Problem B-153, *Fibonacci Quarterly*, 6 (1968), 401.

[499] K.G. Recke, Problem B-157, *Fibonacci Quarterly*, 7 (1969), 108.

[500] N.R. Robbins, Fibonacci and Lucas Numbers of the Forms $w^2 - 1, w^3 \pm 1$, *Fibonacci Quarterly*, 19 (1981), 369–373.

[501] J.W. Roche, Fibonacci and Approximating Roots, *Mathematics Teacher*, 92 (1999), 523.

[502] J.F. Rigby, Equilateral Triangles and the Golden Ratio, *Mathematical Gazette*, 72 (1988), 27–30.

[503] N. Routledge, Matrices and Fibonacci Numbers, *Mathematical Spectrum*, 22 (1989–1990), 63.

[504] I.D. Ruggles, Problem B-1, *Fibonacci Quarterly*, 1(1) (1963), 73.

[505] I.D. Ruggles, Some Fibonacci Results Using Fibonacci-type Sequences, *Fibonacci Quarterly*, 1(2) (1963), 75–80.

[506] I.D. Ruggles and V.E. Hoggatt, Jr., A Primer for the Fibonacci Numbers–Part III, *Fibonacci Quarterly*, 1(3) (1963), 61–65.

[507] I.D. Ruggles and V.E. Hoggatt, Jr., A Primer for the Fibonacci Numbers–Part IV, *Fibonacci Quarterly*, 1(4) (1963), 65–71.

[508] G.E. Runion, *The Golden Section and Related Curiosa*, Scott, Foresman, Glenview, Ill., 1972.

[509] S. Sahib, Solution to Problem B-464, *Fibonacci Quarterly*, 20 (1982), 370–371.

[510] J. Satterly, Meet Mr. Tau, *School Science and Mathematics*, 56 (1956), 731–741.

[511] J. Satterly, Meet Mr. Tau Again, *School Science and Mathematics*, 56 (1956), 150–151.

[512] P. Schub, A Minor Fibonacci Curiosity, *Scripta Mathematica*, 16 (1950), 214.

[513] P. Schub, A Minor Fibonacci Curiosity, *Scripta Mathematica*, 17 (1970), 214.

[514] J.A. Scott, A Fractal Curve Suggested by the MA Crest, *Mathematical Gazette*, 80 (1996), 236–237.

[515] R.A. Seamons, Problem B-89, *Fibonacci Quarterly*, 4 (1966), 190.

[516] R.A. Seamons, Problem B-107, *Fibonacci Quarterly*, 5 (1967), 107.

[517] J. Sedlacek, On the Skeletons of a Graph or Digraph, *Proceedings of the Calgary International Conference of Combinatorial Structures and Their Applications*, Gordon & Breach, New York, 387–391.

[518] H.J. Seiffert, Solution to Problem B-964, *Fibonacci Quarterly*, 42 (2004), 184.

[519] J. Shallit, Problem B-311, *Fibonacci Quarterly*, 14 (1976), 287.

[520] J. Shallit, Problem B-423, *Fibonacci Quarterly*, 18 (1980), 85.

[521] A.G. Shannon, Problem B-167, *Fibonacci Quarterly*, 7 (1969), 331.

[522] A.G. Shannon, Problem B-382, *Fibonacci Quarterly*, 17 (1979), 282–283.

[523] B. Sharpe, On Sums $F_x^2 \pm F_y^2$, *Fibonacci Quarterly*, 3 (1965), 63.

[524] W.E. Sharpe, Golden Weaves, *Mathematical Gazette*, 62 (1978), 42–44.

[525] D.L. Silverman, Problem B-41, *Fibonacci Quarterly*, 2 (1964), 155.

[526] D.L. Silverman, Problem 567, *Journal of Recreational Mathematics*, 9 (1976–1977), 208.

[527] P. Singh, The So-called Fibonacci Numbers in Ancient and Medieval India, *Historia Mathematica*, 12 (1985), 229–244.

[528] S. Singh, The Beast 666, *Journal of Recreational Mathematics*, 21 (1989), 244.

[529] S. Singh, Solution to Problem B-443, *Fibonacci Quarterly*, 20 (1981), 90.

[530] C. Singleton, Problem 2568, *Journal of Recreational Mathematics*, 39 (2001), 311.

[531] C. Singleton, Touching Circles, *Journal of Recreational Mathematics*, 35 (2006), 74–75.

[532] K.J. Smith, *The Nature of Mathematics*, 7th edition, Brooks/Cole, Pacific Grove, Calif., 1995.

[533] F. Soddy, The Kiss Precise, *Nature*, 137 (1936), 1021.

[534] A. Sofo, Generalizations of a Radical Identity, *Mathematical Gazette*, 83 (1999), 274–276.

[535] L. Somer, Solution to Problem B-382, *Fibonacci Quarterly*, 17 (1979), 282–283.

[536] W. Squire, Problem H-83, *Fibonacci Quarterly*, 6 (1968), 54–55.

[537] F.S. Stancliff, A Curious Property of F_{11}, *Scripta Mathematica*, 19 (1953), 126.

[538] P. Stănică, Solution to Problem B-911, *Fibonacci Quarterly*, 39 (2001), 468.

[539] T.E. Stanley, Soution to Problem B-203, *Fibonacci Quarterly*, 9 (1971), 547–548.

[540] E.P. Starke, Solution to Problem 621, *American Mathematical Monthly*, 43 (1936), 307–308.

[541] H. Steinhaus, *Mathematical Snapshots*, 3rd edition, Oxford University Press, New York, 1969.

[542] H. Steinhaus, *One Hundred Problems in Elementary Mathematics*, Dover, New York, 1979.

[543] M. Stephen, The Mysterious Number PHI, *Mathematics Teacher*, 49 (1956), 200–204.

[544] F. Stern, Problem B-374, *Fibonacci Quarterly*, 17 (1979), 93.

[545] R. Stong, Solution to Problem 11258, *American Mathematical Monthly*, 115 (2008), 949–950.

[546] A. Struyk, One Curiosum Leads to Another, *Scripta Mathematica*, 17 (1970), 230.

[547] M.N.S. Swamy, Problem B-83, *Fibonacci Quarterly*, 4 (1966), 375.

[548] M.N.S. Swamy, Problem H-69, *Fibonacci Quarterly*, 5 (1967), 163–165.

[549] M.N.S. Swamy, Problem H-127, *Fibonacci Quarterly*, 6 (1968), 51.

[550] M.N.S. Swamy, Solution to Problem H-93, *Fibonacci Quarterly*, 6 (1968), 145–148.

[551] M.N.S. Swamy, Problem H-150, *Fibonacci Quarterly*, 8 (1970), 391–392.

[552] M.N.S. Swamy, Problem H-157, *Fibonacci Quarterly*, 9 (1971), 65–67.

[553] M.N.S. Swamy, Problem H-158, *Fibonacci Quarterly*, 9 (1971), 67–69.

[554] S.B. Tadlock, Products of Odds, *Fibonacci Quarterly*, 3 (1965), 54–56.

[555] M.H. Tallman, Problem H-23, *Fibonacci Quarterly*, 1(3) (1963), 47.

[556] L. Taylor, Problem B-460, *Fibonacci Quarterly*, 19 (1981), 466.

[557] L. Taylor, Problem B-461, *Fibonacci Quarterly*, 19 (1981), 466.

[558] D. Thoro, Reciprocals of Generalized Fibonacci Numbers, *Fibonacci Quarterly*, 1(4) (1963), 30.

[559] D. Thoro, The Golden Ratio: Computational Considerations, *Fibonacci Quarterly*, 1(3) (1963), 53–59.

[560] P. Tomkins, *Mysteries of the Mexican Pyramids*, Harper & Row, New York, 1976.

[561] C.W. Trigg, Geometric Proof of a Result of Lehmer's, *Fibonacci Quarterly*, 11 (1973), 593–540.

[562] M.R. Turner, Certain Congruence Properties (modulo 100) of Fibonacci Numbers, *Fibonacci Quarterly*, 12 (1974), 87–91.

[563] H.L. Umansky, Pythagorean Triangles from Recurrent Series, *Scripta Mathematica*, 22 (1956), 88.

[564] H.L. Umansky, Curiosa, Zero Determinants, *Scripta Mathematica*, 22 (1956), 88.

[565] H.L. Umansky, Letter to the Editor, *Fibonacci Quarterly*, 8 (1970), 89.

[566] H.L. Umansky, Problem B-233, *Fibonacci Quarterly*, 10 (1972), 329–330.

[567] H.L. Umansky and M. Tallman, Problem H-101, *Fibonacci Quarterly*, 6 (1968), 259–260.

[568] Z. Usiskin, Problem B-265, *Fibonacci Quarterly*, 12 (1974), 315.

[569] Z. Usiskin, Problem B-266, *Fibonacci Quarterly*, 12 (1974), 315–316.

[570] S. Vajda, Some Summation Formulas for Binomial Coefficients, *Mathematical Gazette*, 34 (1950), 211–212.

[571] S. Vajda, *Fibonacci and Lucas Numbers, and the Golden Section*, John Wiley & Sons, Inc., New York, 1989.

[572] J. Vinson, The Relation of the Period Modulo *m* to the Rank of Appartition of *m* in the Fibonacci Sequence, *Fibonacci Quarterly*, 1(2) (1963), 37–45.

[573] N.N. Vorobev, *Fibonacci Numbers*, Pergamon Press, New York, 1961.

[574] M. Wachtel, Problem B-391, *Fibonacci Quarterly*, 16 (1978), 562.

[575] R.C. Walker, An Electrical Network and the Golden Section, *Mathematical Gazette*, 59 (1975), 162–165.

[576] C.R. Wall, Problem B-32, *Fibonacci Quarterly*, 2 (1964), 72.

[577] C.R. Wall, Problem B-40, *Fibonacci Quarterly*, 2 (1964), 155.

[578] C.R. Wall, Problem B-55, *Fibonacci Quarterly*, 2 (1964), 324.

[579] C.R. Wall, Problem B-56, *Fibonacci Quarterly*, 2 (1964), 325.

[580] C.R. Wall, Problem B-45, *Fibonacci Quarterly*, 3 (1965), 76.

[581] C.R. Wall, Problem B-127, *Fibonacci Quarterly*, 5 (1967), 465.

[582] C.R. Wall, Problem B-131, *Fibonacci Quarterly*, 5 (1967), 466.

[583] C.R. Wall, Problem B-141, *Fibonacci Quarterly*, 6 (1968), 186.

[584] C.R. Wall, On Triangular Fibonacci Numbers, *Fibonacci Quarterly*, 23 (1985), 77–79.

[585] D.D. Wall, Fibonacci Series Modulo *m*, *American Mathematical Monthly*, 67 (1960), 525–532.

[586] M.J. Wasteels, Quelques Propriétés des Nombres de Fibonacci, *Mathesis*, 2 (1902), 60–63.

[587] L. Weinstein, A Divisibility Property of Fibonacci Numbers, *Fibonacci Quarterly*, 4 (1966), 83–84.

[588] J. Wessner, Solution to Problem B-63, *Fibonacci Quarterly*, 3 (1965), 239–240.

[589] J. Wessner, Solution to Problem B-88, *Fibonacci Quarterly*, 4 (1966), 190.

[590] J. Wessner, Solution to Problem B-127, *Fibonacci Quarterly*, 6 (1968), 294–295.

[591] R.E. Whitney, Solution to Problem H-48, *Fibonacci Quarterly*, 3 (1965), 303.

[592] R.E. Whitney, Extensions of Recurrence Relations, *Fibonacci Quarterly*, 4 (1966), 37–42.

[593] R.E. Whitney, Compositions of Recursive Formulae, *Fibonacci Quarterly*, 4 (1966), 363–366.

[594] R.E. Whitney, Problem H-188, *Fibonacci Quarterly*, 9 (1971), 513.

[595] D. William, A Fibonacci Sum, *Mathematical Gazette*, 69 (1985), 29–31.

[596] William Lowell Putnam Mathematical Competition, *Mathematics Magazine*, 81 (2008), 72–77.

[597] H.C. Williams, *Edouard Lucas and Primality Testing*, John Wiley & Sons, Inc., New York, 1998.

[598] J. Wlodarski, The Golden Ratio in an Electrical Network, *Fibonacci Quarterly*, 9 (1971), 188, 194.

[599] J. Wlodarski, The Balmer Series and the Fibonacci Numbers, *Fibonacci Quarterly*, 11 (1973), 526, 540.

[600] J.E. Woko, A Pascal-like Triangle for $\alpha^n + \beta^n$, *Mathematical Gazette*, 81 (1997), 75–79.

[601] C.K. Wong and T.W. Maddocks, A Generalized Pascal's Triangle, *Fibonacci Quarterly*, 13 (1975), 134–136.

[602] J. Woodlum, Problem B-119, *Fibonacci Quarterly*, 6 (1968), 187.

[603] *The World Book Encyclopedia*, Volume 20, World Book, Chicago, Ill., 1982.

[604] G. Wulczyn, Solution to Problem B-256, *Fibonacci Quarterly*, 12 (1974), 221.

[605] G. Wulczyn, Problem H-210, *Fibonacci Quarterly*, 12 (1974), 401–402.

[606] G. Wulczyn, Problem B-342, *Fibonacci Quarterly*, 14 (1976), 470.

[607] G. Wulczyn, Problem B-355, *Fibonacci Quarterly*, 16 (1978), 186.

[608] G. Wulczyn, Problem B-384, *Fibonacci Quarterly*, 17 (1979), 283.

[609] G. Wulczyn, Problem B-443, *Fibonacci Quarterly*, 19 (1981), 87.

[610] G. Wulczyn, Problem B-464, *Fibonacci Quarterly*, 19 (1981), 466.

[611] M. Wunderlich, On the Non-existence of Fibonacci Squares, *Mathematics of Computation*, 17 (1963), 455.

[612] M. Wunderlich, Another Proof of the Infinite Primes Theorem, *American Mathematical Monthly*, 72 (1965), 305.

[613] M. Yodder, Solution to Problem B-165, *Fibonacci Quarterly*, 8 (1970), 112.

[614] D. Zeitlin, Power Identities for Sequences Defined by $W_{n+2} = dW_{n+1} - cW_n$, *Fibonacci Quarterly*, 3 (1965), 241–256.

[615] D. Zeitlin, Solution to Problem H-64, *Fibonacci Quarterly*, 5 (1967), 74–75.

[616] D. Zeitlin, Solution to Problem B-277, *Fibonacci Quarterly*, 13 (1975), 96.

[617] D. Zeitlin and F.D. Parker, Problem H-14, *Fibonacci Quarterly*, 1(2) (1963), 54.

[618] M.J. Zerger, The Golden State – Illinois, *Journal of Recreational Mathematics*, 24 (1992), 24–26.

[619] M.J. Zerger, Vol. 89 (and 11 to Centennial), *Mathematics Teacher*, 89 (1996), 3, 26.

[620] Z. Zhusheng, Problem 43.7, *Mathematical Spectrum*, 43 (2010–2011), 92.

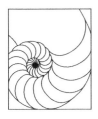

SOLUTIONS TO ODD-NUMBERED EXERCISES

EXERCISES 2

1. 2.

3. $F_{-n} = (-1)^{n+1} F_n$.

5. $L_{-n} = (-1)^n L_n$.

7. $20 + 19 + 15 + 5 + 1 = 60 = 17 + 13 + 11 + 9 + 7 + 3$.

9. We claim that $g_n = c^{F_n}$, where $n \geq 0$. Since $g_0 = c^{F_0}$ and $g_{n+1} = c^{F_n} \cdot c^{F_{n-1}} = c^{F_{n+1}}$, by PMI, the formula is true for $n \geq 0$.

11. $\displaystyle\sum_{i=1}^{n} L_i = L_{n+2} - 3$.

13. $\displaystyle\sum_{i=1}^{n} L_i^2 = L_n L_{n+1} - 2$.

15. $F_3 + F_5 = 2 + 5 = 7 = L_4; F_6 + F_8 = 8 + 21 = 29 = L_7$.

17. The proof follows by PMI.

19. We have $a_n = a_{n-1} + a_{n-2} + 1$, where $a_1 = 0 = a_2$. Letting $b_n = a_n + 1$, $b_n = b_{n-1} + b_{n-2}$, where $b_1 = 1 = b_2$. So $a_n = b_n - 1 = F_n - 1$, where $n \geq 1$.

21. (Brown) Suppose $F_h < F_i < F_j < F_k$ are in AP, so $F_i - F_h = F_k - F_j = d$ (say). Then $d = F_i - F_h < F_i$, whereas $d = F_k - F_j \geq F_k - F_{k-1} = F_{k-2} \geq F_i$, a contradiction.

23. (DeLeon) Since $F_n < x < F_{n+1}$ and $F_{n+1} < y < F_{n+2}$, $F_n + F_{n+1} < x + y < F_{n+1} + F_{n+2}$; that is, $F_{n+2} < x + y < F_{n+3}$, a contradiction.

Fibonacci and Lucas Numbers with Applications, Volume One, Second Edition. Thomas Koshy.
© 2018 John Wiley & Sons, Inc. Published 2018 by John Wiley & Sons, Inc.

25. $a_n = \dfrac{k^n + k - 2}{(k-1)k^n}$, $n \geq 0$.

27. $\dfrac{1}{k-1}$.

EXERCISES 3

1. Since $b_1 = 1$, $b_2 = 2$, and $b_n = b_{n-1} + b_{n-2}$, it follows that $b_n = F_{n+1}$, where $n \geq 1$.
3. Yes.
5. No.
7. $l_n = F_n$.
9. $2F_n - 1$.
11. 10, 15.
13. 14, 25.

EXERCISES 4

1. 00000, 10000, 01000, 00100, 10100, 00010, 10010, 01010, 00001, 10001, 01001, 00101, 10101.
3. $a_1 = 1, a_2 = 2, a_n = a_{n-1} + a_{n-2}, n \geq 3$.
5. $b_1 = 1, b_2 = 2, b_n = b_{n-1} + b_{n-2}, n \geq 3$.
7. 4, 7.
9. We have $x_1 = 1/F_1$ and $x_2 = 1/F_2$. By the iterative formula,
$$x_{n+1} = \frac{x_{n-1}x_n^2 - x_n x_{n-1}^2}{x_n^2 - x_{n-1}^2} = \frac{x_{n-1}x_n}{x_{n-1} + x_n} = \frac{1}{F_n + F_{n-1}} = \frac{1}{F_n}. \text{ So, by PMI,}$$
   the formula works for $n \geq 1$.
11. $a_1 = 2, a_2 = 3, a_n = a_{n-1} + a_{n-2}, n \geq 3$.
13. None; $\{\{1\}, \emptyset\}$; $\{\{1\}, \emptyset\}, \{\{2\}, \emptyset\}, \{\{1,2\}, \emptyset\}$.
15. $S_n = F_{t_n+2} - F_{t_{n-1}+2}$, where $t_k = k(k+1)/2$.

EXERCISES 5

1. The result is true when $n = 1$ and $n = 2$. Assume it works for k. Then
$$\sum_{i=1}^{k+1} F_{2i-1} = \sum_{i=1}^{k} F_{2i-1} + F_{2k+1} = F_{2k} + F_{2k+1} = F_{2k+2}. \text{ So by PMI, the result}$$
   works for $n \geq 1$.

3. $\displaystyle\sum_{i=1}^{n} L_i = \sum_{i=1}^{n}(L_{i+2} - L_{i+1}) = L_{n+2} - L_2 = L_{n+2} - 3.$

5. $\displaystyle\sum_{i=1}^{n} L_{2i} = \sum_{i=1}^{2n} L_i - \sum_{i=1}^{n} L_{2i-1} = (L_{2n+2} - 3) - (L_{2n} - 2) = (L_{2n+2} - L_{2n}) - 1$
$$= L_{2n+1} - 1.$$

7. The result is true when $n = 1$. Assume it is true for $k \geq 1$. Then
$$\sum_{i=1}^{k+1} L_i^2 = \sum_{i=1}^{k} L_i^2 + L_{k+1}^2 = (L_k L_{k+1} - 2) + L_{k+1}^2 = L_{k+1}L_{k+2} - 2.$$ Thus,
by PMI, the formula works for all $n \geq 1$.

9. $L_3 L_1 - L_2^2 = 4 \cdot 1 - 9 \neq (-1)^2$.

11. $v_1 = \alpha + \beta = 1; v_2 = \alpha^2 + \beta^2 = (\alpha + \beta)^2 - 2\alpha\beta = 3$; and $v_{n-1} + v_{n-2} = \alpha^{n-2}(1 + \alpha) + \beta^{n-2}(1 + \beta) = \alpha^{n-2}\alpha^2 + \beta^{n-2}\beta^2 = v_n$.

13. The result is true when $n = 1$. Assume it works for all positive
integers $\leq k$. Then $\sqrt{5}F_k = \sqrt{5}(\alpha + \beta)F_k = \sqrt{5}(\alpha F_k + +\beta F_k) = (\alpha^{k+1} - \beta^{k+1}) + (\alpha\beta)(\alpha^{k-1} - \beta^{k-1}) = (\alpha^{k+1} - \beta^{k+1}) - \sqrt{5}F_{k-1}$; that is,
$\sqrt{5}(F_k + F_{k-1}) = \alpha^{k+1} - \beta^{k+1}$. So $F_{k+1} = (\alpha^{k+1} - \beta^{k+1})/\sqrt{5}$. So the result
is true for all $n \geq 1$.

15. Clearly, $L_1 = F_1$. Conversely, let $L_n = F_n$. Then $\sqrt{5}(\alpha^n + \beta^n) = \alpha^n - \beta^n$.
This implies $\alpha^n = (\alpha/\beta)\beta^n$. So $\alpha^{n-1} - \beta^{n-1} = 0$; that is, $F_{n-1} = 0$, so $n = 1$.

17. $x^2 - (L_n + 2k)x + kL_n + k^2 + (-1)^n = 0$.

19. $F_{n-1}x^2 - (F_{n+1} - F_{n-1})x - F_{n+1} = (x + 1)(F_{n-1}x - F_{n+1}) = 0$; so $x = -1$,
F_{n+1}/F_{n-1}.

21. Let S denote the sum. Since $\alpha^i = \alpha F_i + F_{i-1}, \displaystyle\sum_{i=0}^{n} \alpha^i = \alpha\sum_{i=0}^{n} F_i + \sum_{i=0}^{n} F_{i-1}$. That
is, $(\alpha^{n+1} - 1)/(\alpha - 1) = \alpha S + 1 + 0 + (S - F_n)$; that is, $(\alpha F_{n+1} + F_n - 1)/(\alpha - 1) = (\alpha + 1)S + 1 - F_n$. This yields the desired formula.

23. $F_{n+5} = F_{n+4} + F_{n+3} = 2F_{n+3} + F_{n+2} = 3F_{n+2} + 2F_{n+1} = 5F_{n+1} + 3F_n$.

25. $5(F_{n-1}F_{n+1} - F_n^2) = (\alpha^{n-1} - \beta^{n-1})(\alpha^{n+1} - \beta^{n+1}) - (\alpha^n - \beta^n)^2 = 5(-1)^n$. This
gives Cassini's formula.

27. $\sqrt{5}F_{2n} = (\alpha^{2n} - \beta^{2n}) = (\alpha^n - \beta^n)(\alpha^n + \beta^n) = \sqrt{5}F_n L_n$. This yields the given
identity.

29. $F_{n+1}^2 - F_{n-1}^2 = (F_{n+1} - F_{n-1})(F_{n+1} + F_{n-1}) = F_n L_n = F_{2n}$.

31. $F_{n+2} - F_{n-2} = (F_{n+1} + F_n) - (F_n - F_{n-1}) = F_{n+1} + F_{n-1} = L_n$.

33. $F_{n+1}^2 - F_n^2 = (F_{n+1} + F_n)(F_{n+1} - F_n) = F_{n+2}F_{n-1}$.

35. $L_n^2 + L_{n+1}^2 = (\alpha^n + \beta^n)^2 + (\alpha^{n+1} + \beta^{n+1})^2 = \alpha^{2n} + \beta^{2n} + \alpha^{2n+2} + \beta^{2n+2} = (\alpha - \beta)(\alpha^{2n+1} - \beta^{2n+1}) = 5F_{2n+1}$.

37. $L_n^2 - 4(-1)^n = (\alpha^n + \beta^n)^2 - 4(\alpha\beta)^n = (\alpha^n - \beta^n)^2 = 5F_n^2$.

39. $L_{2n} + 2(-1)^n = \alpha^{2n} + \beta^{2n} + 2(\alpha\beta)^n = (\alpha^n + \beta^n)^2 = L_n^2$.

41. By Binet's formula, $L_{n+2} - L_{n-2} = (\alpha^{n+2} + \beta^{n+2}) - (\alpha^{n-2} + \beta^{n-2}) = (\alpha - \beta)(\alpha^n - \beta^n) = 5F_n$.

43. Since $\sqrt{5}(L_n + F_n) = 2(\alpha^{n+1} - \beta^{n+1})$ by Binet's formulas, the result now follows.

45. The result follows by the identities $F_{n+1}^2 + F_n^2 = F_{2n+1}$ and $F_{n+1}^2 - F_{n-1}^2 = F_{2n}$.

47. The inequality works when $n = 3$ and $n = 4$. Assume it is true for n and $n + 1$, where $n \geq 3$. Then $F_{n+2} + F_{n+3} < 2^n + 2^{n+1}$, so $F_{n+4} < 2^{n+1} + 2^{n+1}$; that is, $F_{n+4} < 2^{n+2}$. The result now follows by PMI.

49. Since $6\alpha^2 = 6(1 + \alpha) > 15 = 5L_2$ and $6\alpha^3 = 6(\alpha + \alpha^2) > 24 > 5L_3$, the result is true when $n = 2$ and $n = 3$. Assume it works for n and $n + 1$. Then $6\alpha^{n+2} = 6(\alpha^n + \alpha^{n+1}) > 5(L_n + L_{n+1}) = 5L_{n+2}$. The result now follows by PMI.

51. $L_{-n} = \alpha^{-n} + \beta^{-n} = (-\beta)^n + (-\alpha)^n = (-1)^n L_n$.

53. LHS $= \alpha^n(\alpha^n + \alpha^{-n}) = \alpha^n[\alpha^n + (-\beta)^n] =$ RHS.

55. LHS $= (\alpha^{2m+n} + \beta^{2m+n}) - (\alpha\beta)^m(\alpha^n + \beta^n) = (\alpha^m - \beta^m)(\alpha^{m+n} - \beta^{m+n}) = 5F_m F_{m+n} =$ RHS.

57. $\sqrt{5}$LHS $= (\alpha^{2m+n} - \beta^{2m+n}) + (\alpha\beta)^m(\alpha^n - \beta^n) = (\alpha^m + \beta^m)(\alpha^{m+n} - \beta^{m+n}) = \sqrt{5}L_m F_{m+n}$. The given result now follows.

59. $\sqrt{5}$LHS $= (\alpha^{m+n} - \beta^{m+n}) - (\alpha^{m-n} - \beta^{m-n}) = \alpha^m[\alpha^n - (-\beta)^n] - \beta^m[\beta^n - (-\alpha)^n]$. This equals $(\alpha^m - \beta^m)(\alpha^n + \beta^n)$ if n is odd, and $(\alpha^m + \beta^m)(\alpha^n - \beta^n)$ otherwise. This yields the given result.

61. $L_{3n} = \alpha^{3n} + \beta^{3n} = (\alpha^n + \beta^n)(\alpha^{2n} - \alpha^n\beta^n + \beta^{2n}) =$ RHS.

63. LHS $= F_n(F_n + F_{n-1}) - F_{n-1}(F_n - F_{n-1}) = F_n^2 + F_{n-1}^2 = F_{2n-1} =$ RHS.

65. LHS $= F_{n+1}(F_{n+3} + F_{n+1}) - F_{n+2}(F_{n+1} + F_{n-1}) = F_{n+1}(F_{n+2} + 2F_{n+1}) - F_{n+2}(F_{n+1} + F_{n-1}) = 2F_{n+1}^2 - (F_{n+1} + F_n)(F_{n+1} - F_n) = F_{n+1}^2 + F_n^2 = F_{2n+1} =$ RHS.

67. LHS $= F_{n+2}(F_{n+2} - F_n) - F_n(F_{n+2} - F_n) - F_{n+2}F_n = F_{n+1}(F_{n+2} - F_n) - F_{n+2}F_n = F_{n+1}^2 - F_{n+2}F_n = (-1)^n =$ RHS.

69. $L_n^3 = (L_{n+1} - L_{n-1})^3 = L_{n+1}^3 - 3L_{n+1}L_{n-1}(L_{n+1} - L_{n-1}) - L_{n-1}^3$. So $L_n^3 + 2L_{n-1}^3 + 6L_{n-1}L_{n+1}^2 = (L_{n+1} + L_{n-1})^3 = (5F_n)^3 = 125F_n^3$.

71. This follows by Exercise 5.70 and the identity $L_{n+2}^2 = L_{n+1}^3 + L_n^3 + 3L_{n+2}L_{n+1}L_n$.

73. LHS $= (F_{n+1}^3 - F_{n-1}^3) + F_n^3 = F_n[F_{n+1}(F_{n+1} + F_{n-1}) + F_{n-1}^2] + F_n^3 = F_n(F_{n+1}L_n + F_{n-1}^2) + F_n^3 = F_{n+1}F_{2n} + F_n(F_{n-1}^2 + F_n^2) = F_{n+1}F_{2n} + F_nF_{2n-1} = F_{3n} =$ RHS.

75. LHS $= \alpha^m(\alpha^n + \beta^n) + \alpha^{m-1}(\alpha^{n-1} + \beta^{n-1}) = \alpha^{m+n} + \alpha^{m+n-2} = \alpha^n(\alpha^m + \beta^m) + \alpha^{n-1}(\alpha^{m-1} + \beta^{m-1}) =$ RHS.

77. $5(\text{LHS}) = (\alpha^m - \beta^m)(\alpha^{n+k} - \beta^{n+k}) - (\alpha^k - \beta^k)(\alpha^{n+m} - \beta^{n+m}) = (\alpha\beta)^k(\alpha^{m-k} - \beta^{m-k})(\alpha^n - \beta^n)$. This gives the desired identity.

79. By Cassini's formula, $F_{2n}F_{2n+4} = F_{2n+2}^2 - 1 = (F_{2n+2} - 1)(F_{2n+2} + 1)$. So $F_{2n}F_{2n+2}F_{2n+4} = (F_{2n+2} - 1)F_{2n+2}(F_{2n+2} + 1)$.

81. Follows by the identities in Exercises 5.73 and 5.74.

83. The formula works when $n = 1$ and $n = 2$. Suppose it works for $n - 1$. Using Theorem 5.4, $\displaystyle\sum_{k=1}^{2n-1}(2n - k)F_k^2 = (F_{2n-2} + F_{2n-1})^2 = F_{2n}^2$. Thus, by PMI, the formula is true for all $n \geq 1$.

85. (Khan) Let n be odd. Then sum $= F_2F_2 + F_2(F_4 - F_2) + (F_4 - F_2)F_4 + \cdots + F_{n-1}(F_{n+1} - F_{n-1}) + (F_{n+1} - F_{n-1})F_{n+1} = F_{n+1}^2$. On the other hand, let n be even. Then sum $= F_1(F_3 - F_1) + (F_3 - F_1)F_3 + F_3(F_5 - F_3) + \cdots + F_{n-1}(F_{n+1} - F_{n-1})F_{n+1} = F_{n+1}^2 - 1$.

87. (Carlitz) $x^{2n} - L_nx^n + (-1)^n = x^{2n} - (\alpha^n + \beta^n)x^n + (\alpha\beta)^n = (x^n - \alpha^n)(x^n - \beta^n)$. Since $(x - \alpha)|(x^n - \alpha^n), (x - \beta)|(x^n - \beta^n)$, and $x^2 - x - 1 = (x - \alpha)(x - \beta)$, the desired result follows.

89. (Wulczyn) LHS $= (\alpha^{2n} + \beta^{2n}) + (\alpha^2 + \beta^2)(\alpha\beta)^{n-1} = (\alpha^{n+1} + \beta^{n+1})(\alpha^{n-1} + \beta^{n-1}) = L_{n+1}L_{n-1}$.

91. The result follows by Exercise 5.69.

93. (Zeitlin) Adding the two Binet's formulas, $2\alpha^n = L_n + \sqrt{5}F_n$; that is, $(1 + \sqrt{5})^n = 2^{n-1}L_n + \sqrt{5}(2^{n-1}F_n)$. So $a_n = 2^{n-1}L_n$ and $b_n = 2^{n-1}F_n$. Thus $2^{n-1}|a_n$ and $2^{n-1}|b_n$.

95. (Bruckman) Since $F_{2n+1}F_{2n-1} - F_{2n}^2 = 1$, $2F_{2n+1}F_{2n-1} - 1 = 2F_{2n}^2 + 1 = F_{2n}^2 + F_{2n+1}F_{2n-1}$. So if $2F_{2n+1}F_{2n-1} - 1$ is a prime, then so are $2F_{2n}^2 + 1$ and $F_{2n}^2 + F_{2n+1}F_{2n-1}$.

97. (Homer) LHS $= \displaystyle\sum_{k=1}^{n}g_{k+2}F_k + \sum_{k=1}^{n}g_{k+1}F_k - \sum_{k=1}^{n}g_kF_k = \sum_{k=3}^{n}(F_{k-2} + F_{k-1} - F_k)g_k + g_{n+2}F_n + g_{n+1}(F_n + F_{n-1}) + g_2F_1 - g_1F_1 - g_2F_2 = g_{n+2}F_n + g_{n+1}F_{n+1} - g_1$.

99. (Yodder) The formula works when $n = 1$ and $n = 2$. Assume it is true for all positive integers $\leq n$, where $n \geq 2$. Let n be odd. Then $f[(7 \cdot 2^n + 1)/3] = f[(7 \cdot 2^{n-1} - 1)/3] + f[(7 \cdot 2^{n-1} + 2)/3] = f[(7 \cdot 2^{n-1} - 1)/3] + f[(7 \cdot 2^{n-2} + 1)/3] = L_n + L_{n-1} = L_{n+1}$. Similarly, when n is even, $f[(7 \cdot 2^n - 1)/3] = L_{n+1}$. Thus, by the strong version of PMI, the result is true for all $n \geq 1$.

101. $S_n = \displaystyle\sum_{i=1}^{n(n+1)/2}F_{2i-1} - \sum_{i=1}^{n(n-1)/2}F_{2i-1} = F_{n(n+1)} - F_{n(n-1)}$.

103. Area $= \sqrt{3}(F_{n+1} + F_{n-1})F_n/4 = \sqrt{3}L_nF_n/4 = \sqrt{3}F_{2n}/4$.

105. By the ratio test, the series converges. Let l be its limit. Then $\dfrac{l}{2} = \displaystyle\sum_{n=1}^{\infty} \dfrac{F_n}{2^n}$.

But

$$l = \sum_{n=0}^{\infty} \frac{F_{n+3} - F_{n+2}}{2^n} = 8 \sum_{n=0}^{\infty} \frac{F_{n+3}}{2^{n+3}} - 4 \sum_{n=0}^{\infty} \frac{F_{n+2}}{2^{n+2}}$$

$$= 8 \left(\sum_{n=1}^{\infty} \frac{F_n}{2^n} - \frac{F_1}{2} - \frac{F_2}{4} \right) - 4 \left(\sum_{n=1}^{\infty} \frac{F_n}{2^n} - \frac{F_1}{2} \right)$$

$$= 8 \left(\frac{l}{2} - \frac{3}{4} \right) - 4 \left(l - \frac{1}{2} \right).$$

This yields $l = 4$. (We can obtain the same result using the power series for $\{F_{n+1}\}$.)

107. Let P_n denote the product with upper limit n. Then

$$\lim_{n\to\infty} P_n = \lim_{n\to\infty} \prod_{k=1}^{n} \frac{(L_{2k} + L_{2k-1})L_{2k+2}}{(L_{2k+2} + L_{2k+1})L_{2k}} = \lim_{n\to\infty} \prod_{k=1}^{n} \frac{L_{2k+1}L_{2k+2}}{L_{2k+3}L_{2k}}$$

$$= \frac{L_3}{L_2} \cdot \lim_{n\to\infty} \frac{L_{2n+2}}{L_{2n+3}} = \frac{4}{3\alpha}.$$

EXERCISES 6

1. RHS $= 4F_n F_{n-1} + F_{n-2}^2 = 4F_n F_{n-1} + (F_n - F_{n-1})^2 = (F_n + F_{n-1})^2 = F_{n+1}^2 =$ LHS.

3. RHS $= 4F_{n-1}^2 + 4F_{n-1}(F_n - F_{n-1}) + (F_n - F_{n-1})^2 = (F_n + F_{n-1})^2 = F_{n+1}^2 =$ LHS.

5. RHS $= 8F_n F_{n-1} + (2F_{n-1} - F_n)^2 = 4F_n F_{n-1} + 4F_{n-1}^2 + F_n^2 = 4F_{n-1}(F_n + F_{n-1}) + F_n^2 = 4F_{n+1}F_{n-1} + F_n^2 = 4[F_n^2 + (-1)^n] + F_n^2 = 5F_n^2 + 4(-1)^n = L_n^2 =$ LHS.

7. RHS $= 4F_n^2 + 4F_{n-1}^2 - 4F_{n-2}^2 + (F_{n-1} - F_{n-2})^2 = 4F_n^2 + 5F_{n-1}^2 - 3F_{n-2}^2 - 2F_{n-1}F_{n-2} = 4F_n^2 + 5F_{n-1}^2 - 3(F_n - F_{n-1})^2 - 2F_{n-1}(F_n - F_{n-1}) = F_n^2 + 4F_{n-1}^2 + 4F_n F_{n-1} = (F_n + 2F_{n-1})^2 = L_n^2 =$ LHS.

9. This follows by the identity $(x + y)^3 = x^3 + y^3 + 3xy(x + y)$.

EXERCISES 7

1. $A_n = 2F_{n-2} + 3F_{n-1} = 2F_n + F_{n-1} = F_n + F_{n+1} = F_{n+2}$.

3. $-\sqrt{5}$.

5. $\lim\limits_{n\to\infty} \dfrac{G_n}{L_n} = \lim\limits_{n\to\infty} \dfrac{c\alpha^n - d\beta^n}{\alpha^n + \beta^n} = c.$

7. $\text{Sum} = \sum\limits_{i=1}^{n}(G_{i+2} - G_{i+1}) = G_{n+2} - G_2 = G_{n+2} - b.$

9. $\text{Sum} = \sum\limits_{i=1}^{2n} G_i - \sum\limits_{i=1}^{n} G_{2i-1} = (G_{2n+2} - b) - (G_{2n} + a - b) = G_{2n+2} - G_{2n} - a = G_{2n+1} - a.$

11. Since $\sum\limits_{i=1}^{10} F_{i+j} = 11F_{j+7}$, we have

$$\sum_{i=1}^{10} G_{k+i} = \sum_{i=1}^{10}(aF_{k+i-2} + bF_{k+i-1}) = a\sum_{i=1}^{10} F_{k+i-2} + b\sum_{i=1}^{10} F_{k+i-1}$$

$$= a(11K_{k-2+7}) + b(11F_{k-1+7}) = 11(aF_{k+5} + bF_{k+6}) = 11G_{k+7}.$$

13. Let $S_i = \sum\limits_{j=1}^{i} G_j = G_{i+2} - b.$ Then

$$\sum_{i=1}^{n-1} S_i = \sum_{i=1}^{n-1} G_{i+2} - (n-1)b = S_{n+1} - a - nb = G_{n+3} - a - (n+1)b;$$

$$\sum_{i=1}^{n-1} iG_i = nS_n - \sum_{i=1}^{n-1} S_i = n(G_{n+2} - b) - [G_{n+3} - a - (n+1)b]$$

$$= nG_{n+2} - G_{n+3} + a + b.$$

15. (Milsom) The formula works when $n = 1$. Assume it works for $k \geq 1$. Then

$$\sum_{i=1}^{k+1} F_i G_{3i} = \sum_{i=1}^{k} F_i G_{3i} + F_{k+1}G_{3k+3} = F_{k+1}(F_k G_{2k+1} + G_{3k+3}).$$

Since $G_{m+n} = F_{n-1}G_m + F_n G_{m+1}$, this implies

$$\sum_{i=1}^{k+1} F_i G_{3i} = F_{k+1}[F_k(G_{2k+1} + G_{2k+2}) + F_{k+1}G_{2k+3}]$$

$$= F_{k+1}[(F_{k+2} - F_{k+1})(G_{2k+1} + G_{2k+2}) + F_{k+1}G_{2k+3}]$$

$$= F_{k+1}F_{k+2}(G_{2k+1} + G_{2k+2}) = F_{k+1}F_{k+2}G_{2k+3}.$$

Thus, by PMI, the result is true for all $n \geq 1$.

17. (Peck) We have $(x + y + z)^n = \sum\limits_{i,j,k\geq 0} \dbinom{n}{i,j,k} x^i y^j z^k$, where $\dbinom{n}{i,j,k} = \dfrac{n!}{i!j!k!}$ and $i + j + k = n$. Let $x = 1$, $y = \alpha$, and $z = -\alpha^2$. Then $x + y + z = 0$. So

$$\sum_{i,j,k\geq0} \binom{n}{i,j,k} \alpha^j(-\alpha^2)^k = 0; \text{ that is,}$$

$$\sum_{i,j,k\geq0} \binom{n}{i,j,k}(-1)^k\alpha^{j+2k} = 0. \tag{1}$$

$$\text{Similarly, } \sum_{i,j,k\geq0} \binom{n}{i,j,k}(-1)^k\beta^{j+2k} = 0. \tag{2}$$

Now multiply (1) by c, (2) by d, and then subtract. Dividing the result by $\alpha - \beta$ gives the desired result.

19. LHS $= (c\alpha^{n+k} - d\beta^{n+k})(c\alpha^{n-k} - d\beta^{n-k}) = c^2\alpha^{2n} + d^2\beta^{2n} - cd(\alpha\beta)^{n-k}(\alpha^{2k} + \beta^{2k}) = 5(\alpha^{2n} + \beta^{2n}) - \mu(-1)^{n-k}L_{2k} = 5L_{2n} - \mu(-1)^{n-k}L_{2k}$.

21. $5(\text{LHS}) = (c\alpha^n - d\beta^n)^2 + (c\alpha^{n-1} - d\beta^{n-1})^2 = c^2\alpha^{2n-2}(1 + \alpha^2) + d^2\beta^{2n-2}(1 + \beta^2) = \sqrt{5}(c^2\alpha^{2n-1} - d^2\beta^{2n-1})$. Similarly, $\sqrt{5}(\text{RHS}) = c^2\alpha^{2n-1} - d^2\beta^{2n-1}$. Thus LHS = RHS.

23. Follows by changing n to $m+n$ in Exercise 7.22.

25. $5(\text{LHS}) = (-1)^{n+k-1}\mu L_{m-n+2k} + (-1)^n\mu L_{m-n} = 5(\text{RHS})$; so LHS = RHS.

27. $5(\text{RHS}) = 5L_{2n-6} - 2\mu(-1)^{n-3} + 20L_{2n-3} - 4\mu(-1)^{n-2} = 5L_{2n-6} + 20L_{2n-3} - 2\mu(-1)^n = 5L_{2n} - 2\mu(-1)^n = 5(\text{LHS})$; so LHS = RHS.

29. $5(\text{LHS}) = 5L_{2n} + 5L_{2n+6} = 10L_{2n+2} + 10L_{2n+4} = 5(\text{RHS})$; so LHS = RHS.

31. Since $(x - y)^2 + (2xy)^2 = (x + y)^2$, the result follows with $x = G_m$ and $y = G_n$.

33. This follows by Candido's identity.

35. LHS $= 5(aF_{n+r-2} + bF_{n+r-1})^2 + 5(aF_{n-r-2} + bF_{n-r-1})^2 = 5a^2(F_{n-2+r}^2 + F_{n-2-r}^2) + b^2(F_{n-1+r}^2 + F_{n-1-r}^2) + 2ab(F_{n-2+r}F_{n-1+r} + F_{n-2-r}F_{n-1+r}) = a^2[L_{2n-4}L_{2r} - 4(-1)^{n-2+r}] + b^2[L_{2n-2}L_{2r} - 4(-1)^{n-1+r}] + 2ab[L_{2n-3}L_{2r} - 2(-1)^{n-2+r}] = (a^2L_{2n-4} + b^2L_{2n-2} + 2abL_{2n-3})L_{2r} - 4a^2(-1)^{n+r} + 4b^2(-1)^{n+r} - 4ab(-1)^{n+r} = (a^2L_{2n-4} + 2abL_{2n-3} + b^2L_{2n-2})L_{2r} - 4\mu(-1)^{n+r} = \text{RHS}$.

37. Let $x = G_{n+2}$ and $y = G_{n+1}$. Then $x^2 - y^2 = (x + y)(x - y) = G_{n+3}G_n$, $2xy = 2G_{n+1}G_{n+2}$, and $x^2 + y^2 = G_{n+2}^2 + G_{n+1}^2$. Since $(x^2 - y^2)^2 + 4x^2y^2 = (x^2 + y^2)^2$, the result follows.

39. Let $x = G_n$ and $y = G_{n-1}$. Then $x^2 + xy + y^2 = G_n^2 + G_nG_{n-1} + G_{n-1}^2 = G_{n-1}(G_{n-1} + G_n) + G_n^2 = G_{n-1}G_{n+1} + G_n^2 = 2G_n^2 + \mu(-1)^n$. The given result now follows by the identity $(x + y)^5 - x^5 - y^5 = 5xy(x + y)(x^2 + xy + y^2)$.

41. Let $x = G_{n-1}$ and $y = G_n$. Then $x^2 + xy + y^2 = 2G_n^2 + \mu(-1)^n$. The result now follows from the identity $x^4 + y^4 + (x + y)^4 = 2(x^2 + xy + y^2)^2$.

43. Let $x = G_{n-1}$ and $y = G_n$. Then $x^2 + xy + y^2 = 2G_n^2 + \mu(-1)^n$ and $x^4 + 3x^3y + 4x^2y^2 + y^4 = G_{n-1}^4 + 3G_{n-1}^3G_n + 4G_{n-1}^2G_n^2 + 3G_{n-1}G_n^3 + G_n^4 = G_{n-1}^4 + G_n^4 + 3G_{n-1}G_n(G_{n-1}^2 + G_n^2) + 4G_{n-1}^2G_n^2 = G_{n-1}^4 + G_n^4 + 3G_{n-1}G_n[(3a - b)G_{2n-1} - \mu F_{2n-1}] + 4G_{n-1}^2G_n^2$. The result now follows from the identity $x^8 + y^8 + (x + y)^8 + 2(x^2 + xy + y^2)^4 + 8x^2y^2(x^4 + 3x^3y + 4x^2y^2 + 3xy^3 + y^4)$.

45. (Lind) $\sum_{i=1}^{n} iG_i = nG_{n+2} - G_{n+3} + a + b$ and $\sum_{i=1}^{n} G_i = G_{n+2} - b$. So

$$A_n = \frac{nG_{n+2} - G_{n+3} + a + b}{G_{n+2} - b}. \text{ Then } A_{n+1} - A_n = \frac{(n+1)G_{n+3} - G_{n+4} + a + b}{G_{n+3} - b}$$

$$-\frac{nG_{n+2} - G_{n+3} + a + b}{G_{n+2} - b}. \text{ Since } \lim_{n \to \infty} \frac{G_{n+k}}{G_n} = \alpha^k, \text{ it follows that } \lim_{n \to \infty}(A_{n+1} - A_n)$$

$$= \frac{(n+1) - \alpha + 0}{1 - 0} - \frac{n - \alpha + 0}{1 - 0} = 1.$$

47. (Bruckman) Since $A_n = C_{n+2} - C_{n+1} - C_n$, we have

$$A_n = H_{n+2}K_0 + H_{n+1}K_1 - H_{n+1}K_0$$

$$+ \sum_{i=0}^{n} H_i(K_{n-i+2} - K_{n-i+1} - K_{n-i})$$

$$= H_{n+2}K_0 + H_{n+1}K_1 - H_{n+1}K_0 = H_{n+1}K_1 + H_nK_0 \qquad (3)$$

$$A_{n+1}A_{n-1} - A_n^2 = (H_{n+2}K_1 + H_{n+1}K_0)(H_nK_1 + H_{n-1}K_0)$$

$$- (H_{n+1}K_1 + H_nK_0)^2$$

$$= (H_{n+2}H_n - H_{n+1}^2)K_1^2 + (H_{n+2}H_{n-1} - H_{n+1}H_n)K_0K_1 +$$

$$(H_{n+1}H_{n-1} - H_n^2)K_0^2$$

But $H_{n+2}H_n - H_{n+1}^2 = (-1)^n\mu$. So $A_{n+1}A_{n-1} - A_n^2 = (-1)^n\mu(K_0^2 + K_0K_1 - K_1^2) = (-1)^n\mu(K_2K_0 - K_1^2) = (-1)^n\mu\nu$. So the characteristic of the sequence $\{A_n\}$ is $\mu\nu$.

It remains to show that $\{A_n\}$ is gibonacci. From (3), $A_{n-1} + A_{n-2} = (H_nK_1 + H_{n-1}K_0) + (H_{n-1}K_1 + H_{n-2}K_0) = H_{n+1}K_1 + H_nK_0 = A_n$. Thus $\{A_n\}$ is a gibonacci sequence.

49. Follows from Exercise 7.48 with $p = L_n, q = L_{n+1}, r = L_{n+2},$ and $s = L_{n+3}$.

51. Clearly, $x_1 = 1/G_1$ and $x_2 = 1/G_2$. Assume the formula works for $n - 1$
and n. Then $x_{n+1} = \dfrac{x_{n_1}x_n}{x_{n-1} + x_n} = \dfrac{1/G_{n-1} \cdot 1/G_n}{1/G_{n-1} + 1/G_n} = \dfrac{1}{G_{n+1}}.$ Thus, by PMI,
it works for all $n \geq 1$.

53. $G_{m+n} + G_{m-n} = G_m[F_{n+1} + (-1)^nF_{n-1}] + G_{m-1}F_n[1 - (-1)^n].$ This yields the given result.

55. This follows by Exercises 7.55 and 7.56.

57. (Swamy) $\sum_{k=1}^{n} a_{2k-1} = \sum_{k=1}^{2n} a_k - \sum_{k=1}^{n} a_k = (a_{4n+1} - a) - (a_{2n+1} - a) = a_{4n+1} - a_{2n+1}.$

51. $G_{4m+1} + a = G_{4m+1} + G_1 = G_{(2m+1)+2m} + G_{(2m+1)-2m} = G_{2m+1}L_{2m},$ by Exercise 7.55.

61. $G_{4m+3} + a = G_{4m+3} + G_1 = G_{(2m+2)+(2m+1)} + G_{(2m+2)-(2m+1)} = (G_{2m+3} + G_{2m+1})F_{2m+1},$ by Exercise 7.55.

63. By Exercise 7.54, $G_{4m+1} - a = G_{4m+1} - G_1 = G_{(2m+1)+2m} - G_{(2m+1)-2m} =$
    $(G_{2m+2} + G_{2m})F_{2m}$.

65. By Exercise 7.54, $G_{4m+3} - a = G_{4m+3} - G_1 = G_{(2m+2)+(2m+1)} - G_{(2m+2)-(2m+1)}$
    $= G_{2m+2}L_{2m+1}$.

67. $(G_{4m+1} - a, G_{4m+2} - b) = ((G_{2m+2} + G_{2m})F_{2m}, (G_{2m+3} + G_{2m+1})F_{2m}) = F_{2m}$.

EXERCISES 8

1. 28,657.

3. 28,657.

5. 144.

7. 610.

9. 1364.

11. 47.

13. 123.

15. 521.

17. 4181.

19. 4181.

21. 1364.

23. 4181.

25. $L_{n+1} = \alpha L_n + 1/2 + \theta$, where $0 \le \theta < 1$, so $\lim\limits_{n\to\infty} \dfrac{L_{n+1}}{L_n} = \alpha$.

27. $L_{n+1} = \dfrac{L_n + 1 + \sqrt{5L_n^2 - 2L_n + 1}}{2} + \theta$, where $0 \le \theta < 1$. So $\lim\limits_{n\to\infty} \dfrac{L_{n+1}}{L_n} = \alpha$.

29. $\lfloor (17711 + 1/2)/\alpha \rfloor = 10{,}946$.

31. 24,476.

33. $\lim\limits_{n\to\infty} \dfrac{U_n}{F_{n+1}} = \lim\limits_{n\to\infty} \sqrt{\left(\dfrac{F_n}{F_{n+1}}\right)^2 + \left(\dfrac{F_{n-2}}{F_{n-1}} \cdot \dfrac{F_{n-1}}{F_n} \cdot \dfrac{F_n}{F_{n+1}}\right)^2} = \sqrt{\dfrac{1}{\alpha^2} + \dfrac{1}{\alpha^6}} =$
    $\dfrac{\sqrt{\alpha^4 + 1}}{\alpha^2} = \dfrac{\sqrt{3}\alpha}{\alpha^2} = -\sqrt{3}\beta$.

35. Follows by the Pythagorean theorem.

37. $\lim\limits_{n\to\infty} \dfrac{x_n}{y_n} = \lim\limits_{n\to\infty} \sqrt{\dfrac{G_n^2 + G_{n-2}^2}{G_{n-1}^2 + G_{n-3}^2}} = \lim\limits_{n\to\infty} \dfrac{G_n}{G_{n-1}} \sqrt{\dfrac{1 + (G_{n-2}/G_n)^2}{1 + (G_{n-3}/G_{n-1})^2}}$
    $= \alpha\sqrt{\dfrac{1 + 1/\alpha^2}{1 + 1/\alpha^2}} = \alpha$.

EXERCISES 9

1. 8.
3. 4.
5. 8.
7. 4.
9. $8 = 42 \cdot 1024 - 43 \cdot 1000.$
11. $4 = (-85) \cdot 2076 + 164 \cdot 1076.$
13. $8 = (-71) \cdot 1976 + 79 \cdot 1776.$
15. $4 = (-97) \cdot 3076 + 151 \cdot 1976.$
17. By the division algorithm, $a = bq + r$, where q is an integer and $0 \le r < b$. Since $d' = (b, r), d'|b$ and $d'|r$. Therefore, $d'|a$. Thus $d'|a$ and $d'|b$. So $d'|d$.
19. We have

$$L_{n+1} = 1 \cdot L_n \quad + L_{n-1}$$
$$L_n = 1 \cdot L_{n-1} + L_{n-2}$$
$$\vdots$$
$$4 = 1 \cdot 3 \quad + \boxed{1}$$
$$3 = 3 \cdot 1 \quad + 0.$$

So $L_{n+1}L_n = \sum_{i=2}^{n} L_i^2 + 3 \cdot 1^2 = \sum_{i=1}^{n} L_i^2 + 3 - 1.$ Thus $\sum_{i=1}^{n} L_i^2 = L_{n+1}L_n - 2.$

EXERCISES 10

1. $F_7 = 13, F_{21} = 10,946,$ and $13|10946.$
3. $(F_{12}, F_{18}) = (144, 2584) = 8 = F_6 = F_{(12,18)}.$
5. $(F_{144}, F_{1925}) = F_{(144,1925)} = F_1 = 1.$
7. $18|46368,$ so $L_6|F_{24}.$
9. $(F_{144}, F_{440}) = F_{(144,440)} = F_8 = 21.$
11. Since $\sqrt{5}F_{mn} = \alpha^{mn} - \beta^{mn} = (\alpha^m - \beta^m)[\alpha^{(n-1)m} + \alpha^{(n-2)m}\beta^m - \alpha^{(n-2)m}\beta^{2m} + \cdots + \beta^{(n-1)m}], F_m|F_{mn}.$
13. (Leveque) Suppose $(F_n, F_{n+1}) = d > 1$. Then, by the WOP, there is such a least positive integer m. Since $(F_1, F_2) = 1, m > 1$. If $(F_m, F_{m+1}) = d$, then $(F_{m-1}, F_m) = d$. This contradicts the choice of m. So $(F_n, F_{n+1}) = 1$ for all $n \ge 1$.
15. $5|10,$ but $L_5 \nmid L_{10}.$
17. $F_4 F_8 \nmid F_{32}.$

19. $[F_8, F_{12}] = [21, 144] = 1008 \neq F_{24} = F_{[8,12]}$.

21. $(F_n, L_n) = 1$ or 2.

23. 5.

25. 11.

27. 5.

29. 11.

31. $37 | F_{19}$.

33. $F_3 | F_{3n}$; that is, $2 | F_{3n}$.

35. Suppose $3 | F_n$; that is, $F_4 | F_n$. So $4 | n$. Conversely, let $4 | n$. Then $F_4 | F_n$; that is, $3 | F_n$.

37. $5 | n$ if and only if $F_5 | F_n$, that is, if and only if $5 | F_n$.

39. $L_{3n} = \alpha^{3n} + \beta^{3n} = (1 + 2\alpha)^n + (1 + 2\beta)^n = \sum_{i=0}^{n} \binom{n}{i} [(2\alpha)^i + (2\beta)^i]$

$= \sum_{i=0}^{n} \binom{n}{i} 2^i L_i$. Since $L_0 = 2$, it follows that $2 | L_{3n}$.

41. Let $(F_n, L_n) = 2$. Then $2 | F_n$, so $3 | n$. Conversely, suppose $3 | n$. Let $n = 3m$. Then $2 | F_{3m}$ and $2 | L_{3m}$. So $2 | (F_{3m}, L_{3m})$. Thus $(F_n, L_n) = (F_{3m}, L_{3m}) = 2$.

43. Since $L_{i-j} = (-1)^j (F_{i+1} L_j - F_i L_{j+1})$, $L_{(2k-1)n} = L_{2kn-n} = (-1)^n (F_{2kn+1} L_n - F_{2kn} L_{n+1})$. Since $F_{2n} | F_{2kn}$, $F_n L_n | F_{2kn}$. So $L_n |$ RHS; that is, $L_n | L_{(2k-1)n}$.

45. The result is true when $n = 1$. Assume it is true for all positive integers $n \leq k$. Since $F_{m(k+1)} = F_{mk+m} = F_{mk-1} F_m + F_{mk} F_{m+1}$, it follows that $F_m | F_{m(k+1)}$. So, by the strong version of PMI, the result holds for all $n \geq 1$.

47. LHS $= (21 + 34) F_7 = 715$, where $n = 8$. RHS $= [21, 34] + (-1)^8 (21, 34) = 714 + 1 = 715 =$ LHS.

49. (Freeman) The identity is true when $n = 2$. Assume it is true for an arbitrary integer $k \geq 2$. Then

$$
\begin{aligned}
F_{2k+1} &= F_{2k} + F_{2k-1} = F_k L_k + F_{k+1} L_{k+2} - L_k L_{k+1} \\
&= F_k L_k + (F_{k+2} - F_k) L_{k+2} - L_k L_{k+1} \\
&= F_k L_k + F_{k+2}(L_{k+3} - L_{k+1}) - F_k L_{k+2} - L_k L_{k+1} \\
&= F_{k+2} L_{k+3} - F_{k+2} L_{k+1} - F_k L_{k+1} - (L_{k+2} - L_{k+1}) L_{k+1} \\
&= F_{k+2} L_{k+3} - L_{k+1} L_{k+2} + L_{k+1}(L_{k+1} - F_{k+2} - F_k) \\
&= F_{k+2} L_{k+3} - L_{k+1} L_{k+2} + L_{k+1} \cdot 0 \\
&= F_{k+2} L_{k+3} - L_{k+1} L_{k+2}.
\end{aligned}
$$

Thus, by PMI, the result follows.

51. $n = 5$, so LHS $= (11 + 18) F_6 = 232 = [11, 18] + (11, 18) F_9 =$ RHS.

53. LHS $= (72 + 116) F_7 = 2444 = [72, 116] + (72, 116) F_{11} =$ RHS.

55. Follows since $(F_m, F_n) = F_{(m,n)}$ and $(a, b) = (a, a + b) = (b, a + b)$.

57. (Carlitz) Suppose $F_k | L_n$. Let $n = mk + r$, where $0 \le r < k$. Since
$\alpha^n + \beta^n = \alpha^r(\alpha^{mk} - \beta^{mk}) + \beta^{mk}(\alpha^r + \beta^r)$, $F_k | \beta^{mk} L_r$; so $F_k | L_r$. Since
$L_r = F_{r-1} + F_{r+1}$, it follows that $L_r < F_{r+2}$ for $r > 2$. Hence we need only
consider the case $F_{r+1} | L_r$. Then this implies $F_{r+1} | F_{r-1}$, which is impossible
for $r \ge 2$. Thus $F_k \nmid L_n$ for $k > 4$.

59. $F_{4n} - 1 = F_{4n} - F_2 = F_{(2n+1)+(2n-1)} - F_{(2n+1)-(2n-1)} = F_{2n+1} L_{2n-1}$.

61. $F_{4n+2} + 1 = F_{4n+2} + F_2 = F_{(2n+2)+2n} - F_{(2n+2)-2n} = F_{2n+2} L_{2n}$.

63. $F_{4n+3} + 1 = F_{4n+3} + F_1 = F_{(2n+2)+(2n+1)} - F_{(2n+2)-(2n+1)} = F_{2n+1} L_{2n+2}$.

65. LHS $= (F_{2n} L_{2n+1}, F_{2n+2} L_{2n+1}) = L_{2n+1}(F_{2n}, F_{2n+2}) = L_{2n+1} = $ RHS.

67. LHS $= (F_{4n} - 1, F_{4n+1} + 1) = (F_{2n+1} L_{2n-1}, F_{2n+1} L_{2n}) = F_{2n+1}(L_{2n-1}, L_{2n}) = F_{2n+1} = $ RHS.

69. LHS $= (F_{4n+2} - 1, F_{4n+3} + 1) = (F_{2n} L_{2n+2}, F_{2n+1} L_{2n+2}) = L_{2n+2}(F_{2n}, F_{2n+1}) = L_{2n+2} = $ RHS.

71. Since $g_0 = 0, g_1 = 12$, and $g_{n+2} = 7g_{n+1} - g_n$, the proof follows by the
strong version of PMI.

73. (Stanley) $F_{kn+2r} = F_{2r-1} F_{kn} + F_{2r} F_{kn+1}$ and $F_{kn-2r} = F_{2r-1} F_{kn} - F_{2r} F_{kn-1}$.
Therefore,

$$F_{kn-2r} + F_{kn} + F_{kn+2r} = (2F_{2r-1} + 1)F_{kn} + (F_{kn+1} - F_{kn-1})F_{2r}$$
$$= (2F_{2r-1} + 1)F_{kn} + F_{kn} F_{2r} = (L_{2r} + 1)F_{kn}.$$

75. (Lord) When $k = 1, h = 5$ and $5 | F_5$. So the statement is true when $k = 1$.
Assume it is true for k. Letting $x = \alpha^h$ and $y = \beta^h$ in the identity $x^5 - y^5 = (x - y)(x^4 + x^3 y + x^2 y^2 + xy^3 + y^4)$, we get $F_{5h} = F_h(L_{4h} - L_{2h} + 1)$.
But $L_{4h} - L_{2h} + 1 = (5F_{2h}^2 + 2) - (5h^2 - 2) + 1 \equiv 0 \pmod 5$. So $F_{5h} \equiv 0 \pmod{5h}$. Thus the statement is true for all $k \ge 1$.

EXERCISES 11

1. $F_{11} = \displaystyle\sum_{i=0}^{2} \binom{10 - i}{i} = 89$.

3. $\displaystyle\sum_{i=0}^{n} \binom{n}{i} L_i = \sum_{i=0}^{n} \binom{n}{i} \alpha^i + \sum_{i=0}^{n} \binom{n}{i} \beta^i = (1 + \alpha)^n + (1 + \beta)^n = \alpha^{2n} + \beta^{2n} = L_{2n}$.

5. $(\alpha - \beta)$LHS $= \displaystyle\sum_{i=0}^{n} \binom{n}{i} \alpha^{i+j} - \sum_{i=0}^{n} \binom{n}{i} \beta^{i+j} = \alpha^j(1 + \alpha)^n - \beta^j(1 + \beta)^n$

$= \alpha^{2n+j} - \beta^{2n+j}$. This gives the desired identity.

7. $(\alpha - \beta)$LHS $= \displaystyle\sum_{i=0}^{n} \binom{n}{i}(-1)^i \alpha^{i+j} - \sum_{i=0}^{n} \binom{n}{i}(-1)^i \beta^{i+j} = \alpha^j \sum_{i=0}^{n} \binom{n}{i}(-\alpha)^i$

$- \beta^j \displaystyle\sum_{i=0}^{n} \binom{n}{i}(-\beta)^i = \alpha^j(1 - \alpha)^n - \beta^j(1 - \beta)^n = \alpha^j \beta^n - \beta^j \alpha^n =$

$(-1)^{j+1}(\alpha^{n-j} - \beta^{n-j})$. This yields the given result.

9. When $n = 4$, LHS $= 5 \sum_{i=0}^{4} \binom{4}{i} F_i^2 = 175 = \sum_{i=0}^{4} \binom{4}{i} L_{2i} =$ RHS; and when $n = 5$, LHS $= 625 =$ RHS.

11. $\sum_{i=0}^{n} \binom{n}{i} L_{2i} = 5 \sum_{i=0}^{n} \binom{n}{i} F_i^2 + 2 \sum_{i=0}^{n} \binom{n}{i} (-1)^i = 5 \sum_{i=0}^{n} \binom{n}{i} F_i^2$.

13. $\sum_{i=0}^{n} \binom{n}{i} L_i^2 = \sum_{i=0}^{n} \binom{n}{i} L_{2i} + 2 \sum_{i=0}^{n} \binom{n}{i} (-1)^i = \sum_{i=0}^{n} \binom{n}{i} L_{2i}$.

15. $2^n L_n = (1 + \sqrt{5})^n + (1 - \sqrt{5})^n = 2 \sum_{i=0}^{\lfloor n/2 \rfloor} \binom{n}{2i} 5^i$. This yields the result.

17. $\sqrt{5} \cdot$ LHS $= \sum_{i=0}^{n} \binom{n}{i} (-2)^i (c\alpha^i - d\beta^i) = c(1 - 2\alpha)^n - d(1 - 2\beta)^n$

$= (\sqrt{5})^n [c(-1)^n - d]$. This gives the desired identity.

19. Since $(-1)^i (\alpha^{2i} - \beta^{2i}) = (-\alpha^2)^i - (-\beta^2)^i$, $\sqrt{5} \cdot$ LHS $= \sum_{i=0}^{n} \binom{n}{i} [(-\alpha^2)^i -$

$(-\beta^2)^i] = (1 - \alpha^2)^n - (1 - \beta^2)^n = (-1)^n (\alpha^n - \beta^n)$. This gives the identity.

21. $\sqrt{5} \cdot$ LHS $= \sum_{i=0}^{n} \binom{n}{i} (-1)^i (c\alpha^i - d\beta^i) = c \sum_{i=0}^{n} \binom{n}{i} (-\alpha)^i - d \sum_{i=0}^{n} \binom{n}{i} (-\beta)^i =$

$c(1 - \alpha)^n - d(1 - \beta)^n = c\beta^n - d\alpha^n = (-1)^n (c\alpha^{-n} - d\beta^{-n})$. This gives the identity.

23. $\sqrt{5} \cdot$ LHS $= \sum_{i=0}^{n} \binom{n}{i} (-1)^i (c\alpha^{i+j} - d\beta^{i+j}) = c\alpha^j \sum_{i=0}^{n} \binom{n}{i} (-\alpha)^i$

$- d\beta^j \sum_{i=0}^{n} \binom{n}{i} (-\beta)^i = c\alpha^j (1 - \alpha)^n - d\beta^j (1 - \beta)^n = c\alpha^j \beta^n - d\beta^j \alpha^n$

$= (-1)^n (c\alpha^{j-n} - d\beta^{j-n})$. This yields the identity.

25. $\sqrt{5} \cdot$ LHS $= \sum_{i=0}^{n} \binom{n}{i} (\alpha^{k+2i} - \beta^{2k+j}) = \alpha^k (1 + \alpha^2)^n - \beta^k (1 + \beta^2)^n = [\alpha^{n+k} -$

$(-1)^n \beta^{n+k}] (\sqrt{5})^n$. This gives the desired result.

27. $5^n = (2\alpha - 1)^{2n} = \sum_{i=0}^{2n} \binom{2n}{i} (2\alpha)^i (-1)^{2n-i} = \sum_{i=0}^{2n} \binom{2n}{i} (-2\alpha)^i$. Similarly,

$5^n = (1 - 2\beta)^{2n} = \sum_{i=0}^{2n} \binom{2n}{i} (-2\beta)^i$. Adding the two, $2 \cdot 5^n = \sum_{i=0}^{2n} \binom{2n}{i} (-2)^i L_i$.

This yields the given result.

29. (Swamy) $\sqrt{5} \cdot$ LHS $= \sum_{k=0}^{n} \binom{n}{k} \alpha^{4mk} - \sum_{k=0}^{n} \binom{n}{k} \beta^{4mk} = (1 + \alpha^{4m})^n - (1 + \beta^{4m})^n$

$= \alpha^{2mn} (\alpha^{2m} + \alpha^{-2m})^n - \beta^{2mn} (\beta^{2m} + \beta^{-2m})^n = (\alpha^{2m} + \beta^{2m})^n (\alpha^{2mn} - \beta^{2mn}) =$
$\sqrt{5} \cdot$ RHS. This gives the desired result.

EXERCISES 12

1. $L_8 = \displaystyle\sum_{i=0}^{4} \frac{8-i}{i}\binom{8-i}{i} = 1 + 8 + 20 + 16 + 2 = 47.$

3. $r^5 + s^5 = \displaystyle\sum_{i=0}^{2} A(5,i)p^{5-2i}q^i = p^5 + 5p^3 q + 5pq^2.$

5. Since $r + s = p$ and $r^2 + s^2 = p^2 + 2q$, the formula works when $n = 1$ and $n = 2$. Now assume it works for all positive integers $\le k$. Notice that $r^{k+1} + s^{k+1} = p(r^k + s^k) + q(r^{k-1} + s^{k-1}).$
   Let $n = k + 1$. Then

$$\text{RHS} = p \sum_{i=0}^{\lfloor k/2 \rfloor} \frac{k}{k-i}\binom{k-i}{i} p^{k-2i} q^i$$

$$+ q \sum_{i=0}^{\lfloor (k-1)/2 \rfloor} \frac{k-1}{k-1-i}\binom{k-1-i}{i} p^{k-1-2i} q^i$$

$$= \sum_{i=0}^{\lfloor k/2 \rfloor} \frac{k}{k-i}\binom{k-i}{i} p^{k+1-2i} q^i$$

$$\sum_{j=0}^{\lfloor (k+1)/2 \rfloor} \frac{k-1}{k-j}\binom{k-j}{j-1} p^{k+1-2j} q^j. \qquad (4)$$

Let k be even. Then, after some basic algebra, equation (4) yields

$$\text{RHS} = \sum_{i=0}^{k/2} \left[\frac{k}{k-i}\binom{k-i}{i} + \frac{k-1}{k-i}\binom{k-i}{i-1} \right] p^{k+1-2i} q^i$$

$$= \sum_{i=0}^{k/2} \frac{(k+1)(k-i)!}{(k+1-2i)!\,i!} p^{k+1-2i} q^i$$

$$= \sum_{i=0}^{k/2} \frac{k+1}{k+1-i}\binom{k+1-i}{i} p^{k+1-2i} q^i$$

$$= \sum_{i=0}^{\lfloor (k+1)/2 \rfloor} \frac{k+1}{k+1-i}\binom{k+1-i}{i} p^{k+1-2i} q^i. \qquad (5)$$

Similarly, equation (4) leads to equation (5) when k is odd. Thus, by the strong version of PMI, the formula works for all $n \ge 1$.

7. $F_{10} = \displaystyle\sum_{i=0}^{5} \binom{9-i}{i} = 1 + 8 + 21 + 20 + 5 = 55.$

9. $r + s = p, r - s = \Delta = \sqrt{p^2 + 4q}$. Then $32(r^5 - s^5) = (p + \Delta)^5 - (p - \Delta)^5 = 32(p^4 + 3p^3 q + q^2)\Delta$. This gives the required result.

11. Clearly, the formula works when $n = 1$ and $n = 2$. Now assume it works for all positive integers $\leq k$. Notice that $r^{k+1} - s^{k+1} = p(r^k - s^k) + q(r^{k-1} - s^{k-1})$. Let $n = k + 1$. Then

$$\text{RHS} = p\Delta \sum_{i=0}^{\lfloor k/2 \rfloor} \binom{k-i-1}{i} p^{k-2i} q^i + q\Delta \sum_{i=0}^{\lfloor (k-1)/2 \rfloor} \binom{k-i-2}{i} p^{k-2i-1} q^i$$

$$= \Delta \sum_{i=0}^{\lfloor k/2 \rfloor} \binom{k-i-1}{i} p^{k+1-2i} q^i + \Delta \sum_{j=0}^{\lfloor (k+1)/2 \rfloor} \binom{k-1-j}{j-1} p^{k+1-2j} q^j. \quad (6)$$

Let k be even. Then, after some basic algebra, equation (6) yields

$$\text{RHS} = \Delta \sum_{i=0}^{k/2} \left[\binom{k-i-1}{i} + \binom{k-i-1}{i-1} \right] p^{k+1-2i} q^i$$

$$= \Delta \sum_{i=0}^{k/2} \binom{k-i}{i} p^{k+1-2i} q^i$$

$$= \Delta \sum_{i=0}^{\lfloor (k+1)/2 \rfloor} \binom{k-i}{i} p^{k+1-2i} q^i. \quad (7)$$

Similarly, equation (6) leads to equation (7) when k is odd. Thus, by the strong version of PMI, the formula works for all $n \geq 1$.

13. $\displaystyle\sum_{j=0}^{3} C(6-j,j) = \sum_{j=0}^{3} \binom{6-j}{j} + \sum_{j=0}^{3} \binom{5-j}{j} = (1+5+6+1) + (0+1+3+1)$

   $= 18.$

15. $\displaystyle\sum_{k=1}^{n} C(k,j) = \sum_{k=1}^{n} \binom{k}{j} + \sum_{k=1}^{n} \binom{k-1}{j-1} = \binom{n+1}{j+1} + \binom{n}{j} = C(n+1, j+1).$

17. $C(n, n-2) = \dbinom{n}{n-2} + \dbinom{n-1}{n-3} = \dbinom{n}{2} + \dbinom{n-1}{2} = (n-1)^2.$

19. $\displaystyle\sum_{j=0}^{3} B(6-j,j) = \sum_{j=0}^{3} \binom{6-j}{j} + \sum_{j=0}^{3} \binom{5-j}{j} = (1+5+6+1) + (1+4+3)$

   $= 21.$

21. $D(n, 2) = \dbinom{n}{2} + \dbinom{n-1}{2} = (n-1)^2.$

23. $A(n, n) = 1, A(n, 0) = F_{2n-1}, n \geq 0; A(n, j) = A(n-1, j) + A(n-1, j-1),$
   $n > j.$

25. Let S_n denote the nth row sum, where $S_0 = 1$. Then $S_n = 1 + \displaystyle\sum_{k=0}^{n-1} F_{2n-2k} =$

   $1 + \displaystyle\sum_{k=1}^{n} F_{2k} = 1 + (F_{2n+1} - 1) = F_{2n+1}.$

27. (Bruckman) Let D_n denote the nth diagonal sum, where $D_0 = 1 = D_1$, and

$$D_n = \begin{cases} \displaystyle\sum_{k=1}^{n/2} F_{4k} + 1 & \text{if } n \text{ is even} \\ \displaystyle\sum_{k=1}^{(n+1)/2} F_{4k-2} & \text{otherwise.} \end{cases} \quad \text{Let } n = 2m.$$

Then $D_{2m} = \displaystyle\sum_{k=1}^{m} F_{4k} + 1 = F_1 + \sum_{k=1}^{m} (F_{4k+1} - F_{4k-1}) = \sum_{k=0}^{2m} (-1)^i F_{2i+1}$.

On the other hand, let $n = 2m + 1$. Then $D_{2m+1} = \displaystyle\sum_{k=1}^{m+1} F_{4k-2} = \sum_{k=1}^{m+1} (F_{4k-1} -$

$F_{4k-3}) = \displaystyle\sum_{i=0}^{2m+1} (-1)^{i+1} F_{2i+1}$. Combining the two cases, $D_n = \displaystyle\sum_{k=0}^{n} (-1)^{n-i} F_{2i+1} =$

$\displaystyle\sum_{k=0}^{n} (-1)^{n-i} (F_{i+1}^2 + F_i^2) = \sum_{k=0}^{n} [(-1)^{n-i} (F_{i+1}^2 - (-1)^{n-i-1} F_i^2)] = F_{n+1}^2 - 0 = F_{n+1}^2$.

29. Let $n = 2m + 1$. Then the rising diagonal sum is $\displaystyle\sum_{i=0}^{m} F_{4i+3}$. Using PMI, it follows that this sum equals $F_{2m+2} F_{2m+3}$. On the other hand, let $n = 2m$. Then the diagonal sum equals $\displaystyle\sum_{i=0}^{m} F_{4i+1} = F_{2m+1} F_{2m+2} = F_{n+1} F_{n+2}$.

31. $S_0(a, b) = a, S_1(a, b) = a + b; S_n(a, b) = S_{n-1}(a, b) + S_{n-2}(a, b), n \geq 2$.

33. Since $S_0(a, b) = a = aF_1 + bF_0$ and $S_1(a, b) = a + b = aF_2 + bF_1$, the result is true when $n = 0$ and $n = 1$. Now assume it is true for all nonnegative integers $\leq k$, where $k \geq 1$. Then $S_{k+1}(a, b) = S_k(a, b) + S_{k-1}(a, b) = (aF_{k+1} + bF_k) + (aF_k + bF_{k-1}) = aF_{k+2} + bF_{k+1}$. Thus, by the strong version of PMI, the result is true for all $n \geq 0$.

35. $T_n(a, b) = T_{n-1}(a, b) + T_{n-2}(a, b)$, where $T_0(a, b) = a$ and $T_1(a, b) = a - b$. The result now follows by the strong version of PMI, as in Exercise 13.33.

EXERCISES 13

1. Yes.
3. Yes.
5. No.
7. Yes.
9. $a_n = 2(-1)^n + 2^n, n \geq 0$.
11. $a_n = 3(-2)^n + 2 \cdot 3^n, n \geq 0$.
13. $a_n = F_{n+2}, n \geq 0$.
15. $L_n = \alpha^n + \beta^n, n \geq 0$.
17. General solution of the homogeneous part: $a_n = A\alpha^n + B\beta^n$. Look for a particular solution of the form $a_n = c$; this gives $c = 1$. So the general

solution for the recurrence is $a_n = A\alpha^n + B\beta^n + 1$. Using the initial conditions, we get $a_n = 2F_{n-1} + 1$.

19. General solution of the homogeneous part: $a_n = A\alpha^n + B\beta^n$. The particular solution is of the form $a_n = Km^n$. Then $Km^{n+2} = Km^{n+1} + Km^n + m^n$, so

$$K = \frac{1}{m^2 - m - 1}. \text{ Thus } a_n = A\alpha^n + B\beta^n + \frac{m^n}{m^2 - m - 1}, A \text{ and } B \text{ are to be}$$

determined using the initial conditions.

21. Characteristic polynomial: $x^3 - 2x^2 - 2x + 1 = (x - \alpha^2)(x - \beta^2)(x + 1)$; so the general solution is $a_n = A\alpha^{2n} + b\beta^{2n} + C(-1)^n$. Using the initial conditions, we get $A = 6/5 = B$ and $C = 8/5$. This yields $a_n = L_n^2 + F_n^2$.

23. $F_{n+3}^2 = 2F_{n+2}^2 + 2F_{n+1}^2 - F_n^2$; so the desired recurrence is $a_{n+3} = 2a_{n+2} + 2a_{n+1} - a_n$, where $n \geq 0$.

25. Let $a_n = L_n^2 + F_n^2$. It follows by Exercises 11.21 and 11.23 that $a_{n+3} = 2a_{n+2} + 2a_{n+1} - a_n$, where $n \geq 0$.

27. $\dfrac{2}{x - 1} - \dfrac{1}{x + 3}$.

29. $\dfrac{2}{1 + 2x} + \dfrac{3}{1 - 3x}$.

31. $\dfrac{1}{2 + 3x} + \dfrac{2x - 1}{x^2 + 1}$.

33. $\dfrac{1 - x}{x^2 + 2} + \dfrac{2x}{x^2 + 3}$.

35. $\dfrac{x - 1}{x^2 + 1} + \dfrac{2x + 1}{x^2 - x + 1}$.

37. $a_n = 2^n, n \geq 0$.

39. $a_n = 2n - 1, n \geq 1$.

41. $a_n = 2^{n+1} - 6n \cdot 2^n, n \geq 0$.

43. $a_n = 5 \cdot 2^n - 3^n, n \geq 0$.

45. $a_n = F_{n+3}, n \geq 0$.

47. $a_n = 3 \cdot 2^n + n \cdot 2^{n+1}, n \geq 0$.

49. $a_n = 3(-2)^n + 2^n - 3^n, n \geq 0$.

51. $a_n = 2^n + 3n \cdot 2^n - 3^n, n \geq 0$.

53. $9a_n = 30 \cdot 2^n + 15(-2)^n - 5 \cdot 3^n + 23(-3)^n, n \geq 0$.

55. $(r - s)(a_n + ca_{n+1}) = (r^{n+1} - s^{n+1}) + c(r^{n+2} - s^{n+2}) = r^{n+2}(c + 1/r) - s^{n+2}(c + 1/s) = r^{n+2}(c - s) - s^{n+3}(c - r) = r^{n+3} - s^{n+3}$. This yields the desired result.

57. Since $f(x) = \dfrac{e^{\alpha x} - e^{\beta x}}{\sqrt{5}}$, $f(-x)\dfrac{e^{-\alpha x} - e^{-\beta x}}{\sqrt{5}} = \dfrac{e^{\beta x} - e^{\alpha x}}{\sqrt{5}e^{(\alpha + \beta)x}} = -\dfrac{f(x)}{e^x}$; so $f(x) = -e^x f(-x)$.

59. By Exercise 13.58, $g(x) = e^{\alpha x} + e^{\beta x}$, so $g(-x) = e^{-\alpha x} - e^{-\beta x} = \dfrac{e^{\alpha x} + e^{\beta x}}{e^x} = \dfrac{g(x)}{e^x}$. Thus $g(x) = e^x g(-x)$.

61. Differentiating $\dfrac{1}{1 - x - x^2} = \displaystyle\sum_{n=0}^{\infty} F_{n+1} x^n$ with respect to x, we get

$\dfrac{1 + 2x}{(1 - x - x^2)^2} = \displaystyle\sum_{n=1}^{\infty} n F_{n+1} x^{n-1}$. This yields the given result.

63. (Hansen) Let $\Delta = 1 - x - x^2$. Then $\displaystyle\sum_{m=0}^{\infty} (L_m L_n + L_{m-1} L_{n-1}) x^m$

$$= L_n \sum_{m=0}^{\infty} L_m x^m + L_{n-1} \sum_{m=0}^{\infty} L_m x^m = L_n \left(\frac{2 - x}{\Delta} \right) + L_{n-1} \left(\frac{-1 + 3x}{\Delta} \right)$$

$$= \frac{[L_n + (L_n - L_{n-1})] + [2L_{n-1} + (L_{n-1} - L_n)] x}{\Delta}$$

$$= \frac{L_n + L_{n-1} x}{\Delta} + \frac{L_{n-2} + L_{n-3} x}{\Delta}$$

$$= \sum_{m=0}^{\infty} (L_{m+n} + L_{m+n-2}) x^m.$$

Thus $L_m L_n + L_{m-1} L_{n-1} = L_{m+n} + L_{m+n-2} = 5 F_{m+n-1}$.

65. Let $A(t) = \dfrac{e^{\alpha t} - e^{\beta t}}{\alpha - \beta} = B(t)$. Then

$$\sum_{n=0}^{\infty} \left[\sum_{k=0}^{n} \binom{n}{k} F_k F_{n-k} \right] \frac{t^n}{n!} = \frac{e^{2\alpha t} + e^{2\beta t} - 2e^{(\alpha+\beta)t}}{(\alpha - \beta)^2} = \sum_{n=0}^{\infty} \left(\frac{2^n L_n - 2}{5} \right) \frac{t^n}{n!}.$$

Equating the coefficients of the corresponding terms gives the desired result.

67. (Padilla) Since $(1 - x - x^2) A_n(x) = F_n x^{n+2} + F_{n+1} x^{n+1} - x$, the desired result follows.

69. $(1 - x - x^2) B(x) = -x^2 \displaystyle\sum_{n=0}^{\infty} \frac{F_n}{n!} x^n - \sum_{n=0}^{\infty} \frac{(n + 1) F_{n+1}}{(n + 1)!} x^{n+1} + e^x$. So $\sqrt{5}(1 - x -$

$x^2) B(x) = \sqrt{5} e^x - x^2 (e^{\alpha x} - e^{\beta x}) - x(\alpha e^{\alpha x} - \beta e^{\beta x})$.

71. $\displaystyle\sum_{m=0}^{\infty} L_{m+n} x^m = \sum_{m=0}^{\infty} (\alpha^{m+n} + \beta^{m+n}) x^m = \alpha^n \sum_{m=0}^{\infty} \alpha^m x^m + \beta^n \sum_{m=0}^{\infty} \beta^m x^m$

$$= \frac{\alpha^n}{1 - \alpha x} + \frac{\beta^n}{1 - \beta x}$$

$$= \frac{(\alpha^n + \beta^n) + (\alpha^{n-1} + \beta^{n-1}) x}{1 - x - x^2} = \frac{L_n + L_{n-1} x}{1 - x - x^2}.$$

73. (Carlitz) Let $C(x) = \displaystyle\sum_{n=0}^{\infty} C_n x^n$, where $C_0 = 0$. Then

$$(1 - x - x^2)C(x) = C_1 x + (C_2 - C_1)x^2 + \sum_{n=3}^{\infty} F_n x^n = F_1 x + F_2 x^2 + \sum_{n=3}^{\infty} F_n x^n$$

$$= \sum_{n=0}^{\infty} F_n x^n = \frac{x}{1 - x - x^2}$$

$$C(x) = \frac{x}{(1 - x - x^2)^2} = x \sum_{i=0}^{\infty}(i + 1)(x + x^2)^i$$

$$= \sum_{i=0}^{\infty}(i + 1)x^{i+1}(1 + x)^i = \sum_{i=0}^{\infty}(i + 1)x^{i+1} \sum_{j=0}^{i} \binom{i}{j} x^j$$

$$= \sum_{n=0}^{\infty} \left[\sum_{i=0}^{n}(i + 1)\binom{i}{n - i} \right] x^{n+1};$$

$$C_{n+1} = \sum_{i=0}^{n}(i + 1)\binom{i}{n - i} = \sum_{i=0}^{\lfloor n/2 \rfloor}(n - i + 1)\binom{n - i}{i}.$$

75. Let $A(t) = e^{\alpha t} + e^{\beta t}$ and $B(t) = e^t$. Then $A(t)B(t) = \displaystyle\sum_{n=0}^{\infty} \left[\sum_{k=0}^{n} \binom{n}{k} L_k \right] \frac{t^n}{n!}$.

That is, $\displaystyle\sum_{n=0}^{\infty} L_{2n} \frac{t^n}{n!} = e^{\alpha^2 t} + e^{\beta^2 t} = e^{(\alpha + 1)t} + e^{(\beta + 1)t} = \sum_{n=0}^{\infty} \left[\sum_{k=0}^{n} \binom{n}{k} L_k \right] \frac{t^n}{n!}$.

So $L_{2n} = \displaystyle\sum_{k=0}^{n} \binom{n}{k} L_k$.

77. Let $A(t) = \dfrac{e^{\alpha^2 t} - e^{\beta^2 t}}{\alpha - \beta}$ and $B(t) = e^{-t}$. Then $\dfrac{e^{(\alpha^2 - 1)t} - e^{(\beta^2 - 1)t}}{\alpha - \beta} =$

$\displaystyle\sum_{n=0}^{\infty} \left[\sum_{k=0}^{n}(-1)^{n-k} \binom{n}{k} F_{2k} \right] \frac{t^n}{n!}$. That is, $\dfrac{e^{\alpha t} - e^{\beta t}}{\alpha - \beta} = \displaystyle\sum_{n=0}^{\infty} \left[\sum_{k=0}^{n}(-1)^{n-k} \binom{n}{k} F_{2k} \right] \frac{t^n}{n!}$.

So $F_n = \displaystyle\sum_{k=0}^{n}(-1)^{n-k} \binom{n}{k} F_{2k}$.

79. (Church and Bicknell) Let $A(t) = \dfrac{e^{\alpha^m t} - e^{\beta^m t}}{\alpha - \beta}$ and $B(t) = e^{\alpha^m t} + e^{\beta^m t}$. This

yields $\displaystyle\sum_{n=0}^{\infty} 2^n F_{mn} \frac{t^n}{n!} = \dfrac{e^{2\alpha^m t} - e^{2\beta^m t}}{\alpha - \beta} = \sum_{n=0}^{\infty} \left[\sum_{k=0}^{n} \binom{n}{k} F_{mk} L_{mn-mk} \right] \frac{t^n}{n!}$. The

identity now follows.

81. (Church and Bicknell) Let $A(t) = e^{\alpha^m t} + e^{\beta^m t} = B(t)$. This yields $(e^{\alpha^m t} +$

$$e^{\beta^m t})^2 = \sum_{n=0}^{\infty} \left[\sum_{k=0}^{n} \binom{n}{k} L_{mk} L_{mn-mk} \right] \frac{t^n}{n!}.$$ That is, $\sum_{n=0}^{\infty} (2^n L_{mn} + 2L_m^n) \frac{t^n}{n!} =$

$$e^{2\alpha^m t} + e^{2\beta^m t} + 2e^{(\alpha^m + \beta^m)t} = \sum_{n=0}^{\infty} \left[\sum_{k=0}^{n} \binom{n}{k} L_{mk} L_{mn-mk} \right] \frac{t^n}{n!}.$$ The given

identity now follows.

83. (Church and Bicknell)

$$\sum_{n=0}^{\infty} F_{mn} \frac{t^n}{n!} = \frac{e^{\alpha^m t} - e^{\beta^m t}}{\alpha - \beta} = \frac{e^{(\alpha F_m + F_{m-1})t} - e^{(\beta F_m + F_{m-1})t}}{\alpha - \beta}$$

$$= e^{F_{m-1}t} \left(\frac{e^{\alpha F_m t} - e^{\beta F_m t}}{\alpha - \beta} \right) = \sum_{n=0}^{\infty} \left[\sum_{k=0}^{n} \binom{n}{k} F_{m-1}^{n-k} F_m^k F_k \right] \frac{t^n}{n!}.$$

The desired identity now follows by equating the coefficients of $t^n/n!$.

85. The series $\sum_{k=0}^{\infty} F_k x^k = \dfrac{x}{1 - x - x^2}$ converges when $|x| < 1/\alpha$. Letting

$x = 1/2$ yields the desired result.

EXERCISES 14

1. Consider a $1 \times 2n$ board. It has F_{2n+1} tilings. Exactly one of them contains all dominoes. So there are $F_{2n+1} - 1$ tilings with at least one square each. Now pivot on the location of the last square. Since the board is of even length, each tiling must contain an even number of squares. So the last square must occupy an even-numbered cell, say, $2k + 2$, where $0 \le k \le n - 1$: subtiling ⎵⎵⎵⎵⎵⎵□⎵⎵⎵⎵⎵⎵. There are F_{2k+2} tilings with the last

   length $2k+1$ dominoes

   square in cell $2k + 2$. So the total number of tilings with at least one square equals $\sum_{k=0}^{n-1} F_{2k+2}$, that is, $\sum_{k=1}^{n} F_{2k}$. This count, coupled with the initial one, yields the given result.

3. Consider a $1 \times 2n$ board. Use breakability at cell n. See Example 14.3.

5. The possible pairs (i, j) are $(0, 0)$, $(0, 1)$, $(0, 2)$, $(0, 3)$, $(1, 0)$, $(1, 1)$, $(1, 2)$, $(2, 0)$, $(2, 1)$, and $(3, 0)$. Correspondingly, the desired sum equals $21 = F_8$.

7. Consider a circular board with $m + n$ cells. Suppose it is breakable at cell m. Then cells 1 through m form a $1 \times m$ board; it can be tiled in F_{m+1} ways. The remaining cells $m + 1$ through $m + n$ can be glued end to end to form an n-bracelet. There are L_n such bracelets. So the number of such tilings equals $F_{m+1}L_n$.

Suppose the tiling is *not* breakable at cell m. Then cells 1 through $m - 1$ form a $1 \times (m - 1)$ board; and cells $m + 2$ through $m + n$ an $(n - 1)$-bracelet. There are $F_m L_{n-1}$ such tilings. So the total equals $F_{m+1} L_n + F_m L_{n-1}$. Equating the two counts gives the desired identity.

9. Consider two pairs of two boards each. One pair consists of a board B_1 of length $n + 1$ and the other B_2 of length $n - 1$. The second pair consists of two boards of length n. The first pair can be tiled in $F_{n+2} F_n$ different ways A, and the second in F_n^2 ways (B). Now try to establish a bijection between the tilings A and tilings B by shifting cell 1 from B_1 to B_2. Count the matchable tilings and use PMI.

11. Consider a $1 \times (n - 1)$ board; and two circular boards, one with $n + 1$ cells and the other with $n - 1$ cells. Let S denote the set of tilings of the linear board, and T the set of $(n + 1)$-bracelets or $(n - 1)$-bracelets. Then $|S| = F_n$ and $|T| = L_{n+1} + L_{n-1}$. Now establish a one-to-five correspondence between S and T.

13. Using the identities $F_{2a} = F_a L_a$, $F_{a+2} - F_{a-2} = L_a$, and $F_{a+b} = F_{a+1} F_b + F_a F_{b-1}$, we have

$$\begin{aligned}
\text{RHS} &= F_{n+2}(F_{n+2}^2 - F_n^2) - F_n^2(F_{n+2} - F_{n-2}) = F_{n+2} L_{n+1} F_{n+1} - F_n^2 L_n \\
&= F_{n+2} F_{2n+2} - F_n F_{2n} = F_{2n+2}(F_{n+1} + F_n) - F_n F_{2n} \\
&= F_{2n+2} F_{n+1} + F_n(F_{2n+2} - F_{2n}) \\
&= F_{2n+2} F_{n+1} + F_{2n+1} F_n = F_{(2n+1)+(n+1)} = F_{3n+2}.
\end{aligned}$$

15. Using the identities $L_a = F_{a+1} + F_{a-1} = F_{a+2} - F_{a-2}$, $F_{2a-2} = F_{2a+2} - 2F_{2a+1} + F_{2a}$, $2F_a - F_{a-2} = F_{a+1}$, $F_{a+1} - 2F_{a-2} = F_{a-1} + F_{a-3} = L_{a-2}$, and $L_{a+b} = F_{a+1} L_b + F_a L_{b-1}$, we have

$$\begin{aligned}
\text{RHS} &= F_{n+2}(F_{n+2}^2 - F_n^2) - 2F_n^2(F_{n+2} - F_{n-2}) + F_{n-2}(F_n^2 - F_{n-2}^2) \\
&= F_{n+2} F_{2n+2} - 2F_n F_{2n} + F_{n-2} F_{2n-2} \\
&= F_{2n+2}(F_{n+1} + F_n) - F_n F_{2n} + F_{n-2}(F_{2n+2} - 2F_{2n+1} + F_{2n}) \\
&= F_{2n+2} L_{n-1} + F_{2n+2} F_{n+1} - 2F_{2n+1} F_{n-2} - F_{2n} F_{n+1} \\
&= F_{2n+2} L_{n-1} + F_{2n+1} F_{n+1} - 2F_{2n+1} F_{n-2} = F_{2n+2} L_{n-1} + F_{2n+1} L_{n-2} \\
&= L_{(2n+1)+(n-1)} = L_{3n}.
\end{aligned}$$

17. $\text{LHS} = (L_{n+1} - L_{n-1}) + (F_n + F_{n-1}) = L_{n+1} - F_{n-2} + F_{n-1} = \text{RHS}$.

19. We have $i(T_{n,k}) = L_{n+k} + F_{n-3} F_k$. Now partition the family of independent subsets of the vertex set into two disjoint subsets, those that contain c_n and those that do not. There are $i(P_{n-3}) i(P_k) = F_{n-1} F_{k+2}$ independent subsets that contain c_n, and $i(P_{n-1+k}) = F_{n+k+1}$ independent subsets that do not. Equating the two counts yields the result.

21. $t_{n,n-k-3} = L_n + F_{n-3-(n-k-3)}F_{n-k-3} = L_n + F_{n-k-3}F_k = t_{n,k}.$

23. Using the formula $\sum\limits_{k=0}^{n} F_k F_{n-k} = nL_n - F_n$, the result follows.

25. The formula works when $n = 3$. Assume it is true for an arbitrary odd integer $n \geq 3$. Then

$$\sum_{k=0}^{n-1}(-1)^k t_{n+2,k} = \sum_{k=0}^{n-3}(-1)^k t_{n+2,k} + (-1)^{n-2}t_{n+2,n-2} + (-1)^{n-1}t_{n+2,n-1}$$

$$= \sum_{k=0}^{n-3}(-1)^k (t_{n,k} + t_{n+1,k}) - (L_{n+2} + F_{n-2}) + L_{n+2}$$

$$= \sum_{k=0}^{n-3}(-1)^k t_{n,k} + \sum_{k=0}^{n-2}(-1)^k t_{n+1,k} + t_{n+1,n-2} - F_{n-2}$$

$$= 2F_n + 0 + L_{n+1} - F_{n-2} = 2F_{n+2}.$$

Thus, by PMI, the formula works for all odd integers ≥ 3.

EXERCISES 15

1. Since $H(1, 1) = 1 = H(2, 1)$ and $H(n, 1) = H(n - 1, 1) + H(n - 2, 1)$,
   $H(n, 1) = F_n.$

3. $H(n, n - 1) = H(n - 1, n - 2) + H(n - 2, n - 3) = H(n - 1, 1) + H(n - 2, 1) = H(n, 1) = F_n.$

5. $L_{n+2} \equiv (-1)^{j-1}L_{n-2j} \pmod 5$. Let $n = 2(m - 1)$ and $j = m - 1$. Then $L_{2m} \equiv (-1)^{m-2}L_0 \equiv 2(-1)^m \pmod 5.$

7. Let U, V, W, and X be as in the figure below. Then $U = A + B$, $V = C + D$, $W = D - C$, and $X = B - A$.

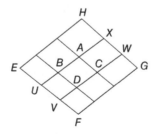

Then $E = V - U = C + D - A - B$, $F = U + V = A + B + C + D$, $G = X + W = B + D - A - C$, and $H = W - X = A + D - B - C.$

9. Subtracting equation (15.10) from equation (15.9), $H(n-1,j) + H(n-1,j-1) - H(n-2,j-1) = F_n$; that is, $H(n,j) + H(n,j-1) - H(n-1,j-1) = F_{n+1}$.

11. By equation (15.10), $H(n,j) + H(n-2,j-1) = F_{n+1}$. So $H(n-2,j) + H(n-4,j-1) = F_{n-1}$. Subtracting, $H(n,j) - H(n-4,j-2) = F_{n+1} - F_{n-1} = F_n$.

EXERCISES 16

1. Yes.

3. Yes.

5. $\dfrac{AC}{CB} = \alpha$; that is, $\dfrac{1-CB}{CB} = \alpha$. So $\dfrac{1}{CB} = \alpha + 1 = \alpha^2$. Thus $BC = 1/\alpha^2$ and $AC = \alpha \cdot AC = 1/\alpha$.

7. $t^2 + t - 1 = 0$.

9. β.

11. Let x_n denote the sum with n radicals. Then $x_n = 1$ if n is odd; and $x_n = 0$ otherwise. So the sequence $\{x_n\}$ does not converge.

13. Let $a/b = c/d = k$. Then $b/a = 1/k = d/c$.

15. Let $a/b = c/d = k$. Then $(a-b)/b = (bk-b)/b = k - 1 = (dk-d)/d = (c-d)/d$.

17. $\dfrac{\text{Sum of the triangular faces}}{\text{Base area}} = \dfrac{4(2b \cdot a)/2}{(2b)^2} = a/b = \alpha$; so $a = b\alpha$.

19. $1 + 1/\alpha = (\alpha + 1)/\alpha = \alpha^2/\alpha = \alpha$.

21. $\alpha^2 = \alpha + 1$, so $\alpha^n = \alpha^{n-1} + \alpha^{n-2}$, where $n \geq 2$.

23. LHS $= \dfrac{1/\alpha}{1 - 1/\alpha} = \dfrac{1}{\alpha - 1} = \alpha$.

25. LHS $= \dfrac{1/\alpha^2}{1 - 1/\alpha^2} = \dfrac{1}{\alpha^2 - 1} = 1/\alpha = -\beta$.

27. LHS is an infinite geometric series with first term $1/\alpha$ and common ratio $1/\alpha^2 < 1$. So LHS $= \dfrac{1/\alpha}{1 - 1/\alpha^2} = 1$.

29. Letting $x = 1/\alpha^2$ in $\dfrac{x}{(1-x)^2} = \sum\limits_{n=1}^{\infty} nx^n$ yields given result.

31. $\alpha\sqrt{3 - \alpha} = \sqrt{3\alpha^2 - \alpha^3} = \sqrt{3(\alpha+1) - \alpha(\alpha+1)} = \sqrt{\alpha + 2}$.

33. $\alpha + 2 = \dfrac{5 + \sqrt{5}}{2} = \dfrac{10 + 2\sqrt{5}}{4}$; so $\sqrt{\alpha + 2} = \dfrac{\sqrt{10 + 2\sqrt{5}}}{2}$.

35. $\cos^2 \pi/10 = \dfrac{1 + \cos \pi/5}{2} = \dfrac{\alpha + 2}{4}$. So $\cos \pi/10 = \dfrac{\sqrt{\alpha + 2}}{2}$.

37. $v + \dfrac{1}{v^2} = v + \dfrac{1}{v+1} = v + \dfrac{v-1}{v} = \dfrac{v^2-1}{v} + 1 = 1 + 1 = 2$.

39. $\lim\limits_{n\to\infty} \dfrac{L_n}{L_{n+1}} = \lim\limits_{n\to\infty} \dfrac{\alpha^n[1+(\beta/\alpha)^n]}{\alpha^{n+1}[1+(\beta/\alpha)^{n+1}]} = 1/\alpha.$

41. $\lim\limits_{n\to\infty} \dfrac{G_{n+1}}{G_n} = \lim\limits_{n\to\infty} \dfrac{c\alpha^{n+1}-d\beta^{n+1}}{c\alpha^n-d\beta^n} = \alpha.$

43. $\pi\alpha.$

45. $\dfrac{\pi\alpha^2\sqrt{3-\alpha}}{24}.$

47. Since $\star$ is associative, the equations $(x \star y) \star z = a + b(a + cxy + z + czx + cyz) + b^2(x+y) + c^2xyz + caz$ and $x \star (y \star z) = a + b(a + x + cyz + cxy + czx) + b^2(y+z) + cax + c^2xyz)$ are equal; so $b^2(x-z) + b(z-x) + (z-x) = 0$. But x, y, and z are arbitrary, so $b^2 - b - 1 = 0$. Thus $b = \alpha$ or β.

49. (Alexanderson) Let $p(x) = x^n - xF_n - F_{n-1}$, $g(x) = x^2 - x - 1$, and $h(x) = x^{n-2} + x^{n-3} + 2x^{n-4} + \cdots + F_k x^{n-k-1} + \cdots + F_{n-2}x + F_{n-1}$. Then $p(x) = g(x)h(x)$. When $x \ge 0$, $h(x) > 0$. When $x > \alpha$, $g(x), h(x) > 0$; so $p(x) > 0$. When $0 \le x < \alpha$, $g(\alpha) < 0$; so $p(\alpha) < 0$.
Suppose $x_k > \alpha$. Since $p(x) > 0$, $x_k^n > x_k F_n + F_{n-1} = x_{k+1}^n$, so $x_k > x_{k+1}$. In addition, $x_{k+1}^n = x_k F_n + F_{n-1} > \alpha F_n + F_{n-1} = \alpha^n$. So $x_{k+1} > \alpha$; Therefore, $x_k > x_{k+1} > \alpha$. Thus $x_0 > \alpha$ implies $x_0 > x_1 > x_2 > \cdots > \alpha$. Similarly, if $0 \le x_0 < \alpha$, then $0 \le x_0 < x_1 < x_2 < \cdots < \alpha$.
Thus, in both cases, the sequence $\{x_k\}$ is monotonic and bounded, so x_k converges to a limit l as $k \to \infty$: $l = \sqrt[n]{F_{n-1}+lF_n}$. Since $l^2 - lF_n - F_{n-1} = 0$, l is the positive zero of $p(x)$. Since α is the unique positive root of $p(x)$, it follows that $l = \lim\limits_{k\to\infty} x_k = \alpha.$

51. Since $|\beta| < 1$, sum $= \dfrac{1}{1-|\beta|} = \dfrac{1}{1+\beta} = \dfrac{1}{\beta^2} = \alpha^2.$

53. $t = \dfrac{x^{t+1}}{t+1}\Big|_0^1 = \dfrac{1}{t+1}$. Then $t^2 + t - 1 = 0$; so $t = \alpha$ or β.

55. $\lim\limits_{n\to\infty} \dfrac{F_{n+k}}{L_n} = \lim\limits_{n\to\infty} \dfrac{\alpha^{n+k}[1-(\beta/\alpha)^{n+k}]}{\sqrt{5}\alpha^n[1+(\beta/\alpha)^n]} = \dfrac{\alpha^k}{\sqrt{5}}.$

57. (Ford) By PMI, $a_n = F_n + k(F_{n+1}-1)$. So $\lim\limits_{n\to\infty} \dfrac{a_n}{F_n} = 1 + k\alpha.$

59. By PMI, $b_n = L_n + kF_{n-1}$. Since $\dfrac{F_{n-1}}{L_n} = \dfrac{F_{n-1}}{F_n} \cdot \dfrac{F_n}{L_n}$, it follows that the desired limit is $1 + k/(\sqrt{5}\alpha)$.

61. (Lord) Sum $= \alpha^{-2n} \sum\limits_{i=0}^{n}(\alpha^3)^i = \alpha^{-2n}(1+\alpha^3)^n = \alpha^{-2n}(2\alpha^2)^n = 2^n.$

63. (Ford) Assume $x_n \ne -1$ for every n. By PMI, $x_n = \dfrac{x_0 F_{n-1} + F_n}{x_0 F_n + F_{n+1}}$. So x_n is defined when $x_0 \ne -F_{n+1}/F_n, n \ge 1$. Then $\lim\limits_{n\to\infty} x_n = \dfrac{1}{\alpha} \cdot \dfrac{x_0+\alpha}{x_0+\alpha} = \dfrac{1}{\alpha} = -\beta.$

65. $\{u, v\}$ is a basis of V; so V is two-dimensional.

67. If follows by Exercise 16.66 that $r = \alpha, \beta$.

69. By Exercise 16.68, $F_n = a\alpha^n + b\beta^n$. It follows by the initial conditions $F_1 = 1 = F_2$, $a = 1/(\alpha - \beta) = -b$. Thus $F_n = (\alpha^n - \beta^n)/(\alpha - \beta)$.

71. Let $K_n = G_{n+2}^4 - G_n G_{n+1} G_{n+3} G_{n+4}$. Then $K_{n+1} = G_{n+3}^4 - G_{n+1} G_{n+2} G_{n+4} G_{n+5} = (2G_{n+1} + G_n)^4 - G_{n+1}(G_{n+1} + G_n)(3G_{n+1} + 2G_n)(5G_{n+1} + 3G_n) = G_{n+1}^4 - 2G_{n+1}^3 G_n - G_{n+1}^2 G_n^2 + 2G_{n+1} G_n^3 + G_n^4 = (G_{n+1} + G_n)^4 - G_n G_{n+1}(2G_{n+1} + G_n)(3G_{n+1} + 2G_n) = G_{n+2}^4 - G_n G_{n+1} G_{n+3} G_{n+4} = K_n$. This implies K_n is a constant K.

73. By Exercise 16.72, $K = (b^2 - ba - a^2)^2$. When $K = 0$, $b^2 - ba - a^2$. This implies $b/a = \alpha$.

EXERCISES 17

1. Since $\triangle ABC$ is a golden triangle, $\dfrac{AB}{AC} = \dfrac{BC}{AC} = \alpha$. Since $\triangle ABC \sim \triangle ADC$, $\dfrac{AB}{AC} = \dfrac{AD}{CD}$; so $\dfrac{AD}{CD} = \dfrac{AC}{CD} = \alpha$, $\triangle CAD$ is a golden triangle.

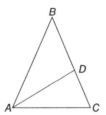

3. Since $\triangle ABC$ is golden, $\dfrac{BC}{AC} = \alpha$ and $\triangle CAD$ is also golden. Since $\triangle ABD$ is isosceles, $AD = AC = BD$. Let h be the length of the altitude from A to $\overline{BC}$. Then $\dfrac{\text{Area } ABC}{\text{Area } BDA} = \dfrac{1/2 \cdot BC \cdot h}{1/2 \cdot BD \cdot h} = \dfrac{BC}{BD} = \dfrac{BC}{AC} = \alpha$.

5. $\dfrac{\text{Area } ABC}{\text{Area } BDA} = \alpha$; so $\dfrac{BC}{BD} = \alpha$. Then $\dfrac{\text{Area } ABC}{\text{Area } CDA} = \dfrac{1/2 \cdot BC \cdot h}{1/2 \cdot CD \cdot h} = \dfrac{BC}{CD} = \dfrac{BC}{BC - BD} = \dfrac{1}{1 - 1/\alpha} = \alpha^2$.

7. $\dfrac{AB}{BC} = \dfrac{BC}{BE} = \dfrac{l}{w} = k$. Let $\dfrac{\text{Area } ABCD}{\text{Area } AEFD} = k$; that is, $\dfrac{AB \cdot BC}{AE \cdot BC} = k$, so $\dfrac{AB}{AE} = k$. Thus $\dfrac{l}{AE} = k$, so $AE = w$. Since $ABCD$ and $BCFE$ are similar, $\dfrac{AB}{BC} = \dfrac{BC}{CF}$; that is, $\dfrac{l}{w} = \dfrac{w}{l - w}$, so $l/w = k = \alpha$.

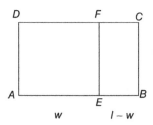

Conversely, let $k = l/w = \alpha$. Then $\dfrac{AB}{BC} = \dfrac{BC}{CF} = \alpha$, so $\dfrac{BC}{AB - AE} = \alpha$.

That is, $\dfrac{1}{\alpha - AE/BC} = \alpha$. So $AE = BC = w$. Then $\dfrac{\text{Area } ABCD}{\text{Area } AEFD} =$

$\dfrac{AB \cdot BC}{AE \cdot BC} = \dfrac{AB}{AE} = \dfrac{AB}{BC} = k = \alpha$.

In both cases, $AE = w$, so $AEFD$ is a square.

9. Using Figure 17.36, since both are golden rectangles, $\dfrac{FG}{BG} = \dfrac{BG}{BC} = \alpha$. Since

$\triangle FPG \sim \triangle BPC$, $\alpha = \dfrac{FG}{BG} = \dfrac{FP}{GP}$. Since $\triangle BPG \sim \triangle BPC$, $\alpha = \dfrac{BG}{BC} = \dfrac{BP}{CP}$;

so $\dfrac{FP}{GP} = \dfrac{BP}{CP} = \alpha$.

11. Since $BGHC$ is a golden rectangle, $\dfrac{BG}{BC} = \alpha$; that is, $\dfrac{AC}{BC} = \alpha$. Then

$\dfrac{AB}{BG} = \dfrac{AC + BC}{BG} = \dfrac{AC + BC}{AC} = 1 + \dfrac{BC}{AC} = 1 + 1/\alpha = \alpha$; so $ABGF$ is a

golden rectangle.

13. $\angle SPQ = 180° - (\angle APS + \angle BPQ) = 180° - (45° + 45°) = 90°$. So $PQRS$ is

a rectangle. Since $\triangle APS \sim \triangle BPQ$, $\dfrac{AP}{BP} = \dfrac{PS}{PQ}$. But $\dfrac{AP}{BP} = \alpha$, so $\dfrac{PS}{PQ} = \alpha$.

Thus $PQRS$ is a golden rectangle.

15. Shorter side $= (-1)^{n+1}(bF_n - aF_{n-1})$ and longer side $= (-1)^n(bF_{n-1} - aF_{n-2})$.

17. $GC^2 = GH^2 + HC^2 = (\omega - 1)^2 + (1 - \theta)^2 = (1/\theta - 1)^2 + (1 - \theta)^2 = \theta^4(1 + \theta^2) = \theta(1 - \theta) + (1 - \theta)^2 = 1 - \theta = \theta^3$; so $GC = \theta\sqrt{\theta}$.

19. $AG^2 = AE^2 + EG^2 = 1 + \theta^2 = 1 + 1/\omega^2 = \omega^3/\omega^2 = \omega$; so $AG = \sqrt{\omega}$.

21. $\dfrac{AE}{AI} = \dfrac{\omega - EB}{\theta} = \dfrac{\omega - (\omega - 1)}{1/\omega} = \omega$. So $AEGI$ is a supergolden rectangle.

23. $\omega^3 - \omega^2 = 1$; so $\omega - 1 = 1/\omega^2 = \theta/\omega$, as desired.

EXERCISES 18

1. (Prielipp) $(L_{n-1}L_{n+2})^2 + (2L_nL_{n+1})^2 = [(L_{n+1} - L_n)(L_{n+1} + L_n)]^2 + (2L_nL_{n+1})^2 = (L_{n+1}^2 - L_n^2)^2 + (2L_nL_{n+1})^2 = (L_{n+1}^2 + L_n^2)^2 = (L_{2n} + L_{2n+2})^2$.

3. $(G_{n-1}G_{n+2})^2 + (2G_nG_{n+1})^2 = [(G_{n+1} - G_n)(G_{n+1} + G_n)]^2 + (2G_nG_{n+1})^2 = (G_{n+1}^2 - G_n^2)^2 + (2G_nG_{n+1})^2 = (G_{n+1}^2 + G_n^2)^2$. So $\triangle ABC$ is a Pythagorean triangle.

5. $\triangle CQR$ and $\triangle CDR$ are isosceles triangles with $CQ = CR$ and $CR = DR$, respectively. Thus $CQ = CR = DR$. Continuing like this, we get $CQ = CR = DR = DS = ES = ET = AT = AP$.
   Since the point of intersection of any two diagonals originating at adjacent vertices of a regular pentagon divides each diagonal in the golden ratio, it follows that $DQ = \alpha DR$, so $RQ = DQ - DR = (\alpha - 1)DR$. Similarly, $ST = (\alpha - 1)SD$. So $QR = ST$. Continuing like this, we get $PQ = QR = RS = ST = TP$. Thus $PQRST$ is a regular pentagon.

7. By Heron's formula, area $= \sqrt{s(s - a)(s - b)(s - c)}$, where $2s = 2AP + TP = 2(a/\alpha^2) + a/\alpha^3 = 2a/\alpha$. So $s = a/\alpha$. Then area $=$
   $\sqrt{a/\alpha(a/\alpha - a/\alpha^2)(a/\alpha - a/\alpha^2)(a/\alpha - a/\alpha^3)} = \dfrac{a^2}{\alpha^4}\sqrt{(\alpha - 1)(\alpha - 1)(\alpha^2 - 1)} = \dfrac{a^2(\alpha - 1)\sqrt{\alpha}}{\alpha^4}$.

9. Area $\triangle CDS = 1/2 \cdot CS \cdot h = \dfrac{1}{2} \cdot \dfrac{a}{\alpha} \cdot \dfrac{a\sqrt{\alpha + 2}}{2\alpha^2} = \dfrac{a^2\sqrt{\alpha + 2}}{4\alpha^3}$.

11. Area $SPRD = 2(\text{area} \triangle RDS) = 2 \cdot \dfrac{1}{2} \cdot \dfrac{a}{\alpha^3} \cdot \dfrac{a\sqrt{\alpha + 2}}{2\alpha^2} = \dfrac{a^2\sqrt{\alpha + 2}}{2\alpha^5}$.

13. $\alpha^3:1$.

15. $SR = a/\alpha^3$, $TQ = (a/\alpha^3)\alpha = a/\alpha^2$, and $ZY = SR/\alpha^3 = a/\alpha^6$; so $2s = 2PV + ZV = 2(SR/\alpha^2) + a/\alpha^6 = 2a/\alpha^5 + a\alpha^6 = 2a(\alpha + 1)/\alpha^6 = 2a/\alpha^4$; so $s = a/\alpha^4$. By Heron's formula, area $\triangle PVZ =$
   $\sqrt{\dfrac{a}{\alpha^4}\left(\dfrac{a}{\alpha^4} - \dfrac{a}{\alpha^5}\right)\left(\dfrac{a}{\alpha^4} - \dfrac{a}{\alpha^5}\right)\left(\dfrac{a}{\alpha^4} - \dfrac{a}{\alpha^6}\right)} = \dfrac{a^2}{\alpha^{10}}\sqrt{(\alpha - 1)(\alpha - 1)(\alpha^2 - 1)} = \dfrac{a^2(\alpha - 1)\sqrt{\alpha}}{\alpha^{10}}$.

17. Area $\triangle DRS = \dfrac{1}{2} \cdot RS \cdot h = \dfrac{1}{2} \cdot \dfrac{a}{\alpha^3} \cdot \dfrac{a}{\alpha} \cos \pi/5 = \dfrac{a^2 \cos \pi/5}{2\alpha^4} = \dfrac{a^2}{4\alpha^3}$. So

desired area = area $PQRST + 5(\text{area } \triangle DRS) =$

$$\left\{ \dfrac{5a^2}{\alpha\sqrt{3-\alpha}} - \dfrac{5a^2[\alpha^2 + (\alpha-1)\sqrt{\alpha-1}]}{4\alpha^4} \right\} + \dfrac{5a^2}{4\alpha^3} =$$

$$\dfrac{5a^2[\alpha^3 - \sqrt{3-\alpha} - (\alpha-1)\sqrt{3\alpha-4}]}{4\alpha^4\sqrt{3-\alpha}}.$$

19. By the Pythagorean theorem, $(ar^2)^2 = a^2 + (ar)^2$. Then $r^4 = r^2 + 1$, so $r^2 = \alpha$ and hence $r = \sqrt{\alpha}$.

21. Since $\triangle GOK$ is isosceles, $GK = GO = r$. $\triangle GLD$ and $\triangle GOK$ are similar and golden; so $\dfrac{GD}{LD} = \dfrac{GK}{OK} = \alpha$; that is, $\dfrac{GD}{LD} = \dfrac{r}{OK} = \alpha$. Therefore, $GD = \alpha LD$. But $LD = OD = r$, so $GD = r\alpha$. Since $AD = GD$, $AD = r\alpha$.

23. $x^5 - 1 = (x-1)(x^4 + x^3 + x^2 + x + 1)$. The equation $x^4 + x^3 + x^2 + x + 1 = 0$ can be written as $(x^2 + 1/x^2) + (x + 1/x) + 1 = 0$. Letting $y = x + 1/x$, this becomes $y^2 + y - 1 = 0$, so $y = -\alpha, -\beta$. When $y = -\alpha, x^2 + \alpha x + 1 = 0$, so $x = \dfrac{-\alpha \pm \sqrt{\alpha^2 - 4}}{2}$. Similarly, $y = -\beta$ yields $x = \dfrac{-\beta \pm \sqrt{\beta^2 - 4}}{2}$. Thus the five solutions are 1, $\dfrac{-\alpha \pm \sqrt{\alpha^2 - 4}}{2}$, and $\dfrac{-\beta \pm \sqrt{\beta^2 - 4}}{2}$.

25. $AB^2 = \left(\dfrac{\alpha-1}{2} + \dfrac{\alpha}{2}\right)^2 + \left(\dfrac{\alpha\sqrt{3-\alpha}}{2} - \dfrac{\sqrt{3-\alpha}}{2}\right)^2 =$

$\dfrac{(2\alpha-1)^2 + (3-\alpha)(\alpha-1)^2}{4} = 3 - \alpha$; so $AB = \sqrt{3-\alpha}$.

27. $AE = \sqrt{3-\alpha}$, $BD = \alpha\sqrt{3-\alpha}$, and $h = \dfrac{\alpha}{2} + \dfrac{\alpha-1}{2} = \dfrac{2\alpha-1}{2}$.

Area $ABDE = \dfrac{1}{2}(AE + BD)h = \dfrac{1}{2}\sqrt{3-\alpha} + \alpha\sqrt{3-\alpha} \cdot \dfrac{2\alpha-1}{2} =$

$\dfrac{(3\alpha-1)\sqrt{3-\alpha}}{4}$.

29. $PQ^2 = \left(\dfrac{2\beta+1}{2} + \dfrac{\beta}{2}\right)^2 + \left(\dfrac{\beta\sqrt{3-\alpha}}{2} + \dfrac{\beta\sqrt{3-\alpha}}{2}\right)^2$; so $PQ = \dfrac{\sqrt{19\beta+13}}{2}$.

$AP^2 = \left(\dfrac{2\beta+1}{2} + \dfrac{\alpha}{2}\right)^2 + \left(\dfrac{3\sqrt{3-\alpha}}{2} + \dfrac{\sqrt{3-\alpha}}{2}\right)^2$; so $AP = \dfrac{\sqrt{9\beta+21}}{2}$.

$QC^2 = \left(1 + \dfrac{\beta}{2}\right)^2 + \dfrac{\beta^4(3-\alpha)}{4}$; so $QC = 3 + 4\beta$.

31. $1 : 1/\alpha : 1/\alpha^2 : 1/\alpha^3$.

33. $\triangle ACE \sim \triangle AED$. So $\dfrac{AE}{AC} = \dfrac{AD}{AE} = x$ (say). Since $AE = AB$ and $AC = BD$,

$\dfrac{AD}{AE} = \dfrac{AB + BD}{AE} = \dfrac{AE + BD}{AE} = 1 + \dfrac{AC}{AE}$; that is, $x = 1 + 1/x$, so $x = \alpha$.

35. $F_2 = 2 \sum\limits_{k=0}^{1} (-1)^k \cos^{1-k} \pi/5 \sin^k \pi/10 = 1.$

$F_4 = 2^3 \sum\limits_{k=0}^{3} (-1)^k \cos^{3-k} \pi/5 \sin^k \pi/10 = \alpha^3 + \alpha^2\beta + \alpha\beta^2 + \beta^3$

$$= (\alpha + \beta)^3 - 2\alpha\beta(\alpha + \beta) = 3.$$

37. (Hoggatt) By equations (18.4) and (18.5), we have

$$F_n = \frac{2^{n+2}}{5}(\cos^n \pi/5 \sin \pi/5 \sin 3\pi/5$$

$$+ \cos^n 3\pi/5 \sin 3\pi/5 \sin 9\pi/5)$$

$$(-1)^n F_n = \frac{2^{n+2}}{5}(\cos^n 2\pi/5 \sin 2\pi/5 \sin 6\pi/5$$

$$+ \cos^n 4\pi/5 \sin 4\pi/5 \sin 12\pi/5)$$

$$F_n + (-1)^n F_n = \frac{2^{n+2}}{5} \sum\limits_{k=1}^{4} \cos^n k\pi/5 \sin k\pi/5 \sin 3k\pi/5.$$

This yields the desired result.

39. The asymptotes are $y = \pm\dfrac{x}{\sqrt{\alpha}}$. Solving the equations $y = \pm\dfrac{x}{\sqrt{\alpha}}$ and $y^2 = 4ax$, we get the given points.

41. $PQ = \sqrt{(a\alpha^2 - a\beta^2)^2 + (2a\alpha - 2a\beta)^2} = \sqrt{5a^2(\alpha - \beta)^2} = 5a.$

43. $x - \beta y + a\beta^2 = 0, x - \alpha y - 4a + a\alpha = 0.$

45. The slopes of the tangents at P and Q are $1/\alpha$ and $1/\beta$, respectively. Then

$\tan\theta = \left| \dfrac{1/\alpha - 1/\beta}{1 + 1/\alpha \cdot 1/\beta} \right| = \dfrac{\alpha - \beta}{0} = \infty$; so $\theta = \pi/2$.

47. The slopes of the normals at P and Q are $-\alpha$ and $-\beta$, respectively. Then

$\tan\theta = \left| \dfrac{-\alpha + \beta}{1 + \alpha\beta} \right| = \dfrac{\alpha - \beta}{0} = \infty$; so $\theta = \pi/2$.

49. $SQ = \sqrt{(a - a\beta^2)^2 + (0 - 2a\beta)^2} = \sqrt{5}a|\beta|$ and

$QR = \sqrt{(a\beta^2 - 0)^2 + (2a\beta + 2a)^2} = \sqrt{5}a\beta^2$. So $SQ : QR = |\beta| : \beta^2 = \alpha : 1.$

EXERCISES 19

1. $1 + \cfrac{1}{2 + \cfrac{1}{5 + \cfrac{1}{3}}}.$

3. $\dfrac{52}{23}$.

5. $1, \dfrac{3}{2}, \dfrac{10}{7}, \dfrac{43}{30}, \dfrac{225}{157}$.

7. $C_4 = \dfrac{p_4}{q_4} = \dfrac{5 \cdot 43 + 10}{5 \cdot 30 + 7} = \dfrac{225}{157}$; $C_5 = \dfrac{p_5}{q_5} = \dfrac{6 \cdot 225 + 43}{6 \cdot 157 + 30} = \dfrac{1393}{972}$.

9. Since $C_0 = 1 = F_2/F_1$ and $C_1 = 2 = F_3/F_2$, the formula works for $n = 0$ and $n = 1$. Assume it is true for all integers $0 \le k \le n$, where $n \ge 1$. Then $p_n = 1 \cdot p_{n-1} + p_{n-2} = F_n + F_{n-1} = F_{n+1}$. Similarly, $q_n = F_n$. Thus $C_n = p_n/q_n = F_{n+1}/F_n$. Thus, by PMI, the formula works for all $n \ge 0$.

11. $C_n = \dfrac{F_{n+1}}{F_n}$. So $C_n - C_{n-1} = \dfrac{F_{n+1}}{F_n} - \dfrac{F_n}{F_{n-1}} = \dfrac{F_{n+1}F_{n-1} - F_n^2}{F_n F_{n-1}} = \dfrac{(-1)^n}{F_n F_{n-1}}$. Thus $\lim\limits_{n \to \infty} (C_n - C_{n-1}) = 0$.

13. $x = L_n + \dfrac{1}{x}$, so $x = \dfrac{L_n + \sqrt{L_n^2 + 4}}{2}$.

EXERCISES 20

1. $F_n Q + F_{n-1} I = F_n \begin{bmatrix} 1 & 1 \\ 1 & 0 \end{bmatrix} + F_{n-1} \begin{bmatrix} 1 & 0 \\ 0 & 1 \end{bmatrix} = \begin{bmatrix} F_{n+1} & F_n \\ F_n & F_{n-1} \end{bmatrix} = Q^n$.

3. LHS $= \sum\limits_{i=0}^{n+1} Q^i - \sum\limits_{i=1}^{n} Q^i = Q^{n+1} - Q^0 = Q^{n+1} - I$.

5. Follows by equating the corresponding elements in the last two matrices in Corollary 20.2.

7. $F_{m+n+1} = F_{m+(n+1)} = F_{m+1}F_{n+1} + F_m F_n$ and $F_{m+n-1} = F_{(m-1)+n} = F_m F_n + F_{m-1}F_{n-1}$. Subtracting, we get the desired identity.

9. Since $m, n \ge 2$, $mn > m + n - 1$. Then $F_{mn} > F_{m+n-1} = F_m F_n + F_{m-1}F_{n-1} > F_m F_n$.

11. Add identities (20.2) and (20.3): $2F_{m+n} = F_m(F_{n-1} + F_{n+1}) + F_n(F_{m-1} + F_{m+1}) = F_m L_n + F_n L_m$.

13. Using Exercise 20.11, $2F_{m-n} = F_m L_{-n} + F_{-n} L_m = (-1)^n F_m L_n + (-1)^{n+1} F_n L_m = (-1)^n (F_m L_n - F_n L_m)$.

15. By Binet's formula, LHS $= 6(\alpha^{m+n} + \beta^{m+n}) + 4(\alpha^m \beta^n + \alpha^n \beta^m) = 6L_{m+n} + 4(-1)^n(\alpha^{m-n} + \beta^{m-n}) = 6L_{m+n} + 4(-1)^n L_{m-n} = $ RHS.

17. This follows by changing m to $-m$ in the identity in Exercise 20.5.

19. Using Exercises 20.5 and 20.17, $F_{m+n} - (-1)^n F_{m-n} = (F_{m+1} + F_{m-1})F_n = L_m F_n$.

21. By Exercises 20.5 and 20.17, $F_{m+n} + F_{m-n} = F_m F_{n-1}[1 + (-1)^n] + F_n[F_{m+1} - (-1)^n F_{m-1}]$. This yields the desired result.

23. By Exercises 20.8 and 20.18, LHS $= F_{m+1}L_n[1 + (-1)^n] + F_m[L_{n-1} - (-1)^n L_{n+1}] =$

$$\begin{cases} F_m(L_{n-1} + L_{n+1}) & \text{if } n \text{ is odd} \\ 2F_{m+1}L_n + F_m(L_{n-1} - L_{n+1}) & \text{otherwise} \end{cases} = \begin{cases} 5F_m F_n & \text{if } n \text{ is odd} \\ L_m F_n & \text{otherwise} \end{cases} = \text{RHS}.$$

25. This follows by Exercises 20.21 and 20.22.

27. By Exercises 20.8 and 20.18, $L_{a+b} - (-1)^b L_{a-b} = 5F_a F_b$. Letting $a = n - m$ and $b = m + 1$; and $a = n - m - 1$ and $b = m$, this yields $L_{n+1} + (-1)^m L_{n-2m-1} = 5F_{n-m}F_{m+1}$ and $L_{n-1} - (-1)^m L_{n-2m-1} = 5F_{n-m-1}F_m$, respectively. Subtracting, we get $L_n + 2(-1)^m L_{n-2m-1} = 5(F_{n-m}F_{m+1} - F_{n-m-1}F_m)$. This gives the desired result.

29. This follows by Exercise 20.17.

31. The result is true when $n = 1$. Now assume it is true for an arbitrary integer k. Then $M^k = M^k \cdot M = \begin{bmatrix} F_{2k-1} & F_{2k} \\ F_{2k} & F_{2k+1} \end{bmatrix} \begin{bmatrix} 1 & 1 \\ 1 & 2 \end{bmatrix} = \begin{bmatrix} F_{2k+1} & F_{2k+2} \\ F_{2k+2} & F_{2k+3} \end{bmatrix}$. Thus, by PMI, the result is true for every $n \geq 1$.

33. (Rabinowitz) $4A_{2n} = \begin{bmatrix} 1 & 3 \\ 3 & 1 \end{bmatrix} A_n^2 - \begin{bmatrix} 2 & 2 \\ 2 & 2 \end{bmatrix} A_n A_{n+1} + \begin{bmatrix} 0 & 2 \\ 2 & 0 \end{bmatrix} A_{n+1}^2$.

35. $|M - \lambda I| = \lambda^2 - \alpha \lambda - \beta$.

37. $\dfrac{\beta}{\sqrt{5}} \begin{bmatrix} -\alpha & -1 \\ -1 & \alpha \end{bmatrix}$.

39. Follows by PMI.

41. Follows by PMI.

43. Follows by PMI.

45. By Cramer's rule, $y = \dfrac{\begin{bmatrix} G_n & G_{n+1} \\ G_{n+1} & G_{n+2} \end{bmatrix}}{\begin{bmatrix} G_n & G_{n-1} \\ G_{n+1} & G_n \end{bmatrix}} = \dfrac{G_n G_{n+2} - G_{n+1}^2}{G_n^2 - G_{n-1}G_{n+1}}$. But $y = 1$. So

$G_n G_{n+2} - G_{n+1}^2 = -(G_{n-1}G_{n+1} - G_n^2)$. Let $q_n = G_{n+1}G_{n-1} - G_n^2$, where $q_1 = G_2 G_0 - G_1^2 = b(b - a) - a^2 = -\mu$. Then $q_n = (-1)^{n+1}q_1 = (-1)^n \mu$; that is, $G_{n+1}G_{n-1} - G_n^2 = (-1)^n \mu$.

47. $\begin{bmatrix} F_{n+1} \\ F_n \end{bmatrix}$.

49. $\begin{bmatrix} F_n \\ -F_{n+1} \end{bmatrix}$ if n is odd; and $\begin{bmatrix} -F_n \\ F_{n+1} \end{bmatrix}$ otherwise.

51. $V_m Q^{n+1} = (L_{m+1}, L_m) \begin{bmatrix} F_{n+2} & F_{n+1} \\ F_{n+1} & F_n \end{bmatrix} = (L_{m+n+2}, L_{m+n+1}) = V_{m+n+1}$.

53. $(a + e - b - d)(e + j - h - f) - (b + f - c - e)(d + h - g - e)$.

55. $\lambda(P) = \begin{vmatrix} 1 & 1 & 2 \\ 1 & 2 & 3 \\ 2 & 2 & 2 \end{vmatrix} - \begin{vmatrix} 0 & 0 & 1 \\ 0 & 1 & 2 \\ 1 & 1 & 1 \end{vmatrix} = -1.$

57. $\lambda R = L_{n+1}L_{n-1} - L_n^2 = 5(-1)^{n+1}.$

59. $G_{n+k} + G_{n-k} - 2G_n.$

61. Notice that $F_{n+1}^2 - F_{n-1}F_n + 2F_nF_{n+1} = F_{n+1}(F_{n+1} + F_n) + F_n(F_{n+1} - F_{n-1}) = F_{n+1}F_{n+2} + F_n^2 = F_{n+2}(F_{n+2} - F_n) + F_n^2 = F_{n+2}^2 - F_n(F_{n+2} - F_n) = F_{n+2}^2 - F_nF_{n+1}$, and $2F_{n+1}^2 + 2F_nF_{n+1} = 2F_{n+1}(F_{n+1} + F_n) = 2F_{n+1}F_{n+2}.$
The given formula works when $n = 1$. Now assume it works for $n \geq 1$. Then

$$P^{n+1} = P^n \cdot P = \begin{bmatrix} F_{n-1}^2 & F_{n-1}F_n & F_n^2 \\ 2F_{n-1}F_n & F_{n+1}^2 - F_{n-1}F_n & 2F_nF_{n+1} \\ F_n^2 & F_nF_{n+1} & F_{n+1}^2 \end{bmatrix} \begin{bmatrix} 0 & 0 & 1 \\ 0 & 1 & 2 \\ 1 & 1 & 1 \end{bmatrix}$$

$$= \begin{bmatrix} F_n^2 & F_{n-1}F_n + F_n^2 & F_{n-1}^2 + 2F_{n-1}F_n + F_n^2 \\ 2F_nF_{n+1} & F_{n+1}^2 - F_{n-1}F_n + 2F_nF_{n+1} & 2F_{n+1}^2 + 2F_nF_{n+1} \\ F_{n+1}^2 & F_nF_{n+1} + F_{n+1}^2 & F_n^2 + 2F_nF_{n+1} + F_{n+1}^2 \end{bmatrix}$$

$$= \begin{bmatrix} F_n^2 & F_nF_{n+1} & F_{n+1}^2 \\ 2F_nF_{n+1} & F_{n+2}^2 - F_nF_{n+1} & 2F_{n+1}F_{n+2} \\ F_{n+1}^2 & F_{n+1}F_{n+2} & F_{n+2}^2 \end{bmatrix}.$$

Thus, by PMI, the formula works for all $n \geq 1$.

63. LHS $= F_n(F_n - F_{n-1}) + F_{n+1}(F_{n+1} - F_n) - 2F_nF_{n-1} = F_nF_{n-2} + F_{n-1}(F_{n+1} - F_n) - F_nF_{n-1} = F_nF_{n-2} + F_{n-1}^2 - F_nF_{n-1} = F_nF_{n-2} - F_{n-1}(F_n - F_{n-1}) = F_nF_{n-2} - F_{n-1}F_{n-2} = F_{n-2}(F_n - F_{n-1}) = F_{n-2}^2.$

65. $\sum_{i=1}^{2n} F_{k+i} = \sum_{i=1}^{k+2n} F_i - \sum_{i=1}^{k} F_i = (F_{k+2n+2} - 1) - (F_{k+2} - 1) = F_{k+2n+2} - F_{k+2} = L_{k+n+2}F_n$, by Exercise 20.22. This gives the desired result.

67. Recall that $F_{n+2} - F_{n-2} = L_n$, $L_{n+2} - L_{n-2} = 5F_n$, and $L_{n+1}F_{2n} + L_nF_{2n-1} = L_{3n}$. In addition, we will need the following facts:
   1) Let $S_n = L_{n+2}^3 - L_{n+4}L_{n+2}L_n$. Then $S_n = 3S_{n-2} - S_{n-4}.$
   2) $L_{n+3}^3 - L_{n+3}L_{n+2} - L_{n-1}L_{n-3} = 15F_{2n}.$
The given identity now follows by these properties, after some tedious algebra.

69. (Singh) The given identity can be rewritten as $F_{m+2n}F_{m+n} - F_{m-n}F_{m-2n} - L_{3n}F_m^2 = 2L_n(F_{m+n}F_{m-n} - F_m^2)$. By Theorem 5.11, RHS $= 2L_n(-1)^{m+n+1}F_n^2 = 2(-1)^mL_nF_n^2$. Using the identity $L_{a+b} - (-1)^bL_{a-b} = 5F_aF_b$, we have

$$5(F_{m+2n}F_{m+n} - F_{m-n}F_{m-2n}) = [L_{2m+3n} + (-1)^mL_n] - [L_{2m-3n} - (-1)^mL_n]$$

$$= (L_{2m+3n} - L_{2m-3n}) + 2(-1)^mL_n$$

$$= L_{2m}L_{3n} + 2(-1)^mL_n$$

$$= L_{3n}[5F_m^2 + 2(-1)^m] + 2(-1)^m L_n$$

$$= 5L_{3n}F_m^2 + 2(-1)^m(L_{3n} + L_n)$$

$$= 5L_{3n}F_m^2 + 10(-1)^m F_{2n}F_n$$

$$= 5L_{3n}F_m^2 + 10(-1)^m L_n F_n^2.$$

So LHS $= 2(-1)^m L_n F_n^2 =$ RHS.

71. (Herring et al.) Since $L_n^2 = L_{2n} + 2(-1)^n$ and $F_{2n} = F_n L_n$, it follows by PMI

that $\displaystyle\sum_{k=1}^{n} \frac{F_{2k-1}}{L_{2k} + 1} = \frac{F_{2n}}{L_{2n} + 1}$. Consequently, since $\displaystyle\lim_{n\to\infty} \frac{L_n}{F_n} = \sqrt{5}$, it follows

that $\displaystyle\sum_{k=1}^{\infty} \frac{F_{2k-1}}{L_{2k} + 1} = \lim_{n\to\infty} \frac{F_{2n}}{L_{2n} + 1} = \frac{1}{\sqrt{5}}$, as desired.

73. (Kwong) Since $L_{m+n} + (-1)^n L_{m-n} = L_m L_n$ and $L_{m+n} - (-1)^n L_{m-n} = 5F_m F_n$, we have $L_{2k} + 2 = L_{2k-1}^2$ and $L_{2k} - 2 = 5F_{2k-1}^2$. So, when

$k \geq 2$, $5F_{2k-1}^2(L_{2k-1}^2 - 1) = (L_{2k} - 2)(L_{2k} + 1) = L_{2k}^2 - L_{2k-1}^2$. Then

$$\frac{F_{2k-1}^2}{L_{2k}^2 - 1} = \frac{L_{2k}^2 - L_{2k-1}^2}{5(L_{2k-1}^2 - 1)(L_{2k}^2 - 1)} = \frac{1}{5}\left(\frac{1}{L_{2k-1}^2 - 1} - \frac{1}{L_{2k}^2 - 1}\right). \text{ So}$$

$$\sum_{k=1}^{\infty} \frac{F_{2k-1}^2}{L_{2k}^2 - 1} = \frac{1}{L_2^2 - 1} + \frac{1}{5} \cdot \frac{1}{L_2^2 - 1} = \frac{3}{20}.$$

EXERCISES 21

1. The characteristic polynomial is $|Q - \lambda I| = \lambda^2 - \lambda - 1$. So the eigenvalues are α and β.

3. 11111 11121 11211 12111 12121.

5. 11112 11212 12112.

7. 1121 1211.

9. Count the number of paths of length n from v_1 to v_2 in two different ways. By Corollary 21.1, there are F_n such paths. Now consider an arbitrary path P of length n from v_1 to v_2. Suppose P ends at v_1 after m steps. There are F_{m+1} such subpaths. Each remaining path is of length $n - m$; there are F_{n-m} such subpaths. So there are $F_{m+1}F_{n-m}$ such paths P. On the other hand, suppose P ends at v_2 after m steps. Similarly, there are $F_m F_{n-m-1}$ such paths. Thus there is a total of $F_{m+1}F_{n-m} + F_m F_{n-m-1}$ paths P. Equating the two counts, the result follows.

11.

$$\sum_{i=1}^{n} L_{2i-1} = 2\sum_{i=1}^{n} F_{2i} - \sum_{i=1}^{n} F_{2i-1}$$

$$= 2\left(\begin{array}{l}\text{number of paths of even} \\ \text{length } \leq 2n \text{ from } v_1 \text{ to } v_2\end{array}\right)$$

$$- \left(\begin{array}{c} \text{number of closed paths of even} \\ \text{length } \leq 2n \text{ from } v_2 \text{ to } v_2 \end{array} \right)$$

$$= 2(F_{2n+1} - 1) - F_{2n} = L_{2n} - 2.$$

13. It follows from Q^{2i} that there are $\sum\limits_{i=1}^{n} F_{2i}$ paths of even length $\leq 2n$ from v_1 to v_2. There are $F_{2n+1} - 1$ such paths.

EXERCISES 22

1. $L_n L_{n+2} - L_{n+1}^2 = 5(-1)^n$.

3. (Finkelstein) Let $r = L_{6d}/L_{3d}$ and $s = L_{6d+1} - rL_{3d+1}$. Then, by PMI, $L_{n+6d} = rL_{n+3d} + sL_n$ for all n. In particular, let $n = a$, $a + d$, and $a + 2d$. Then the rows of the determinant are linearly dependent, so the determinant is zero.

5. Since $G_n = aF_{n-2} + bF_{n-1}$, the determinant is zero by Exercise 22.3.

7. (Jaiswal) Consider the determinant $D = \begin{vmatrix} G_p & G_{p+m} & G_{p+m+n} \\ G_q & G_{q+m} & G_{q+m+n} \\ G_r & G_{r+m} & G_{r+m+n} \end{vmatrix}$. Since

$G_{k+m+n} = G_{k+m}F_{n+1} + G_{k+m-1}F_n$, it follows that

$$D = F_{n+1} \begin{vmatrix} G_p & G_{p+m} & G_{p+m} \\ G_q & G_{q+m} & G_{q+m} \\ G_r & G_{r+m} & G_{r+m} \end{vmatrix} + F_n \begin{vmatrix} G_p & G_{p+m} & G_{p+m-1} \\ G_q & G_{q+m} & G_{q+m-1} \\ G_r & G_{r+m} & G_{r+m-1} \end{vmatrix}$$

$$= F_n \begin{vmatrix} G_p & G_{p+m} & G_{p+m-1} \\ G_q & G_{q+m} & G_{q+m-1} \\ G_r & G_{r+m} & G_{r+m-1} \end{vmatrix} = F_n \begin{vmatrix} G_p & G_{p+m-2} & G_{p+m-1} \\ G_q & G_{q+m-2} & G_{q+m-1} \\ G_r & G_{r+m-2} & G_{r+m-1} \end{vmatrix}$$

$$= F_n \begin{vmatrix} G_p & G_{p+m-2} & G_{p+m-3} \\ G_q & G_{q+m-2} & G_{q+m-3} \\ G_r & G_{r+m-2} & G_{r+m-3} \end{vmatrix}.$$

Thus, by alternately subtracting columns 2 and 3 from one another, the process can be continued to decrease the subscripts. After a certain stage, when m is even, columns 1 and 2 would become identical; and when m is odd, columns 1 and 3 would become identical. In either case, $D = 0$. The given determinant Δ can be written as the sum of eight determinants. Using the facts that $D = 0$ and a determinant vanishes when two columns are identical, $\Delta = \begin{vmatrix} G_p & G_{p+m} & k \\ G_q & G_{q+m} & k \\ G_r & G_{r+m} & k \end{vmatrix} + \cdots + \cdots = \Delta_1 + \Delta_2 + \Delta_3$ (say). Since

$$G_{m-1}G_n - G_m G_{n-1} = (-1)^{m-1}\mu F_{n-m},$$

$$\Delta_1 = kF_m \begin{vmatrix} G_p & G_{p-1} & 1 \\ G_q & G_{q-1} & 1 \\ G_r & G_{r-1} & 1 \end{vmatrix} = k\mu F_m[(-1)^{r-1}F_{q-r} + (-1)^{p-1}F_{r-p} + (-1)^{q-1}F_{p-q}].$$

So $\Delta = k\mu F_m[(-1)^q F_{r-q} - (-1)^p F_{r-p} + (-1)^p F_{q-p}][F_m - F_{m+n} + (-1)^m F_n].$

9. (Jaiswal) Since $\begin{vmatrix} a & b & c & d \\ b & a & d & c \\ c & d & a & b \\ d & c & b & a \end{vmatrix} = [(a+b)^2 - (c+d)^2][(a-b)^2 - (c-d)^2],$

the given determinant $\Delta = [(G_{n+3} + G_{n+2})^2 - (G_{n+1} + G_n)^2][(G_{n+3} - G_{n+2})^2 - (G_{n+1} - G_n)^2] = (G_{n+4}^2 - G_{n+2}^2)(G_{n+1}^2 - G_{n-1}^2)$. But $(G_{m+1}^2 - G_{m-1}^2) = aG_{2m-2} + bG_{2m-1}$. So $\Delta = (aG_{2n+4} + bG_{2n+5})(aG_{2n-2} + bG_{2n-1})$.

11. By Theorem 22.7 with $k = 2, m = 1, r = 0$, and $a_n = L_n$, $D =$

$$(-1)^{n \cdot 2 \cdot 3/2} A_2(L_0) = (-1)^{3n} \begin{vmatrix} L_0^2 & L_1^2 & L_2^2 \\ L_1^2 & L_2^2 & L_3^2 \\ L_2^2 & L_3^2 & L_4^2 \end{vmatrix} = (-1)^n \begin{vmatrix} 4 & 1 & 9 \\ 1 & 9 & 16 \\ 9 & 16 & 49 \end{vmatrix} = 250(-1)^n.$$

13. By Theorem 22.7 with $k = 3, m = 1, r = 0$, and $a_n = L_n$, $D = (-1)^{n \cdot 3 \cdot 4/2} A_3(L_0)$

$$= \begin{vmatrix} L_0^3 & L_1^3 & L_2^3 & L_3^3 \\ L_1^3 & L_2^3 & L_3^3 & L_4^3 \\ L_2^3 & L_3^3 & L_4^3 & L_5^3 \\ L_3^3 & L_4^3 & L_5^3 & L_6^3 \end{vmatrix} = \begin{vmatrix} 4 & 1 & 27 & 64 \\ 1 & 27 & 64 & 343 \\ 27 & 64 & 343 & 1331 \\ 343 & 1331 & 5832 & 24389 \end{vmatrix} = 0.$$

15. $g_{n+2}(x) = 2xg_{n+1}(x) + g_n(x), n \geq 0$.

17. (Brown) $Q^n = \begin{bmatrix} F_{n+1} & F_n \\ F_n & F_{n-1} \end{bmatrix}; e^{Q^n} = \begin{bmatrix} \sum\limits_{k=0}^{\infty} \dfrac{F_{nk+1}}{k!} & \sum\limits_{k=0}^{\infty} \dfrac{F_{nk}}{k!} \\ \sum\limits_{k=0}^{\infty} \dfrac{F_{nk}}{k!} & \sum\limits_{k=0}^{\infty} \dfrac{F_{nk-1}}{k!} \end{bmatrix}$. We have

$\sum\limits_{k=0}^{\infty} \dfrac{F_{nk}}{k!} x^k = \dfrac{e^{\alpha^n x} - e^{\beta^n x}}{\alpha - \beta}$ and $\sum\limits_{k=0}^{\infty} \dfrac{L_{nk}}{k!} x^k = e^{\alpha^n x} + e^{\beta^n x}$. Since $L_a = F_{a+1} +$

F_{a-1}, then $\sum\limits_{k=0}^{\infty} \dfrac{F_{nk+1} + F_{nk-1}}{k!} = e^{\alpha^n} + e^{\beta^n}$; that is,

$$\sum\limits_{k=0}^{\infty} \dfrac{F_{nk+1}}{k!} = e^{\alpha^n} + e^{\beta^n} - \sum\limits_{k=0}^{\infty} \dfrac{F_{nk-1}}{k!}. \tag{8}$$

We also have

$$\sum\limits_{k=0}^{\infty} \dfrac{F_{nk+1}}{k!} = \sum\limits_{k=0}^{\infty} \dfrac{F_{nk}}{k!} + \sum\limits_{k=0}^{\infty} \dfrac{F_{nk-1}}{k!} = \dfrac{e^{\alpha^n} - e^{\beta^n}}{\alpha - \beta} + \sum\limits_{k=0}^{\infty} \dfrac{F_{nk-1}}{k!}. \tag{9}$$

From equations (8) and (9),

$$\sum_{k=0}^{\infty} \frac{F_{nk+1}}{k!} = \frac{1}{2}\left[(e^{\alpha^n} + e^{\beta^n}) + \frac{e^{\alpha^n} - e^{\beta^n}}{\alpha - \beta}\right] \quad \text{and}$$

$$\sum_{k=0}^{\infty} \frac{F_{nk-1}}{k!} = \frac{1}{2}\left[(e^{\alpha^n} + e^{\beta^n}) - \frac{e^{\alpha^n} - e^{\beta^n}}{\alpha - \beta}\right].$$

Thus

$$|e^{Q_n}| = \left(\sum_{k=0}^{\infty} \frac{F_{nk+1}}{k!}\right)\left(\sum_{k=0}^{\infty} \frac{F_{nk-1}}{k!}\right) - \left(\sum_{k=0}^{\infty} \frac{F_{nk}}{k!}\right)^2$$

$$= \frac{1}{4}\left[(e^{\alpha^n} + e^{\beta^n})^2 - \left(\frac{e^{\alpha^n} - e^{\beta^n}}{\alpha - \beta}\right)^2\right] - \left(\frac{e^{\alpha^n} - e^{\beta^n}}{\alpha - \beta}\right)^2$$

$$= e^{\alpha^n + \beta^n} = e^{L_n}.$$

19. (Parker) Let D_n denote the given determinant. Expanding it by the last column, $D_n = a_n D_{n-1} + b_{n-1} D_{n-2}$. Then $g_n = g_{n-1} + g_{n-2}$, where $g_1 = 1$ and $g_2 = 2$; so $g_n = F_{n+1}$.

21. (Parker) Let D_n denote the given determinant. Then $D_1 = a + b$ and $D_2 = a^2 + ab + b^2$. Expanding D_n by row 1, $D_n = (a+b)D_{n-1} - abD_{n-2}$. Solving this second-order LHRWCCs, we get

$$D_n = \begin{cases} (n+1)a^n & \text{if } a = b \\ (a^{n+1} - b^{n+1})/(a-b) & \text{otherwise.} \end{cases}$$

EXERCISES 23

1. $F_6 13^5 + F_5 13^6 = 8 \cdot 13^5 + 5 \cdot 13^6 \equiv 134 + 48 \equiv 1 \pmod{181}$; $L_6 13^5 + L_5 13^6 = 18 \cdot 13^5 + 11 \cdot 13^6 \equiv 30 + 178 \equiv 27 \equiv 1 + 2 \cdot 13 \pmod{181}$.

3. $F_4 - 11F_5 = 3 - 11 \cdot 5 \equiv 79 \equiv (-1)^5 11^5 \pmod{131}$.

5. Let $F_n \equiv 0 \pmod 3$. Then $3|F_n$; that is, $F_4|F_n$. So $n \equiv 0 \pmod 4$. Conversely, let $n \equiv 0 \pmod 4$. Then $4|n$, so $F_4|F_n$; that is, $3|F_n$. So $F_n \equiv 0 \pmod 3$.

7. $F_n \equiv 0 \pmod 5$ if and only if $5|F_n$; that is, if and only if $F_5|F_n$. Thus $F_n \equiv 0 \pmod 5$ if and only if $n \equiv 0 \pmod 5$.

9. Since $5F_n^2 = L_n^2 - 4(-1)^n$, $F_n^2 \equiv L_n^2 \pmod 4$.

11. Since $2L_{m+n} = L_m L_n + 5F_m F_n$, $2L_{m+n} \equiv L_m L_n \pmod 5$.

13. $L_{(2k-1)n} = \alpha^{(2k-1)n} + \beta^{(2k-1)n} = (\alpha^n + \beta^n)[\alpha^{(2k-2)n} - \alpha^{(2k-3)n}\beta^n + \cdots + \beta^{(2k-2)n}] \equiv 0 \pmod{L_n}$.

15. Since $F_{m+n} = F_{m-1}F_n + F_m F_{n+1}$, $F_{n+24} = F_{23}F_n + F_{24}F_{n+1}$, where $F_{23} = 28{,}657 \equiv 1 \pmod 9$ and $F_{24} = 46{,}368 \equiv 0 \pmod 9$. So $F_{n+24} \equiv F_n \pmod 9$.

17. Since $F_3|F_{3n}, F_{3n} \equiv 0 \pmod 2$.

19. Since $F_5|F_{5n}, F_{5n} \equiv 0 \pmod 5$.

21. Since $2|n$ and $3\nmid n$, n is of the form $6k+2$ or $6k+4$.
    Case 1. Let $n = 6k+2$. Then $L_n = L_{6k+2} = F_{6k+1}L_2 + F_{6k}L_1$. Since $6|6k$, $8|F_{6k}$; so $F_{6k} \equiv 0 \pmod 4$. So $L_n \equiv F_{6k+1} \cdot 3 + 0 \equiv 3F_{6k+1} \pmod 4$. But $F_{6k+1} \equiv 1 \pmod 4$. Thus $L_n \equiv 3 \pmod 4$.
    Case 2. Let $n = 6k+4$. Then $L_n = L_{6k+4} = L_{(6k+1)+3} = F_{6k+2}L_3 + F_{6k+1}L_2 \equiv F_{6k+2} \cdot 0 + 1 \cdot 3 \equiv 3 \pmod 4$. Thus in both cases, $L_n \equiv 3 \pmod 4$.

23. By Exercise 10.39, $2|L_{3n}$; so the congruence follows.

25. Let $(F_n, L_n) = 2$. Then $2|F_n$; that is, $F_3|F_n$. So $n \equiv 0 \pmod 3$. Conversely, let $n = 3m$. Then $2|F_{3m}$, and by Exercise 23.23, $2|L_{3n}$ also. So $2|(F_{3m}, L_{3m})$. Suppose $(F_{3m}, L_{3m}) = d > 2$ for all n. Then $F_{3m+3} = F_{3m}F_4 + F_{3m-1}F_3 = 3F_{3m} + 2F_{3m-1}$ and $L_{3m+3} = F_{3m}L_4 + L_{3m-1}L_3 = 7F_{3m} + 4F_{3m-1}$. Then $d|F_{3m+3}$ implies $d|2F_{3m-1}$. This is a contradiction, since $d > 2$ and $(F_{3m}, F_{3m-1}) = 1$. Thus $(F_n, L_n) = (F_{3m}, L_{3m}) = 2$.

27. Since $5(F_n^2 + F_{n-2}^2) = 3L_{2n-2} - 4(-1)^n$ by Exercise 5.34, the congruence follows.

29. Follows by the identity $5F_nF_{n+1} = L_{2n+1} - (-1)^n$.

31. Follows by the identity $L_{m+n}^2 + 5F_{m-n}^2 = L_{2m}L_{2n}$.

33. (Wessner) The result is true when $n = 1$. Assume it is true for every $i \leq k$, where $k \geq 1$. Then $2^{k+1}F_k = 2(2^k F_k + 2 \cdot 2^{k-1}F_{k-1}) \equiv 2[2^k + 2 \cdot 2(k-1)] \equiv 2(k-4) \equiv 2(k+1) \pmod 5$. By PMI, the result now follows.

35. (Bruckman) We will use the identity $F_{5m} = 25F_m^5 + 25(-1)^m F_m^3 + 5F_m$ and PMI, where $m \geq 0$. The result is true when $n = 0$ and $n = 1$. Assume it is true for k: $F_{5k} \equiv 5^k \pmod{5^{k+3}}$, so $F_m = m(1 + 125a)$ for some integer a, where $m = 5^k$. By the preceding identity, $F_{5m} = 5^2m^5(1 + 125a)^5 - 5^2m^3(1 + 125a)^3 + 5m(1 + 125a) \equiv 5^2m^5 - 5^2m^3 + 5m \pmod{5^4m}$. Since $k \geq 1$, $5|m$; so $5^2|m^2$. Hence $5^4m|5^2m^3$. So $F_{5m} \equiv 5m \pmod{5^4m}$; that is, $F_{5k+1} \equiv 5^{k+1} \pmod{5^{k+4}}$. Thus, by PMI, the result is true for all $n \geq 0$.

37. (Prielipp) Since $L_n^2 = 5F_n^2 + 4(-1)^n, (L_n^2)^2 = (5F_n^2)^2 + 8(-1)^n(5F_n^2) + 4^2$. This yields the given congruence.

39. $\displaystyle\sum_{i=1}^{20} F_{n+i} = \sum_{i=1}^{n+20} F_i - \sum_{i=1}^{n} F_i = F_{n+22} - F_{n+2} = (F_nF_{21} + F_{n+1}F_{22}) - (F_n + F_{n+1}) = F_n(F_{21} - 1) + F_{n+1}(F_{22} - 1) \equiv F_n \cdot 0 + F_{n+1} \cdot 0 \equiv 0 \pmod{F_{10}}$, since $F_{21} \equiv F_{22} \equiv 1 \pmod{55}$.

41. Follows by PMI.

43. (Milsom) RHS $= (\alpha\beta)^{n(n-1)}(\alpha^n + \beta^n) = \alpha^n + \beta^n = L_n = $ LHS.

45. (Prielipp) Using the identities $L_{2n} = 5F_n^2 + 2(-1)^n$, $F_{2n} = F_nL_n$, and $L_n^2 - F_n^2 = 4F_n^2 + 4(-1)^n$, we have

$$\text{LHS} = 25F_{2n}^2 + 10 - [5F_n^2 + 2(-1)^n]^2 + 6 - 6(-1)^n[5F_n^2 + 2(-1)^n]$$

$$= 25F_n^2[(L_n^2 - F_n^2) - 2(-1)^n] = 50F_n^2[(2F_n^2) + (-1)^n].$$

This yields the desired congruence.

47. 5, 89, 11, 199.

49. Let $5|n$. Then $5|F_n$, so $F_n = 5m$ for some integer m. Then $F_{F_n} = F_{5m}$. Since $5|F_{5m}$, it follows that $F_{F_n} \equiv 0 \pmod 5$. Conversely, let $F_{F_n} \equiv 0 \pmod 5$. Then $5|F_n$, so $n \equiv 0 \pmod 5$.

51. $F_{33} = F_{32} + F_{31} \equiv 9 + 9 \equiv 8 \pmod{10}$.

53. $F_{703} = F_{11 \cdot 60 + 43} \equiv F_{43} \equiv 7 \pmod{10}$.

55. $L_{45} \equiv L_9 \equiv 6 \pmod{10}$; so $L_{93} = L_{12 \cdot 7 + 9} \equiv L_9 \equiv 6 \pmod{10}$.

57. $L_{25} = L_{24} + L_{23} \equiv 7 + 2 \equiv 1 \pmod 8$.

59. $L_{i-2} = L_i - L_{i-1} \equiv 3 - 4 \equiv 7 \pmod 8$.

61. The result is true when $n = 0$. Assume it is true for $k \geq 0$. Then $F_{6(k+1)+1} = F_{6k+7} = F_{6k+4} + 2F_{6k+5} = \cdots = 5F_{6k+1} + 8F_{6k+2} \equiv 5 \cdot 1 + 0 \equiv 1 \pmod 4$.
Thus the result follows by PMI.

63. $F_{6n-1} = F_{6n+1} - F_{6n} \equiv 1 - 0 \equiv 1 \pmod 4$.

65. $L_{6n+2} = F_{6n+1} + F_{6n+3} = 3F_{6n+1} + F_{6n} \equiv 3 \cdot 1 + 0 \equiv 3 \pmod 4$, since $F_6|F_{6n}$.

67. By Binet's formula, RHS $= L_{12n+9} - L_{4n+3} + L_{4n+3} = L_{12n+9} = $ LHS.

69. Using the identity $L_{m+n} = F_{m-1}L_n + F_m L_{n-1}$, $L_{4n+2} = F_{4n-1}L_2 + F_{4n}L_1 = 3F_{4n-1} + F_{4n} \equiv 0 \pmod 3$.

71. By Exercises 23.69 and 23.70, $L_{4n+1} = L_{4n+2} - L_{4n} \equiv 0 - \pm 1 \equiv \pm 1 \pmod 3$.

73. Follows by Exercise 23.21.

75. $L_{m+2k} \equiv (-1)L_m \equiv (-1)^2 L_{m-2k} \equiv (-1)^3 L_{m-4k} \equiv \cdots \equiv (-1)^{\lfloor m/4 \rfloor} L_{m-2\lfloor m/4 \rfloor k}$ $\pmod{L_k}$. Let $k = 2$ and $m = 4n - 2$. Then $L_{4n+2} = L_{(4n-2)+2 \cdot 2} \equiv (-1)^{n-1}L_{4n-2-4\lfloor (4n-2)/4 \rfloor} \equiv (-1)^{n-1}L_2 \equiv 0 \pmod 3$.

77. (Kramer and Hoggatt)

$$(\alpha - \beta)\text{LHS} = \alpha^{5^{n+1}} - \beta^{5^{n+1}} = \alpha^{5^n \cdot 5} - \beta^{5^n \cdot 5}$$

$$= (\alpha^{5^n} - \beta^{5^n})(\alpha^{5^n \cdot 4} + \alpha^{5^n \cdot 3}\beta^{5^n} + \alpha^{5^n \cdot 2}\beta^{5^n \cdot 2} + \alpha^{5^n}\beta^{5^n \cdot 3} + \beta^{5^n \cdot 4})$$

$$\text{LHS} = F_{5^n}[(\alpha^{5^n \cdot 4} + \beta^{5^n \cdot 4}) + (-1)^{5^n}(\alpha^{5^n \cdot 2} + \beta^{5^n \cdot 2}) + (-1)^{5^n \cdot 2}]$$

$$= F_{5^n}(L_{4 \cdot 5^n} - L_{2 \cdot 5^n} + 1) = \text{RHS}.$$

79. (Turner) By Exercise 12.14, $2^{n-1}F_n \equiv n + 5n(n-1)(n-2)/6 \pmod{25}$, so $2^{60k-1}F_{60k} \equiv 60k + 50k(60k-1)(60k-2) \equiv 10k \pmod{25}$. Since $2^{20} \equiv 1 \pmod{25}$, it follows that $F_{60k} \equiv 20k \pmod{25}$. Since $6|60k$, $F_6|F_{60k}$; that is, $8|F_{60k}$. So $F_{60k} \equiv 0 \equiv 20k \pmod 4$. Combining the two congruences yields the desired result. (Follows from Exercise 23.81 also.)

81. By the addition formula, $F_{60k+n} = F_{60k}F_{n-1} + F_{60k+1}F_n \equiv 20kF_{n-1} + (60k + 1)F_n \pmod{100}$, by Exercises 23.79 and 23.80.

83. (Peck) Follows since $F_{(n+2)k} - F_{nk} = L_k F_{(n+1)k}$, where k is odd.

85. (Zeitlin) Since $F_{(n+2)k} - L_k F_{(n+1)k} + (-1)^k F_{nk} = 0$, $F_{(n+2)k} - 2F_{(n+1)k} + (-1)^k F_{nk} = (L_k - 2)F_{(n+1)k}$. So when n is even, $F_{(n+2)k} + F_{nk} \equiv 2F_{(n+1)k} \pmod{(L_k - 2)}$.

87. Follows since $L_{(2m+1)(4n+1)} - L_{2m+1} = 5F_{(2m+1)2n}F_{(2m+1)(2n+1)}$.

89. (Prielipp) The statement is true when $n = 0$ and $n = 1$. Assume it is true for all nonnegative integers $\leq k$. Then $F_{3k-2} + F_{3k+1} + F_{k+4} \equiv 0 \pmod 3$. But $6F_{3k-1} + 4F_{3k-2} + 3F_{3k+1} = F_{3k+4}$, so $F_{3k-2} + F_{3k+1} \equiv F_{3k+4} \pmod 3$; that is, $F_{3k+4} + F_{k+4} \equiv F_{3k+4} \pmod 3$. Thus, by PMI, the result follows.

91. (Somer) The result is true when $n = 2$. Assume it works for an arbitrary integer $k \geq 2$. Since $L_m^2 = L_{2m} + 2(-1)^m$, $L_{2k+1} = L_{2k}^2 - 2(-1)^{2k} \equiv 7^2 - 2 \equiv 7 \pmod{10}$. Thus, by PMI, the result follows.

93. (Wulczyn) The result is true when $n = 1$. Assume it works for $k \geq 1$. Since $L_{2m}^2 = L_{4m} + 2$, $L_{3 \cdot 2^{k+1}} = L_{3 \cdot 2^k}^2 - 2 \equiv 2 + 2^{2k+4} + 2^{4k+4} \pmod{2^{4k+5}} \equiv 2 + 2^{2k+4} \pmod{2^{2k+6}}$. Thus, by PMI, the congruence works for all $n \geq 1$.

EXERCISES 24

1. Letting $x = 1/2$ in the series $\sum_{i=1}^{\infty} F_i x^{i-1} = \dfrac{1}{1 - x - x^2}$ gives the desired result.

3. $u_n = A\alpha^n + B\beta^n$, where $A = \dfrac{b - a\beta}{\alpha - \beta}$ and $B = \dfrac{a\alpha - \beta}{\alpha - \beta}$.

5. By Exercise 24.4, $\sum_{i=0}^{\infty} \dfrac{u_i}{k^{i+1}} = \dfrac{a(k - 1) + b}{(k - \alpha)(k - \beta)}$. Let $a = 0$ and $b = 1$. Desired

   sum $= \dfrac{1}{(k - \alpha)(k - \beta)} = \dfrac{1}{k^2 - k - 1}$.

7. Differentiating the power series $\sum_{i=0}^{\infty} L_i x^i = \dfrac{2 - x}{1 - x - x^2}$, we get $\sum_{i=1}^{\infty} i L_i x^{i-1} = \dfrac{1 + 4x - x^2}{(1 - x - x^2)^2}$. Letting $x = 1/2$ yields the given result.

9. Letting $x = 1/2$ in Exercise 24.8 gives the result.

11. Since $\beta < 0$, $1 + \beta^2 > 0$, and $\beta^{2k-1} < 0$. So $0 > \beta^{2k-1}(1 + \beta^2)$; that is, $\alpha^{2k-1} - \alpha^{2k-1} > \beta^{2k-1} + \beta^{2k+1}$. Then $\alpha^{2k-1} - \beta^{2k-1} > -\beta(\alpha^{2k} - \beta^{2k})$; thus $F_{2k-1}/F_{2k} > -\beta$, so $-m/n \notin (\beta, -\beta)$.

13. $F_{2k}/F_{2k-1} \geq 1 > -\beta$, $F_{2k}/F_{2k-1} \notin (\beta, -\beta)$.

15. Yes, by Exercise 24.14.

17. Solving the characteristic equation $t^2 - at - b = 0$, $t = r$ or s. So the general solution is $U_n = Rr^n + Ss^n$, where R and S are to be determined. The two initial conditions yield the linear system $R + S = c$ and $Rr + Ss = d$.

   Solving $R = \dfrac{c}{2} + \dfrac{2d - ca}{2\sqrt{a^2 + 4b}} = P$ and $S = \dfrac{c}{2} - \dfrac{2d - ca}{2\sqrt{a^2 + 4b}} = Q$. So $U_n = Pr^n + Qs^n, n \geq 0$.

19. Letting $a = b = d = 1, B = -10$, and $c = 0$ in Theorem 24.2 gives the result.

21. (Pond) Since $\lim\limits_{n\to\infty} \dfrac{a_{n+1}}{a_n} = \lim\limits_{n\to\infty} \dfrac{1/F_{n+1}}{1/F_n} = 1/\alpha < 1$, the series converges by d'Alembert's test.

23. (Lindstrom) Let $S_n = \sum\limits_{i=1}^{n} \dfrac{1}{F_i}$. Then $240S_{13} = 240 + 240 + 120 + 80 +$

$48 + 30 + \dfrac{240}{13} + \dfrac{240}{21} + \dfrac{240}{34} + \dfrac{240}{55} + \dfrac{240}{89} + \dfrac{240}{144} + \dfrac{240}{233} > 803$; so

$S > S_{13} > 803/240$.

25. (Peck) Since $\alpha^{n+1} = \alpha F_{n+1} + F_n$, sum $= \sum\limits_{n=1}^{\infty} \dfrac{1}{\alpha^{n+1}} = \dfrac{1/\alpha^2}{1-\alpha} = 1$.

27. (Graham) $\sum\limits_{n=1}^{\infty} \dfrac{(-1)^{n+1}}{F_n F_{n+1} F_{n+2}} = \sum\limits_{n=1}^{\infty} \left(\dfrac{1}{F_{n+1}} - \dfrac{F_{n+1}}{F_n F_{n+2}} \right) =$

$\sum\limits_{n=1}^{\infty} \left(\dfrac{1}{F_{n+1}} - \dfrac{F_{n+2} - F_n}{F_n F_{n+2}} \right) = \sum\limits_{n=1}^{\infty} \left(\dfrac{1}{F_{n+1}} - \dfrac{1}{F_n} + \dfrac{1}{F_{n+2}} \right) = \sum\limits_{n=1}^{\infty} \left(\dfrac{1}{F_{n+1}} - \dfrac{1}{F_n} \right)$

$+ \sum\limits_{n=1}^{\infty} \left(\dfrac{1}{F_{n+2}} \right) = -1 + \left(\sum\limits_{n=1}^{\infty} \dfrac{1}{F_n} - 2 \right) = -3 + \sum\limits_{n=1}^{\infty} \dfrac{1}{F_n}$. The desired result

now follows.

29. Sum $= \lim\limits_{n\to\infty} \sum\limits_{k=2}^{n} \left[\left(\dfrac{1}{G_{n-1}} - \dfrac{1}{G_n} \right) + \left(\dfrac{1}{G_n} - \dfrac{1}{G_{n+1}} \right) \right] =$

$\lim\limits_{n\to\infty} \left[\left(\dfrac{1}{G_1} - \dfrac{1}{G_n} \right) + \left(\dfrac{1}{G_2} - \dfrac{1}{G_{n+1}} \right) \right] = \dfrac{1}{a} + \dfrac{1}{b}$.

31. RHS $= 1 + \sum\limits_{i=2}^{n} \dfrac{F_{i+1}F_{i-1} - F_i^2}{F_i F_{i-1}} = 1 + \sum\limits_{i=2}^{n} \left(\dfrac{F_{i+1}}{F_i} - \dfrac{F_i}{F_{i-1}} \right) =$

$1 + \left(\dfrac{F_{n+1}}{F_n} - \dfrac{F_2}{F_1} \right) = \dfrac{F_{n+1}}{F_n} =$ LHS.

33. (Carlitz) Since $F_{2n+1} = F_{n+1}L_{n+2} - F_{n+2}L_n$, $\sum\limits_{n=1}^{m} \dfrac{F_{2n+1}}{L_n L_{n+1} L_{n+2}} =$

$\sum\limits_{n=1}^{m} \left(\dfrac{F_{n+1}}{L_n L_{n+1}} - \dfrac{F_{n+2}}{L_{n+1} L_{n+2}} \right) = \dfrac{F_2}{L_1 L_2} - \dfrac{F_{m+2}}{L_{m+1} L_{m+2}}$. Since $\lim\limits_{m\to\infty} \dfrac{F_{m+2}}{L_{m+1} L_{m+2}} = 0$,

the result follows.

35. (Carlitz) LHS $= \sqrt{5} \sum\limits_{n=0}^{\infty} \dfrac{(-1)^n}{\alpha^{2(2n+1)} - \beta^{2(2n+1)}} = \sqrt{5} \sum\limits_{n=0}^{\infty} \dfrac{(-1)^n}{\alpha^{2(2n+1)}} \cdot \dfrac{1}{1 - \alpha^{-4(2n+1)}}$

$= \sqrt{5} \sum\limits_{n=0}^{\infty} (-1)^n \sum\limits_{r=0}^{\infty} \alpha^{-2(2r+1)(2n+1)} = \sqrt{5} \sum\limits_{r=0}^{\infty} \dfrac{\alpha^{-2(2r+1)}}{1 + \alpha^{-4(2r+1)}} =$

$\sqrt{5} \sum\limits_{r=0}^{\infty} \dfrac{1}{\alpha^{2(2r+1)} + \beta^{2(2r+1)}} =$ RHS.

37. LHS $= \lim\limits_{n\to\infty} \sum\limits_{k=1}^{n} \left(\dfrac{1}{F_k F_{k+1}} - \dfrac{1}{F_{k+1} F_{k+2}} \right) = \lim\limits_{n\to\infty} \left(\dfrac{1}{F_1 F_2} - \dfrac{1}{F_{n+1} F_{n+2}} \right) =$
$1 - 0 = 1.$

39. $\sum\limits_{k=1}^{n} \left(\dfrac{1}{F_k F_{k+1} F_{k+2}} - \dfrac{1}{F_{k+1} F_{k+2} F_{k+3}} \right) = \sum\limits_{k=1}^{n} \dfrac{F_{k+3} - F_k}{F_k F_{k+1} F_{k+2} F_{k+3}}$. That is,

$\dfrac{1}{1 \cdot 1 \cdot 2} - \dfrac{1}{F_{n+1} F_{n+2} F_{n+3}} = 2 \sum\limits_{k=1}^{n} \dfrac{F_{k+1}}{F_k F_{k+1} F_{k+2} F_{k+3}}$. So $2 \sum\limits_{k=1}^{\infty} \dfrac{F_{k+1}}{F_k F_{k+1} F_{k+2} F_{k+3}}$

$= \dfrac{1}{2} - 0 = \dfrac{1}{2}$. The given result follows now.

41. Let $S_n = \sum\limits_{k=1}^{n} \dfrac{F_{k+1}}{F_k F_{k+3}} = \dfrac{1}{2} \sum\limits_{k=1}^{n} \left(\dfrac{1}{F_k} - \dfrac{1}{F_{k+3}} \right) =$

$\dfrac{1}{2} \left(\dfrac{1}{F_1} + \dfrac{1}{F_2} + \dfrac{1}{F_3} - \dfrac{1}{F_{n+1}} - \dfrac{1}{F_{n+2}} - \dfrac{1}{F_{n+3}} \right) =$

$\dfrac{1}{2} \left(\dfrac{5}{2} - \dfrac{1}{F_{n+1}} - \dfrac{1}{F_{n+2}} - \dfrac{1}{F_{n+3}} \right)$. So $\lim\limits_{n\to\infty} S_n = 5/4.$

43. (Mana) Let $\dfrac{1-x}{1-3x+x^2} = \sum\limits_{i=0}^{\infty} C_i x^i$. This series converges for $|x| < r$,

where r is the zero of $1 - 3x + x^2$ with the least absolute value, namely, β^2.

So $1 - x = (1 - 3x + x^2) \sum\limits_{i=0}^{\infty} C_i x^i$. Equating the coefficients of like terms,

$C_0 = 1, C_1 = 2$, and $C_{n+2} - 3C_{n+1} + C_n = 0$ for $n \geq 2$. This implies
$C_n = F_{2n+1}.$

45. Notice that $\beta < F_k/L_k < -\beta$. Using equations (24.16) and (24.17), LHS $=$

$\dfrac{F_k/L_k}{1 - F_k/L_k - F_k^2/L_k^2} = \dfrac{F_k L_k}{L_k^2 - L_k F_k - F_k^2} = \dfrac{F_{2k}}{L_k^2 - L_k F_k - F_k^2} = $ RHS.

EXERCISES 25

1. Let $S_j = \sum\limits_{k=1}^{j} F_{2k} = F_{2j+1} - 1$. Then

$$\text{LHS} = \sum\limits_{k=1}^{n} F_{2k} + 2 \sum\limits_{k=2}^{n} F_{2k} + \cdots + 2 \sum\limits_{k=n}^{n} F_{2k}$$

$$= 2[S_n + (S_n - S_1) + \cdots + (S_n - S_{n-1})]$$

$$= 2[nS_n - \sum\limits_{j=1}^{n-1} S_j] = 2[n(F_{2n+1} - 1) - (F_{2n} - 1) - n + 1]$$

$$= 2(nF_{2n+1} - F_{2n}) = \text{RHS}.$$

3. Let $S_j = \sum_{k=1}^{j} L_{2k-1} = L_{2j} - 1$. Then, as in Exercise 25.1,

$$\text{LHS} = (2n-1)S_n - 2\sum_{j=1}^{n-1} S_j = (2n-1)S_n - 2\sum_{j=1}^{n-1}(L_{2j} - 2)$$

$$= (2n-1)(L_{2n} - 2) - 2[(L_{2n-1} - 1) - 2(n-1)]$$

$$= (2n-1)L_{2n} - 2L_{2n-1} = \text{RHS}.$$

5. Let $S_j = \sum_{k=1}^{j} L_{2k} = L_{2j+1} - 1$. Then, as in Exercise 25.1,

$$\text{LHS} = 2\left[nS_n - \sum_{j=1}^{n-1} S_j\right] = 2\left[n(L_{2n+1} - 1) - \sum_{j=1}^{n-1}(L_{2j+1} - 1)\right]$$

$$= 2[n(L_{2n+1} - 1) - (L_{2n} - 3) + (n-1)]$$

$$= 2(nL_{2n+1} - L_{2n} + 2) = \text{RHS}.$$

7. $\text{LHS} = (a-d)\sum_{i=1}^{n} L_i + d\sum_{i=1}^{n} iL_i = (a-d)(L_{n+2} - 3) + d(nL_{n+2} - L_{n+3} + 4)$

$$= (a + nd - d)L_{n+2} - d(L_{n+3} - 7) - 3a = \text{RHS}.$$

9. $\text{LHS} = (a-d)\sum_{i=1}^{n} F_i^2 + d\sum_{i=1}^{n} iF_i^2 = (a-d)F_nF_{n+1} + d(nF_nF_{n+1} - F_n^2 + v)$

$$= (a + nd - d)F_nF_{n+1} - d(F_n^2 - v) = \text{RHS}.$$

11. Let $S_n = \sum_{i=1}^{n} iL_i^2$ and $S_n^* = \sum_{i=1}^{n}(n-i+1)L_i^2$. Then $S_n + S_n^* = (n+1)\sum_{i=1}^{n} L_i^2 =$
$(n+1)(L_nL_{n+1} - 2)$. So $S_n^* = (n+1)(L_nL_{n+1} - 2) - (nL_nL_{n+1} - L_n^2 + v) =$
$L_nL_{n+1} + L_n^2 - 2(n+1) - v = L_nL_{n+2} - 2(n+1) - v$.

13. Let $S = \sum_{i=1}^{n}[a + (i-1)d]L_i^2$ and $S^* = \sum_{i=1}^{n}[a + (n-i)d]L_i^2$. Then $S + S^* =$
$[2a + (n-1)d]\sum_{i=1}^{n} L_i^2 = [2a + (n-1)d](L_nL_{n+1} - 2)$. So $S^* = [2a +$
$(n-1)d](L_nL_{n+1} - 2) - (a + nd - d)(L_nL_{n+1} - v) + d(L_n^2 - 2n - v) =$
$a(L_nL_{n+1} - 2) + d(L_n^2 - 2n - v)$.

15. Let $S_j = \sum_{k=1}^{j} G_k = G_{j+2} - b$. Then

$$\text{Sum} = \sum_{i=1}^{n} G_i + \sum_{i=2}^{n} G_i + \cdots + \sum_{i=n}^{n} G_i$$

$$= S_n + (S_n - S_1) + \cdots + (S_n - S_{n-1})$$

$$= nS_n - \sum_{j=1}^{n-1} S_j = n(G_{n+2} - b) - \sum_{i=1}^{n}(G_{i+2} - b)$$

$$= n(G_{n+2} - b) - (G_{n+3} - b - a - b) + (n-1)b$$

$$= nG_{n+2} - G_{n+3} + a + b.$$

17. $\displaystyle\sum_{i=1}^{n} G_{2i-1} = \sum_{i=1}^{n} G_{2i} - \sum_{i=1}^{n} G_{2i-2} = G_{2n} - G_0 = G_{2n} + a - b.$

19. Let $S_j = \displaystyle\sum_{k=1}^{j} G_{2k-1} = G_{2j} + a - b$. Then

$$\text{Sum} = S_n + 2(S_n - S_1) + \cdots + 2(S_n - S_{n-1})$$

$$= S_n + 2(n-1)S_n + \cdots + 2(S_n - S_{n-1})$$

$$= S_n + 2(n-1)S_n - 2\sum_{i=1}^{n-1} S_i = (2n-1)S_n - 2\sum_{i=1}^{n-1}(G_{2i} + a - b)$$

$$= (2n-1)S_n - 2\sum_{i=1}^{n-1}[G_{2i} - 2(n-1)(a-b)]$$

$$= (2n-1)(G_{2n} + a - b) - 2(G_{2n-1} - a) - 2(n-1)(a-b)$$

$$= (2n-1)G_{2n} - 2G_{2n-1} + 3a - b.$$

21. $\text{Sum} = \displaystyle\sum_{i=1}^{10} i^2 F_i = 121F_{12} - 23F_{14} + 2F_{16} - 8 = 121 \cdot 144 - 23 \cdot 377 +$
$2 \cdot 987 - 8 = 10{,}719.$

23. $\text{Sum} = \displaystyle\sum_{i=1}^{5} i^3 F_i = 216F_7 - 127F_9 + 42F_{11} - 6F_{13} + 50 = 216 \cdot 13 - 127 \cdot 34 +$
$42 \cdot 89 - 6 \cdot 233 + 50 = 880.$

25. $\text{Sum} = \displaystyle\sum_{i=1}^{5} i^4 F_i = 1296F_7 - 1105F_9 + 590F_{11} - 180F_{13} + 24F_{15} - 416 =$
$4072.$

27. $\text{LHS} = F_1 + 4F_2 + 9F_3 + 16F_4 + 25F_5 + 36F_6 = 484 = 49F_8 - 15F_{10} +$
$2F_{12} - 8 = \text{RHS}.$

29. $\text{LHS} = F_1 + 8F_2 + 27F_3 + 64F_4 + 125F_5 + 216F_6 = 2608 = 343F_8 -$
$169F_{10} + 48F_{12} - 6F_{14} + 50 = \text{RHS}.$

31. $\text{LHS} = L_1 + 16L_2 + 81L_3 + 256L_4 + 625L_5 + 1296L_6 = 32{,}368 = 2401L_8 -$
$1695L_{10} + 770L_{12} - 204L_{14} + 24L_{16} - 930 = \text{RHS}.$

33. Let $S_j = \displaystyle\sum_{t=1}^{j} L_t = L_{j+2} - 3$; and $A_i = \displaystyle\sum_{j=1}^{i} S_j = \sum_{j=1}^{i}(L_{j+2} - 3) = \sum_{j=1}^{2+i} L_j - 4 - 3i =$
$L_{4+i} - 3i - 7$, so $A_{n-1} = L_{n+3} - 3n - 4.$ Using the technique of staggered

addition, we have

$$\sum_{i=1}^{n-1}(2i+1)S_i = 3\sum_{i=1}^{n-1}S_i + 2\sum_{i=2}^{n-1}S_i + 2\sum_{i=3}^{n-1}S_i + \cdots + 2\sum_{i=n-1}^{n-1}S_i$$

$$= 3A_{n-1} + 2(A_{n-1}-A_1) + 2(A_{n-1}-A_2) + \cdots + 2(A_{n-1}-A_{n-2})$$

$$= [3+2(n-2)]A_{n-1} - 2\sum_{i=1}^{n-2}A_i$$

$$= (2n-1)A_{n-1} - 2\sum_{i=1}^{n-2}(L_{4+i}-3i-7)$$

$$= (2n-1)A_{n-1} - 2\left(\sum_{i=1}^{n+2}L_i - \sum_{i=1}^{4}L_i\right) + (n-2)(3n+11)$$

$$= (2n-1)A_{n-1} - 2(L_{n+4}-3) + 30 + (n-2)(3n+11)$$

$$= (2n-1)(L_{n+3}-3n-4) - 2L_{n+4} + (n-2)(3n+11) + 36;$$

$$\sum_{i=1}^{n}i^2 L_i = \sum_{i=1}^{n}L_i + 3\sum_{i=2}^{n}L_i + 5\sum_{i=3}^{n}L_i + \cdots + (2n-1)\sum_{i=n}^{n}L_i$$

$$= \sum_{i=1}^{n}(2i-1)S_n - \sum_{i=1}^{n-1}(2i+1)S_i$$

$$= n^2(L_{n+2}-3) - \big[(2n-1)(L_{n+3}-3n-4)$$

$$\qquad -2L_{n+4} + (n-2)(3n+11) + 36\big]$$

$$= n^2 L_{n+2} - (2n-1)L_{n+3} + 2L_{n+4} - 18$$

$$= n^2 L_{n+2} + (2n-1)(L_{n+2}-L_{n+4}) + 2L_{n+4} - 18$$

$$= (n-1)^2 L_{n+2} - (2n-3)L_{n+4} - 18$$

$$= (n+1)^2 L_{n+2} - 2L_{n+2} - (2n-3)L_{n+4} - 18$$

$$= (n+1)^2 L_{n+2} - (2n+3)L_{n+4} + (6L_{n+4}-2L_{n+2}) - 18$$

$$= (n+1)^2 L_{n+2} - (2n+3)L_{n+4} + 2L_{n+6} - 18.$$

EXERCISES 26

1. Let $G_n = L_n$. Then $\mu = -5$ and $G_{n+1}(G_n + G_{n+2}) = L_{n+1}(L_n + L_{n+2}) = L_{2n+1} + L_{2n+3}$. This yields the desired reult.

3. Let θ_n = RHS. Then $\tan\theta_n = \dfrac{1/L_{2n} + 1/L_{2n+2}}{1 - 1/L_{2n} \cdot 1/L_{2n+2}} = \dfrac{L_{2n} + L_{2n+2}}{L_{2n}L_{2n+2} - 1} =$

$\dfrac{5F_{2n+1}}{5F_{2n+1}^2} = \dfrac{1}{F_{2n+1}} =$ RHS.

5. Let $2\theta = \tan^{-1} 2$. Then $\tan 2\theta = 2$; that is, $\dfrac{2\tan\theta}{1 - \tan^2\theta} = 2$. Solving, $\tan\theta = -\alpha, -\beta$. Since $\tan\theta \geq 0$, $\tan\theta = -\beta$. So $\theta = \tan^{-1}(-\beta) = \dfrac{1}{2}\tan^{-1} 2$.

7. This follows, since the tangent of the LHS is infinite.

9. (Peck) We have $\tan^{-1} 1/F_{2n} = \tan^{-1} 1/F_{2n+2} + \tan^{-1} 1/F_{2n+1}$. By Exercise 26.7, this implies $\tan^{-1} 1/F_{2n} = (\pi/2 - \tan^{-1} F_{2n+2}) + (\pi/2 - \tan^{-1} F_{2n+1}) = \pi - \tan^{-1} F_{2n+2} - \tan^{-1} F_{2n+1}$. This gives the desired result.

EXERCISES 27

1. $43 = 34 + 8 + 1$.

3. $137 = 89 + 34 + 8 + 5 + 1$.

5. $43 = 29 + 11 + 3$.

7. $137 = 123 + 11 + 3$.

9. $43 = 34 + 8 + 1$.

1	2	3	5	8	13	21	34
49*	98	147	245	392*	637	1029	1666*

$43 \cdot 49 = 49 + 392 + 1666 = 2107$.

11. $111 = 89 + 21 + 1$.

1	2	3	5	8	13	21	34	55	89
121*	242	363	605	968	1573	2541*	4114	6655	10769*

$111 \cdot 121 = 121 + 2541 + 10769 = 13{,}431$.

13. $43 = 29 + 11 + 3$

1	3	4	7	11	18	29
49	147*	196	343	539*	882	1421*

$43 \cdot 49 = 147 + 539 + 1421 = 2107$.

15. $111 = 76 + 29 + 4 + 2$.

2	1	3	4	7	11	18	29	47	76
242*	121	363	484*	847	1331	2178	3509*	5687	9196*

$111 \cdot 121 = 242 + 484 + 3509 + 9196 = 13{,}431$.

17. Let n be a positive integer and L_m the largest Lucas number $\leq n$. Then $n = L_m + n_1$, where $n_1 \leq L_m$. Let L_{m_1} be the largest Lucas number $\leq n_1$. Then $n = L_m + L_{m_1} + n_2$, where $n_2 \leq L_{m_1}$. Continuing like this, we get $n = L_m + L_{m_1} + L_{m_2} + \cdots$, where $n \geq L_m > L_{m_1} > L_{m_2} > \cdots$. Since this decreasing sequence of positive integers terminates, the desired result follows.

EXERCISES 29

1. $U_5 = F_5 = 5$, $U_6 = F_8 = 21$, $X_4 = L_3 = 4$, and $X_7 = L_{13} = 521$. LHS $= 5U_5 U_6 = 525 = X_7 - (-1)^{F_5} X_4 = $ RHS.

3. $X_5 X_6 = L_5 L_8 = 11 \cdot 47 = 517 - 4 = X_7 + (-1)^{F_5} X_4 = $ RHS.

5. LHS $= 5V_5 V_6 = 5F_{11} F_{18} = 5 \cdot 89 \cdot 2584 = 1{,}149{,}880 = 1{,}149{,}851 + 29 = L_{29} + L_7 = W_7 - (-1)^{L_5} W_4 = $ RHS.

7. Since $W_k = \alpha^{L_k} + \beta^{L_k}$, $W_n W_{n+1} = \left(\alpha^{L_{n+2}} + \beta^{L_{n+2}}\right) + \left(\alpha^{L_n}\beta^{L_{n+1}} + \alpha^{L_{n+1}}\beta^{L_n}\right)$
$= W_{n+2} + \left[\alpha^{L_{n+1}}\left(-\alpha^{-L_n}\right) + \beta^{L_{n+1}}\left(-\beta^{-L_n}\right)\right] = W_{n+2} + (-1)^{L_n}\left(\alpha^{L_{n+1}-L_n}\right.$
$\left. + \beta^{L_{n+1}-L_n}\right) = W_{n+2} + (-1)^{L_n}\left(\alpha^{L_{n-1}} + \beta^{L_{n-1}}\right) = W_{n+2} + (-1)^{L_n} W_{n-1}$.

9. LHS $= 5U_4 U_6 = 5F_3 F_8 = 5 \cdot 2 \cdot 21 = 210 = 199 + 11 = L_{11} + L_5 = W_5 - (-1)^3 X_5 = $ RHS.

11. $\sqrt{5}U_n = \alpha^{F_n} - \beta^{F_n}$; so
$$5U_{n-1} U_{n+1} = \left(\alpha^{F_{n-1}+F_{n+1}} + \beta^{F_{n-1}+F_{n+1}}\right) - \left(\alpha^{F_{n-1}}\beta^{F_{n+1}} + \alpha^{F_{n+1}}\beta^{F_{n-1}}\right)$$
$$= \left(\alpha^{L_n} + \beta^{L_n}\right) - \left[\alpha^{F_{n+1}}\left(-\alpha^{-F_{n-1}}\right) + \beta^{F_{n+1}}\left(-\beta^{-F_{n-1}}\right)\right]$$
$$= W_n - (-1)^{F_{n-1}}\left(\alpha^{F_{n+1}-F_{n-1}} + \beta^{F_{n+1}-F_{n-1}}\right)$$
$$= W_n - (-1)^{F_{n-1}}\left(\alpha^{F_n} + \beta^{F_n}\right) = W_n - (-1)^{F_{n-1}} X_n.$$

13. $2L_{13} = \sqrt{5L_{12}^2 - 20(-1)^{12}} + L_{12} = \sqrt{5 \cdot 322^2 - 20} + 322 = 1042$; so $L_{13} = 521$.

15. $2L_{n+1} = 5F_n + L_n = \sqrt{5L_n^2 - 20(-1)^n} + L_n$; so $L_{n+1} = \left[\sqrt{5L_n^2 - 20(-1)^n} + L_n\right]/2$.

17. $L_{3n} = \alpha^{3n} + \beta^{3n} = (1 + 2\alpha)^n + (1 + \beta)^{2n} = \sum_{i=0}^{n} \binom{n}{i}\left[(2\alpha)^i + (2\beta)^i\right] = \sum_{i=0}^{n} \binom{n}{i} 2^i L_i$. Since $L_0 = 2$, it follows that $2 | L_{3n}$. Conversely, let $2 | L_n$. Since $F_{2n} = F_n L_n$, this implies $2 | F_{2n}$; that is, $F_3 | F_{2n}$. So $3 | n$, since $3 \nmid 2$.

19. 233.

21. 24,476.

EXERCISES 30

1. Let $z = a + bi$. Then $\|z\| = a^2 + b^2 \geq 0$.

3. Let $w = a + bi$ and $z = c + di$. Since $wz = (ac - bd) + (ad + bc)i$, $\|wz\| = (ac - bd)^2 + (ad + bc)^2 = (a^2 + b^2)(c^2 + d^2) = \|w\| \cdot \|z\|$.

5. The result is true when $n = 1$ and $n = 2$. Assume it is true for all positive integers $k \leq n$. Then $f_{n+1} = f_n + f_{n-1} = (F_n + iF_{n-1}) + (F_{n-1} + iF_{n-2}) = (F_n + F_{n-1}) + i(F_{n-1} + F_{n-2}) = F_{n+1} + iF_n$. Thus the result is true for all $n \geq 1$.

7. Let $z = a + bi$. Then $\bar{z} = a - bi$, so $\|\bar{z}\| = a^2 + b^2 = \|z\|$.

9. $f_{-10} = F_{-10} + iF_{-11} = (-1)^{11}F_{10} + i(-1)^{12}F_{11} = -55 + 89i$; and $l_{-10} = L_{-10} + iL_{-11} = (-1)^{10}L_{10} + i(-1)^{11}L_{11} = 123 - 199i$.

11. $2m - 1 = 7$ and $2n - 1 = 21$, so $(2m-1)|(2n-1)$. $F_7 = 13, F_3F_{11} - F_4F_{10} = 2 \cdot 89 - 3 \cdot 55 = 13$, so $F_7|(F_3F_{11} - F_4F_{10})$.

13. LHS $= (F_{n+1} + iF_n)(F_{n-1} + iF_{n-2}) - (F_n + iF_{n-1})^2 = (F_{n+1}F_{n-1} - F_n^2) - (F_nF_{n-2} - F_{n-1}^2) + i(F_{n+1}F_{n-2} - F_nF_{n-1}) = 2(-1)^n + i(-1)^{n-1} = (2 - i)(-1)^n$.

15. LHS $= (L_{n+1} + iL_n) + (L_{n-1} + iL_{n-2}) = (L_{n+1} + L_{n-1}) + i(L_n + L_{n-2}) = 5(F_n + iF_{n-1}) = 5f_n$.

17. LHS $= (F_{n+2} + iF_{n+1}) - (F_{n-2} + iF_{n-3}) = (F_{n+2} - F_{n-2}) + i(F_{n+1} - F_{n-3}) = L_n + iL_{n-1} = l_n$.

19. LHS $= (F_n + iF_{n-1})^2 + (F_{n+1} + iF_n)^2 = (F_n^2 + F_{n+1}^2) - (F_{n-1}^2 + F_n^2) + 2i(F_nF_{n-1} + F_{n+1}F_n) = F_{2n+1} - F_{2n-1} + 2iF_{2n} = (1 + 2i)F_{2n} = $ RHS.

21. $f_n l_n = (F_n + iF_{n-1})(L_n + iL_{n-1}) = (F_{2n} - F_{2n-2}) + i(F_nL_{n-1} + F_{n-1}L_n) = F_{2n-1} + 2iF_{2n-1} = (1 + 2i)F_{2n-1} = $ RHS.

23. LHS $= (L_{n-1} + iL_{n-2})(L_{n+1} + iL_n) - (L_n + iL_{n-1})^2 = (L_{n+1}L_{n-1} - L_n^2) - (L_nL_{n-2} - L_{n-1}^2) + i(L_{n+1}L_{n-2} - L_nL_{n-1}) = 10(-1)^{n+1} - 5i(-1)^{n-1} = 5(2 - i)(-1)^{n+1} = $ RHS.

25. LHS $= (L_n + iL_{n-1})^2 - 5(F_n + iF_{n-1})^2 = (L_n^2 - 5F_n^2) - (L_{n-1}^2 - 5F_{n-1}^2) + 2i(L_nL_{n-1} - 5F_nF_{n-1}) = 8(-1)^n + 4i(-1)^{n-1} = 4(2 - i)(-1)^n = $ RHS.

27. LHS $= \displaystyle\sum_{k=1}^{n}(F_{2k-1} + iF_{2k-2}) = F_{2n} + i(F_{2n-1} - 1) = f_{2n} - i = $ RHS.

29. LHS $= \displaystyle\sum_{k=1}^{2n}(-1)^k(F_k + iF_{k-1}) = \sum_{k=1}^{2n}(-1)^kF_k + i\sum_{k=1}^{2n}(-1)^kF_{k-1} = \sum_{k=1}^{n}F_{2k} - \sum_{k=1}^{n}F_{2k-1} + i\left(\sum_{k=1}^{n}F_{2k-1} - \sum_{k=1}^{n-1}F_{2k}\right) = (F_{2n+1} - 1) - F_{2n} + i[F_{2n} - (F_{2n-1} - 1)] = (F_{2n-1} - 1) + i(F_{2n-2} + 1) = (F_{2n-1} + iF_{2n-2}) + i - 1 = f_{2n-1} + i - 1 = $ RHS.

31. $C_n\overline{C_n} = (F_n + iF_{n+1})(F_n - iF_{n+1}) = F_n^2 + F_{n+1}^2 = F_{2n+1}$.

33. $l(0) = l(w)$ implies $2 = \alpha^w + \beta^w$ and $l(z + w) = l(z)$ implies $\alpha^{w+z} + \beta^{w+z} = \alpha^z + \beta^z$. So $\alpha^{w+z} + \beta^z(2 - \alpha^w) = \alpha^z + \beta^z$. Then $\alpha^w(\alpha^z - \beta^z) + 2\beta^z = \alpha^z + \beta^z$, so $\alpha^w(\alpha^z - \beta^z) = \alpha^z - \beta^z$. Thus $\alpha^w = 1$, so $\mathrm{Re}(w) = 0$. Let $w = 0 + yi$. Then $\alpha^{yi} = 0$, which is possible only if $y = 0$. Thus $w = 0$, a contradiction.

35. $5(\mathrm{RHS}) = (\alpha^{z+1} - \beta^{z+1}) + (\alpha^z - \beta^z) = \alpha^z(\alpha + 1) - \beta^z(\beta + 1) = \alpha^{z+2} - \beta^{z+2} = 5f(z + 2)$. This yields the identity.

37. $\mathrm{LHS} = (\alpha^z + \beta^z)^2 - (\alpha^z - \beta^z)^2 = 4(\alpha\beta)^z = 4(-1)^z = 4e^{\pi z i} = \mathrm{RHS}$.

39. $l(-z) = \alpha^{-z} + \beta^{-z} = \dfrac{\alpha^z + \beta^z}{(\alpha\beta)^z} = \dfrac{l(z)}{(-1)^z} = l(z)e^{\pi z i}$.

41. $5(\mathrm{RHS}) = (\alpha^z - \beta^z)(\alpha^{w+1} - \beta^{w+1}) + (\alpha^{z-1} - \beta^{z-1})(\alpha^w - \beta^w) = \alpha^{z+w}(\alpha + \alpha^{-1}) + \beta^{z+w}(\beta + \beta^{-1}) - \alpha^{z-1}\beta^w(\alpha\beta + 1) - \alpha^w\beta^{z-1}(\alpha\beta + 1) = (\alpha - \beta)(\alpha^{z+w} - \beta^{z+w})$. So $\mathrm{RHS} = f(z + w) = \mathrm{LHS}$.

43. Let $w = 0$ and $z = n$ in equation (30.1). Then $F_n = \dfrac{1}{\sqrt{5}}\displaystyle\sum_{k=0}^{\infty}\dfrac{(\ln^k\alpha - \ln^k\beta)n^k}{k!}$.

33. $f(0) = 0$ so $f(n)$ implies $2 = a^0 + K^0$ and $K = a$ ... so ...

...

35. $3f(MLS) = 6m^{x/n} - 2g^{...}$ ...

(i) LHS $= ...$

...

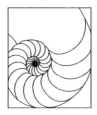

Index

PURE AND APPLIED MATHEMATICS

A Wiley Series of Texts, Monographs, and Tracts

Founded by RICHARD COURANT

Editors Emeriti: MYRON B. ALLEN III, PETER HILTON, HARRY HOCHSTADT, ERWIN KREYSZIG, PETER LAX, JOHN TOLAND

*Now available in a lower priced paperback edition in the Wiley Classics Library
†Now available in paperback

*Now available in a lower priced paperback edition in the Wiley Classics Library
†Now available in paperback

*Now available in a lower priced paperback edition in the Wiley Classics Library
†Now available in paperback

ROITMAN—Introduction to Modern Set Theory

ROSSI—Theorems, Corollaries, Lemmas, and Methods of Proof

*RUDIN—Fourier Analysis on Groups

SENDOV—The Averaged Moduli of Smoothness: Applications in Numerical Methods and Approximations

SENDOV and POPOV—The Averaged Moduli of Smoothness

SEWELL—The Numerical Solution of Ordinary and Partial Differential Equations, Second Edition

SEWELL—Computational Methods of Linear Algebra, Second Edition

SHICK—Topology: Point-Set and Geometric

SHISKOWSKI and FRINKLE—Principles of Linear Algebra With *Maple*™

SHISKOWSKI and FRINKLE—Principles of Linear Algebra With *Mathematica*®

*SIEGEL—Topics in Complex Function Theory

  Volume 1—Elliptic Functions and Uniformization Theory

  Volume 2—Automorphic Functions and Abelian Integrals

  Volume 3—Abelian Functions and Modular Functions of Several Variables

SMITH and ROMANOWSKA—Post-Modern Algebra

ŠOLÍN–Partial Differential Equations and the Finite Element Method

STADE—Fourier Analysis

STAHL and STENSON—Introduction to Topology and Geometry, Second Edition

STAHL—Real Analysis, Second Edition

STAKGOLD and HOLST—Green's Functions and Boundary Value Problems, Third Edition

STANOYEVITCH—Introduction to Numerical Ordinary and Partial Differential Equations Using MATLAB®

*STOKER—Differential Geometry

*STOKER—Nonlinear Vibrations in Mechanical and Electrical Systems

*STOKER—Water Waves: The Mathematical Theory with Applications

WATKINS—Fundamentals of Matrix Computations, Third Edition

WESSELING—An Introduction to Multigrid Methods

†WHITHAM—Linear and Nonlinear Waves

ZAUDERER—Partial Differential Equations of Applied Mathematics, Third Edition

*Now available in a lower priced paperback edition in the Wiley Classics Library
†Now available in paperback

Printed and bound by CPI Group (UK) Ltd, Croydon, CR0 4YY

16/04/2025

14658370-0002